LONDON MATHEMATICAL SOCIETY STUDENT TEXTS

Managing Editor: Ian J. Leary,
Mathematical Sciences, University of Southampton, UK

London Mathematical Society Student Texts 92

The Block Theory of Finite Group Algebras

Volume II

MARKUS LINCKELMANN

City, University of London

CAMBRIDGE
UNIVERSITY PRESS

CAMBRIDGE
UNIVERSITY PRESS

University Printing House, Cambridge CB2 8BS, United Kingdom

One Liberty Plaza, 20th Floor, New York, NY 10006, USA

477 Williamstown Road, Port Melbourne, VIC 3207, Australia

314-321, 3rd Floor, Plot 3, Splendor Forum, Jasola District Centre, New Delhi - 110025, India

79 Anson Road, #06-04/06, Singapore 079906

Cambridge University Press is part of the University of Cambridge.

It furthers the University's mission by disseminating knowledge in the pursuit of
education, learning and research at the highest international levels of excellence.

www.cambridge.org
Information on this title: www.cambridge.org/9781108441803
DOI: 10.1017/9781108349307

© Markus Linckelmann 2018

First published 2018

A catalogue record for this publication is available from the British Library

ISBN – 2 Volume Set 978-1-108-44190-2 Paperback
ISBN – Volume I 978-1-108-42591-9 Hardback
ISBN – Volume I 978-1-108-44183-4 Paperback
ISBN – Volume II 978-1-108-42590-2 Hardback
ISBN – Volume II 978-1-108-44180-3 Paperback

Contents

Volume II

Volume I

Introduction

Group representation theory investigates the structural connections between groups and mathematical objects admitting them as automorphism groups. Its most basic instance is the action of a group G on a set M, which is equivalent to a group homomorphism from G to the symmetric group S_M of all permutations of the set M, thus representing G as an automorphism group of the set M. Classical representation theory, developed during the last decade of the 19th century by Frobenius and Schur, investigates the representations of a finite group G as linear automorphism groups of complex vector spaces, or equivalently, modules over the complex group algebra $\mathbb{C}G$. Modular representation theory, initiated by Brauer in the 1930s, considers finite group actions on vector spaces over fields with positive characteristic, and more generally, on modules over complete discrete valuation rings as a link between different characteristics. Integral representation theory considers representations of groups over rings of algebraic integers, with applications in number theory. Topologists have extensively studied the automorphism groups of classifying spaces of groups in connection with K-theory and transformation groups. Methods from homotopy theory and homological algebra have shaped the area significantly.

Within modular representation theory, viewed as the theory of module categories of finite group algebras over complete discrete valuation rings, the starting point of block theory is the decomposition of finite group algebras into indecomposable direct algebra factors, called *block algebras*. The block algebras of a finite group algebra are investigated individually, bearing in mind that the module category of an algebra is the direct sum of the module categories of its blocks. Block theory seeks to gain insight into which way the structure theory of finite groups and the representation theory of block algebras inform each other.

Few algebras are expected to arise as block algebras of finite groups. Narrowing down the pool of possible block algebras with essentially representation

1

theoretic methods has been very successful for blocks of finite and tame representation type, but remains a major challenge beyond those cases. Here is a sample list of properties that have to be satisfied by an algebra B which arises as a block algebra of a finite group algebra over a complete discrete valuation ring $\mathcal{O}$ with residue field k of prime characteristic p and field of fractions K of characteristic 0.

- B is symmetric.
- $K \otimes_{\mathcal{O}} B$ is semisimple.
- The canonical map $Z(B) \to Z(k \otimes_{\mathcal{O}} B)$ is surjective.
- B is separably equivalent to $\mathcal{O}P$ for some finite p-group P.
- The Cartan matrix of $k \otimes_{\mathcal{O}} B$ is positive definite, its determinant is a power of p, and its largest elementary divisor is the smallest power of p that annihilates all homomorphism spaces in the $\mathcal{O}$-stable category $\underline{\mathrm{mod}}(B)$.
- The decomposition map from the Grothendieck group of finitely generated $K \otimes_{\mathcal{O}} B$-modules to the Grothendieck group of finitely generated $k \otimes_{\mathcal{O}} B$-modules is surjective.
- B is defined over a finite extension of the p-adic integers $\mathbb{Z}_p$, and $Z(B)$ is defined over $\mathbb{Z}_{(p)}$.
- $k \otimes_{\mathcal{O}} B$ is defined over a finite field $\mathbb{F}_q$, where q is a power of p, and $Z(k \otimes_{\mathcal{O}} B)$ is defined over $\mathbb{F}_p$.

The dominant feature of block theory is the dichotomy of invariants associated with block algebras. Block algebras of finite groups have all the usual 'global' invariants associated with algebras – module categories, derived and stable categories, cohomological invariants including Hochschild cohomology, and numerical invariants such as the numbers of ordinary and modular irreducible characters. Due to their provenance from finite groups, block algebras have further 'local' invariants which cannot be, in general, associated with arbitrary algebras. The prominent conjectures that drive block theory revolve around the interplay between 'global' and 'local' invariants. Source algebras of blocks capture invariants from both worlds, and in an ideal scenario, the above mentioned conjectures would be obtained as a consequence of a classification of the source algebras of blocks with a fixed defect group. In this generality, this has been achieved in two cases, namely for blocks with cyclic and Klein four defect groups. The local structure of a block algebra B includes the following invariants.

- A defect group P of B.
- A fusion system $\mathcal{F}$ of B on P.

- A class $\alpha \in H^2(\mathcal{F}^c; k^{\times})$ such that α restricts on $\mathrm{Aut}_{\mathcal{F}}(Q)$ to the Külshammer–Puig class α_Q, for any Q belonging to the category $\mathcal{F}^c$ of $\mathcal{F}$-centric subgroups in P.
- The number of weights of $(\mathcal{F}, \alpha)$.

A sample list for the global structure of B includes the following invariants of B as an algebra, as well as their relationship with the local invariants.

- The numbers $|\mathrm{Irr}_K(B)|$ and $|\mathrm{IBr}_k(B)|$ of isomorphism classes of simple $K \otimes_{\mathcal{O}} B$-modules and $k \otimes_{\mathcal{O}} B$-modules, respectively, with their heights.
- The $\mathcal{O}$-stable module category $\underline{\mathrm{mod}}(B)$ and its dimension as a triangulated category.
- The bounded derived category $D^b(B)$ and its dimension as a triangulated category.
- The module category $\mathrm{mod}(B)$ as an abelian category, structure and Loewy lengths of projective indecomposable B-modules.
- The generalised decomposition matrix of B.

All of the above local and global invariants can be calculated, at least in principle, from the source algebras of B, and hence methods to determine source algebras are a major theme in this book.

Volume I introduces the broader context and many of the methods that are fundamental to modular group representation theory. Chapter 1 provides background on algebras and introduces some of the main players in this book – group algebras, twisted group algebras as well as category algebras, for the sake of giving a broader picture. Chapter 2 switches the focus from algebras to module categories and functors. Chapter 3 develops the classical representation theory of finite groups – that is, representations over complex vector spaces – just far enough to prove Burnside's $p^a q^b$-Theorem and describe Brauer's characterisation of characters. Turning to modular representation theory, Chapter 4 handles the general theory of algebras over discrete valuation rings. Chapter 5 combines this material with group actions, leading to Green's theory of vertices and sources, Puig's notion of pointed groups, and further fundamental module theoretic results on special classes of modules, as well as Green's Indecomposability Theorem.

In Volume II, the core theme of this book takes centre stage. Chapter 6 develops in a systematic way block theory, including Brauer's three main theorems, some Clifford Theory, the work of Alperin and Broué on Brauer pairs and Puig's notion of source algebras. Chapter 7 describes modules over finite p-groups, with an emphasis on endopermutation modules. This is followed by another core chapter on local structure, containing in particular a brief introduction to fusion systems, connections between characters and local structure, the structure theory of nilpotent blocks and their extensions. Chapter 9 on isometries illustrates the interaction between the concepts introduced up to this point. Applications in subsequent chapters include the structure theory of blocks with cyclic or Klein four defect groups.

Along the way, some of the fundamental conjectures alluded to above will be described. Alperin's weight conjecture predicts that the number of isomorphism classes of simple modules of a block algebra should be determined by its local invariants. Of a more structural nature, Broué's abelian defect group conjecture would offer, if true, some explanation for these numerical coincidences at least in the case of blocks with abelian defect groups. The finiteness conjectures of Donovan, Feit and Puig predict that once a defect group is fixed, there are only 'finitely many blocks' with certain properties. These conjectures are known to hold for blocks of various classes of finite simple groups and their extensions. Complemented by rapidly evolving reduction techniques, this points to the possibility of proving parts of these conjectures by invoking the classification of finite simple groups. A lot more work seems to be needed to provide the understanding that would transform mystery into insight.

The representation theory of finite group algebras draws significantly on methods from areas including ring theory, category theory, and homological algebra. Rather than giving systematic introductions to those areas, we develop background material as we go along, trying not to lose sight of the actual topic.

Acknowledgements. The author wishes to thank David Craven, Charles Eaton, Adam Glesser, Radha Kessar, Justin Lynd, Michal Stolorz, Michael Livesey, Lleonard Rubio y Degrassi and Benjamin Sambale for reading, commenting on, and correcting parts of various preliminary versions. Special thanks go to Dave Benson, Morty Harris, Sejong Park, as well as to Robert Boltje and the members of his research seminar at UCSC, for long lists of very detailed comments and suggestions.

6

Blocks and Source Algebras

The prominent conjectures in block theory can be divided into three classes. The *finiteness conjectures* predict that for a fixed finite p-group P, there should be only finitely many 'different' block algebras with defect groups isomorphic to P; this includes the finiteness conjectures of Donovan and Puig. There are *counting conjectures* that express certain numerical invariants of a block algebra in terms of invariants associated with the defect groups, such as Alperin's weight conjecture. Last but not least, there are *structural conjectures*, trying to predict structural information about block algebras, their module and derived module categories in terms of 'local' information. This includes Broué's abelian defect conjecture, for instance. The line between the latter two classes of conjectures is fluid in that, whenever possible, one seeks to give structural explanations for the numerical coincidences observed. We will describe these conjectures as we go along in this chapter.

Besides classical material – including Brauer's three main theorems – this chapter introduces and investigates source algebras of blocks. This notion, due to Puig and very briefly touched upon in 5.6.12 above in the context of G-algebras, is fundamental for the general structure theory of block algebras of finite groups.

Throughout this chapter we denote by $\mathcal{O}$ a complete discrete valuation ring having a residue field k of prime characteristic p and a quotient field K. If not stated otherwise, we allow the case $K = \mathcal{O} = k$ in order to ensure that statements over $\mathcal{O}$ can be applied over k if necessary. Many results will require the field k to be 'large enough' to be a splitting field for the blocks under consideration. For an individual block there is always a finite field that is large enough. When it comes to statements and conjectures for all blocks, one can choose k algebraically closed, so that k is large enough for all blocks. If $\mathrm{char}(K) = 0$, then the field K cannot be chosen to be algebraically closed, however, so one cannot fix K to be large enough for all blocks.

5

6.1 Blocks of group algebras and defect groups

Let G be a finite group. Decompose the group algebra $\mathcal{O}G$ as a direct product

$$\mathcal{O}G = B_1 \times B_2 \times \cdots \times B_r$$

of indecomposable $\mathcal{O}$-algebras B_i. By 1.7.7, each block algebra B_i is of the form $B_i = \mathcal{O}Gb_i$ for a unique primitive idempotent $b_i \in Z(\mathcal{O}G)$, and the set $\{b_i\}_{1 \leq i \leq r}$ is a primitive decomposition of 1 in $Z(\mathcal{O}G)$. The canonical map $\mathcal{O}G \to kG$ sends $Z(\mathcal{O}G)$ onto $Z(kG)$, and hence by 4.7.1, the image $\bar{b}$ in kG of any primitive idempotent $b \in Z(\mathcal{O}G)$ remains primitive in $Z(kG)$. Thus $\bar{B}_i = k \otimes_{\mathcal{O}} B_i = kG\bar{b}$ remains an indecomposable k-algebra, or equivalently, the decomposition of kG as a direct product of its block algebras is equal to

$$kG = \bar{B}_1 \times \bar{B}_2 \times \cdots \times \bar{B}_r.$$

The different parametrisations of blocks of $\mathcal{O}G$ either by primitive idempotents in $Z(\mathcal{O}G)$ or by indecomposable algebra factors of $\mathcal{O}G$ or by indecomposable direct bimodule summands of $\mathcal{O}G$ all have their merits depending on the circumstances.

A block algebra of a finite group is more than just an algebra – the presence of the group gives rise to invariants that cannot, a priori, be read off the algebra structure alone. We recall the definition and basic properties of defect groups of primitive G-algebras specialised to block algebras:

Definition 6.1.1 Let G be a finite group, let b be a block of $\mathcal{O}G$, and let $B = \mathcal{O}Gb$ be the correponding block algebra. A *defect group of b* or of B is a minimal subgroup P of G with the property $b \in (\mathcal{O}G)_P^G$, or equivalently, such that $b = \mathrm{Tr}_P^G(c)$ for some $c \in (\mathcal{O}G)^P$.

Theorem 6.1.2 *Let G be a finite group, b be a block of $\mathcal{O}G$ and P a defect group of b.*

(i) *P is a p-subgroups of G.*

(ii) *For any subgroup H of G such that $b \in (\mathcal{O}G)_H^G$ there is $x \in G$ such that $P \subseteq {}^xH$.*

(iii) *The defect groups of b form a G-conjugacy class of p-subgroups of G.*

Proof These statements hold more generally for primitive G-algebras and are proved in 5.6.5.   □

Since the defect groups of a block are all conjugate, they have the same order.

Definition 6.1.3 Let G be a finite group, and let b be a block of $\mathcal{O}G$. The *defect of b* is the unique integer $d = d(b)$ such that p^d is the order of the defect groups of b.

The defect groups of a block of $\mathcal{O}G$ need not be Sylow p-subgroups of G, but there is at least one block whose defect groups are the Sylow p-subgroups. Since 1 is the unique idempotent in $\mathcal{O}$, it follows that the augmentation homomorphism $\mathcal{O}G \to \mathcal{O}$ maps all blocks but one to zero.

Definition 6.1.4 Let G be a finite group. The *principal block of $\mathcal{O}G$* is the unique block b_0 of $\mathcal{O}G$ such that $\eta(b_0) = 1$, where $\eta : \mathcal{O}G \to \mathcal{O}$ is the augmentation homomorphism of $\mathcal{O}G$.

Theorem 6.1.5 *Let G be a finite group. The defect groups of the principal block of $\mathcal{O}G$ are the Sylow p-subgroups of G.*

Proof Denote by b is the principal block of $\mathcal{O}G$; that is, b is not in the kernel of the augmentation homomorphism $\eta : \mathcal{O}G \to \mathcal{O}$. Let P be a defect group of b. Let $c \in (\mathcal{O}G)^P$ such that $b = \mathrm{Tr}_P^G(c)$. Since $\eta(^x c) = \eta(c)$ for any $x \in G$ it follows that $1 = \eta(b) = \eta(\mathrm{Tr}_P^G(c)) = |G : P|\eta(c)$, which is only possible if $|G : P|$ is invertible in $\mathcal{O}$, hence if $|G : P|$ is prime to p, or equivalently, if P is a Sylow p-subgroup of G. $\square$

Defect groups do not change upon replacing $\mathcal{O}$ by its residue field k.

Theorem 6.1.6 *Let G be a finite group and let b be a block of $\mathcal{O}G$. Then the image $\bar{b}$ of b in kG is a block of kG, and moreover b and $\bar{b}$ have the same defect groups.*

Proof The centre $Z(\mathcal{O}G)$ is $\mathcal{O}$-free, having as a basis the G-conjugacy class sums in $\mathcal{O}G$. Similarly, $Z(kG)$ has as a k-basis the G-conjugacy class sums in kG. Thus the canonical map $Z(\mathcal{O}G) \to Z(kG)$ is surjective. The lifting theorem for idempotents implies that either $\bar{b} = 0$ or $\bar{b}$ is a block of kG. The case $\bar{b} = 0$ cannot occur because the kernel of the canonical map $Z(\mathcal{O}G) \to Z(kG)$ is $J(\mathcal{O})Z(\mathcal{O}G)$, which is contained in the radical of $Z(\mathcal{O}G)$, hence contains no idempotent. Let P be a defect group of b. Then $b \in (\mathcal{O}G)_P^G$, hence $\bar{b} \in (kG)_P^G$. Thus P contains a defect group Q of $\bar{b}$. If Q is any defect group of $\bar{b}$ then $\bar{b} \in (kG)_Q^G$, and hence $b \in (\mathcal{O}G)_Q^G + J(\mathcal{O})Z(\mathcal{O}G)$. Since b is primitive in $Z(\mathcal{O}G)$, Rosenberg's Lemma 4.4.8 implies that b is contained in one of the two ideals $(\mathcal{O}G)_Q^G$ or $J(\mathcal{O})Z(\mathcal{O}G)$. However, the ideal $J(\mathcal{O})Z(\mathcal{O}G)$ contains no idempotent, and hence $b \in (\mathcal{O}G)_Q^G$. This shows that Q contains a defect group of b, whence the result. $\square$

Let G be a finite group and M an $\mathcal{O}G$-module. By 1.7.9 we have a decomposition $M = \oplus_b bM$, of M as a direct sum of $\mathcal{O}G$-modules, where in the sum b runs over the set of blocks of $\mathcal{O}G$. Each summand bM has the property that b acts as the identity on bM, and hence bM can be viewed as an $\mathcal{O}Gb$-module. Conversely, every $\mathcal{O}Gb$-module can be viewed as an $\mathcal{O}G$-module via the projection $\mathcal{O}G \rightarrow \mathcal{O}Gb$; that is, with $x \in G$ acting on $m \in M$ as xbm. In particular, if M is indecomposable, then there is a unique block b of $\mathcal{O}G$ such that $bM = M$. In that case we have $b'M = \{0\}$ for all blocks $b' \neq b$ because any two different blocks of $\mathcal{O}G$ are orthogonal. An indecomposable $\mathcal{O}Gb$-module remains indecomposable when viewed as an $\mathcal{O}G$-module, and hence has a vertex and a source.

Theorem 6.1.7 *Let G be a finite group, b be a block of $\mathcal{O}G$, and P be a defect group of b. Set $B = \mathcal{O}Gb$. Every B-module is relatively P-projective. In particular, every indecomposable B-module has a vertex contained in P, and the $\mathcal{O}$-rank of any finitely generated $\mathcal{O}$-free B-module is divisible by the p-part of $|G : P|$.*

Proof We are going to show that every B-module M is relatively P-projective by applying Higman's criterion. Since M is a B-module, it follows that $bm = m$ for all $m \in M$. Write $b = \mathrm{Tr}_P^G(c)$ for some $c \in (\mathcal{O}G)^P$. Define $\varphi : M \rightarrow M$ by setting $\varphi(m) = cm$ for all $m \in M$. Since c commutes with all elements in P we have $\varphi \in \mathrm{End}_{\mathcal{O}P}(M)$. For all $m \in M$ we have

$$\mathrm{Tr}_P^G(\varphi)(m) = \sum_{x \in [G/P]} (^x\varphi)(m) = \sum_{x \in [G/P]} x\varphi(x^{-1}m)$$

$$= \sum_{x \in [G/P]} xcx^{-1}m = \mathrm{Tr}_P^G(c)m = bm = m$$

and thus $\mathrm{Id}_M = \mathrm{Tr}_P^G(\varphi)$. Higman's criterion implies that M is relatively P-projective. In particular, if M is indecomposable, then P contains a vertex of M. The last statement follows from Theorem 5.12.13. $\square$

Remark 6.1.8 Let G be a finite group. The group algebra $\mathcal{O}G$ is a Hopf algebra, with comultiplication induced by the diagonal map $G \rightarrow G \times G$ and antipode induced by taking inverses in G. The antipode permutes the blocks of $\mathcal{O}G$, but the comultiplication is in general not compatible with the block decomposition of $\mathcal{O}G$, and hence block algebras need not inherit a Hopf algebra structure. In particular, the tensor product of two modules belonging to a block of $\mathcal{O}G$ may contain summands in different blocks. The interaction between the block decomposition and the Hopf algebra structure is complicated in general, but one can make this work sometimes to one's advantage: starting with tensor products of modules belonging to well-understood blocks may reveal properties of other blocks.

Let G be a finite group and B a block algebra of $\mathcal{O}G$. It is an open question whether the $\mathcal{O}$-algebra structure of B determines the defect groups of B. There are no known examples of block algebras having equivalent module categories but nonisomorphic defect groups. It is expected that once a defect group is fixed, there should be only 'finitely many possibilities' for blocks with that defect group. The following conjecture is an instance for this expectation:

Conjecture 6.1.9 (Donovan) *Suppose that k is algebraically closed. For a given finite p-group P there are only finitely many equivalence classes of module categories of block algebras over k with defect groups isomorphic to P.*

Remark 6.1.10 Donovan's conjecture can be formulated for block algebras over $\mathcal{O}$. It is not known whether the Morita equivalence class of a block of kG determines that of the unique corresponding block of $\mathcal{O}G$, not even up to finitely many possibilities. In particular, it is not known whether Donovan's conjecture over k implies the analogous version over $\mathcal{O}$. There is another issue to be aware of. Having chosen k algebraically closed ensures that k is a splitting field for all finite group algebras, while for a fixed ring $\mathcal{O}$ of characteristic zero, the quotient field K cannot be chosen to be algebraically closed, hence will not be a splitting field for all finite groups. A canonical choice for $\mathcal{O}$ as a suitable ring for formulating Donovan's conjecture would be the ring of Witt vectors of an algebraic closure of $\mathbb{F}_p$.

Donovan's conjecture holds for blocks with a cyclic defect group (by work of Janusz [91] and Kupisch [117]), for blocks with a Klein four defect group (by work of Erdmann [66]); for blocks with a dihedral or semi-dihedral defect group, and for 'most' cases of blocks with a generalised quaternion defect group (again by work of Erdmann [67]). This conjecture has further been verified for blocks of certain families of finite groups, such as finite p-solvable groups ([110]), symmetric groups ([203]), alternating groups ([86]), and certain classes of blocks finite groups of Lie type (e.g. [93], [87], [88]). Using the classification of finite simple groups, Koshitani has shown in [107] that Donovan's conjecture holds for principal 3-blocks with an abelian defect group, and Eaton, Kessar, Külshammer and Sambale have shown in [63] that Donovan's conjecture holds for all 2-blocks with an elementary abelian 2-group as a defect group. Since Cartan matrices are invariant under Morita equivalences, Donovan's conjecture would imply the following conjecture.

Conjecture 6.1.11 (Weak Donovan Conjecture) *Suppose that k is algebraically closed. For a given finite p-group P there are only finitely many integers that are entries of Cartan matrices of blocks with defect group isomorphic to P. Equivalently, the entries of the Cartan matrix of a block with defect groups isomorphic to P are bounded by a function depending only on P.*

This weaker version of Donovan's conjecture has been reduced to blocks of quasi-simple finite groups by Düvel [61]. Kessar showed in [94] that the 'gap' between the weak form and the original form of Donovan's conjecture can be formulated as a rationality conjecture; see 6.12.12 below.

6.2 Characterisations of defect groups

There are many ways of characterising defect groups, reflecting the different ways of regarding blocks as idempotents, algebras, or as bimodules. For G a finite group and b a block of $\mathcal{O}G$, the block algebra $B = \mathcal{O}Gb$ is an indecomposable direct factor of $\mathcal{O}G$ as an algebra as well as an indecomposable direct summand of $\mathcal{O}G$ as an $\mathcal{O}G$-$\mathcal{O}G$-bimodule. Via our standard convention, B can be regarded as an indecomposable $\mathcal{O}(G \times G)$-module, with $(x, y) \in G \times G$ acting on $a \in B$ by xay^{-1}.

Theorem 6.2.1 *Let G be a finite group, let b be a block of $\mathcal{O}G$ and let P be a subgroup of G. Set $B = \mathcal{O}Gb$. The following are equivalent.*

   (i) *P is a minimal subgroup such that $b \in (\mathcal{O}G)_P^G$; that is, P is a defect group of b.*
  (ii) *P is a maximal p-subgroup of G such that $\mathrm{Br}_P(b) \neq 0$.*
 (iii) *P is a p-subgroup of G satisfying $b \in (\mathcal{O}G)_P^G$ and $\mathrm{Br}_P(b) \neq 0$.*
  (iv) *ΔP is a vertex of B as an $\mathcal{O}(G \times G)$-module.*
   (v) *P is a minimal subgroup such that the map $B \otimes_{\mathcal{O}P} B \to B$ induced by multiplication in B splits as a homomorphism of B-B-bimodules.*
  (vi) *P is a minimal subgroup such that B is isomorphic to a direct summand of $B \otimes_{\mathcal{O}P} B$ as a B-B-bimodule.*
 (vii) *P is a maximal p-subgroup such that $\mathcal{O}P$ is isomorphic to a direct summand of B as an $\mathcal{O}P$-$\mathcal{O}P$-bimodule.*
(viii) *P is a p-subgroup such that $\mathcal{O}P$ is isomorphic to a direct summand of B as an $\mathcal{O}P$-$\mathcal{O}P$-bimodule and such that B is isomorphic to a direct summand of $B \otimes_{\mathcal{O}P} B$ as a B-B-bimodule.*

The fact that B must have a vertex of the form ΔP for some p-subgroup P of G follows already from 5.11.6. The equivalence of the statements (i), (ii),

(iii) in 6.2.1 follows from the more general result 5.6.7 on G-algebras. In order to address the remaining statements, we prove a lemma dealing with the type of tensor product appearing in statement (v) of 6.2.1. Even though B is an $\mathcal{O}$-algebra, the tensor product $B \otimes_{\mathcal{O}P} B$ is not, in general, an algebra, because the image of $\mathcal{O}P$ in B need not be in the centre of B. But $B \otimes_{\mathcal{O}P} B$ is certainly a B-B-bimodule, or a $\mathcal{O}(G \times G)$-module, with $(x, y) \in G \times G$ acting an $B \otimes_{\mathcal{O}P} B$ by sending $c \otimes c'$ to $xc \otimes c'y^{-1}$. Since b is a central idempotent, $B \otimes_{\mathcal{O}P} B$ is a direct summand of $\mathcal{O}G \otimes_{\mathcal{O}P} \mathcal{O}G$ as an $\mathcal{O}(G \times G)$-module. The following lemma generalises 5.11.6.

Lemma 6.2.2 *Let G be a finite group and let P be a subgroup of G. We have an isomorphism of $\mathcal{O}(G \times G)$-modules $\mathcal{O}G \otimes_{\mathcal{O}P} \mathcal{O}G \cong \mathrm{Ind}_{\Delta P}^{G \times G}(\mathcal{O})$ mapping $x \otimes y$ to $(x, y^{-1}) \otimes 1$ for all $x, y \in G$.*

Proof The statement makes sense: we have $\mathrm{Ind}_{\Delta P}^{G \times G}(\mathcal{O}) = \mathcal{O}(G \times G) \otimes_{\mathcal{O}\Delta P} \mathcal{O}$, and for any $x, y \in G$ and $u \in P$, the image of $xu \otimes y$ is $(xu, y^{-1}) \otimes 1 = (x, y^{-1}u^{-1})(u, u) \otimes 1 = x, y^{-1}u^{-1} \otimes (u, u)1 = (x, (uy)^{-1}) \otimes 1$, and that is equal to the image of $x \otimes uy$. An element $(g, h) \in G \times G$ acts on $\mathcal{O}G \otimes_{\mathcal{O}P} \mathcal{O}G$ by sending $x \otimes y$ to $gx \otimes yh^{-1}$, and it acts on $\mathrm{Ind}_{\Delta P}^{G \times G}(\mathcal{O})$ by sending $(x, y^{-1}) \otimes 1$ to $(gx, hy^{-1}) \otimes 1$, which is the image of $gx \otimes yh^{-1}$. Thus the given map is a homomorphism of $\mathcal{O}(G \times G)$-modules. Its inverse sends $(x, y^{-1}) \otimes 1$ to $x \otimes y$. $\square$

The next result will imply the equivalence of the statements (i), (iv), (v), (vi) in 6.2.1; it holds not just for blocks but arbitrary central idempotents.

Proposition 6.2.3 *Let G be a finite group, let b be an idempotent in $Z(\mathcal{O}G)$ and let P be a subgroup of G. Set $B = \mathcal{O}Gb$. The following are equivalent.*

(i) We have $b \in (\mathcal{O}G)_P^G$.
(ii) B is relatively ΔP-projective as an $\mathcal{O}(G \times G)$-module.
(iii) The map $\mu : B \otimes_{\mathcal{O}P} B \to B$ induced by multiplication in B splits as a homomorphism of B-B-bimodules.

Proof We show first that (i) implies (iii). Suppose (i) holds. Then $b = \mathrm{Tr}_P^G(c)$ for some $c \in (\mathcal{O}G)^P$. Since b is an idempotent we have $b = b^2 = \mathrm{Tr}_P^G(c)b = \mathrm{Tr}_P^G(cb)$. Thus, after replacing c by cb we may assume that c belongs to B. Set $\sigma(b) = \sum_{y \in [G/P]} yb \otimes cy^{-1}$. This expression is independent of the choice of $[G/P]$; indeed, if $y' = yu$ for some $y, y' \in G$ and some $u \in P$ then $y'b \otimes c(y')^{-1} = yub \otimes cu^{-1}y^{-1} = ybuu^{-1} \otimes cy^{-1} = yb \otimes cy^{-1}$ because u commutes with both b, c and the tensor product is taken over $\mathcal{O}P$. We show next that $a\sigma(b) = \sigma(b)a$ for all $a \in \mathcal{O}Gb$. Let $x \in G$. If y runs over $[G/P]$ then xy runs over a set of representatives of the right P-cosets in G as well, and so

$$x\sigma(b) = \sum_{y\in[G/P]} xyb \otimes cy^{-1} = \sum_{y\in[G/P]} xyb \otimes cy^{-1}x^{-1}x = \sigma(b)x.$$

It follows that $a\sigma(b) = \sigma(b)a$ for all $a \in B$. Thus we can define a homomorphism of $\mathcal{O}(G \times G)$-modules $\sigma : B \to B \otimes_{\mathcal{O}P} B$ by setting $\sigma(a) = a\sigma(b)$ for all $a \in B$. We check that σ is a section as claimed: for any $a \in B$ we have

$$\mu(\sigma(a)) = \mu\Big(\sum_{y\in[G/P]} ayb \otimes cy^{-1} \Big) = \sum_{y\in[G/P]} aybcy^{-1} = a\mathrm{Tr}_P^G(c) = ab = a.$$

Thus (i) implies (iii). We show next that (iii) implies (ii). Suppose that (iii) holds. Then B is a direct summand of $B \otimes_{\mathcal{O}P} B$ as an $\mathcal{O}(G \times G)$-module, thus of $\mathcal{O}G \otimes_{\mathcal{O}P} \mathcal{O}G \cong \mathrm{Ind}_{\Delta P}^{G\times G}(\mathcal{O})$, by 6.2.2. Hence B is relatively $\mathcal{O}\Delta P$-projective. This shows that (ii) is a consequence of (iii). Suppose finally that (ii) holds. Since $\Delta P \subseteq G \times P$ it follows that B is relatively $G \times P$-projective as an $\mathcal{O}(G \times G)$-module. This means that there is $\varphi \in \mathrm{End}_{\mathcal{O}(G\times P)}(B)$ such that $\mathrm{Id}_B = \mathrm{Tr}_{G\times P}^{G\times G}(\varphi)$. Now the $\mathcal{O}(G \times P)$-endomorphism φ of B is in particular an endomorphism as left B-module, hence is given by right multiplication with an element in B; say $\varphi(a) = ac$ for all $a \in B$. But φ commutes also with the right action of $\mathcal{O}P$ on B, so $\varphi(au) = \varphi(a)u$ for all $u \in P$. This is equivalent to $auc = acu$ for all $a \in B$ and all $u \in P$. In particular, considering this for $a = b$ yields that $c \in B^P$. Now if x runs over a set of representatives of G/P in G then $(1, x)$ runs over a set of representatives of $G \times G/G \times P$ in $G \times G$. Thus $\mathrm{Id}_B = \mathrm{Tr}_{G\times P}^{G\times G}(\varphi) = \sum_{x\in[G/P]} {}^{(1,x)}\varphi$. For any $x \in G$ and any $a \in B$ we have $({}^{(1,x)}\varphi)(a) = (1, x)\varphi((1, x^{-1})a) = \varphi(ax)x^{-1} = axcx^{-1}$. It follows that $a = \mathrm{Id}_B(a) = a\mathrm{Tr}_P^G(c)$. But that is only possible if $\mathrm{Tr}_P^G(c)$ is the unit element in B, hence if $\mathrm{Tr}_P^G(c) = b$. Thus (ii) implies (i). $\square$

Proof of Theorem 6.2.1 The equivalence of (i), (ii), (iii) is a particular case of 5.6.7. The equivalence of (iii) and (vii) follows from 5.8.8. The equivalence of the statements (i), (iv), (v), is a direct consequence of 6.2.3. The equivalence of (v) and (vi) follows from 2.6.13. Statement (viii) follows from combining the (equivalent) statements (vi) and (vii). If (viii) holds, then $\mathcal{O}P$ is isomorphic to a direct summand of B, as an $\mathcal{O}P$-$\mathcal{O}P$-bimodule. Thus $B(P)$ is nonzero, or equivalently $\mathrm{Br}_P(b) \neq 0$. Moreover, if (viii) holds, then B is isomorphic to a direct summand of $B \otimes_{\mathcal{O}P} B$ as a B-B-bimodule, and hence 6.2.3 implies that $b \in (\mathcal{O}G)_P^G$. Thus (viii) implies (iii). $\square$

Corollary 6.2.4 *Let G be a finite group, b a block of $\mathcal{O}G$, and P a defect group of b. Set $B = \mathcal{O}Gb$. The block algebra B and the defect group algebra $\mathcal{O}P$ are separably equivalent.*

Proof Set $M = B$, viewed as a B-$\mathcal{O}P$-bimodule. Then $M^* \cong B$ viewed as an $\mathcal{O}P$-B-bimodule, by the symmetry of B. By 6.2.1 (v) or (viii), the B-B-bimodule B is isomorphic to a direct summand of $M \otimes_{\mathcal{O}P} M^*$. By 6.2.1 (vii) or (viii), the $\mathcal{O}P$-$\mathcal{O}P$-bimodule $\mathcal{O}P$ is isomorphic to a direct summand of $B \cong M^* \otimes_B M$. The result follows. □

It is not clear whether $\mathcal{O}P$ being separably equivalent to B characterises the isomorphism class of P. In other words, it is not clear whether two finite p-groups with separably equivalent group algebras have to be isomorphic. If true, this would imply that Morita equivalent blocks have isomorphic defect groups.

Proposition 6.2.5 *Let G be a finite group, let b be a block of $\mathcal{O}G$ and let P be a minimal subgroup of G such that $b \in (\mathcal{O}G)_P^G$.*

(i) We have $\mathrm{Br}_P(b) \neq 0$.

(ii) For any subgroup Q such that $\mathrm{Br}_Q(b) \neq 0$ there is $x \in G$ such that $Q \subseteq {}^xP$.

Proof This is 5.6.6 applied to the primitive G-algebra $\mathcal{O}Gb$. □

The next result shows that defect groups of any block of a finite group G contain the largest normal p-subgroup $O_p(G)$ of G.

Theorem 6.2.6 *Let G be a finite group, P a normal p-subgroup of G and b a block of $\mathcal{O}G$.*

(i) Any defect group of b contains P.

(ii) We have $b \in (\mathcal{O}C_G(P))^G$.

(iii) Let e be a block of $\mathcal{O}C_G(P)$ such that $eb = e$, and denote by H the stabiliser of e in G. Then e remains a block of $\mathcal{O}H$, and b is equal to the sum of all different G-conjugates of e; equivalently, $b = \mathrm{Tr}_H^G(e)$. Moreover, the $\mathcal{O}Gb$-$\mathcal{O}He$-bimodule $\mathcal{O}Ge$ and its dual $e\mathcal{O}G$ induce a Morita equivalence between $\mathcal{O}Gb$ and $\mathcal{O}Hc$.

(iv) Let i be a primitive idempotent i in $\mathcal{O}C_G(P)$. Then i remains primitive in $(\mathcal{O}G)^P$ and satisfies $\mathrm{Br}_P(i) \neq 0$.

Proof The canonical map $\mathcal{O}G \to kG$ maps $(\mathcal{O}C_G(P))^G$ onto $(kC_G(P))^G$, and hence we may assume that $\mathcal{O} = k$. Write $kG = kC_G(P) \oplus k[G \setminus C_G(P)]$. Since P is normal in G, so is $C_G(P)$, and hence this direct sum decomposition is stable with respect to the action of G by conjugation. Thus $Z(kG) = (kG)^G = (kC_G(P))^G \oplus (\ker(\mathrm{Br}_P))^G$. It follows from 5.4.4 that $(\ker(\mathrm{Br}_P))^G$ is a nilpotent ideal in $Z(kG)$. In particular, $(\ker(\mathrm{Br}_P))^G$ contains no block idempotent, whence (i). If we write $b = c + d$ with $c \in (kC_G(P))^G$ and $d \in (\ker(\mathrm{Br}_P))^G$ then, for any integer $n > 0$, we have $b^{p^n} = c^{p^n} + d^{p^n}$, and since d is nilpotent, we get that $b = c^{p^n}$ for n large enough, which proves (ii). Since $(kC_G(P))^G \subseteq Z(kC_G(P))$

this implies that b is a sum of blocks of $kC_G(P)$. If e is a block of $kC_G(P)$ then either $eb = 0$ or $eb = e$. Note that ${}^x e$ is also a block of $kC_G(P)$ for any $x \in G$. Since $b = {}^x b$ we get that if $eb = e$ then also ${}^x eb = {}^x e$. Therefore, if c is the sum of all G-conjugates of e, then $cb = c$, and $c = \mathrm{Tr}_H^G(e)$. But then $b = cb + (b - cb)$ which forces $b - cb = 0$ by the primitivity of b in $Z(kG)$. Thus $b = cb = c$. Note that $b \in \mathcal{O}Gbe\mathcal{O}Gb$ and $e \in \mathcal{O}Gb$. It follows from 2.8.7 that $\mathcal{O}Gb$ and $\mathcal{O}He$ are Morita equivalent via the bimodules as stated. This proves (iii). Similarly, we have $(kG)^P = kC_G(P) \oplus (\ker(\mathrm{Br}_P))$ and $\ker(\mathrm{Br}_P) \subseteq J((kG)^P)$ by 5.4.4. Thus a primitive idempotent in $kC_G(P)$ remains primitive in $(kG)^P$ by 4.7.10. Since i belongs to $kC_G(P)$, it follows that $\mathrm{Br}_P(i) = i \neq 0$ as stated. □

We will see in 6.8.3 that the Morita equivalence in 6.2.6 (iii) preserves defect groups and that this is a special case of a more general equivalence.

Corollary 6.2.7 *Let G be a finite group having a p-subgroup P such that $G = PC_G(P)$. The sets block idempotents of $\mathcal{O}C_G(P)$ and of $\mathcal{O}G$ are equal.*

Proof Since $G = PC_G(P)$ it follows that $(\mathcal{O}C_G(P))^G = Z(\mathcal{O}C_G(P))$, and hence the result is an immediate consequence of 6.2.6 (ii). □

Corollary 6.2.8 *Let G be a finite group having a normal p-subgroup P such that $C_G(P) = Z(P)$. Then the group algebra $\mathcal{O}G$ has a unique block.*

Proof The algebra $\mathcal{O}C_G(P) = \mathcal{O}Z(P)$ is local, hence has a unique block. By 6.2.6, every block idempotent of $\mathcal{O}G$ is contained in $\mathcal{O}C_G(P)$. The result follows. □

Using the classification of finite simple groups, Harris has shown in [79] that for p odd the finite groups with a unique p-block are as in this corollary, while for $p = 2$ the group algebras $\mathcal{O}M_{22}$ and $\mathcal{O}M_{24}$ are indecomposable (and, more precisely, M_{22}, M_{24} are exactly the possible components of finite groups having a unique 2-block). We note the following special case for future reference:

Corollary 6.2.9 *Let P be a finite p-group, and let E be a p'-subgroup of $\mathrm{Aut}(P)$. Then $\mathcal{O}(P \rtimes E)$ is a primitive interior P-algebra; in particular, $\mathcal{O}(P \rtimes E)$ has a unique block.*

Proof Since E is a p'-subgroup of $\mathrm{Aut}(P)$, an elementary argument implies that $C_{P \rtimes E}(P) = Z(P)$. Since $\mathcal{O}Z(P)$ is local, it follows from 6.2.6 that 1 remains primitive in $(\mathcal{O}(P \rtimes E))^{\Delta P}$. The second statement of 6.2.9 is also a special case of 6.2.8. □

By combining earlier results, the order of a defect group can be read off the $\mathcal{O}$-stable category of a block.

Proposition 6.2.10 *Let G be a finite group, B a block algebra of $\mathcal{O}G$, and P a defect group of B. Suppose that* char(K) $= 0$, *and let $\pi \in \mathcal{O}$ such that $J(\mathcal{O}) = \pi\mathcal{O}$. Let d be the smallest positive integer such that π^d annihilates $\underline{\mathrm{Hom}}_B(U, V)$ for any two finitely generated B-modules U, V. Then $\pi^d\mathcal{O} = |P|\mathcal{O}$.*

Proof The integer d is an invariant of separable equivalences. Since B and $\mathcal{O}P$ are separably equivalent, we may assume that $G = P$ and $B = \mathcal{O}P$. If P is non-trivial, the result follows from 4.13.20. If P is trivial, then so is its $\mathcal{O}$-stable category, and hence $d = 0$ in that case. □

Corollary 6.2.11 *Let G be a finite group, B a block algebra of $\mathcal{O}G$, and P a defect group of B. Suppose that* char(K) $= 0$. *Then $|P|$ is the smallest positive integer such that $|P| \cdot \underline{\mathrm{End}}_{B\otimes_\mathcal{O} B^{op}}(B) = \{0\}$.*

Proof Let d be the smallest positive integer such that π^d annihilates $\underline{\mathrm{End}}_{B\otimes_\mathcal{O} B^{op}}(B)$. By 4.13.16, π^d annihilates $\underline{\mathrm{Hom}}_B(U, V)$ for all finitely generated B-modules. It follows from 6.2.10 that $|P|$ divides π^d in $\mathcal{O}$. By 6.2.1, ΔP is a vertex of B as an $\mathcal{O}(G \times G)$-module. It follows from 4.13.22 that $|P|$ annihilates $\underline{\mathrm{End}}_{B\otimes_\mathcal{O} B^{op}}(B)$. Thus $|P|\mathcal{O} = \pi^d\mathcal{O}$, which implies the result. □

6.3 Brauer pairs for finite group algebras

Brauer considered pairs (Q, e) consisting of a p-subgroup of a finite group G and a block e of $kQC_G(Q)$ with defect group Q. Alperin and Broué extended this in [5, 3.1], laying the foundations for associating fusion systems to blocks in work of Puig in the early 1990s; see §8.5 below. The concept of Brauer pairs was further extended to p-permutation G-algebras and related to local pointed groups in work of Broué and Puig [40]. We have developed this material in this generality in §5.8; we review this here specialised to finite group algebras, allowing for some redundancy for convenience. By 6.2.7 the blocks of $kC_G(Q)$ and of $k\mathcal{Q}C_G(Q)$ correspond bijectively to each other.

Definition 6.3.1 Let G be a finite group, let b be a block of $\mathcal{O}G$, and set $B = \mathcal{O}Gb$. A *Brauer pair* on $\mathcal{O}G$ is a pair (P, e) consisting of a p-subgroup P of G and a block e of $kC_G(P)$. A (G, b)-*Brauer pair* or B-*Brauer pair* is a Brauer pair (P, e) on $\mathcal{O}G$ with the additional property $\mathrm{Br}_P(b)e = e$.

Thus a (G, b)-Brauer pair (P, e) is a (G, B)-Brauer pair in the sense of 5.9.1 specialised to the case where $B = \mathcal{O}Gb$ and where we identify

$B(P) = kC_G(P)\mathrm{Br}_P(b)$. As mentioned in the general case, if $\mathrm{Br}_P(b) \neq 0$, then P is contained in a defect group of b and $\mathrm{Br}_P(b)$ is a sum of block idempotents of $kC_G(P)$. Thus the existence of a (G, b)-Brauer pair (P, e) implies that P is contained in a defect group of b.

The set of Brauer pairs on $\mathcal{O}G$ is a G-set with respect to the action of G by conjugation: if (P, e) is a Brauer pair on $\mathcal{O}G$ and if $x \in G$ then clearly the pair $^x(P, e) = (^xP, \, ^xe)$ is a Brauer pair, and if (P, e) is a (G, b)-Brauer pair, so is $^x(P, e)$. The set of Brauer pairs on a group algebra $\mathcal{O}G$ is partially ordered; the inclusion of Brauer pairs in 5.9.8 specialises to this case as follows.

Definition 6.3.2 Let G be a finite group, let (P, e), (Q, f) be Brauer pairs on $\mathcal{O}G$. We write $(Q, f) \leq (P, e)$ and say that (Q, f) *is contained in* (P, e) if $Q \leq P$ and if there is a primitive idempotent $i \in (\mathcal{O}G)^P$ satisfying $\mathrm{Br}_P(i)e \neq 0$ and $\mathrm{Br}_Q(i)f \neq 0$.

Theorem 6.3.3 ([5, Theorem 3.4], [40, 1.8]) *Let G be a finite group, let (P, e) be a Brauer pair on $\mathcal{O}G$ and let Q be a subgroup of P. There is a unique block f of $kC_G(Q)$ such that $(Q, f) \leq (P, e)$.*

Proof This is a special case of 5.9.8. $\qquad\square$

In this way the set of Brauer pairs becomes a partially ordered set. The partial order is compatible with the G-action by conjugation, so this is a partially ordered G-set (or G-poset, for short). If Q is normal in P, then the inclusion of Brauer pairs can be characterised as follows (this is the approach in [5]):

Proposition 6.3.4 *Let G be a finite group, let (P, e) be a Brauer pair on $\mathcal{O}G$ and let Q be a normal subgroup of P. The unique block f of $kC_G(Q)$ such that $(Q, f) \leq (P, e)$ is the unique P-stable block of $kC_G(Q)$ satisfying $\mathrm{Br}_P(f)e = e$.*

Proof This is a special case of 5.9.9. $\qquad\square$

The poset of all Brauer pairs on $\mathcal{O}G$ is not connected in general; we will see that the connected components are in bijection with the blocks of $\mathcal{O}G$ that hence every connected component is also a G-poset.

Proposition 6.3.5 *Let G be a finite group and let (P, e) be a Brauer pair on $\mathcal{O}G$. There is a unique block b of $\mathcal{O}G$ such that (P, e) is a (G, b)-Brauer pair, and this is the unique block satisfying $(1, b) \leq (P, e)$.*

Proof Since $1_{\mathcal{O}G} = \sum_{b \in \mathcal{B}} b$, where $\mathcal{B}$ is the set of blocks of $\mathcal{O}G$, we get $1_{kC_G(P)} = \sum_{b \in \mathcal{B}} \mathrm{Br}_P(b)$, and each $\mathrm{Br}_P(b)$ is either zero or a sum of blocks of $kC_G(P)$. Thus there is a unique block b of $\mathcal{O}G$ such that $be = e$ and such that

$b'e = 0$ for every other block b' of $\mathcal{O}G$. Clearly $(1, b) \leq (P, e)$, and this characterises b by 6.3.3. □

Proposition 6.3.6 *Let G be a finite group, let b be a block of $\mathcal{O}G$ and let (P, e), (Q, f) be Brauer pairs on $\mathcal{O}G$ such that $(Q, f) \leq (P, e)$. Then (P, e) is a (G, b)-Brauer pair if and only if (Q, f) is a (G, b)-Brauer pair.*

Proof Let i be a primitive idempotent in $(\mathcal{O}G)^P$ such that $\mathrm{Br}_P(i)e \neq 0$. The primitivity of i implies that i belongs to a unique block algebra of $\mathcal{O}G$. We have $\mathrm{Br}_Q(i)f \neq 0$ by the definition of the inclusion of Brauer pairs. If (P, e) is a b-Brauer pair then $\mathrm{Br}_P(b)e = e$, so necessarily $i \in (\mathcal{O}Gb)^P$. Since $\mathrm{Br}_P(i)f \neq 0$ this implies $\mathrm{Br}_P(b)f \neq 0$, hence (Q, f) is a (G, b)-Brauer pair. Conversely, if (Q, f) is a (G, b)-Brauer pair, then a similar reasoning shows that $i \in (\mathcal{O}Gb)^P$ and hence that (P, e) is a (G, b)-Brauer pair. □

Theorem 6.3.7 *Let G be a finite group, let b be a block of $\mathcal{O}G$ and let (P, e) be a maximal (G, b)-Brauer pair. Then P is a defect group of b and for any (G, b)-Brauer pair (Q, f) there is $x \in G$ such that $^x(Q, f) \leq (P, e)$. In particular, all maximal (G, b)-Brauer pairs are G-conjugate.*

Proof This is a special case of 5.9.11. □

Remark 6.3.8 Let G be a finite group and H a subgroup of G. Let Q be a p-subgroup of H such that $C_G(Q) \subseteq H$, and let f be a block of $kC_G(Q)$. Then (Q, f) is a Brauer pair on $\mathcal{O}H$ and on $\mathcal{O}G$. Thus (Q, f) determines unique blocks c of $\mathcal{O}H$ and b of $\mathcal{O}G$ such that (Q, f) is an (H, c)-Brauer pair and a (G, b)-Brauer pair. We say in that case that b *corresponds to* c. The block b is in that case uniquely determined by c. Indeed, if (P, e) is a maximal (H, c)-Brauer pair such that $(Q, f) \leq (P, e)$, then $C_G(P) \subseteq H$, and hence (P, e) is also a Brauer pair on $\mathcal{O}G$. It follows from 6.3.6 that (Q, f) and (P, e) determine the same block b of $\mathcal{O}G$. Since the maximal (H, c)-Brauer pairs are H-conjugate, it follows that c determines b in this way uniquely. In general, c is not uniquely determined by b. Brauer's first main Theorem 6.7.6 below considers a situation where this correspondence does yield a bijection between certain subsets of blocks of $\mathcal{O}G$ and $\mathcal{O}H$.

The following result is a generalisation to Brauer pairs of an elementary group theoretic fact: if Q is a subgroup of a Sylow p-subgroup P of a finite group G, then $C_P(Q)$ need not be a Sylow p-subgroup of $C_G(Q)$, but for some G-conjugate of Q contained in P this will be true; similarly for normalisers. We will reformulate this later in the context of *fusion systems of blocks* in terms of *fully centralised* and *fully normalised* subgroups; see 8.5.3 below.

Theorem 6.3.9 *Let G be a finite group, b a block of G and (P, e) a maximal (G, b)-Brauer pair. Let Q be a subgroup of P and f the block of $kC_G(Q)$ such that $(Q, f) \le (P, e)$.*

(i) *$C_P(Q)$ is contained in a defect group of $kC_G(Q)f$, and there is $x \in G$ such that $^x(Q, f) \le (P, e)$ and such that $C_P(^xQ)$ is a defect group of $kC_G(^xQ)^xf$.*

(ii) *$N_P(Q)$ is contained in a defect group of $kN_G(Q, f)f$, and there is $x \in G$ such that $^x(Q, f) \le (P, e)$ and such that $N_P(^xQ)$ is a defect group of $kN_G(^xQ, ^xf)^xf$.*

Proof Let i be a primitive idempotent in $(\mathcal{O}Gb)^P$ such that $\mathrm{Br}_P(i)e = \mathrm{Br}_P(i) \ne 0$. Then, by 6.3.3, f is the unique block of $kC_G(Q)$ such that $\mathrm{Br}_Q(i) \in kC_G(Q)e_Q$. Since $\mathrm{Br}_P(i) \ne 0$ we also have $\mathrm{Br}_{C_P(Q)}(\mathrm{Br}_Q(i)) = \mathrm{Br}_{QC_P(Q)}(i) \ne 0$. Thus $\mathrm{Br}_{C_P(Q)}(e_Q) \ne 0$, which shows that $C_P(Q)$ is contained in a defect group of e_Q as a block of $kC_G(Q)$. Let (R, g) be a maximal $(C_G(Q), f)$-Brauer pair. The subgroup R of $C_G(Q)$ is a defect group of $kC_G(Q)f$ and g is a block of $C_{C_G(Q)}(R) = C_G(QR)$, hence (QR, g) is a Brauer pair on $\mathcal{O}G$. Moreover, $g = g\mathrm{Br}_R(f)$, hence $(Q, f) \le (QR, g)$, which shows in particular that (QR, g) is in fact a (G, b)-Brauer pair. Thus there is $x \in G$ such that $^x(QR, g) \le (P, e)$. Since R centralises Q we have $^xR \le C_P(^xQ)$, hence equality by the first part of the argument. This proves (i). For (ii), note first that the statement makes sense: f remains a block of $kN_G(Q, f)$ by 6.2.6, and $N_P(Q) \le N_G(Q, f)$ by the uniqueness of the inclusion of Brauer pairs. Arguing as before, we have $\mathrm{Br}_{N_P(Q)}(\mathrm{Br}_Q(i)) = \mathrm{Br}_{QN_P(Q)}(i) \ne 0$, and hence $N_P(Q)$ is contained in a defect group of $kN_G(Q, f)f$. Let now (R, g) be a maximal $(N_G(Q, f), f)$-Brauer pair. The subgroup R of $N_G(Q, f)$ is a defect group of $kN_G(Q, f)f$ and g is a block of $C_{N_G(Q,f)}(R)$ satisfying $g = g\mathrm{Br}_R(f)$. Note that $C_G(QR) \le C_{N_G(Q,f)}(R) \le N_G(QR)$. Thus g is a sum of blocks of $kC_G(QR)$, and any such block g_1 will still satisfy $g_1 = g_1\mathrm{Br}_R(f)$. As before, this shows that (QR, g_1) is a (G, b)-Brauer pair containing (Q, f), and thus there is $x \in G$ such that $^x(QR, g_1) \le (P, e)$. Since R normalises Q, it follows that $^xR \le N_P(^xQ)$, and hence equality by the first part of the argument. $\square$

Corollary 6.3.10 *Let G be a finite group, b a block of G and (P, e) a maximal (G, b)-Brauer pair. Then $Z(P)$ is a defect group of $kC_G(P)e$, and P is a defect group of $kN_G(P, e)e$.*

Proof This is an immediate consequence of 6.3.9. $\square$

Corollary 6.3.11 *Let G be a finite group, b a block of G and (P, e) a maximal (G, b)-Brauer pair. Let Q be a subgroup of $Z(P)$ and let e_Q the block of $kC_G(Q)$*

such that $(Q, e_Q) \leq (P, e)$. *Then* P *is a defect group of* e_Q *as a block of* $kC_G(Q)$ *and of* $kN_G(Q, e_Q)$.

Proof Since Q is a subgroup of $Z(P)$ we have $C_P(Q) = N_P(Q) = P$. By 6.3.9, P is contained in a defect group of e_Q as a block of $kC_G(Q)$ and of $kN_G(Q, e_Q)$. Again by 6.3.9, the defect groups of $kC_G(Q)e_Q$ and $kN_G(Q)e_Q$ have order at most $|P|$. The result follows. □

Corollary 6.3.12 *Let G be a finite group, b a block of G and (P, e) a maximal (G, b)-Brauer pair. Suppose that P is abelian. Let Q be a subgroup of P and let e_Q the block of $kC_G(Q)$ such that $(Q, e_Q) \leq (P, e)$. Then P is a defect group of e_Q as a block of $kC_G(Q)$ and of $kN_G(Q, e_Q)$.*

Proof This is a special case of 6.3.11. □

We will prove slightly more general statements in 8.5.3, 8.5.4, and 8.5.7 in terms of fusion systems of blocks.

Definition 6.3.13 Let G be a finite group, b a block of $\mathcal{O}G$ and P a defect group of b. We say that b *is a block of principal type* if $\mathrm{Br}_Q(b)$ is a block of $kC_G(Q)$ for every subgroup Q of P.

If b is a block of principal type of $\mathcal{O}G$, then the G-poset of (G, b)-Brauer pairs is isomorphic to the G-poset of p-subgroups Q of G satisfying $\mathrm{Br}_Q(b) \neq 0$, and we have $N_G(Q, e) = N_G(Q)$ for any (G, b)-Brauer pair (Q, e). Principal blocks are of principal type:

Theorem 6.3.14 (Brauer's Third Main Theorem) *Let G be a finite group and b the principal block of $\mathcal{O}G$. Then, for any p-subgroup Q of G, $\mathrm{Br}_Q(b)$ is the principal block of $kC_G(Q)$. Equivalently, if (Q, e) is a (G, b)-Brauer pair then e is the principal block of $kC_G(Q)$. In particular, the principal block is a block of principal type.*

Proof We may assume $\mathcal{O} = k$. Let Q be a p-subgroup of G. Denote by $\eta_G : kG \to k$ and $\eta_{C_G(Q)} : kC_G(Q) \to k$ the augmentation maps. Note that $\ker(\mathrm{Br}_Q) \subseteq \ker(\eta_G)$. Thus η_G restricted to $(kG)^Q$ is equal to the composition of the Brauer homomorphism $\mathrm{Br}_Q : (kG)^Q \to kC_G(Q)$ followed by the augmentation map $\eta_{C_G(Q)}$. Since b is the principal block we have $\eta_G(b) \neq 0$ hence $\eta_{C_G(Q)}(\mathrm{Br}_Q(b)) \neq 0$. This means that the principal block of $kC_G(Q)$ occurs in $\mathrm{Br}_Q(b)$. For P a Sylow p-subgroup of G all blocks occurring in $\mathrm{Br}_P(b)$ are $N_G(P)$-conjugate. Since the principal block of $kC_G(P)$ is $N_G(P)$-stable it follows that $e_P = \mathrm{Br}_P(b)$ is the principal block of $C_G(P)$. If (Q, e) is a (G, b)-Brauer pair such that $(Q, e) \leq (P, e_P)$ then, on one hand, e is uniquely determined by (P, e_P) as follows: let $i \in (\mathcal{O}G)^P$ be a primitive idempotent such that $\mathrm{Br}_P(i)$ is a primitive idempotent in $kC_G(P)$ corresponding to the trivial $kC_G(P)$-module, or equivalently, satisfying $\eta_{C_G(P)}(\mathrm{Br}_P(i)) \neq 0$. By the uniqueness of the inclusion of Brauer pairs, we have $\mathrm{Br}_Q(i) \in kC_G(Q)e$. We also have $\eta_{C_G(Q)}(\mathrm{Br}_Q(i)) \neq 0$, and so $\mathrm{Br}_Q(i)$ is an idempotent in $kC_G(Q)$ in which appears a primitive idempotent corresponding to the trivial $kC_G(Q)$-module. This forces e to be the principal block of $kC_G(Q)$. Then, for any $x \in G$, the block ^xe is the principal block of $kC_G(^xQ)$ because the isomorphism $kC_G(Q) \cong kC_G(^xQ)$ given by conjugation with x preserves augmentation maps. Since any (G, b)-Brauer pair is conjugate to a subpair of (P, e_P) the result follows.   □

6.4 Source algebras of blocks

Source algebras of blocks, introduced by Puig in [169], provide a notion of equivalence between blocks that is strong enough to preserve not only the equivalence class of the module category of a block but also invariants that are not known to be determined by the module category, such as the defect groups of a block, vertices and sources of modules, and generalised decomposition numbers. In addition, source algebras tend to be 'smaller' than the block algebras themselves, with some key properties that hold for principal blocks: source algebras satisfy a version of Brauer's Third Main Theorem (a fact that follows from 5.9.5), and as a consequence, one can read off the fusion systems of a block from its source algebras, while the block algebra itself may be 'too big' for that. The theme of fusion systems of source algebras will be developed in §8.7. As we have seen in 5.6.12, source algebras can be defined more generally for primitive G-algebras. When specialised to block algebras, this definition reads as follows.

Definition 6.4.1 (cf. [169, 3.2]) Let G be a finite group, let b be a block of $\mathcal{O}G$ and let P be a defect group of b. A primitive idempotent $i \in (\mathcal{O}Gb)^P$ satisfying $\mathrm{Br}_P(i) \neq 0$ is called a *source idempotent of* b. If $i \in (\mathcal{O}Gb)^P$ is a source idempotent, the interior P-algebra $i\mathcal{O}Gi$, with structural homomorphism sending $u \in P$ to $ui = iui = iu$ is called a *source algebra of* b. The $\mathcal{O}(G \times P)$-module $\mathcal{O}Gi$ is called a *source module of* b.

The following conjecture is another instance of the expectation that block algebras should be determined by their defect groups up to finitely many possibilities.

Conjecture 6.4.2 (Puig's Conjecture, 1982) *For a given finite p-group P there are only finitely many isomorphism classes of interior P-algebras that occur as source algebras of blocks of finite groups with P as a defect group.*

Puig's conjecture is known to hold in full generality when P is cyclic ([127]) or a Klein four group ([51]). The Klein four defect case requires the classification of finite simple groups. There are many partial results whereby one restricts the attention to blocks of certain classes of finite groups rather than all finite groups. For instance, Puig's conjecture has been verified for the blocks of symmetric groups, alternating groups, their central extensions, and various classes of finite groups of Lie type. By 5.6.13 or 6.4.6 below, the module category of a block algebra is equivalent to that of any of its source algebras. Thus Puig's conjecture implies Donovan's conjecture 6.1.9.

The map sending a source idempotent i to $\mathcal{O}Gi$ induces a bijection between local points of P on $\mathcal{O}Gb$ and isomorphism classes of direct summands with diagonal vertex ΔP of the $\mathcal{O}(G \times P)$-module $\mathcal{O}Gb$; this is an easy consequence 5.1.15 and Higman's criterion. The connection between source algebras and source modules is given via the isomorphism $i\mathcal{O}Gi \cong \mathrm{End}_{\mathcal{O}G}(\mathcal{O}Gi)^{op}$ sending $a \in i\mathcal{O}Gi$ to right multiplication by a on $\mathcal{O}Gi$. By its very construction, the source algebra $i\mathcal{O}Gi$ contains an image Pi of the defect group P as part of its structure, and the previous algebra isomorphism is an isomorphism of interior P-algebras. The source idempotent i is not unique, but we will see that the isomorphism class of $i\mathcal{O}Gi$ does not depend on the choice of i, up to precomposing the structural homomorphism by an automorphism of P. The condition $\mathrm{Br}_P(i) \neq 0$ implies that $\mathrm{Br}_P(i)$ is a primitive idempotent in $kC_G(P)$ and by 6.3.3, for each subgroup Q of P there is a unique block e_Q of $kC_G(Q)$ such that $\mathrm{Br}_Q(i) \in kC_G(Q)e_Q$. In many arguments the fact that i is primitive is irrelevant – what matters is that i determines a unique block of $kC_G(Q)$ for each subgroup Q of P, and the following terminology captures this.

Definition 6.4.3 ([136]) Let G be a finite group, b be a block of $\mathcal{O}G$ and P be a defect group of b. An idempotent $i \in (\mathcal{O}Gb)^P$ satisfying $\mathrm{Br}_P(i) \neq 0$ is called an *almost source idempotent of b* if for any subgroup Q of P there is a unique block e_Q of $kC_G(Q)$ such that $\mathrm{Br}_Q(i)e_Q = \mathrm{Br}_Q(i)$. The interior P-algebra $i\mathcal{O}Gi$, with structural homomorphism sending $u \in P$ to $ui = iui = iu$ is called an *almost source algebra of b*.

Replacing a block algebra $\mathcal{O}Gb$ by one of its almost source algebras $i\mathcal{O}Gi$ replicates a fundamental property of principal blocks from Brauer's Third Main Theorem 6.3.14: we will see in 8.7.3 that after possibly replacing a subgroup Q of a defect group P by a suitable conjugate, the algebra $(i\mathcal{O}Gi)(Q) \cong \mathrm{Br}_Q(i)kC_G(Q)\mathrm{Br}_Q(i)$ is Morita equivalent to $kC_G(Q)e_Q$, hence in particular indecomposable. The block idempotent b itself is an almost source idempotent if and only if b is a block of principal type. As a left or right $\mathcal{O}P$-module, an almost source algebra $i\mathcal{O}Gi$ of a block with a defect group P is projective since it is a direct summand of $\mathcal{O}G$ as an $\mathcal{O}P$-$\mathcal{O}P$-bimodule. In particular, the

$\mathcal{O}$-rank of $i\mathcal{O}Gi$ is divisible by $|P|$. The source algebras of a block are essentially unique:

Theorem 6.4.4 *Let G be a finite group, let b be a block of $\mathcal{O}G$, let P be a defect group of b and let i, j be two source idempotents in $(\mathcal{O}Gb)^P$. Then there is $x \in N_G(P)$ and $c \in ((\mathcal{O}Gb)^P)^\times$ such that $j = {}^{cx}i$. In particular, there is an $\mathcal{O}$-algebra isomorphism $i\mathcal{O}Gi \cong j\mathcal{O}Gj$ sending iai to ${}^{cx}(iai) = j({}^{cx}a)j$ for any $a \in \mathcal{O}Gb$. This isomorphism sends ui to $({}^xu)j$ for any $u \in P$.*

Proof Let γ, δ be the local points of P on $\mathcal{O}Gb$ containing i and j, respectively. Then, by 5.5.16, there is $x \in N_G(P)$ such that $\delta = {}^x\gamma$. Thus ${}^xi$, j belong both to the same local point δ of P on $\mathcal{O}Gb$, hence they are conjugate in $(\mathcal{O}Gb)^P$, say ${}^{cx}i = j$ for some $c \in ((\mathcal{O}Gb)^P)^\times$. The result follows.   □

Remark 6.4.5 The isomorphism $i\mathcal{O}Gi \cong j\mathcal{O}Gj$ in the statement of 6.4.4 falls short of being an isomorphism of interior P-algebras, because it sends ui to ${}^xuj$ (instead of uj) for any $u \in P$. Thus the source algebras are not unique, up to isomorphism, as interior P-algebras. But 6.4.4 shows what the ambiguity is: if B, B' are two source algebras of $\mathcal{O}Gb$ with structural homomorphisms $\sigma : P \to B^\times$ and $\sigma' : P \to (B')^\times$ then there is an algebra isomorphism $\beta : B \cong B'$ and a group automorphism τ of P such that $\sigma' \circ \tau = \beta \circ \sigma$. In other words, if B is some source algebra with structural homomorphism $\sigma : P \to B^\times$ then any other source algebra is isomorphic, as interior P-algebra, to the interior P-algebra B with structural homomorphism $\sigma \circ \tau : P \to B^\times$ for some $\tau \in \mathrm{Aut}(P)$. In particular, the number of isomorphism classes of interior P-algebras occurring as source algebras of a given block b is bounded by $\mathrm{Aut}(P)$.

We turn now to showing that the block theoretic invariants mentioned at the beginning of this section can indeed be read off the source algebras of a block. The module categories of a block algebra and of any of its source algebras are equivalent. This holds more generally for almost source algebras; we restate this for future reference.

Theorem 6.4.6 ([169, 3.5]) *Let G be a finite group, let b be a block of $\mathcal{O}G$, let P be a defect group of b and let i be an idempotent in $(\mathcal{O}Gb)^P$ such that $\mathrm{Br}_P(i) \neq 0$. The algebras $\mathcal{O}Gb$ and $i\mathcal{O}Gi$ are Morita equivalent. More precisely, there is an equivalence of categories $\mathrm{Mod}(\mathcal{O}Gb) \cong \mathrm{Mod}(i\mathcal{O}Gi)$ sending an $\mathcal{O}Gb$-module M to the $i\mathcal{O}Gi$-module iM and sending a homomorphism of $\mathcal{O}Gb$-modules $\varphi : M \to M'$ to the homomorphism of $i\mathcal{O}Gi$-modules $iM \to iM'$ obtained from restricting φ to iM.*

Proof This is a special case of 5.6.13.   □

By Proposition 2.8.20, Morita equivalences are compatible with tensor products of algebras, and hence this Theorem implies in particular that left multiplication by i on an $\mathcal{O}Gb\text{-}\mathcal{O}P$-bimodule yields an equivalence between the category of $\mathcal{O}Gb\text{-}\mathcal{O}P$-bimodules and the category of $i\mathcal{O}Gi\text{-}\mathcal{O}P$-bimodules. The relative separability of block algebras with respect to their defect group algebras carries also over to almost source algebras.

Theorem 6.4.7 *Let G be a finite group, b be a block of $\mathcal{O}G$, P be a defect group of b and i be an idempotent in $(\mathcal{O}Gb)^P$ such that $\mathrm{Br}_P(i) \neq 0$. Set $A = i\mathcal{O}Gi$. The canonical map $A \otimes_{\mathcal{O}P} A \to A$ induced by multiplication in A splits as homomorphism of A-A-bimodules. In particular, A is isomorphic to a direct summand of $A \otimes_{\mathcal{O}P} A$ as A-A-bimodule, and every A-module is relatively $\mathcal{O}P$-projective.*

Proof Let I be a primitive decomposition of b in $(\mathcal{O}Gb)^P$. Thus $(\mathcal{O}Gb)^P = \oplus_{j \in I}(\mathcal{O}Gb)^P j$, hence $(\mathcal{O}Gb)^G_P = \sum_{j \in I} \mathrm{Tr}^G_P((\mathcal{O}Gb)^P j)$. This is a sum of left ideals in $Z(\mathcal{O}Gb)$. Since $Z(\mathcal{O}Gb)$ is commutative, this is in fact a sum of ideals. As P is a defect group, this sum of ideals contains b. By Rosenberg's Lemma, one of these ideals contains b. Thus $b = \mathrm{Tr}^G_P(cj)$ for some $j \in I$ and some $c \in (\mathcal{O}Gb)^P$. Then j belongs to a local point of P by the minimality of P subject to $b \in (\mathcal{O}Gb)^G_P$. Since all local points of P on $\mathcal{O}Gb$ are G-conjugate we may choose c such that $b = \mathrm{Tr}^G_P(ci)$. One checks that the map sending $a \in \mathcal{O}G$ to $\sum_{x \in [G/P]} axci \otimes ix^{-1}$ in $\mathcal{O}Gi \otimes_{\mathcal{O}P} i\mathcal{O}G$ is a section for the canonical $\mathcal{O}G$-$\mathcal{O}G$-bimodule homomorphism $\mathcal{O}Gi \otimes_{\mathcal{O}G} i\mathcal{O}G \to \mathcal{O}Gb$ given by multiplication in $\mathcal{O}G$. Multiplying this homomorphism by i on the left and right yields the required section as stated. Thus A is isomorphic to a direct summand of $A \otimes_{\mathcal{O}P} A$ as an A-A-bimodule. $\qquad\square$

Corollary 6.4.8 *Let G be a finite group, b a block of $\mathcal{O}G$, P a defect group of b and i an idempotent in $(\mathcal{O}Gb)^P$ such that $\mathrm{Br}_P(i) \neq 0$. Set $A = i\mathcal{O}Gi$. An A-module U is projective if and only if its restriction to $\mathcal{O}P$ is projective.*

Proof If U is projective as an A-module then it is projective as an $\mathcal{O}P$-module because A is projective as an $\mathcal{O}P$-module. By 6.4.7, A is a direct summand of $A \otimes_{\mathcal{O}P} A$ as A-A-bimodule; tensoring by $- \otimes_A U$ shows that the A-module U is isomorphic to a direct summand of the A-module $A \otimes_{\mathcal{O}P} U$. Thus if U is projective as an $\mathcal{O}P$-module then the A-module $A \otimes_{\mathcal{O}P} U$ is projective, and hence so is its direct summand U. $\qquad\square$

The source algebras of a block of $\mathcal{O}G$ are determined by those of the corresponding block of kG; this is analogous to the unique lifting property of p-permutation kG-modules to p-permutation $\mathcal{O}G$-modules (cf. 5.11.2).

Theorem 6.4.9 ([170, 7.8]) *Let G be a finite group, b a block of $\mathcal{O}G$ with defect group P and i an idempotent in $(\mathcal{O}Gb)^P$ such that $\mathrm{Br}_P(i) \neq 0$. Set $A = i\mathcal{O}Gi$. Let B be an interior P-algebra such that B has a $P \times P$-stable $\mathcal{O}$-basis and such that $k \otimes_{\mathcal{O}} A \cong k \otimes_{\mathcal{O}} B$ as interior P-algebras over k. Then $A \cong B$ as interior P-algebras over $\mathcal{O}$.*

Proof Since A, B are permutation $\mathcal{O}(P \times P)$-modules such that $k \otimes_{\mathcal{O}} A \cong k \otimes_{\mathcal{O}} B$, the uniqueness property 5.11.2 implies that an isomorphism of interior P-algebras $k \otimes_{\mathcal{O}} A \cong k \otimes_{\mathcal{O}} B$ lifts to an isomorphism of $\mathcal{O}(P \times P)$-modules $A \cong B$. But A is also relatively $\mathcal{O}P$-separable, by 6.4.7, and thus the lifting Theorem 4.8.2 applies and yields an isomorphism of interior P-algebras $A \cong B$.   □

One can prove 6.4.9 also without using 4.8.2.

Alternative proof of 6.4.9 Set $M = \mathcal{O}Gi \otimes_{\mathcal{O}P} B$, viewed as $\mathcal{O}G$-B-bimodule. Set $\bar{A} = k \otimes_{\mathcal{O}} A$, $\bar{B} = k \otimes_{\mathcal{O}} B$, $\bar{M} = k \otimes M$ and denote by $\bar{b}$, $\bar{i}$ the canonical images of b, i in kG, respectively. By a standard adjunction we have $\mathrm{End}_{B^{\mathrm{op}}}(M) \cong \mathrm{Hom}_{\mathcal{O}(G \times P)}(\mathcal{O}Gi, M)$. By the assumptions, $\mathcal{O}Gi$ and M are p-permutation $\mathcal{O}(G \times P)$-modules. Thus, using the same adjunction over k, we get from 5.10.2 that the canonical algebra homomorphism $\mathrm{End}_{B^{\mathrm{op}}}(M) \to \mathrm{End}_{\bar{B}^{\mathrm{op}}}(\bar{M})$ is surjective. It follows from 6.4.7 that $kG\bar{i}$ is isomorphic to a direct summand of $\bar{M}$ as $kG\bar{b}$-$\bar{B}$-bimodule. Such a summand corresponds to an idempotent $\bar{e}$ in $\mathrm{End}_{\bar{B}^{\mathrm{op}}}(\bar{M})$, and since $kG\bar{i}$ induces a Morita equivalence between $kG\bar{b}$ and $\bar{A} \cong \bar{B}$, we get that the structural map $kGb \to \mathrm{End}_{\bar{B}^{\mathrm{op}}}(\bar{M})$ induces an isomorphism $kGb \cong \bar{e} \cdot \mathrm{End}_{\bar{B}^{\mathrm{op}}}(\bar{M}) \cdot \bar{e}$. The surjectivity mentioned above yields an idempotent $e \in \mathrm{End}_{B^{\mathrm{op}}}(M)$ lifting $\bar{e}$, and then Nakayama's Lemma implies that the structural algebra homomorphism $\mathcal{O}Gb \to e \cdot \mathrm{End}_{B^{\mathrm{op}}}(M) \cdot e$ is again an isomorphism. The idempotent e corresponds to a direct summand X of M as $\mathcal{O}Gb$-B-bimodule such that $k \otimes_{\mathcal{O}} X \cong kGi$. Thus X induces a Morita equivalence between $\mathcal{O}Gb$ and B. Composing this with the Morita equivalence between A and $\mathcal{O}Gb$ implies that iX induces a Morita equivalence between A and B. More precisely, since $k \otimes_{\mathcal{O}} iX \cong ikGi$ we have $iX \cong B$ as right B-module, and hence $A \cong \mathrm{End}_{B^{\mathrm{op}}}(iX) \cong B$, whence the result.   □

Vertices and sources can be read off almost source algebras; this generalises 6.4.8.

Theorem 6.4.10 ([126, 6.3]) *Let G be a finite group, b a block of $\mathcal{O}G$ with defect group P and $i \in (\mathcal{O}Gb)^P$ an almost source idempotent. Set $A = i\mathcal{O}Gi$. Let M be an indecomposable $\mathcal{O}Gb$-module and let Q be a minimal subgroup of P such that the A-module iM is isomorphic to a direct summand of $A \otimes_{\mathcal{O}Q} V$*

for some $\mathcal{O}Q$-module V. Then Q is a vertex of M. Moreover, the module V can be chosen to be indecomposable, and then V is an $\mathcal{O}Q$-source of M. The module V can further be chosen to be isomorphic to a direct summand of iM as an $\mathcal{O}Q$-module; in particular, M is isomorphic to a direct summand of $\mathcal{O}Gi \otimes_{\mathcal{O}Q} iM$.

Proof The A-module iM is indecomposable since it corresponds to the indecomposable $\mathcal{O}Gb$-module M via the canonical Morita equivalence between $\mathcal{O}Gb$ and A from 6.4.6. Thus iM is isomorphic to a direct summand of $A \otimes_{\mathcal{O}Q} V$ for some indecomposable $\mathcal{O}Q$-module V. It follows from 2.6.12 that V can be chosen to be a direct summand of iM as an $\mathcal{O}Q$-module. The minimality of Q implies that V has vertex Q. Choosing V as a direct summand of iM implies that V is a direct summand of $\mathrm{Res}_Q^G(M)$; in particular, Q is contained in a vertex of M. Since M is a direct summand of $\mathcal{O}Gi \otimes_{\mathcal{O}Q} V$, hence of $\mathrm{Ind}_Q^G(V)$, it follows that Q is a vertex and V a source of M. □

In view of 6.4.10, we extend the notion of vertices and sources to indecomposable modules over almost source algebras. Let G be a finite group, b a block of $\mathcal{O}G$, P a defect group of b, and $i \in (\mathcal{O}Gb)^P$ an almost source idempotent. Set $A = i\mathcal{O}Gi$. Let M be an indecomposable A-module. then a minimal subgroup Q of P such that M is isomorphic to a direct summand of $A \otimes_{\mathcal{O}Q} V$ for some indecomposable direct summand V of $\mathrm{Res}_Q(M)$ is called a *vertex of M*, and V is called an $\mathcal{O}Q$-*source* of M. By 6.4.10 the pair (Q, V) is then also a vertex-source pair of the indecomposable $\mathcal{O}Gb$-module $\mathcal{O}Gi \otimes_A M$.

We will see in 6.13.10 below that the matrix of generalised decomposition numbers associated with a block is an invariant of its source algebras. The following observation is useful for bounding the dimension of source algebras in some cases.

Proposition 6.4.11 *Let G be a finite group and b a block of kG with a nontrivial defect group P. Let $i \in (kGb)^P$ be a source idempotent and set $A = ikGi$. Let S be a simple A-module. Then $\mathrm{Res}_{kP}^A(S)$ has no nonzero projective direct summand.*

Proof The idempotent i is primitive in $(kGb)^P$, and hence kGi is indecomposable as a kG-kP-bimodule. It is also nonprojective as a bimodule since P is nontrivial. By 6.4.6 and the remarks after that theorem, left multiplication by i is a Morita equivalence, and thus the A-$\mathcal{O}P$-bimodule A is indecomposable and nonprojective. The result is therefore a special case of 4.14.9. Alternatively, S remains semisimple after extending k, and hence we may assume that k is a splitting field. Then the structural map $A \to \mathrm{End}_k(S)$ is surjective, hence induces a surjective map from A_1^P onto $\mathrm{End}_k(S)_1^P$. The idempotent $i = 1_A$ is primitive in A^P, hence A^P is a local algebra. Since $\mathrm{Br}_P(i) \neq 0$, the ideal A_1^P is a proper ideal, thus contained in $J(A^P)$. Its image $\mathrm{End}_k(S)_1^P$ is therefore contained in the radical of $\mathrm{End}_{kP}(S)$. In particular, $\mathrm{End}_k(S)_1^P$ contains no idempotent. Higman's criterion implies that as a kP-module, S has no nonzero projective summand. $\qquad\square$

One of the reasons for introducing almost source idempotents is the technical issue that source idempotents are not always mapped to source idempotents under Brauer homomorphisms – we postpone a more detailed investigation of this issue to §8.7. We state two observations in which source idempotents are preserved under the Brauer homomorphism.

Proposition 6.4.12 *Let G be a finite group, b a block of $\mathcal{O}G$ with a defect group P, and let $i \in (\mathcal{O}Gb)^P$ be an almost source idempotent. Set $A = i\mathcal{O}Gi$ and $j = \mathrm{Br}_P(i)$. Denote by e the unique block of $kC_G(P)$ satisfying $je \neq 0$. Then j is an almost source idempotent of $kC_G(P)$. Moreover, if i is a source idempotent of $\mathcal{O}Gb$, then j is a source idempotent of $kC_G(P)e$. In particular, the algebras $A(P)$ and $kC_G(P)e$ are Morita equivalent.*

Proof By 6.3.10, $Z(P)$ is a defect group of $kC_G(P)$. Since the conjugation action by $Z(P)$ on $C_G(P)$ is trivial, we have $\mathrm{Br}_{Z(P)}(j) = j \neq 0$ and $\mathrm{Br}_{Z(P)}(e) = e$. Thus j is an almost source idempotent of $kC_G(P)e$. The Brauer homomorphism Br_P maps $(\mathcal{O}G)^P$ onto $kC_G(P) = (kC_G(P))^{Z(P)}$. Thus if i is primitive in $(\mathcal{O}Gb)^P$, then j is primitive in $kC_G(P)e = (kC_G(P)e)^{Z(P)}$, thus a source idempotent. The last statement follows from 6.4.6. $\qquad\square$

Proposition 6.4.13 (cf. [109, Lemma]) *Let G be a finite group, b a block of $\mathcal{O}G$ with a defect group P, and let $i \in (\mathcal{O}Gb)^P$ be a source idempotent (resp. almost source idempotent). Set $A = i\mathcal{O}Gi$. Let Q be a subgroup of $Z(P)$ and denote by f the unique block of $kC_G(Q)$ satisfying $\mathrm{Br}_Q(i)f \neq 0$. Then P is a defect group of $kC_G(Q)f$, the idempotent $\mathrm{Br}_Q(i)$ is a source idempotent (resp. almost source idempotent) in $(kC_G(Q)f)^P$, and $A(Q)$ is a source algebra (resp. almost source algebra) of $kC_G(Q)f$. In particular, $kC_G(Q)f$ and $A(Q)$ are Morita equivalent.*

Proof By 6.3.11, P is a defect group of $kC_G(Q)f$. Since Q is a subgroup of $Z(P)$, we have $P \leq C_G(Q)$, and hence the Brauer homomorphism Br_Q maps $(\mathcal{O}G)^P$ onto $(kC_G(Q))^P$. Thus if i is primitive in $(\mathcal{O}Gb)^P$, then $\mathrm{Br}_Q(i)$ is a primitive idempotent in $(kC_G(Q)f)^P$. Moreover, we have $\mathrm{Br}_P(\mathrm{Br}_Q(i)) = \mathrm{Br}_P(i) \neq 0$, which implies that $\mathrm{Br}_Q(i)$ is an almost source idempotent of $kC_G(Q)f$. The last statement follows from 6.4.6. $\qquad\square$

Corollary 6.4.14 *Let G be a finite group, and let b a block of kG with an abelian defect group P. Let $i \in (\mathcal{O}Gb)^P$ be a source idempotent, and let Q be a subgroup of P. Denote by e the unique block of $kC_G(Q)$ satisfying $\mathrm{Br}_Q(i)e \neq 0$. Then P is a defect group of e and $\mathrm{Br}_Q(i)$ is a source idempotent of e. In particular, $kC_G(Q)f$ and $A(Q)$ are Morita equivalent.*

Proof This is a special case of 6.4.13. □

See Proposition 8.7.2 and 8.7.3 for further statements on almost source idempotents.

Theorem 6.4.15 ([109, Theorem]) *Let G be a finite group, let b be a block of $\mathcal{O}G$ and let (P, e) be a maximal (G, b)-Brauer pair. Set $H = N_G(P, e)$. For any subgroup Q of P denote by e_Q and f_Q the unique blocks of $kC_G(Q)$ and $kC_H(Q)$ satisfying $(Q, e_Q) \leq (P, e)$ and $(Q, f_Q) \leq (P, e)$, respectively. Let f be a primitive idempotent in $(\mathcal{O}Gb)^H$ such that $\mathrm{Br}_P(f) = e$ and set $X = \mathcal{O}Gf$. Then, as an $\mathcal{O}(G \times H)$-module, X is indecomposable with vertex ΔP, and for any subgroup Q of $Z(P)$ the $k(C_G(Q) \times C_H(Q))$-module $e_Q X(\Delta Q)f_Q$ is up to isomorphism the unique indecomposable direct summand of $e_Q kC_G(Q)f_Q$ with vertex ΔP.*

Proof Let $\hat{e}$ be the block of $\mathcal{O}C_G(P)$ that lifts the block e of $kC_G(P)$. Then $\hat{e}$ is still a block of $\mathcal{O}H$, with (P, e) as unique maximal Brauer pair. Let $j \in (\mathcal{O}H\hat{e})^P$ be a source idempotent of $\hat{e}$ as block of $\mathcal{O}H$. By 6.8.3, j remains a source idempotent of the block $c = \mathrm{Tr}_H^{N_G(P)}(\hat{e})$ of $\mathcal{O}N_G(P)$ corresponding to b through the Brauer correspondence. Thus by 6.15.1, the idempotent $i = jf$ is a source idempotent of the block b in $(\mathcal{O}Gb)^P$, and since f was chosen such that $\mathrm{Br}_P(f) = e$ we have $\mathrm{Br}_P(i)e \neq 0$. Let Q be a subgroup of $Z(P)$. By 6.4.13, $i_Q = \mathrm{Br}_Q(i)$ is a source idempotent of the block e_Q, and $j_Q = \mathrm{Br}_Q(j)$ is a source idempotent of the block f_Q. Since $i = jf = fj$ we have $i_Q = \mathrm{Br}_Q(f)j_Q$, and this is therefore in particular a primitive idempotent in $(kC_G(Q)e_Q)^P$. Since $X = \mathcal{O}Gf$ we have $X(\Delta Q) = kC_G(Q)\mathrm{Br}_Q(f)$, and therefore

$$e_Q X(\Delta Q)j_Q = e_Q kC_G(Q)\mathrm{Br}_Q(f)j_Q = e_Q kC_G(Q)i_Q.$$

As $i_Q \in kC_G(Q)e_Q$ this implies in particular that $e_Q X(\Delta Q)j_Q$ is nonzero. Since i_Q is primitive in $(kC_G(Q)e_Q)^P$, it follows that the $(kC_G(Q)e_Q, kP)$-bimodule $e_Q kC_G(Q)i_Q$ is indecomposable. Moreover, kP is isomorphic to a subalgebra of the source algebra $j_Q kC_H(Q)j_Q$ via multiplication by j_Q, and hence $e_Q X(\Delta Q)j_Q$ is indecomposable as a $(kC_G(Q)e_Q, j_Q kC_H(Q)j_Q)$-bimodule. By 6.4.13, the block algebra $kC_H(Q)f_Q$ and its source algebra $j_Q kC_H(Q)j_Q$ are Morita equivalent, which implies that $e_Q X(\Delta Q)f_Q$ is indecomposable as a $k(C_G(Q) \times C_H(Q))$-module. Since X is a direct summand of $\mathcal{O}Gb$ as $\mathcal{O}(G \times H)$-module, $X(\Delta Q)$ is a direct summand of $kC_G(Q)\mathrm{Br}_Q(b)$ as $k(C_G(Q) \times C_H(Q))$-module, and hence $e_Q X(\Delta Q) f_Q$ is a direct summand of $e_Q k C_G(Q) f_Q$. As noted in 6.7.4, the $\mathcal{O}(G \times H)$-module X is indecomposable with ΔP as a vertex. By 6.3.4 we have $e = e\mathrm{Br}_P(e_Q) = e\mathrm{Br}_P(f_Q) = e\mathrm{Br}_P(f)$. Thus, if we denote by $\bar{f}$ the canonical image of f in $(kG)^H$, we get $e\mathrm{Br}_P(e_Q \bar{f} f_Q) = e \neq 0$, so that $\mathrm{Br}_P(e_Q \bar{f} f_Q) \neq 0$, hence $(e_Q X(\Delta Q) f_Q)(\Delta P) \neq 0$, which implies that ΔP is a vertex of $e_Q X(\Delta Q) f_Q$. For the last part we observe that the $k(C_G(Q) \times C_H(Q))$-module $e_Q k C_G(Q) f_Q$ is a direct summand of $kC_G(Q) f_Q = \mathrm{Ind}_{C_H(Q) \times C_H(Q)}^{C_G(Q) \times C_H(Q)}(kC_H(Q) f_Q)$. Moreover, the $k(C_H(Q) \times C_H(Q))$-module $kC_H(Q) f_Q$ is indecomposable with ΔP as a vertex, and the normaliser of ΔP in $C_G(Q) \times C_H(Q)$ is contained in $C_H(Q) \times C_H(Q)$. Thus the Green correspondence implies that the $k(C_G(Q) \times C_H(Q))$-module $kC_G(Q) f_Q$ has exactly one indecomposable direct summand with ΔP as vertex, up to isomorphism. The result follows. □

Proposition 6.4.16 *Let $\mathcal{O}'$ be a complete discrete valuation ring containing $\mathcal{O}$ having the same residue field k as $\mathcal{O}$. Let G be a finite group, b a block of $\mathcal{O}G$, P a defect group of $\mathcal{O}Gb$ and $i \in (\mathcal{O}Gb)^P$ a source idempotent of $\mathcal{O}Gb$. Then b is a block of $\mathcal{O}'G$, P is a defect group of $\mathcal{O}'Gb$, and i is a source idempotent of $\mathcal{O}'Gb$. In particular, if A is a source algebra of $\mathcal{O}Gb$, then $\mathcal{O}' \otimes_\mathcal{O} A$ is a source algebra of $\mathcal{O}'Gb$.*

Proof The canonical map $Z(\mathcal{O}G) \to Z(kG)$ is surjective, and we have an obvious isomorphism $Z(\mathcal{O}'G) \cong \mathcal{O}' \otimes_\mathcal{O} Z(\mathcal{O}G)$. It follows from 4.7.3 applied to $Z(\mathcal{O}G)$ that b remains a block of $\mathcal{O}'G$. The fact that $\mathcal{O}Gb$ and $\mathcal{O}'Gb$ have the same defect groups is an immediate consequence of 6.1.6. Similarly, the map $(\mathcal{O}G)^P \to (kG)^P$ is surjective, and hence this map sends local points of P on $\mathcal{O}G$ to local points of P on kG. It follows from 4.7.3 applied to the algebra $(\mathcal{O}Gb)^P$ that i remains primitive in $(\mathcal{O}'Gb)^P$, whence the result. □

The following result yields a sufficient criterion for when an almost source algebra is in fact a source algebra.

Proposition 6.4.17 *Let G be a finite group, B a block algebra of kG, P a defect group of B and $i \in B^P$ an idempotent satisfying $\mathrm{Br}_P(i) \neq 0$. If iS is indecomposable as a kP-module for every simple B-module S, then i is primitive in B^P, or equivalently, $ikGi$ is a source algebra of B.*

Proof Since $\mathrm{Br}_P(i) \neq 0$, it follows that there is a primitive idempotent $j \in (ikGi)^P = (iBi)^P$ satisfying $\mathrm{Br}_P(j) \neq 0$; that is, j is a source idempotent of

B. By 6.4.6 we have $jS \neq \{0\}$ for every simple B-module S. If $j' = i - j$ is nonzero, then j and j' are orthogonal idempotents in B whose sum is i. Since $J(B)$ contains no idempotent, it follows that there exists a simple B-module S such that $j'S \neq \{0\}$. But then $iS = jS \oplus j'S$ is a direct sum of two nonzero kP-modules. The result follows. $\qquad\square$

6.5 Characters in blocks

Definition 6.5.1 Let G be a finite group and b a central idempotent of $\mathcal{O}G$. Set $B = \mathcal{O}Gb$. Suppose that the quotient field K of $\mathcal{O}$ has characteristic 0. Let $\chi \in \mathrm{Irr}_K(G)$, $\varphi \in \mathrm{IBr}_k(G)$, and $\Phi \in \mathrm{IPr}_{\mathcal{O}}(G)$. We say that χ *belongs to* b if $\chi(b) = \chi(1)$, or equivalently, if b acts as identity on a KG-module with χ as character. We say that φ *belongs to* b if the image $\bar{b}$ of b in kG acts as identity on a simple kG-module with Brauer character φ. Finally, we say that Φ belongs to b if b acts as identity on a projective indecomposable $\mathcal{O}G$-module with Φ as character. We define the following notation:

$$\mathrm{Irr}(B) = \mathrm{Irr}_K(G, b) = \{\chi \in \mathrm{Irr}_K(G) \mid \chi \text{ belongs to } b\},$$

$$\mathrm{IBr}(B) = \mathrm{IBr}_k(G, b) = \{\varphi \in \mathrm{IBr}_k(G) \mid \varphi \text{ belongs to } b\},$$

$$\mathrm{IPr}(B) = \mathrm{IPr}\mathcal{O}(G, b) = \{\Phi \in \mathrm{IPr}_{\mathcal{O}}(G) \mid \Phi \text{ belongs to } b\}.$$

For any $\chi \in \mathrm{Irr}_K(G)$ set

$$e(\chi) = \frac{\chi(1)}{|G|} \sum_{x \in G} \chi(x^{-1})x.$$

With this notation, $\chi \in \mathrm{Irr}_K(G)$ belongs to b if and only if $\chi(xb) = \chi(x)$ for all $x \in G$. Indeed, if b acts as identity on a KG-module V with χ as its character, then x and xb act in the same way on V. This suggests how to extend this terminology to class functions: we say that $\psi \in \mathrm{Cl}_K(G)$ *belongs to* b if $\psi(xb) = \psi(x)$ for all $x \in G$. If K is a splitting field for KG, and if $\psi = \sum_{\chi \in \mathrm{Irr}_K(G)} a_\chi \chi$ for some coefficients $a_\chi \in K$, then ψ belongs to b if and only if χ belongs to b for any $\chi \in \mathrm{Irr}_K(G)$ such that $a_\chi \neq 0$, or equivalently, if and only if ψ is a K-linear combination of $\mathrm{Irr}_K(G, b)$. This is an immediate consequence of the fact that $\mathrm{Irr}_K(G)$ is a K-basis of $\mathrm{Cl}_K(G)$. Similarly, a function in $\mathrm{Cl}_K(G_{p'})$ *belongs to* b if it is a K-linear combination of $\mathrm{IBr}_k(G, b)$. Since central idempotents from kG lift uniquely to central idempotents in $\mathcal{O}G$, we may extend this terminology further: we say that a class function in $\mathrm{Cl}_K(G)$ or $\mathrm{Cl}_K(G_{p'})$ *belongs to a central idempotent* b *in* kG if it belongs to the unique central idempotent in $\mathcal{O}G$ that lifts b.

Proposition 6.5.2 *Let G be a finite group. Suppose that K has characteristic zero.*

(i) If χ, $\chi' \in \mathrm{Cl}_K(G)$ belong to different blocks of $\mathcal{O}G$, then $\langle \chi, \chi' \rangle_G = 0$.

(ii) If φ, $\varphi' \in \mathrm{Cl}_K(G_{p'})$ belong to different blocks of $\mathcal{O}G$, then $\langle \varphi, \varphi' \rangle'_G = 0$.

Proof We may assume that K and k are splitting fields for G. Statement (i) is a trivial consequence of the orthogonality relations. By Brauer's reciprocity 5.14.8, the class function φ is a K-linear combination of characters of projective indecomposable $\mathcal{O}Gb$-modules, restricted to $G_{p'}$, where b is the block to which φ belongs. But φ' belongs to a block b' different from b, hence is a linear combination of elements in $\mathrm{IBr}_k(G, b')$ by Theorem 5.13.13. Thus (ii) follows from Brauer's reciprocity 5.14.8.   □

If K is large enough, then by 3.3.1 the elements $e(\chi)$ in 6.5.1, with $\chi \in \mathrm{Irr}_K(G)$, are the primitive idempotents in $Z(KG)$. Their connection with block idempotents is as follows.

Theorem 6.5.3 *Let G be a finite group. Suppose that the quotient field K of $\mathcal{O}$ is a splitting field of characteristic zero for KG. Let b be a block of $\mathcal{O}G$. Then*

$$b = \sum_{\chi \in \mathrm{Irr}_K(G,b)} e(\chi)$$

and the character β of $\mathcal{O}Gb$ as an $\mathcal{O}(G \times G)$-module is given by

$$\beta(x, y) = \sum_{\chi \in \mathrm{Irr}_K(G,b)} \chi(x)\chi(y^{-1})$$

for all $(x, y) \in G \times G$. Moreover, $\mathrm{Irr}_K(G)$ is the disjoint union of the sets $\mathrm{Irr}_K(G, b)$, and $\mathrm{IBr}_k(G)$ is the disjoint union of the sets $\mathrm{IBr}_k(G, b)$, with b running over the blocks of $\mathcal{O}G$. In particular, we have

$$\mathbb{Z}\mathrm{Irr}_K(G) = \oplus_b \mathbb{Z}\mathrm{Irr}_K(G, b)$$

$$\mathbb{Z}\mathrm{IBr}_k(G) = \oplus_b \mathbb{Z}\mathrm{IBr}_k(G, b)$$

with b running over the blocks of $\mathcal{O}G$.

Proof We have $\mathcal{O}G = \prod_b \mathcal{O}Gb$, where b runs over the blocks of $\mathcal{O}G$. Tensoring with $K \otimes_{\mathcal{O}} -$ yields an isomorphism of K-algebras

$$KG = \prod_b KGb,$$

where b runs again over the blocks of $\mathcal{O}G$. Note that the blocks b are still idempotents in $Z(KG)$, but not necessarily primitive. The set of idempotents $e(\chi)$, with $\chi \in \mathrm{Irr}_K(G)$, is a primitive decomposition of 1 in $Z(KG)$; thus we have

$$KG = \prod_{\chi \in \mathrm{Irr}_K(G)} KGe(\chi).$$

Thus, for any block b of $\mathcal{O}G$ we have

$$KGb = \prod_{\chi \in \mathrm{Irr}_K(G,b)} KGe(\chi)$$

for some unique subset $\mathrm{Irr}_K(G, b)$ of $\mathrm{Irr}_K(G)$. Clearly $\mathrm{Irr}_K(G)$ is the disjoint union of these subsets. Since the $e(\chi)$ are pairwise orthogonal we have $be(\chi) = e(\chi)$ if $\chi \in \mathrm{Irr}_K(G, b)$ and $be(\chi) = 0$, otherwise. Thus, if V is a simple KG-module with χ as character, then $e(\chi)$ acts as identity on V, and hence b acts as identity on V if and only if $be(\chi) = e(\chi)$, in which case we get $\chi(b) = \chi(1)$. The formula for β follows from 3.3.12. Similarly, by 1.7.9, for any simple kG-module there is exactly one block of kG, hence, by 6.1.6, exactly one block of $\mathcal{O}G$, which acts as identity on that simple module. The result follows. $\square$

Theorem 6.5.4 *Let G be a finite group and b a block of $\mathcal{O}G$. Write $b = \sum_{x \in G} \lambda_x x$ with coefficients $\lambda_x \in \mathcal{O}$. If $x \in G$ is not a p'-element, then $\lambda_x = 0$.*

Proof We may assume that k is large enough for G. Similarly, we may assume that K is large enough and has characteristic zero. (The equal characteristic case follows then because in that case, K would merely be an extension field of k.) The formula for b from 6.5.3 implies that the coefficient λ_x is equal to $\sum_{\chi \in \mathrm{Irr}_K(G,b)} \frac{\chi(1)}{|G|} \chi(x^{-1})$. Again by 6.5.3 this expression is equal to $\frac{1}{|G|}\beta(1, x)$, where β is the character of $\mathcal{O}Gb$ as an $\mathcal{O}(G \times G)$-module. Since $\mathcal{O}Gb$ is projective as a right $\mathcal{O}Gb$-module, it follows from 5.12.14 that $\beta(1, x) = 0$ unless possibly x is a p'-element. $\square$

Corollary 6.5.5 *Let G be a finite group. Suppose that K has characteristic zero. Let W be a complete discrete valuation ring contained in $\mathcal{O}$ having k as a residue field, and let η be a primitive $|G|_{p'}$ root of unity in K. Let b be a block of $\mathcal{O}G$. Write $b = \sum_{x \in G} \lambda_x x$ with coefficients $\lambda_x \in \mathcal{O}$. Then, for any $x \in G$, we have $\lambda \in \mathbb{Q}[\eta] \cap W$.*

Proof We have $\lambda_x \in W$ by 6.4.16. If x is not a p'-element, then by 6.5.4 we have $\lambda_x = 0$, so there is nothing to prove. If x is a p'-element, then its order divides $|G|_{p'}$, and hence $\chi(x) \in \mathbb{Z}[\eta]$ for all $\chi \in \mathrm{Irr}(G, b)$. It follows that $e(\chi) \in \mathbb{Q}[\eta]$ for all such χ, and hence the formula $b = \sum_{\chi \in \mathrm{Irr}_K(G,b)} e(\chi)$ from 6.5.3 implies that $\lambda_x \in \mathbb{Q}[\eta]$. $\square$

Proposition 6.5.6 *Let G be a finite group and b a block of $\mathcal{O}G$. Set $B = \mathcal{O}Gb$. Let $\chi \in \mathrm{Irr}(B)$. Suppose that K and k are splitting fields for G and that $\mathrm{char}(K) = 0$.*

(i) The map $\omega_\chi : Z(\mathcal{O}G) \to \mathcal{O}$ sending $z \in Z(\mathcal{O}G)$ to $\frac{\chi(z)}{\chi(1)}$ is a surjective algebra homomorphism that sends all blocks different from b to zero and that induces a surjective algebra homomorphism $Z(B) \to \mathcal{O}$.

(ii) We have $J(Z(B)) = \{z \in Z(B) \mid \frac{\chi(z)}{\chi(1)} \in J(\mathcal{O})\}$ and $Z(B)^\times = \{z \in Z(B) \mid \frac{\chi(z)}{\chi(1)} \in \mathcal{O}^\times\}$. In particular, we have $\chi(z) \neq 0$ for any $\in Z(B)^\times$.

(iii) The map ω_χ extends to the unique algebra homomorphism $Z(KG) \to K$ sending $e(\chi)$ to 1 and $e(\chi')$ to 0 for any $\chi' \in \mathrm{Irr}_K(G)$ such that $\chi' \neq \chi$.

Proof By Corollary 3.5.9, we have $\frac{\chi(z)}{\chi(1)} \in \mathcal{O}$ for any $z \in Z(\mathcal{O}G)$. By 3.5.5, the map sending $z \in Z(\mathcal{O}G)$ to $\frac{\chi(z)}{\chi(1)}$ in $\mathcal{O}$ is a unitary, hence surjective, algebra homomorphism. Since χ belongs to b, we have $\chi(b') = 0$ for any block idempotent $b' \neq b$, and hence ω_χ restricts to a surjective algebra homomorphism $Z(B) \to \mathcal{O}$. This proves (i). The algebra $Z(B)$ is local, hence every element in $Z(B)$ is either invertible or in the radical. Thus the kernel of the homomorphism $Z(B) \to \mathcal{O}$ is contained in $J(Z(B))$. This implies that $J(Z(B))$ is equal to the inverse image in $Z(B)$ of $J(\mathcal{O})$, and that $Z(B)^\times$ is the inverse image of $\mathcal{O}^\times$, whence (ii). Statement (iii) has been proved as part of 3.5.5. □

The following description of blocks of characters is used essentially as a definition of blocks in [32, §9]. It implies that the partition of $\mathrm{Irr}_K(G)$ into p-blocks can be read off the character table of G, where K is a splitting field of characteristic zero for the finite group G.

Theorem 6.5.7 *Let G be a finite group. Suppose that K and k are splitting fields for G and that $\mathrm{char}(K) = 0$. Two irreducible characters $\chi, \psi \in \mathrm{Irr}_K(G)$ belong to the same block of $\mathcal{O}G$ if and only if $\frac{\chi(z)}{\chi(1)} - \frac{\psi(z)}{\psi(1)} \in J(\mathcal{O})$ for all $z \in Z(\mathcal{O}G)$.*

Proof By 6.5.6, the map sending $z \in Z(\mathcal{O}G)$ to $\frac{\chi(z)}{\chi(1)}$ in $\mathcal{O}$ is a surjective algebra homomorphism. This homomorphism sends every block of $\mathcal{O}G$ to zero except the unique block b to which χ belongs. The algebra $Z(kG\bar{b})$ is local, hence admits only a unique surjective algebra homomorphism $Z(kG\bar{b}) \to k$. Thus χ, and ψ belong to the same block b of $\mathcal{O}G$ if and only if the maps sending $z \in Z(\mathcal{O}G)$ to $\frac{\chi(z)}{\chi(1)}$ and $\frac{\psi(z)}{\psi(1)}$ induce the same map from $Z(kG)$ to k, whence the result. □

Theorem 6.5.8 *Let G be a finite group and b a block of G. Set $B = \mathcal{O}Gb$. Suppose that K and k are splitting fields for G and that $\mathrm{char}(K) = 0$.*

(i) *The algebra $Z(B)$ is split local.*

(ii) *We have $Z(B) \subseteq \sum_{\chi \in \mathrm{Irr}(B)} \mathcal{O} \cdot e(\chi)$.*

(iii) *We have $J(Z(B)) \subseteq \sum_{\chi \in \mathrm{Irr}(B)} J(\mathcal{O}) \cdot e(\chi)$.*

(iv) *We have $Z(B)^\times \subseteq \sum_{\chi \in \mathrm{Irr}(B)} \mathcal{O}^\times \cdot e(\chi)$.*

(v) *For $\chi \in \mathrm{Irr}(B)$ let $\lambda_\chi \in \mathcal{O}$ such that $\sum_{\chi \in \mathrm{Irr}(B)} \lambda_\chi e(\chi) \in Z(B)^\times$. Then $\lambda_\chi \in \mathcal{O}^\times$ and the image of λ_χ in $k^\times$ does not depend on χ.*

Proof Statement (i) follows from Proposition 1.14.11. We have $Z(K \otimes_\mathcal{O} B) = \prod_{\chi \in \mathrm{Irr}(B)} Ke(\chi)$. Let $a \in Z(B)$, and write $a = \sum_{\chi \in \mathrm{Irr}(B)} \lambda_\chi e(\chi)$ with uniquely determined $\lambda_\chi \in K$. By Theorem 3.5.5 we have $ae(\chi) = \frac{\chi(a)}{\chi(1)} e(\chi)$, hence $\lambda_\chi = \frac{\chi(a)}{\chi(1)}$i, and this is in $\mathcal{O}$ by Proposition 6.5.6 or Corollary 3.5.9. This proves (ii). If $a \in J(Z(B))$, then some power of a is in $J(\mathcal{O})Z(B)$ by 4.1.8. Thus some power of λ_χ is in $J(\mathcal{O})$, which forces $\lambda_\chi \in J(\mathcal{O})$. This proves (iii). Since $Z(B)$ is split local, it follows that $Z(B)^\times = \mathcal{O}^\times \cdot (1 + J(Z(B)))$, and hence any element in $Z(B)^\times$ can be written in the form $\sum_{\chi \in \mathrm{Irr}(B)} (\mu + \mu_\chi) e(\chi)$ for some $\mu \in \mathcal{O}^\times$ and some $\mu_\chi \in \mathcal{O}$ such that $\sum_{\chi \in \mathrm{Irr}(B)} \mu_\chi e(\chi) \in J(Z(B))$. Then $\mu_\chi \in J(\mathcal{O})$ by (iii). Thus $\mu + \mu_\chi \in \mathcal{O}^\times$, and the image of this element in k depends only on μ. This proves (iv) and (v). $\qquad\square$

Proposition 6.5.9 *Let G be a finite group, u a p-element in G, and j an idempotent in $(\mathcal{O}G)^{\langle u \rangle}$. Let $\chi \in \mathbb{Z}\mathrm{Irr}_K(G)$. We have $\chi(uj) - \chi(j) \in J(\mathcal{O})$.*

Proof By 5.12.16, the value $\chi(uj)$ depends only on $\mathrm{Br}_{\langle u \rangle}(j)$. By 5.4.12, χ maps the kernel of $\mathrm{Br}_{\langle u \rangle}(j)$ to $J(\mathcal{O})$. Thus we may replace j by an idempotent j' in $\mathcal{O}C_G(u)$ lifting $\mathrm{Br}_{\langle u \rangle}(j)$. Then both uj' and j' are contained in $\mathcal{O}C_G(u)$, so we may replace χ by an irreducible character η of $C_G(u)$. Since $u - 1$ belongs to the radical of $Z(\mathcal{O}C_G(u))$, it follows from 3.5.5 and 6.5.6 that $\eta(uj') - \eta(j') = \eta((u-1)j') = \frac{\eta(u-1)}{\eta(1)} \eta(j') \in J(\mathcal{O})$, whence the result. $\qquad\square$

The results developed earlier on the decomposition map $d_G : \mathbb{Z}\mathrm{Irr}_K(G) \to \mathbb{Z}\mathrm{IBr}_k(G)$ of a finite group G are compatible with the block decomposition of group algebras. The following obvious fact is a restatement of some earlier comments on class functions and blocks in this Section.

Proposition 6.5.10 *Let G be a finite group. Let $\chi \in \mathrm{Irr}_K(G)$ and $\varphi \in \mathrm{IBr}_k(G)$. If $d_\varphi^\chi \neq 0$, then χ and φ belong to the same block of $\mathcal{O}G$.*

Proof Let b be the block of $\mathcal{O}G$ to which χ belongs. Then b acts as identity on any $\mathcal{O}$-free $\mathcal{O}G$-module Y having χ as character. Thus the image $\bar{b}$ of b in kG acts as identity on any composition factor of $k \otimes_\mathcal{O} Y$. The result follows. $\qquad\square$

Thus d_G induces a group homomorphism

$$d_{G,b} : \mathbb{Z}\mathrm{Irr}_K(G, b) \to \mathbb{Z}\mathrm{IBr}_k(G, b),$$

and d_G is the sum of the maps $d_{G,b}$, the sum taken over the set of blocks b of $\mathcal{O}G$. Extending our earlier notation, we set

$$L^0(G, b) = \ker(d_{G,b}).$$

That is, $L^0(G, b)$ consists of all generalised characters in $\mathbb{Z}\mathrm{Irr}_K(G, b)$ that vanish on all p'-elements of G. Clearly $L^0(G) = \oplus_b L^0(G, b)$, the direct sum taken over the set of blocks of $\mathcal{O}G$. We denote further by $\mathrm{Pr}_\mathcal{O}(G, b)$ the subgroup of $\mathbb{Z}\mathrm{Irr}_K(G, b)$ generated by the characters of finitely generated projective $\mathcal{O}Gb$-modules, and by $\mathrm{Pr}_k(G, b)$ the subgroup of $\mathbb{Z}\mathrm{IBr}_k(G, b)$ generated by the Brauer characters of finitely generated $kG\bar{b}$-modules, where $\bar{b}$ is the canonical image of b in kG. The scalar product on $\mathbb{Z}\mathrm{Irr}_K(G)$ restricts to a scalar product on $\mathbb{Z}\mathrm{Irr}_K(G, b)$, for any block b of $\mathcal{O}G$. The direct sum decomposition $\mathbb{Z}\mathrm{Irr}_K(G) = \oplus_b \mathbb{Z}\mathrm{Irr}_K(G, b)$, with b running over the set of blocks of $\mathcal{O}G$, is clearly a direct sum of pairwise perpendicular subgroups.

Theorem 6.5.11 *Let G be a finite group and b a block of $\mathcal{O}G$. Suppose that the quotient field K of $\mathcal{O}$ is a splitting field of characteristic zero for KGb. Let D be the decomposition matrix of $\mathcal{O}Gb$ and let C be the Cartan matrix of $\mathcal{O}Gb$. The following hold.*

(i) *The decomposition map $d_{G,b} : \mathbb{Z}\mathrm{Irr}_K(G, b) \to \mathbb{Z}\mathrm{IBr}_k(G, b)$ is surjective and induces an isomorphism $\mathrm{Pr}_\mathcal{O}(G, b) \cong \mathrm{Pr}_k(G, b)$.*
(ii) *The matrix C is positive definite, we have ${}^t D \cdot D = C$, and $\det(C) > 0$.*
(iii) *We have $L^0(G, b)^\perp = \mathrm{Pr}_\mathcal{O}(G, b)$; equivalently, $\mathrm{Pr}_\mathcal{O}(G, b)$ consists of all generalised characters in $\mathbb{Z}\mathrm{Irr}_K(G, b)$ that vanish on all p-singular elements of G.*

Proof We may assume that K is a splitting field for KG. Statement (i) is an immediate consequence of the surjectivity of the map d_G in 5.14.1, together with the obvious fact that d_G is the sum of the maps $d_{G,b}$ mentioned before. Since the decomposition matrix and the Cartan matrix of $\mathcal{O}G$ are obtained blockwise, by piecing together the decomposition and Cartan matrices of the blocks of $\mathcal{O}G$, statement (ii) follows from 5.14.2. Finally, (iii) follows from 4.17.3. □

Proposition 6.5.12 *Let G be a finite group, b a block of $\mathcal{O}G$ and P a defect group of b. Let $\chi \in \mathbb{Z}\mathrm{Irr}_K(G, b)$ and $\varphi \in \mathbb{Z}\mathrm{IBr}_k(G, b)$. The p-part of $|G : P|$ divides $\chi(1)$ and $\varphi(1)$.*

Proof This is an immediate consequence of the last statement of Theorem 6.1.7, applied over $\mathcal{O}$ and over k. □

With the notation of the previous theorem, extending scalars yields a direct sum of perpendicular $\mathbb{Q}$-vector spaces

$$\mathbb{Q} \otimes_\mathbb{Z} \mathbb{Z}\mathrm{Irr}_K(G, b) = \mathbb{Q} \otimes_\mathbb{Z} L^0(G, b) \oplus \mathbb{Q} \otimes_\mathbb{Z} \mathrm{Pr}_\mathcal{O}(G, b).$$

The decomposition map $d_{G,b}$ coincides with the canonical projection of $\mathbb{Q} \otimes_\mathbb{Z} \mathbb{Z}\mathrm{Irr}_K(G, b)$ onto the second summand $\mathbb{Q} \otimes_\mathbb{Z} \mathrm{Pr}_\mathcal{O}(G, b)$. Since the decomposition map $d_{G,b}$ is surjective with kernel $L^0(G, b)$, mapping $\mathrm{Pr}_\mathcal{O}(G, b)$ isomorphically to $\mathrm{Pr}_k(G, \bar{b})$, we have an isomorphism of abelian groups

$$\mathbb{Z}\mathrm{Irr}_K(G, b)/(L^0(G, b) \oplus \mathrm{Pr}_\mathcal{O}(G, b)) \cong \mathbb{Z}\mathrm{IBr}_k(G, b)/\mathrm{Pr}_k(G, \bar{b}).$$

The order of this group is the determinant of the Cartan matrix of b, and the orders of the cyclic direct factors of this group are the elementary divisors of the Cartan matrix of b.

Theorem 6.5.13 *Let G be a finite group and b a block of $\mathcal{O}G$. Suppose that K is a splitting field for KGb. Denote by a and d the integers such that p^a is the order of a Sylow p-subgroup of G and such that p^d is the order of a defect group of b. Let $\chi \in \mathbb{Z}\mathrm{Irr}_K(G, b)$ or $\chi \in \mathbb{Z}\mathrm{IBr}_K(G, b)$. Denote by χ^0 the class function on G defined, for $x \in G$, by $\chi^0(x) = \chi(x)$ if x is a p'-element, and $\chi^0(x) = 0$, otherwise.*

(i) We have $\chi^0 \in \mathbb{Z}\mathrm{Irr}_K(G, b)$ if and only if p^a divides $\chi(1)$.
(ii) We have $p^d \chi^0 \in \mathrm{Pr}_\mathcal{O}(G, b)$.

Proof The surjectivity of the decomposition map in 6.5.11 implies that every element in $\mathbb{Z}\mathrm{IBr}_k(G, b)$ is the restriction to $G_{p'}$ of an element in $\mathbb{Z}\mathrm{Irr}_K(G, b)$, and hence it suffices to show the result for $\chi \in \mathbb{Z}\mathrm{Irr}_K(G, b)$. Suppose first that χ is an element in $\mathbb{Z}\mathrm{Irr}_K(G, b)$ with the property that χ^0 is again a generalised character. Then $\mathrm{Res}_P^G(\chi^0)$ is a generalised character of a Sylow p-subgroup S of G. But χ^0 vanishes on p-singular elements, hence $\mathrm{Res}_S^G(\chi^0)$ is a multiple of the regular character of S, which forces that p^a divides $\chi(1) = \chi^0(1)$. For the other implication in (i) we use Brauer's characterisation of characters. Any elementary subgroup of G with respect to a prime not necessarily equal to p can be written as a direct product $R \times E$, for some p-subgroup R of G and some p'-subgroup E of $C_G(R)$. Since p^a divides $\chi(1)$, the restriction to R of χ^0 is an integer multiple of the regular character of R, and the restriction of χ^0 to E is equal to that of χ, hence also a generalised character of E. Since $\mathrm{Res}_{R \times E}^G(\chi^0)$ vanishes, by construction, outside of E it follows that $\mathrm{Res}_{R \times E}^G(\chi^0)$ is the product of the restrictions of χ^0 to R and E, hence a generalised character.

Thus, by 3.7.2, χ^0 is a generalised character. Since χ and χ^0 coincide on p'-elements we have $d_G(\chi^0) = d_G(\chi) \in \mathbb{Z}\mathrm{IBr}_k(G, b)$. Since moreover d_G is compatible with the block decomposition of $\mathcal{O}G$ it follows that χ^0 belongs indeed to $\mathbb{Z}\mathrm{Irr}_K(G, b)$. This shows (i). Setting $\psi = p^d \chi$, we get that $\psi^0 = p^d \chi^0$. Since p^{d-a} divides $\chi(1)$ by 6.5.12, we get that p^a divides $\psi(1)$, and hence ψ^0 is a generalised character associated with b, by (i). But ψ^0 also vanishes on p-singular elements, hence belongs to $\mathrm{Pr}_{\mathcal{O}}(G, b)$, by 4.17.3 (ii). $\qquad\square$

Remark 6.5.14 With the notation from the previous theorem, we have $\chi^0 = (t_G^1 \circ d_G^1)(\chi)$, where d_G^1 and t_G^1 are as in 5.15.1 and 5.15.13, respectively.

Corollary 6.5.15 *Let G be a finite group, b a block of kG, and P a defect group of b. Suppose that k is a splitting field for kGb. The elementary divisors of the Cartan matrix C of b divide $|P|$. In particular, $\det(C)$ is a power of p.*

Proof The second statement in 6.5.13 implies that the abelian group $\mathbb{Z}\mathrm{IBr}_k(G, b)/\mathrm{Pr}_k(G, b)$ is annihilated by $|P|$, whence the result. $\qquad\square$

There are more precise statements regarding the elementary divisors of the Cartan matrix of a block; see e.g. [152, Chapter 5, Theorem 11.6]. We conclude this section with some consequences, from [106], of the material developed so far which we will be used later, notably in the Section 10.7. The character of a block algebra with the conjugation action by G detects the numbers of irreducible characters and Brauer characters as follows.

Proposition 6.5.16 ([106, Lemma 4.4]) *Let G be a finite group and B a block algebra of $\mathcal{O}G$. Suppose that the quotient field K of $\mathcal{O}$ is a splitting field of characteristic zero for $K \otimes_{\mathcal{O}} B$. Denote by γ the character of B as an $\mathcal{O}G$-module, with G acting by conjugation on B. For any $x \in G$, we have*

$$\gamma(x) = \sum_{\chi \in \mathrm{Irr}(B)} \chi(x)\chi(x^{-1}).$$

Moreover, we have

$$|\mathrm{Irr}_K(B)| = \langle \gamma, 1 \rangle,$$

$$|\mathrm{IBr}_k(B)| = \langle \gamma, 1 \rangle',$$

where 1 denotes the trivial character of G.

Proof The character γ is obtained from restricting the character β in 6.5.3 of B as an $\mathcal{O}(G \times G)$-module to the diagonal subgroup of $G \times G$; that is, after identifying that diagonal subgroup with G, we have $\gamma(x) = \sum_{\chi \in \mathrm{Irr}(B)} \chi(x)\chi(x^{-1})$ for all $x \in G$ as stated. Thus $\langle \gamma, 1 \rangle = \sum_{\chi \in \mathrm{Irr}_K(B)} \langle \chi, \chi \rangle$. This sum is equal to

$|\mathrm{Irr}_K(B)|$ by the orthogonality relations, whence the first equality. The restriction of $\chi \in \mathrm{Irr}(B)$ to $G_{p'}$ is equal to $\sum_{\varphi \in \mathrm{IBr}(B)} d_\varphi^\chi \varphi$. By 5.14.4, if $\Phi \in \mathrm{IPr}_{\mathcal{O}}(G)$ is the character of a projective cover of a simple $\mathcal{O}G$-module with Brauer character φ, then $\Phi = \sum_{\chi \in \mathrm{Irr}_K(G)} d_\varphi^\chi \chi$. Combining these facts yields $\langle \gamma, 1 \rangle' = \sum_\chi \langle \chi, \chi \rangle' = \sum_{\chi, \varphi} d_\varphi^\chi \langle \varphi, \chi \rangle' = \sum_\varphi \langle \varphi, d_\varphi^\chi \chi \rangle' = \sum_{\Phi, \varphi} \langle \varphi, \Phi \rangle'$, where in the sums χ, φ, Φ run over $\mathrm{Irr}(B)$, $\mathrm{IBr}(B)$, $\mathrm{IPr}(B)$, respectively. Brauer's reciprocity 5.14.8 implies that in the last sum, the summands are zero unless Φ is the character of a projective cover of a simple $\mathcal{O}G$-module with Brauer character φ, in which case this summand is 1. The second equality follows. □

We have the following subtle generalisation of Theorem 6.2.6 (ii), due to Knörr and Robinson, which in addition to some of the above material makes use of Thompson's $A \times B$-Lemma.

Theorem 6.5.17 ([106, Lemma 3.1]) *Let G be a finite group. Let m be a nonnegative integer, and for $0 \le i \le m$ let Q_i be a p-subgroup of G such that Q_i is a proper subgroup of Q_{i+1} for $0 \le i \le m - 1$. Set $N = \cap_{i=0}^m N_G(Q_i)$. Every block idempotent of $\mathcal{O}N$ is contained in $\mathcal{O}C_G(Q_m)$.*

Proof We have $C_G(Q_m) \le N$, so the statement makes sense. If $m = 0$, then $N = N_G(Q_0)$, and hence in that case the result follows from 6.2.6 (ii). Assume that $m > 0$. The issue is that Q_m need not normalise all of the Q_i, and hence Q_m need not be contained in N, so we cannot directly apply 6.2.6 (ii). Since N and the subgroups Q_i are contained in $N_G(Q_m)$, we may replace G by $N_G(Q_m)$; that is, we may assume that Q_m is normal in G. Then $N = \cap_{i=0}^{m-1} N_G(Q_i)$, because Q_m is already normal in G. Arguing by induction, every block idempotent of $\mathcal{O}N$ is contained in $\mathcal{O}C_G(Q_{m-1})$. Setting $U = O_p(C_G(Q_{m-1}))$, it follows from 6.2.6 (ii) that every block of $\mathcal{O}N$ is contained in $\mathcal{O}C_G(U)$. Note that U contains $C_{Q_m}(Q_{m-1})$ because Q_m is normal in G. Let b be a block idempotent in $\mathcal{O}N$. Write $b = \sum_{x \in N} \lambda_x x$ for some coefficients $\lambda_x \in \mathcal{O}$. Let $y \in N$ such that $\lambda_y \ne 0$. It follows from 6.5.4 that y is a p'-element. By the induction hypothesis, y centralises Q_{m-1}. The group $\langle y \rangle \times Q_{m-1}$ acts in the obvious way on Q_m (using yet again that Q_m is normal in G). By the above, y acts trivially on $C_{Q_m}(Q_{m-1})$. But then Thompson's $A \times B$-Lemma [76, Theorem 5.3.4] implies that y acts trivially on Q_m, or equivalently, $y \in C_G(Q_m)$. The result follows. □

If Q is a normal p-subgroup of a finite group G, then by Corollary 1.11.8, the kernel of the canonical algebra homomorphism $kG \to kG/Q$ is contained in the radical of kG, hence contains no idempotent. In particular, if b is a block of kG, then the image $\bar{b}$ of b in kG/Q is a central idempotent – but $\bar{b}$ need no longer be a block of kG/Q. Nonetheless, the passage from b to $\bar{b}$ commutes with Brauer homomorphisms in the following sense.

Proposition 6.5.18 ([106, Lemma 3.5]) *Let G be a finite group, Q a normal p-subgroup of G and R a p-subgroup of G containing Q. Set $\bar{G} = G/Q$, $\bar{R} = R/Q$, and for $a \in kG$ denote by $\bar{a}$ the image of a in $k\bar{G}$. Let b be an idempotent in $Z(kG)$. We have*

$$\overline{\mathrm{Br}_R(b)} = \mathrm{Br}_{\bar{R}}(\bar{b}).$$

Proof Write $b = \sum_{x \in G} \lambda_x x$ with coefficients $\lambda_x \in k$. Let $y \in G$ such that $\lambda_y \neq 0$. It follows from 6.5.4 that y is a p'-element. It follows from 6.2.6 (ii) that $y \in C_G(Q)$. Suppose that the image $\bar{y}$ of y in $\bar{G}$ belongs to $C_{\bar{G}}(\bar{R})$. Then $[R, y] \subseteq Q$. Since y centralises Q, it follows that $[R, y, y] = \{1\}$. A standard result on coprime group actions [76, Theorem 5.3.6] implies that $[R, y] = \{1\}$, or equivalently, that $y \in C_G(R)$. This shows that $\mathrm{Br}_{\bar{R}}(\bar{b})$ is the image of $\mathrm{Br}_R(b)$ as stated.   □

Remark 6.5.19 Decomposition matrices, Cartan matrices, and decomposition maps for blocks can be detected at the source algebra level because they are invariant under Morita equivalences. Let G be a finite group, b a block of $\mathcal{O}G$, P a defect group of b, and $i \in (\mathcal{O}Gb)^P$ a source idempotent. Set $A = i\mathcal{O}Gi$. Suppose that K, k are splitting fields for G and that $\mathrm{char}(K) = 0$. The standard Morita equivalence 6.4.6 between $\mathcal{O}Gb$ and A implies that if X is a simple KGb-module, then iX is a simple $K \otimes_{\mathcal{O}} A$-module. The map sending the character χ of X to the character, denoted $i\chi$, of iX, induces a bijection $\mathrm{Irr}_K(G, b) \cong \mathrm{Irr}_K(A)$. This bijection induces an isomorphism $\mathbb{Z}\mathrm{Irr}_K(G, b) \cong \mathbb{Z}\mathrm{Irr}_K(A) = R_K(A)$. Since this isomorphism is induced by a Morita equivalence, it induces isomorphisms $\mathrm{Pr}_{\mathcal{O}}(G, b) \cong \mathrm{Pr}_{\mathcal{O}}(A)$ and $L^0(G, b) \cong L^0(A)$. We have an obvious commutative diagram

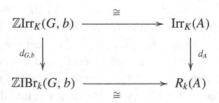

where $d_A : R_K(A) \to R_k(A)$ is the decomposition map as in 4.17.1.

6.6 Blocks with a central defect group

The main result of this section shows that if a block b of $\mathcal{O}G$ has a defect group P that is contained in $Z(G)$ and if k is large enough, then the source algebras of $\mathcal{O}Gb$ are isomorphic to $\mathcal{O}P$. In particular, $\mathcal{O}Gb$ is Morita equivalent to $\mathcal{O}P$. We start with a special case. A block b of $\mathcal{O}G$ has defect zero if and only if it has the trivial group $\{1\}$ as its defect group. The following theorem is a list of characterisations of blocks of defect zero.

Theorem 6.6.1 *Let G be a finite group and b a block of kG. Set $B = kGb$. The following are equivalent.*

  (i) The block b has defect zero.
  (ii) The algebra B has a projective simple module.
 (iii) The algebra B is simple.

In particular, the map sending a simple kG-module S to the unique block of kG to which S belongs induces a bijection between the isomorphism classes of simple projective kG-modules and the defect zero blocks of kG.

Proof If (i) holds, then every finite-dimensional B-module is relatively 1-projective, hence projective. Thus (i) implies (ii). Suppose that (ii) holds. Then B has a projective simple module. Since B is symmetric, this simple module is also injective. Thus 1.14.7 implies that B is simple, and hence (ii) implies (iii). If (iii) holds, then 1.16.21 implies that B is separable, hence isomorphic to a direct summand of $B \otimes_k B$ as a B-B-bimodule. It follows from 6.2.1 that b has defect zero, so (iii) implies (i). The last statement follows. $\square$

There are many more characterisations of defect zero blocks in the situation where k and K are splitting fields (some of which do not require this hypothesis; see [35]).

Theorem 6.6.2 *Let G be a finite group, b be a block of $\mathcal{O}G$ and $\chi \in \mathrm{Irr}_K(G, b)$. Denote by $\bar{b}$ the image of b in kG. Suppose that k is a splitting field for $kG\bar{b}$ and that K is a splitting field for KGb such that $\mathrm{char}(K) = 0$. Then the following are equivalent:*

  (i) The block b has defect zero.
  (ii) The algebra $kG\bar{b}$ has a projective simple module.
 (iii) We have $kG\bar{b} \cong M_n(k)$ for some positive integer n.
 (iv) We have $\mathcal{O}Gb \cong M_n(\mathcal{O})$ for some positive integer n.
  (v) We have $KGb \cong M_n(K)$ for some positive integer n.
 (vi) $\mathrm{Irr}_K(G, b)] = \{\chi\}$.
 (vii) We have $b = e(\chi)$.
(viii) The order of a Sylow p-subgroup of G divides $\chi(1)$.
 (ix) The integer $\frac{|G|}{\chi(1)}$ is prime to p.
  (x) The decomposition matrix of $\mathcal{O}Gb$ is (1).

40 *Blocks and Source Algebras*

(xi) *The Cartan matrix of $k G \bar{b}$ is (1).*
(xii) *We have $\chi(x) = 0$ for any p-singular element $x \in G$.*
(xiii) *We have $\chi(u) = 0$ for any nontrivial p-element $u \in G$.*
(xiv) *χ is the character of a projective $\mathcal{O}Gb$-module.*

Proof The equivalence of (i), (ii), (iii) follows from 6.6.1. The implication (iii) $\Rightarrow$ (iv) follows from 4.7.12. The implication (iv) $\Rightarrow$ (v) is trivial. The equivalences (v) $\Leftrightarrow$ (vi) $\Leftrightarrow$ (vii) and (viii) $\Leftrightarrow$ (ix) are obvious. If (viii) holds, then $e(\chi) = \frac{\chi(1)}{|G|} \sum_{x \in G} \chi(x^{-1}) x$ belongs to $\mathcal{O}G$, hence is a block idempotent. Thus (viii) $\Rightarrow$ (vii). The decomposition map $d : \mathbb{Z}\mathrm{Irr}_K(G, b) \to \mathbb{Z}\mathrm{IBr}_k(G, b)$ is surjective; thus, if b has χ as its unique irreducible character then $\mathbb{Z}\mathrm{Irr}_K(G, b) \cong \mathbb{Z}$ with χ as generator, this forces $d(\chi) = \varphi$, with φ the unique irreducible Brauer character of b, and hence the decomposition matrix D of $\mathcal{O}Gb$ is equal to $D = (1)$. This proves (vi) $\Rightarrow$ (x). Since $D^t \cdot D$ yields the Cartan matrix of $k G \bar{b}$ we get (x) $\Rightarrow$ (xi). If the Cartan matrix of $k G \bar{b}$ is (1) then $k G \bar{b}$ has a unique (up to isomorphism) simple $k G \bar{b}$ module V, which is also the unique (up to isomorphism) projective indecomposable $k G \bar{b}$-module. Thus the structural homomorphism $k G \bar{b} \to \mathrm{End}_k(V)$ is surjective (because V is simple) and injective (because V is the unique projective indecomposable), which shows (xi) $\Rightarrow$ (iii). If (iv) holds then χ is the character of the unique projective indecomposable $\mathcal{O}Gb$-module U, hence χ vanishes outside p-regular elements by 5.12.14. This proves (iv) $\Rightarrow$ (xii) and (iv) $\Rightarrow$ (xiv). The implication (xii) $\Rightarrow$ (xiii) is trivial. Let S be a Sylow p-subgroup of G. The number $\langle \mathrm{Res}_S^G(\chi), 1_S \rangle_S$ is an integer, and if (xiii) holds then this integer is equal to $\frac{\chi(1)}{|S|} \sum_{x \in S} \chi(x) = \frac{\chi(1)}{|S|}$, which shows the implication (xiii) $\Rightarrow$ (viii). Suppose that (iii) holds; that is, $k G \bar{b} \cong \mathrm{End}_k(V)$, where V is as before the unique (up to isomorphism) simple $k G \bar{b}$-module. Since V is also projective, Higman's criterion 2.6.2 implies that $\mathrm{Id}_V \in (\mathrm{End}_k(V))_1^G$. Through the isomorphism $k G \bar{b} \cong \mathrm{End}_k(V)$ this translates to $b \in (k G \bar{b})_1^G$, and hence b is a block of defect zero. This shows the implication (iii) $\Rightarrow$ (ii). If (xiv) holds then so does (xii) by 5.12.14, which completes the proof. $\square$

The proof of the equivalence of the statements (i), (ii), (iii), (iv) and (v) above does not require K to be a splitting field. In fact, it shows that if b has defect zero and k is a splitting field for $k G \bar{b}$, then K is automatically a splitting field for KGb. This is used in the next observation, showing that the source algebras of defect zero blocks are trivial:

Proposition 6.6.3 *Let G be a finite group and b a block of $\mathcal{O}G$ of defect zero. Suppose that k is a splitting field for b. The source algebras of $\mathcal{O}Gb$ are isomorphic to the trivial $\mathcal{O}$-algebra $\mathcal{O}$.*

Proof By 6.6.2, we have $\mathcal{O}Gb \cong M_n(\mathcal{O})$ for some positive integer n. A source idempotent of $\mathcal{O}Gb$ corresponds to a primitive idempotent j in $M_n(\mathcal{O})$, and clearly $jM_n(\mathcal{O})j \cong \mathcal{O}$. □

We have encountered projective simple modules for twisted finite group algebras in the context of multiplicity modules. Projective simple modules play a crucial role in the context of Alperin's weight conjecture 1.15.12 for finite groups and 6.10.2 below for blocks. The existence of a projective simple module has the following implication.

Proposition 6.6.4 *Suppose that k is algebraically closed. Let G be a finite group, $\alpha \in H^2(G; k^\times)$. If $k_\alpha G$ has a projective simple module, then $O_p(G) = \{1\}$.*

Proof By 1.2.18 there is an extension G' of G by a finite central p'-subgroup Z of G' and an idempotent e in $Z(kG')$ such that $kG'e \cong k_\alpha G$. Thus $kG'e$ has a projective simple module, or equivalently, kG' has a block of defect zero. It follows from 6.2.6 that $O_p(G')$ is trivial. By elementary group theory, the image of $O_p(G')$ in G is equal to $O_p(G)$, and hence $O_p(G)$ is trivial as well. □

Proposition 6.6.5 *Let G be a finite group and P a p-subgroup of $Z(G)$. The canonical algebra homomorphism $\mathcal{O}G \to \mathcal{O}G/P$ induces a bijection between the blocks of $\mathcal{O}G$ with P as a defect group and the blocks of defect zero of $\mathcal{O}G/P$. In particular, if B is a block algebra of $\mathcal{O}G$ with the central defect group P, then $k \otimes_{\mathcal{O}} B$ has a unique isomorphism class of simple modules.*

Proof Let b be a block of $\mathcal{O}G$ with defect group P. Set $B = \mathcal{O}Gb$. For $a \in \mathcal{O}G$ denote by $\bar{a}$ the image of a in $\mathcal{O}G/P$. Set $\bar{G} = G/P$ and $\bar{B} = \mathcal{O}\bar{G}\bar{b}$. Since P is central in G, we have $B^P = B$. Moreover, if $a \in B$, then the image of $\mathrm{Tr}_P^G(a)$ in $\bar{B}$ is $\mathrm{Tr}_1^{\bar{G}}(\bar{a})$. Thus the canonical homomorphism $\mathcal{O}G \to \mathcal{O}\bar{G}$ maps B_P^G onto $\bar{B}_1^{\bar{G}}$, and so $\bar{b}$ is a defect zero block of $\mathcal{O}\bar{G}$, or equivalently, $\bar{B}$ is a matrix algebra over $\mathcal{O}$. The kernel of the canonical map $\mathcal{O}G \to \mathcal{O}\bar{G}$ is the ideal generated by the augmentation ideal of $\mathcal{O}P$, hence contained in $J(\mathcal{O}G)$. Thus the result follows from 4.7.5 applied to the ideals R_P^G and $\bar{B}_1^{\bar{G}}$ in $Z(B)$ and $Z(\bar{B})$, respectively. The last statement follows from 6.6.1. □

The structure of blocks with a central defect group is described in the next theorem.

Theorem 6.6.6 *Let G be a finite group and b a block of $\mathcal{O}G$ with a defect group P contained in $Z(G)$. Set $B = \mathcal{O}Gb$. Suppose that k is a splitting field for B. Then the following hold.*

(i) *As an interior P-algebra, $\mathcal{O}P$ is a source algebra of B.*

(ii) *Every primitive idempotent in B is a source idempotent, there is a unique conjugacy class of source idempotents and a unique isomorphism class of source modules for B.*

(iii) *As an $\mathcal{O}$ algebra, B is isomorphic to a matrix algebra over $\mathcal{O}P$.*

(iv) *Let U be a projective indecomposable B-module. The p-part of $\mathrm{rk}_{\mathcal{O}}(U)$ is equal to the p-part of $|G|$.*

(v) *The algebra $k \otimes_{\mathcal{O}} B$ has a unique isomorphism class of simple modules. If K is a splitting field for B, then B has $|P|$ irreducible characters.*

Proof If $P = 1$, then all statements follow from 6.6.2. For the general case, set $\bar{G} = G/P$, denote by $\bar{b}$ the image of b in $\mathcal{O}\bar{G}$, and set $\bar{B} = \mathcal{O}\bar{G}\bar{b}$. By 6.6.5, $\bar{B}$ is a defect zero block of $\mathcal{O}\bar{G}$, hence a matrix algebra over $\mathcal{O}$ by the assumptions on k being large enough. If $i \in B^P = B$ is a primitive idempotent, then its image $\bar{i}$ is a primitive idempotent in the matrix algebra $\bar{B}$, which in particular shows the uniqueness statements in (ii) about source idempotents and modules, because a matrix algebra has a unique point. We have $\bar{i}\mathcal{O}\bar{G}\bar{i} \cong \mathcal{O}$. It follows that $i\mathcal{O}Gi = \mathcal{O}Pi + I(\mathcal{O}P)i\mathcal{O}Gi$, hence $i\mathcal{O}Gi = \mathcal{O}Pi$ by Nakayama's Lemma. But $i\mathcal{O}Gi$ is also projective as a left $\mathcal{O}P$-module and hence $i\mathcal{O}Gi \cong \mathcal{O}P$. Since B is Morita equivalent to the local algebra $\mathcal{O}P$ we have $B \cong M_n(\mathcal{O}P)$ for some positive integer by 2.8.9, which proves (iii). For (iv), we may choose $U = \mathcal{O}Gi$. The $\bar{B}$-module $\bar{U} = \bar{B}\bar{i}$ is a projective indecomposable module for the defect zero block $\bar{B}$. It follows from 6.6.2 and 3.5.7 that its rank has the same p-part as $|\bar{G}|$. Since U is projective as an $\mathcal{O}P$-module, we have $\mathrm{rk}_{\mathcal{O}}(U) = |P| \cdot \mathrm{rk}_{\mathcal{O}}(\bar{U})$, whence (v). The last statement is an immediate consequence of the fact that B is Morita equivalent to its source algebra $\mathcal{O}P$. $\qquad\square$

Corollary 6.6.7 *Let G be a finite group and b a block of $\mathcal{O}G$ with a defect group P such that $G = PC_G(P)$. Then b is a block of $\mathcal{O}C_G(P)$ having $Z(P)$ as a defect group, every primitive idempotent $i \in \mathcal{O}C_G(P)b$ is a source idempotent of b as block of $\mathcal{O}G$, there is a unique conjugacy class of source idempotents in $(\mathcal{O}Gb)^P$ and a unique isomorphism class of source modules for b. We have an isomorphism of interior P-algebras $i\mathcal{O}Gi \cong \mathcal{O}P$. In particular, there is a positive integer n such that $\mathcal{O}Gb \cong M_n(\mathcal{O}P)$ as $\mathcal{O}$-algebras.*

Proof By the assumptions on G we have $G/P \cong C_G(P)/Z(P)$. By 6.2.7, b is a block of $\mathcal{O}C_G(P)$. Thus $b = \mathrm{Tr}_P^G(c) = \mathrm{Tr}_{Z(P)}^{C_G(P)}(c)$ for some $c \in \mathcal{O}C_G(P)$. Thus $Z(P)$ contains a defect group of b as a block of $\mathcal{O}C_G(P)$, and since $Z(P)$ is normal in $C_G(P)$ it is equal to a defect group of b as a block of $\mathcal{O}C_G(P)$. If i is a primitive idempotent in $\mathcal{O}C_G(P)b$, then $\mathrm{Br}_P(i)$ is the image of i in $kC_G(P)\bar{b}$, hence nonzero, and i is primitive in $(\mathcal{O}Gb)^P$ by 6.2.6 (iii). This shows that i

is a source idempotent for b as a block of $\mathcal{O}G$ and that any source idempotent is conjugate to i. By 6.6.6 we have $i\mathcal{O}C_G(P)i \cong \mathcal{O}Z(P)$. Since $G = PC_G(P)$ it follows that the structural map $\mathcal{O}P \to i\mathcal{O}Gi$ is surjective. But $i\mathcal{O}Gi$ is projective as an $\mathcal{O}P$-module, so this map is also injective. Since $\mathcal{O}Gb$ is Morita equivalent to the local algebra $\mathcal{O}P$ it follows from 2.8.9 that $\mathcal{O}Gb \cong M_n(\mathcal{O}P)$. $\qquad\square$

Corollary 6.6.8 *Let G be a finite group, b a block of $\mathcal{O}G$ with a defect group P, and let $i \in (\mathcal{O}Gb)^P$ be a source idempotent of b. Set $A = i\mathcal{O}Gi$. Denote by e the block of $kC_G(P)$ satisfying $\mathrm{Br}_P(i)e \neq 0$. Then $\mathrm{Br}_P(i)$ is a source idempotent of e, the interior $Z(P)$-algebra $A(P)$ is a source algebra of e, and we have an isomorphism $A(P) \cong kZ(P)$ as interior $Z(P)$-algebras.*

Proof By 6.3.10, $Z(P)$ is a defect group of $kC_G(P)e$. Since Br_P maps $(\mathcal{O}G)^P$ onto $kC_G(P)$, the idempotent $j = \mathrm{Br}_P(i)$ is primitive in $kC_G(P)e$. Thus 6.6.6 implies that j is a source idempotent of $kC_G(P)$ and that the source algebra $A(P) = jkC_G(P)j$ is isomorphic to $kZ(P)$ as an interior $Z(P)$-algebra. $\qquad\square$

Blocks with a central defect group are a special case of blocks with a normal defect group, which will be described in §6.14.

6.7 Brauer's First Main Theorem

The following theorem, known as the Brauer correspondence, describes a bijection between blocks of a finite groups G with a given defect group P and blocks of any subgroups of G containing $N_G(P)$ with P as defect group.

Theorem 6.7.1 (Brauer correspondence) *Let G be a finite group and let P be a p-subgroup of G. Let H be a subgroup of G containing $N_G(P)$. For any block b of $\mathcal{O}G$ with P as a defect group there is a unique block c of H with P as defect group such that $\mathrm{Br}_P(b) = \mathrm{Br}_P(c)$, and this correspondence defines a bijection between the sets of blocks of $\mathcal{O}G$ and of $\mathcal{O}H$ with P as a defect group.*

Proof We have $C_G(P) \subseteq N_G(P) \subseteq H$, so $\mathrm{Br}_P(b)$ and $\mathrm{Br}_P(c)$ are both idempotents in $kC_G(P) = kC_H(P)$. By 5.4.5 we have $\mathrm{Br}_P((\mathcal{O}G)_P^G) = (kC_G(P))_P^{N_G(P)}$. It follows from 4.7.19 that Br_P induces a bijection between primitive idempotents in $(\mathcal{O}G)_P^G$ not contained in $\ker(\mathrm{Br}_P)$ and primitive idempotents in the ideal $(kC_G(P))_P^{N_G(P)}$ of the algebra $(kC_G(P))^{N_G(P)}$. The same argument applies to H, and so we get a bijection as required. $\qquad\square$

The Brauer correspondence can be interpreted as a particular case of the Green correspondence:

Theorem 6.7.2 *Let G be a finite group, P a p-subgroup of G and b a block of $\mathcal{O}G$ with P as a defect group. Let H be a subgroup of G containing $N_G(P)$ and c the block of $\mathcal{O}H$ having P as a defect group and satisfying $\mathrm{Br}_P(b) = \mathrm{Br}_P(c)$. Set $B = \mathcal{O}Gb$ and $C = \mathcal{O}Hc$.*

(i) *The $\mathcal{O}(H \times H)$-module C is the Green correspondent of the $\mathcal{O}(G \times G)$-module B.*

(ii) *As a C-C-bimodule, C is isomorphic to a direct summand of cBc.*

(iii) *Any indecomposable direct summand of cBc as an $\mathcal{O}(H \times H)$-module that is not isomorphic to C has a vertex of the form $^{(1,y)}\Delta Q$ for some proper subgroup Q of P and some element $y \in G$ such that ^yQ is contained in P.*

Proof The group $H \times H$ contains the normaliser N in $G \times G$ of the diagonal subgroup ΔP. Since $\mathrm{Br}_P(b) = \mathrm{Br}_P(c)$ we have $B(P) \cong C(P)$ as $\mathcal{O}N$-modules. Both B and C are trivial source modules over $\mathcal{O}(G \times G)$ and $\mathcal{O}(H \times H)$, respectively. The characterisation in 5.10.5 of the Green correspondents of trivial source modules, together with the fact that trivial source modules over k lift uniquely, up to isomorphism, to trivial source modules over $\mathcal{O}$ in 5.10.2, implies that C is the Green correspondent of B with respect to $H \times H$. This proves (i). In particular, C is isomorphic to a direct summand of B as an $\mathcal{O}H$-$\mathcal{O}H$-bimodule. Since c acts as identity on C by left or right multiplication, it follows that C is isomorphic to a direct summand of cBc as a C-C-bimodule. This shows (ii). The Green correspondence implies that any other indecomposable direct summand Y of cBc as an $\mathcal{O}(H \times H)$-module has a vertex in $^{(x,y)}\Delta P \cap H \times H$ for some $(x, y) \in (G \times G) \setminus (H \times H)$, so at least one of x, y does not belong to H. Since c acts as identity on both sides of cBc, it follows that Y has a vertex contained in $P \times P$. Thus we may choose (x, y) such that a vertex of Y is contained in $^{(x,y)}\Delta P \cap P \times P$. Since at least one of x, y is not in G, hence not in $N_G(P)$, it follows that Y has a vertex of the form $^{(x,y)}\Delta Q$ for some proper subgroup Q of P such that ^xQ and ^yQ are contained in P. After replacing Q by ^xQ and y by yx^{-1} we obtain statement (iii).   □

The compatibility of the Green correspondence with the Brauer correspondence extends to modules in the following way:

Theorem 6.7.3 *Let G be a finite group, b a block of $\mathcal{O}G$, M an indecomposable $\mathcal{O}Gb$-module with vertex Q. The Green correspondent $f(M)$ belongs to a block c of $\mathcal{O}N_G(Q)$ satisfying $\mathrm{Br}_Q(b)\bar{c} = \bar{c}$, where $\bar{c}$ is the canonical image of c in $kN_G(Q)$.*

Proof Since any vertex of M is contained in a defect group of b we have $\mathrm{Br}_Q(b) \neq 0$. Thus $\mathrm{Br}_Q(b)$ is a sum of block idempotents of $kN_G(Q)$, and hence there is a central idempotent d in $Z(\mathcal{O}N_G(Q))$ that lifts $\mathrm{Br}_Q(b)$. Then $b = bd + (b - bd)$ is a sum of two orthogonal idempotents in $(\mathcal{O}G)^{N_G(Q)}$, and we have $b - bd \in \ker(\mathrm{Br}_Q)$. It follows that $M = bM = bdM \oplus (b - bd)M$ is a decomposition of M as a direct sum of $\mathcal{O}N_G(Q)$-modules. Moreover, since $b - bd \in \ker(\mathrm{Br}_Q)$, no summand of $\mathrm{Res}_Q^{N_G(Q)}((b - bd)M)$ has Q as a vertex, by Higman's criterion. Thus the Green correspondent $f(M)$, which is an indecomposable direct summand of $\mathrm{Res}_{N_G(Q)}^G(M)$ with vertex Q must be a direct summand of $bdM = dM$, whence the result. $\qquad\square$

Source idempotents of blocks and their Brauer correspondents are related as follows:

Proposition 6.7.4 ([68, 4.10]) *Let G be a finite group and b a block of $\mathcal{O}G$ with maximal (G, b)-Brauer pair (P, e). Set $H = N_G(P, e)$ and let c be the block of $\mathcal{O}H$ that lifts e. Let $j \in (\mathcal{O}Hc)^P$ be a source idempotent of c as block of $\mathcal{O}H$. There is a unique point β of H on $\mathcal{O}Gb$ such that for $f \in \beta$ we have $\mathrm{Br}_P(f) = e$. For any $f \in \beta$, the element $i = fj$ is a source idempotent of the block b of $\mathcal{O}G$, and the $\mathcal{O}(G \times H)$-module $\mathcal{O}Gf$ is indecomposable with vertex ΔP.*

Proof As an $\mathcal{O}(H \times H)$-module, $\mathcal{O}Hc$ is indecomposable with vertex ΔP and trivial source. Since $\mathrm{Br}_P(b)e = e$ there is a primitive idempotent $f \in (\mathcal{O}Gb)^H$ such that $\mathrm{Br}_P(f)e = e$. The $\mathcal{O}(G \times H)$-module $\mathcal{O}Gf$ is indecomposable because f is primitive in $(\mathcal{O}Gb)^H$. We have $\mathrm{Br}_P(f) \neq 0$, and hence ΔP is contained in a vertex of $\mathcal{O}Gf$ as an $\mathcal{O}(G \times H)$-module. Then $\mathrm{Br}_P(i) = \mathrm{Br}_P(fj) = \mathrm{Br}_P(f)\mathrm{Br}_P(j) = e\mathrm{Br}_P(j) \neq 0$ because $\mathrm{Br}_P(j) \in kC_G(P)e$. The $\mathcal{O}(H \times P)$-module $\mathcal{O}Hj$ is indecomposable and has vertex ΔP because $j \in (\mathcal{O}Hc)^P$ is primitive local. As an $\mathcal{O}(G \times P)$-module, $\mathcal{O}Gi$ is isomorphic to a direct summand of $\mathcal{O}Gj = \mathrm{Ind}_{H \times P}^{G \times P}(\mathcal{O}Hj)$. By the Green correspondence, $\mathcal{O}Gj$ has up to isomorphism, a unique indecomposable direct summand with vertex ΔP, and every other summand has a vertex of order strictly smaller than $|P|$. Thus $\mathcal{O}Gi$ has up to isomorphism, a unique indecomposable direct summand with vertex ΔP, and every other summand has a vertex of order strictly smaller than $|P|$. But $\mathcal{O}Gf$ is also isomorphic to a direct summand of $\mathrm{Ind}_{G \times P}^{G \times H}(W)$ for some indecomposable $\mathcal{O}(G \times P)$-module W with vertex ΔP; in particular, ΔP contains a vertex of $\mathcal{O}Gf$, hence is a vertex of $\mathcal{O}Gf$. Moreover, $\mathcal{O}Gi$ is isomorphic to a direct summand of $\mathrm{Res}_{G \times P}^{G \times H}\mathrm{Ind}_{G \times P}^{G \times H}(W)$. By Mackey's formula, this module is a direct sum of the form $\oplus_{x \in [H/P]}{}^{(1,x)}W$. Since H normalises P, all summands have a vertex of order $|P|$, thus all indecomposable direct summands of $\mathcal{O}Gi$ have a vertex of order $|P|$. Combining these observations shows that $\mathcal{O}Gi$ is indecomposable as an $\mathcal{O}(G \times P)$-module, and hence i is primitive in $(\mathcal{O}Gb)^P$. $\square$

This can be reformulated in terms of source modules as follows.

Proposition 6.7.5 ([7, Theorem 5]) *Let G be a finite group and b a block of $\mathcal{O}G$ with maximal (G, b)-Brauer pair (P, e). Set $H = N_G(P, e)$ and let c be the block of $\mathcal{O}H$ that lifts e.*

  (i) *Up to isomorphism, there is a unique indecomposable direct summand X of $\mathrm{Res}_{G \times H}^{G \times G}(\mathcal{O}Gb)$ with vertex ΔP such that X is isomorphic to a direct summand of $\mathrm{Ind}_{H \times H}^{G \times H}(\mathcal{O}Hc)$.*
  (ii) *If N is a source module for the block c of $\mathcal{O}H$, then $M = X \otimes_{\mathcal{O}H} N$ is a source module for the block b of $\mathcal{O}G$.*
  (iii) *The $\mathcal{O}(H \times P)$-module N is isomorphic to a direct summand of $\mathrm{Res}_{H \times P}^{G \times P}(M)$, and every other summand of $\mathrm{Res}_{H \times P}^{G \times P}(M)$ has a vertex of order strictly smaller than $|P|$.*

Proof The existence and uniqueness of X follows from the Green correspondence. Using the notation of 6.7.4 we have $X \cong \mathcal{O}Gf$ and $N \cong \mathcal{O}Hj$, hence $M = X \otimes_{\mathcal{O}H} N \cong \mathcal{O}Gf \otimes_{\mathcal{O}H} \mathcal{O}Hj \cong \mathcal{O}Gfj$. By 6.7.4, this bimodule is indecomposable, whence the statements (i) and (ii). Statement (iii) follows from the Green correspondence. $\square$

The Brauer correspondence applies in particular to $H = N_G(P)$, in which case some of the previous statements can be made more precise. This is the content of Brauer's First Main Theorem, which summarises in particular the considerations in this section.

Theorem 6.7.6 (Brauer's First Main Theorem) *Let G be a finite group, let P be a p-subgroup of G, and let b be a block of $\mathcal{O}G$ with P as a defect group.*

  (i) *There is a unique block c of $\mathcal{O}N_G(P)$ with P as a defect group such that $\mathrm{Br}_P(b) = \mathrm{Br}_P(c)$, and this correspondence defines a bijection between the sets of blocks of $\mathcal{O}G$ and of $\mathcal{O}N_G(P)$ with P as a defect group.*
  (ii) *The block c of $\mathcal{O}N_G(P)$ is contained in $(\mathcal{O}C_G(P))^{N_G(P)}$.*
  (iii) *If e is a block of $\mathcal{O}C_G(P)$ satisfying $ec = e$, then $c = \mathrm{Tr}_{N_G(P,e)}^{N_G(P)}(e)$; that is, c is the sum of an $N_G(P)$-conjugacy class of blocks of $\mathcal{O}C_G(P)$. In particular, $\mathcal{O}N_G(P)e$ and its dual induce a Morita equivalence between $\mathcal{O}N_G(P)c$ and $\mathcal{O}N_G(P, e)e$.*
  (iv) *The image $\bar{e}$ of e in $\mathcal{O}C_G(P)/Z(P)$ is a block of defect zero.*
  (v) *If k is large enough then $|N_G(P, e) : PC_G(P)|$ is prime to p.*

Statement (iii) can be made more precise: not only are the blocks $\mathcal{O}N_G(P)c$ and $\mathcal{O}N_G(P, e)e$ Morita equivalent, but they have in fact isomorphic source algebras. By 6.2.6 (iv), if i is a primitive idempotent in $\mathcal{O}C_G(P)e$, then i is a source idempotent in both $\mathcal{O}N_G(P, e)e$ and $\mathcal{O}N_G(P)c$, and we have an equality of source algebras $i\mathcal{O}N_G(P, e)i = i\mathcal{O}N_G(P)i$. This is a special case of 6.8.3 below.

Proof of Theorem 6.7.6 Statement (i) is the particular case of 6.7.1 applied to $H = N_G(P)$. Statement (ii) and (iii) restate 6.2.6. To prove (iv), observe that, by 6.2.6, we have $c = \mathrm{Tr}_P^{N_G(P,e)}(z)$ for some $z \in \mathcal{O}C_G(P)$. Thus $e = ec = \mathrm{Tr}_P^{N_G(P,e)}(ez)$. Since $PC_G(P)$ is normal in $N_G(P, e)$ we get that $e = \mathrm{Tr}_P^{PC_G(P)}(y)$ for some $y \in \mathcal{O}C_G(P)$. Now $C_G(P)/Z(P) \cong PC_G(P)/P$ and so $e = \mathrm{Tr}_{Z(P)}^{C_G(P)}(y)$. Therefore $\bar{e} = \mathrm{Tr}_1^{C_G(P)/Z(P)}(\bar{y})$, where $\bar{e}, \bar{y}$ are the images of e, y in $\mathcal{O}C_G(P)/Z(P)$. This shows that $\bar{e}$ is a sum of defect zero blocks of $\mathcal{O}C_G(P)/Z(P)$. This is in fact a single block: the canonical map sends $(\mathcal{O}C_G(P))^{Z(P)} = \mathcal{O}C_G(P)$ onto $\mathcal{O}C_G(P)/Z(P)$, hence this map sends $(\mathcal{O}C_G(P))_{Z(P)}^{C_G(P)}$ onto $(\mathcal{O}C_G(P)/Z(P))_1^{C_G(P)/Z(P)}$. Thus $\bar{e}$ remains primitive in $Z(\mathcal{O}C_G(P)/Z(P))$, proving (iv). For (v) we write $e = \mathrm{Tr}_P^{N_G(P,e)}(ez) = \mathrm{Tr}_{PC_G(P)}^{N_G(P,e)}(\mathrm{Tr}_P^{PC_G(P)}(ez))$. The element $\mathrm{Tr}_P^{PC_G(P)}(ez)$ belongs to the local algebra $Z(\mathcal{O}C_G(P)e)$ which is split as k is large enough. Thus $\mathrm{Tr}_P^{PC_G(P)}(ez) = \lambda e + r$ for some $\lambda \in \mathcal{O}$ and some $r \in J(Z(\mathcal{O}C_G(P)e))$. It follows that

$$e = \mathrm{Tr}_{PC_G(P)}^{N_G(P,e)}(\lambda e + r) = \lambda|N_G(P, e) : PC_G(P)| + \mathrm{Tr}_{PC_G(P)}^{N_G(P,e)}(r).$$

In this expression the second term is in $J(Z(\mathcal{O}C_G(P)e)))$, but e is not. Thus in particular $|N_G(P, e) : PC_G(P)|$ is not in $J(\mathcal{O})$, which is equivalent to saying that $|N_G(P, e) : PC_G(P)|$ is prime to p. $\square$

Definition 6.7.7 Let G be a finite group, let b be a block of $\mathcal{O}G$ and let P be a defect group of b. The unique block c of $\mathcal{O}N_G(P)$ with P as defect group satisfying $\mathrm{Br}_P(b) = \mathrm{Br}_P(c)$ is called the *Brauer correspondent* of b. If e is a block of $\mathcal{O}C_G(P)$ satisfying $ec = e$ then the group $E = N_G(P, e)/PC_G(P)$ is called the *inertial quotient of b*.

Remark 6.7.8 The inertial quotient E is unique up to isomorphism, since all pairs (P, e) consisting of a defect group P of b and a block e of $\mathcal{O}C_G(P)$ satisfying $ec = e$ are G-conjugate. We identify E to a subgroup of $\mathrm{Out}(P) = \mathrm{Aut}(P)/\mathrm{Inn}(P)$, via the map sending $x \in N_G(P, e)$ to the class modulo inner automorphisms of the automorphism of P given by conjugation with x. If P is abelian then $P \subseteq C_G(P)$ and hence $E = N_G(P, e)/C_G(P)$ is a group of automorphisms of P; thus we may consider the semidirect product $P \rtimes E$. In general, the action of E on P is defined only up to inner automorphisms. If k is large enough, then E is a p'-group. In that case the inverse image L of E in $\mathrm{Aut}(P)$ has $\mathrm{Inn}(P)$ as a normal Sylow p-subgroup. The Schur–Zassenhaus Theorem implies that $\mathrm{Inn}(P)$ has a complement in L that is unique up to conjugation by an element in $\mathrm{Inn}(P)$, and that maps isomorphically onto E. In other words, the canonical map $E \rightarrow \mathrm{Out}(P)$ lifts to a map $E \rightarrow \mathrm{Aut}(P)$, uniquely up to conjugation by an element in $\mathrm{Inn}(P)$. By choosing such a lift me may again consider the corresponding semidirect product $P \rtimes E$. Whenever we use this notation, such as in 6.14.1 below, we implicitly assert that the concepts or results do not depend on this choice. Note that the second statement of 6.7.6 implies that $\mathrm{Br}_P(c)$ is just the canonical image $\bar{c}$ of c in $kC_G(P)^{N_G(P)}$.

Example 6.7.9 Suppose p is an odd prime. Set $G = SL_2(p)$. The order of G is equal to $p(p+1)(p-1)$. Thus the group $P = \left\{ \left(\begin{smallmatrix} 1 & b \\ 0 & 1 \end{smallmatrix} \right) \middle| b \in \mathbb{F}_p \right\}$ is a Sylow p-subgroup of G and $N_G(P) = P \rtimes E$ where $E = \left\{ \left(\begin{smallmatrix} a & 0 \\ 0 & a^{-1} \end{smallmatrix} \right) \middle| a \in \mathbb{F}_p^{\times} \right\}$. Set $t = \left(\begin{smallmatrix} -1 & 0 \\ 0 & -1 \end{smallmatrix} \right)$. Then $T = \{1, t\}$ is the unique subgroup of order 2 of E, and we have $C_G(P) = P \times T$. Then $\mathcal{O}C_G(P) \cong \mathcal{O}P \otimes_{\mathcal{O}} \mathcal{O}T$ has two blocks corresponding to the two irreducible characters of T with values in $\mathcal{O}$; explicitly, the two blocks of $\mathcal{O}C_G(P)$ are $e_0 = \frac{1}{2}(1+t)$ and $e_1 = \frac{1}{2}(1-t)$. Since t is actually in the centre of $N_G(P)$ it follows from 6.7.6 (ii) that $c_0 = e_0$ and $c_1 = e_1$. Thus 6.7.6 (i) implies that $\mathcal{O}G$ has exactly two blocks b_0, b_1 with P as defect group, and every other block has defect zero.

The following conjecture, due to Broué, relates the derived category of a block with an abelian defect group to that of its Brauer correspondent.

Conjecture 6.7.10 (Broué's abelian defect group conjecture, [38, 6.1]) *Let G be a finite group, b a block of $\mathcal{O}G$, P a defect group of b, and denote by c the block of $\mathcal{O}N_G(P)$ corresponding to b under the Brauer correspondence. If P is abelian, then there is an equivalence of bounded derived categories $D^b(\mathrm{mod}(\mathcal{O}Gb)) \cong D^b(\mathrm{mod}(\mathcal{O}N_G(P)c))$.*

This is stated in [38, 6.1] with the assumption that K is large enough. For a refinement of this conjecture to *splendid derived equivalences* see 9.7.6 below. At the level of characters, a derived equivalence between block algebras over $\mathcal{O}$ induces a *perfect isometry* between the blocks; we will come back to this theme in chapter 9. Broué's abelian defect group conjecture holds if P is cyclic or a Klein four group. We will consider these two cases in detail in the chapters 11 and 12. The cyclic case was the first nontrivial derived equivalence that was shown to exist between block algebras in work of Rickard [185]. Broué's abelian defect group conjecture holds for blocks with abelian defect groups of symmetric groups (Chuang–Rouquier [48], using earlier work of Chuang–Kessar [47]), blocks with abelian defect groups of alternating groups (Marcuş [148], using [48] and Clifford Theory [147]). There is a long (and growing) list of special cases, notably including many blocks of sporadic simple groups and their central extensions, to which many authors have contributed to date.

Brauer's First Main Theorem, in conjunction with earlier results, can be used to rule out certain p-subgroups of a finite group G to occur as defect groups of a block of $\mathcal{O}G$. We denote by $O_p(G)$ the largest normal p-subgroup of G. A necessary condition for a p-subgroup P of G to occur as a defect group of a block is that P must be the largest normal p-subgroup of its normaliser.

Theorem 6.7.11 *Let G be a finite group, let b be a block of $\mathcal{O}G$ and let P be a defect group of b. Then $P = O_p(N_G(P))$.*

Proof By 6.7.6 there is a block c of $\mathcal{O}N_G(P)$ having P as defect group. Since P is normal in $N_G(P)$ we have $P \subseteq O_p(N_G(P))$. By 6.2.6, $O_p(N_G(P))$ is contained in every defect group of every block of $\mathcal{O}N_G(P)$. Thus $P = O_p(N_G(P))$. $\square$

The following theorem, due to Green, shows that any defect group is the intersection of two Sylow subgroups.

Theorem 6.7.12 (Green) *Let G be a finite group and let P be a defect group of a block b of $\mathcal{O}G$. For any Sylow p-subgroup S of G containing P there is an element $x \in C_G(P)$ such that $P = S \cap {}^x S$.*

Proof Let S be a Sylow p-subgroup of G containing P. By 5.6.7, the diagonal subgroup $\Delta P = \{(u, u) | u \in P\}$ of $G \times G$ is a vertex of $\mathcal{O}Gb$ as an $\mathcal{O}(G \times G)$-module. Thus $\mathrm{Res}_{S \times S}^{G \times G}(\mathcal{O}Gb)$ has a direct summand with vertex ΔP. Now $\mathcal{O}Gb$ is a direct summand of $\mathcal{O}G \cong \mathrm{Ind}_{\Delta G}^{G \times G}(\mathcal{O})$ as $\mathcal{O}(G \times G)$-modules. Hence every indecomposable summand of $\mathrm{Res}_{S \times S}^{G \times G}(\mathcal{O}Gb)$ is isomorphic to a direct summand of $\mathrm{Res}_{S \times S}^{G \times G}\mathrm{Ind}_{\Delta G}^{G \times G}(\mathcal{O})$. Any $(S \times S)$-ΔG-double coset in $G \times G$ has a representative of the form $(1, y)$ for some $y \in G$. Then $(S \times S) \cap {}^{(1,y)}\Delta G = \{(u, {}^y u) \mid u \in S \cap {}^{y^{-1}} S\}$; denote this group by $\Delta_y(S)$. Thus Mackey's formula yields

$$\mathrm{Res}_{S \times S}^{G \times G}(\mathrm{Ind}_{\Delta G}^{G \times G}(\mathcal{O})) \cong \oplus_y \mathrm{Ind}_{\Delta_y(S)}^{S \times S}(\mathcal{O})$$

with y running over a suitable subset of G. Now any of the summands $\mathrm{Ind}_{\Delta_y(S)}^{S \times S}(\mathcal{O})$ is indecomposable, with vertex $\Delta_y(S)$. Thus this direct sum has a summand with vertex ΔP if one of the $\Delta_y(S)$ is conjugate in $S \times S$ to ΔP.

That is, there is an element x in S such that $\Delta P = {}^{(1,x)}\Delta_y(S) = \Delta_{xy}(S)$, which is equivalent to $P = S \cap {}^{(xy)^{-1}}S$ and $u = {}^{xy}u$ for all $u \in P$.   □

One can use Theorem 6.7.12 to give an alternative proof of 6.2.6 (i): a normal p-subgroup of G is contained in any Sylow p-subgroup of G, hence in any intersection of two Sylow p-subgroups and hence in any defect group of a block of $\mathcal{O}G$ by 6.7.12.

Theorem 6.7.13 (Brauer [30]) *Let G be a finite group and b a block of $\mathcal{O}G$. Suppose that k is large enough. The p-part of $\mathrm{rk}_{\mathcal{O}}(\mathcal{O}Gb)$ is equal to p^{2a-d}, where p^a is the order of a Sylow p-subgroup of G and p^d is the order of a defect group of b.*

Proof We may assume $\mathcal{O} = k$. Let P be a defect group of b. As an $\mathcal{O}(G \times G)$-module, $\mathcal{O}Gb$ is indecomposable, and has ΔP as a vertex. Note that p^{2a-d} is the p-part of $|G \times G : \Delta P|$. By 5.12.15, it suffices to show the result for the Green correspondent of kGb. By 5.10.5, this Green correspondent is $(kGb)(P) = kC_G(P)c$, where $c = \mathrm{Br}_P(b)$, viewed as a module for $kN_{G \times G}(\Delta P)$. Brauer's first main theorem implies that $kC_G(P)c$ is the product of the $N_G(P)$-conjugacy class of a block $kC_G(P)e$ with a central defect group $Z(P)$. Thus the p-part of $\dim_k(kC_G(P)e$ is equal to that of $|Z(P)| \cdot |C_G(P)/Z(P)|^2$. There are $|N_G(P) : N_G(P, e)|$ blocks in the $N_G(P)$-conjugacy class of e, so the p-part of $\dim_k(kC_G(P)c)$ is equal to the p-part of $|Z(P)| \cdot |C_G(P)/Z(P)|^2 \cdot |N_G(P) : N_G(P, e)|$. This is equal to the p-part of $|N_G(P)/PC_G(P)| \cdot |C_G(P)| \cdot |C_G(P)/Z(P)|$, where we use that $|N_G(P, e)/PC_G(P)|$ is prime to p, thanks to the assumption on k being large enough. Using $C_G(P)/Z(P) \cong PC_G(P)/P$, this is in turn equal to the p-part of $|N_G(P)/P| \cdot |C_G(P)|$. We have $N_{G \times G}(\Delta P) = \Delta N_G(P)(C_G(P) \times C_G(P))/\Delta P$ and $\Delta N_G(P) \cap (C_G(P) \times C_G(P)) = \Delta C_G(P)$. Thus $|N_{G \times G}(\Delta P)/\Delta P| = |N_G(P)/P| \cdot |C_G(P)|$. The result follows.   □

Remark 6.7.14 With the notation of 6.7.13, as an $\mathcal{O}(G \times G)$-module, $\mathcal{O}Gb$ is indecomposable, with vertex ΔP, and trivial source. Thus, by 5.10.5, $kC_G(P)/Z(P)\bar{c}$ is the multiplicity module of $\mathcal{O}Gb$, where $\bar{c}$ is the image in $kC_G(P)/Z(P)$ of $c = \mathrm{Br}_P(b)$. This is a projective indecomposable module over $kN_{G \times G}(\Delta P)/\Delta P$. This module is in fact simple because $kC_G(P)/Z(P)\bar{c}$ is a direct product of matrix algebras, permuted transitively by $N_G(P)$. The proof of 6.7.13 is a special case of a more general result on modules with simple multiplicity modules; see 6.11.13.

The last statement in this Section, due to Puig, is a useful criterion for when the structural map of an interior G-algebra restricts to a split injective bimodule homomorphism on a block of G.

Proposition 6.7.15 ([176, Proposition 3.8]) *Let G be a finite group, b a block of $\mathcal{O}G$, P a defect group of b, and B an interior G-algebra. Suppose that k is large enough and that the conjugation action of P on B stabilises an $\mathcal{O}$-basis of B, and that $\mathrm{Br}_{\Delta P}(b) \cdot B(\Delta P) \cdot \mathrm{Br}_{\Delta P}(b)$ is projective as a left or right $kZ(P)$-module. Then the map $\alpha : \mathcal{O}Gb \to B$ induced by the structural homomorphism $G \to B^{\times}$ is split injective as a homomorphism of $\mathcal{O}Gb$-$\mathcal{O}Gb$-bimodules.*

Proof We follow the proof given in [142, 6.3], which plays this back to a special case of 5.10.8. After replacing B by $b \cdot B \cdot b$ we may assume that $B(\Delta P)$ is projective as a left or right $kZ(P)$-module. As an $\mathcal{O}(G \times G)$-module, $\mathcal{O}Gb$ has vertex ΔP and trivial source. By the assumptions, B is a permutation $\mathcal{O}\Delta P$-module. We have $N_{G \times G}(\Delta P) = (C_G(P) \times C_G(P)) \cdot N_{\Delta G}(\Delta P)$. Set $N = N_{G \times G}(\Delta P)/\Delta P$. Denote by Z the image of $Z(P) \times \{1\}$ in N; this is equal to the image of $\{1\} \times Z(P)$, normal in N, and canonically isomorphic to $Z(P)$. Consider the induced map $\alpha(\Delta P) : kC_G(P)\mathrm{Br}_{\Delta P}(b) \to B(\Delta P)$. Since k is large enough, if e is a block of $kC_G(P)$ occurring in $\mathrm{Br}_{\Delta P}(b)$, then $kC_G(P)/Z(P)\bar{e}$ is a matrix algebra, where $\bar{e}$ is the canonical image of e in $kC_G(P)/Z(P)$. Thus $kC_G(P)/Z(P)\bar{e}$ is simple as a module over $k(C_G(P) \times C_G(P))$. By Brauer's First Main Theorem, the blocks e arising in this way are permuted transitively by $N_G(P)$, it follows that $k \otimes_{kZ} kC_G(P)\mathrm{Br}_{\Delta P}(b) \cong kC_G(P)/Z(P)c$ is a simple kN-module, where c is the image of $\mathrm{Br}_{\Delta P}(b)$ in $kC_G(P)/Z(P)$, or equivalently, c is the sum of the $\bar{e}$ as above. By the assumptions, $A(\Delta P)$ is projective as a kZ-module, and hence the obvious composition of algebra homomorphisms $kZ(P) \to kC_G(P)\mathrm{Br}_{\Delta P} \to B(\Delta P)$ is injective. Thus $kC_G(P)\mathrm{Br}_{\Delta P}(b)$ has a summand isomorphic to kZ, as a kZ-module, which is mapped injectively into $B(\Delta P)$ by $\alpha(\Delta P)$. The result follows from the implication (ii) $\Rightarrow$ (i) in 5.10.8. $\qquad\square$

In order to illustrate typical applications of this proposition, we present an alternative proof of 6.4.7.

Alternative proof of 6.4.7 By 2.6.13 it suffices to show that A is isomorphic to a direct summand to $A \otimes_{\mathcal{O}P} A \cong \mathrm{End}_{\mathcal{O}P^{\mathrm{op}}}(A)$. Consider $\mathrm{End}_{\mathcal{O}P^{\mathrm{op}}}(\mathcal{O}Gi)$ as an interior G-algebra via the left multiplication by G on $\mathcal{O}Gi$. As an $\mathcal{O}Gb$-$\mathcal{O}Gb$-bimodule, this is isomorphic to $\mathcal{O}Gi \otimes_{\mathcal{O}P} i\mathcal{O}Gb$, and hence this is a permutation $\mathcal{O}(P \times P)$ module that is projective as a left $\mathcal{O}P$-module and as a right $\mathcal{O}P$-module. The hypothesis $\mathrm{Br}_P(i) \neq 0$ implies that $i\mathcal{O}Gi$ has a direct summand isomorphic to $\mathcal{O}P$ as an $\mathcal{O}P$-$\mathcal{O}P$-bimodule. Thus $(\mathrm{End}_{\mathcal{O}P^{\mathrm{op}}}(\mathcal{O}Gi))(\Delta P)$ is nonzero and projective as a left or right $kZ(P)$-module. It follows from 6.7.15 that the homomorphism $\mathcal{O}Gb \to \mathrm{End}_{\mathcal{O}P^{\mathrm{op}}}(\mathcal{O}Gi)$ is split injective as a homomorphism of $\mathcal{O}Gb$-$\mathcal{O}Gb$-bimodules. Since multiplying by i on the left

and right is a Morita equivalence by 6.4.6, it follows that the structural map
$A \to \text{End}_{\mathcal{O}P^{\text{op}}}(A)$ is split injective as a homomorphism of A-A-bimodules, and
hence A is isomorphic to a direct summand of $A \otimes_{\mathcal{O}P} A$. Tensoring over A with
an A-module M shows that $M \cong A \otimes_A M$ is isomorphic to a direct summand of
$A \otimes_{\mathcal{O}P} A \otimes_A \otimes_A M \cong A \otimes_{\mathcal{O}P} M$, hence M is relatively $\mathcal{O}P$-projective. $\square$

6.8 Clifford Theory

Let N be a normal subgroup of a finite group G. Clifford's Theorem 1.9.9 states
that the restriction of a simple kG-module to kN is semisimple, and that the
isotypic components of this restriction are permuted transitively by G. Some of
the main results in this section are block theoretic reduction techniques going
back to work of Fong and Reynolds. Clifford Theory has become the generic
term for results describing relationships between blocks of a finite group and
blocks of its normal subgroups.

Definition 6.8.1 Let G be a finite group and N a normal subgroup of G. Let b
be a block of $\mathcal{O}G$ and c a block of $\mathcal{O}N$. We say that b *covers* c if $bc \neq 0$.

Conjugation with any element $x \in G$ induces an algebra automorphism of
$\mathcal{O}N$, hence of $Z(\mathcal{O}N)$, and thus, in particular, conjugation by x permutes the
set of blocks of $\mathcal{O}N$. The blocks of $\mathcal{O}N$ covered by a fixed block of $\mathcal{O}G$ form a
G-conjugacy class of blocks of $\mathcal{O}N$:

Proposition 6.8.2 *Let G be a finite group, N a normal subgroup of G and b a
block of $\mathcal{O}G$.*

(i) *The set of blocks c of $\mathcal{O}N$ satisfying $bc \neq 0$ is a G-conjugacy class of
blocks of $\mathcal{O}N$.*

(ii) *If there is a unique block c of $\mathcal{O}N$ satisfying $bc \neq 0$, then c is G-stable and
$bc = b$.*

(iii) *If $b \in \mathcal{O}N$, then b is a G-conjugacy class sum of blocks of $\mathcal{O}N$.*

Proof Let c be a block of $\mathcal{O}N$ such that $bc \neq 0$. Since b is invariant under conjugation with elements in G we have $b({}^x c) = {}^x(bc) \neq 0$ for any $x \in G$. Let d be
the sum of the different G-conjugates of c. Then $dc = c$, and d is G-stable. Thus
d is an idempotent in $(\mathcal{O}N)^G \subseteq Z(\mathcal{O}G)$, and hence d is a sum of blocks of $\mathcal{O}G$.
We have $bd \neq 0$ since $bdc = bc \neq 0$. It follows that $bd = b$. If c' is any block
of $\mathcal{O}N$ satisfying $bc' \neq 0$, then also $dc' \neq 0$, and hence c' is conjugate, in G, to
c. This shows (i). The statements (ii) and (iii) are immediate consequences of
(i). $\square$

The following result implies that in order to describe the structure of the source algebras of a block b of $\mathcal{O}G$, we can always assume that for any normal subgroup N of G, the block b covers a unique block c of $\mathcal{O}N$, or equivalently, that c is G-stable and $bc = b$. In the context of character theory, this is known as Fong's first reduction.

Theorem 6.8.3 *Let G be a finite group, let N be a normal subgroup of G and let c be a block of $\mathcal{O}N$. Denote by H the stabiliser of c in G. Let b be a block of $\mathcal{O}G$ satisfying $bc \neq 0$.*

(i) *There is a unique block d of $\mathcal{O}Hc$ such that $b = \mathrm{Tr}_H^G(d)$, and then $bd = d = dc$.*
(ii) *We have $d\mathcal{O}Gd = \mathcal{O}Hd$ and $\mathcal{O}Gb = \mathcal{O}Hd\mathcal{O}H$; in particular, we have a Morita equivalence $\mathrm{Mod}(\mathcal{O}Gb) \cong \mathrm{Mod}(\mathcal{O}Hd)$ induced by the $\mathcal{O}Gb$-$\mathcal{O}Hd$-bimodule $\mathcal{O}Gd$ and its dual $d\mathcal{O}G$.*
(iii) *If P is a defect group of d, then P is a defect group of b, every source idempotent $i \in (\mathcal{O}Hd)^P$ of d is a source idempotent of b in $(\mathcal{O}Gb)^P$, and we have an equality of source algebras $i\mathcal{O}Gi = i\mathcal{O}Hi$.*

We state a part of the proof as a separate lemma, for future reference.

Lemma 6.8.4 *With the notation of 6.8.3, the element $\mathrm{Tr}_H^G(c)$ is an idempotent in $Z(\mathcal{O}G)$ and the trace map Tr_H^G induces an algebra isomorphism*

$$\mathrm{Tr}_H^G : Z(\mathcal{O}Hc) \cong Z(\mathcal{O}G\mathrm{Tr}_H^G(c))$$

with inverse given by multiplication with c.

Proof The different G-conjugates of c are all blocks of $\mathcal{O}N$ because N is normal in G. Thus they are pairwise orthogonal. If $x \in G \setminus H$, then $c \neq {}^xc$, hence $c({}^xc) = 0$, or explicitly, $cxcx^{-1} = 0$. Multiplying by x on the right yields $cxc = 0$. This holds for all $x \in G \setminus H$, and hence $c\mathcal{O}Gc = c\mathcal{O}Hc = \mathcal{O}Hc$. In particular, if $z \in Z(\mathcal{O}G)$, then $zc = czc \in Z(\mathcal{O}Hc)$. The trace $\mathrm{Tr}_H^G(c)$ is the sum of all different G-conjugates of c and hence an idempotent in $Z(\mathcal{O}G)$. Let $y \in Z(\mathcal{O}Hc)$. Then $y = yc$. Thus $\mathrm{Tr}_H^G(y)c = \sum_{x \in [G/H]} xycx^{-1}c$. The only summand that does not vanish in this sum is the one with $x \in H$, and this summand yields y. Thus $\mathrm{Tr}_H^G(y)c = y$. Conversely, let $z \in Z(\mathcal{O}G)\mathrm{Tr}_H^G(c)$. Then $z = z\mathrm{Tr}_H^G(c) = \mathrm{Tr}_H^G(zc)$. As noted above, we have $zc \in Z(\mathcal{O}Hc)$, and hence Tr_H^G and multiplication by c induce $\mathcal{O}$-linear isomorphisms between $Z(\mathcal{O}Hc)$ and $Z(\mathcal{O}G)\mathrm{Tr}_H^G(c)$. In order to see that these maps are algebra isomorphisms it suffices to show that multiplication by c is multiplicative. If $z_1, z_2 \in Z(\mathcal{O}G\mathrm{Tr}_H^G(c))$, then $(z_1c)(z_2c) = z_1z_2c$ because z_2 is central, whence the multiplicativity. $\square$

Proof of Theorem 6.8.3 It follows from 6.8.4 that Tr_H^G and multiplication by c induce a bijection between the set of blocks b of $\mathcal{O}G$ satisfying $b\mathrm{Tr}_H^G(c) = b$ and the set of blocks d of $\mathcal{O}H$ satisfying $dc = d$. If b is a block of $\mathcal{O}G$ satisfying $bc \neq 0$, then $b(^x c) = {}^x c$ for any $x \in G$, and so the condition $bc \neq 0$ is equivalent to $b\mathrm{Tr}_H^G(c) = b$. In particular, by the previous arguments, for every block b of $\mathcal{O}G$ satisfying $bc \neq 0$ there is a unique block d of $\mathcal{O}Hc$ such that $\mathrm{Tr}_H^G(d) = b$. Since d belongs to $\mathcal{O}Hc$, it follows that $dc = d$. Thus the different G-conjugates of d by representatives of G/H of are pairwise orthogonal, and hence $d = d\mathrm{Tr}_H^G(d) = bd$. This proves (i). The equality $b = \mathrm{Tr}_H^G(d)$ implies that $b \in \mathcal{O}Gd\mathcal{O}G$, and $d = dc$ implies that $d\mathcal{O}Gd = dc\mathcal{O}Gdc = dc\mathcal{O}Hdc = \mathcal{O}Hd$. Note that $\mathcal{O}Gd = b\mathcal{O}Gd$ and $d\mathcal{O}Gb = d\mathcal{O}G$ since $d \in \mathcal{O}Gb$. It follows from 2.8.7 that the bimodules $\mathcal{O}Gd$ and $d\mathcal{O}G$ induce a Morita equivalence. This proves (ii). Let P be a defect group of d. That is, there is $u \in (\mathcal{O}Hd)^P$ such that $d = \mathrm{Tr}_P^H(u)$. Then $b = \mathrm{Tr}_H^G(d) = \mathrm{Tr}_P^G(u)$, and therefore P contains a defect group R of b. If $v \in (\mathcal{O}Gb)^R$ such that $b = \mathrm{Tr}_R^G(v) = \sum_{x \in [H\backslash G/R]} \mathrm{Tr}_{H\cap {}^x R}^H({}^x v)$ then $d = db = d\mathrm{Tr}_R^G(v) = \sum_{x \in [H\backslash G/R]} \mathrm{Tr}_{H\cap {}^x R}^H(d({}^x v))$. Rosenberg's Lemma 4.4.8 implies that $d \in (\mathcal{O}Hd)_{H\cap {}^x R}^H$, so $|P| \leq |R|$ and hence $P = R$. If $i \in (\mathcal{O}Hd)^P$ is a source idempotent for c then $i = id$, hence $i\mathcal{O}Gi = id\mathcal{O}Gdi = i\mathcal{O}Hi$ by (ii). In particular, i remains primitive in $(i\mathcal{O}Gi)^P = (i\mathcal{O}Hi)^P$, so i remains a source idempotent of b. This proves (iii). $\qquad\square$

Corollary 6.8.5 *Let G be a finite group and b a block of $\mathcal{O}G$. Let P be a defect group of $\mathcal{O}Gb$. There is a subgroup H of G containing P and a block c of $\mathcal{O}H$ having P as a defect group such that $\mathcal{O}Hc$ and $\mathcal{O}Gb$ have isomorphic source algebras and such that for any normal subgroup N of H there is an H-stable block e_N of $\mathcal{O}N$ satisfying $e_N c = c$.*

Proof Applying repeatedly 6.8.3 yields a subgroup H and a block c of $\mathcal{O}H$ such that for any normal subgroup N of H there is an H-stable block e_N of $\mathcal{O}N$ satisfying $e_N c = c$, and such that a defect group and source algebra of c is a defect group and source algebra of b, respectively. Since the defect groups of b are G-conjugate, after possibly replacing H by a G-conjugate, we may choose H and c in such a way that P is a defect group of c $\qquad\square$

Proposition 6.8.6 *Let G be a finite group and N a normal subgroup of G. Let b a block of $\mathcal{O}G$ and c a block of $\mathcal{O}N$. We have $bc \neq 0$ if and only if $\mathcal{O}Nc$ is isomorphic to a direct summand of $\mathcal{O}Gb$ as an $\mathcal{O}(N \times N)$-module.*

Proof Suppose that $bc \neq 0$. We have $\mathcal{O}G = \oplus_{x \in [G/N]} x\mathcal{O}N$ as right $\mathcal{O}N$-modules. Since N is normal in G, this is also a direct sum decomposition of $\mathcal{O}G$ as left $\mathcal{O}N$-modules, hence as $\mathcal{O}(N \times N)$-modules. Multiplying by c on the right yields a direct sum decomposition

$$\mathcal{O}Gc = \oplus_{x \in [G/N]} x\mathcal{O}Nc.$$

As an $\mathcal{O}(N \times N)$-module, any of the summands $x\mathcal{O}Nc$ is obtained from $\mathcal{O}Nc$ by restriction along the algebra automorphism of $\mathcal{O}(N \times N)$ given by conjugation with $(x, 1)$ on $N \times N$; that is, $x\mathcal{O}Nc \cong {}^{(x,1)}\mathcal{O}Nc$. In particular, the summands $x\mathcal{O}Nc$ are all indecomposable as $\mathcal{O}(N \times N)$-modules. Since $bc \neq 0$, it follows that $\mathcal{O}Gbc$ is a nonzero direct summand of $\mathcal{O}Gb$ and of $\mathcal{O}Gc$, as an $\mathcal{O}(N \times N)$-module. The Krull–Schmidt Theorem implies that there is $x \in [G/N]$ such that $x\mathcal{O}Nc$ is isomorphic to a direct summand of $\mathcal{O}Gb$ as an $\mathcal{O}(N \times N)$-module. Since $\mathcal{O}Gb$ is invariant under left multiplication by x^{-1}, it follows that $\mathcal{O}Nc$ is isomorphic to a direct summand of $\mathcal{O}Gb$ as an $\mathcal{O}(N \times N)$-module. Conversely, if $\mathcal{O}Nc$ is isomorphic to a direct summand of $\mathcal{O}Gb$, then right multiplication by c on $\mathcal{O}Gb$ is nonzero because it is the identity on $\mathcal{O}Nc$. In particular, $bc \neq 0$, whence the result. □

Corollary 6.8.7 *Let G be a finite group and N a normal subgroup of G. Let c be a G-stable block of $\mathcal{O}N$, and let b be a block of $\mathcal{O}G$ such that $bc = b$. Multiplication by b induces an injective algebra homomorphism $\mathcal{O}Nc \to \mathcal{O}Gb$ that is split injective as a homomorphism of $\mathcal{O}(N \times N)$-modules.*

Proof By 6.8.6, the $\mathcal{O}(N \times N)$-module $\mathcal{O}Nc$ is isomorphic to a direct summand of $\mathcal{O}Gb$ as an $\mathcal{O}(N \times N)$-module. The result follows from 2.6.18. □

Proposition 6.8.8 *Let G be a finite group, and let N be a normal subgroup of G. Let c be a G-stable block of $\mathcal{O}N$, and let b be a block of $\mathcal{O}G$ such that $bc = b$. We have $J(\mathcal{O}Nc)\mathcal{O}Gb \subseteq J(\mathcal{O}Gb)$. If N contains a defect group of $\mathcal{O}Gb$, then $J(\mathcal{O}Nc)\mathcal{O}Gb = J(\mathcal{O}Gb)$.*

Proof We may assume that $\mathcal{O} = k$. Since N is normal in G, we have $J(kNc)kGb = kGbJ(kNc)$. Thus $J(kNc)kGb$ is a nilpotent ideal in kGb, hence contained in $J(kGb)$. As a kNc-module, the quotient $kGb/J(kNc)kGb$ is semisimple. If N contains a defect group of kGb, then $kGb/J(kNc)kGb$ is a relatively N-projective kGb-module. It follows from 2.6.5 that $kGb/J(kNc)kGb$ is semisimple as a kGb-module, and hence that $J(kGb) \subseteq J(kNc)kGb$, implying the result. □

Theorem 6.8.9 *Let G be a finite group, and let N be a normal subgroup of G.*

(i) *Let b be a block of $\mathcal{O}G$ and c a block of $\mathcal{O}N$ such that $bc \neq 0$. For any defect group Q of c there is a defect group P of b such that $Q = P \cap N$.*

(ii) *Let b be a block of $\mathcal{O}G$. For any defect group P of b there is a P-stable block c of $\mathcal{O}N$ such that $bc \neq 0$, and then $P \cap N$ is a defect group of c.*

(iii) Let c be a G-stable block of $\mathcal{O}N$ such that $Z(\mathcal{O}Nc)$ is split. There is a block b such that $bc = b$ and such that for any defect group P of b, the image PN/N of P in G/N is a Sylow p-subgroup of G/N.

Note that the first two statements do not say that if P is a defect group of b then $P \cap N$ is a defect group of any block c of $\mathcal{O}N$ satisfying $bc \neq 0$, but that we always can replace either c or P by a suitable G-conjugate with this property. The reason is that while all defect groups of b are G-conjugate, their intersections with N need not all be N-conjugate.

Proof of Theorem 6.8.9 By 6.8.3, in order to prove (i), we may assume that the block c of $\mathcal{O}N$ is G-stable (after possibly replacing G by the stabiliser H of c in G). Then c is a sum of blocks of $\mathcal{O}G$, and thus, if $bc \neq 0$ then in fact $bc = b$. By 6.8.6, the $\mathcal{O}(N \times N)$-module $\mathcal{O}Nc$ is isomorphic to a direct summand of $\text{Res}^{G \times G}_{N \times N}(\mathcal{O}Gb)$. Since ΔQ is a vertex of $\mathcal{O}Nc$ it follows that some $G \times G$-conjugate of ΔQ is contained in ΔP, hence Q is contained in a G-conjugate of P. After replacing P by a suitable G-conjugate we may therefore assume that $Q \subseteq P \cap N$. But we also have $\text{Br}_P(b) \neq 0$, hence $\text{Br}_{P \cap N}(b) \neq 0$. Since $b = cb$ this forces $\text{Br}_{P \cap N}(c) \neq 0$ and so $P \cap N$ is contained in some defect group of c. This implies the equality $Q = P \cap N$, whence (i). The group G acts transitively on the set of blocks c of $\mathcal{O}N$ satisfying $bc \neq 0$. If d is the sum of all different G-conjugates of these c, then $bd = b$. Thus if P fixes no block c of $\mathcal{O}N$ satisfying $bc \neq 0$, then $\text{Br}_P(d) = 0$, hence $\text{Br}_P(b) = 0$, contradicting the fact that P is a defect group of b. Thus there is a P-stable block c of $\mathcal{O}N$ satisfying $bc \neq 0$ and satisfying $\text{Br}_P(c) \neq 0$. Thus also $\text{Br}_{P \cap N}(c) \neq 0$, which shows that $P \cap N$ is contained in a defect group Q of c as a block of $\mathcal{O}N$. But Q is also contained in a defect group of b by (i), hence contained in $P' \cap N$ for some defect group P' of b. Since P and P' are conjugate and N is normal in G, it follows that $|P \cap N| = |P' \cap N|$, implying that $Q = P \cap N$. This shows (ii). For (iii) we consider $\mathcal{O}Nc$ as a G-algebra. Since c is a block of $\mathcal{O}N$ it follows that c is primitive in $(\mathcal{O}Nc)^G$; that is, $\mathcal{O}Nc$ is a primitive G-algebra. Let P be a defect group of $\mathcal{O}Nc$ as a G-algebra. That is, P is a minimal subgroup of G for which there exists an element $a \in (\mathcal{O}Nc)^P$ satisfying $c = \text{Tr}_P^G(a)$. We have $\text{Tr}_P^{PN}(a) \in Z(\mathcal{O}Nc)$, and $Z(\mathcal{O}Nc)$ is split local by the assumptions. Thus $\text{Tr}_P^{PN}(a) = \lambda c + r$ for some $\lambda \in \mathcal{O}$ and some $r \in J(Z(\mathcal{O}Nc))$. Since conjugation by G stabilises $J(Z(\mathcal{O}Nc))$, it follows that $c = \text{Tr}_P^G(a) = \text{Tr}_{PN}^G(\lambda c + r) = |G : PN|c + s$ for some $s \in J(Z(\mathcal{O}Nc))$. As c is an idempotent, this forces $|G : PN|$ to be invertible in $\mathcal{O}$, hence prime to p. We have $\text{Br}_P(c) \neq 0$, and hence there is at least one block b of $\mathcal{O}G$ satisfying $bc = b$ and $\text{Br}_P(b) \neq 0$. The result follows. $\square$

Corollary 6.8.10 *Let G be a finite group, and let N be a normal subgroup of G. Let c be a G-stable block of $\mathcal{O}N$ such that $Z(\mathcal{O}Nc)$ is split. Suppose that c* remains a block of $\mathcal{O}G$. If P is a defect group of $\mathcal{O}Gc$, then PN/N is a Sylow p-subgroup of G/N.

Proof This follows from 6.8.9, since the assumptions imply that the block $b = c$ is the unique block of $\mathcal{O}G$ covering c. □

We consider the special case where c is a block of a normal subgroup N of p-power index. By 6.8.3 there is no loss of generality if we assume that c is G-stable.

Proposition 6.8.11 *Let G be a finite group and N a normal subgroup such that G/N is a p-group. Let c be a block of $\mathcal{O}N$. Then there is a unique block b of $\mathcal{O}G$ that covers c, and we have $b = \mathrm{Tr}_H^G(c)$, where H is the stabiliser in G of c. In particular, if c is G-stable, then c remains a block of $\mathcal{O}G$. Assume that c is G-stable, denote by P a defect group of $\mathcal{O}Gc$, and set $Q = P \cap N$. The following hold.*

  (i) *We have $G = NP$, and Q is a defect group of $\mathcal{O}Nc$.*
 (ii) *There is a block f of $kC_N(Q)$ such that (Q, f) is a maximal (N, c)-Brauer pair and such that (Q, f) is P-stable, or equivalently, such that $P \leq N_G(Q, f)$.*

Proof Since N contains all p'-elements of G, it follows from 6.5.4 that every block of $\mathcal{O}G$ is contained in $(\mathcal{O}N)^G$. Thus if b is a block of $\mathcal{O}G$ covering c, then $bc = c$, and 6.8.3 (i) implies that $b = \mathrm{Tr}_H^G(c)$, where H is the stabiliser of c in G. In particular, if c is G-stable, then $b = c$. Suppose now that c is G-stable, and let P and Q be as in the statement. By 6.8.10, the image of P in G/N is a Sylow p-subgroup in G/N, hence equal to G/N by the assumptions. This shows that $G = NP$. By 6.8.9, some G-conjugate of P intersected with N is a defect group of $\mathcal{O}Nc$. Since $G = NP$, the G-conjugates of P are in fact N-conjugates, and hence P itself intersected with N yields a defect group of $\mathcal{O}Nc$, proving (i). Since Q is normal in the defect group P of c as a block of $\mathcal{O}G$, we have $0 \neq \mathrm{Br}_P(c) = \mathrm{Br}_P(\mathrm{Br}_Q(c))$. Write $\mathrm{Br}_Q(c) = \sum_{f \in I} f$, with f running over the set I of blocks of $kC_N(Q)$ satisfying $\mathrm{Br}_Q(c)f = f$. The group P normalises Q, N, hence also $C_N(Q)$. It follows that P permutes the set I. If P has no fixed point in I, then $\mathrm{Br}_Q(c)$ is a sum of nontrivial P-orbits in I, which yields the contradiction $\mathrm{Br}_P(c) = 0$. Thus P has a fixed point in I, which proves (ii). □

Proposition 6.8.12 ([114, Proposition 2.1]) *Let G be a finite group, let M and N be normal subgroups of G such that $N \leq M$ and such that M/N is a p-group. Let b be a block of $\mathcal{O}G$ and c a block of $\mathcal{O}N$ such that $bc \neq 0$ and such that $Z(\mathcal{O}Nc)$ is split. Suppose that N contains a defect group of b. Let H be the stabiliser of c in G. Then $H \cap M = N$. Equivalently, $H/N \cap O_p(G/N) = \{1\}$.*

Proof By 6.8.3 there is a unique block d of $\mathcal{O}Hc$ such that $b = \mathrm{Tr}_H^G(d)$, and a defect group of d is a defect group of b, hence contained in N by the assumptions. Set $L = H \cap M$. Then N is normal in L of p-power index, and L stabilises c. Thus $\mathcal{O}N$ contains any block of $\mathcal{O}L$, and hence c remains a block of $\mathcal{O}L$. It follows from 6.8.10 that if R is a defect group of $\mathcal{O}Lc$, then RN/N is a Sylow p-subgroup of L/N. Note that L is normal in H and that $cd = d$. By 6.8.9, there is a defect group P of $\mathcal{O}Hd$ that contains R. Since $P \subseteq N$ by the assumptions, it follows that $L = N$. If M is the inverse image of $O_p(G/N)$ in G, then M is the unique maximal normal subgroup of G containing N as a normal subgroup of p-power index, whence the equivalence with the last statement. $\square$

Fong's second reduction theorem replaces $O_{p'}(G)$ by a p'-subgroup of $Z(G)$. The following slightly more general result, also known as Fong–Reynolds reduction, is a key tool in the situation where the normal subgroup N of a finite group G has a block c of defect zero, hence is a matrix algebra, provided that $\mathcal{O}$ is large enough. In view of 6.8.3, there is no loss of generality in assuming that the block c of the normal subgroup N is G-stable.

Theorem 6.8.13 ([72], [184])   *Let G be a finite group, N a normal subgroup of G and c a G-stable block of defect zero of $\mathcal{O}N$. Set $S = \mathcal{O}Nc$ and suppose that k is a splitting field for S. Set $L = G/N$. For any $x \in G$ there is s_x in $S^\times$ such that $^x t = s_x t (s_x)^{-1}$ for all $t \in S$ and such that $s_x s_y = s_{xy}$ if at least one of x, y is in N. Then the 2-cocycle $\alpha \in Z^2(G; \mathcal{O}^\times)$ defined by $s_x s_y = \alpha(x, y)^{-1} s_{xy}$ for x, $y \in G$ depends only on the images of x, y in L and induces a 2-cocycle, still denoted α, in $Z^2(L; \mathcal{O}^\times)$, and we have an isomorphism of $\mathcal{O}$-algebras*

$$\mathcal{O}Gc \cong S \otimes_\mathcal{O} \mathcal{O}_\alpha L$$

sending xb to $s_x \otimes \bar{x}$, where $x \in G$ and $\bar{x}$ is the image of x in L.

Proof Let R be a set of representatives of G/N in G. Then $\mathcal{O}G = \bigoplus_{x \in \mathcal{R}} x\mathcal{O}N$, and since c is contained in $\mathcal{O}N$ we get that $\mathcal{O}Gc = \bigoplus_{x \in \mathcal{R}} x\mathcal{O}Nc$ as a right $\mathcal{O}N$-module. In particular, $\mathrm{rk}_\mathcal{O}(\mathcal{O}Gc) = |L|\mathrm{rk}_\mathcal{O}(\mathcal{O}Nc)$. By 4.1.5, for $x \in \mathcal{R}$ there is $s_x \in S^\times$ such that $^x t = s_x t (s_x)^{-1}$ for all $t \in S$. For $x \in \mathcal{R}$ and $y \in N$ set $s_{xy} = s_x yc$. Then, for all $x \in G$ we have $^x t = s_x t (s_x)^{-1}$, and for $x \in G$, $y \in N$ we have $s_{xy} = s_x yc = s_x s_y$. Since $s_y = yc$ we have $(s_y)^x = (s_x)^{-1} s_y s_x$, and hence $s_{yx} = s_{xy^x} = s_x y^x c = s_x s_x^{-1} y s_x c = y s_x c = s_y s_x$. Thus α induces a 2-cocycle on L as stated. The map $x \mapsto s_x \otimes \bar{x}$ is a surjective linear map. It is also multiplicative: for x, $y \in G$, the image of xy under this map is $s_{xy} \otimes \bar{x}\bar{y} = \alpha(x, y)s_x s_y \otimes \alpha(x, y)^{-1}\bar{x}\bar{y} = s_x s_y \otimes \bar{x}\bar{y}$, which is the product of the images of x and y. Thus this map induces a surjective algebra homomorphism $\mathcal{O}G \to S \otimes_\mathcal{O} \mathcal{O}_\alpha L$, which sends c to the unit element, hence which induces a surjective algebra homomorphism $\mathcal{O}Gc \to S \otimes_{\mathcal{O}} \mathcal{O}_\alpha L$. Both sides have the same $\mathcal{O}$-rank, and hence this map is an isomorphism. □

The 2-cocycle α in 6.8.13 is inverse to that defined by the action of G on S. One of the key arguments used in the proof of the Fong–Reynolds reduction is that every algebra automorphism of a defect zero block is an inner automorphism, thanks to the Skolem–Noether Theorem. For blocks of $\mathcal{O}N$ with nontrivial defect groups this may not be true in general, and leads to considering the subgroup of G consisting of those elements in G that induce an inner automorphism of $\mathcal{O}Nc$. This leads to the following result, due to Dade.

Theorem 6.8.14 ([54]) *Let G be a finite group, N a normal subgroup of G and c a G-stable block of $\mathcal{O}N$. Denote by H the subgroup of all elements $x \in G$ such that conjugation by x induces an inner automorphism of $\mathcal{O}Nc$ as an $\mathcal{O}$-algebra. Then H is a normal subgroup of G containing N, and the following hold.*

 (i) Every block b of $\mathcal{O}G$ satisfying $bc = b$ is contained in $\mathcal{O}H$.
 (ii) The map sending a block d of $\mathcal{O}Hc$ to its G-conjugacy class sum is a bijection between G-conjugacy classes of blocks of $\mathcal{O}Hc$ and blocks of $\mathcal{O}Gc$.
(iii) If $H = N$, then c is a block of $\mathcal{O}G$, and if P is a defect group of $\mathcal{O}Gc$ and $Z(\mathcal{O}Hc)$ is split, then PN/N is a Sylow p-subgroup of G/N.

We collect the technicalities for the proof of this theorem in the following lemma.

Lemma 6.8.15 ([54]) *Let G be a finite group, N a normal subgroup and c a G-stable block of $\mathcal{O}N$. Set $A = \mathcal{O}Gc$ and $B = \mathcal{O}Nc$. Denote by H the subgroup of G consisting of all $x \in G$ such that conjugation by x induces an inner automorphism of B as an $\mathcal{O}$-algebra.*

 (i) We have $A^N = \oplus_{x \in [G/N]} (Bx)^N$.
 (ii) Let $x \in G$. Then $(Bx)^N$ contains an element in $(A^N)^\times$ if and only if $x \in H$.
(iii) Let $x \in G$. Then $(A^N)(Bx)^N = (Bx)^N(A^N)$; in particular, this set is a twosided ideal in A^N.
(iv) Let $x \in G$. We have $(A^N)(Bx)^N \subseteq J(A^N)$ if and only if $x \in G \setminus H$.

Proof We have $\mathcal{O}G = \oplus_{x \in [G/H]} \mathcal{O}Nx$. Since N is normal, this is a decomposition of $\mathcal{O}G$ as a direct sum of $\mathcal{O}N$-$\mathcal{O}N$-bimodules. Thus multiplication by c yields a decomposition $A = \oplus_{x \in [G/H]} Bx$ of A as a direct sum of B-B-bimodules. In particular, each summand in this decomposition is stable under the conjugation action of N, and hence taking fixed points with respect to this action yields the equality in (i). Let $x \in G$. If $u \in (Bx)^N \cap (A^N)^\times$, then $ux^{-1} \in B^\times$, hence $xu^{-1} \in B^\times$ acts as conjugation by x on B whence $x \in H$. Conversely, if $x \in H$, then there is $v \in B^\times$ such that vx^{-1} centralises B, hence $vx^{-1} \in (Bx^{-1})^N \cap (A^N)\times$, proving (ii). Let $x \in G$ and $v \in A^N$. Since N is normal in G, we have $xvx^{-1} \in A^N$, hence xvx^{-1} commutes with all elements in B. Let $u \in B$ such that $ux \in (Bx)^N$. Then $uxv = u(xvx^{-1})x = (xvx^{-1})ux \in A^N(Bx)^N$. This shows $(Bx)^N A^N \subseteq A^N(Bx)^N$, and a similar argument shows the reverse inclusion, whence (iii). For (iv) observe first that $(Bx)^N(Bx^{-1})^N \subseteq B^N = Z(B)$. This inclusion is an equality if and only if $(Bx)^N$ contains an invertible element in A^N because $Z(B)$ is local. Thus if $x \in G \setminus H$, then $(Bx)^N(Bx^{-1})^N \subseteq J(Z(B))$. Since x^{-1} is a power of x it follows from (iii) that some power of the ideal $A^N(Bx)^N$ is contained in $A^N J(Z(B)) \subseteq J(A^N)$. It follows that $A^N(Bx)^N \subseteq J(A^N)$. If $x \in H$, then $A^N(Bx)^N$ contains an invertible element in A^N, hence is not contained in $J(A^N)$. This completes the proof. $\square$

Proof of Theorem 6.8.14 Let $x \in H$. Then there is $u \in (\mathcal{O}Hc)^\times$ such that $^x h = {}^u h$ for all $h \in \mathcal{O}Hc$. For any $y \in G$ we have $^{(yxy^{-1})}h = {}^{yx}(y^{-1}hy) = {}^y(uy^{-1}hyu^{-1}) = yuy^{-1}hyu^{-1}y^{-1} = {}^{(yuy^{-1})}h$, and $yuy^{-1} \in \mathcal{O}Nc$ since N is normal and c is G-stable. This shows that H is normal in G, and clearly H contains N. Let b be a block of $\mathcal{O}Hc$. Then b is in particular a central idempotent in $(\mathcal{O}Gc)^N$. By 6.8.15 we have $(\mathcal{O}Gc)^N = (\mathcal{O}Hc)^N + J((\mathcal{O}Gc)^N)$, hence $(\mathcal{O}Hc)^N$ contains a conjugate of any idempotent in $(\mathcal{O}Gc)^N$. Since b is central, it follows that $(\mathcal{O}Hc)^N$ contains b, and hence b is a sum of G-conjugate blocks of $\mathcal{O}Hc$, proving (i) and (ii). If $H = N$, then (ii) implies that c remains a block of $\mathcal{O}G$. In that case it follows from 6.8.10 that the image in G/N of a defect group of $\mathcal{O}Gc$ is a Sylow p-subgroup of G/N. $\square$

6.9 The Harris–Knörr correspondence

The Harris–Knörr correspondence is a correspondence between the sets of the covering blocks of two blocks that correspond to each other via the Brauer correspondence.

Theorem 6.9.1 ([80]) *Let G be a finite group, N a normal subgroup of G, and let Q be a p-subgroup of N. Set $H = N_G(Q)$ and $L = H \cap N = N_N(Q)$. Then L is a normal subgroup of H. Let d be a block of $\mathcal{O}N$ with Q as a defect group. Let e be the block of $\mathcal{O}L$ with Q as a defect group that corresponds to d via the Brauer correspondence. Let c be a block of $\mathcal{O}H$ covering e, and let R be a defect group of c such that $R \cap L = Q$. Then $N_G(R) \le H$. Denote by b the unique block of $\mathcal{O}G$ with defect group R corresponding to c through the Brauer* *correspondence. The map sending c to b as above induces a bijection between the blocks of $\mathcal{O}H$ covering e and the blocks b of $\mathcal{O}G$ covering d.*

Using the fact that the Brauer correspondence is a special case of the Green correspondence, we will prove this theorem as a consequence of a result, due to Alperin [2], which combines the Green correspondence and Clifford Theory.

Definition 6.9.2 ([2]) Let G be a finite group and let U be an indecomposable $\mathcal{O}G$-module. Let N be a normal subgroup of G and let V be an indecomposable $\mathcal{O}N$-module. We say that U *covers* V if V is isomorphic to a direct summand of $\mathrm{Res}_N^G(U)$ and if U has a vertex R such that $R \cap N$ is a vertex of V.

This definition extends the notion of covered blocks in 6.8.1. That is, if N is a normal subgroup of a finite group G and if b is a block of $\mathcal{O}G$ that covers a block c of $\mathcal{O}N$ in the sense of 6.8.1, then the $\mathcal{O}(G \times G)$-module $\mathcal{O}Gb$ covers the $\mathcal{O}(N \times N)$-module $\mathcal{O}Nc$ in the sense of 6.9.2. Indeed, if Q is a defect group of c, then by 6.8.9 there is a defect group P of b such that $Q = P \cap N$. By 6.2.1 the diagonal subgroups ΔP and ΔQ are vertices of the $\mathcal{O}(G \times G)$-module $\mathcal{O}Gb$ and the $\mathcal{O}(H \times H)$-module $\mathcal{O}Hc$, respectively. Moreover, by 6.8.6, the $\mathcal{O}(N \times N)$-module $\mathcal{O}Nc$ is isomorphic to a direct summand of $\mathrm{Res}_{N \times N}^{G \times G}(\mathcal{O}Gb)$.

Theorem 6.9.3 ([2]) *Let G be a finite group, N a normal subgroup of G, and let Q be a p-subgroup of N. Let H be a subgroup of G containing $N_G(Q)$ and set $L = H \cap N$. Then L is a normal subgroup of H containing $N_N(Q)$. Let V be an indecomposable $\mathcal{O}N$-module with vertex Q, and let V' be an indecomposable $\mathcal{O}L$-module with vertex Q that is a Green correspondent of V. Let R be a p-subgroup of G such that $R \cap L = Q$. Then $N_G(R) \le H$. Let U be an indecomposable $\mathcal{O}G$-module with vertex R and let U' be an indecomposable $\mathcal{O}H$-module with vertex R that is a Green correspondent of U. Then U covers V if and only if U' covers V'. The correspondence sending U' to U induces a bijection between the isomorphism classes of indecomposable $\mathcal{O}H$-modules that cover V' and the isomorphism classes of indecomposable $\mathcal{O}G$-modules that cover V.*

Proof Since N is normal in G, it follows trivially that $L = H \cap N$ is normal in H. We have $N_G(R) \le N_G(R \cap N) = N_G(Q) \le H$, and $N_N(Q) = N_G(Q) \cap N \le H \cap N = L$. In what follows we will use without further comment some basic statements on vertices of modules, notably 5.1.5 and 5.1.6.

Suppose that U' covers V'. We need to show that U covers V. The $\mathcal{O}G$-module U has R as a vertex, and the $\mathcal{O}N$-module V has $Q = R \cap N$ as a vertex. Thus, in order to show that U covers V, it suffices to show that V is isomorphic to a direct summand of $\mathrm{Res}_N^G(U)$. By the assumptions, V' is a summand of

$\text{Res}_L^H(U')$ and V is a summand of $\text{Res}_N^G(U)$. Since also V' is a summand of $\text{Res}_L^N(V)$, it follows that V' is a summand of $\text{Res}_L^G(U) = \text{Res}_L^N(\text{Res}_N^G(U))$. Thus there is an indecomposable summand W of $\text{Res}_N^G(U)$ such that V' is a summand of $\text{Res}_L^N(W)$. Now V' has Q as a vertex, and $N_N(Q) \leq L$. The Burry–Carlson–Puig Theorem 5.5.20 implies that Q is a vertex of W. But then W is the Green correspondent of V', hence $W \cong V$. This shows that V is a summand of $\text{Res}_N^G(U)$ with vertex $Q = R \cap N$, hence that U covers V.

Conversely, suppose that U covers V. We need to show that U' covers V'. Since R is a vertex of U' and $Q = R \cap L$ is a vertex of V', it remains to show that V' is a summand of $\text{Res}_L^H(U')$. We consider the intermediate group HN between G and N. Let Y be an indecomposable $\mathcal{O}HN$-module corresponding to U through the Green correspondence. Then U' is also a Green correspondent of Y, by the transitivity 5.2.6 of the Green correspondence.

We show first that V is a summand of $\text{Res}_N^{HN}(Y)$. Since U and Y are Green correspondents, it follows that U is a summand of $\text{Ind}_{HN}^G(Y)$. By the assumption on U, the module V is a summand $\text{Res}_N^G(U)$, hence of

$$\text{Res}_N^G(\text{Ind}_{HN}^G(Y)) \cong \oplus_x {}^xY,$$

where x runs over some subset of G. Each summand ${}^xY$ is an indecomposable module for the group algebra of $xHNx^{-1}$ with vertex xRx^{-1}. Therefore each indecomposable summand of ${}^xY$ as an $\mathcal{O}N$-module has a vertex contained in $N \cap xRx^{-1} = xQx^{-1}$. Thus if V is a summand of ${}^xY$, then xQx^{-1} and Q must be conjugate in N, and hence there is $n \in N$ such that $xn \in N_G(Q) \leq H$, or equivalently, $x \in HN$. But in that case we obviously have that ${}^xY \cong Y$. This shows that V is a summand of $\text{Res}_N^{HN}(Y)$.

Again by the Green correspondence, Y is a summand of $\text{Ind}_H^{HN}(U')$, so V is a summand of $\text{Res}_N^{HN}(\text{Ind}_H^{HN}(U')) \cong \text{Ind}_L^N(\text{Res}_L^H(U'))$, where we have used Mackey's formula. Thus there is an indecomposable summand Z of $\text{Res}_L^H(U')$ such that V is a summand of $\text{Ind}_L^N(Z)$. Since R is a vertex of U', it follows that Z has a vertex contained in $L \cap hRh^{-1} = hQh^{-1}$ for some $h \in H$, where we use that H normalises L. Since the summand V of $\text{Ind}_L^N(Z)$ has Q as a vertex, it follows that Z has hQh^{-1} as a vertex. Then hQh^{-1} is also a vertex of V. Thus the vertices Q and hQh^{-1} of V are conjugate in N. Let $n \in N$ such that $Q = nhQh^{-1}n^{-1}$. Then $nh \in N_G(Q) \leq H$, hence $n \in H \cap N = L$. It follows that ${}^nZ \cong Z$ has Q as a vertex. Thus Z is a Green correspondent of V, and hence $Z \cong V'$. Since Z was chosen as a summand of $\text{Res}_L^H(U')$, it follows that V' is a summand of $\text{Res}_L^H(U')$.

It remains to show that the correspondence $U' \mapsto U$, with U and U' as in the statement, induces a bijection on isomorphism classes. Let U', U_1' be indecomposable $\mathcal{O}H$-modules and let U, U_1 be indecomposable $\mathcal{O}G$-modules.

Suppose that U' and U have a common vertex R satisfying $R \cap N = Q$. By the above, we have $N_G(R) \leq H$. Suppose that U' is the Green correspondent of U with respect to R. Similarly, suppose that U_1' and U_1 have a common vertex R_1 satisfying $R_1 \cap N = Q$. As before, we have $N_G(R_1) \leq H$. Suppose that U_1' is the Green correspondent of U_1 with respect to R_1. We need to show that $U \cong U_1$ if and only if $U' \cong U_1'$. Suppose that $U' \cong U_1'$. Then both R, R_1 are vertices of U', hence they are H-conjugate. Thus U and U_1 are summands of $\operatorname{Ind}_H^G(U')$ which both have R as a vertex, and hence $U \cong U_1$ by the uniqueness properties of the Green correspondence. Conversely, suppose that $U \cong U_1$. Then R, R_1 are both vertices of U, hence they are G-conjugate. Let $x \in G$ such that $R_1 = xRx^{-1}$. Then $Q = R_1 \cap N = xRx^{-1} \cap N = x(R \cap N)x^{-1} = xQx^{-1}$, hence $x \in N_G(Q) \leq H$. It follows that R, R_1 are conjugate in H, and hence they are both vertices of U'. Thus U' and U_1' are summands of $\operatorname{Res}_H^G(U)$ with R_1 as a vertex. As before, the uniqueness properties of the Green correspondence imply that $U' \cong U_1'$. This completes the proof. $\square$

Proof of Theorem 6.9.1 Theorem 6.9.1 follows from 6.9.3 applied with $G \times G$, $N \times N$, $H \times H$, ΔR, ΔQ, $\mathcal{O}Gb$, $\mathcal{O}Hc$, $\mathcal{O}Nd$, $\mathcal{O}Le$, instead of G, N, H, R, Q, U, U', V, V', respectively. To see this, note that $N_{G \times G}(\Delta Q) \leq N_G(Q) \times N_G(Q) \leq H \times H$. Clearly $N \times N$ is normal in $G \times G$. Let b be a block of $\mathcal{O}G$ that covers the block d of $\mathcal{O}N$. By 6.8.9, b has a defect group R such that $R \cap N = Q$. As mentioned above, the $\mathcal{O}(G \times G)$-module $\mathcal{O}Gb$ has vertex ΔR and covers the $\mathcal{O}(N \times N)$-module $\mathcal{O}Nd$ in the sense of 6.9.2. Its Green correspondent for $H \times H$ is of the form $\mathcal{O}Hc$, where c is the Brauer correspondent of b. The result follows from 6.9.3. $\square$

6.10 Weights of blocks

We assume in this section that k is algebraically closed. For any finite-dimensional k-algebra A we denote by $\ell(A)$ the number of isomorphism classes of simple A-modules, and by $w(A)$ the number of isomorphism classes of simple projective A-modules. Let G be a finite group and b a block of kG. Let (Q, e) be a (G, b)-Brauer pair. Then e is a block of $kC_G(Q)$, hence remains a block of $kN_G(Q, e)$. The canonical image $\bar{e}$ of e in $k(N_G(Q, e)/Q)$ is a central idempotent, but need not be a block of $k(N_G(Q, e)/Q)$. We use the following terminology for the projective simple modules over the corresponding factor algebra $k(N_G(Q, e)/Q)\bar{e}$.

Definition 6.10.1 Let G be a finite group, b a block of $\mathcal{O}G$, and set $B = \mathcal{O}Gb$. Let (Q, e) be a (G, b)-Brauer pair. Denote by $\bar{e}$ the image of e in $k(N_G(Q, e)/Q)$. A simple projective $k(N_G(Q, e)/Q)\bar{e}$-module is called a *weight of the block*

b or a *weight of the block algebra B*, or, for short, a (G, b)-*weight* or a *B-weight*.

The version 1.15.12 of Alperin's weight conjecture for group algebras has the following refinement for block algebras.

Conjecture 6.10.2 (Alperin's Weight Conjecture for blocks [3]) *Let G be a finite group and b a block of kG. Set $B = kGb$. We have*

$$\ell(B) = \sum_{(Q,e)} w(k(N_G(Q, e)/Q)\bar{e}),$$

where (Q, e) runs over a set of representatives of the G-conjugacy classes of (G, b)-Brauer pairs, and where $\bar{e}$ is the canonical image of e in $k(N_G(Q, e)/Q)$.

The sum on the right side is clearly independent of the choice of conjugacy class representatives of Brauer pairs. Alperin's weight conjecture holds trivially if B has defect zero. In that case, both sides are 1, since on the right side, the only nonzero term arises for $(Q, e) = (1, b)$. If B has a nontrivial defect group, then the summand indexed by $(1, b)$ is zero, since B has no projective simple module in that case.

Alperin's weight conjecture holds for blocks with cyclic defect groups (Dade [53]), 2-blocks with Klein four, generalised dihedral, quaternion, semi-dihedral defect groups (Brauer [28], [29], Olsson [161]), blocks of symmetric groups and general linear groups (Alperin [3], Alperin and Fong [6], An [8]), blocks of finite groups of Lie type in defining characteristic (Cabanes [44]), blocks of finite p-solvable groups (Okuyama [158]), and many more cases. Knörr and Robinson [106] reformulated Alperin's weight conjecture in terms of alternating sums indexed over chains of p-subgroups. This led Dade to an array of conjectures taking into account further invariants. Alperin's weight conjecture has been reduced to a statement on nonabelian finite simple groups in work of Späth [208], extending work of Navarro and Tiep [157] on the nonblockwise version. In order to relate Alperin's weight conjecture for blocks to its group theoretic version 1.15.12, we reformulate the right side in the conjecture as a sum indexed by conjugacy classes of p-subgroups rather than Brauer pairs.

Proposition 6.10.3 *Let G be a finite group and b a block of kG. For any p-subgroup Q of G, set $b_Q = \mathrm{Br}_Q(b)$, and denote by $\bar{b}_Q$ the image of b_Q in $N_G(Q)/Q$. We have*

$$\sum_{(Q,e)} w(k(N_G(Q, e)/Q)\bar{e}) = \sum_{Q} w(k(N_G(Q)/Q)\bar{b}_Q),$$

where in the left sum (Q, e) runs over a set of representatives of the G-conjugacy classes of (G, b)-Brauer pairs, and where in the right sum Q runs over a set of representatives of the conjugacy classes of p-subgroups of G.

Proof Both sides are clearly independent of choices of conjugacy class representatives. Let Q be a p-subgroup of G. If the term in the right sum indexed by a conjugate of Q is nonzero, then $\mathrm{Br}_Q(b) \neq 0$, and hence Q is contained in a defect group of P. Suppose that there is a projective simple $k(N_G(Q)/Q)\bar{b}_Q$-module S. Then S is a simple $kN_G(Q)b_Q$-module, and hence there is a unique block c of $kN_G(Q)b_Q$ to which S belongs, when regarded as a $kN_G(Q)b_Q$-module. By 6.2.6, we have $c = \mathrm{Tr}_{N_G(Q,e)}^{N_G(Q)}(e)$ for some block e of $kC_G(Q)$, and the blocks $kC_G(Q)c$ and $kN_G(Q, e)e$ are canonically Morita equivalent. Since a Morita equivalence preserves the properties of being projective and simple, it follows that S corresponds to a unique projective simple $k(N_G(Q, e)/Q)\bar{e}$-module S'. Since c is a block of $kN_G(Q)b_Q$, hence a sum of blocks of $kC_G(Q)b_Q$, it follows that (Q, e) is a (G, b)-Brauer pair. The block e is not uniquely determined by c, but it is unique up to conjugation by an element in $N_G(Q)$. By reversing the argument, starting with a simple projective $kN_G(Q, e)/Q\bar{e}$-module, one sees that every (G, b)-weight arises, up to isomorphism, in this way. The result follows. $\square$

Corollary 6.10.4 *The block theoretic version 6.10.2 of Alperin's weight conjecture implies the group theoretic version 1.15.12.*

Proof Let G be a finite group. The set of blocks $\mathcal{B}$ of kG is a primitive decomposition of 1 in $Z(kG)$. Thus, for Q a p-subgroup of G, the set $\{\mathrm{Br}_Q(b) | b \in \mathcal{B}, \mathrm{Br}_Q(b) \neq 0\}$ is a (not necessarily primitive) decomposition of 1 in $Z(N_G(Q))$. The reformulation of Alperin's weight conjecture in 6.10.3 implies that the isomorphism classes of p-weights of kG are partitioned according to the isomorphism classes of (G, b)-weights, with b running over $\mathcal{B}$. Since the set of isomorphism classes of simple kG-modules is also partitioned in terms of the sets of isomorphism classes of simple kGb-modules, with $b \in \mathcal{B}$, the result follows. $\square$

It is not known whether the group theoretic version of Alperin's weight conjecture implies the block theoretic version.

Remark 6.10.5 Okuyama's result 5.10.6 can be rephrased as saying that every simple kG-module with a trivial source gives rise to a p-weight of G.

Proposition 6.10.6 ([114, Proposition 2.2]) *Let G be a finite group, let M and N be normal subgroups of G such that $N \leq M$ and such that M/N is a p-group. Let S be a projective simple kG-module, and let T be a simple summand* *of* $\mathrm{Res}_N^G(S)$. *Then* $^xT \not\cong T$ *for any* $x \in M \setminus N$. *In particular, if* G *stabilises the isomorphism class of* T, *then* $O_p(G/N) = \{1\}$.

Proof The block b of kG to which S belongs has defect zero, and so does the block c of kN to which T belongs. Thus the stabiliser of c in G is equal to the stabiliser of the isomorphism class of T. It follows from 6.8.12 that no element in $M \setminus N$ stabilises T. Applied to the inverse image of $O_p(G/N)$ in G, this yields the last statement.   □

Proposition 6.10.7 *Let* G *be a finite group, let* M *and* N *be normal subgroups of* G *such that* $N \leq M$ *and such that* M/N *is a* p-*group. Let* S *be a projective simple* kG-*module, and let* T *be a simple summand of* $\mathrm{Res}_N^G(S)$. *The* kM-*module* $\mathrm{Ind}_N^M(T)$ *is projective, simple, and isomorphic to a direct summand of* $\mathrm{Res}_M^G(S)$.

Proof By 6.10.6 no element in $M \setminus N$ stabilises T. It follows from 5.12.1 that $\mathrm{Ind}_N^M(T)$ is indecomposable, and also projective, as T is projective. The inclusion $T \rightarrow \mathrm{Res}_N^G(S) = \mathrm{Res}_N^M\mathrm{Res}_M^G(S)$ corresponds via Frobenius' reciprocity to a nonzero map $\mathrm{Ind}_N^M(T) \rightarrow \mathrm{Res}_M^G(S)$. The right side is a semisimple projective module, hence so is the image of the nonzero map $\mathrm{Ind}_N^M(T) \rightarrow \mathrm{Res}_M^G(S)$. It follows that this map splits; that is, the image of this map is isomorphic to a direct summand of $\mathrm{Ind}_N^M(T)$. The fact that $\mathrm{Ind}_N^M(T)$ is indecomposable forces that this map is split injective, implying the result.   □

We have noted in 1.15.14 that there are some structural restrictions for a finite group to have a projective simple module. The next result shows that giving rise to a weight yields strong restrictions on Brauer pairs.

Theorem 6.10.8 ([3]) *Let* G *be a finite group,* b *a block of* kG, *and* (Q, e) *a* (G, b)-*Brauer pair. Denote by* $\bar{e}$ *the image of* e *in* $k(N_G(Q, e)/Q)$. *Suppose that there is a projective simple* $k(N_G(Q, e)/Q)\bar{e}$-*module. Then the following hold.*

   (i) The group $Z(Q)$ *is a defect group of* $kC_G(Q)e$.
  (ii) For any normal subgroup M *of* $N_G(Q, e)$ *containing* Q *we have* $O_p(N_G(Q, e)/M) = \{1\}$.
 (iii) We have $O_p(N_G(Q, e)/QC_G(Q)) = \{1\}$ *and* $O_p(N_G(Q, e)) = Q$.

Proof Let S be a simple projective $k(N_G(Q, e)/Q)\bar{e}$-module. The group $C_G(Q)/Z(Q)$, identified to its obvious image in $N_G(Q, e)/Q$, is a normal subgroup of $N_G(Q, e)/Q$. Thus the restriction of S to $kC_G(Q)/Z(Q)\bar{e}$ is projective and semisimple. Therefore, if T is a simple summand of this restriction, then T belongs to a defect zero block of $kC_G(Q)/Z(Q)$ which is a summand of $\bar{e}$. It follows from 6.6.5 that this defect zero block is the canonical image of a block of $kC_G(Q)e$ with defect group $Z(Q)$. Since e is already a block of $kC_G(Q)$, this implies that e is a block of $kC_G(Q)$ with $Z(Q)$ as its defect group and that $\bar{e}$ is a defect zero block of $kC_G(Q)/Z(Q)$. In particular, this proves (i). By definition, $N_G(Q, e)$ stabilises the block e of $kC_G(Q)$, hence the defect zero block $\bar{e}$ of $kC_G(Q)/Z(Q)$, and hence the isomorphism class of the unique simple $kC_G(Q)e$-module. It follows from 6.10.6 applied to $N_G(Q, e)/Q$ and the normal subgroup M/Q that $O_p(N_G(Q, e)/M)$ is trivial, whence (ii). Statement (iii) follows from (ii) applied to $M = QC_G(Q)$ and $M = Q$, respectively. The latter case follows also from 1.15.14. $\qquad\square$

Remark 6.10.9 In the terminology of fusion systems of blocks – developed in Section 8.5 below – the statements (i) and (ii) of this Theorem can be rephrased as follows: denoting by $\mathcal{F}$ the fusion system of b on P with respect to some maximal Brauer pair containing (Q, e), if $k(N_G(Q, e_Q)/Q)\bar{e}$ has a weight, then Q is an $\mathcal{F}$-centric radical subgroup of P.

Corollary 6.10.10 *Let G be a finite group and b a block of kG. Suppose that b has an abelian defect group P. Set $H = N_G(P)$ and denote by c the block of kH that is the Brauer correspondent of b. Then Alperin's weight conjecture holds for b if and only if $\ell(kGb) = \ell(kHc)$.*

Proof Let e be a block of $kC_G(P)$ such that (P, e) is a maximal (G, b)-Brauer pair. Let (Q, f) be a (G, b)-Brauer pair such that $(Q, f) \le (P, e)$. By 6.3.12, P is a defect group of $kC_G(Q, f)$. It follows from 6.10.8 that if $kN_G(Q, f)/Q\bar{f}$ has a projective simple module, then $(Q, f) = (P, e)$. Thus Alperin's weight conjecture for kGb is equivalent to $\ell(kGb) = w(kN_G(P, e)/P\bar{e})$. The group P acts trivially on every simple $kN_G(P, e)$-module, hence P is contained in every vertex of a simple $kN_G(P, e)$-module. Every simple $kN_G(P, e)e$-module has a vertex contained in the defect group P of e, hence has P as its vertex. Thus every simple $kN_G(P, e)e$-module is the inflation to $kN_G(P, e)$ of a projective simple $kN_G(P, e)/P\bar{e}$-module. It follows that $w(kN_G(P, e)/P\bar{e}) = \ell(kN_G(P, e)e)$. The Morita equivalence from 6.7.6 between $kN_G(P, e)e$ and kHc implies that $w(kN_G(P, e)/P\bar{e}) = \ell(kHc)$, whence the result. $\qquad\square$

Remark 6.10.11 Since the number of isomorphism classes of simple modules is invariant under derived equivalences, it follows from 6.10.10 that Broué's abelian defect group conjecture 6.7.10 implies Alperin's weight conjecture for blocks with abelian defect groups.

A (G, b)-weight of $kN_G(Q, e)/Q$ corresponds to a p-weight of G via the Morita equivalence from 6.8.3 between $kN_G(Q, e)e$ and $kN_G(Q)c$, where $c = \text{Tr}_{N_G(Q,e)}^{N_G(Q)}(e)$, from 6.8.3. The following result sheds some light on the Green correspondents of weights.

Proposition 6.10.12 ([3, Lemma 1]) *Let G be a finite group Q a p-subgroup of G, S a simple projective $kN_G(Q)/Q$-module. Then S has Q as a vertex, when viewed as a $kN_G(Q)$-module. Let U be an indecomposable kG-module with vertex Q that is a Green correspondent of S viewed as a $kN_G(Q)$-module. Then U is isomorphic to a direct summand of $\mathrm{Ind}_P^G(k)$, where P is a Sylow p-subgroup of G.*

Proof By 5.1.8, Q is a vertex of S as a $kN_G(Q)$-module. We may choose P such that $R = N_P(Q)$ is a Sylow p-subgroup of $L = N_G(Q)$. Since R/Q is a p-group, there is a nonzero kR/Q-homomorphism $k \to \mathrm{Res}_{R/Q}^{L/Q}(S)$. Frobenius' reciprocity yields a nonzero homomorphism $\mathrm{Ind}_{R/Q}^{L/Q}(k) \to S$. This homomorphism is surjective as S is simple, and it is split as S is projective. Thus S is isomorphic to a direct summand of $\mathrm{Ind}_R^L(k)$, when viewed as a kL-module. Mackey's formula implies that $\mathrm{Ind}_R^L(k)$ is isomorphic to a direct summand of $\mathrm{Res}_L^G\mathrm{Ind}_P^G(k)$. The corollary 5.5.19 to the Burry–Carlson–Puig Theorem implies that U is isomorphic to a direct summand of $\mathrm{Ind}_P^G(k)$ as stated. □

6.11 Heights of characters

We assume in this section that K has characteristic 0. The height of an irreducible character of a finite group is an arithmetic invariant that shows up in many conjectures involving characters in blocks.

Definition 6.11.1 Let G be a finite group, B a block algebra of $\mathcal{O}G$ and P a defect group of b. Let χ be an element in $\mathbb{Z}\mathrm{Irr}(B)$ or in $\mathbb{Z}\mathrm{IBr}(B)$. Let a and d be the nonnegative integers such that p^a is the order of a Sylow p-subgroup of G and such that p^d is the order of P. The *height of* χ is the integer, denoted $\mathrm{ht}(\chi)$, such that $p^{a-d+\mathrm{ht}(\chi)}$ is the exact power of p dividing $\chi(1)$.

Since p^{a-d} is the p-part of $|G:P|$, it follows from 6.5.12 that the height $\mathrm{ht}(\chi)$ is a nonnegative integer. We will see that every block has at least one character of height zero. One of the key applications of this fact is a theorem of Brauer and Feit which bounds the number of irreducible characters in a block in terms of the defect groups.

Example 6.11.2 Let P be a finite p-group. Suppose that K is a splitting field of P. By 3.5.7, the degree of any $\chi \in \mathrm{Irr}_K(P)$ divides $|P|$, hence is a power of p. Since $\mathcal{O}P$ has a unique block, having P itself as a defect group, it follows that $\chi(1) = p^{\mathrm{ht}(\chi)}$. The height zero characters of P are thus exactly the characters of degree 1.

Definition 6.11.3 Let G be a finite group. Suppose that K is a splitting field for G, and let $\chi \in \mathrm{Irr}_K(G)$. The *defect of* χ is the nonnegative integer denoted $d(\chi)$ such that $p^{d(\chi)}$ is the exact power of p dividing $\frac{|G|}{\chi(1)}$.

Proposition 6.11.4 *Let G be a finite group, B a block algebra of $\mathcal{O}G$, d the integer such that p^d is the order of the defect groups of B, and let $\chi \in \mathrm{Irr}(B)$. Suppose that K is a splitting field for G.*

(i) We have $d = \mathrm{ht}(\chi) + d(\chi)$; in particular, $d(\chi) \le d$ and $\mathrm{ht}(\chi) \le d$.
(ii) If $d > 0$ then $d(\chi) > 0$ and $\mathrm{ht}(\chi) < d$.

Proof Let p^a be the order of a Sylow p-subgroup of G. Then $p^a = p^{a-d+\mathrm{ht}(\chi)}p^{d-\mathrm{ht}(\chi)}$. Since p^a is the exact power of p dividing $|G| = \frac{|G|}{\chi(1)}\chi(1)$ and $p^{a-d+\mathrm{ht}(\chi)}$ is the exact power of p dividing $\chi(1)$ it follows that $p^{d-\mathrm{ht}(\chi)}$ is the exact power of p dividing $\frac{|G|}{\chi(1)}$, whence $d(\chi) = d - \mathrm{ht}(\chi)$, proving (i). If $d(\chi) = 0$ then $\chi(1)$ is divisible by p^a, hence belongs to a block of defect zero, by 6.6.2. Statement (ii) follows. $\square$

Proposition 6.11.5 *Let G be a finite group, B a block algebra of $\mathcal{O}G$, P a defect group of B and $\chi \in \mathrm{Irr}(B)$. The following are equivalent.*

(i) We have $\mathrm{ht}(\chi) = 0$.
(ii) We have $\frac{|G|}{|P|\chi(1)} \in \mathbb{Z}$.
(iii) The rational number $\frac{|G|}{|P|\chi(1)}$ is an integer that is coprime to p.
(iv) We have $\frac{|G|}{|P|\chi(1)} \in \mathcal{O}$.
(v) We have $\frac{|G|}{|P|\chi(1)} \in \mathcal{O}^\times$.
(vi) We have $\frac{\chi(1)|P|}{|G|} \in \mathcal{O}^\times$.

Proof Note that $\frac{|G|}{\chi(1)}$ is a rational integer, by 3.5.7. We have $\mathrm{ht}(\chi) = 0$ if and only if the defect of χ is the unique integer d such that $p^d = |P|$, whence the equivalence of (i) with any of (ii), (iii), (iv), (v). The equivalence of (v) and (vi) holds trivially. $\square$

The following result on trivial source modules will be used in the existence proof of height zero characters in blocks.

Theorem 6.11.6 *Let G be a finite group, B a block algebra of $\mathcal{O}G$ and P a defect group of B. Let $i \in B^P$ be a source idempotent of B.*

(i) The p-part of $\mathrm{rk}_\mathcal{O}(Bi)$ is equal to the p-part of $|G|$.
(ii) There is an indecomposable B-module U with vertex P and trivial source $\mathcal{O}$ such that the p-part of $|G : P|$ is equal to the p-part of $\mathrm{rk}_\mathcal{O}(U)$.

Proof Since $i \in B$, we have $Bi = \mathcal{O}Gi$. Let e be the block of $kC_G(P)$ such that $\mathrm{Br}_P(i) \in kC_G(P)e$. Denote by $\hat{e}$ the unique block of $\mathcal{O}C_G(P)$ that lifts e, and let j be a primitive idempotent in $\mathcal{O}C_G(P)\hat{e}$ that lifts $\mathrm{Br}_P(i)$. Note that j belongs to $(\mathcal{O}G)^P$, so we may choose i such that $j = i + j'$ for some idempotent $j' \in \ker(\mathrm{Br}_P)$. Then $\mathcal{O}Gj = \mathcal{O}Gi \oplus \mathcal{O}Gj'$. All indecomposable summands of $\mathcal{O}Gj'$ as an $\mathcal{O}(G \times P)$-module have vertices of order strictly smaller than $|P|$. It follows from Theorem 5.12.13 that the $\mathcal{O}$-rank of $\mathcal{O}Gj'$ is divisible by $p|G|$. It suffices therefore to show that the p-part of $|G|$ is the p-part of $\mathrm{rk}_{\mathcal{O}}(\mathcal{O}Gj) = \mathrm{Ind}_{C_G(P)}^G \mathcal{O}C_G(P)j$. Clearly $\mathrm{rk}_{\mathcal{O}}(\mathcal{O}Gj) = |G : C_G(P)|\mathrm{rk}_{\mathcal{O}}(\mathcal{O}C_G(P)j)$. Now $\mathcal{O}C_G(P)j$ is a projective indecomposable module for the block $\mathcal{O}C_G(P)\hat{e}$ with defect group $Z(P)$. By Theorem 6.6.6, its p-part is equal to that of $|C_G(P)|$. This proves (i). Note that $\mathcal{O}Gi$ is projective as a right $\mathcal{O}P$-module. It follows from (i) that the p-part of $\mathcal{O}Gi \otimes_{\mathcal{O}P} \mathcal{O}$ is equal to the p-part of $|G : P|$. The $\mathcal{O}G$-module $\mathcal{O}Gi \otimes_{\mathcal{O}P} \mathcal{O}$ is a direct summand of $\mathcal{O}G \otimes_{\mathcal{O}P} \mathcal{O} = \mathrm{Ind}_P^G(\mathcal{O})$, so every indecomposable direct summand of $\mathcal{O}Gi \otimes_{\mathcal{O}P} \mathcal{O}$ is a trivial source module with a vertex R contained in P. Since the p-part of $|G : R|$ divides the rank of any trivial source module with vertex R, it follows that $\mathcal{O}Gi \otimes_{\mathcal{O}P} \mathcal{O}$ has an indecomposable direct summand U with vertex P such that the p-parts of $|G : P|$ and $\mathrm{rk}_{\mathcal{O}}(U)$ are equal. This proves (ii). $\square$

Theorem 6.11.7 *Let G be a finite group, B a block algebra of $\mathcal{O}G$ and P a defect group of B. Then there are $\chi \in \mathrm{Irr}(B)$ and $\varphi \in \mathrm{IBr}(B)$ such that χ and φ have height zero and such that the decomposition number d_φ^χ is prime to p.*

Proof Let U be a trivial source B-module as in the statement of Theorem 6.11.6 (ii). Then some irreducible constituent χ of the character of an U has the property that the p-part of $|G : P|$ is the exact power of p dividing $\chi(1)$. The decomposition map applied to χ yields in particular $\chi(1) = \sum_\varphi d_\varphi^\chi \varphi(1)$, where φ runs over the set $\mathrm{IBr}_k(G, b)$. Since by 6.5.12, $\varphi(1)$ is divisible by the p-part of $|G : P|$ there must be a summand $d_\varphi^\chi \varphi$ such that d_φ^χ is prime to p and such that the p-part of $|G : P|$ is the exact power of p dividing $\varphi(1)$. $\square$

Blocks of defect zero are characterised as the blocks with a unique irreducible character. There is no such characterisation of blocks with a unique irreducible Brauer character, or equivalently, with a unique isomorphism class of simple modules over k. Classes of blocks with a unique irreducible Brauer character include *nilpotent blocks*, which are precisely the blocks that are Morita equivalent to their defect group algebras – see Section 8.11 for details. Using 6.5.13 and 6.11.7 it is possible to list a few general properties of blocks with a unique irreducible Brauer character:

Theorem 6.11.8 *Let G be a finite group, B be a block algebra of $\mathcal{O}G$ and P a defect group of B. Suppose that k is a splitting field for G. Suppose that B has a unique irreducible Brauer character φ. Denote by Φ the character of a projective indecomposable B-module. The unique entry of the Cartan matrix of B is equal to $|P|$, and we have*

$$\mathrm{Res}^G_{G_{p'}}(\Phi) = |P| \cdot \varphi.$$

In particular, the basic algebras of B are split local of $\mathcal{O}$-rank $|P|$.

Proof Any projective indecomposable B-module is isomorphic to $Bj = \mathcal{O}Gj$ for some primitive idempotent j in B. Moreover, an idempotent in $\mathcal{O}G$ is primitive if and only of its image in kG is primitive. Thus we may replace $\mathcal{O}$ by any extension ring having k as a residue field, and hence we may assume that K is also a splitting field for all subgroups of G. Set $\bar{B} = k \otimes_{\mathcal{O}} B$. By 5.14.10 we have $\mathrm{Res}^G_{G_{p'}}(\Phi) = c \cdot \varphi$, where c is the unique entry of the Cartan matrix of B. It follows from 6.5.13 (ii) that $|P| \cdot \varphi = d \cdot \mathrm{Res}^G_{G_{p'}}(\Phi)$ for some integer d. Thus $cd = |P|$, hence c divides $|P|$. By 6.11.7, the p-part of $|G : P|$ is the exact p-power dividing $\varphi(1)$, and hence $|P|$ is the smallest positive integer with the property that $|P| \cdot \varphi$ is a multiple of $\mathrm{Res}^G_{G_{p'}}(\Phi)$. This shows $c = |P|$. The basic algebras of B are split, by the assumption on k, and local, by the assumption $\ell(B) = 1$. The statement on their rank follows from the fact that the unique (up to isomorphism) projective indecomposable $\bar{B}$-module has exactly $|P|$ composition factors, all isomorphic to the unique (up to isomorphism) simple $\bar{B}$-module. □

Heights of characters can be read off the source algebras (cf. [164], [191]). We will prove this based on the following result.

Proposition 6.11.9 *Let G be a finite group, B a block algebra of $\mathcal{O}G$, and P a defect group of B. Let $i \in B^P$ be a source idempotent. Suppose that k is a splitting field for all subgroups of G.*

(i) Let $c \in iB^Pi$. We have $\mathrm{Tr}^G_P(c) \in Z(B)^\times$ if and only if $c \in (iB^Pi)^\times$.
(ii) We have $\mathrm{Tr}^G_P(J(iB^Pi)) \subseteq J(Z(B))$.
(iii) We have $\mathrm{Tr}^G_P(i) \in Z(B)^\times$.

Proof Note first that (i) is a consequence of (ii) and (iii). Indeed, the algebra iB^Pi is local because i is primitive in B^P, hence $(iB^Pi)^\times = \mathcal{O}^\times(i + J(iB^Pi))$ and any element c in iB^Pi that is not invertible belongs to $J(iB^Pi)$. Since k is a splitting field for the subgroups, the source idempotent i remains primitive in $(\mathcal{O}G)^P$ upon replacing $\mathcal{O}$ by an extension with the same residue field k; thus we may assume that K is a splitting field for G. Let $\chi \in \mathrm{Irr}(B)$ such that $\mathrm{ht}(\chi) = 0$; that is, such that $\frac{\chi(1)}{|G:P|}$ is an invertible element in $\mathcal{O}$. Let $y \in J((i\mathcal{O}Gi)^P)$. Then $\chi(y) \in J(\mathcal{O})$. Thus $\frac{\chi(\mathrm{Tr}_P^G(y))}{\chi(1)} = \frac{|G:P|}{\chi(1)}\chi(y)$ is an element in $J(\mathcal{O})$. Statement (ii) follows from the characterisation of $J(Z(B))$ in 6.5.6. Set $b = 1_B$. By 5.6.13 there are $c, d \in B^P$ such that $b = \mathrm{Tr}_P^G(cid)$. Let $\chi \in \mathrm{Irr}(B)$. Since χ is a central function and since i is an idempotent, we have $\chi(cid) = \chi(idci)$. The algebra iB^Pi is local, and hence $idci = \lambda i + y$ for some $\lambda \in \mathcal{O}$ and some $y \in J(iB^Pi)$. It follows that

$$\chi(1) = \chi(b) = \chi(\mathrm{Tr}_P^G(cid)) = |G:P|\chi(cid) = \chi(\mathrm{Tr}_P^G(idci))$$
$$= \lambda\chi(\mathrm{Tr}_P^G(i)) + \chi(\mathrm{Tr}_P^G(y)).$$

Dividing by $\chi(1)$ yields $1 = \lambda\frac{\chi(\mathrm{Tr}_P^G(i))}{\chi(1)} + \frac{\chi(\mathrm{Tr}_P^G(y))}{\chi(1)}$. It follows from (ii) and 6.5.6 that the second summand belongs to $J(\mathcal{O})$. Thus the first summand belongs to $\mathcal{O}^\times$, and hence $\frac{\chi(\mathrm{Tr}_P^G(i))}{\chi(1)} \in \mathcal{O}^\times$. Thus $\mathrm{Tr}_P^G(i) \in Z(B)^\times$ by 6.5.6. $\square$

Corollary 6.11.10 *With the notation and hypotheses of 6.11.9, for any $\chi \in \mathrm{Irr}(B)$ and any $u \in Z(P)$ we have $\chi(ui) \neq 0$.*

Proof Let $u \in Z(P)$. Then $ui \in (iB^Pi)^\times$, and hence $\mathrm{Tr}_P^G(ui) \in Z(B)^\times$ by 6.11.9. Let $\chi \in \mathrm{Irr}(B)$. It follows from 6.5.6 that $0 \neq \chi(\mathrm{Tr}_P^G(ui)) = |G:P|\chi(ui)$, whence the result. $\square$

See [164] and [213, 9.3] for characterisations in terms of multiplicity modules of primitive interior G-algebras A with the property that $\mathrm{Tr}_P^G(i) \in (A^G)^\times$ for all defect groups P and all source idempotents i.

Proposition 6.11.11 ([164], [191]) *Let G be a finite group, B a block algebra of $\mathcal{O}G$, P a defect group of b, and $i \in B^P$ a source idempotent. Suppose that K, k are splitting fields for all subgroups of G. Let $\chi \in \mathrm{Irr}_K(B)$. Then $\mathrm{ht}(\chi)$ is the largest nonnegative integer such that $p^{\mathrm{ht}(\chi)}$ divides $\chi(i)$ in $\mathcal{O}$. In particular, $\mathrm{ht}(\chi) = 0$ if and only if $\chi(i)$ is prime to p.*

Proof It follows from 6.11.9 that $\mathrm{Tr}_P^G(i) \in Z(B)^\times$. Then 6.5.6 implies that $\frac{\chi(\mathrm{Tr}_P^G(i))}{\chi(1)} \in \mathcal{O}^\times$. We have $\chi(\mathrm{Tr}_P^G(i)) = |G:P|\chi(i)$. Thus the p-part of $\chi(i)$ is equal to the p-part of $\frac{\chi(1)}{|G:P|}$, which is equal to $p^{\mathrm{ht}(\chi)}$ by the definition of heights of characters. $\square$

Corollary 6.11.12 *With the notation and hypotheses of 6.11.11, let $\chi \in \mathrm{Irr}_K(B)$ such that $\mathrm{ht}(\chi) = 0$. Then for any element $u \in P$ we have $\chi(ui) \notin J(\mathcal{O})$. In particular, $\chi(ui) \neq 0$.*

Proof Let $u \in P$. By 6.5.9 we have $\chi(ui) - \chi(i) \in J(\mathcal{O})$. By 6.11.11 we have $\chi(i) \notin J(\mathcal{O})$. Thus $\chi(ui) \notin J(\mathcal{O})$. □

Theorem 6.11.13 ([104, Theorem 4.5]) *Let G be a finite group and let U be an $\mathcal{O}$-free indecomposable $\mathcal{O}Gb$-module with vertex P and an absolutely indecomposable $\mathcal{O}P$-source Y. Suppose that the multiplicity module M of U with respect to Y is simple. Then $\mathrm{rk}_{\mathcal{O}}(U)_p = |G : P|$ if and only if $\mathrm{rk}_{\mathcal{O}}(Y)$ is prime to p.*

Proof By the assumptions and 5.7.9, the multiplicity module M is a simple and projective module for the twisted group algebra $k_\alpha T/P$, where T is the stabiliser in $N_G(P)$ of a source Y of U. Thus M belongs to a defect zero block of a central p'-extension of T/P, and hence $\dim_k(M)_p = |T : P|_p$. It follows from 5.12.15 that $\mathrm{rk}_{\mathcal{O}}(U)_p = |G : P|$ holds if and only if $\mathrm{rk}_{\mathcal{O}}(Y)$ is prime to p. □

Corollary 6.11.14 ([104, Corollary 4.5]) *Let G be a finite group, B a block algebra of $\mathcal{O}G$ and P a defect group of B. Let $\chi \in \mathrm{Irr}_K(B)$ and let U be an $\mathcal{O}$-free B-module with character χ and absolutely indecomposable sources. Then χ has height zero if and only if P is a vertex of U and the sources of U have rank prime to p.*

Proof Note that U is necessarily indecomposable, so has a vertex and a source. By 6.1.7, P contains a vertex Q of U, and by 5.12.13, $\chi(1)$ is divisible by $|G : Q|_p$. If χ has height zero, then $\chi(1)_p = |G : P|_p$, so necessarily P is a vertex of any module with character χ in that case. The result follows from 6.11.13. □

Let P be a finite p-group and K a splitting field of characteristic of P. By 3.3.14 there exists an irreducible character in $\mathrm{Irr}_K(P)$ whose degree is a positive power of p if and only if P is nonabelian. Brauer's Height Zero Conjecture suggests that a far reaching generalisation of this fact should be true:

Conjecture 6.11.15 (Brauer's Height Zero Conjecture [26]) *Let G be a finite group, B a block algebra of $\mathcal{O}G$, and P a defect group of B. Suppose that K is a splitting field for G. Then P is abelian if and only if all characters in $\mathrm{Irr}_K(B)$ have height zero.*

One direction of this conjecture has been proved by Kessar and Malle, using the classification of finite simple groups.

Theorem 6.11.16 ([102]) *Let G be a finite group, B a block algebra of $\mathcal{O}G$ and P a defect group of b. Suppose that K is a splitting field for G. If P is abelian, then all characters in $\mathrm{Irr}_K(B)$ have height zero.*

Eaton and Moretó extended Brauer's Height Zero Conjecture as follows. For B a block algebra of $\mathcal{O}G$, denote by $\mathrm{mh}(B)$ the smallest positive height of some $\chi \in \mathrm{Irr}(B)$, with the convention $\mathrm{mh}(B) = \infty$ if all $\chi \in \mathrm{Irr}(B)$ have height zero.

Conjecture 6.11.17 ([64, Conjecture A]) *Let G be a finite group, B a block algebra of $\mathcal{O}G$, and P a defect group of B. Suppose that K contains a primitive $|G|^3$-th root of unity. We have $\mathrm{mh}(B) = \mathrm{mh}(\mathcal{O}P)$.*

Eaton and Moretó proved in [64] that this conjecture holds if G is a general linear group with defining characteristic p, or a symmetric or sporadic group. They proved further that if G is p-solvable, or if Dade's projective conjecture holds for every subquotient of G, then $\mathrm{mh}(\mathcal{O}P) \leq \mathrm{mh}(B)$.

Exercise 6.11.18 Let G be a finite group, H a subgroup, and let $\psi \in \mathrm{Irr}_K(H)$. Set $\chi = \mathrm{Ind}_H^G(\psi)$. Suppose that $\chi \in \mathrm{Irr}_K(G)$. Show that $d(\chi) = d(\psi)$.

6.12 On the number of irreducible characters in a block

Donovan's conjecture 6.1.9 would imply that the number of irreducible characters in a block is bounded in terms of the order of the defect groups. The following theorem, due to Brauer and Feit, shows that there is such a bound. The proof combines the existence of height zero characters in blocks with earlier results, notably 6.5.13. We assume that K has characteristic zero and that K and k are splitting fields for all blocks that arise in this section.

Theorem 6.12.1 ([31]) *Let G be a finite group, B a block algebra of $\mathcal{O}G$, and P a defect group of B. We have*

$$|\mathrm{Irr}_K(B)| \leq \sum_{\chi \in \mathrm{Irr}_K(B)} p^{2\mathrm{ht}(\chi)} \leq \frac{1}{4}|P|^2 + 1.$$

In particular, $|\mathrm{Irr}_K(B)|$ and $|\mathrm{IBr}_k(B)|$ are bounded by a function depending only on $|P|$.

Proof The first inequality is obvious, since every summand in the sum in the middle is a positive integer. Let d and a be the integers such that $|P| = p^d$ and $|G|_p = p^a$. For $\chi \in \mathrm{Irr}_K(B)$, denote by $\hat{\chi}$ the class function on G defined by $\hat{\chi}(x) = |P|\chi(x)$ if x is a p'-element in G and $\hat{\chi}(x) = 0$ if the order of x is divisible by p. By 6.5.13 (ii), the class function $\hat{\chi}$ is a generalised character in $\mathrm{Pr}_{\mathcal{O}}(B)$. Write $\hat{\chi} = \sum_{\psi \in \mathrm{Irr}_K(B)} a(\chi, \psi)\psi$ for some integers $a(\chi, \psi)$. Note that $a(\chi, \psi) = \langle \hat{\chi}, \psi \rangle = \langle \chi, \hat{\psi} \rangle = a(\psi, \chi)$. We have $\langle \hat{\chi}, \hat{\chi} \rangle = \sum_{\psi \in \mathrm{Irr}_K(B)} a(\chi, \psi)^2$

and $\langle \hat{\chi}, \hat{\chi} \rangle = |P| \langle \chi, \hat{\chi} \rangle$, whence

$$\sum_{\psi \in \mathrm{Irr}_K(B)} a(\chi, \psi)^2 = |P| a(\chi, \chi).$$

We have

$$a(\chi, \psi) = \langle \hat{\chi}, \psi \rangle = \frac{1}{|G|} \sum_{x \in G} \hat{\chi}(x) \psi(x^{-1}).$$

The summands with x of order divisible by p vanish, and the remaining summands depend only on the conjugacy class of x. Denote by $\mathcal{R}$ a set of representatives of the conjugacy classes of p'-elements in G. The above expression yields

$$a(\chi, \psi) = \frac{\chi(1)|P|}{|G|} \sum_{x \in \mathcal{R}} \frac{\chi(C_x)}{\chi(1)} \psi(x^{-1}),$$

where C_x is the sum in $\mathcal{O}G$ of the conjugacy class of x. Since $a(\chi, \psi) = a(\psi, \chi)$, we also have

$$a(\chi, \psi) = \frac{\psi(1)|P|}{|G|} \sum_{x \in \mathcal{R}} \frac{\psi(C_x)}{\psi(1)} \chi(x^{-1}).$$

Suppose that χ has height zero. Then $\frac{\chi(1)|P|}{|G|} \in \mathcal{O}^\times$, by 6.11.5. In that case, p^a is the highest power of p dividing $\hat{\chi}(1)$, and hence 6.5.13 (i) implies that $\frac{1}{p}\hat{\chi}$ is not a generalised character, or equivalently, that there is $\psi \in \mathrm{Irr}_K(B)$ such that $a(\chi, \psi)$ is prime to p. The two previous equations imply that such a character ψ has then also height zero and satisfies $\sum_{x \in \mathcal{R}} \frac{\psi(C_x)}{\psi(1)} \chi(x^{-1}) \in \mathcal{O}^\times$. By 6.5.7 the image of $\omega_\psi(C_x) = \frac{\psi(C_x)}{\psi(1)}$ in k does not depend on $\psi \in \mathrm{Irr}_K(B)$. Thus $\sum_{x \in \mathcal{R}} \frac{\psi(C_x)}{\psi(1)} \chi(x^{-1}) \in \mathcal{O}^\times$ for any $\psi \in \mathrm{Irr}_K(B)$. The last equation for $a(\chi, \psi)$ implies that if $\chi \in \mathrm{Irr}_K(B)$ has height zero, then the exact power of p dividing $a(\chi, \psi)$ is equal to $p^{\mathrm{ht}(\psi)}$, for any $\psi \in \mathrm{Irr}_K(B)$ (and this is also the exact power of p dividing $a(\psi, \chi)$). In particular, $a(\chi, \psi)^2 \geq p^{2\mathrm{ht}(\psi)} > 0$. Since $\mathrm{ht}(\chi) = 0$, we have $p^{2\mathrm{ht}(\chi)} = 1$. It follows that

$$|P| a(\chi, \chi) = \sum_{\psi \in \mathrm{Irr}_K(B)} a(\chi, \psi)^2 \geq a(\chi, \chi)^2 + \sum_{\psi \in \mathrm{Irr}_K(B)} p^{2\mathrm{ht}(\psi)} - 1$$

or equivalently,

$$|P| a(\chi, \chi) - a(\chi, \chi)^2 + 1 \geq \sum_{\psi \in \mathrm{Irr}_K(B)} p^{2\mathrm{ht}(\psi)}.$$

If r, s are any two real numbers, then $(\frac{r}{2} - s)^2 = \frac{1}{4}r^2 - rs + s^2 \geq 0$, hence $\frac{r^2}{4} \geq rs - s^2$. Applied to $r = |P|$ and $s = a(\chi, \chi)$, the previous inequality yields the second inequality in the statement, whence the result.   □

The number $|\mathrm{Irr}_K(B)|$ is equal to $\dim_k(Z(\bar{B}))$, where $\bar{B} = k \otimes_{\mathcal{O}} B$. The centre $Z(\bar{B})$ can be interpreted as the Hochschild cohomology of $\bar{B}$ in degree zero. Using the above theorem of Brauer and Feit, it is shown in [100] that $\dim_k(HH^n(\bar{B}))$ is bounded in terms of the defect groups of B for all integers $n \geq 0$. The bound for $|\mathrm{Irr}_K(B)|$ stated above is not best possible; one can for instance use the second inequality to improve this bound if there exists a character with positive height in B. From the point of view of finiteness conjectures such as Donovan's, the key point of this theorem is that there is a bound in terms of the defect group at all. Brauer has conjectured a much sharper bound:

Conjecture 6.12.2 (Brauer's $k(B)$-conjecture) *Let G be a finite group, B a block algebra of $\mathcal{O}G$, and P a defect group of B. Then $|\mathrm{Irr}_K(B)| \leq |P|$.*

For blocks of p-solvable groups, Brauer's $k(B)$-conjecture reduces to Nagao's $k(GV)$-problem from [151]. This is the special case of Brauer's conjecture above where P is elementary abelian and normal of p'-index in G. The $k(GV)$-problem has been shown to have an affirmative answer by Gluck, Magaard, Riese and Schmid, making use of earlier work of Knörr, Gow, Robinson and Thompson, amongst others. We mention without proof the following result of Sambale.

Theorem 6.12.3 ([202, Theorem 5]) *Let G be a finite group and B a block algebra of $\mathcal{O}G$ with Cartan matrix C. We have*

$$|\mathrm{Irr}_K(B)| \leq \frac{\det(C) - 1}{\ell(B)} + \ell(B) \leq \det(C).$$

The *sectional rank* of a finite p-group P is the largest integer $s(P)$ for which there exists a subgroup Q and a normal subgroup R of Q such that Q/R is elementary abelian of order $p^{s(P)}$.

Conjecture 6.12.4 (Malle and Robinson [146]) *Let G be a finite group, B a block algebra of kG, and P a defect group of B. Then $\ell(B) \leq p^{s(P)}$.*

Malle and Robinson showed that this conjecture holds for blocks of finite p-solvable groups, symmetric and alternating groups as well as their covering groups, quasisimple groups of Lie type in defining characteristic, unipotent blocks of quasisimple groups of classical Lie type in nondefining characteristic, and many more cases. This conjecture would imply that $p^{s(P)}(|\mathrm{Irr}_K(P)| - 1)$ is an upper bound for $|\mathrm{Irr}_K(G, b)|$, which can be less than the bound in 6.12.1.

Conjecture 6.12.5 (Olsson) *Let G be a finite group, B a block algebra of $\mathcal{O}G$, and P a defect group of B. Then the number of height zero characters in $\mathrm{Irr}_K(B)$ is at most $|P : [P, P]|$.*

Olsson's conjecture is known to hold for blocks of finite p-solvable groups, symmetric and alternating groups. See [84] for further results and references.

McKay conjectured that given a finite group G and a Sylow p-subgroup P, then the groups G and $N_G(P)$ should have the same number of irreducible characters of degree prime to p. For $p = 2$ this has been proved by Malle and Späth, using the classification of finite simple groups:

Theorem 6.12.6 (Malle-Späth (2015)) *Let G be a finite group, and let S be a Sylow 2-subgroup of G. Then G and $N_G(S)$ have the same number of K-valued irreducible characters of odd degree.*

McKay's conjecture has the following block theoretic version:

Conjecture 6.12.7 (Alperin–McKay) *Let G be a finite group and b a block of $\mathcal{O}G$ with defect group P. Let c be the block of $\mathcal{O}N_G(P)$ corresponding to B. Then $\mathcal{O}Gb$ and $\mathcal{O}N_G(P)c$ have the same number of irreducible characters of height zero.*

The McKay conjecture follows from the Alperin–McKay conjecture by running over all blocks of kG having a Sylow p-subgroup of G as a defect group. It follows from [113] and the positive solution of the kGV-problem that the Alperin–McKay conjecture implies Olsson's conjecture. The Alperin–McKay conjecture has been refined by Isaacs and Navarro [90], taking into account certain congruences of character degrees. Another refinement, due to Navarro [154], takes into account Galois actions on irreducible characters. Yet another refinement due to Turull [217] takes into account fields of values of characters and local Schur indices over the p-adic number field. The Alperin–McKay conjecture (and in many cases some of its refinements) holds for blocks with cyclic defect groups (as before a consequence of Dade's work), blocks of p-solvable groups (proved independently by Dade and by Okuyama and Wajima), blocks of symmetric groups and their covers (Fong, Olsson, Michler), blocks with a trivial intersection defect group (An, Eaton), certain finite groups of Lie type (Lehrer, Michler, Olsson, Späth). Uno [219] has formulated a generalisation of the McKay conjecture, combining conjectures of Dade and Isaacs–Navarro. See also [9, §2]. Murai described in [150] properties of a minimal counterexample to the Alperin–McKay conjecture. Späth has reduced in [207] the Alperin–McKay conjecture to a statement on nonabelian finite simple groups.

Knörr and Robinson [106] reformulated Alperin's weight conjecture 6.10.2 in terms of alternating sums indexed by chains of p-subgroups, counting ordinary irreducible characters rather that Brauer characters. We use the following notation. For G a finite group and $\sigma = Q_0 < Q_1 < \cdots < Q_m$ a nonempty chain of p-subgroups of G we denote by $N_G(\sigma)$ the stabiliser of σ in G; that is, $N_G(\sigma)$ is the intersection of the normalisers $N_G(Q_i)$, $0 \le i \le m$. If b is a block of $\mathcal{O}G$, we denote by b_σ the sum of blocks of $\mathcal{O}G_\sigma$ corresponding to b. The following conjecture has been shown by Knörr and Robinson in [106] to be equivalent to Alperin's weight conjecture.

Conjecture 6.12.8 *Let G be a finite group and b a block of $\mathcal{O}G$ with a nontrivial defect group. We have*

$$|\mathrm{Irr}_K(G, b)| = \sum_\sigma (-1)^{|\sigma|} |\mathrm{Irr}_K(G_\sigma, b_\sigma)|,$$

where σ runs over a set of representatives of the G-conjugacy classes of nonempty chains of nontrivial p-subgroups of G.

If $\sigma = Q_0 < Q_1 < \cdots < Q_m$ is a nonempty chain of p-subgroups of a finite group G, then $N_G(\sigma)$ contains $C_G(Q_m)$, and by Theorem 6.5.17, the blocks occurring in b_σ are conjugacy class sums of blocks e of $\mathcal{O}G_\sigma$ such that $(Q_m, \bar{e})$ is a (G, b)-Brauer pair, where $\bar{e}$ is the image in $kC_G(Q_m)$ of e. It is shown in [106] that the sum on the right side remains unchanged if instead of taking the sum over all chains of nontrivial subgroups, one takes the sum over certain partially ordered sets of *normal* or *radical* chains. We will come back to this in Section 10.7, where we give in particular a proof of the fact that Alperin's weight conjecture 6.10.2 and the Knörr–Robinson conjecture 6.12.8 are indeed equivalent.

Dade's *ordinary conjecture* announced in [58], is obtained from the Knörr–Robinson version of Alperin's weight conjecture by specifying defects of characters on both sides. For G a finite group, b a block or a sum of blocks of $\mathcal{O}G$ and d a nonnegative integer, we denote by $\mathrm{Irr}_d(G, b)$ the set of $\chi \in \mathrm{Irr}_K(G, b)$ such that $d(\chi) = d$. For the remainder of this section we assume that K is a splitting field for all finite groups involved.

Conjecture 6.12.9 (Dade's ordinary conjecture) *Let G be a finite group G such that $O_p(G) = \{1\}$, and let b be a block of $\mathcal{O}G$ with a nontrivial defect group. For any positive integer d we have*

$$|\mathrm{Irr}_d(G, b)| = \sum_\sigma (-1)^{|\sigma|} |\mathrm{Irr}_d(G_\sigma, b_\sigma)|,$$

where σ runs over a set of representatives of the G-conjugacy classes of nonempty chains of nontrivial p-subgroups of G.

Dade's *projective conjecture*, announced in [59], is stronger in that it allows for the group G to have a nontrivial normal p-subgroup. Eaton showed in [62] that Dade's projective conjecture is equivalent to a conjecture due to Robinson [189, Conjecture 4.1]. To describe Robinson's conjecture, we will use the following notation. For G a finite group, b a block (or a sum of blocks) of $\mathcal{O}G$, d a positive integer, and $\lambda \in \mathrm{Irr}_K(O_p(Z(G)))$, we denote by $\mathbf{k}_d(G, b, \lambda)$ the number of $\chi \in \mathrm{Irr}_K(G, b)$ such that $d = d(\chi)$ and such that $\langle \mathrm{Res}^G_{O_p(Z(G))}(\chi), \lambda \rangle \neq 0$. In addition, if $\sigma = Q_0 < Q_1 < \cdots < Q_n$ is a nonempty chain of p-subgroups Q_i of G, then we denote by $w_d(G, b, \lambda, \sigma)$ the number of $\chi \in \mathrm{Irr}_K(G, b)$ such that $d = d(\chi)$, $\langle \mathrm{Res}^G_{O_p(Z(G))}(\chi), \lambda \rangle \neq 0$, and such that χ can be afforded by a relatively $\mathcal{O}Q_0$-projective $\mathcal{O}G$-module. For σ as before, we denote by G_σ the stabiliser of σ in G; that is, G_σ is the intersection of the normalisers in G of the Q_i. If $0 \leq i \leq n$, we denote by σ_i the chain $Q_0 < Q_1 < \cdots < Q_i$, and we say that σ is a *radical chain* if $Q_i = O_p(G_{\sigma_i})$ for $0 \leq i \leq n$. We set $|\sigma| = n$, the *length of σ*. Chains of length zero are p-subgroups of G.

Conjecture 6.12.10 (Robinson's conjecture) *Let G be a finite group G and b a block of $\mathcal{O}G$. For every integer $d \geq 0$, and every $\lambda \in \mathrm{Irr}_K(O_p(Z(G)))$ we have*

$$\mathbf{k}_d(G, b, \lambda) = \sum_\sigma (-1)^{|\sigma|} w_d(G_\sigma, b_\sigma, \lambda, \sigma),$$

where in the sum σ runs over a set of representatives of the G-conjugacy classes of radical chains of nontrivial p-subgroups of G, and where b_σ is the sum of all blocks of $\mathcal{O}G_\sigma$ that correspond to b.

The Brauer–Feit bound on the number of irreducible characters of a block is one of the ingredients that goes into establishing a connection between Donovan's conjecture and rationality properties of blocks. In order to describe this, we need the following notation. Let G be a finite group. If τ is an automorphism of the field k then τ extends to a ring automorphism, abusively still denoted by τ, of the finite group algebra kG, sending $\sum_{x \in G} \lambda_x x$ to $\sum_{x \in G} \tau(\lambda_x)x$. If B is a block algebra of kG then so is its image $\tau(B)$ under this ring automorphism. Since τ commutes with the relative trace map Tr^G_P, it follows that B and $\tau(B)$ have the same defect groups. But unless τ is trivial, it is not a k-algebra automorphism of kG. Thus, although B and $\tau(B)$ are isomorphic as rings, they may not be isomorphic as k-algebras – they need not even be Morita equivalent. We denote by $m(G, B)$ the number of pairwise Morita inequivalent blocks of kG of the form $\tau^t(B)$, where τ is the Frobenius automorphism $\lambda \mapsto \lambda^p$ of k and where

t runs over the set of nonnegative integers. Certain quantum complete intersections provided the first example of ring isomorphic but not Morita equivalent block algebras over k; this is due to Benson and Kessar [16].

Conjecture 6.12.11 (Kessar's rationality conjecture) *Suppose that k is algebraically closed. For any finite p-group P there is a positive integer $m(P)$, depending only on the isomorphism class of P, such that for any finite group G and any block B of kG with defect groups isomorphic to P, we have $m(G, B) \leq m(P)$.*

Principal blocks are defined over $\mathbb{F}_p$, and hence Kessar's rationality conjecture holds for principal blocks. Donovan's conjecture for principal blocks is therefore reduced to principal blocks of finite simple groups by Düvel's results in [61]. In conjunction with the following result, due to Kessar, this suggests a strategy how to go about Donovan's conjecture.

Theorem 6.12.12 ([94, Theorem 1.4]) *The Weak Donovan Conjecture 6.1.11 and Kessar's rationality conjecture 6.12.11 are together equivalent to Donovan's conjecture 6.1.9.*

Proof Donovan's conjecture 6.1.9 implies both conjectures 6.1.11 and 6.12.11. Suppose conversely that the conjectures 6.1.11 and 6.12.11 hold. Let P be a finite p-group. Let m be a positive integer such that $m(G, B) \leq m$ for any finite group G and any block B of kG with defect groups isomorphic to P. Set $M = m!$. Denote by τ the automorphism of k sending $\lambda \in k$ to λ^p. Let t be a positive integer such that $t \leq m$ and such that B and $\tau^t(B)$ are Morita equivalent. Let e be an idempotent in B such that eBe is a basic algebra of B. Then $\tau^t(e)kG\tau^t(e) = \tau^t(eBe)$ is a basic algebra of $\tau^t(B)$. Since B and $\tau^t(B)$ are Morita equivalent, they have isomorphic basic algebras. Let $\psi : eBe \cong \tau^t(eBe)$ be a k-algebra isomorphism. Set $\Phi = \psi^{-1} \circ \tau^t$, where τ^t denotes here the restriction of τ^t to eBe. Then Φ is a ring automorphism of eBe. Setting $q = p^t$, it follows that Φ satisfies $\Phi(\lambda x) = \lambda^q \Phi(x)$ for all $x \in eBe$ and all $\lambda \in k$. By 1.14.17, the Φ-fixed point set $(eBe)^\Phi$ in eBe is an $\mathbb{F}_q$-subalgebra of eBe satisfying $eBe = k \otimes_{\mathbb{F}_q} (eBe)^\Phi$. Since t divides $M = m!$, this shows that for any basic algebra A of a block B of kG with a defect group isomorphic to P, there is an $\mathbb{F}_{p^M}$-algebra C such that $A \cong k \otimes_{\mathbb{F}_{p^M}} C$. The dimension of such a basic algebra is bounded in terms of its Cartan matrix, thus in terms of the largest entry as well as the size of the matrix. The largest entry is bounded in terms of P since we assume that the Weak Donovan conjecture holds. The size of the Cartan matrix is bounded in terms of P thanks to the Brauer–Feit Theorem 6.12.1. Thus $\dim_k(A)$ is bounded in terms of a function depending only on P.

But then the number of possible 3-dimensional matrices of multiplicative structure constants of A is bounded in terms of P, and this is equivalent to Donovan's conjecture 6.1.9. □

Calculating numerical invariants of blocks such as their numbers of isomorphism classes of simple modules or irreducible characters remains one of the great challenges in block theory. A wealth of results calculating block invariants for 'small' defect groups can be found in Sambale [201].

6.13 Brauer's Second Main Theorem

We assume in this section that the quotient field K of $\mathcal{O}$ has characteristic 0. We assume further that K and k are splitting fields for all finite groups and their subgroups in this section. The generalised decomposition numbers $\chi(u_\epsilon)$ considered in 5.15.3 are compatible with the block decomposition:

Theorem 6.13.1 *Let G be a finite group and b a block of $\mathcal{O}G$. Let $\chi \in$ Irr$_K(G, b)$, let u be a p-element in G, let ϵ be a local point of $\langle u \rangle$ on $\mathcal{O}G$ and $\varphi = \varphi_\epsilon \in \mathrm{IBr}_k(C_G(u))$. If $\chi(u_\epsilon) \neq 0$, then φ belongs to a block e of $kC_G(u)$ such that (u, e) is a (G, b)-Brauer element.*

Proof Let $j \in \epsilon$. If χ belongs to the block b, then $\chi(x) = \chi(xb)$ for all $x \in G$, and hence $\chi(uj) = \chi(ubj)$. By 5.12.16 this value is zero if $\mathrm{Br}_u(bj) = 0$. Thus, if $\chi(u_\epsilon) \neq 0$, then $\mathrm{Br}_u(bj) = \mathrm{Br}_u(b)\mathrm{Br}_u(j) \neq 0$, which means that the unique block e of $kC_G(u)$ to which $\mathrm{Br}_u(j)$ belongs must satisfy $\mathrm{Br}_u(b)e = e$. □

The following equivalent reformulation of Brauer's Second Main Theorem is from [36, A.2.1].

Theorem 6.13.2 *Let G be a finite group and b a block of $\mathcal{O}G$. Let u be a p-element in G. Let $c \in Z(\mathcal{O}C_G(u))$ be the unique central idempotent whose image $\bar{c}$ in $Z(kC_G(u))$ is equal to $\mathrm{Br}_u(b)$.*

(i) If $\chi \in \mathrm{Cl}_K(G)$ belongs to b, then $d_G^u(\chi) \in \bar{\mathrm{Cl}}_K(C_G(u)_{p'})$ belongs to c.
(ii) If $\varphi \in \mathrm{Cl}_K(C_G(u))_{p'}$ belongs to c, then $t_G^u(\varphi) \in \mathrm{Cl}_K(G)$ belongs to b.

Proof Note that $\bar{c}$ is the sum of all blocks e of $kC_G(u)$ such that (u, e) is a (G, b)-Brauer element. It follows that (i) is equivalent to 6.13.1. Suppose that $\chi \in \mathrm{Irr}_K(G)$ does not belong to b. By 5.15.14 we have $\langle \chi, t_G^u(\varphi) \rangle_G = \langle d_G^u(\chi), \varphi \rangle'_{C_G(u)}$. By 6.13.1 or by (i), none of the constituents of $d_G^u(\chi)$ belongs to a block that arises in a decomposition of c, and hence by 6.5.2 this scalar product is zero. This proves (ii). □

Definition 6.13.3　Let G be a finite group. Let b a block of $\mathcal{O}G$ and (u, e) a (G, b)-Brauer element. Denote by $\hat{e}$ the unique block of $\mathcal{O}C_G(u)$ that lifts e. For $\chi \in \text{Cl}_K(G)$ define a class function $d_G^{(u,e)}(\chi)$ in $\text{Cl}_K(C_G(u))_{p'}$ by setting

$$d_G^{(u,e)}(\chi)(s) = \chi(us\hat{e})$$

for all $s \in C_G(u)_{p'}$.

With this notation, statement (i) of 6.13.2 can be reformulated as follows:

Corollary 6.13.4　*Let G be a finite group. Let b be a block of $\mathcal{O}G$ and u a p-element in G. Let $\chi \in \text{Cl}_K(G)$ be a class function that belongs to b. Then*

$$d^u(\chi) = \sum_e d_G^{(u,e)}(\chi),$$

where e runs over all blocks of $kC_G(u)\text{Br}_u(b)$.

Proof　If c is the idempotent in $Z(\mathcal{O}C_G(u))$ that lifts $\text{Br}_u(b)$, then $c = \sum_e \hat{e}$, where e runs over the blocks of $kC_G(u)\text{Br}_u(b)$. By 6.13.2 (i), if χ belongs to b, then for any p'-element in $C_G(u)$ we have $\chi(us) = \chi(usb) = \chi(usc) = \sum_e \chi(us\hat{e})$, where e is as in the statement and where we use 5.12.16 for the second equality.　□

We use Brauer's Second Main Theorem to reorganise the generalised decomposition numbers in a way that is compatible with the block decomposition.

Definition 6.13.5　Let G be a finite group and b a block of $\mathcal{O}G$. With the notation as in 6.13.1, for any b-Brauer element (u, e), any $\varphi \in \text{IBr}_k(C_G(u), e)$ and any $\chi \in \text{Irr}_K(G, b)$ write

$$d_{\chi,\varphi}^{(u,e)} = \chi(u_\epsilon),$$

where ϵ is the unique local point of $\langle u \rangle$ on $\mathcal{O}Gb$ such that $\varphi_\epsilon = \varphi$. The *generalised decomposition matrix of the block b* is the matrix

$$D_b = (d_{\chi,\varphi}^{(u,e)})$$

such that the rows of D_b are indexed by the set of irreducible characters $\text{Irr}_K(G, b)$ belonging to b and such that the columns are indexed by triples (u, e, φ), with (u, e) running over a set of representatives of the G-conjugacy classes of b-Brauer elements and $\varphi \in \text{IBr}_k(C_G(u), e)$. The entries of the matrix D_b are called *generalised decomposition numbers of b*.

Proposition 6.13.6 *Let G be a finite group. Let (u, e) be a (G, b)-Brauer element and $\chi \in \mathrm{Irr}_K(G, b)$. We have*

$$d_G^{(u,e)}(\chi) = \sum_{\varphi \in \mathrm{IBr}_k(C_G(u), e)} d_{\chi,\varphi}^{(u,e)} \varphi.$$

Proof This follows immediately from the definitions 6.13.3 and 6.13.5. □

With the above notation, the columns in D_b indexed by $(1, b, \varphi)$, where $\varphi \in \mathrm{IBr}_k(G, b)$, correspond to the ordinary decomposition matrix of the block algebra $\mathcal{O}Gb$ as defined earlier. The ordinary decomposition matrix multiplied by its transpose yields the Cartan matrix. We will show now that the generalised decomposition matrix multiplied with its transpose yields a block diagonal matrix whose blocks are the Cartan matrices of the b-Brauer elements, implying in particular that D_b is a square non singular matrix. We introduce the following terminology:

Definition 6.13.7 Let G be a finite group and b a block of $\mathcal{O}G$. The *generalised Cartan matrix of b* is the block diagonal matrix C_b whose blocks, denoted $C_{(u,e)}$, are indexed by a system of representatives of the G-conjugacy classes of b-Brauer elements (u, e), such that $C_{(u,e)}$ is the Cartan matrix of $kC_G(u)e$.

The ordinary Cartan matrix of b appears in C_b as the block corresponding to the b-Brauer element $(1, b)$. Note that C_b is non singular because Cartan matrices of blocks are non singular. The matrices C_b, D_b are defined up to a choice of an order on the labelling sets of the rows and columns, and while there is no canonical choice, one choice always gives rise to a dual choice as follows. If (u, e) is a b-Brauer element then so is (u^{-1}, e) because $C_G(u) = C_G(u^{-1})$, and thus if (u, e) runs over a set of representatives of the G-conjugacy classes of b-Brauer elements then so does (u^{-1}, e).

Theorem 6.13.8 *Let G be a finite group and b a block of $\mathcal{O}G$. For any two b-Brauer elements (u, e), (v, f) and any $\varphi \in \mathrm{IBr}_k(C_G(u), e)$, $\psi \in \mathrm{IBr}_k(C_G(v), f)$ we have*

$$\sum_{\chi \in \mathrm{Irr}_K(G,b)} d_{\chi,\varphi}^{(u,e)} \cdot d_{\chi,\psi}^{(v^{-1},f)} = \begin{cases} 0 & \text{if } u \text{ and } v \text{ are not conjugate} \\ c(\varphi, \psi) & \text{if } u = v \end{cases}$$

where $c(\varphi, \psi)$ is the Cartan number of $kC_G(u)e$ associated with φ, ψ. Moreover, if we denote by C_b, D_b the generalised Cartan matrix and the generalised decomposition matrix of b, respectively, then D_b is a square matrix and we have $^TD_b \cdot D_b = C_b$ where ^TD_b is the transpose of D_b with rows reindexed dually to the indexing of the columns of D_b. In particular, D_b is non singular.

Proof Let $i \in (\mathcal{O}G)^{\langle u \rangle}$ and $j \in (\mathcal{O}G)^{\langle v \rangle}$ be primitive idempotents such that $\bar{i} = \mathrm{Br}_u(i)$ and $\bar{j} = \mathrm{Br}_v(j)$ are primitive idempotents in $kC_G(u)e$ and $kC_G(v)f$, respectively, with the property that $kC_G(u)\bar{i}$ is a projective cover of a simple $kC_G(u)$-module with Brauer character φ, and $kC_G(v)j$ is a projective cover of a simple $kC_G(v)$-module with Brauer character ψ. In the sums that follow the index χ runs over $\mathrm{Irr}_K(G, b)$. The product of the row of $^T D_b$ indexed by (u, e, φ) and the column of D_b indexed by (v^{-1}, f, ψ) is equal to the sum

$$\sum_\chi d^{(u,e)}_{\chi,\varphi} \cdot d^{(v^{-1},f)}_{\chi,\psi} = \sum_\chi \chi(ui)\chi(v^{-1}j).$$

The character Ψ of $\mathcal{O}Gb$ as left $\mathcal{O}(G \times G)$-module is given by $\Psi(x, y) = \sum_\chi \chi(x)\chi(y^{-1})$. Thus the previous sum is equal to the trace of (u, v) on $i\mathcal{O}Gj$, with u acting by left multiplication and v acting with right multiplication by v^{-1}. Now $i\mathcal{O}Gj$ is a permutation module for the cyclic group $R = \langle(u, v)\rangle$ generated by (u, v). By 3.1.16 and 5.8.1, the trace of (u, v) on $i\mathcal{O}Gj$ is the dimension of $(i\mathcal{O}Gj)(R)$. If u and v are not conjugate then $(i\mathcal{O}Gj)(R) = \{0\}$ by 5.8.9. In the case $u = v$ we have $(i\mathcal{O}Gj)(R) = \bar{i}kC_G(u)\bar{j}$. The dimension of this space is the Cartan invariant $c(\varphi, \psi)$ as claimed, where the notation is as in 5.14.10. The result follows. $\qquad\square$

Remark 6.13.9 If K is a subfield of $\mathbb{C}$ then $d^{(u,e)}_{\chi,\varphi}$ is the complex conjugate of $d^{(u^{-1},e)}_{\chi,\varphi}$. In that case the equation $^T D_b \cdot D_b = C_b$ in the previous theorem is equivalent to

$$^t \bar{D}_b \cdot D_b = C_b,$$

where $^t \bar{D}_b$ is the complex conjugate of the transpose of D_b.

Remark 6.13.10 Generalised decomposition numbers can be read off the source algebras. Let G be a finite group, b a block of $\mathcal{O}G$ with a defect group P and let $i \in (\mathcal{O}Gb)^P$ be a source idempotent. Denote by γ the local point of P on $\mathcal{O}Gb$ containing γ. By 5.15.3 and 6.13.5, the matrix of generalised decomposition numbers of a block b of $\mathcal{O}G$ is the matrix

$$D_b = (\chi(u_\epsilon))_{\chi,u_\epsilon},$$

where χ runs over $\mathrm{Irr}_K(G, b)$ and u_ϵ over a system of representatives of the G-conjugacy classes of local pointed elements on $\mathcal{O}Gb$, and where $\chi(u_\epsilon) = \chi(ui_\epsilon)$ for some $i_\epsilon \in \epsilon$. Any such u_ϵ can be chosen such that $u_\epsilon \in P_\gamma$ for some fixed choice of a defect pointed group P_γ of b, and hence $i_\epsilon \in \epsilon$ can be chosen in $i\mathcal{O}Gi$. Therefore $\chi(u_\epsilon)$ is the character value of the character of the simple $K \otimes_\mathcal{O} i\mathcal{O}Gi$-module corresponding to χ, evaluated at the element ui_ϵ in the source algebra $i\mathcal{O}Gi$.

The compatibility of local pointed elements with the block decomposition of a finite group algebra $\mathcal{O}G$ has the following immediate consequence, generalising 5.15.6.

Theorem 6.13.11 *Let G be a finite group and B a block algebra of $\mathcal{O}G$. Denote by $\mathcal{R}$ a set of representatives of the G-conjugacy classes of local pointed elements on B. For any local pointed element u_ϵ on $\mathcal{O}G$ denote by $\bar{u}_\epsilon$ the image of uj in $B/[B, B]$ for some $j \in \epsilon$. Then $\bar{u}_\epsilon$ is independent of the choice of j in ϵ, and the set $\{\bar{u}_\epsilon | u_\epsilon \in \mathcal{R}\}$ is an $\mathcal{O}$-basis of $B/[B, B]$ that is independent of the choice of $\mathcal{R}$.*

Proof This follows from the corresponding statement for the group algebra $\mathcal{O}G$ in 5.15.6 and the fact that $\mathcal{O}G/[\mathcal{O}G, \mathcal{O}G]$ is a module over $Z(\mathcal{O}G)$, hence decomposes according to the block decomposition of $\mathcal{O}G$. $\qquad\square$

The following theorem is a blockwise version of 1.15.11.

Theorem 6.13.12 (Brauer) *Let G be a finite group and B a block algebra of $\mathcal{O}G$. Denote by $\mathcal{U}$ a set of representatives of the G-conjugacy classes of B-Brauer elements. We have*

$$|\mathrm{Irr}_K(B)| = \sum_{(u,e)\in\mathcal{U}} \ell(C_G(u), e).$$

Proof The terms on the right side count the number of simple modules of the algebras $kC_G(u)e$, with $(u, e) \in \mathcal{U}$. Every conjugacy class of local pointed elements on B belongs to exactly one of the conjugacy class of B-Brauer elements. The simple $kc_G(u)e$-modules are parametrised by the set of local points ϵ of $\langle u \rangle$ on B such that $\mathrm{Br}_u(\epsilon)e \neq 0$. For fixed u, two local pointed elements u_ϵ, $u_{\epsilon'}$ are G-conjugate if and only if they are $C_G(u)$-conjugate, hence if and only if $\epsilon' = \epsilon$. Thus the right side is equal to the number of G-conjugacy classes of local pointed elements on B, and this is the number of columns of the generalised decomposition matrix D of B. The left side in the stated equation is equal to the number of rows of D. Since D is a square matrix, the result follows. $\qquad\square$

6.14 Blocks with a normal defect group

We show that the source algebras of blocks with a normal defect group P and inertial quotient E are twisted semidirect group algebras of the form $\mathcal{O}_\alpha(P \rtimes E)$. We assume that the residue field $k = \mathcal{O}/J(\mathcal{O})$ of prime characteristic p of $\mathcal{O}$ is a perfect field. Then, by 4.3.9 there is a canonical group isomorphism $\mathcal{O}^\times \cong k^\times \times (1 + J(\mathcal{O}))$ through which we can view $H^2(E; k^\times)$ as a subgroup of $H^2(E, \mathcal{O}^\times)$. Let G be a finite group, b a block of $\mathcal{O}G$ and P a defect group. Let e be a block of $\mathcal{O}C_G(P)$ such that $(P, \bar{e})$ is a maximal (G, b)-Brauer pair, where $\bar{e}$ is the canonical image of e in $kC_G(P)$. If k is large enough then, by 6.7.6, the inertial quotient $E = N_G(P, e)/PC_G(P)$ is a p'-group, and by 6.7.8, the action of E on P modulo inner automorphisms lifts uniquely, up to an inner automorphism of P, to an action of E on P. The corresponding semidirect product of such a lift is denoted by $P \rtimes E$. Any element in $H^2(E; k^\times)$ can therefore be considered as element in $H^2(P \rtimes E; k^\times)$ via the canonical surjection $P \rtimes E \to E$. If the defect group P is normal in G then, by 6.2.6, b is contained in $\mathcal{O}C_G(P)$ and more precisely, $b = \mathrm{Tr}^G_{N_G(P,e)}(e)$. In particular, e remains a block of $\mathcal{O}N_G(P, e)$, and by 6.8.3, the blocks $\mathcal{O}Gb$ and $\mathcal{O}N_G(P, e)e$ have isomorphic source algebras, and as an $\mathcal{O}$-algebra, $\mathcal{O}Gb$ is isomorphic to a matrix algebra over $\mathcal{O}N_G(P, e)$. Thus, when investigating the structure of a block with a normal defect group P we may assume that b remains a block of $\mathcal{O}C_G(P)$.

Theorem 6.14.1 ([112, Theorem A]) *Let G be a finite group and b a block of $\mathcal{O}G$ with a defect group P such that P is normal in G. Let e be a block of $\mathcal{O}C_G(P)$ such that $be = e$. Suppose that k is large enough and set $E = N_G(P, e)/PC_G(P)$. Let i in $\mathcal{O}C_G(P)e$ be a primitive idempotent. Then i is a source idempotent of b as block of $\mathcal{O}G$, and there is $\alpha \in H^2(E; k^\times)$ such that we have an isomorphism of interior P-algebras*

$$i\mathcal{O}Gi \cong \mathcal{O}_\alpha(P \rtimes E)$$

which restricts to an isomorphism of interior P-algebras

$$i\mathcal{O}PC_G(P)i \cong \mathcal{O}P.$$

As an $\mathcal{O}$-algebra, $\mathcal{O}Gb$ is isomorphic to a matrix algebra over $\mathcal{O}_\alpha(P \rtimes E)$; in particular, the algebras $\mathcal{O}Gb$ and $\mathcal{O}_\alpha(P \rtimes E)$ are Morita equivalent.

Proof By the remarks preceding the statement of this theorem, we may assume that $G = N_G(P, e)$, and hence that $b = e$. Set $C = PC_G(P)$. With this notation we have $E = G/C$. For any $\bar{x} \in E$ let $x \in G$ be a representative. Since the algebra $\mathcal{O}_\alpha(P \rtimes E)$ has a $P \times P$-stable $\mathcal{O}$-basis, in order to determine the structure of the source algebras of b, we may assume by 6.4.9 that $\mathcal{O} = k$. Consider the decomposition $kG = \oplus_{\bar{x} \in E} kCx$. This is a decomposition of kG as a direct sum of kC-kC-bimodules because C is normal in G. Let i be a primitive idempotent in $kC_G(P)e$. By 6.2.6, the idempotent i is a source idempotent for $b = e$ as block of kC and of kG. By 6.6.7 we have an isomorphism $ikCi \cong kP$ as interior P-algebra. Since $kC_G(P)e$ has a unique point, for any $x \in G$ there is $u_x \in (kC_G(P)e)^\times$ such that $xix^{-1} = (u_x)^{-1}iu_x$. Equivalently, $u_x x$ commutes with i, hence $s_x = u_x xi$ is an invertible element in $ikGi$ which acts as x on the image of P in $ikCi$. By comparing dimensions we get a decomposition $ikGi = \oplus_{\bar{x} \in E} ikCis_x$. For $\bar{x}, \bar{y} \in E$ the elements $s_x s_y$ and s_{xy} act in the same way, as xy, on $ikCi \cong kP$. Thus $s_x s_y = \alpha(x, y)s_{xy}$ for some

$\alpha(x, y) \in Z(kP)^{\times}$. This yields a class $\bar{\alpha}$ in $H^2(E; Z(kP)^{\times})$; other words, $ikGi$ is a twisted crossed product of $ikCi$ by E. Now $Z(kP)^{\times} = k^{\times} \times (1 + J(Z(kP)))$. Since the group $1 + J(Z(kP))$ has a p-power exponent and since E is a p'-group, the group $H^2(E; 1 + J(Z(kP)))$ is trivial, and so we may choose a 2-cocycle α representing the class $\bar{\alpha}$ with values in $k^{\times}$. The isomorphism $ikGi \cong k_{\alpha}(P \rtimes E)$ follows. For the last statement, reverting to general $\mathcal{O}$, choose a a primitive decomposition J of b in $\mathcal{O}C_G(P)e$. Then $\mathcal{O}Gb \cong \mathrm{End}_{\mathcal{O}Gb}(\mathcal{O}Gb)^{\mathrm{op}} \cong \mathrm{End}_{\mathcal{O}Gb}(\oplus_{j \in J} \mathcal{O}Gj)$. Since the elements in J are all conjugate this is a matrix algebra over $i\mathcal{O}Gi$. $\square$

Theorem 6.14.1 implies that $\mathcal{O}_{\alpha}(P \rtimes E)$ is a primitive interior P-algebra; for $\alpha = 1$, this was shown in 6.2.9. It is possible to identify the class $\alpha \in H^2(E; k^{\times})$ more precisely. Suppose that G is a finite group and b a block of kG with a defect group P that is normal in G. Set $C = C_G(P)$. Let e be a block of $kC_G(P)$ such that $be = e$. Set $H = N_G(P, e)$ and $\bar{C} = C/Z(P)$. Then H acts by conjugation on kCe, hence on the matrix algebra $S = k\bar{C}\bar{e}$, where $\bar{e}$ is the image of e in $k\bar{C}$. The Skolem–Noether Theorem implies that for $x \in H$ there is $s_x \in S^{\times}$ such that $^x s = s_x s(s_x)^{-1}$ for all $s \in S$. This yields a 2-cocycle β in $Z^2(H; k^{\times})$ defined by $s_x s_y = \beta(x, y)s_{xy}$, for $x, y \in H$. This 2-cocycle can be chosen in a particular way: if $y \in PC$, choose for s_x the image in $k\bar{C}\bar{e}$ of ye under the maps induced by the canonical homomorphisms $PC \to PC/P \cong C/Z(P)$. The consequence of this choice of β is that the values $\beta(x, y)$, with $x, y \in H$, depend only on the images of x, y in $E = H/PC$, and hence determine a 2-cocycle in $Z^2(E; k^{\times})$. We denote its image in $H^2(E; k^{\times})$ abusively again by β, and we denote by β^{-1} the class obtained from inverting β; that is, as 2-cocycles, we have $\beta^{-1}(x, y) = \beta(x, y)^{-1}$, for $x, y \in E$.

Theorem 6.14.2 ([174, 14.6]) *With the notation above, the class $\alpha = \beta^{-1}$ satisfies the conclusion of Theorem 6.14.1.*

Proof By 5.3.8, the algebra $\mathcal{O}_{\alpha}(P \rtimes E)$ has a $P \times P$-stable $\mathcal{O}$-basis. Therefore, by 6.4.9 we may assume that $\mathcal{O} = k$. We may further assume that $G = H = N_G(P, e)$, so that $b = e$. Set $\bar{G} = G/P$ and $C = C_G(P)$. We identify PC/P and $C/Z(P)$ whenever expedient, and keep the notation preceding the statement of 6.14.2; in particular, $S = k(PC/P)\bar{b}$. As a block of kPC/P, $\bar{b}$ has defect zero, and PC/P is a normal subgroup of $\bar{G}$. By 6.8.13 we may choose $s_x \in S^{\times}$ in such a way that s_x acts as $x \in G$ on S and such that $s_x = \bar{x}\bar{b}$ if $x \in PC$ and $\bar{x} = xP \in PC/P$. Set $\alpha = \beta^{-1}$. By 6.8.13, there is an algebra isomorphism

$$k\bar{G}\bar{b} \cong S \otimes_k k_{\alpha}E$$

sending $x \in G$ to $s_x \otimes \bar{x}$, where $\bar{x}$ is the image of x in $k_{\alpha}E$. A primitive idempotent i in $kCPb$ is a source idempotent of b, and its image $\bar{i}$ in S is a primitive idempotent in S. Thus $\bar{i}S\bar{i} \cong k$, hence $\bar{i}kG\bar{i} \cong k_\alpha E$. By 5.3.8, the kP-kP-bimodule structure of $k_\alpha(P \rtimes E)$ does not depend on α. Thus, by 6.14.1, we have an isomorphism of kP-kP-bimodules $ikGi \cong k_\alpha(P \rtimes E)$. Since $ikGi$ is relatively kP-separable, it follows from 4.8.2 that there is a homomorphism of interior P-algebras $ikGi \to k_\alpha(P \rtimes E)$ that lifts the isomorphism $\bar{i}kG\bar{i} \cong k_\alpha E$. The image of this homomorphism contains both P and an inverse image of E, and hence this is an algebra isomorphism as both sides have the same dimension.   □

See also [7] for an alternative proof of 6.14.2.

6.15 The bimodule structure and rank of a source algebra

We describe some aspects of the $\mathcal{O}P$-$\mathcal{O}P$-bimodule structure of a source algebra A of a block with defect group P, obtained from combining results of previous section. In particular, we show that $|P|$ is the highest power of p that divides the $\mathcal{O}$-rank of A. More precise results relating the bimodule structure of source algebras to fusion systems will be given in §8.7. We assume that k is perfect, and hence we have a canonical identification $\mathcal{O}^\times = k^\times \times (1 + J(\mathcal{O}))$, through which we consider $H^2(E; k^\times)$ as a subgroup of $H^2(E; \mathcal{O}^\times)$, for any finite group E. The key result of this section is the fact that the source algebra B of the Brauer correspondent of a block embeds canonically into the source algebra A of that block in such a way that B becomes a direct summand of A as a B-B-bimodule, and such that every indecomposable summand of a complement has strictly smaller vertices.

Theorem 6.15.1 (cf. [174, 14.6]) *Let G be a finite group, b a block of $\mathcal{O}G$, (P, e) a maximal (G, b)-Brauer pair, and f a primitive idempotent in $(\mathcal{O}Gb)^{N_G(P,e)}$ such that $\mathrm{Br}_P(f) = e$. Set $H = N_G(P)$, and denote by c the block of $\mathcal{O}H$ corresponding to b. Let j be a source idempotent of $\mathcal{O}Hc$ such that $\mathrm{Br}_P(j)e \neq 0$. Then $i = fj$ is a source idempotent of $\mathcal{O}Gb$. Set $A = i\mathcal{O}Gi$ and $B = j\mathcal{O}Hj$.*

(i) *Multiplication by f induces an injective algebra homomorphism $B \to A$ which is split injective as a B-B-bimodule homomorphism.*

(ii) *If Y is a complement of the image of B in A as a B-B-bimodule, then every indecomposable direct summand of Y as an $\mathcal{O}(P \times P)$-module has a vertex of the form $\{(u, \varphi(u)) | u \in Q\}$ for some proper subgroup Q of P and some injective group homomorphism $\varphi : Q \to P$. In particular we have $Y(\Delta P) = \{0\}$ and $\frac{\mathrm{rk}_\mathcal{O}(A)}{|P|} \equiv \frac{\mathrm{rk}_\mathcal{O}(B)}{|P|} \pmod{p}$.*

(iii) *Multiplication in A induces a split surjective A-A-bimodule homomorphism $A \otimes_B A \to A$, and if X is an A-A-bimodule satisfying $A \otimes_B A \cong A \oplus X$, then $X(\Delta P) = \{0\}$.*

(iv) *Set $E = N_G(P, e)/PC_G(P)$ and assume that k is large enough. Then $B \cong \mathcal{O}_\alpha(P \rtimes E)$ for some $\alpha \in H^2(E; k^\times)$, and we have $\frac{\text{rk}_{\mathcal{O}}(A)}{|P|} \equiv |E| \pmod{p}$; in particular, $\frac{\text{rk}_{\mathcal{O}}(A)}{|P|}$ is not divisible by p.*

(v) *Suppose that k and K are splitting fields for all subgroups of G and that char(K) = 0. There are $\chi \in \text{Irr}_K(G, b)$ and $\varphi \in \text{IBr}_k(G, b)$ such that the three integers $\chi(i)$, $\varphi(i)$, and d_φ^χ are all coprime to p.*

Proof The fact that $i = fj$ is a source idempotent is proved in 6.7.4. Since $\mathcal{O}Hj$ is isomorphic to a direct summand of $\text{Res}_{H \times P}^{G \times P}(\mathcal{O}Gi)$ it follows that $B \cong \text{End}_{\mathcal{O}H}(\mathcal{O}Hj)^{\text{op}}$ is isomorphic to a direct summand of $A \cong \text{End}_{\mathcal{O}G}(\mathcal{O}Gi)^{\text{op}}$ as a B-B-bimodule. This shows (i). It follows from 6.7.2 (iii) that Y is as stated in (ii). Thus every indecomposable direct summand of Y is annihilated by $\text{Br}_{\Delta P}$ and has $\mathcal{O}$-rank divisible by $p|P|$, whence (ii). By 6.4.7, the canonical bimodule homomorphism $A \otimes_{\mathcal{O}P} A \to A$ has a section $\rho : A \to A \otimes_{\mathcal{O}P} A$. Composing ρ with the canonical map $A \otimes_{\mathcal{O}P} A \to A \otimes_B A$ yields a section of the canonical map $A \otimes_B A \to A$. Similarly, B is a direct summand of $B \otimes_{\mathcal{O}P} B$, and hence $A \otimes_B A$ is a direct summand of $A \otimes_{\mathcal{O}P} A$. It follows from (ii) that we have B-B-bimodule isomorphisms $A \otimes_B A \cong (B \oplus Y) \otimes_B (B \oplus Y) \cong B \oplus Y \oplus Y \oplus Y \otimes_B Y$ and that $Y(\Delta P) = \{0\}$. By the above, $Y \otimes_B Y$ is a direct summand of $Y \otimes_{\mathcal{O}P} Y$, hence $(Y \otimes_B Y)(\Delta P) = \{0\}$. Thus $(A \otimes_B A)(\Delta P) \cong B(\Delta P) \cong A(\Delta P)$, implying (iii). If k is large enough, then the structure of B follows from 6.14.1, and hence statement (iii) follows from (ii). For the proof of (iv), let J be a primitive decomposition of 1 in A. Then $\text{rk}_{\mathcal{O}}(A) = \sum_{j \in J} \text{rk}_{\mathcal{O}}(Aj)$. Each Aj is a projective indecomposable A-module. If $\varphi \in \text{IBr}_k(G, b)$, then the multiplicity of a projective cover if a simple A-module corresponding to φ is equal to $\varphi(i)$, where we use that $i = 1_A$. Denoting by Φ_φ the character of a projective indecomposable $\mathcal{O}Gb$-module which is a projective cover of a simple module with Brauer character φ, we get that $\text{rk}_{\mathcal{O}}(A) = \sum_{\varphi \in \text{IBr}_k(G, b)} \Phi_\varphi(i)\varphi(i)$. Each Aj remains projective as an $\mathcal{O}P$-module, hence has a rank divisible by $|P|$. Thus $\Phi_\varphi(i)$ is divisible by $|P|$. By (iii), the integer $\frac{\text{rk}_{\mathcal{O}}(A)}{|P|} = \sum_{\varphi \in \text{IBr}_k(G,b)} \frac{\Phi_\varphi(i)}{|P|}\varphi(i)$ is not divisible by p. Thus at least one summand is not divisible by p; in, particular, there is $\varphi \in \text{IBr}_k(G, b)$ such that $\varphi(i)$ is not divisible by p. The surjectivity of the decomposition map 6.5.11 implies that $\varphi(i)$ is a $\mathbb{Z}$-linear combination of the $\chi(i)$, where $\chi \in \text{Irr}_K(G, b)$, and hence at least one of the $\chi(i)$ is not divisible by p. Writing $\chi = \sum_{\varphi \in \text{IBr}(G,b)} d_\varphi^\chi \varphi$ implies that one can choose χ, φ such that none of $\chi(i)$, d_φ^χ, $\varphi(i)$, is divisible by p. $\qquad\square$

The next result is a version of 6.7.15 for almost source algebras.

Proposition 6.15.2 *Let G be a finite group, b a block of $\mathcal{O}G$, and P a defect group of b. Let $i \in (\mathcal{O}Gb)^{\Delta P}$ be an almost source idempotent and set $A = i\mathcal{O}Gi$. Suppose that k is a splitting field for $A(\Delta P)$. Let B be an interior P-algebra over $\mathcal{O}$ having a finite $P \times P$-stable $\mathcal{O}$-basis such that B is projective as a left $\mathcal{O}P$-module and as a right $\mathcal{O}P$-module. Suppose that $B(\Delta P) \neq \{0\}$. Let $\alpha : A \to B$ be a homomorphism of interior P-algebras. Consider B as an A-A-bimodule via α. Then α is split injective as a homomorphism of A-A-bimodules.*

Proof We play this back to 6.7.15. Let e be the block of $kC_G(P)$ such that $\mathrm{Br}_{\Delta P}(i)e \neq 0$. It follows from 6.4.12 that the algebras $A(\Delta P)$ and $kC_G(P)e$ are Morita equivalent, and hence k is a splitting field for $kC_G(P)e$. Set $C = \mathrm{End}_{B^{\mathrm{op}}}(\mathcal{O}Gi \otimes_A B)$, viewed as an interior G-algebra with structural homomorphism given by left multiplication by G on $\mathcal{O}Gi \otimes_A B$. Consider the structural homomorphism $\mathcal{O}Gb \to C = \mathrm{End}_{B^{\mathrm{op}}}(\mathcal{O}Gi \otimes_A B)$. Since A is isomorphic to a direct summand of $A \otimes_{\mathcal{O}P} A$, the $\mathcal{O}Gb$-B-bimodule $\mathcal{O}Gi \otimes_A B$ is isomorphic to a direct summand of $\mathcal{O}Gi \otimes_{\mathcal{O}P} B$. By a standard adjunction, we have $\mathrm{End}_{B^{\mathrm{op}}}(\mathcal{O}Gi \otimes_{\mathcal{O}P} B) \cong \mathrm{Hom}_{\mathcal{O}P^{\mathrm{op}}}(\mathcal{O}Gi, \mathcal{O}Gi \otimes_{\mathcal{O}P} B) \cong \mathcal{O}Gi \otimes_{\mathcal{O}P} B \otimes_{\mathcal{O}P} i\mathcal{O}G$. This bimodule is projective as a left $\mathcal{O}P$-module and as a right $\mathcal{O}P$-module, and it has a $P \times P$-stable basis, by the hypotheses on B. Thus the interior G-algebra C has a $P \times P$-stable $\mathcal{O}$-basis and is projective as a left and right $\mathcal{O}P$-module. It follows from 5.8.8 that $C(\Delta P)$ is projective as a left or right $kZ(P)$-module. Moreover, writing $\mathcal{O}Gi = A \oplus (1 - i)\mathcal{O}Gi$ implies that B is a direct summand of $\mathcal{O}Gi \otimes_A B$ as an $\mathcal{O}P$-B-bimodule, and hence C has a direct summand isomorphic to B as an $\mathcal{O}P$-$\mathcal{O}P$-bimodule. It follows that $C(\Delta P)$ is nonzero. Therefore, by 6.7.15, the algebra homomorphism $\mathcal{O}Gb \to C$ is split injective as a homomorphism of $\mathcal{O}Gb$-$\mathcal{O}Gb$-bimodules. Multiplying this homomorphism by i on both sides and using the standard Morita equivalence between the categories of $\mathcal{O}Gb$-$\mathcal{O}Gb$-bimodules and of A-A-bimodules obtained from 6.4.6 we get that the homomorphism $A \to \mathrm{End}_{B^{\mathrm{op}}}(A \otimes_A B) \cong \mathrm{End}_{B^{\mathrm{op}}}(B) \cong B$ is split injective as a homomorphism of A-A-bimodules. Since this homomorphism is clearly equal to α, the result follows. $\square$

6.16 On automorphisms of blocks and source algebras

Let P be a finite p-group, and let A be an interior P-algebra. We denote by $\mathrm{Aut}_P(A)$ the group of algebra automorphisms of A which fix the image of P in A elementwise. We denote by $\mathrm{Out}_P(A)$ the image of $\mathrm{Aut}_P(A)$ in $\mathrm{Out}(A)$. If α, $\beta \in \mathrm{Aut}_P(A)$ represent the same class in $\mathrm{Out}(A)$, then $\alpha = \gamma \circ \beta$, where γ is an inner automorphism of A, hence given by conjugation with an element $c \in A^\times$. Since both α and β fix the image of P in A, the same is true for conjugation by c; that is, we have $c \in A^P$. Thus $\mathrm{Out}_P(G)$ is isomorphic to the quotient of $\mathrm{Aut}_P(A)$ by the group $\mathrm{Inn}_P(A)$ of inner automorphisms of A induced by conjugation with elements in $(A^P)^\times$. Since P acts by inner automorphisms on A, we have $Z(A) \subseteq A^P$. We have an obvious exact sequence of groups

$$1 \longrightarrow Z(A)^\times \longrightarrow (A^P)^\times \longrightarrow \mathrm{Aut}_P(A) \longrightarrow \mathrm{Out}_P(A) \longrightarrow 1$$

and the image of the map $(A^P)^\times \to \mathrm{Aut}_P(A)$ is $\mathrm{Inn}_P(A)$.

Theorem 6.16.1 ([174, 14.9]) *Let A be a source algebra of a block of a finite group with defect group P and inertial quotient E. Suppose that k is large enough. There is a canonical injective group homomorphism $\mathrm{Out}_P(A) \to \mathrm{Hom}(E; k^\times)$. In particular, $\mathrm{Out}_P(A)$ is a p'-group.*

Let E be a p'-subgroup of $\mathrm{Aut}(P)$, and let $\alpha \in H^2(E; k^\times)$. Any $\zeta \in \mathrm{Hom}(E, k^\times)$ induces an automorphism of $\mathcal{O}_\alpha(P \rtimes E)$ sending ux to $\zeta(x)ux$, where $u \in P$ and $x \in E$. Theorem 6.16.1 follows from the next result, which shows in particular, that the group of automorphisms of $\mathcal{O}_\alpha(P \rtimes E)$ obtained as above is isomorphic to $\mathrm{Out}_P(\mathcal{O}_\alpha(P \rtimes E))$.

Lemma 6.16.2 *With the notation from 6.16.1, assume that k is large enough and let $B = \mathcal{O}_\alpha(P \rtimes E)$ be the subalgebra of A as in 6.15.1. Let $\tau \in \mathrm{Aut}_P(A)$. Then there is $\sigma \in \mathrm{Aut}_P(A)$ and $\zeta \in \mathrm{Hom}(E; k^\times)$ with the following properties.*

(i) The classes of τ and σ in $\mathrm{Out}_P(A)$ are equal.

(ii) The automorphism σ stabilises the subalgebra B of A, and we have $\sigma(ux) = \zeta(x)ux$ for all $u \in P$, $x \in E$.

(iii) The map sending $\zeta \in \mathrm{Hom}(E, k^\times)$ to the automorphism of B sending ux to $\zeta(x)ux$ for $u \in P$ and $x \in E$ induces a group isomorphism $\mathrm{Hom}(E; k^\times) \cong \mathrm{Out}_P(B)$.

(iv) The map sending τ to the restriction of σ to B induces an injective group homomorphism $\mathrm{Out}_P(A) \to \mathrm{Out}_P(B)$.

Proof Since τ fixes the image of P in A, for any $x \in E$, the elements x and $\tau(x)$ act in the same way on P, and hence $x^{-1}\tau(x) \in (A^P)^\times = k^\times \times (1+J(AP))$, where we use that A^P is a split local algebra by the assumptions. Thus $x^{-1}\tau(x) = \zeta(x)(1+r(x))$ for a uniquely determined $\zeta(x) \in k^\times$ and $r(x) \in J(A^P)$. The component $\zeta(x)$ in $k^\times$ is invariant under any conjugation that stabilises A^P. We use this to verify that $\zeta : E \to k^\times$ is a group homomorphism. Let $x, y \in E$. The component of $x^{-1}\tau(x)y^{-1}\tau(y)$ in $k^\times$ is $\zeta(x)\zeta(y)$, and this is also the component of the conjugate $(yx^{-1}\tau(x)y^{-1})(\tau(y)y^{-1})$. The component in $k^\times$

of the first factor coincides with that of $x^{-1}\tau(x)$, so that the component in $k^\times$ of the product is that of $(x^{-1}\tau(x))(\tau(y)y^{-1})$, which in turn is equal to that of $y^{-1}x^{-1}\tau(x)\tau(y)$, hence to $\zeta(xy)$. Thus ζ is a group homomorphism. For any $x \in E$ we have $x^{-1}\tau(x)\zeta(x)^{-1} \in 1 + J(A^P)$. Thus $s = \sum_{x \in E} x^{-1}\tau(x)\zeta(x)^{-1} \in |E| + J(A^P)$. Since $|E|$ is prime to p, we have $s \in (A^P)^\times$. For any $y \in E$ we have $s\tau(y)\zeta(y)^{-1} = \sum_{x \in E} x^{-1}\tau(xy)\zeta(xy)^{-1} = y\sum_{x \in E} y^{-1}x^{-1}\tau(xy)\zeta(xy)^{-1} = ys$. Thus s is an invertible element in A^P satisfying $s\tau(x)s^{-1} = \zeta(x)x$ for all $x \in E$, and hence the automorphism σ defined by $\sigma(a) = s\tau(a)s^{-1}$ satisfies (i) and (ii). It follows from (ii) that the map sending ζ to the automorphism $ux \mapsto \zeta(x)ux$ in (iii) induces a surjective group homomorphism $\mathrm{Hom}(E; k^\times) \to \mathrm{Out}_P(B)$. We need to show that this is also injective. Let $\zeta \in \mathrm{Hom}(E, k^\times)$ such that the automorphism $ux \mapsto \zeta(x)ux$ of B is inner; that is, such that there is an element $d \in 1 + J(B^P)$ satisfying $\zeta(x)ux = d^{-1}uxd$. Multiplying by the inverse of x in $B^\times$ on the right and by d in the left implies $(^x d) = \zeta(x)d$. Since the action of x on B preserves $1 + J(B^P)$, this forces $\zeta(x) = 1$ for all $x \in E$, whence (iii). In order to show (iv), let $\zeta \in \mathrm{Hom}(E, k^\times)$ and let $\sigma \in \mathrm{Aut}_P(A)$ such that $\sigma(ux) = \zeta(x)ux$ for all $u \in P$ and $x \in E$. For the map in (iv) to be well-defined and injective, we need to show that σ is inner on A if and only of its restriction to B is inner. If σ is inner on A, then by 2.8.16, $A_\sigma \cong A$ as A-A-bimodules, hence as B-B-bimodules. Thus B and B_σ are indecomposable direct summands of A, and since σ acts as identity on P, we have $B_\sigma(\Delta P) = B(\Delta P) \neq \{0\}$. By 6.15.1, A has up to isomorphism a unique indecomposable B-B-bimodule summand with this property. It follows that $B_\sigma \cong B$, and hence σ restricted to B is an inner automorphism of B, again by 2.8.16. Conversely, if σ induces an inner automorphism on B, then $A_\sigma \cong A$ as A-B-bimodules. By 6.15.1, A is isomorphic to a direct summand of $A \otimes_B A$, and any other summand vanishes under applying $\mathrm{Br}_{\Delta P}$. It follows that A_σ is isomorphic to a direct summand of $A_\sigma \otimes_B A \cong A \otimes_B A$, and hence $A_\sigma \cong A$ as A-A-bimodules, implying that σ is inner. This completes the proof. $\qquad\square$

Proof of Theorem 6.16.1 This follows from combining the statements in 6.16.2. $\qquad\square$

Corollary 6.16.3 *Let G be finite group, b a block of $\mathcal{O}G$, and (P, e) a maximal (G, b)-Brauer pair. Let $\varphi \in \mathrm{Aut}(G)$. Suppose that φ fixes P elementwise and that the induced automorphism on $kC_G(P)$ fixes e. Then the automorphism of $\mathcal{O}G$ induced by φ fixes b, and the image of φ in $\mathrm{Out}(\mathcal{O}Gb)$ has order prime to p. In particular, if the order of φ in $\mathrm{Out}(G)$ is a power of p, then the automorphism of $\mathcal{O}Gb$ induced by φ is inner and given by conjugation with an element in $((\mathcal{O}Gb)^P)^\times$.*

Proof We use the same letter φ to denote the algebra automorphism of $\mathcal{O}G$ induced by φ. Since φ fixes e, it fixes b and stabilises the unique local point γ of P on $\mathcal{O}Gb$ associated with e. Let $i \in \gamma$. Then $\varphi(i) \in \gamma$, hence $\varphi(i) = cic^{-1}$ for some $c \in ((\mathcal{O}Gb)^P)^\times$. Define ψ by setting $\psi(a) = c^{-1}\varphi(a)c$ for all $a \in \mathcal{O}Gb$. Thus ψ and φ represent the same class in $\mathrm{Out}(\mathcal{O}Gb)$. The restriction of ψ to the source algebra $A = i\mathcal{O}Gi$ belongs to $\mathrm{Aut}_P(A)$, and hence its order in $\mathrm{Out}(A)$ is prime to p, by 6.16.1. Since A and $\mathcal{O}Gb$ are Morita equivalent, it follows from 2.8.18 that ψ has order prime to p in $\mathrm{Out}(\mathcal{O}Gb)$, whence the result. $\qquad\square$

We sketch an alternative proof of 6.16.1 using Theorem 5.7.11.

Alternative proof of Theorem 6.16.1 Let G be a finite group, b a block of $\mathcal{O}G$, (P, e) a maximal (G, b)-Brauer pair and $i \in (\mathcal{O}Gb)^P$ a source idempotent such that $\mathrm{Br}_P(i)e \neq 0$. Set $E = N_G(P, e)/PC_G(P)$. Let f be a primitive idempotent in $(\mathcal{O}Gb)^{N_G(P,e)}$ satisfying $\mathrm{Br}_P(f) = e$ as in 6.15.1. Let $j \in \mathcal{O}C_G(P, e)$ be a primitive idempotent; by 6.14.1 this is a source idempotent of the block $\hat{e}$ of $\mathcal{O}N_G(P, e)$ that lifts e. By 6.15.1, we may choose notation such that $jf = i$. Set $A = i\mathcal{O}Gi$. Let $\tau \in \mathrm{Aut}_P(A)$. Then A_τ is an indecomposable A-A-bimodule. As an A-$\mathcal{O}P$-module, we have $A_\tau = A$ because τ fixes the image of P in A element-wise; in particular, A_τ remains indecomposable as an A-$\mathcal{O}P$-bimodule. Through the standard Morita equivalences between block algebras and their source algebras, the A-A-bimodule A_τ corresponds to the necessarily indecomposable $\mathcal{O}(G \times G)$-module $\mathcal{O}Gi_\tau \otimes_A i\mathcal{O}G$. The isomorphism class of this module is determined by its restriction to $\mathcal{O}(G \times N_G(P))$ via the Green correspondence, hence by the $\mathcal{O}(G \times N_G(P, e))$-module summand $\mathcal{O}Gi_\tau \otimes_A i\mathcal{O}Gf$. Multiplying this module by i on the left and j on the right yields that the isomorphism class of the A-A-bimodule A_τ is determined by its restriction to an A-$\mathcal{O}_\alpha(P \rtimes E)$-bimodule. Note that since $A_\tau = A$ is indecomposable as an A-$\mathcal{O}P$-bimodule, it follows that the $\mathcal{O}(G \times N_G(P, e))$-module $\mathcal{O}Gi_\tau \otimes_A i\mathcal{O}Gf$ remains indecomposable upon restriction to $\mathcal{O}(G \times PC_G(P))$. Since $j\mathcal{O}PC_G(P)j \cong \mathcal{O}P$ and since τ fixes P, it follows that $\mathcal{O}Gi_\tau \otimes_A i\mathcal{O}Gf$ is isomorphic to $\mathcal{O}Gbf$ as an $\mathcal{O}(G \times PC_G(P))$-module. Since E is a p'-group, it follows from 5.7.11, applied with the groups $G \times N_G(P, e)$ and the normal subgroup $G \times PC_G(P)$, that the extensions of $\mathcal{O}Gf$ from an $\mathcal{O}Gb$-$\mathcal{O}PC_G(P)$-bimodule to an $\mathcal{O}Gb$-$\mathcal{O}N_G(P, e)$-bimodule are parametrised by the group $\mathrm{Hom}(E, k^\times)$. Let $\alpha \in H^2(E; k^\times)$ such that $j\mathcal{O}N_G(P, e)j \cong \mathcal{O}_\alpha(P \rtimes E)$; cf. 6.14.1. The standard Morita equivalences imply that the extensions of the A-$\mathcal{O}P$-bimodule A to an A-$\mathcal{O}_\alpha(P \rtimes E)$-bimodule structure are parametrised by $\mathrm{Hom}(E, k^\times)$. The result follows. $\qquad\square$

Remark 6.16.4 If k is algebraically closed, then the outer automorphism group $\mathrm{Out}(A)$ of a finite-dimensional k-algebra is an algebraic group. By contrast, if $\mathcal{O}$ is a p-adic ring with a finite residue field k and A is an $\mathcal{O}$-free $\mathcal{O}$-algebra such that $K \otimes_{\mathcal{O}} A$ is a finite-dimensional separable K-algebra, then $\mathrm{Out}(A)$ is a finite group. See [52, §55] for a proof, as well as [140], [20] for further results on automorphisms and Picard groups of block algebras.

Exercise 6.16.5 Let G be a finite group, B a block algebra of $\mathcal{O}G$ with defect group P and source algebra $A = iBi$ for some source idempotent $i \in B^P$. Combine the Theorems 6.16.1, 6.4.9, 5.11.2 and their proofs to show the following statements.

(a) The group $\mathrm{Out}_P(A)$ is canonically isomorphic to the subgroup of $\mathrm{Pic}(B)$ given by trivial source bimodules that are isomorphic to a direct summand of $Bi \otimes_{\mathcal{O}P} iB$.

(b) The canonical algebra homomorphism $A \to \bar{A} = k \otimes_{\mathcal{O}} A$ induces an isomorphism $\mathrm{Out}_P(A) \cong \mathrm{Out}_P(\bar{A})$.

The connection between $\mathrm{Out}_P(A)$ and Morita equivalences given by trivial source bimodules will be investigated further in Section 9.7; see in particular Theorem 9.7.4.

6.17 On the centre of block algebras

Let B be a block algebra and A a source algebra of B. Since A and B are canonically Morita equivalent, there is an isomorphism of centres $Z(B) \cong Z(A)$. This isomorphism is compatible with the Brauer construction.

Proposition 6.17.1 ([96, 3.3]) *Let G be a finite group, B a block algebra of kG, P a defect group of b and let i be a source idempotent in B^P. Suppose that k is a splitting field for all subgroups of G. Set $A = iBi = ikGi$. Multiplication by i induces an isomorphism of the centres $Z(B) \cong Z(A)$. This isomorphism maps $Z^{\mathrm{pr}}(B)$ onto $Z^{\mathrm{pr}}(A)$ and $Z(B) \cap \ker(\mathrm{Br}_P)$ onto $Z(A) \cap \ker(\mathrm{Br}_P)$.*

Proof Multiplication by i induces a Morita equivalence between B and A. Any Morita equivalence preserves centres and projective ideals (and this does not require as yet the hypothesis on k being large enough). Since i commutes with P, multiplication by i maps $Z(B) \cap \ker(\mathrm{Br}_P)$ into $Z(A) \cap \ker(\mathrm{Br}_P)$. If $z \in Z(B)$ such that $\mathrm{Br}_P(iz) = 0$, then $\mathrm{Br}_P(\mathrm{Tr}_P^G(i)z) = \mathrm{Br}_P(\mathrm{Tr}_P^G(iz)) = \mathrm{Tr}_P^{N_G(P)}(\mathrm{Br}_P(iz)) = 0$, where we use the formula 5.4.5. By 6.11.9, $\mathrm{Tr}_P^G(i)$ is invertible (this is where we use that k is large enough), and since Br_P is an algebra homomorphism, it follows that $\mathrm{Br}_P(z) = 0$. $\qquad\square$

Theorem 6.17.2 ([37, Prop. III (1.1)]) *Let G be a finite group, b a block of kG, (P, e) a maximal (G, b)-Brauer pair, and $E = N_G(P, e)/PC_G(P)$ the corresponding inertial quotient. Suppose that k is a splitting field for all subgroups of G. Set $B = kGb$. We have an algebra isomorphism $\mathrm{Br}_P(Z(B)) \cong (kZ(P))^E$.*

Proof Since $Z(B) = B_P^G$ it follows from 5.4.5 that $\mathrm{Br}_P(B) = (kC_G(P) \, \mathrm{Br}_P(b))_P^{N_G(P)}$. Since $\mathrm{Br}_P(b)$ is the sum of the different $N_G(P)$-conjugates of e, this is isomorphic to $(kC_G(P)e)_P^{N_G(P,e)}$. As a block of $kPC_G(P)$, e has defect group P, and hence $Z(kPC_G(P)e) = (kPC_G(P))_P^{PC_G(P)}$. Note that $Z(kC_G(P)) = Z(kPC_G(P)) \cap kC_G(P)$. This implies that $Z(kC_G(P)e) = (kC_G(P))_P^{PC_G(P)}$. Thus $(kC_G(P)e)_P^{N_G(P,e)} = (Z(kC_G(P)e))_{PC_G(P)}^{N_G(P,e)} = (Z(kC_G(P)e))^E$, where we use that E is a p'-group. As a block of $kC_G(P)$, e has the central defect group $Z(P)$, and so $Z(kC_G(P)e) = kZ(P)e \cong kZ(P)$. The result follows. $\qquad\square$

Corollary 6.17.3 *Let G be a finite group, b a block of kG, P a defect group of b and let i be a source idempotent in $(kGb)^P$. Denote by E the associated inertial quotient. Suppose that k is a splitting field for all subgroups of G. Set $A = ikGi$. Then $\mathrm{Br}_P(Z(A)) \cong (kZ(P))^E$.*

Proof This follows from combining 6.17.1 and 6.17.2. $\qquad\square$

Proposition 6.17.4 ([222]) *Let G be a finite group and b a block of kG. We have*

$$Z(kGb) = \sum_{(Q,e)} \mathrm{Tr}_Q^G(kC_G(Q)e)b$$

where (Q, e) runs over a system of representatives of the G-conjugacy classes of (G, b)-Brauer pairs.

We state one technical step separately.

Lemma 6.17.5 *Let G be a finite group and b a block of kG. Let Q be a p-subgroup of G. We have $(kGb)_Q^G \subseteq \sum_{(R,f)} \mathrm{Tr}_R^G(kC_G(R)f)b$, where (R, f) runs over all (G, b)-Brauer pairs satisfying $R \leq Q$.*

Proof We argue by induction over Q. For $Q = 1$ this is trivial, since both sides are equal to $(kGb)_1^G$. Suppose that Q is nontrivial. Using the fact that $kC_G(Q)(\mathrm{Br}_Q(b)b - b) \in \ker(\mathrm{Br}_Q)$, we have $(kGb)^Q = kC_G(Q)\mathrm{Br}_Q(b)b + \sum_{R<Q}(kGb)_R^Q$. Thus $(kGb)_Q^G = \mathrm{Tr}_Q^G(kC_G(Q)\mathrm{Br}_Q(b)b) + \sum_{R<Q}(kGb)_R^G$. Now $\mathrm{Br}_Q(b)$ is a sum of blocks e of $kC_G(Q)$ such that (Q, e) is a (G, b)-Brauer

pair, and the second sum is contained in the right side by induction, whence the result. $\qquad\square$

Proof of Proposition 6.17.4 We have $Z(kGb) = (kGb)_P^G$, where P is a defect group of b. Thus 6.17.5 yields $Z(kGb) = \sum_{(Q,e)} \mathrm{Tr}_Q^G(kC_G(Q)e)b$, where (Q, e) runs over all (G, b)-Brauer pairs. Since the space $\mathrm{Tr}_Q^G(kC_G(Q)e)b$ is invariant under conjugation with any $x \in G$, the sum on the right side does not change if we sum over a set of representatives of the G-conjugacy classes of (G, b)-Brauer pairs. $\qquad\square$

Definition 6.17.6 Let G be a finite group and (Q, e) a Brauer pair on kG. We set

$$m(Q, e) = \dim_k(\mathrm{Tr}_Q^{N_G(Q,e)}(kC_G(Q)e)).$$

The sum in 6.17.4 is in fact a direct sum.

Proposition 6.17.7 *Let G be a finite group and b a block of kG. Let $\mathcal{S}$ be a set of representatives of the G-conjugacy classes of (G, b)-Brauer pairs. We have*

$$Z(kGb) = \oplus_{(Q,e)\in\mathcal{S}} \mathrm{Tr}_Q^G(kC_G(Q)e)b.$$

For any (G, b)-Brauer pair (Q, e), the trace map $\mathrm{Tr}_{N_G(Q,e)}^G$ and multiplication by b induce linear isomorphisms

$$\mathrm{Tr}_Q^{N_G(Q,e)}(kC_G(Q)e) \cong \mathrm{Tr}_Q^G(kC_G(Q)e) \cong \mathrm{Tr}_Q^G(kC_G(Q)e)b.$$

In particular, we have

$$\dim_k(Z(kGb)) = \sum_{(Q,e)\in\mathcal{S}} m(Q, e).$$

Proof The starting point of the proof is the vector space decomposition

$$Z(kG) = \oplus_{Q\in\mathcal{R}} \mathrm{Tr}_Q^G(kC_G(Q))$$

from 5.4.7, where $\mathcal{R}$ is a set of representatives of the conjugacy classes of p-subgroups of G. By 5.4.6, each summand $\mathrm{Tr}_Q^G(kC_G(Q))$ is isomorphic to $\mathrm{Tr}_Q^{N_G(Q)}(kC_G(Q))$. The latter space is isomorphic to the direct sum of the spaces $\mathrm{Tr}_Q^{N_G(Q,e)}(kC_G(Q)e)$, where e runs over a set of representatives of the $N_G(Q)$-conjugacy classes of blocks of $kC_G(Q)$. Two blocks e and e' of $kC_G(Q)$ belong to the same $N_G(Q)$-conjugacy class if and only if the pairs (Q, e) and (Q, e') are G-conjugate. Thus if Q runs over $\mathcal{R}$ and for each Q, e runs over a set of representatives of the $N_G(Q)$-conjugacy classes of blocks of $kC_G(Q)$, then (Q, e) runs over a set of representatives of the G-conjugacy classes of Brauer pairs on G. It follows that $Z(kGb)$ is equal, as a vector space, to the direct sum

$\oplus_{(Q,e)} \mathrm{Tr}_Q^G(kC_G(Q)e)$, with (Q, e) running over a set of representatives of the G-conjugacy classes of Brauer pairs on kG. Thus $\dim_k(Z(kG)) = \sum_{(Q,e)} m(Q, e)$, where (Q, e) runs over a set of representatives of the G-conjugacy classes of Brauer pairs on kG. Note that the dimension of $\mathrm{Tr}_Q^G(kC_G(Q)e)b$ is at most $m(Q, e)$, because this space is a homomorphic image of $\mathrm{Tr}_Q^{N_G(Q,e)}(kC_G(Q)e)$. It follows from 6.17.4 that $\dim_k(Z(kGb)) \geq \sum_{(Q,e)\in\mathcal{S}} m(Q, e)$. Taking the sum over all blocks and comparing dimensions shows that this must be an equality, and it moreover shows that the spaces $\mathrm{Tr}_Q^{N_G(Q,e)}(kC_G(Q)e)$, $\mathrm{Tr}_Q^G(kC_G(Q)e)$ and $\mathrm{Tr}_Q^G(kC_G(Q)e)b$ all have the same dimension $m(Q, e)$. □

The bases of $Z(\mathcal{O}G)$ and $\mathcal{O}G/[\mathcal{O}G, \mathcal{O}G]$ as well their dual bases considered in 5.15.6, 5.15.18 and 5.15.19 are compatible with the block decomposition of $\mathcal{O}G$, because each local pointed element u_ϵ on $\mathcal{O}G$ determines a unique block b of $\mathcal{O}G$ such that $\epsilon \cdot b \neq 0$. We briefly restate this in order to review the notation needed for the rationality result 6.17.9 below. For any local pointed element u_ϵ on $\mathcal{O}G$, we denote as in 5.15.2 by φ_ϵ the irreducible Brauer character of $C_G(u)$ corresponding to ϵ; that is, if $j \in \epsilon$, then $kC_G(u)\mathrm{Br}_u(j)$ is a projective cover of a simple $kC_G(u)$-module with Brauer character φ_ϵ. For any p-element u in G and any class function φ on $C_G(u)_{p'}$, we denote by $t_G^u(\varphi)$ the class function on G defined by $t_G^u(\varphi)(x) = 0$ if the p-part of x is not conjugate to u, and by $t_G^u(\varphi)(us) = \varphi(s)$ for any $s \in C_G(u)_{p'}$.

Lemma 6.17.8 *Let G be a finite group and B a block algebra of $\mathcal{O}G$. Let $\mathcal{R}$ be a set of representatives of the conjugacy classes of local pointed elements on B. For any local pointed element u_ϵ on B, denote by $\bar{u}_\epsilon$ the image of uj in $B/[B, B]$, where $j \in \epsilon$, and denote by $z(u_\epsilon)$ the image in $Z(B)$ of $t_G^u(\varphi_\epsilon)$ under the canonical isomorphism $(B/[B, B])^* \cong Z(B)$.*

(i) *The set $\{\bar{u}_\epsilon\}_{u_\epsilon \in \mathcal{R}}$ is an $\mathcal{O}$-basis of $B/[B, B]$.*
(ii) *The set $\{t_G^u(\varphi_\epsilon)\}_{u_\epsilon \in \mathcal{R}}$ is an $\mathcal{O}$-basis of $(B/[B, B])^*$.*
(iii) *The set $\{z(u_\epsilon)\}_{u_\epsilon \in \mathcal{R}}$ is an $\mathcal{O}$-basis of $Z(B)$.*

Proof Statement (i) is a restatement of 6.13.11. By 5.15.18, the set in (ii) is the dual basis of the basis in (i). Statement (iii) follows from 5.15.19. □

The following result of Cliff, Plesken and Weiss exploits these bases in order to prove a rationality property of centres of block algebras.

Theorem 6.17.9 ([49, (5.1) Theorem]) *Let G be a finite group and B a block algebra of $\mathcal{O}G$. The set $\{z(u_\epsilon)\}$, with u_ϵ running over a set of representatives of the G-conjugacy classes of local pointed elements on B, is an $\mathcal{O}$-basis of $Z(B)$.*

The multiplicative constants of this basis are rational numbers with denominators prime to p.

Proof By 6.17.8, the set $\{z(u_\epsilon)\}$, with u_ϵ running over a set of representatives of the G-conjugacy classes of local pointed elements on B, is an $\mathcal{O}$-basis of $Z(B)$. For u_ϵ a local pointed element on B, write

$$z(u_\epsilon) = \sum_{\chi \in \mathrm{Irr}(B)} a(u_\epsilon, \chi) e(\chi)$$

with coefficients $a(u_\epsilon, \chi) \in \mathcal{O}$, where $e(\chi)$ is the primitive idempotent in $Z(K \otimes_\mathcal{O} B)$ corresponding to χ. We observe first that in order to show that the multiplicative constants of the basis $\{z(u_\epsilon)\}$ are contained in a subfield L of K, it suffices to show that the numbers $a(u_\epsilon, \chi)$ are contained in L. Indeed, since the set of the $e(\chi)$, $\chi \in \mathrm{Irr}(B)$, and the set $\{z(u_\epsilon)\}$ are both bases of $Z(K \otimes_\mathcal{O} B)$, it follows that the matrix of coefficients $a(u_\epsilon, \chi)$ is invertible. If this matrix has coefficients in a subfield L, then the same is true for its inverse, and hence each $e(\chi)$ is an L-linear combination of the $z(u_\epsilon)$. We start by showing that the $a(u_\epsilon, \chi)$ are contained in $L = \mathbb{Q}[\zeta]$ for some root of unity ζ of order a sufficiently high power of p.

Let $\chi \in \mathrm{Irr}(B)$. Denote by $\omega : Z(K \otimes_\mathcal{O} B) \to K$ the associated central character as in 6.5.6. Let u_ϵ be a local pointed element on B, and set $\varphi = \varphi_\epsilon$. In the following calculation, we use the explicit description of $z(u_\epsilon)$ from 5.15.19. Note that $C_G(u) = C_G(u^{-1})$. We have

$$a(u_\epsilon, \chi) = \omega(z(u_\epsilon)) = \sum_s \varphi(s^{-1}) \frac{\chi(c(u^{-1}s))}{\chi(1)}$$

$$= \sum_s \varphi(s^{-1}) \frac{|G|}{|C_G(us)|\chi(1)} \chi(u^{-1}s).$$

In the sum, s runs over a set of representatives of the p'-conjugacy classes in $C_G(u)$. Thus, if we let run s over all p'-elements in $C_G(u)$, we need to adjust by $|C_G(u) : C_G(us)|$. By the definition of generalised decomposition numbers, we have

$$\chi(u^{-1}s) = \sum_\psi d^{u^{-1}}_{\chi, \psi} \psi(s)$$

where ψ runs over $\mathrm{IBr}_k(C_G(u))$, and by 5.15.7 the numbers $d^{u^{-1}}_{\chi, \psi}$ are in $\mathbb{Z}[\zeta]$ for some root of unity ζ of p-power order. Thus we get

$$a(u_\epsilon, \chi) = \frac{|G|}{|C_G(u)|\chi(1)} \sum_\psi d^{u^{-1}}_{\chi, \psi} \sum_s \varphi(s^{-1})\psi(s)$$

where s runs over all p'-elements in $C_G(u)$. By 5.14.11 the last sum yields coefficients of the inverse of the Cartan matrix of $kC_G(u)$, and these coefficients are rational. Thus $a(u_\epsilon, \chi) \in \mathbb{Q}[\zeta]$. By the argument at the beginning of this proof, it follows that the multiplicative constants of the basis $\{z(u_\epsilon)\}$ are contained in $\mathbb{Q}[\zeta]$.

Since a Brauer character of a simple kG-module takes values in $Q[\zeta']$ for some p'-root of unity ζ', it follows from the explicit description in 5.15.19 that the multiplicative constants of the basis $\{z(u_\epsilon)\}$ are contained in $\mathbb{Q}[\zeta']$. But this forces the multiplicative constants to be rational. These multiplicative constants are also in $\mathcal{O}$, and hence their denominators are prime to p. □

Corollary 6.17.10 *Let G be a finite group and B a block algebra of kG. Suppose that k is algebraically closed. Then $Z(B)$ has an $\mathbb{F}_p$-subalgebra C such that $Z(B) = k \otimes_{\mathbb{F}_p} C$.*

Proof The image in $Z(B)$ of the basis considered in 6.17.9 has multiplicative constants in the subfield generated by the images in k of rational numbers with p'-denominators, and that subfield is $\mathbb{F}_p$. Thus the $\mathbb{F}_p$-subalgebra C generated by this basis has the property as stated. □

Donovan's conjecture would imply that there are only finitely many k-algebras that arise as centres of blocks with a fixed defect. The above results imply that this is indeed the case.

Corollary 6.17.11 ([95, Corollary 1.3]) *Suppose that k is algebraically closed. Let d be a positive integer and set $m = \frac{1}{4}p^{2d} + 1$. There are at most p^{m^3} isomorphism classes of commutative k-algebras that occur as centres of blocks of defect d.*

Proof The Brauer–Feit Theorem 6.12.1 implies that the dimension of the centre of a block of defect d is at most $\frac{1}{4}p^{2d} + 1$. By 6.17.10, there is a basis with multiplicative constants in $\mathbb{F}_p$. There are at most p^{m^3} possibilities for 3-dimensional matrices of multiplicative constants in $\mathbb{F}_p$. □

Remark 6.17.12 Despite being statements about k-algebras, the proofs of 6.17.10 and 6.17.11 require some character theory, as do the alternative proofs due to Kessar [95], using isotypies; see §9.6 below.

7

Modules over Finite p-Groups

Green's theory of vertices and sources of indecomposable modules over group algebras puts the spotlight on representations of finite p-groups over p-local rings. There are only very few finite p-groups whose indecomposable modules can be classified over a field of characteristic p. We describe the indecomposable modules over cyclic p-groups and the Klein four group, as needed for later chapters on blocks with cyclic or Klein four defect groups. The structure theory of blocks with cyclic or Klein four defect relies on these descriptions. In general the module categories of finite p-groups are beyond reach, and this is one of the major hurdles in block theory.

Dade introduced in the late 1970s the class of *endopermutation modules* over finite p-group algebras that turned out to be of fundamental relevance for the structure theory of block algebras. We will describe Dade's classification of endopermutation modules over abelian p-groups, needed for Dade's Fusion Splitting Theorem, one of the Clifford theoretic cornerstones in block theory, as well as for Puig's Gluing Theorem, which is a key ingredient for certain stable equivalences between block algebras. Combined work of Alperin, Bouc, Carlson, and Thévenaz culminated in the classification of endopermutation modules for arbitrary finite p-groups in [24], nearly three decades after these modules were introduced by Dade. This classification is beyond the scope of this book; we refer to [214] for an overview and references.

We will use without further reference some basic results on finite p-group algebras from earlier sections. The group algebra kP of a finite p-group P over a field k of characteristic p is split local by 1.11.1 and symmetric by 2.11.2, with unique minimal one-dimensional ideal $\mathrm{soc}(kP) = k \cdot \sum_{y \in P} y$. In particular, every finitely generated projective kP-module is free. By 1.11.2 and 1.11.3, or also 2.5.3, the element $\sum_{y \in P} y$ annihilates any indecomposable nonprojective kP-module, and if V is a free kP-module, then the dimension of $(\sum_{y \in P} y)V = V_1^P$ is equal to the rank of V as a kP-module. Throughout this chapter p is a prime and $\mathcal{O}$ is a complete local principal ideal domain having a residue field k of characteristic p. If not stated otherwise, we allow the case $\mathcal{O} = k$.

7.1 Modules over cyclic *p*-groups

If P is a finite cyclic p-group, then kP has only finitely many isomorphism classes of finitely generated indecomposable modules, and these are classified as follows.

Theorem 7.1.1 *Let P be a finite cyclic p-group. Set $q = |P|$. For any integer i such that $1 \le i \le q$ set $V_i = kP/J(kP)^i$, viewed as a left kP-module.*

(i) *For $1 \le i \le q$ the kP-module V_i is indecomposable, and $\dim_k(V_i) = i$.*

(ii) *We have $V_1 \simeq k$ and $V_q \cong kP$.*

(iii) *Every finitely generated indecomposable kP-module is isomorphic to V_i for a uniquely determined i.*

(iv) *Every submodule of kP is equal to $J(kP)^i$ for a unique integer i such that $0 \le i \le q$, with the convention $J(kP)^0 = kP$; in particular, kP has only finitely many submodules.*

(v) *Any finitely generated indecomposable kP-module V has a unique composition series, and, setting $i = \dim_k(V)$, this is equal to*

$$V \supset J(kP)V \supset J(kP)^2 V \supset \cdots \supset J(kP)^{i-1} V \supset \{0\}.$$

(vi) *For $1 \le i \le q$ we have $V_i \cong J(kP)^{q-i}$.*

(vii) *All finitely generated kP-modules are selfdual, and all finitely generated indecomposable kP-modules are absolutely indecomposable.*

The most notable feature of this classification is that kP is of *finite representation type*; that is, there are only finitely many isomorphism classes of finitely generated indecomposable kP-modules for P a finite cyclic p-group. No noncyclic finite p-group has this property. The proof we present here uses the fact that kP can be identified with a quotient of the polynomial ring $k[x]$, which is a principal ideal domain. The theorem follows from the classification of finitely generated modules over principal ideal domains. Another proof, due to Nakayama, will be given in the section on serial algebras in the chapter on blocks with cyclic defect groups.

Lemma 7.1.2 *Let P be a finite p-group with generator y. The map sending the variable x to the element $y - 1$ in the group algebra kP induces a surjective algebra homomorphism $k[x] \to kP$ whose kernel is the ideal $(x^{|P|})$. In particular, this induces an algebra isomorphism $k[x]/(x^{|P|}) \cong kP$.*

Proof Since char$(k) = p$, we have $(y-1)^{|P|} = y^{|P|} - 1 = 1 - 1 = 0$ in kP. Thus the map sending x to $y-1$ induces a surjective algebra homomorphism $k[x] \to kP$ whose kernel contains $(x^{|P|})$. The image of the set of monomials $\{1, x, x^2, \ldots, x^{|P|-1}\}$ in $k[x]/(x^{|P|})$ is a k-basis. It follows that $\dim_k(k[x]/(x^{|P|})) = |P| = \dim_k(kP)$ which completes the proof. $\square$

Proof of 7.1.1 The k-algebra $k[x]$ is in particular a principal ideal ring. By the classification of finitely generated modules over principal ideal domains, any finite-dimensional indecomposable $k[x]$-module is isomorphic to $k[x]/(f^i)$ for some monic irreducible polynomial $f \in k[x]$ and some positive integer i. Any indecomposable $k[x]/(x^{|P|})$-module can be viewed as a $k[x]$-module via the algebra homomorphism $k[x] \to k[x]/(x^{|P|})$, and becomes an indecomposable $k[x]$-module that is annihilated by $(x^{|P|})$. But then an indecomposable $k[x]$-module of the form $k[x]/(f^i)$ with f and i as before is annihilated by $x^{|P|}$ if and only if $x^{|P|} \in (f^i)$. Since f is irreducible and monic this forces $f = x$ and $1 \le i \le |P|$. Since $J(kP) = (y-1)kP$ for any generator y of P, the isomorphism $k[x]/(x^{|P|}) \cong kP$ from the previous lemma implies the statements (i), (ii) and (iii) of the theorem. The inverse image in $k[x]$ of a submodule of $k[x]/(x^{|P|})$ is an ideal (f) for some polynomial f such that $(x^{|P|}) \subseteq (f)$, or equivalently, such that f divides $x^{|P|}$ in $k[x]$. The divisors of $x^{|P|}$ are, up to associates, exactly the monomial x^i with $0 \le i \le |P|$, hence every submodule of $k[x]/(x^{|P|})$ is of the form $(x^i)/(x^{|P|})$ for some i such that $0 \le i \le |P|$. Using the isomorphism $k[x]/(x^{|P|}) \cong kP$ as before proves (iv). This in turn implies that kP has a unique composition series, namely the series $kP \supset J(kP) \supset J(kP)^2 \supset \cdots \supset J(kP)^{|P|-1} \supset \{0\}$. Any finitely generated indecomposable kP-module is a quotient of kP, whence (v). Since every submodule of kP has a simple socle, hence is indecomposable, combining the previous statements yields (vi). The isomorphism class of a finitely generated indecomposable kP-module V is determined by its dimension, and hence V is selfdual. Moreover, V is a quotient of kP by a uniquely determined submodule of kP, and hence $\text{End}_{kP}(V)$ is a quotient of $\text{End}_{kP}(kP) \cong kP$. In particular, $\text{End}_{kP}(V)$ is split local. This proves (vii). $\square$

Statement (v) of 7.1.1 says that every indecomposable kP-module is *uniserial* if P is cyclic. Again, no noncyclic finite p-group has this property. The next result is a very special case of a theorem of Chouinard.

Theorem 7.1.3 *Let P be a nontrivial cyclic finite p-group and Q a nontrivial subgroup of P. A finitely generated kP-module V is projective if and only if the kQ-module $\text{Res}_Q^P(V)$ is projective.*

Proof Let V be a finite-dimensional kP-module and let Q be a nontrivial subgroup of P such that $\mathrm{Res}_Q^P(V)$ is projective. In order to show that V is projective, we argue by induction over the index $|P : Q|$, hence we may assume that Q is the maximal subgroup of P. Let y be a generator of P. Then y^p is a generator of Q. We may assume that V is indecomposable, hence $V \cong J(kP)^i = kP(y-1)^i$ for some i. Since $\mathrm{Res}_Q^P(V)$ is projective and Q nontrivial, the dimension of V is divisible by p, and hence $i = pm$ for some integer $m \geq 0$. Thus $(y-1)^i = (y^p - 1)^m$, and $y^p - 1 \in kQ$. It follows that $\mathrm{Res}_Q^P(V) \cong \bigoplus_{z \in [P/Q]} zkQ(y^p - 1)^m$, which is isomorphic to p copies of $kQ(y^p - 1)^m$. It also follows from 7.1.1 that this module is projective if either $m = 0$ or $m \geq |Q|$. But then either $i = |P|$, in which case V is zero, or $i = 0$, in which case $V \cong kP$, and hence V is projective as a kP-module in both cases. The converse is trivial. $\qquad\square$

Lemma 7.1.4 *Let P be a finite cyclic p-group. Set $q = |P|$, and let s be a generator of P. Set $x = s - 1$. Let U be a finitely generated projective kP-module and let $u \in U$. Let i be an integer such that $0 \leq i \leq q$. We have $x^i u = 0$ if and only if there exists $v \in U$ such that $u = x^{q-i}v$. In particular, we have $\mathrm{soc}(U) = x^{q-1}U$ and $x^{q-1} = \sum_{y \in P} y$.*

Proof We have $J(kP)^i = x^i kP$. It follows from 7.1.1 that the annihilator of $J(kP)^i$ in kP is exactly $J(kP)^{q-i}$; applied to $i = 1$ this implies that $\mathrm{soc}(kP) = kx^{q-1} = k\sum_{u \in P} u$. Therefore x^{q-1} and $\sum_{y \in P} y$ differ at most by a nonzero scalar. Comparing the coefficients at 1 of these two elements of kP shows that this scalar is equal to 1. The result holds for kP instead of U. Since a projective kP-module is free, the result holds for any finitely generated projective kP-module. $\qquad\square$

Any kP-endomorphism of kP arises as given by right multiplication with an element in kP. An arrow labelled by an element in kP in the following description of projective resolutions is to be understood as the map induced by multiplication with that element.

Proposition 7.1.5 *Let P be a finite cyclic p-group. Set $q = |P|$. Denote by s a generator of P, and set $x = s - 1$. Let i be an integer such that $1 \leq i \leq q - 1$. A minimal projective resolution of V_i has the form*

$$\cdots \xrightarrow{x^i} kP \xrightarrow{x^{q-1}} kP \xrightarrow{x^i} kP \xrightarrow{x^{q-i}} V_i \longrightarrow 0.$$

We have $\Omega(V_i) \cong V_{q-i}$ and $\Omega^2(V_i) \cong V_i$. The kP-module V_i has period 2, unless $p = 2$ and $i = \frac{q}{2}$, in which case V_i has period 1. In particular, the trivial kP-module $k \cong V_1$ has period 2, except if $|P| = 2$, in which case it has period 1.

Proof Since V_i is uniserial, it has a projective cover of the form $kP \to V_i$. Comparing dimensions yields $\Omega(V_i) \cong V_{q-i}$. The endomorphism $kP \to kP$ given by multiplication with x^{q-i} has as image $J(kP)^{q-i} \cong V_i$ and as kernel $J(kP)^i \cong V_{q-i}$. Similarly for the endomorphism given by multiplication with x^i. The result follows. $\square$

Proposition 7.1.6 *Let P be a finite cyclic p-group of order q. Let i, j be positive integers such that i, $j \leq q$ and let V_i, V_j be indecomposable kP-modules of dimension i, j, respectively. We have $\dim_k(\mathrm{Hom}_{kP}(V_i, V_j)) = \min\{i, j\}$.*

Proof After dualising if necessary we may assume that $i \leq j$. Identify V_i, V_j to the unique submodules of kP of dimension i, j, respectively. Every homomorphism from V_i to V_j extends to an endomorphism of kP, since kP is injective. Since $i \leq j$, every endomorphism of kP sends V_i to V_j. The space $\mathrm{End}_{kP}(kP)$ has dimension q, and the annihilator of V_i in $\mathrm{End}_{kP}(kP)$ has dimension $q - i$. Thus $\mathrm{Hom}_{kP}(V_i, V_j)$ has dimension i. The result follows. $\square$

Proposition 7.1.7 *Let P be a finite cyclic p-group of order q. Let i, j be positive integers such that $i + j \leq q$ and let V_i, V_j be indecomposable kP-modules of dimension i, j, respectively. Then $\underline{\mathrm{Hom}}_{kP}(V_i, V_j) \cong \mathrm{Hom}_{kP}(V_i, V_j)$; that is, no nonzero homomorphism from V_i to V_j factors through a projective module. Moreover, $V_i \otimes_k V_j$ has no nonzero projective direct summand.*

Proof If $\alpha : V_i \to V_j$ factors through a projective kP-module, it factors through a projective cover $\pi : kP \to V_j$ of V_j, say $\alpha = \pi \circ \gamma$ for some $\gamma : V_i \to kP$. Then $\mathrm{Im}(\gamma)$ has length at most i, and $\ker(\pi)$ has length $q - j \geq i$. Thus $\ker(\pi)$ contains $\mathrm{Im}(\gamma)$, or equivalently, $\alpha = \pi \circ \gamma = 0$. This shows the first statement. Since all kP-modules are selfdual, we have $\mathrm{Hom}_k(V_i, V_j) \cong V_i \otimes_k V_j$. Write $V_i \otimes_k V_j = W \oplus Y$ such that Y is projective and W has no nonzero projective summand. Then $\mathrm{Hom}_{kP}(V_i, V_j) \cong W^P \oplus Y^P$, and we have $Y^P = Y_1^P$ and $W_1^P = \{0\}$. Thus 2.13.11 implies that $\underline{\mathrm{Hom}}_{kP}(V_i, V_j) \cong W^P$. This forces $Y^P = \{0\}$, and hence $Y = \{0\}$. $\square$

Proposition 7.1.8 *Let P be a finite cyclic p-group and let U be a finitely generated indecomposable kP-module such that $\underline{\mathrm{End}}_{kP}(U) \cong k$. Then U is isomorphic to k or $\Omega(k)$.*

Proof It follows from 7.1.5 that after possibly replacing U by $\Omega(U)$, we may assume that U has length $i \leq \frac{|P|}{2}$. Then by 7.1.7 we have $\underline{\mathrm{End}}_{kP}(U) \cong \mathrm{End}_{kP}(U)$, which has dimension i. Thus $i = 1$ and the result follows. $\square$

Proposition 7.1.9 *Let P be a finite cyclic p-group. Set $q = |P|$. Denote by s a generator of P, and set $x = s - 1$. For any kP-module V, any $\psi \in \text{End}_k(V)$ and any $v \in V$ we have $\text{Tr}_1^P(\psi)(v) = \sum_{i=0}^{q-1} sx^{q-i-1}\psi(x^i v)$.*

Proof The map Tr_1^P depends only on the canonical symmetrising form σ : $kP \to k$ sending 1 to 1 and $u \in P \setminus \{1\}$ to 0, but not on the choice of a basis together with its dual basis with respect to σ in kP. It suffices therefore to show that $\{x^i | 0 \leq i \leq q - 1\}$ is a k-basis of kP that is dual to the basis $\{sx^{q-i-1} | 0 \leq i \leq q - 1\}$ with respect to σ. Using the last statement of 7.1.4, we have $\sigma(sx^{q-i-1}x^i) = \sigma(sx^{q-1}) = \sigma(s \sum_{u \in P} u) = 1$, where $0 \leq i \leq q - 1$. If $0 \leq i, j \leq q - 1$ such that $i \neq j$, then $\sigma(sx^{q-i-1}x^j) = \sigma(sx^{q+j-i-1})$. If $j > i$, then this expression is zero as $q + j - i - 1 \geq q$ in that case. If $j < i$, then $m = q + j - i - 1 \leq q - 2$, and hence the coefficient of $s^{-1} = s^{q-1}$ in x^m is zero. It follows that the coefficient of 1 of sx^m is zero, or equivalently, $\sigma(sx^m) = 0$ as required. $\qquad\square$

Proposition 7.1.10 *Let P be a finite cyclic p-group, V a finitely generated kP-module, and T a one-dimensional submodule of V. Let s be a generator of P, and set $x = s - 1$. Suppose that there exists $\psi \in \text{End}_k(V)$ such that $\psi(T) \subseteq T$ and such that $\text{Tr}_1^P(\psi)(v) = xv$ for all $v \in V$. Then either T is a direct summand of V or T is a submodule of a projective indecomposable summand of V.*

Proof Set $q = |P|$. Let $t \in T$ such that $T = kt$. Note that $xT = \{0\}$. Suppose that T is not a direct summand of V. Then $T \subseteq \text{rad}(V) = xV$, hence $t = xv$ for some $v \in V$. If $q = 2$ then $kPv \cong kP$ is projective indecomposable, hence also injective and therefore a direct summand of V containing T. Assume that $q > 2$. Since T is simple, we have $xT = \{0\}$, hence $x^2 v = 0$. Using 7.1.9 we have $t = xv = \text{Tr}_1^P(\psi)(v) = \sum_{i=0}^{q-1} sx^{q-i-1}\psi(x^i v)$. This sum has at most two nonzero summands, namely for i equal to 0 or 1. The summand for $i = 1$ is $sx^{q-2}\psi(xv)$, which is zero because $\psi(xv) = \psi(t) \in T$ and $q - 2 > 0$, hence x^{q-2} annihilates T. Thus $t = sx^{q-1}\psi(v)$, which shows that T is the socle of the projective indecomposable module $kPs\psi(v) \cong kP$. This submodule is also injective, hence a direct summand. $\qquad\square$

Vertices of indecomposable modules for cyclic p-groups are uniquely determined by the dimension of the module.

Proposition 7.1.11 *Let P be a finite cyclic p-group and Q a subgroup of P. Let U be a finitely generated indecomposable kP-module.*

(i) *The module U is relatively Q-projective if and only if $|P : Q|$ divides $\dim_k(U)$.*

(ii) The subgroup Q of P is the unique vertex of U if and only if $|P : Q|$ is the exact power of p dividing $\dim_k(U)$.

(iii) The group P is a vertex of U if and only if $\dim_k(U)$ is prime to p.

Proof We will use 7.1.1, applied to both P and Q, without further reference. We may assume that P is nontrivial. If U is relatively Q-projective, then Green's Indecomposability Theorem 5.12.3 implies that $U \cong \operatorname{Ind}_Q^P(W)$ for some indecomposable kQ-module W. Thus $|P : Q|$ divides $\dim_k(U)$ if U is relatively Q-projective. Note that $\operatorname{Res}_Q^P(\operatorname{Ind}_Q^P(W))$ is a direct sum of modules ismorphic to W, because Q is normal in P and because all conjugates of W are isomorphic to W since they have the same dimension. Thus there are $|Q|$ isomorphism classes of modules of the form $\operatorname{Ind}_Q^P(W)$, with W an indecomposable $\mathcal{O}Q$-module. There are also $|Q|$ isomorphism classes of indecomposable kP-modules of dimension divisible by $|P : Q|$, whence (i). Statement (ii) follows from (i), and (iii) is a special case of (ii).   □

The fact that $\dim_k(U)$ is divisible by $|P : Q|$ if U is relatively Q-projective holds for arbitrary finite p-groups (as a consequence of Theorem 5.12.13), but the converse in (i) above is specific to cyclic p-groups.

7.2 Modules over a Klein four group

Although a Klein four group has infinitely many isomorphism classes of indecomposable modules in characteristic 2, these can still be classified. This is one of the few 2-groups having what is called *tame representation type* in characteristic 2. We assume in this section that k is algebraically closed of characteristic 2. Let V_4 be a Klein four group and write V_4 as a direct product $V_4 = \langle s \rangle \times \langle t \rangle$ for some involutions s and t in V_4.

Theorem 7.2.1 ([13], [83]) *The following list is a complete set of representatives of the isomorphism classes of finitely generated indecomposable kV_4-modules.*

(i) The four-dimensional projective indecomposable kV_4-module.

(ii) The Heller translates $\Omega^n(k)$ of the trivial kV_4-module, with $n \in \mathbb{Z}$; these represent all odd-dimensional indecomposable kV_4-modules. They satisfy $(\Omega^n(k))^* \cong \Omega^{-n}(k)$ and $\dim_k(\Omega^n(k)) = \dim_k(\Omega^{-n}(k)) = 2n+1$. The trivial module is the only odd-dimensional self-dual indecomposable kV_4-module.

(iii) Let n be a positive integer, denote by I the $n \times n$-identity matrix and for $\lambda \in k$, denote by J_λ the $n \times n$-matrix in Jordan canonical form, with λ

in the diagonal, 1 *above the diagonal, and zero everywhere else. For any* $(\lambda, \mu) \in k^2 \setminus \{(0, 0)\}$, *there is a 2n-dimensional indecomposable nonprojective* kV_4-*module* $V_{2n}(\lambda, \mu)$, *such that the actions of s and t are represented by the* $2n \times 2n$-*matrices, written as* 2×2-*matrices of* $n \times n$-*blocks,*

$$s \mapsto \begin{pmatrix} I & I \\ 0 & I \end{pmatrix} \quad t \mapsto \begin{pmatrix} I & J_{\mu\lambda^{-1}} \\ 0 & I \end{pmatrix}$$

if $\lambda \neq 0$, *or*

$$s \mapsto \begin{pmatrix} I & J_{\lambda\mu^{-1}} \\ 0 & I \end{pmatrix} \quad t \mapsto \begin{pmatrix} I & I \\ 0 & I \end{pmatrix}$$

if $\mu \neq 0$, *and both descriptions are isomorphic if both* λ, μ *are nonzero. More precisely, we have* $V_{2n}(\lambda, \mu) \cong V_{2n}(\lambda', \mu')$ *if and only if* $(\lambda', \mu') = (\kappa\lambda, \kappa\mu)$ *for some nonzero* κ *in k, and we have* $\Omega(V_{2n}(\lambda, \mu)) \cong V_{2n}(\lambda, \mu)$; *that is, the even-dimensional indecomposable nonprojective* kV_4-*modules are periodic of period 1.*

(iv) *Let* n *be a positive integer and* $(\lambda, \mu) \in k^2 \setminus \{0, 0\}$. *The A-module* $V_{2n}(\lambda, \mu)$ *has a filtration* $\{0\} \subseteq V_2 \subseteq V_4 \subseteq \cdots \subseteq V_{2n-2} \subseteq V_{2n} = V_{2n}(\lambda, \mu)$ *such that* $V_{2i} \cong V_{2i}(\lambda, \mu)$ *and such that* $V_{2n}/V_{2i} \cong V_{2(n-i)}(\lambda, \mu)$ *for* $1 \leq i \leq n - 1$.

In other words, the $2n$-dimensional indecomposable nonprojective kV_4 modules are parametrised by the elements of the projective line $\mathbb{P}^1(k)$, which is by definition the quotient of $k^2 \setminus \{0\}$ by the action of the group $k^\times$ by scalar multiplication. After rescaling, one can always choose at least one of λ, μ to be 1; more precisely, the points of $\mathbb{P}^1(k)$ are exactly represented by the points in k^2 of the form $(\lambda, 1)$, with $\lambda \in k$, and the point $(1, 0)$. The natural action of $GL_2(k)$ on k^2 induces a transitive action on $\mathbb{P}^1(k)$. The group $GL_2(k)$ acts as well on A by algebra automorphisms in such a way that an element $g \in GL_2(k)$ induces the automorphism on A sending x to $\lambda x + \mu y$ and y to $\lambda' x + \mu' y$, where (λ, μ), (λ', μ') are the images under g of the unit vectors in k^2. In this way, $GL_2(k)$ induces an action on the isomorphism classes of even-dimensional indecomposable nonprojective A-modules such that ${}^g(V_{2n}(1, 0)) \cong V_{2n}(\lambda, \mu)$ and ${}^g(V_{2n}(0, 1)) \cong V_{2n}(\lambda', \mu')$.

In order to prove the above theorem, we fix the following notation. Set $A = kV_4$. The algebra A is 4-dimensional, local, commutative and symmetric. Setting $x = s - 1$ and $y = t - 1$, the set $\{1, x, y, xy\}$ is a k-basis of A satisfying $x^2 = y^2 = 0$ and $xy = yx$. It is worth noting that the classification of the isomorphism classes of indecomposable A-modules depends only on these relations and the algebraic closure of k but not on the characteristic of k. The ideal generated by x and y is equal to $J(A)$, which is also equal to the k-span of $\{x, y, xy\}$. We have $J(A)^2 = \mathrm{soc}(A) = kxy$ and $J(A)^3 = \{0\}$. In particular, 4.11.8 applies, which yields the following statement:

Proposition 7.2.2 *Let U be a finitely generated indecomposable nonprojective nonsimple A-module. We have* $\mathrm{rad}(U) = \mathrm{soc}(U)$.

With the notation of 7.2.2, if we set $U_1 = \mathrm{rad}(U)$ and denote by U_0 a vector space complement of U_1 in U, then x, y as above induce linear maps from U_0 to U_1 and annihilate U_1; that is, the action of A is completely described by a pair of linear maps from U_0 to U_1. A description of the isomorphism classes of indecomposable A-modules amounts therefore to classifying indecomposable pairs of matrices. This approach, which goes back to work of Kronecker, further developed by Dieudonné [60], has been used by Heller and Reiner [83] and Bašev [13] to classify indecomposable modules of the Klein four group algebra even when the ground field is not algebraically closed – this is essentially based on the theory of rational canonical forms. See Benson [14, 4.3] for an exposition of this material. We restrict attention to the case where k is algebraically closed. We use 7.2.2 to describe first all odd-dimensional indecomposable A-modules. For U an indecomposable, nonprojective, and nonsimple A-module we set $r_1(U) = \dim_k(U/\mathrm{rad}(U))$ and $r_2(U) = \dim_k(\mathrm{rad}(U))$. We clearly have $\dim_k(U) = r_1(U) + r_2(U)$.

Lemma 7.2.3 *Let U be a finitely generated indecomposable nonprojective nonsimple A-module.*

  (i) *If $r_1(U) > r_2(U)$, then $U \cong \Omega^n(k)$ for some positive integer n. Moreover, in that case we have $n = r_2(U) = r_1(U) - 1$ and $\dim_k(U) = 2n + 1$.*

 (ii) *If $r_1(U) < r_2(U)$ then $U \cong \Omega^{-n}(k)$ for some positive integer n. Moreover, in that case we have $n = r_1(U) = r_2(U) - 1$ and $\dim_k(U) = 2n + 1$.*

(iii) *If $r_1(U) = r_2(U)$ then $\dim_k(U) = \dim_k(\Omega^n(U))$ for any integer n.*

Proof Suppose that $r_1(U) > r_2(U)$. We argue by induction over $\dim_k(U)$. If $\Omega^{-1}(U)$ is simple, then $U \cong \Omega(k)$, hence $r_1(U) = 2$ and $r_2(U) = 1$, which proves the result in this case. Assume that $\Omega^{-1}(U)$ is not simple. Since $r_2(U) = \dim_k(\mathrm{rad}(U)) = \dim_k(\mathrm{soc}(U))$, the module U has an injective envelope I of dimension $4r_2(U)$. Since I is also a projective cover of $\Omega^{-1}(U)$, we have $r_1(\Omega^{-1}(U)) = r_2(U)$. We have $\dim_k(\Omega^{-1}(U)) = \dim_k(I) - \dim_k(U) = 4r_2(U) - (r_1(U) + r_2(U)) = 3r_2(U) - r_1(U) < r_1(U) + r_2(U) = \dim_k(U)$. Finally, we have $r_2(\Omega^{-1}(U)) = \dim_k(\Omega^{-1}(U)) - r_1(\Omega^{-1}(U)) = 3r_2(U) -$

$r_1(U) - r_2(U) = 2r_2(U) - r_1(U) < r_2(U) = r_1(\Omega^{-1}(U))$. Thus $\Omega^{-1}(U)$ satisfies the induction hypothesis and has smaller dimension than U. It follows
that $\Omega^{-1}(U) \cong \Omega^n(k)$, where $n = r_2(\Omega^{-1}(k)) = r_1(\Omega^{-1}(k)) - 1$. Applying Ω
yields $U \cong \Omega^{n+1}(k)$. Moreover, by the above we have $r_2(U) = r_1(\Omega^{-1}(U)) =$
$n + 1$ and $r_1(U) = 2r_2(U) - r_2(\Omega^{-1}(U)) = 2(n + 1) - n = n + 2$. This
shows (i). A similar argument, with the injective envelope of U replaced by
a projective cover, shows (ii). Alternatively, the dual of a module satisfying
the hypothesis in (ii) satisfies the hypothesis, hence the conclusions in (i), and
the conclusions in (ii) follow from dualising those in (i). If $r_1(U) = r_2(U)$,
then a projective cover of U has dimension $4r_1(U) = 2\dim_k(U)$, whence
$\dim_k(\Omega(U)) = 2\dim_k(U) - \dim_k(U) = \dim_k(U)$. Dualising and iterating this
argument shows (iii). □

The Heller translates of the trivial A-module represent therefore exactly
the isomorphism classes of all odd-dimensional indecomposable A-modules,
and any indecomposable nonprojective A-module U of even dimension has the
property $r_1(U) = r_2(U)$.

Lemma 7.2.4 *Let U be a finitely generated indecomposable nonprojective A-
module and let V be an indecomposable nonsimple submodule of U. Then $\Omega(V)$
is isomorphic to a submodule of $\Omega(U)$.*

Proof Since $J(A)^3 = \{0\}$ this is a special case of 4.11.9. □

Lemma 7.2.5 *Let U be an even-dimensional indecomposable nonprojective
A-module. Then U has no submodule isomorphic to $\Omega^m(k)$ for some positive
integer m.*

Proof Suppose that U has a submodule isomorphic to $\Omega^m(k)$ for some positive integer m. Then, for any $n \geq 0$, $\Omega^{m+n}(k)$ is indecomposable nonprojective
and nonsimple. Inductively we get from 7.2.4 that $\Omega^{m+n}(k)$ is isomorphic to
a submodule of $\Omega^n(U)$ for any $n \geq 0$. This is, however, not possible, since
by 7.2.3, the modules $\Omega^{m+n}(k)$ have arbitrary large dimension, whereas the
modules $\Omega^n(U)$ all have dimension $\dim_k(U)$, with n running over the positive
integers. □

Lemma 7.2.6 *Let U be a finitely generated A-module having no projective
direct summand such that $\mathrm{rad}(U) = \mathrm{soc}(U)$ and $r_1(U) = r_2(U)$. If U does not
have an indecomposable direct summand isomorphic to $\Omega^m(k)$ for some positive integer m, then every indecomposable direct summand of U has even
dimension.*

Proof Since $\mathrm{rad}(U) = \mathrm{soc}(U)$, the module U has no trivial direct summand. Any indecomposable odd-dimensional direct summand W of U satisfied $r_1(W) = r_2(W) \pm 1$. Thus if U has an indecomposable odd-dimensional direct summand, then it has a direct summand W satisfying $r_1(W) = r_2(W) + 1$, hence W is isomorphic to $\Omega^m(k)$ for some positive integer m, contradicting the assumptions. $\qquad\square$

Lemma 7.2.7 *Let U be an indecomposable nonprojective even-dimensional A-module and let V be an indecomposable 2-dimensional submodule of U. Then every indecomposable direct summand of U/V has even dimension.*

Proof We show first that $\mathrm{rad}(U/V) = \mathrm{soc}(U/V)$. Let $u \in U$ such that $u + V \in \mathrm{soc}(U/V)$; that is, $xu, yu \in V$. If one of xu, yu is nonzero, then $V' = Au$ is an indecomposable 2-dimensional submodule whose socle coincides with that of V. Thus $V' + V$ is a 3-dimensional submodule with a simple socle, hence isomorphic to $\Omega(k)$. This contradicts 7.2.5. Therefore, $xu = yu = 0$, or equivalently, $u \in \mathrm{soc}(U) = \mathrm{rad}(U)$, which implies that $u + V \in \mathrm{rad}(U/V)$. This shows that $\mathrm{soc}(U/V) \subseteq \mathrm{rad}(U/V)$; the other inclusion holds for all indecomposable nonprojective A-modules. We show next that $r_1(U/V) = r_2(U/V)$. Indeed, since $\mathrm{rad}(U/V)$ is the image of $\mathrm{rad}(U) = \mathrm{soc}(U)$ and since $V \cap \mathrm{soc}(U) = \mathrm{soc}(V)$ is one-dimensional, we get that $r_1(U/V) = r_1(U) - 1$. But then, since $\dim_k(U/V) = \dim_k(U) - 2$ we also get that $r_2(U/V) = r_2(U) - 1 = r_1(U) - 1 = r_1(U/V)$ as claimed. By 7.2.6, if U/V has an odd-dimensional indecomposable direct summand, then U/V has a direct summand W isomorphic to $\Omega^m(k)$ for some positive integer m. Let X be the inverse image of W in U. One easily checks that $r_1(X) = r_2(X) + 1$ and that X has no trivial direct summand. Thus X has a summand isomorphic to $\Omega^t(k)$ for some positive integer t, contradicting again 7.2.5. The Lemma follows. $\qquad\square$

Proof of Theorem 7.2.1 The odd-dimensional indecomposable A-modules are classified in 7.2.3. In order to classify the even-dimensional indecomposable modules, we first observe that it suffices to show that if U is $2n$-dimensional indecomposable nonprojective for some positive integer, then $xU = \mathrm{soc}(U)$ or $yU = \mathrm{soc}(U)$. Indeed, suppose that $xU = \mathrm{soc}(U)$. Let U' be a vector space complement of $\mathrm{soc}(U)$ in U. Since $\dim_k(U') = \dim_k(\mathrm{soc}(U))$, this implies that multiplication by x induces a k-linear isomorphism $U' \cong \mathrm{soc}(U)$. Choose bases such that this isomorphism is represented by the identity matrix I. The matrix representing y is then unique up to conjugacy, hence, after simultaneous change of bases of U', $\mathrm{soc}(U)$, may be chosen to be a matrix J in Jordan canonical form. For U to be indecomposable, J has to consist of a single Jordan block. Thus $J = J_\mu$ for some $\mu \in k$, and hence $U \cong V_{2n}(1, \mu)$. More explicitly, U has a basis that is the union of a basis $\{u_i | 1 \le i \le n\}$ of U' and a basis $\{v_i | 1 \le i \le n\}$ of $\mathrm{soc}(U)$ such that $xu_i = v_i$ for $1 \le i \le n$, $yu_i = \mu v_i + v_{i+1}$ for $1 \le i \le n-1$ and $yu_n = \mu v_n$. For a fixed t, $1 \le t \le n$, the set $\{u_i, v_i | t \le i \le n\}$ is then the basis of a submodule isomorphic to $V_{2(n-t+1)}(1, \mu)$, and thus the modules described in (iii) have a filtration as stated in (iv). A similar argument applies if $yU = \mathrm{soc}(U)$. It remains to show that $xU = \mathrm{soc}(U)$ or $yU = \mathrm{soc}(U)$. Suppose first that U has an indecomposable submodule V of dimension 2. By 7.2.7, U/V is a direct sum of even-dimensional indecomposable modules, hence satisfies $x(U/V) = \mathrm{soc}(U/V)$ or $yU/V = \mathrm{soc}(U/V)$ by induction, which implies that $xU = \mathrm{soc}(U)$ or $yU = \mathrm{soc}(U)$, respectively. Suppose finally that U has no 2-dimensional indecomposable submodule. Arguing by induction, we assume that every indecomposable proper submodule of U has odd dimension. Let m be maximal such that U has a submodule W isomorphic to $\Omega^{-m}(k)$. Since $r_1(W) < r_2(W)$ there is an element $u \in U - \mathrm{rad}(U)$ such that $u \notin W$. By the assumption on U, the dimension of Au is 3, and the intersection $W \cap Au$ is contained in $\mathrm{soc}(U)$, because otherwise $W \cap Au$ would be an indecomposable 2-dimensional module. Set $W' = W + Au$. Suppose first that $W \cap Au$ has dimension 1. Then W' is an odd-dimensional module satisfying $\mathrm{rad}(W') = \mathrm{soc}(W')$ and $r_1(W') = r_2(W') - 1$. By 7.2.5, W' is a direct sum of modules of the form $\Omega^{-t}(k)$, with t positive. The equation $r_1(W') = r_2(W') - 1$ implies that there cannot be two such summands; that is, W' is indecomposable. Since $\dim_k(W') = \dim_k(W) + 2$ we get that $W' \cong \Omega^{-(m+1)}(k)$, contradicting the maximality of m. This contradiction shows that $W \cap Au$ must have dimension 2. This means that $W \cap Au = \mathrm{soc}(Au)$, thus xu and yu are both contained in W, and in particular, $\mathrm{soc}(W') = \mathrm{soc}(W)$. Thus W' has even dimension and satisfies $\mathrm{rad}(W') = \mathrm{soc}(W')$ and $r_1(W') = r_2(W')$. It follows from 7.2.6 that W' is a direct sum of indecomposable even-dimensional modules, which by induction have 2-dimensional submodules. This contradiction shows that the even-dimensional nonprojective indecomposable A-modules are as stated in 7.2.1 (iii). One verifies directly that the 2-dimensional indecomposable modules have period 1. Thus, if U is an even-dimensional nonprojective indecomposable A-module, then both U and $\Omega(U)$ have isomorphic 2-dimensional submodules by 7.2.4. But then $U \cong \Omega(U)$, because as a consequence of the filtration in (iv), U has exactly one isomorphism class of indecomposable 2-dimensional submodules, and hence the isomorphism class of U is determined by that of its 2-dimensional submodules. $\qquad\square$

Theorem 7.2.8 *Let V be a finitely generated indecomposable A-module. Then* $\underline{\mathrm{End}}_A(V) \cong k$ *if and only if* $V \cong \Omega^n(k)$ *for some integer n.*

Proof Since Ω induces a stable equivalence on $\underline{\mathrm{mod}}(A)$, we have $\underline{\mathrm{End}}_A(V) \cong k$ whenever $V \cong \Omega^n(k)$ for some integer. For the converse, let V be a finitely generated indecomposable A-module that is not isomorphic to any of the $\Omega^n(k)$. By 7.2.1, V has even dimension $2m$ for some integer $m \geq 1$. If $m > 1$, then the description of the $2m$-dimensional modules in 7.2.1 (iv) implies that there is an endomorphism β of V having as image a 2-dimensional indecomposable submodule W. This submodule is not contained in $\mathrm{rad}(V)$, and so by 4.13.5 this endomorphism does not factor through a projective kP-module. Using again the structure of V one sees that Id_V and β remain linearly independent in $\underline{\mathrm{End}}_A(V)$ because no nontrivial linear combination yields an endomorphism with image contained in $\mathrm{rad}(V)$. Thus $\dim_k(\underline{\mathrm{End}}_A(V)) \geq 2$. It remains to consider the case $\dim_k(V) = 2$. Denote by γ the endomorphism sending V to its socle. One easily verifies that γ does not factor through the injective envelope kP of V. As before, the images of Id_V and γ in $\underline{\mathrm{End}}_A(V)$ are linearly independent. The result follows. $\square$

The odd-dimensional indecomposable A-modules are Heller translates of the trivial A-module k, and hence they are stable under any algebra automorphism of A because this is true for the trivial A-module and the projective indecomposable A-module A. The even-dimensional indecomposable A-modules that are stable under the automorphism group of order 3 of V_4 are described in the following result.

Proposition 7.2.9 *Let n be a positive integer, V be a $2n$-indecomposable nonprojective A-module, and let ζ be a cube root of unity in $k^\times$. The isomorphism class of V is stable under the automorphism group of order 3 of V_4 if and only if V is isomorphic to either $V_{2n}(\zeta, 1)$ or $V_{2n}(1, \zeta) \cong V_{2n}(\zeta^2, 1)$.*

Proof By 7.2.1, the isomorphism class of an even-dimensional A-module is determined by its 2-dimensional submodules, so we may assume that $n = 1$. Let α be the algebra automorphism of A induced by the 3-automorphism of V_4 given by the 3-cycle $s \mapsto t \mapsto st \mapsto s$. Let $(\lambda, \mu) \in k^2$ and let $\{a, b\}$ be a basis of $V = V_2(\lambda, \mu)$ such that s, t act as $sa = a + \lambda b$, $ta = a + \mu b$ and as identity on b. Then $_\alpha V$ is equal to V as a k-vector space, but with s, t acting by $sa = a + \mu b$, $ta = (\lambda + \mu)b$. Thus $V \cong {}_\alpha V$ if and only if there is a nonzero scalar $\gamma \in k$ such that $(\lambda, \mu, \lambda + \mu) = \gamma(\mu, \lambda + \mu, \lambda)$. Both λ, μ are nonzero in that case because if one is zero so is the other, which would imply that V is a copy of two trivial modules. Thus we may assume $\mu = 1$. The equality $(\lambda, 1, \lambda + 1) = \gamma(1, \lambda + 1, \lambda)$ forces $\gamma = \lambda$, hence $\gamma^2 + \gamma = 1$, implying that γ is a cube root of unity in $k^\times$. The result follows. $\square$

One can classify from here all indecomposable kA_4-modules. Identify V_4 with the Sylow 2-subgroup of A_4. Since V_4 is normal in A_4, the simple kA_4-modules correspond to the simple kC_3-modules, where $C_3 = A_4/V_4$. Thus there are three isomorphism classes of simple kA_4-modules, and they all have dimension 1.

Theorem 7.2.10 *Let* $T_1 = k$, T_2, T_3 *be a set of representatives of the isomorphism classes of simple kA_4-modules. Let ζ be a cube root of unity in $k^\times$. The following list is a complete set of representatives of the isomorphism classes of finitely generated indecomposable kA_4-modules.*

(i) *The three projective indecomposable kA_4-modules P_1, P_2, P_3 corresponding to T_1, T_2, T_3, respectively. For $i \in \{1, 2, 3\}$ we have $\mathrm{Res}_{V_4}^{A_4}(P_i) \cong kV_4$, and the Loewy series of P_i is determined by*

$$\mathrm{rad}(P_i)/\mathrm{soc}(P_i) \cong T_j \oplus T_l$$

where $\{i, j, l\} = \{1, 2, 3\}$. This decomposition is unique, and P_i has exactly two quotients and two submodules of dimension 2.

(ii) *The Heller translates $\Omega^n(T_i)$ of the simple kA_4-modules T_i, with $i \in \{1, 2, 3\}$ and $n \in \mathbb{Z}$. They satisfy $\mathrm{Res}_{V_4}^{A_4}(\Omega^n(T_i)) \cong \Omega^n(k)$ for all i, n as before.*

(iii) *The induced modules $\mathrm{Ind}_{V_4}^{A_4}(V_{2n}(\lambda, \mu))$, with $(\lambda, \mu) \in \mathbb{P}^1(k)$ different from $(\zeta, 1)$, $(1, \zeta)$ viewed as elements in $\mathbb{P}^1(k)$. These modules are periodic of period 1.*

(v) *Suppose that (λ, μ) is equal to either $(\zeta, 1)$ or $(1, \zeta)$. Then the kV_4-module $V_{2n}(\lambda, \mu)$ extends, up to isomorphism, in exactly three ways to an indecomposable kA_4-module. These modules are periodic of period 3.*

Proof Since $|A_4 : V_3| = 3$ is odd, it follows that every finitely generated indecomposable kA_4-module arises as a direct summand of $\mathrm{Ind}_{V_4}^{A_4}(V)$ for some indecomposable kV_4-module. If V is A_4-stable, we will see that $\mathrm{Ind}_{V_4}^{A_4}(V)$ is a direct sum of three pairwise nonisomorphic kA_4 modules corresponding to the three possible extensions of V. If V is not A_4-stable, then $\mathrm{Ind}_{V_4}^{A_4}(V)$ is indecomposable by 5.12.1. The three simple kC_3-modules give rise to three primitive idempotents e_i in kC_3 which remain primitive in kA_4, such that $P_i = kA_4e_i$ is a projective cover of T_i. Since $\dim_k(kA_4) = 12$, it follows that each P_i has dimension 4, hence remains indecomposable projective upon restriction to kV_4. Thus $\mathrm{rad}(P_1)/\mathrm{soc}(P_1)$ is a sum of two simple kA_4-modules, not both T_1 because kA_4 is an indecomposable algebra. Thus $\mathrm{rad}(P_1)/\mathrm{soc}(P_1)$ has a summand T_2 or T_3. Since P_1 is stable under field automorphisms but T_2 and T_3 are exchanged under sending ζ to ζ^2, it follows that $\mathrm{rad}(P_1)/\mathrm{soc}(P_1) = T_2 \oplus T_3$. Tensoring with T_i

implies the analogous statement for P_2, P_3. Since $\text{rad}(P_i)/\text{soc}(P_i)$ is a direct sum of two nonisomorphic simple modules, this direct sum is unique, whence (i). Statement (ii) follows from (i). If (λ, μ) is as in (iii), then $V_{2n}(\lambda, \mu)$ is not A_4-stable. By the initial remarks, this implies that the modules $\text{Ind}_{V_4}^{A_4}(V_{2n}(\lambda, \mu))$ are indecomposable. These modules have period 1 because this is true for the $V_{2n}(\lambda, \mu)$, by 7.2.1 (iii). Statement (iv) is an easy verification; it suffices to construct one extension to A_4 and then tensor with the T_i. Thus the above list is complete. These modules are determined by their 2-dimensional indecomposable submodules, and so they have the same period as the 2-dimensional indecomposable kA_4-modules. See the proof of 7.2.11 for the statement on the period. $\quad\square$

Corollary 7.2.11 *There are six isomorphism classes of indecomposable kA_4-modules of dimension 2. More precisely, with the notation above, if $\{i, j, l\} = \{1, 2, 3\}$ then there is a unique indecomposable kA_4-module T_j^i, up to isomorphism, with composition series T_i, T_j. Moreover, we have $\Omega(T_j^i) \cong T_i^l$. In particular, T_j^i has period 3.*

Proof The 2-dimensional indecomposable kA_4-modules correspond to the three extensions of each of $V_2(\zeta, 1)$, $V_2(1, \zeta)$, where ζ is a cube root of unity in $k^\times$. Alternatively, these modules arise as the two-dimensional quotients or also 2-dimensional submodules of the projective indecomposable kA_4-modules P_i. The isomorphism $\Omega(T_j^i) \cong T_i^l$ follows immediately from the Loewy series of the projective cover P_i of T_i. $\quad\square$

The list of finitely generated indecomposable kA_4-modules with a one-dimensional stable endomorphism ring is as follows.

Theorem 7.2.12 *Let V be a finitely generated indecomposable kA_4-module. Then $\underline{\text{End}}_{kA_4}(V) \cong k$ if and only if either $V \cong \Omega^n(T)$ for some integer n and a simple kA_4-module T, or if $\dim_k(V) = 2$.*

Proof The Heller translates of the simple kA_4-modules have one-dimensional stable endomorphism algebras because the Heller operator is an equivalence on the stable category. If $\dim_k(V) = 2$, then V has two nonisomorphic composition factors, and hence $\text{End}_{kA_4}(V) \cong k$, implying that also $\underline{\text{End}}_{kA_4}(V) \cong k$. If $V \cong \text{Ind}_{V_4}^{A_4}(W)$ for some finitely generated indecomposable kV_4-module W, then W is not A_4-stable. Mackey's formula implies that $\text{Res}_{V_4}^{A_4}(V) \cong W_1 \oplus W_2 \oplus W_3$, where $W_1 \cong W$ and W_2, W_3 are obtained from twisting W with the two 3-automorphisms of V_4. Using the version of Frobenius' reciprocity for stable categories from 2.15.4 we get that $\underline{\text{End}}_{kA_4}(V) \cong \underline{\text{Hom}}_{kV_4}(W, \text{Res}_{V_4}^{A_4}(V)) \cong \oplus_{i=1}^3 \underline{\text{Hom}}_{kV_4}(W, W_i)$. This space has dimension at least 2 because $\text{Hom}_{kV_4}(W, W)$ has dimension greater than one, by 7.2.8. If V has dimension $2n$ with $n \geq 2$, such that $\text{Res}_{V_4}^{A_4}(V)$ remains indecomposable, we follow the reasoning as for kV_4 in the proof of 7.2.8. In that case, V has a submodule V' of dimension $2(n-1)$ that is also a quotient of V by a 2-dimensional submodule, and the map $V \to V' \subseteq V$ yields an endomorphism whose image in $\underline{\text{End}}_{kA_4}(V)$ is nonzero and linearly independent of Id_V, because no nontrivial linear combination of these two maps has image contained in $\text{rad}(V)$. Thus $\underline{\text{End}}_{kA_4}(V)$ has dimension at least 2. $\qquad\square$

Remark 7.2.13 Bondarenko and Drozd showed that kP has tame representation type if and only if $p = 2$ and P is generalised dihedral, semi-dihedral, or quaternion. In particular, all noncyclic finite p-group algebras, for p odd, have *wild representation type*, and this puts these algebras firmly beyond the possibility of classifying their modules. The trichotomy theorem, due to Drozd, states that a finite-dimensional algebra over an algebraically closed field has either finite, tame, or wild representation type.

7.3 Endopermutation modules

Endopermutation modules, introduced by Dade in the late 1970s, are omnipresent in block theory, where they arise as sources of simple modules of nilpotent blocks (described in §8.11) and as sources of bimodules inducing Morita equivalences or stable equivalences of Morita type between blocks (see Chapter 9).

From 7.3.6 onwards, we will assume in this section that k is perfect, since we will be making use of some results lifting group actions on matrix algebras, such as 5.3.1, which require this hypothesis.

Given a finite group P and an $\mathcal{O}P$-module V, we consider the dual $V^* = \text{Hom}_{\mathcal{O}}(V, \mathcal{O})$ as a left $\mathcal{O}P$-module via $(y \cdot \mu)(v) = \mu(y^{-1}v)$, where $y \in P$, $v \in V$ and $\mu \in V^*$. Given two $\mathcal{O}P$-modules V, W, we consider the tensor product $V \otimes_{\mathcal{O}} W$ as an $\mathcal{O}P$-module, with $y \in P$ acting diagonally by $y \cdot (v \otimes w) = yv \otimes yw$. Unless indicated otherwise, the terminology and the results of this and the following section are due to Dade [55].

Definition 7.3.1 Let P be a finite p-group. An *endopermutation $\mathcal{O}P$-module* is a finitely generated $\mathcal{O}$-free $\mathcal{O}P$-module V with the property that $V \otimes_{\mathcal{O}} V^*$ is a permutation $\mathcal{O}P$-module.

This definition depends not just on $\mathcal{O}P$ as an algebra but also on the image of P in $\mathcal{O}P$. This is because the tensor product depends on the comultiplication

$\Delta : \mathcal{O}P \to \mathcal{O}P \otimes_{\mathcal{O}} \mathcal{O}P$, which in turn depends on the choice of a group basis. Since $V \otimes_{\mathcal{O}} V^* \cong \mathrm{End}_{\mathcal{O}}(V)$, by 2.9.4, the module V is an endopermutation $\mathcal{O}P$-module if and only if the algebra $\mathrm{End}_{\mathcal{O}}(V)$ is a permutation P-algebra. Clearly any permutation $\mathcal{O}P$-module is an endopermutation $\mathcal{O}P$-module. The relevance of endopermutation modules in block theory is illustrated by the following example.

Example 7.3.2 Suppose that k is algebraically closed. Let N be a normal p'-subgroup of a finite group G and let c be a block of $\mathcal{O}N$. Let P be a p-subgroup of G that stabilises c. Since N is a p'-group, it follows that c has defect 0 and that the rank of the matrix algebra $\mathcal{O}Nc$ is prime to p. By 5.3.4 the action of P on $\mathcal{O}Nc$ lifts uniquely to a group homomorphism $\sigma : P \to (\mathcal{O}Nc)^{\times}$ such that $\det(\sigma(u)) = 1$ for all $u \in P$, hence $\mathcal{O}Nc \cong \mathrm{End}_{\mathcal{O}}(V)$ for some $\mathcal{O}P$-module V. Since $\mathcal{O}Nc$ is a direct summand of the permutation $\mathcal{O}P$-module $\mathcal{O}N$, it follows from 5.11.2 that $\mathcal{O}Nc$ is a permutation $\mathcal{O}P$-module, and hence that V is an endopermutation $\mathcal{O}P$-module. Using the classification of finite simple groups one can show that not every endopermutation $\mathcal{O}P$-module arises in this way; see Section 7.10 for more details.

The following result is a very special case of much more general results. It illustrates some earlier results and will be a useful tool for inductive arguments.

Proposition 7.3.3 *Let P be a cyclic group of order p and V an indecomposable endopermutation kP-module. Then V is isomorphic to the trivial kP-module k or its Heller translate $\Omega(k)$.*

Proof Up to isomorphism, any indecomposable permutation kP-module is either k or kP. Since indecomposable kP-modules are absolutely indecomposable, it follows from 5.1.13 (iii) that $V \otimes_k V^*$ has a unique summand k and that every other summand is isomorphic to kP. In particular, $\dim_k(V) \equiv \pm 1 \pmod{p}$, hence $\dim_k(V)$ is either 1 or $p - 1$, or equivalently, V is isomorphic to k or $\Omega(k)$. Both k and $\Omega(k)$ are endopermutation kP-modules; for k this is obvious, and for $\Omega(k)$ this follows from the isomorphism $\Omega(k) \otimes_k \Omega(k)^* \cong k \oplus Y$ for some projective kP-module Y. One can prove this also without using 5.1.13. After possibly replacing V by $\Omega(V)$, we may assume that $\dim_k(V) = i \leq \frac{p}{2}$. By 7.1.7 we have $\underline{\mathrm{End}}_{kP}(V) \cong \mathrm{End}_{kP}(V)$, which has dimension i. Thus $V \otimes_k V^* \cong k^i \oplus (kP)^j$ for some $j \geq 0$. Taking dimensions yields $i^2 = i + pj$. In particular, $i^2 - i = i(i - 1)$ is divisible by p, which forces $i = 1$. $\quad\square$

The following proposition, besides listing basic properties of endopermutation modules, contains examples of endopermutation modules that are not permutation modules.

Proposition 7.3.4 *Let P be a finite p-group and let V, W be endopermutation $\mathcal{O}P$-modules.*

(i) *For any subgroup Q of P, the module $\mathrm{Res}_Q^P(V)$ is an endopermutation $\mathcal{O}Q$-module.*

(ii) *Any direct summand of V is an endopermutation $\mathcal{O}P$-module.*

(iii) *The $\mathcal{O}$-dual V^* of V is an endopermutation $\mathcal{O}P$-module.*

(iv) *The tensor product $V \otimes_{\mathcal{O}} W$ is an endopermutation $\mathcal{O}P$-module.*

(v) *For any integer n the module $\Omega^n(V)$ is an endopermutation $\mathcal{O}P$-module.*

(vi) *The kP-module $k \otimes_{\mathcal{O}} V$ is an endopermutation kP-module.*

(vii) *The $\mathcal{O}P$-module V is indecomposable if and only if the kP-module $k \otimes_{\mathcal{O}} V$ is indecomposable, and in that case, V and $k \otimes_{\mathcal{O}} V$ have the same vertices.*

Proof The statements (i) and (ii) are trivial. For (iii) we use the compatibility of the tensor product with duality from 2.9.10 together with the fact that the dual of a permutation $\mathcal{O}P$-module is again a permutation $\mathcal{O}P$-module. For (iv) we use the obvious isomorphism $(V \otimes_{\mathcal{O}} W) \otimes_{\mathcal{O}} (V \otimes_{\mathcal{O}} W)^* \cong V \otimes_{\mathcal{O}} V^* \otimes_{\mathcal{O}} W \otimes_{\mathcal{O}} W^*$, which is a permutation $\mathcal{O}P$-module. In order to show (v), consider an exact sequence of the form

$$0 \longrightarrow \Omega(V) \longrightarrow U \longrightarrow V \longrightarrow 0$$

where U is a finitely generated projective $\mathcal{O}P$-module. All involved modules in this sequence are $\mathcal{O}$-free, and hence this sequence is split as sequence of $\mathcal{O}$-modules. Thus applying the functors $\mathrm{Hom}_{\mathcal{O}}(-, V)$ and $\mathrm{Hom}_{\mathcal{O}}(\Omega(V), -)$ yields two short exact sequence of $\mathcal{O}P$-modules

$$0 \longrightarrow \mathrm{Hom}_{\mathcal{O}}(V, V) \longrightarrow \mathrm{Hom}_{\mathcal{O}}(U, V) \longrightarrow \mathrm{Hom}_{\mathcal{O}}(\Omega(V), V) \longrightarrow 0$$

$$0 \longrightarrow \mathrm{Hom}_{\mathcal{O}}(\Omega(V), \Omega(V)) \longrightarrow \mathrm{Hom}_{\mathcal{O}}(\Omega(V), U) \longrightarrow \mathrm{Hom}_{\mathcal{O}}(\Omega(V), V) \longrightarrow 0.$$

The terms in the middle of this sequence are projective by 5.1.12. The right terms coincide. Thus, by Schanuel's Lemma 1.12.5, there are projective $\mathcal{O}P$-modules Y, Y' such that

$$\mathrm{End}_{\mathcal{O}}(V) \oplus Y \cong \mathrm{End}_{\mathcal{O}}(\Omega(V)) \oplus Y'$$

which shows that V is an endopermutation module if and only if $\Omega(V)$ is so. Iterating this argument shows (v). Set $\bar{V} = k \otimes_{\mathcal{O}} V$. The image in $\bar{V} \otimes_k \bar{V}^*$ of a P-stable basis of $V \otimes_{\mathcal{O}} V^*$ is clearly a P-stable basis, hence $\bar{V}$ is an endopermutation kP-module. If $\bar{V}$ is indecomposable then so is V. Conversely, suppose that V is indecomposable. The isomorphism $\mathrm{End}_{\mathcal{O}}(V) \cong V \otimes_{\mathcal{O}} V^*$ maps $\mathrm{End}_{\mathcal{O}P}(U)$

to the P-fixed points $(V \otimes V^*)P$. If X is a P-stable basis of $V \otimes_{\mathcal{O}} V^*$, then any P-fixed point is a linear combination of the P-orbit sums of X in $V \otimes_{\mathcal{O}} V^*$. Since the analogous statement holds for the image of X in $\bar{V}$, it follows that $k \otimes_{\mathcal{O}} V$ is an endopermutation kP-module, and that the map from $(V \otimes_{\mathcal{O}} V^*)^P$ to $(\bar{V} \otimes_k \bar{V}^*)^P$ is surjective. Thus $\text{End}_{kP}(\bar{V})$ is a quotient of the local algebra $\text{End}_{\mathcal{O}P}(U)$, hence itself local, whence the indecomposabilitiy of $\bar{V}$. For Q a subgroup of P, we have $\text{End}_{\mathcal{O}}(V)(Q) \cong \text{End}_k(k \otimes_{\mathcal{O}} V)(Q)$, because both sides have a basis indexed by the Q-fixed points in X, which implies that V and $k \otimes_{\mathcal{O}} V$ have the same vertices if they are indecomposable. $\qquad\square$

Thus taking restrictions, direct summands, tensor products and Heller translates of endopermutation modules yields again endopermutation modules. There is another trivial construction principle: the inflation functor. For a normal subgroup Q of a finite group P and an $\mathcal{O}P/Q$-module U, we denote by $\text{Inf}^P_{P/Q}(U)$ the $\mathcal{O}P$-module obtained from restricting U along the canonical surjection $P \to P/Q$. That is, as $\mathcal{O}$-module, $\text{Inf}^P_{P/Q}(U)$ is equal to U, with $y \in P$ acting on $u \in U$ as its image yQ in P/Q. We define $\text{Inf}^P_{P/Q}$ in the obvious way on homomorphism; in this way, $\text{Inf}^P_{P/Q}$ becomes an exact functor from $\text{Mod}(\mathcal{O}P/Q)$ to $\text{Mod}(\mathcal{O}P)$.

Proposition 7.3.5 *Let P be a finite p-group and Q a normal subgroup of P. If U is an endopermutation $\mathcal{O}P/Q$-module then $\text{Inf}^P_{P/Q}(U)$ is an endopermutation $\mathcal{O}P$-module.*

Proof This is a consequence of the obvious identity $\text{Inf}^P_{P/Q}(U) \otimes_{\mathcal{O}} \text{Inf}^P_{P/Q}(U^*) = \text{Inf}^P_{P/Q}(U \otimes_{\mathcal{O}} U^*)$. $\qquad\square$

The two propositions above show that taking summands of tensor products of Heller translates of endopermutation modules inflated from normal subgroups. It is a consequence of the classification of endopermutation modules mentioned in the introduction to this chapter – that (with an exception at the prime 2) essentially all endopermutation modules can be constructed in this way, starting with modules of rank one. For the remainder of this section, the field k is assumed to be perfect.

Proposition 7.3.6 *Let P be a finite p-group acting on a matrix algebra $S \cong M_n(\mathcal{O})$ for some positive integer n such that S has a P-stable $\mathcal{O}$-basis. If $S(P)$ is nonzero, then $S(P)$ is a matrix algebra over k.*

Proof We may assume $\mathcal{O} = k$ and that $S(P) \neq \{0\}$. If Z is a central subgroup of order p of P, then $S(Z)$ has a P/Z-stable basis by 5.8.1, namely the image in $S(Z)$ of the Z-fixed points of a P-stable basis of S. Moreover, we have

$S(Z)(P/Z) \cong S(P)$, by 5.8.5. Thus, arguing by induction, we may assume that P is cyclic of order p. Write $S = \mathrm{End}_k(V)$ for some k-vector space V of dimension n. By 5.3.1 there is a group homomorphism $P \to S^\times$ lifting the action of P on S, or equivalently, V has a kP-module structure inducing the action of P on S. Thus V is an endopermutation kP-module having an indecomposable direct summand W of vertex P. Since S is a matrix algebra, left and right multiplication by S on itself yields an isomorphism $\mathrm{End}_k(S) \cong S \otimes_k S^{\mathrm{op}}$. This is clearly an isomorphism as kP-modules. Using that S is a permutation kP-module, we get from 5.8.10 that $(\mathrm{End}_k(S))(P) \cong S(P) \otimes_k S(P)^{\mathrm{op}}$. But we also get from 5.8.6 that $(\mathrm{End}_k(S))(P) \cong \mathrm{End}_k(S(P))$. Thus $S(P)$ is central simple by 1.16.5. Since we know already that $S(P)$ is central simple, it suffices to show that $S(P)$ is split. This is equivalent to showing that $(iSi)(P)$ is split for any primitive idempotent i in S^P whose image in $S(P)$ is nonzero. Thus we may assume that V is indecomposable with vertex P. But then, by 7.3.3, we have $V \cong k$ or $V \cong \Omega(k)$. In the first case we have $S \cong S(P) \cong k$. In the second case, S is isomorphic to $k \oplus Y$ for some projective kP-module Y, which also forces $S(P) \cong k$. $\square$

Proposition 7.3.7 *Let P be a finite p-group and V an endopermutation $\mathcal{O}P$-module. Set $S = \mathrm{End}_\mathcal{O}(V)$ and suppose that $S(P) \neq \{0\}$. Then, for any subgroup Q of P, the following hold.*

(i) *There is up to isomorphism a unique endopermutation $kN_P(Q)/Q$-module W_Q such that $S(Q) \cong \mathrm{End}_k(W_Q)$ as $N_P(Q)/Q$-algebras, and W_Q has an indecomposable direct summand with vertex $N_P(Q)/Q$. In particular, $S(Q)$ is a matrix algebra over k.*

(ii) *There is a unique local point δ of Q on S, and its multiplicity m_δ satisfies $m_\delta^2 = \dim_k(S(Q))$.*

(iii) *The endopermutation $\mathcal{O}Q$-module $\mathrm{Res}_Q^P(V)$ has an indecomposable direct summand with vertex Q, and any two indecomposable direct summands of $\mathrm{Res}_Q^P(V)$ with Q as a vertex are isomorphic.*

(iv) *Denote by V_Q an indecomposable direct summand of $\mathrm{Res}_Q^P(V)$ with vertex Q. Then any direct summand of V with vertex Q is isomorphic to $\mathrm{Ind}_Q^P(V_Q)$.*

(v) *If V has a trivial direct summand, then V is a permutation $\mathcal{O}P$-module.*

Proof By 7.3.6, the algebra $S(Q)$ is a matrix algebra. By 5.3.1 the $N_P(Q)/Q$-structure on $S(Q)$ lifts uniquely to an interior $N_P(Q)/Q$-structure. Hence, if we write $S(Q) = \mathrm{End}_k(W_Q)$ for some k-vector space W_Q, then W_Q inherits a unique structure of $kN_P(Q)/Q$-module. Since $S(Q)$ is a permutation $N_P(Q)/Q$-algebra, W_Q is an endopermutation $kN_P(Q)/Q$-module. Moreover, since $S(P)$ is nonzero, so is $S(N_P(Q)) \cong S(Q)(N_P(Q)/Q)$, which shows that W_Q has an indecomposable direct summand with vertex $N_P(Q)/Q$. This shows (i). Statement (ii) is just a reformulation of the fact that $S(Q)$ is simple, hence has a unique point, which lifts to a unique local point δ with the multiplicity as stated. Statement (iii) is a consequence of (ii), since any local point of Q on $S = \text{End}_{\mathcal{O}}(V)$ corresponds to an isomorphism class of indecomposable direct summands of $\text{Res}_Q^P(V)$ with vertex Q. Statement (iv) follows from the uniqueness of V_Q, up to isomorphism, by (iii). Statement (v) is a special case of (iv). $\qquad\square$

Definition 7.3.8 Let P be a finite p group and let Q, R be subgroups of P such that Q is normal in R. Let V be an endopermutation $\mathcal{O}P$-module having an indecomposable direct summand with vertex P. We denote by $\text{Defres}_{R/Q}^P(V)$ the endopermutation kR/Q-module $\text{Res}_{R/Q}^{N_P(Q)/Q}(W_Q)$, where W_Q is an endopermutation $kN_P(Q)/Q$-module satisfying $S(Q) \cong \text{End}_k(W_Q)$.

Proposition 7.3.9 *Let P be a finite p-group and let V be an endopermutation $\mathcal{O}P$-module having an indecomposable direct summand with vertex P. Let Q be a normal subgroup of P, and set $W = \text{Defres}_{P/Q}^P(V)$. We have a canonical surjective algebra homomorphism $\text{End}_{\mathcal{O}P}(V) \to \text{End}_{kP/Q}(W)$; in particular, if V is indecomposable as an $\mathcal{O}P$-module, then W is indecomposable as a kP/Q-module.*

Proof By 5.8.5, the Brauer homomorphism $\text{Br}_Q : \text{End}_{\mathcal{O}Q}(V) \to \text{End}_{\mathcal{O}}(V)(Q)$ composed with the isomorphism $\text{End}_{\mathcal{O}}(V)(Q) \cong \text{End}_k(W)$ from 7.3.7 maps $\text{End}_{\mathcal{O}P}(V)$ onto $\text{End}_{kP}(W) = \text{End}_{kP/Q}(W)$. If V is indecomposable, then $\text{End}_{\mathcal{O}P}(V)$ is local, hence so is the quotient algebra $\text{End}_{kP/Q}(W)$, implying that W is indecomposable. $\qquad\square$

Besides restriction to subgroups, inflation from quotients by normal subgroups, and deflation to sections, there is one further operation on endopermutation modules: tensor induction from subgroups. Ordinary induction from subgroups, however, need not preserve endopermutation modules. This follows from Proposition 7.3.7 (iii), which implies that the direct sum of two endopermutation modules need not be endopermutation. Indeed, the direct sum $V \oplus W$ of two indecomposable endopermutation $\mathcal{O}P$-modules V, W with vertex P is an endopermutation $\mathcal{O}P$-module if and only if V and W are isomorphic. In general, two endopermutation $\mathcal{O}P$-modules are called *compatible* if their direct sum is an endopermutation module.

Proposition 7.3.10 *Let P be a finite p-group and V be an indecomposable endopermutation $\mathcal{O}P$-module with vertex P. Set $S = \text{End}_{\mathcal{O}}(V)$. Then V is absolutely indecomposable, and the following hold.*

 (i) *We have $S(P) \cong k$ and $\mathrm{rk}_{\mathcal{O}}(V) \equiv \pm 1 \pmod{p}$.*
 (ii) *We have $V \otimes_{\mathcal{O}} V^* \cong \mathcal{O} \oplus X$, where every indecomposable summand of X is of the form $\mathrm{Ind}_Q^P(\mathcal{O})$, for some proper subgroup Q of P.*
(iii) *For any finitely generated $\mathcal{O}$-free indecomposable $\mathcal{O}P$-module U with vertex P there is a unique indecomposable direct summand with vertex P in any decomposition of $V \otimes_{\mathcal{O}} U$ as direct sum of indecomposable $\mathcal{O}P$-modules.*
(iv) *We have $S \otimes_{\mathcal{O}} S^{\mathrm{op}} \cong \mathrm{End}_{\mathcal{O}}(\mathcal{O} \oplus X)$ as interior P-algebras. In particular, P has a unique local point on $S \otimes_{\mathcal{O}} S^{\mathrm{op}}$, and if e belongs to this local point, then $e(S \otimes_{\mathcal{O}} S^{\mathrm{op}})e \cong \mathcal{O}$.*

Proof Since V is indecomposable with vertex P, the algebra $S(P)$ is nonzero local. But $S(P)$ is also a matrix algebra by 7.3.6, and hence $S(P) \cong k$. Since $S(P)$ is a quotient of $S^P = \mathrm{End}_{\mathcal{O}P}(V)$, it follows that V is absolutely indecomposable. By 5.8.1 we have $\mathrm{rk}_{\mathcal{O}}(S) \equiv 1 \pmod{p}$. Since $\mathrm{rk}_{\mathcal{O}}(S) = (\mathrm{rk}_{\mathcal{O}}(V))^2$, statement (i) follows, and (ii) is just a reformulation of the isomorphism $S(P) \cong k$. Using 5.1.11, it follows from (ii) that $V \otimes_{\mathcal{O}} V^* \otimes_{\mathcal{O}} U$ has a summand isomorphic to U, and that every other indecomposable summand has a smaller vertex. Thus $V \otimes_{\mathcal{O}} U$ has at least one indecomposable summand with vertex P. It cannot have more than one, because the same argument applied with V^* would yield at least two summands with vertex P in a decomposition of $V \otimes_{\mathcal{O}} V^* \otimes_{\mathcal{O}} U$. This shows (iii). We have

$$S \otimes_{\mathcal{O}} S^{\mathrm{op}} \cong \mathrm{End}_{\mathcal{O}}(V) \otimes_{\mathcal{O}} \mathrm{End}_{\mathcal{O}}(V^*) \cong \mathrm{End}_{\mathcal{O}}(V \otimes_{\mathcal{O}} V^*) \cong \mathrm{End}_{\mathcal{O}}(\mathcal{O} \oplus X),$$

where the last isomorphism is from (ii). The projection π of $\mathcal{O} \oplus X$ onto $\mathcal{O}$ with kernel X belongs to the unique local point of P on $\mathrm{End}_{\mathcal{O}}(\mathcal{O} \oplus X)$ because any indecomposable direct summand of X has a vertex strictly contained in P. Thus $\pi \circ \mathrm{End}_{\mathcal{O}}(\mathcal{O} \oplus X) \circ \pi \cong \mathcal{O}$. The result follows. $\square$

The tensor product induces an abelian group structure on the set of isomorphism classes of indecomposable $\mathcal{O}P$-modules with vertex P; the unit element is represented by the class of the trivial $\mathcal{O}P$-module and the inverse of the class of V is represented by V^*, by 7.3.10 (ii).

Definition 7.3.11 Let P be a finite p-group. The *Dade group* $D_{\mathcal{O}}(P)$ *of P over* $\mathcal{O}$ is the group of isomorphism classes $[V]$ of indecomposable endopermutation $\mathcal{O}P$-modules V with P as vertex, with product given by

$$[U] \cdot [V] = [W]$$

if U, V, W are three indecomposable endopermutation $\mathcal{O}P$-modules with vertex P such that W is isomorphic to a direct summand of $U \otimes_{\mathcal{O}} V$.

The group $D_\mathcal{O}(P)$ is called *cap group* in [55]. As mentioned before, $D_\mathcal{O}(P)$ is indeed a group, with unit element $[\mathcal{O}]$ and inverses given by $[V]^{-1} = [V^*]$. The class $[\Omega(\mathcal{O})]$ generates a cyclic subgroup in $D_\mathcal{O}(P)$; more precisely, we have $[\Omega(\mathcal{O})]^n = [\Omega^n(\mathcal{O})]$ for any integer n. This subgroup is finite if and only if $\mathcal{O}$ is Ω-periodic, which is known to happen precisely if P is cyclic or if $p = 2$ and P is generalised quaternion. We have $[\Omega(\mathcal{O})] \cdot [V] = [\Omega(V)]$ in $D_\mathcal{O}(P)$. We consider next the canonical map $D_\mathcal{O}(P) \to D_k(P)$.

Proposition 7.3.12 *Let P be a finite p-group. If V, W are endopermutation $\mathcal{O}P$-modules each having at least one indecomposable direct summand with vertex P such that $k \otimes_\mathcal{O} V \cong k \otimes_\mathcal{O} W$ then $V \cong T \otimes_\mathcal{O} W$ for some $\mathcal{O}$-free $\mathcal{O}P$-module T of $\mathcal{O}$-rank 1, which is unique up to isomorphism. In particular, the kernel of the canonical map $D_\mathcal{O}(P) \to D_k(P)$ is equal to $\mathrm{Hom}(P, \mathcal{O}^\times)$.*

Proof Suppose first that V, W are indecomposable with vertex P. Let T be the up to isomorphism unique indecomposable direct summand of the endopermutation $\mathcal{O}P$-module $V \otimes_\mathcal{O} W^*$ having P as a vertex. In terms of the Dade group, this translates to $[T] = [V] \cdot [W]^{-1}$, which implies the uniqueness of T as stated. By 7.3.4, the kP-module $\bar{T} = k \otimes_\mathcal{O} T$ is indecomposable with vertex P. Since $k \otimes_\mathcal{O} V \cong k \otimes_\mathcal{O} W$ it follows that $k \otimes_\mathcal{O} V \otimes_\mathcal{O} W^*$ has a trivial summand k. Thus $\bar{T} \cong k$, and hence T has $\mathcal{O}$-rank 1. Tensoring by W implies that $V \otimes_\mathcal{O} W^* \otimes_\mathcal{O} W$ and $T \otimes_\mathcal{O} W$ have isomorphic summands with vertex P. Note that $T \otimes_\mathcal{O} W$ is indecomposable, as W is so. Since $\mathcal{O}$ is a summand of $W^* \otimes_\mathcal{O} W$, it follows that V is a summand of $V \otimes_\mathcal{O} W^* \otimes_\mathcal{O} W$, and hence $V \cong T \otimes_\mathcal{O} W$. This shows also that the kernel of the canonical map $D_\mathcal{O}(P) \to D_k(P)$ is $\mathrm{Hom}(P, \mathcal{O}^\times)$.

For the case where V, W are not necessarily indecomposable, let V_P, W_P be indecomposable direct summands of V, W, respectively, with vertex P. By the first argument, we have $V_P \cong T \otimes_\mathcal{O} W_P$ for a uniquely determined $\mathcal{O}P$-module of rank 1. For any subgroup Q, $\mathrm{Res}_Q^P(V_P)$ has, up to isomorphism a unique direct summand V_Q with vertex Q, hence $V_Q \cong \mathrm{Res}_Q^P(T) \otimes_\mathcal{O} W_Q$. By 7.3.7 (iv) any direct summand of V with vertex Q is isomorphic to $\mathrm{Ind}_Q^P(V_Q) \cong T \otimes_\mathcal{O} \mathrm{Ind}_Q^P(W_Q)$. The result follows. $\square$

The classification of endopermutation modules implies that the map $D_\mathcal{O}(P) \to D_k(P)$ is surjective; that is, all endopermutation kP-modules lift to endopermutation $\mathcal{O}P$-modules. The previous proposition implies that we have a short exact sequence of abelian groups

$$1 \longrightarrow \mathrm{Hom}(P, \mathcal{O}^\times) \longrightarrow D_\mathcal{O}(P) \longrightarrow D_k(P) \longrightarrow 1.$$

See [214, Theorem 14.2]. If p is odd, then this sequence splits canonically, with the section induced by the map which sends an indecomposable endopermutation kP-module $\bar{V}$ to the unique – up to isomorphism – endopermutation $\mathcal{O}P$-module V with determinant 1 satisfying $k \otimes_{\mathcal{O}} V \cong \bar{V}$. For $p = 2$, the lifts with determinant 1 do not yield a group homomorphism from $D_k(P)$ to $D_{\mathcal{O}}(P)$ in general.

Proposition 7.3.13 *Let P be a finite p-group, V an endopermutation $\mathcal{O}P$-module and set $S = \mathrm{End}_{\mathcal{O}}(V)$. Suppose that $\mathcal{O}$ is unramified; that is, $\mathcal{O}$ contains no nontrivial p-power root of unity. Then the following hold.*

  (i) The values of the character χ of V are in $\mathbb{Z}$.
  (ii) For any $u \in P$ we have $\chi(u)^2 = \dim_k(S(\langle u \rangle))$.
  (iii) For any $u \in P$ we have $\chi(u) = \pm m_\delta$, where m_δ is the multiplicity of the unique local point δ of $\langle u \rangle$ on S.

Proof Let $u \in P$. The value $\chi(u)$ depends only on $\mathrm{Res}^P_{\langle u \rangle}(V)$, and hence in order to prove (i), we may assume that $P = \langle u \rangle$ is cyclic. The trace of u is zero on any indecomposable summand of V with a vertex strictly smaller than Q, and hence we may assume that V is indecomposable with vertex P. (This is also equivalent to $m_\delta = 1$, where δ is the unique local point of $\langle u \rangle = P$ on S.) By 5.3.4, there is a unique endopermutation $\mathcal{O}P$-module W with determinant 1 such that $k \otimes_{\mathcal{O}} W \cong k \otimes_{\mathcal{O}} V$, and by 7.3.12 there is a unique $\mathcal{O}P$-module T of $\mathcal{O}$-rank 1 such that $V \cong T \otimes_{\mathcal{O}} W$. Since $\mathcal{O}$ is unramified, if p is odd, then $T \cong \mathcal{O}$ and hence $V \cong W$. If $p = 2$, then either $T \cong \mathcal{O}$, or u acts as -1 on T. By 7.1.1, the isomorphism class of $k \otimes_{\mathcal{O}} W$ is uniquely determined by its dimension, hence invariant under any group automorphism of P. Thus the isomorphism classes of W, T, and hence of V, are invariant under $\mathrm{Aut}(P)$. Since $\mathrm{Aut}(P)$ is transitive on the set of elements that generate P, it follows that the value $\chi(u)$ does not depend on the choice of a generator of P. Therefore, 3.5.4 implies that $\chi(u) \in \mathbb{Z}$. This proves (i). For (ii), note that $\chi(u)^2$ is the character value of u acting diagonally on $V \otimes_{\mathcal{O}} V^* \cong S$ thanks to (i). This is equal to the number of $\langle u \rangle$-fixed points in a $\langle u \rangle$-stable basis of S, hence equal to $\dim_k(S(\langle u \rangle)) = m_\delta^2$, whence (ii) and (iii). $\qquad\square$

With the notation of 7.3.12, if $V \cong T \otimes_{\mathcal{O}} W$, then the algebras $\mathrm{End}_{\mathcal{O}}(V)$ and $\mathrm{End}_{\mathcal{O}}(W)$ are isomorphic as P-algebras, but as interior P-algebras they are only isomorphic if T is trivial. One way to circumvent uniqueness considerations of this nature is to focus on the P-algebra $S = \mathrm{End}_{\mathcal{O}}(V)$ as a matrix algebra with a P-stable basis, rather than the module V itself.

Proposition 7.3.14 *Let P be a finite p-group acting on a matrix algebra $S \cong M_n(\mathcal{O})$ for some positive integer n in such a way that S has a P-stable basis and such that $S(P) \neq \{0\}$. Suppose that k is algebraically closed. There is an endopermutation $\mathcal{O}P$-module V such that $S \cong \mathrm{End}_{\mathcal{O}}(V)$ as P-algebras.*

Proof By 5.3.1, there is a kP-module W such that $k \otimes_{\mathcal{O}} S \cong \mathrm{End}_k(W)$. In order to show that this action lifts to $\mathcal{O}$, we need to show that the class in $H^2(P; \mathcal{O}^\times)$ determined by the action of P on S is trivial. Since S has a P-stable $\mathcal{O}$-basis, it follows that $k \otimes_{\mathcal{O}} S$ has a P-stable k-basis. In particular, W is an endopermutation kP-module. Since $S(P) \neq \{0\}$, it follows that the kP-module W has an indecomposable direct summand with vertex P. We first show that we may assume that W is indecomposable with vertex P. Denote by $j \in (k \otimes_{\mathcal{O}} S)^P$ an idempotent that is a projection of W onto an indecomposable direct summand W' with vertex P. Since S has a P-stable basis, it follows that the canonical map $S^P \to (k \otimes_{\mathcal{O}} S)^P$ is surjective, and hence j lifts to a primitive idempotent i in S^P. By 5.3.5, the class in $H^2(P; \mathcal{O}^\times)$ determined by the actions of P on S and on iSi are equal. Thus we may replace S by iSi, or equivalently, we may assume that S is a primitive P-algebra and that the endopermutation kP-module W is indecomposable with vertex P. By 7.3.10, the dimension of W is prime to p, hence the $\mathcal{O}$-rank of S is prime to p. By 5.3.6 the action of P on S lifts to a group homomorphism $P \to S^\times$, or equivalently, W lifts to an endopermutation $\mathcal{O}P$-module V such that $S \cong \mathrm{End}_{\mathcal{O}}(V)$ as P-algebras. $\square$

Corollary 7.3.15 *Let G be a finite group, P a Sylow p-subgroup, and S a matrix algebra over $\mathcal{O}$ of finite $\mathcal{O}$-rank. Suppose k is algebraically closed and that G acts on S. Denote by α the class in $H^2(G; \mathcal{O}^\times)$ determined by the action of G on S. If S has a P-stable $\mathcal{O}$-basis such that $S(P) \neq \{0\}$, then $\alpha \in H^2(G; k^\times)$, where we identify $k^\times$ with its canonical preimage in $\mathcal{O}^\times$.*

Proof If S has a P-stable $\mathcal{O}$-basis, then by 7.3.14 the action of P on S lifts to a group homomorphism $P \to S^\times$, or equivalently, the class $\mathrm{res}_P^G(\alpha)$ is trivial. It follows from 4.3.10 that $\alpha \in H^2(G; k^\times)$. $\square$

If Q is a subgroup of a finite p-group, then the restriction functor Res_Q^P induces a group homomorphism $D_{\mathcal{O}}(P) \to D_{\mathcal{O}}(Q)$ and the tensor induction functor Ten_Q^P induces a group homomorphism $D_{\mathcal{O}}(Q) \to D_{\mathcal{O}}(P)$. If Q is a normal subgroup of P, then the inflation functor $\mathrm{Inf}_{P/Q}^P$ induces a group homomorphism $D_{\mathcal{O}}(P/Q) \to D_{\mathcal{O}}(P)$. The next result states that $\mathrm{Defres}_{R/Q}^P$ induces a group homomorphism $D_{\mathcal{O}}(P) \to D_k(R/Q)$ for Q a normal subgroup of a subgroup R of P.

Proposition 7.3.16 *Let P be a finite p-group and Q, R subgroups of P such that Q is normal in R. The map sending an indecomposable endopermutation $\mathcal{O}P$-module V with vertex P to the endopermutation kR/Q-module $\mathrm{Defres}_{R/Q}^{P}(V)$ induces a group homomorphism $D_{\mathcal{O}}(P) \to D_k(R/Q)$.*

Proof Since the Brauer construction applied to permutation modules commutes with tensor products, it follows that the given correspondence becomes a group homomorphism on Dade groups. □

We adopt the notational abuse of denoting the group homomorphisms on Dade groups induced by the four operations Res_{Q}^{P}, $\mathrm{Inf}_{P/Q}^{P}$, Ten_{Q}^{P}, $\mathrm{Defres}_{R/Q}^{P}$ by the same symbols. As mentioned earlier, the direct sum of two endopermutation modules need not be an endopermutation module. As a consequence, if U is an indecomposable $\mathcal{O}G$-module with a vertex P and an endopermutation $\mathcal{O}P$-module V as a source, we cannot in general conclude that $\mathrm{Res}_{P}^{G}(U)$ is an endopermutation $\mathcal{O}P$-module, although it is a direct sum of endopermutation modules. Modules whose restriction to a vertex remain endopermutation modules have some useful lifting properties.

Proposition 7.3.17 *Let G be a finite group, U a finitely generated indecomposable $\mathcal{O}G$-module, P a vertex of U, and let V be a source of U. Suppose that $\mathrm{Res}_{P}^{G}(U)$ is an endopermutation $\mathcal{O}P$-module. Then the kG-module $k \otimes_{\mathcal{O}} U$ is indecomposable, with P as a vertex and $k \otimes_{\mathcal{O}} V$ as a source.*

Proof By the assumption, U is free as an $\mathcal{O}$-module and the algebra $\mathrm{End}_{\mathcal{O}}(U)$ has a permutation basis for the conjugation action of P. Thus the canonical map $\mathrm{End}_{\mathcal{O}P}(U) \to \mathrm{End}_{kP}(k \otimes_{\mathcal{O}} U)$ is surjective. Since P is a vertex of U, Higman's criterion implies that $\mathrm{End}_{\mathcal{O}G}(U) = (\mathrm{End}_{\mathcal{O}}(U))_{P}^{G}$. Similarly, since $k \otimes_{\mathcal{O}} U$ is relatively P-projective, we have $\mathrm{End}_{kG}(k \otimes_{\mathcal{O}} U) = (\mathrm{End}_{k}(k \otimes_{\mathcal{O}} U))_{P}^{G}$. It follows that the canonical map $\mathrm{End}_{\mathcal{O}G}(U) \to \mathrm{End}_{kG}(k \otimes_{\mathcal{O}} U)$ is still surjective, and hence $k \otimes_{\mathcal{O}} U$ is indecomposable with a vertex contained in P. By 7.3.4, $k \otimes_{\mathcal{O}} V$ has P as a vertex. The result follows. □

Modules whose restriction to any p-subgroup is an endopermutation module have been investigated more systematically in work of Urfer [220]. The next statement is a bimodule version of 7.3.10 (iii).

Proposition 7.3.18 *Let G be a finite group, P a p-subgroup of G, V an indecomposable endopermutation $\mathcal{O}P$-module with vertex P, and let Y be a finitely generated indecomposable $\mathcal{O}P$-$\mathcal{O}G$-bimodule that has vertex $\Delta P = \{(u, u) | u \in P\}$ when considered as an $\mathcal{O}(P \times G)$-module. Then the $\mathcal{O}(P \times G)$-module $V \otimes_{\mathcal{O}} Y$, with P acting diagonally on the left and G acting on the right on Y, has exactly one indecomposable direct summand M with vertex ΔP, and*

*any other indecomposable direct summand has a vertex of order smaller than
ΔP. Moreover, a ΔP-source of M is isomorphic to the unique indecomposable
direct summand with vertex ΔP of $V \otimes_{\mathcal{O}} W$.*

Proof Since ΔP is a vertex of Y, there is a finitely generated indecomposable $\mathcal{O}\Delta P$-module W with vertex ΔP such that Y is a direct summand of $\mathrm{Ind}_{\Delta P}^{P \times G}(W)$. Thus $V \otimes_{\mathcal{O}} Y$ is a direct summand of $V \otimes_{\mathcal{O}} \mathrm{Ind}_{\Delta P}^{P \times G}(W) \cong V \otimes_{\mathcal{O}} \mathrm{Ind}_{\Delta P}^{P \times P}(W) \otimes_{\mathcal{O}P} \mathcal{O}G$. Using 2.4.12 and 2.4.13, this module is isomorphic to $\mathrm{Ind}_{\Delta P}^{P \times P}(V \otimes_{\mathcal{O}} W) \otimes_{\mathcal{O}P} \mathcal{O}G \cong \mathrm{Ind}_{\Delta P}^{P \times G}(V \otimes_{\mathcal{O}} W)$. In particular, every indecomposable direct summand of $V \otimes_{\mathcal{O}} Y$ has a vertex contained in ΔP. By 7.3.10 (ii), we have $V^* \otimes_{\mathcal{O}} V \cong \mathcal{O} \oplus U$, where U is a direct sum of permutation $\mathcal{O}P$-modules of the form $\mathrm{Ind}_Q^P(\mathcal{O})$ with Q running over a family of proper subgroups of P. Thus $V^* \otimes_{\mathcal{O}} V \otimes_{\mathcal{O}} Y \cong Y \oplus U \otimes Y$, and every indecomposable summand of $U \otimes_{\mathcal{O}} Y$ has a vertex strictly smaller than ΔP. If every indecomposable direct summand of $V \otimes_{\mathcal{O}} Y$ has a vertex strictly smaller than ΔP, then the same is true for $V^* \otimes_{\mathcal{O}} V \otimes_{\mathcal{O}} Y$, a contradiction. Thus $V \otimes_{\mathcal{O}} Y$ has an indecomposable direct summand M with vertex ΔP. If $V \otimes_{\mathcal{O}} Y$ has two indecomposable summands with vertex ΔP, then the above argument implies that $V^* \otimes_{\mathcal{O}} V \otimes_{\mathcal{O}} Y$ has two summands with vertex ΔP, again a contradiction. By 7.3.10, $V \otimes_{\mathcal{O}} W$ has exactly one indecomposable summand with vertex ΔP, and hence this must be an $\mathcal{O}\Delta P$-source for the bimodule summand M of $V \otimes_{\mathcal{O}} Y$. The result follows. $\qquad\qquad\qquad\qquad\qquad\qquad\qquad\qquad\qquad\qquad\qquad\qquad\square$

Remark 7.3.19 Let P be a finite p-group and V an endopermutation $\mathcal{O}P$-module. Suppose that k has at least three distinct elements. It follows from 5.8.13 that the interior P-algebra $S = \mathrm{End}_{\mathcal{O}}(V)$ has a P-stable $\mathcal{O}$-basis contained in $S^\times = \mathrm{GL}(V)$. It follows further from 5.9.14 that the interior P-algebra $S \otimes_{\mathcal{O}} \mathcal{O}P$ has a P-P-stable $\mathcal{O}$-basis contained in the subgroup $S^\times \otimes P$ of $(S \otimes_{\mathcal{O}} \mathcal{O}P)^\times$.

See [214] for references to further material on endopermutation modules and their classification.

7.4 The Dade group action on interior P-algebras

The Dade group $D_{\mathcal{O}}(P)$ of a finite p-group P acts on the class of primitive interior P-algebras with stable bases, in a way that preserves Morita equivalence classes, the local structure, and the property of being relatively P-separable. This technique, due to Puig, is the main tool for translating Morita equivalences with endopermutation source between block algebras into statements on the algebra structure of associated source algebras. It is further used to follow generalised decomposition numbers through these Morita equivalences. For A an interior P-algebra over $\mathcal{O}$ and V an $\mathcal{O}P$-module, setting $S = \mathrm{End}_{\mathcal{O}}(V)$, we consider $S \otimes_{\mathcal{O}} A$ as an interior P-algebra via the diagonal map $P \to S \otimes_{\mathcal{O}} A$, sending $u \in P$ to $\nu(u) \otimes \sigma(u)$, where $\nu(u)$ is the linear endomorphism of V given by the action of u, and where $\sigma : P \to A^{\times}$ is the structural homomorphism of the interior P-algebra A. We assume in this section that k is perfect. We assume further that the $\mathcal{O}$-algebras in this section are finitely generated as $\mathcal{O}$-modules.

Theorem 7.4.1 (cf. [170, §5]) *Let P be a finite p-group, A a primitive interior P-algebra having a ΔP-stable $\mathcal{O}$-basis, such that $A(\Delta P) \neq \{0\}$, and let V be an indecomposable endopermutation $\mathcal{O}P$-module with vertex P. Set $S = \mathrm{End}_{\mathcal{O}}(V)$.*

(i) *The unit element 1 of $S \otimes_{\mathcal{O}} A$ satisfies $1 = e + (1 - e)$ for some primitive idempotent $e \in (S \otimes_{\mathcal{O}} A)^{\Delta P}$ such that $\mathrm{Br}_{\Delta P}(e) \neq 0$ and $\mathrm{Br}_{\Delta P}(1 - e) = 0$. Moreover, e is unique up to conjugation by an invertible element in $(S \otimes_{\mathcal{O}} A)^{\Delta P}$.*

(ii) *If f is a primitive idempotent in $(S^{\mathrm{op}} \otimes_{\mathcal{O}} e(S \otimes_{\mathcal{O}} A)e)^{\Delta P}$ satisfying $\mathrm{Br}_{\Delta P}(f) \neq 0$, then $f(S^{\mathrm{op}} \otimes_{\mathcal{O}} e(S \otimes_{\mathcal{O}} A)e)f \cong A$ as interior P-algebras.*

(iii) *The interior P-algebra $A' = e(S \otimes_{\mathcal{O}} A)e$ is primitive and Morita equivalent to A via the A'-A-bimodule $M = e(V \otimes_{\mathcal{O}} A)$ and the A-A'-bimodule $N = (A \otimes_{\mathcal{O}} V^*)e$. Moreover, A' has a ΔP-stable $\mathcal{O}$-basis and satisfies $A'(\Delta P) \neq \{0\}$.*

(iv) *If A is relatively $\mathcal{O}P$-separable, then $S \otimes_{\mathcal{O}} A$ and eAe are relatively $\mathcal{O}P$-separable.*

Proof Statement (i) is a special case of Proposition 5.9.3. The idempotent f as in (ii) remains primitive local in $(S^{\mathrm{op}} \otimes_{\mathcal{O}} S \otimes_{\mathcal{O}} A)^{\Delta P}$. If j is primitive local in $(S^{\mathrm{op}} \otimes_{\mathcal{O}} S)^{\Delta P}$, then, by 7.3.10 (iv) we have $j(S^{\mathrm{op}} \otimes_{\mathcal{O}} S)j \cong \mathcal{O}$, hence $(j \otimes 1_A)(S^{\mathrm{op}} \otimes_{\mathcal{O}} S \otimes_{\mathcal{O}} A)(j \otimes 1_A) \cong A$. This implies that $j \otimes 1_A$ belongs to the unique local point of P on $S^{\mathrm{op}} \otimes_{\mathcal{O}} S \otimes_{\mathcal{O}} A$, hence is conjugate to f in $(S^{\mathrm{op}} \otimes_{\mathcal{O}} S \otimes_{\mathcal{O}} A)^{\Delta P}$. This shows (ii). By the choice of e, the interior P-algebra $e(S \otimes_{\mathcal{O}} A)e$ is primitive. Since A and $S \otimes_{\mathcal{O}} A$ are Morita equivalent, the number $\ell(A)$ of points of A is equal to $\ell(S \otimes_{\mathcal{O}} A)$, hence $\ell(A') = \ell(e(S \otimes_{\mathcal{O}} A)e) \leq \ell(A)$. In order to show that A and A' are Morita equivalent, it suffices to show that $\ell(A') = \ell(A)$. By (ii) we have $\ell(A) = \ell(f(S \otimes_{\mathcal{O}} A')f) \leq \ell(S^{\mathrm{op}} \otimes_{\mathcal{O}} A') = \ell(A')$. This shows that A and A' are Morita equivalent via the A'-A-bimodule $M = e(V \otimes_{\mathcal{O}} A)$ and the A-A'-bimodule $N' = f(V^* \otimes_{\mathcal{O}} e(S \otimes_{\mathcal{O}} A)e)$. The bimodule N' is isomorphic to $f(V^* \otimes_{\mathcal{O}} (S \otimes_{\mathcal{O}} A)e) \cong$

$f(V^* \otimes_{\mathcal{O}} V \otimes_{\mathcal{O}} (V^* \otimes_{\mathcal{O}} A)e \cong (V^* \otimes_{\mathcal{O}} A)e$, where we have used the fact that f projects $V^* \otimes_{\mathcal{O}} V \otimes_{\mathcal{O}} A$ onto A. Thus N' is isomorphic to the bimodule N as stated in (iii). Since S and A have ΔP-stable bases, so does $S \otimes_{\mathcal{O}} A$, and hence also A', which is a direct summand of $S \otimes_{\mathcal{O}} A$ as an $\mathcal{O}\Delta P$-module. By (i) we have $\mathrm{Br}_{\Delta P}(e) \neq 0$, hence $A'(\Delta P) \neq \{0\}$, proving (iii). If A is relatively P-separable, then by 5.1.18, the interior P-algebra $S \otimes_{\mathcal{O}} A$ is relatively P-separable. Thus $S \otimes_{\mathcal{O}} A$ is isomorphic to a direct summand of $(S \otimes_{\mathcal{O}} A) \otimes_{\mathcal{O}P}$ $(S \otimes_{\mathcal{O}} A) \to S \otimes_{\mathcal{O}} A$. It follows that $A' = e(S \otimes_{\mathcal{O}} A)e$ is isomorphic to a direct summand of $e(S \otimes_{\mathcal{O}} A) \otimes_{\mathcal{O}P} (S \otimes_{\mathcal{O}} A)e$. As e is primitive, A' is indecomposable as an algebra, hence as a bimodule. Thus A' is isomorphic to a direct summand of $e(S \otimes_{\mathcal{O}} A)e' \otimes_{\mathcal{O}P} e''(S \otimes_{\mathcal{O}} A)e$ for some primitive idempotents e', e'' in $(S \otimes_{\mathcal{O}} A)^{\Delta P}$. Since $\mathrm{Br}_{\Delta P}(e) \neq 0$, the idempotents e', e'' must belong to the unique local point of P on $S \otimes_{\mathcal{O}} A$, hence are conjugate to e in $(S \otimes_{\mathcal{O}} A)^{\Delta P}$. Thus we may choose $e' = e'' = e$. It follows that A' is relatively $\mathcal{O}P$-separable. $\quad\square$

Theorem 7.4.2 (cf. [170, 5.3]) *Let P be a finite p-group, A an interior P-algebra having a ΔP-stable $\mathcal{O}$-basis such that $A(\Delta P) \neq \{0\}$, and let V be an endopermutation $\mathcal{O}P$-module having an indecomposable direct summand with vertex P. Set $S = \mathrm{End}_{\mathcal{O}}(V)$. For any subgroup Q of P, there is a canonical bijection between the sets of local points of Q on A and on $S \otimes_{\mathcal{O}} A$. This bijection preserves multiplicities and inclusion. More precisely, the following hold.*

  (i) *Let Q_δ be a local pointed group on A. Let τ be the unique point of the matrix algebra $S(\Delta Q)$. There is a unique local point δ' of Q on $S \otimes_{\mathcal{O}} A$ satisfying $\tau \otimes \mathrm{Br}_{\Delta Q}(\delta) \subseteq \mathrm{Br}_{\Delta Q}(\delta')$, where we identify $S(Q) \otimes_k A(\Delta Q) = (S \otimes_{\mathcal{O}} A)(\Delta Q)$.*

 (ii) *Let Q_δ and $Q_{\delta'}$ be as in (i). We have $m_{\delta'}^{S \otimes \mathcal{O}A} = m_\delta^A \cdot m(Q)$, where $m(Q) = m_\tau^{S(Q)}$ is the multiplicity of the unique point τ of $S(\Delta Q)$, or equivalently, the positive integer satisfying $m(Q)^2 = \dim_k(S(Q))$.*

(iii) *Let Q_δ, R_ϵ be local pointed groups on A such that $Q_\delta \leq R_\epsilon$, and let δ' and ϵ' be the local points of Q and R on $S \otimes_{\mathcal{O}} A$ corresponding to δ and ϵ, respectively. Then $Q_{\delta'} \leq R_{\epsilon'}$.*

(iv) *If A is primitive and e a primitive local idempotent in $(S \otimes_{\mathcal{O}} A)^{\Delta P}$, then for any local pointed group Q_δ the corresponding point δ' of Q on $S \otimes_{\mathcal{O}} A$ has a representative in $e(S \otimes_{\mathcal{O}} A)e$, and the map sending δ to $\delta' \cap e(S \otimes_{\mathcal{O}} A)e$ is an inclusion preserving bijection between the local pointed groups on A and on $e(S \otimes_{\mathcal{O}} A)e$.*

Proof The Brauer homomorphism $A^{\Delta Q} \to A(\Delta Q)$ induces a multiplicity preserving bijection between the sets of local points of Q on A and of points of $A(\Delta Q)$. Similarly, the points of Q on $S \otimes_{\mathcal{O}} A$ correspond to the points of $(S \otimes_{\mathcal{O}} A)(\Delta Q) \cong S(\Delta Q) \otimes_k A(\Delta Q)$. By the assumptions on V, the algebra $S(\Delta Q)$ is nonzero, hence a matrix algebra by 7.3.6. Thus the points of $A(\Delta Q)$ correspond bijectively to those of $S(\Delta Q) \otimes_k A(\Delta Q)$ as stated in (i), and the multiplicity of a point of $S(\Delta Q) \otimes_k A(\Delta Q)$ is $m(Q)$ times the multiplicity of the corresponding point of $A(\Delta Q)$ as stated in (ii). For (iii) we may assume that R normalises Q_δ, by 5.5.18. The Brauer homomorphism $\mathrm{Br}_{\Delta Q} : A \to A(\Delta Q)$ maps $A^{\Delta R/Q}$ onto $A(\Delta Q)^{\Delta R/Q}$ because A has a ΔP-stable $\mathcal{O}$-basis. Thus $\mathrm{Br}_{\Delta Q}$ sends δ to a point of $A(\Delta Q)$ and ϵ to a point of R/Q on $A(\Delta Q)$, preserving inclusion. Therefore we may assume that $Q = \{1\}$ and $\mathcal{O} = k$. We may replace A by jAj for some $j \in \epsilon$. Then A is a primitive R-algebra. Let $j' \in \epsilon'$. By 7.4.1, the algebras A and $j'(S \otimes_k A)j'$ are Morita equivalent. In particular, the point δ' of $S \otimes_k A$ has a representative in $j'(S \otimes_k A)j'$, or equivalently, $Q_{\delta'} \le R_{\epsilon'}$. This shows (iii). If A is primitive, then $\gamma = \{1_A\}$ is the unique local point of P on A, and e belongs to the corresponding (and hence unique) local point γ' of P on $S \otimes_{\mathcal{O}} A$. Thus, if Q_δ is a local pointed group on A, then $Q_\delta \le P_\gamma$, and hence $Q_{\delta'} \le P_{\gamma'}$, which is equivalent to the existence of a representative of δ' in $e(S \otimes_{\mathcal{O}} A)e$. Thus the map sending δ to $\delta' \cap e(S \otimes_{\mathcal{O}} A)e$ is an injective map from the set of local points of Q on A to the set of local points of Q on $e(S \otimes_{\mathcal{O}} A)e$. This map is also surjective because a primitive local idempotent in $(e(S \otimes_{\mathcal{O}} A)e)^{\Delta Q}$ remains primitive local in $(S \otimes_{\mathcal{O}} A)^{\Delta Q}$. The fact that this bijection preserves the inclusion between local pointed groups follows from (iii). $\qquad\qquad\Box$

There are more precise statements in [170, 5.3] relating fusion in A and $S \otimes_{\mathcal{O}} A$; we will postpone this aspect to the relevant block theoretic applications in subsequent chapters. In order to state the relationship with generalised decomposition numbers, it is convenient to slightly extend the notation from 5.15.2: if G is a finite group, A an interior G-algebra over $\mathcal{O}$ that is $\mathcal{O}$-free of finite rank, u_ϵ a local pointed element on A and χ a symmetric $\mathcal{O}$-linear map from A to the quotient field K of $\mathcal{O}$, then we set

$$\chi(u_\epsilon) = \chi(uj),$$

where $j \in \epsilon$. This expression is independent of the choice of j as χ is symmetric. If χ is the character of a finitely generated $\mathcal{O}$-free A-module U, then $\chi(uj)$ is the value at u of the character of the $\mathcal{O}\langle u \rangle$-module jU.

Theorem 7.4.3 (cf. [170, 1.11, 1.12]) *Let P be a finite p-group, A a primitive interior P-algebra having a ΔP-stable $\mathcal{O}$-basis such that $A(\Delta P) \ne \{0\}$, and let V be an indecomposable endopermutation $\mathcal{O}P$-module with vertex P. Set $S = \mathrm{End}_{\mathcal{O}}(V)$. Let $e \in (S \otimes_{\mathcal{O}} A)^{\Delta P}$ be a primitive idempotent belonging to the* *unique local point γ' of P on $S \otimes_{\mathcal{O}} A$. Let u_ϵ be a local pointed element on A, and let ϵ' be the corresponding local point of $\langle u \rangle$ on $e(S \otimes_{\mathcal{O}} A)e$. Let U be an $\mathcal{O}$-free A-module of finite $\mathcal{O}$-rank, and let $U' = e(V \otimes_{\mathcal{O}} U)$ be the corresponding A'-module. Denote by χ the character of U and by χ' the character of U'. We have*

$$\chi'(u_{\epsilon'}) = \omega(u)\chi(u_\epsilon),$$

where ω is the character of an indecomposable direct summand of $\mathrm{Res}^P_{\langle u \rangle}(V)$ with vertex $\langle u \rangle$. If V is defined over the unramified subring $\mathcal{O}'$ of $\mathcal{O}$ having k as residue field, then $\omega(u) \in \{\pm 1\}$ is equal to the sign of the character χ_V of V evaluated at u.

Proof We have $\chi'(u_{\epsilon'}) = \chi'(uj')$, for some (hence any) $j' \in \epsilon' \cap e(S \otimes_{\mathcal{O}} A)e$. By 5.12.16 this value depends only on $\mathrm{Br}_{\Delta\langle u \rangle}(j')$. It follows from 7.4.2 that we may choose j' such that $\mathrm{Br}_{\Delta\langle u \rangle}(j') = \mathrm{Br}_{\Delta\langle u \rangle}(j_Q) \otimes \mathrm{Br}_{\Delta\langle u \rangle}(j)$ for some primitive idempotent $j_Q \in S^{\Delta Q}$ belonging to the unique local point of Q on S, and some $j \in \epsilon$. Since $j' \in e(S \otimes_{\mathcal{O}} A)e$, we have $\chi'(uj') = \chi''(uj')$, where χ'' is the character of the $S \otimes_{\mathcal{O}} A$-module $V \otimes_{\mathcal{O}} U$. Thus $\chi'(uj') = \chi''(u(j_u \otimes j)) = \chi_V(uj_u)\chi(uj)$. Clearly $j_u V$ is an indecomposable direct summand of $\mathrm{Res}^P_{\langle u \rangle}(U)$ with vertex $\langle u \rangle$. Thus $\chi_V(uj_u) = \omega(u)$, as stated. The last statement follows from 7.3.13. $\qquad\square$

The next result is a sufficient criterion for when the unit element of $S \otimes_{\mathcal{O}} A$ remains primitive in $(S \otimes_{\mathcal{O}} A)^{\Delta P}$.

Proposition 7.4.4 *Let P be a finite p-group, A an interior P-algebra having a finite P-stable $\mathcal{O}$-basis such that $A(\Delta P)$ is a nonzero local k-algebra, and let V be an indecomposable endopermutation $\mathcal{O}P$-module with vertex P. Set $S = \mathrm{End}_{\mathcal{O}}(V)$ and $\bar{V} = k \otimes_{\mathcal{O}} V$. Suppose that for every simple $k \otimes_{\mathcal{O}} A$-module T every indecomposable direct summand of the kP-module $\bar{V} \otimes_k \mathrm{Res}_P(T)$ has vertex P. Then the interior P-algebra $S \otimes_{\mathcal{O}} A$ is primitive.*

Proof The kP-module $\bar{V} = k \otimes_{\mathcal{O}} V$ is indecomposable by 7.3.17, and hence we may assume that $\mathcal{O} = k$. Let e be a primitive idempotent in $(S \otimes_k A)^{\Delta P}$ such that $\mathrm{Br}_{\Delta P}(e) \neq \{0\}$. Since $(S \otimes_k A)(\Delta P) \cong S(\Delta P) \otimes_k A(\Delta P) \cong A(\Delta P)$ is local, it follows that $1 - e \in \ker(\mathrm{Br}_{\Delta P})$. Any simple $S \otimes_k A$-module is of the form $V \otimes_k T$ for some simple A-module T, and by the above, any indecomposable direct summand of $(1 - e)(V \otimes_k T)$ as a kP-module has a vertex strictly smaller than P. By the assumptions, $V \otimes_k T$ has no such summand. Thus $1 - e$ annihilates every simple $S \otimes_k A$-module, hence is contained in $J(S \otimes_k A)$. This forces $e = 1$. $\qquad\square$

Corollary 7.4.5 *Let P be a finite p-group, E a p'-subgroup of* Aut(P), $\alpha \in H^2(E; \mathcal{O}^\times)$, *and let V be an indecomposable endopermutation* $\mathcal{O}P$-*module with vertex P. Set* $S = \text{End}_{\mathcal{O}}(V)$. *The interior P-algebra* $S \otimes_{\mathcal{O}} \mathcal{O}_\alpha(P \rtimes E)$ *is primitive.*

Proof The kP-module $\bar{V} = k \otimes_{\mathcal{O}} V$ is indecomposable by 7.3.17, and hence we may assume that $\mathcal{O} = k$. Let e be a primitive idempotent in $(S \otimes_k k_\alpha(P \rtimes E))^{\Delta P}$ such that $\text{Br}_{\Delta P}(e) \neq \{0\}$. By 5.3.8, $k_\alpha(P \rtimes E)$ is isomorphic to $k(P \rtimes E)$ as a kP-kP-bimodule, hence isomorphic to $\oplus_{y \in E} kPy$. If $y \in E \setminus \{1\}$, then y is an outer automorphism of P because its order is prime to p, hence kPy is not isomorphic to kP as a kP-kP-bimodule and thus $(kPy)(\Delta P) = \{0\}$ by 5.8.8. This shows that $(k_\alpha(P \rtimes E))(\Delta P) \cong kZ(P)$, which is a local algebra. Moreover, P acts trivially on any simple $k_\alpha(P \rtimes T)$-module T, and hence any indecomposable direct summand of $\bar{V} \otimes_k \text{Res}_P(T)$ as a kP-module is isomorphic to $\bar{V}$, thus has vertex P. The result follows from 7.4.4. □

7.5 Endotrivial modules

Definition 7.5.1 Let G be a finite group. An *endotrivial* $\mathcal{O}G$-*module* is a finitely generated $\mathcal{O}$-free $\mathcal{O}G$-module V such that $V \otimes_{\mathcal{O}} V^* \cong \mathcal{O} \oplus X$ for some projective $\mathcal{O}G$-module X.

As in the case of endopermutation modules, the definition of endotrivial modules depends a priori on how the group G is embedded in its group algebra $\mathcal{O}G$. If P is a finite p-group, then any endotrivial $\mathcal{O}P$-module is an endopermutation $\mathcal{O}P$-module, because a projective $\mathcal{O}P$-module is a permutation $\mathcal{O}P$-module. Many of the statements in 7.3.4 remain true within the class of endotrivial modules.

Proposition 7.5.2 *Let G be a finite group and let V, W be endotrivial* $\mathcal{O}G$-*modules.*

(i) *For any subgroup H of G, the module* $\text{Res}_H^G(V)$ *is an endotrivial* $\mathcal{O}H$-*module.*

(ii) *We have* $V \cong V' \oplus U$ *for some absolutely indecomposable endotrivial* $\mathcal{O}G$-*module V' and some projective* $\mathcal{O}G$-*module U.*

(iii) *The* $\mathcal{O}$-*dual V* of V is an endotrivial* $\mathcal{O}G$-*module.*

(iv) *The tensor product* $V \otimes_{\mathcal{O}} W$ *is an endotrivial* $\mathcal{O}G$-*module.*

(v) *For any integer n the module* $\Omega^n(V)$ *is an endotrivial* $\mathcal{O}G$-*module.*

(vi) *If V is indecomposable, then the vertices of V are the Sylow p-subgroups of G.*

Proof The statements (i), (iii), (iv) are trivial. Since $V \otimes_{\mathcal{O}} V^*$ has the trivial module $\mathcal{O}$ as its unique indecomposable direct summand, it follows from 2.9.8 that V has, up to isomorphism, a unique indecomposable nonprojective direct summand V', and then clearly V' is endotrivial. This proves (ii). Statement (v) is proved exactly as 7.3.4 (v). Statement (vi) follows from 5.1.11 and the fact that the trivial summand of $V \otimes_{\mathcal{O}} V^*$ has the Sylow p-subgroups of G as vertices, by 5.1.4. □

The inflation $\mathrm{Inf}^P_{P/Q}(W)$ of an endotrivial $\mathcal{O}P/Q$-module W, with Q a non trivial normal subgroup of a finite p-group P, is not endotrivial because the inflation of a projective module is not projective.

Proposition 7.5.3 *Let P be a finite p-group and let V be a finitely generated $\mathcal{O}$-free $\mathcal{O}P$-module. Then V is an endotrivial $\mathcal{O}P$-module if and only if $k \otimes_{\mathcal{O}} V$ is an endotrivial kP-module.*

Proof Suppose that $\bar{V} = k \otimes_{\mathcal{O}} V$ is an endotrivial kP-module; that is, $\bar{V} \otimes \bar{V}^* \cong k \oplus Y$ for some projective kP-module Y. In particular, the rank of V is prime to p. Thus $V \otimes_{\mathcal{O}} V^* \cong \mathcal{O} \oplus X$ for some $\mathcal{O}P$-module X, by 5.1.13. The Krull–Schmidt Theorem implies that $k \otimes_{\mathcal{O}} X \cong Y$, hence X is projective. This implies that V is endotrivial. The converse is obvious. □

Let P be a finite p-group and V a finitely generated kP-module. By a result of Carlson [45], the module V is endotrivial P if and only of $\underline{\mathrm{End}}_{kP}(V) \cong k$. We show two slightly weaker results characterising endotrivial modules.

Proposition 7.5.4 ([128, Corollary 3.9]) *Let P be a finite p-group and V a finitely generated $\mathcal{O}$-free $\mathcal{O}P$-module. Suppose that $\mathrm{char}(\mathcal{O}) = 0$. Then V is endotrivial if and only $\underline{\mathrm{End}}_{\mathcal{O}P}(V) \cong \mathcal{O}/|P|\mathcal{O}$ and $\underline{\mathrm{End}}_{kP}(k \otimes_{\mathcal{O}} V) \cong k$.*

Proof Suppose that $\underline{\mathrm{End}}_{\mathcal{O}P}(V) \cong \mathcal{O}/|P|\mathcal{O}$ and $\underline{\mathrm{End}}_{kP}(k \otimes_{\mathcal{O}} V) \cong k$. By 4.13.19 we have $V^* \otimes_{\mathcal{O}} V \cong \mathcal{O} \oplus W \oplus X$ for some projective $\mathcal{O}P$-module X and some $\mathcal{O}P$-module W that has no nonzero projective direct summand and that satisfies $W^P = \{0\}$. By 4.13.13, $k \otimes_{\mathcal{O}} W$ has no nonzero projective direct summand as a kP-module. Since $\underline{\mathrm{End}}_{kP}(k \otimes_{\mathcal{O}} V) \cong k$, it follows that $(k \otimes_{\mathcal{O}} W)^P = \{0\}$. This forces $W = \{0\}$, since the trivial kP-module k is, up to isomorphism, the unique simple kP-module. The converse is obvious. □

Proposition 7.5.5 *Let P be a finite p-group and let V be a finitely generated kP-module. Then V is an endotrivial kP-module if and only if $\dim_k(V)$ is prime to p and $\underline{\mathrm{End}}_{kP}(V) \cong k$.*

Proof The canonical isomorphism $\mathrm{End}_k(V) \cong V \otimes_k V^*$ sends $\mathrm{End}_{kP}(V)$ to $(V \otimes_k V^*)^P = \mathrm{soc}(V \otimes_k V^*)$. By 2.13.11 we have $\mathrm{End}^{\mathrm{pr}}_{kP}(V) = (\mathrm{End}_k(V))^P_1$.

Thus the previous isomorphism sends $\mathrm{End}_{kP}^{\mathrm{pr}}(V)$ to $(V \otimes_k V^*)_1^P = (\sum_{y \in P} y) \cdot (V \otimes_k V^*)$, the image of the relative trace map Tr_1^P on $V \otimes_k V^*$. Decompose $V \otimes_k V^* = W \oplus X$ where X is projective and W has no nonzero projective summand. Then $(V \otimes V^*)_1^P = X_1^P$, as $W_1^P = (\sum_{y \in P}) \cdot W = \{0\}$ by 1.11.3. Since X is projective, we have $X_1^P = X^P$ by 1.11.2. It follows that the dimension of $\underline{\mathrm{End}}_{kP}(V)$ is equal to the dimension of $\mathrm{soc}(W)$. Thus $\underline{\mathrm{End}}_{kP}(V) \cong k$ if and only if $\mathrm{soc}(W)$ is simple, or equivalently, if and only if W is a proper submodule of kP. If this is the case, the additional condition that $\dim_k(V)$ is prime to p holds if and only if $W \cong k$, by 5.1.13, hence if and only if V is endotrivial. □

The characterisation of endotrivial kP-modules in 7.5.5 does no longer depend on the embedding of P in kP; we therefore have the following.

Corollary 7.5.6 *Let P be a finite p-group and Q a subgroup of $(kP)^\times$ such that $|Q| = |P|$ and such that $kQ = kP$. Let V be a finitely generated kP-module. Then V is an endotrivial kP-module if and only if V is an endotrivial kQ-module.*

Corollary 7.5.7 *Let P be a finite cyclic p-group. A finitely generated indecomposable kP-module V is endotrivial if and only if V is isomorphic to k or $\Omega(k)$.*

Proof This follows from 7.5.5 in conjunction with 7.1.8. □

Proposition 7.5.8 *Let P be a finite p-group and let V be an indecomposable endopermutation $\mathcal{O}P$-module with P as its vertex. Set $S = \mathrm{End}_{\mathcal{O}}(V)$. Then V is an endotrivial $\mathcal{O}P$-module if and only if $S(Q) \cong k$ for every nontrivial subgroup Q of P.*

Proof Write $S \cong V \otimes_{\mathcal{O}} V^* \cong \mathcal{O} \oplus M$ for some permutation $\mathcal{O}P$-module M. The condition $S(Q) \cong k$ is equivalent to $M(Q) = \{0\}$. By 5.8.2 this holds for every nontrivial subgroup Q of P if and only if M is projective, whence the result. □

Definition 7.5.9 Let P be a finite p-group. We denote by $T_{\mathcal{O}}(P)$ the subgroup of $D_{\mathcal{O}}(P)$ generated by the isomorphism classes of indecomposable endotrivial $\mathcal{O}P$-modules.

Proposition 7.5.10 ([173, 2.1.2]) *Let P be a finite p-group. We have an exact sequence of abelian groups*

$$0 \longrightarrow T_{\mathcal{O}}(P) \longrightarrow D_{\mathcal{O}}(P) \overset{d}{\longrightarrow} \prod_Q D_k(N_P(Q)/Q)$$

where Q runs over the nontrivial subgroups of P and where d is the product of the deflation maps $\mathrm{Defres}^P_{N_P(Q)/Q}$.

Proof The map $T_{\mathcal{O}}(P) \to D_{\mathcal{O}}(P)$ in this sequence is the inclusion map. Given an endotrivial $\mathcal{O}P$-module V and setting $S = \mathrm{End}_{\mathcal{O}}(V)$, we have $S(Q) \cong k$ for any nontrivial subgroup Q of P, and hence the class of V is in the kernel of all deflation maps $\mathrm{Defres}^P_{N_P(Q)/Q}$, hence in the kernel of the map d. Conversely, let V be an indecomposable endopermutation $\mathcal{O}P$-module such that the class of V is in the kernel of d. This means that $\mathrm{Defres}^P_{N_P(Q)/Q}(V) \cong k$ for any nontrivial subgroup Q of P. Setting $S = \mathrm{End}_{\mathcal{O}}(V)$ and denoting by W_Q an endopermutation $kN_P(Q)/Q$-module satisfying $\mathrm{End}_k(W_Q) \cong S(Q)$, it follows that W_Q has a trivial direct summand. In view of 7.5.8, we need to show that W_Q is trivial, or equivalently, that $S(Q) \cong k$ for any nontrivial subgroup Q of P. We proceed by induction over the index of Q in P. For $Q = P$ there is nothing to prove since V is indecomposable. Suppose that Q is a nontrivial proper subgroup of P. By induction we have $S(Q)(R/Q) \cong S(R) \cong k$ for any subgroup R of $N_P(Q)$ properly containing Q. Thus W_Q is an endotrivial $kN_P(Q)/Q$-module by 7.5.8. Since W_Q has a trivial summand, it follows that $W_Q \cong k \oplus Y$ for some projective $kN_P(Q)/Q$-module Y. We need to show that Y is zero. If not, then a projection of W_Q onto Y is an idempotent j in $S(Q)_1^{N_P(Q)/Q} = S(Q)_Q^{N_P(Q)}$. The Brauer homomorphism Br_Q maps S_Q^P onto $S(Q)_Q^{N_P(Q)}$, and hence j lifts to an idempotent in S_Q^P. However, V is indecomposable with vertex P, hence S_Q^P contains no idempotent. This shows that Y is zero. The result follows. $\square$

The above proposition makes no statement about the image of the map d. One obvious condition for a family $(V_Q)_Q$ of endopermutation modules for the quotients $N_P(Q)/Q$ to be in the image of d is that if Q, Q' are conjugate subgroups of P, then V_Q should be conjugate to $V_{Q'}$, up to isomorphism. This condition is, of course, not sufficient. For P abelian, one can describe the image of d precisely; see 7.8.2.

Lemma 7.5.11 *Let P be a finite P and let Q, R be normal subgroups of P.*

(i) *If Q is not contained in R, then* $\mathrm{Inf}^P_{P/R}(T_k(P/R)) \subseteq \ker(\mathrm{Defres}^P_{P/Q})$.

(ii) *If Q is contained in R, then* $\mathrm{Defres}^P_{P/Q}(\mathrm{Inf}^P_{P/R}(W)) \cong \mathrm{Inf}^{P/Q}_{P/R}(W)$ *for any indecomposable endopermutation kP/R-module W with vertex P/R, where we identify P/R with $(P/Q)/(R/Q)$.*

Proof Let V be an endotrivial kP/R-module; that is, $V \otimes_k V^* \cong k \oplus \mathrm{Ind}^{P/R}_1(k)^m$ for some integer m. Set $W = \mathrm{Inf}^P_{P/R}(V)$. Then $W \otimes_k W^* \cong k \oplus \mathrm{Ind}^P_R(k)^m$. Setting $S = \mathrm{End}_k(W) \cong W \otimes_k W^*$, we get that $S(Q) \cong k \oplus (\mathrm{Ind}^P_R(k))(Q)^m$. Suppose that Q is not contained in R. Then Q is not conjugate to a subgroup of R because both Q, R are normal in P. Hence $\operatorname{Ind}_R^P(k)(Q) = \{0\}$ by 5.10.3, or equivalently, the class determined by $\operatorname{Inf}_{P/R}^P(V)$ in $D_k(P)$ belongs to the kernel of $\operatorname{Defres}_{P/Q}^P$. This proves (i); statement (ii) is a trivial consequence of the fact that if Q is a subgroup of R, then Q acts trivially on $\operatorname{Inf}_{P/R}^P(W)$. $\qquad\square$

Endotrivial modules are exactly the modules inducing stable equivalences of Morita type.

Proposition 7.5.12 *Let G be a finite group and let V be a finitely generated $\mathcal{O}$-free $\mathcal{O}G$-module. Then V is an endotrivial $\mathcal{O}G$-module if and only if the functor $V \otimes_{\mathcal{O}} -$ induces an equivalence on the $\mathcal{O}$-stable category $\underline{\operatorname{mod}}(\mathcal{O}G)$. In that case, the functor $V^* \otimes_{\mathcal{O}} -$ induces an inverse of $V \otimes_{\mathcal{O}} -$ on $\underline{\operatorname{mod}}(\mathcal{O}G)$, and this stable equivalence is of Morita type, induced by the bimodule $\operatorname{Ind}_{\Delta G}^{G \times G}(V)$ and its dual.*

Proof If $V \otimes_{\mathcal{O}} -$ induces an equivalence on $\underline{\operatorname{mod}}(\mathcal{O}G)$, then $V^* \otimes_{\mathcal{O}} -$ induces an inverse, since this functor is an adjoint. In particular, $V \otimes_{\mathcal{O}} V^* \otimes_{\mathcal{O}} -$ induces the identity functor on $\underline{\operatorname{mod}}(\mathcal{O}G)$, hence sends $\mathcal{O}$ to $\mathcal{O} \oplus X$ for some projective $\mathcal{O}G$-module, implying that V is endotrivial. If V is an endotrivial $\mathcal{O}G$-module, then $V \otimes_{\mathcal{O}} V^* \cong \mathcal{O} \oplus X$ for some projective $\mathcal{O}G$-module X. The projectivity of X implies that $X \otimes_{\mathcal{O}} -$ induces the zero functor on $\underline{\operatorname{mod}}(\mathcal{O}G)$, and hence $V \otimes_{\mathcal{O}} V^*$ induces the identity on $\underline{\operatorname{mod}}(\mathcal{O}G)$. The functors $V \otimes_{\mathcal{O}} -$ and $\operatorname{Ind}_{\Delta G}^{G \times G}(V) \otimes_{\mathcal{O}G} -$ are isomorphic by 2.4.12. By 2.4.13 we have $\operatorname{Ind}_{\Delta G}^{G \times G}(V) \otimes_{\mathcal{O}G} \operatorname{Ind}_{\Delta G}^{G \times G}(V^*) \cong \operatorname{Ind}_{\Delta G}^{G \times G}(V \otimes_{\mathcal{O}} V^*)$. Thus if $V \otimes_{\mathcal{O}} V^* \cong \mathcal{O} \oplus X$ for some projective $\mathcal{O}G$-module X, then the previous expression is of the form $\operatorname{Ind}_{\Delta G}^{G \times G}(\mathcal{O}) \oplus Z$ for some projective $\mathcal{O}G$-$\mathcal{O}G$-bimodule Z, and by 2.4.5 we have $\operatorname{Ind}_{\Delta G}^{G \times G}(\mathcal{O}) \cong \mathcal{O}G$ as $\mathcal{O}G$-$\mathcal{O}G$-bimodules. The last statement follows. $\qquad\square$

Corollary 7.5.13 *Let G be a finite group. The map sending an endotrivial $\mathcal{O}G$-module V to the bimodule $\operatorname{Ind}_{\Delta G}^{G \times G}(V)$ induces an injective group homomorphism $T_{\mathcal{O}}(G) \to \operatorname{StPic}(\mathcal{O}G)$.*

Proof The fact that the map $V \mapsto \operatorname{Ind}_{\Delta G}^{G \times G}(V)$ induces a group homomorphism $T_{\mathcal{O}}(G) \to \operatorname{StPic}(\mathcal{O}G)$ follows from 7.5.12 and 2.4.13. If V, V' are two nonisomorphic endotrivial $\mathcal{O}G$-modules, then the functors $V \otimes_{\mathcal{O}} -$ and $V' \otimes_{\mathcal{O}} -$ on $\underline{\operatorname{mod}}(\mathcal{O}G)$ are nonisomorphic, since the images of the trivial module $\mathcal{O}$ under these two functors are nonisomorphic in $\underline{\operatorname{mod}}(\mathcal{O}G)$. This shows the injectivity of the group homomorphism as stated. $\qquad\square$

The stable equivalences obtained from endotrivial modules can be factorised through a canonical Morita equivalence and a restriction map as follows.

Lemma 7.5.14 *Let G be a finite group and V an endotrivial $\mathcal{O}G$-module. Set $S = \mathrm{End}_{\mathcal{O}}(V)$. Denote by $\sigma : G \to S^{\times}$ the structural homomorphism and by $\delta :$ $\mathcal{O}G \to S \otimes_{\mathcal{O}} \mathcal{O}G$ the diagonal map, sending $x \in G$ to $\sigma(x) \otimes x$. The restriction functor $\mathrm{Res}_{\delta} : \mathrm{mod}(S \otimes_{\mathcal{O}} \mathcal{O}G) \to \mathrm{mod}(\mathcal{O}G)$ induces a stable equivalence of Morita type.*

Proof The restriction functor Res_{δ} is obtained as a composition of the canonical Morita equivalence $\mathrm{mod}(S \otimes_{\mathcal{O}} \mathcal{O}G) \cong \mathrm{mod}(\mathcal{O}G)$ sending $V \otimes U$ to U, where U is any $\mathcal{O}G$-module, followed by the stable equivalence of Morita type induced by the functor $V \otimes_{\mathcal{O}} -$ on $\mathrm{mod}(\mathcal{O}G)$. $\qquad\square$

7.6 Shifted cyclic subgroups and Dade's Lemma

A module over a finite p-group P whose restrictions to proper subgroups is projective need not be projective. Dade's Lemma shows that projectivity can be characterised by the projectivity of restrictions to certain cyclic subgroups of order p in the group of invertible elements of kP. For any element y of order p in P we have $(y - 1)^p = 0$ in kP. Thus, for any $\lambda \in k^{\times}$, the element $1 + \lambda(y - 1)$ has order p.

Definition 7.6.1 Let P be an elementary abelian finite p-group of rank n and let $\mathcal{Y}$ be a basis of P. A *cyclic shifted subgroup of* $(kP)^{\times}$ is a cyclic subgroup of $(kP)^{\times}$ having a generator of the form $1 + \sum_{y \in \mathcal{Y}} \lambda_y(y - 1)$ for some coefficients $\lambda_y \in k$ not all of which are zero.

Proposition 7.6.2 *Let P be a finite elementary abelian p-group and let H be a shifted cyclic subgroup of $(kP)^{\times}$. Then H has order p, the subalgebra of kP generated by H is isomorphic to the group algebra kH over k, and kP is projective as left and right kH-module.*

Proof Let $\mathcal{Y}$ be a basis of P and suppose that H has a generator of the form $w = 1 + \sum_{y \in \mathcal{Y}} \lambda_y(y - 1)$ for some coefficients $\lambda_y \in k$ not all of which are zero. Any element of $\mathrm{GL}_n(k)$, by letting it act on the basis $\mathcal{Y}$ of P in the natural way, can be viewed as an automorphism of kP, and hence there is an automorphism α of kP sending a fixed element $z \in \mathcal{Y}$ to the generator w of H. Then α induces an isomorphism $k\langle z \rangle \cong kH$, and since kP restricted to $k\langle z \rangle$ is projective, so is kP restricted to kH. $\qquad\square$

Theorem 7.6.3 (Dade [55, Lemma 11.8]) *Let P be a finite elementary abelian p-group. Suppose that k is algebraically closed. A finitely generated kP-module*

U is projective if and only if the restriction of U to any cyclic shifted subgroup in $(kP)^\times$ is projective.

Proof If U is projective then so is its restriction to any cyclic shifted subgroup, by 7.6.2. Suppose conversely that the restriction of U to any cyclic shifted subgroup of $(kP)^\times$ is projective. If P is cyclic, then U is projective by 7.1.3. Suppose that P has rank 2. Write $P = \langle s \rangle \times \langle t \rangle$. Set $x = s - 1$ and $y = t - 1$. Then $x^p = y^p = 0$. The set $\{x^i y^j | 0 \le i, j \le p - 1\}$ is a k-basis of kP. The subset of $x^i y^j$ with at least one of i, j positive spans the radical $J = J(kP)$. As an ideal, J is generated by x and y. An easy verification implies that J^{p-1} is generated by the set $\{x^i y^{p-1-i} | 0 \le i \le p - 1\}$ and that J^p is generated by the set $\{x^i y^{p-i} | 1 \le i \le p - 1\}$. Let $z = \lambda x + \mu y$ be a nonzero linear combination of x and y. Then $z \in J$, hence $zJ^{p-1} \subseteq J^p$. Multiplying the generating set of J^{p-1} by z yields elements whose span contains the generating set of J^p, whence $zJ^{p-1} = J^p$. Suppose now that U is a finitely generated kP-module such that the restriction to any of the cyclic shifted subgroups $\langle 1 + z \rangle$ is projective. If U is zero, there is nothing to prove, so assume that U is nonzero. Consider the chain of submodules $J^p U \subseteq J^{p-1} U \subseteq V$, where $V = \{u \in U | Ju \subseteq J^{p-1} U\}$. The quotients $J^{p-1} U / J^p U$ and $V / J^{p-1} U$ are both annihilated by J. Note that V contains $J^{p-2} U$, hence $V / J^{p-1} U$ is nonzero, as J is a nilpotent ideal and U is nonzero. Multiplication by z induces a k-linear map

$$\zeta : V/J^{p-1}U \to J^{p-1}U/J^pU.$$

We show that ζ is an isomorphism. We start by showing that ζ is injective. Let $v \in V$ such that $v + J^{p-1} \in \ker(\zeta)$. That is, $zv \in J^p U = zJ^{p-1} U$. Thus there is $w \in J^{p-1} U$ such that $zv = zw$, hence such that $z(v - u) = 0$. Since U is projective as a module for $k\langle 1 + z \rangle$, it follows from 7.1.4 that $v - w = z^{p-1} u$ for some $u \in U$. Since $z^{p-1} \in J^{p-1}$ we get that $v = w + z^{p-1} u \in J^{p-1} U$, hence the class $v + J^{p-1} U$ is zero. This shows the injectivity of ζ. Since ζ is injective for any choice of z as above, in order to show the surjectivity it suffices to show this for $z = x$, because then both sides have the same dimension. We will show that $xV = J^{p-1} U$. If not, then xV is a proper submodule of $J^{p-1} U = \sum_{0 \le i \le p-1} x^i y^{p-i+1} U = y^{p-1} U + xJ^{p-2} U$. Since $xJ^{p-2} U \subseteq V$ this means that there is an element $u \in U$ such that $y^{p-1} u \notin xV$. If $x^{p-1} y^{p-1} u \ne 0$ then $kPu \cong kP$, contradicting the assumptions on U. Thus $x^{p-1} y^{p-1} u = 0$. Thus $y^{p-1} u$ is annihilated by x^{p-1}. Again since U is projective upon restriction to $\langle 1 + x \rangle$, there is an element a in U such that $y^{p-1} u = xa$. Then $xya = xy^p u = 0$, and yet again, we get an element $b \in U$ such that $x^{p-1} b = ya$. Thus, both elements xa, ya belong to $J^{p-1} U$, hence $a \in V$, from where we get the contradiction $y^{p-1} u \in xV$. Thus ζ is surjective.

This means that we have constructed a 2-dimensional space of homomorphisms all of whose nonzero elements are isomorphisms. By 1.9.4, there is no such space. This contradiction concludes the proof in case P has rank 2.

Suppose now that P has rank greater than 2. We proceed by induction. Write $P = \langle s_1 \rangle \times \cdots \times \langle s_r \rangle$, with elements s_i of order p in P, and set $x_i = s_i - 1$, for $1 \leq i \leq r$. Set $S = \sum_{3 \leq i \leq r} x_i U$; that is, $S = J(kQ)U$, where $Q = \langle s_3, \ldots, s_r \rangle$. Set $V = U/S$. Let $z = \lambda_1 x_1 + \lambda_2 x_2$ be a nonzero linear combination of x_1, x_2. Choose notation such that $\lambda_2 \neq 0$. Then $P' = \langle 1 + z, s_2, \ldots, s_r \rangle$ is again elementary abelian, hence isomorphic to P, and $kP' = kP$. Moreover, the k-subspace of kP spanned by the set $\{z, x_2, \ldots, x_r\}$ is equal to that spanned by $\{x_1, x_2, \ldots, x_r\}$. By induction, U is free as a module over kR, where $R = \langle z, x_3, \ldots, x_r \rangle$. It follows that S, hence V, is free as a module over $k\langle 1 + z \rangle$. This holds for all nonzero linear combinations z of x_1, x_2, and hence V is free as a module over $k\langle s_1, s_2 \rangle$. Now U is free as a module over $k\langle s_3, \ldots, s_r \rangle$, by induction. One verifies that the inverse image in U of a basis of V as a module over $k\langle s_1, s_2 \rangle$ is a basis of U as a kP-module. $\square$

We have closely followed Dade's original proof, which has the advantage of requiring only basic properties of modules over cyclic p-groups. There are cohomological proofs of this result as part of the much broader theory of cohomology varieties – see Benson [15] for an exposition on this topic.

Proposition 7.6.4 ([92]) *Let P be an elementary abelian p-group of order p^2. If $n = 2m$ for some nonnegative integer m, then $\dim_k(\Omega^n(k)) = mp^2 + 1$. If $n = 2m - 1$ for some positive integer m, then $\dim_k(\Omega^n(k)) = mp^2 - 1$. In particular, the trivial kP-module is not periodic.*

Proof Write $P = kQ \times kR$ for some cyclic subgroups $Q = \langle s \rangle$ and $R = \langle t \rangle$. Set $x = s - 1$ and $y = t - 1$. Consider kP/Q as a kP-module; that is, x annihilates kP/Q. Similarly for kP/R. By 2.4.11, the kP-module $kP/Q \otimes_k kP/R$ is projective, hence isomorphic to kP. By 7.1.5, the projective resolutions of the trivial kP/Q-module and the trivial kP/R-module, inflated to kP yield complexes

$$\cdots \longrightarrow kP/Q \xrightarrow{\ y^{p-1}\ } kP/Q \xrightarrow{\ y\ } kP/Q \longrightarrow 0$$

$$\cdots \longrightarrow kP/R \xrightarrow{\ x^{p-1}\ } kP/R \xrightarrow{\ x\ } kP/R \longrightarrow 0$$

whose homology is k in degree zero. Their tensor product is a projective resolution of the trivial kP-module of the form

$$\cdots \longrightarrow X_2 \xrightarrow{\ \delta_2\ } X_1 \xrightarrow{\ \delta_1\ } X_0 \longrightarrow 0$$

where $X_n = \bigoplus_{i=0}^{n} kP/Q \otimes_k kP/R \cong (kP)^{n+1}$ for $n \geq 0$. Since $J(kP) = xkP + ykP$, the differential of this complex has the property that $\text{Im}(\delta_n) \subseteq \text{rad}(X_{n-1})$, for n positive, and hence this complex is a minimal projective resolution of the trivial kP-module. Using $(n+1)p^2 = \dim_k(X_n) = \dim_k(\Omega^n(k)) + \dim_k(\Omega^{n+1}(k))$, an easy induction yields the result on the dimensions of the Heller translates of k. Since these dimensions are pairwise different as n runs over all nonnegative integers, it follows that k is not periodic. $\quad\square$

The results in [92] are more precise in that they describe the module structure of the Heller translates of k explicitly.

Corollary 7.6.5 *If P is a finite p-group containing an elementary abelian subgroup of order p^2, then the trivial kP-module is not periodic.*

Proof If kP has a periodic trivial module, then so does every subgroup of P. The result follows from 7.6.4. $\quad\square$

Remark 7.6.6 The converse of 7.6.5 holds as well: if P does not contain an elementary abelian subgroup of order p^2, then by some basic group theory, either P is cyclic (in which case we have already observed that k has period at most 2), or $p = 2$ and P is a generalised quaternion group, in which case one can calculate explicitly that k has period 4.

7.7 Endotrivial modules over abelian p-groups

The classification of endotrivial modules over abelian p-groups, due to Dade, is a key step towards the classification of all endopermutation modules over abelian p-groups in the next section.

Theorem 7.7.1 ([55, Theorem 10.1]) *Let P be a finite abelian p-group and let V be an indecomposable endotrivial $\mathcal{O}P$-module. Suppose that k is algebraically closed. There is an $\mathcal{O}P$-module T of $\mathcal{O}$-rank 1 and an integer n such that $V \cong \Omega_P^n(T)$. If P has order at least 3, then T is unique up to isomorphism, and if P is noncyclic, then the integer n is unique. In particular, the group $T_k(P)$ is cyclic with $[\Omega_P(k)]$ as a generator. If $|P| = 2$, then $T_k(P)$ is trivial, if $|P| \geq 3$ and P is cyclic, then $T_k(P) \cong \mathbb{Z}/2\mathbb{Z}$, and if P is noncyclic, then $T_k(P) \cong \mathbb{Z}$.*

Proof By 7.3.12 we may assume that $\mathcal{O} = k$. The statement on the order of $\Omega(k)$ in the Dade group follows from 7.1.5 and 7.6.4. If P is cyclic, 7.7.1 follows from 7.5.7.

We consider next the case where P has rank 2. Write $P = Q \times R$ with cyclic groups $Q = \langle g \rangle$ and $R = \langle h \rangle$ of order a and b, respectively, where we choose notation such that $a \geq b \geq 2$. Set $x = g - 1$ and $y = h - 1$. Let V be an indecomposable endotrivial kP-module. Arguing by induction over $|P|$, we may assume that the restriction of V to any proper subgroup of P is endotrivial. We need to show that $\Omega^n(V) \cong k$ for some integer n. We consider the radical and socle series of V as a kR-module. For $i \geq 0$ we set

$$s^i(V) = \mathrm{soc}^i(\mathrm{Res}_R^P(V)) = \{v \in V \mid y^i v = 0\},$$

$$r^i(V) = \mathrm{rad}^i(\mathrm{Res}_R^P(V)) = \{y^i v \mid v \in V\}.$$

Since P is abelian, these are kP-submodules of V. We set

$$t(V) = \dim_k(\mathrm{Hom}_{kQ}(V/s^{b-1}(V), k)).$$

Since every indecomposable kQ-module is uniserial, hence has a unique simple quotient, the number $t(V)$ is equal to the number t of indecomposable direct summands Y_j in a decomposition

$$V/s^{b-1}(V) = \oplus_{j=1}^t Y_j$$

of $V/s^{b-1}(V)$ as a direct sum of indecomposable kQ-modules. Since y annihilates $V/s^{b-1}(V)$, this is a direct sum decomposition as kP-modules. We choose now V amongst its Heller translates in such a way that $t(V)$ is minimal; that is, such that $t(V) \leq t(\Omega^n(V))$ for all integers n. It suffices to show that V is trivial. We proceed in a series of steps. For the first step, note that we have $r^1(V) \subseteq s^{b-1}(V)$.

(1) As a kP-module, $s^{b-1}/r^1(V)$ is a one-dimensional direct summand of $V/r^1(V)$. Moreover, amongst all Heller translates of V with this property, we may choose V such that $V/r^1(V)$ has no nonzero projective summand as a kQ-module.

Since V is endotrivial, so is its restriction to R. Since R is cyclic, we have $\mathrm{Res}_R^P(V) \cong \Omega_R^\epsilon(k) \oplus Y$ for some $\epsilon \in \{0, 1\}$ and some projective kR-module Y. Thus $r^1(V) = yV \cong y\Omega_R^\epsilon(V) + yY$ has codimension 1 in $s^{b-1}(Y) \cong \Omega_R^\epsilon(k) \oplus yY$. In order to prove (2), it suffices to show that $s^{b-1}/r^1(V)$ is a direct summand of $V/r^1(V)$ as a kQ-module, because y annihilates $V/r^1(V)$, and so any kQ-submodule of $V/r^1(V)$ is a kP-submodule. Arguing by contradiction, suppose that $s^{b-1}/r^1(V)$ is not a direct summand of $V/r^1(V)$ as a kQ-module. We introduce the following notation which we will also use at a later stage. Since V is indecomposable and satisfies $\underline{\mathrm{End}}_{kP}(V) \cong k$, the nilpotent endomorphism of

V given by multiplication with x factors through a projective kP-module. Thus there is $\psi \in \text{End}_k(V)$ satisfying $\text{Tr}_1^P(\psi)(v) = xv$ for all $v \in V$. Set $\varphi = \text{Tr}_1^R(\psi)$. Then φ is a kR-endomorphism of V such that $\text{Tr}_1^Q(\varphi) = \text{Tr}_1^P(\psi)$, which is the endomorphism of V equal to multiplication by x. Note that φ preserves the submodules $s^{b-1}(V)$, $r^1(V)$. In particular, φ induces a k-linear map $\bar\varphi$ on $V/r^1(V)$ which preserves $s^{b-1}(V)/r^1(V)$ and which has the property that $\text{Tr}_1^Q(\bar\varphi)$ is equal to the endomorphism of $V/r^1(V)$ induced by left multiplication with x. By 7.1.10, $s^{b-1}(V)/r^1(V)$ is contained in a projective indecomposable direct summand Y_1 of $V/r^1(V)$ as a kQ-module. This means that $V/r^1(V)$ is a nonsplit extension of $V/s^{b-1}(V)$ by a the one-dimensional submodule $s^{b-1}(V)/r^1(V)$, and hence the number t of indecomposable direct summands of $V/r^1(V)$ in a decomposition of this kQ-module is equal to the number $t(V)$ of indecomposable direct summands in a decomposition of the kQ-module $V/s^{b-1}(V)$. Then $(kQ)^t$ is a projective cover of $V/r^1(V)$ as a kQ-module. Multiplication by y^{b-1} on kP induces a surjective map $kP \to kQ$ which coincides with the map induced by the canonical projection $P = Q \times R \to Q$. In this way, all kQ-modules and homomorphisms can be viewed as kP-modules and homomorphisms. Thus we have a commutative diagram of kP-modules with exact rows (using Nakayama's Lemma for the surjectivity of the right map in the first row) of the form

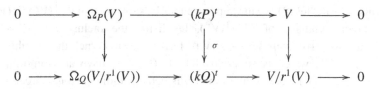

where σ is induced by multiplication with y^{b-1}. Since Ω_Q sends a projective kQ-module to zero, we get that $\Omega_Q(V/r^1(V))$ has at most $t-1$ indecomposable direct summands as a kQ-module. The left vertical map $\Omega_P(V) \to \Omega_Q(V/r^1(V))$ in this diagram is induced by multiplication with y^{b-1}, and hence induces an injective map $\Omega_P(V)/s^{b-1}(\Omega_P(V)) \to \Omega_Q(V/r^1(V))$. Thus $\Omega_P(V)/s^{b-1}(\Omega_P(V))$ has at most $t-1$ summands in a decomposition as a kQ-module, hence $t(\Omega_P(V)) \le t-1$, contradicting our choice of V. This proves that the one-dimensional kP-module $s^{b-1}/r^1(V)$ is a direct summand of $V/r^1(V)$. This shows also that $V/r^1(V)$ is a direct sum of $t+1$ indecomposable kQ-modules. If $V/r^1(V)$ has a nonzero projective summand as a kQ-module, then $\Omega_Q(V/r^1(V))$ is a direct sum of at most t indecomposable kQ-modules none of which is projective, hence the same is true for the submodule $\Omega_P(V)/s^{b-1}(\Omega_P(V))$. In particular, $\Omega_P(V)$ satisfies the same minimality assumption as V, hence $\Omega_P(V)/r^1(\Omega_P(V))$ is a direct sum of

$\Omega_P(V)/s^{b-1}(\Omega_P(V))$ and a one-dimensional module, so has no nonzero projective summand as a kQ-module. Therefore, after possibly replacing V by $\Omega_P(V)$, we may assume that $V/r^1(V)$ has no nonzero projective summand. This proves (1). We show now that the minimality assumption on $t(V)$ passes to the k-dual V^* of V.

(2) We have $t(V^*) \leq t(\Omega^n(V^*))$ for all integers n.

Dualising the inclusion $r^{b-1}(V) \to V$ yields a surjective map $V^* \to (r^{b-1}(V))^*$. Its kernel consists of all $v^* \in V^*$ that annihilate $r^{b-1}(V) = y^{b-1}V$, hence of all v^* that satisfy $y^{b-1} \cdot v^*$. Thus this kernel is equal to $s^{b-1}(V^*)$. This shows that we have an isomorphism $V^*/s^{b-1}(V^*) \cong r^{b-1}(V)^*$. Multiplication by y^{b-1} induces an isomorphism $V/s^{b-1}C \cong r^{b-1}(V)$. Since indecomposable kQ-modules are uniserial and selfdual, we get that $t(V^*) = t(V)$. Thus, for any integer n, we get $t(V^*) = t(V) \leq t(\Omega_P^{-n}(V)) = t((\Omega_P^{-n}(V))^*) = t(\Omega_P^n(V^*))$ as required. This proves (2). In particular, we can apply statement (1) to V^* instead of V. We will use this to show the following statement.

(3) As a kP-module, $r^{b-1}(V)$ is a direct summand of $s^1(V)$ of codimension one.

Since V^* satisfies (2), we may apply (1) to V^*, showing that $s^{b-1}(V^*)/r^1(V^*)$ is a direct summand of $V^*/r^1(V^*)$. Dualising the inclusion $s^1(V) \to V$ yields a surjective map $V^* \to s^1(V)^*$ having as a kernel the annihilator of $s^1(V)$ in V^*, which is equal to $r^1(V^*)$. Thus we have an isomorphism $V^*/r^1(V^*) \cong s^1(V)^*$. Under this isomorphism, the one-dimensional direct summand $s^{b-1}(V^*)/r^1(V^*)$ goes to the annihilator W of $r^{b-1}(V)$ in $s^1(V)^*$. Dualising again yields a split surjective map $s^1(V) \to W^*$ with kernel $r^{b-1}(V)$. This shows (3).

(4) We have $\mathrm{Res}_R^P(V) \cong k \oplus Y$ for some projective kR-module Y.

In order to prove (4), we argue by contradiction. Suppose that (4) does not hold. As before, since R is cyclic and V an endotrivial kP-module, we have $\mathrm{Res}_R^P(V) \cong \Omega_R(k) \oplus Y$ for some projective kR-module Y, and we also have $b \geq 3$, because if $b = 2$ then $\Omega_R(k) \cong k$. In particular, the composition length of $\Omega_R(k)$ is at least 2. This particular shape of $\mathrm{Res}_R^P(V)$ implies also that y does not annihilate $s^{b-1}(V)/r^2(V)$, a fact we will use later. We consider the inclusions of kP-modules

$$r^2(V) \subseteq r^1(s^{b-1}(V)) \subseteq r^1(V) \subseteq s^{b-1}(V).$$

Multiplication by y induces a kP-isomorphism

$$V/r^1(V) \cong r^1(V)/r^2(V)$$

which restricts to a kP-isomorphism

$$s^{b-1}(V)/r^1(V) \cong r^1(s^{b-1}(V))/r^2(V).$$

Thus, from (1), we get an isomorphism

$$r^1(V)/r^2(V) \cong r^1(s^{b-1}(V))/r^2(V) \oplus W_1$$

for some kP-submodule W_1 of $r^1(V)/r^2(V)$. Multiplication by y^{b-2} induces an isomorphism

$$s^{b-1}(V)/r^1(s^{b-1}(V)) \cong s^1(V)$$

which induces an isomorphism

$$r^1(V)/r^1(s^{b-1}(V)) \cong r^{b-1}(V).$$

Thus (3) implies that

$$s^{b-1}(V)/r^1(s^{b-1}(V)) = r^1(V)/r^1(s^{b-1}(V)) \oplus \bar{W}_2$$

for some one-dimensional kP-module $\bar{W}_2$. Note that $r^2(V) \subseteq r^1(s^{b-1}(V))$. Denote by W_2 the inverse image of $\bar{W}_2$ in $s^{b-1}(V)/r^2(V)$. Then

$$s^{b-1}(V)/r^2(V) = r^1(V)/r^2(V) + W_2 = W_1 + W_2,$$

where the second equality uses the fact that W_2 contains the one-dimensional complement $r^1(s^{b-1}(V))$ of W_1 in $r^1(V)/r^2(V)$. In particular, the kQ-module W_2 has dimension 2. Thus either x annihilates W_2 or W_2 is indecomposable of dimension 2. We show that the latter is impossible. Indeed, if W_2 is indecomposable, then its socle is one-dimensional, hence equal to

$$xW_2 = r^1(s^{b-1}(V))/r^2(V).$$

The kR-endomorphism φ of V considered in the proof of statement (1) leaves the submodules $r^1(V)$, $r^1(s^{b-1}(V))$, $r^2(V)$ invariant, hence induces a k-linear endomorphism $\bar{\varphi}$ on $s^{b-1}(V)/r^2(V)$ preserving xW_2 with the property that $\mathrm{Tr}_1^Q(\bar{\varphi})$ is equal to multiplication by x. By 7.1.10, since xW_2 is not a direct summand of W_2, it is the socle of a projective indecomposable direct summand of $s^{b-1}(V)/r^2(V) = W_1 + W_2$, hence of W_1, because W_2 has dimension 2. This however contradicts the choice of V in (1). Thus x annihilates W_2, or equivalently, the generator g of Q acts as identity on W_2. But all the arguments and constructions up to this point depend only on R and not on the choice of the complement Q of R in P. Thus gh acts as identity on W_2. But then $h = g^{-1}gh$

acts as identity on W_2, or equivalently, y annihilates W_2. Since y also annihilates W_1, it follows that y annihilates $s^{b-1}(V)/r^2(V)$, contradicting an observation made at the beginning of the proof of (4). This concludes the proof of (4).

(5) We have $V \cong k$.

By (3) we have $S^1(V) = r^{b-1}(V) \oplus X$ for some kP-module X. By (4) we have $\text{Res}_R^P(V) = k \oplus Y$ for some projective kR-module Y. It follows that $\dim_k(X) = 1$ and that $s^{b-1}(V) = r^1(V) \oplus X$. Moreover, X is a complement to $s^1(Y)$ in $k \oplus s^1(Y) = k \oplus r^{b-1}(Y)$. From (1) we have $V/r^1(V) = s^{b-1}(V)/r^1(V) \oplus \bar{W}$ for some kP-module $\bar{W}$. Denoting by W the inverse image of $\bar{W}$ in V we get that $V = s^{b-1}(V) + W = X + r^1(V) + W$. Nakayama's Lemma yields $V = X + W$; comparing dimensions yields $V = X \oplus W$. Since V was chosen indecomposable, this implies $W = \{0\}$, hence $V = X$ is one-dimensional. This proves 7.7.1 for P of rank 2.

We assume now that P has rank at least 3. Arguing by induction, we may assume that the result holds for any proper subgroup of P. Arguing by contradiction, we may assume that $\Omega^n(V) \not\cong k$ for any integer n.

(6) Let Q be a maximal subgroup of P. There is an integer n such that $\text{Res}_Q^P(\Omega_P^n(V)) \cong k \oplus Y$ for some nonzero projective kQ-module Y.

Since the result holds for Q, we have $\Omega_Q^n(V) \cong k$ for some integer n. Then $\text{Res}_Q^P(\Omega_P^n(V)) \cong k \oplus Y$ for some projective kQ-module Y, and our assumption on $\Omega_P^n(V) \not\cong k$ forces Y to be nonzero. This proves (6).

(7) The group P is elementary abelian.

Suppose not. Write $P = Z \times R$, where $Z = \langle g \rangle$ is cyclic of order $q \geq p^2$. Set $Z' = \langle g^p \rangle$; this is the unique maximal subgroup of Z. Set $Q = Z' \times R$; this is a maximal subgroup of P. By (6) there is an integer n such that $\text{Res}_Q^P(\Omega_P^n(V)) \cong k \oplus Y$ for some nonzero projective kQ-module Y. Let w be a nonzero element in $\text{soc}(kR)$; the annihilator of w in kR is thus $J(kR)$, and hence the annihilator of w in kQ is $kQJ(kR)$. Denote by M the annihilator in $\Omega_P^n(V)$ of w. Then $\Omega_P^n(V)/M \cong Y/J(kR)Y$. As a kZ'-module, this is projective, hence free. Therefore, by 7.1.3, $\Omega_P^n(V)/M$ is projective as a kZ-module. Let $a \in \Omega_P^n(V)$ such that the image $\bar{a}$ in $\Omega_P^n(V)/M$ generates a projective summand as a kZ-module; that is, such that $kZ\bar{a} \cong kZ$. Then $a \notin M$, so $wa \neq 0$. The projectivity of $kZ\bar{a}$ implies that $(g-1)^{q-1}wa \neq 0$. Since $(g-1)^{q-1}w$ generates the socle of kP we have $kPa \cong kP$. Then kPa is a direct summand of $\Omega_P^n(V)$, contradicting the indecomposability of this module. This shows (7).

The integer n in (6) is unique (as a consequence of 7.6.4) and does not depend on the choice of the group basis P of kP or of the choice of the maximal subgroup of P. Indeed, if P' is a subgroup of $(kP)^\times$ isomorphic to P such that $kP' = kP$ and if Q' is a maximal subgroup of P', there is a group isomorphism $P \cong P'$ mapping Q to Q'; this induces an algebra automorphism α of kP mapping P to P' and Q to Q'. Restriction along α is an exact functor which sends the trivial kQ-module to the trivial kQ'-module, projective kQ-modules to projective kQ'-modules and hence Heller translates of the trivial kQ-module to the corresponding Heller translates of the trivial kQ'-module. Since P has rank at least 3, we can write $P = \langle s \rangle \times \langle t \rangle \times R$ for some elements s, t of order p. Set $U = \Omega_P^n(V)/N$, where now N is the annihilator of $\mathrm{soc}(kR)$ in $\Omega_P^n(V)$. Set $S = \langle s \rangle \times \langle t \rangle$. We first show that U has no nonzero projective summand as a kS-module. Indeed, if it did, we could find an element $u \in \Omega_P^n(V)$ such that the image $\bar{u}$ in U has the property $kS\bar{u} \cong kS$. Thus $\mathrm{soc}(kS)$ does not annihilate $\bar{u}$, hence as before, $\mathrm{soc}(kP) = \mathrm{soc}(kS)\mathrm{soc}(kR)$ does not annihilate u, hence generates a projective summand $kPu \cong kP$ in $\Omega_P^n(V)$, a contradiction. Thus U has indeed no nonzero projective direct summand as a kS-module. Let $z = \lambda(s - 1) + \mu(t - 1)$ be a nonzero linear combination of $s - 1, t - 1$. If $\lambda \neq 0$, set $P' = \langle 1 + z \rangle \times \langle t \rangle \times R$; otherwise set $P' = \langle s \rangle \times \langle 1 + z \rangle \times R$. In both cases, P' is a group basis of kP and $Q = \langle 1 + z \rangle \times R$ is a maximal subgroup of P'. Thus, by (6), we have $\mathrm{Res}_Q^{P'}(\Omega_P^n(V)) \cong k \oplus Y$ for some projective kQ-module Y. Thus $\mathrm{Res}_Q^{P'}(U) \cong Y/J(kR)Y$ is projective as a $k\langle 1 + z \rangle$-module. But then U is projective, by 7.6.3. This is only possible if $U = \{0\}$, hence $Y = \{0\}$, forcing $\Omega_P^n(V)$ to have dimension 1. This concludes the proof of 7.7.1. □

7.8 Endopermutation modules over abelian p-groups

The main result of this section is Dade's classification of the endopermutation modules of abelian p-groups. We assume in this section that k is algebraically closed.

Theorem 7.8.1 ([55, Theorem 12.5]) *Let P be a finite abelian p-group and let U be an indecomposable endopermutation $\mathcal{O}P$-module having P as its vertex. There is an $\mathcal{O}P$-module T of $\mathcal{O}$-rank 1, unique up to isomorphism, and for any subgroup Q of index at least 3 in P there is an integer $n_Q \geq 0$ such that U is isomorphic to a direct summand of the tensor product*

$$T \otimes_{\mathcal{O}} (\otimes_Q \mathrm{Inf}_{P/Q}^P(\Omega_{P/Q}^{n_Q}(\mathcal{O})))$$

where in the tensor product, Q runs over all subgroups of P of index at least 3. Moreover, n_Q is unique whenever P/Q is not cyclic. In particular, the Dade group $D_{\mathcal{O}}(P)$ is isomorphic to a direct product

$$D_{\mathcal{O}}(P) \cong \operatorname{Hom}(P, \mathcal{O}^\times) \times \prod_Q T_k(P/Q)$$

where as before Q runs over the subgroups of P of index at least 3, and the canonical map $D_{\mathcal{O}}(P) \to D_k(P)$ is surjective with kernel $\operatorname{Hom}(P, \mathcal{O}^\times)$.

Proof By 7.3.12 we may assume that $\mathcal{O} = k$. Let $\mathcal{C}$ be an upwardly closed set of subgroups of P; that is, if $Q \le R \le P$ and if Q belongs to $\mathcal{C}$, then R belongs to $\mathcal{C}$. Set

$$D(\mathcal{C}) = \cap_{Q \in \mathcal{C}} \ker(\operatorname{Defres}_{P/Q}^P).$$

We will show by induction over the number of subgroups in $\mathcal{C}$ that we have

$$D_k(P) = D(\mathcal{C}) \times \prod_{Q \in \mathcal{C}} \operatorname{Inf}_{P/Q}^P(T_k(P/Q)).$$

If $\mathcal{C}$ consists of a single subgroup of P, then $\mathcal{C} = \{P\}$, since $\mathcal{C}$ is upwardly closed. In that case we have $D(\mathcal{C}) = D_k(P)$ and $T_k(P/P)$ is trivial, so the above decomposition holds trivially. Suppose now that $\mathcal{C}$ contains at least two subgroups of P. Let R be a minimal subgroup of P such that R belongs to $\mathcal{C}$. Set $\mathcal{C}' = \mathcal{C} \setminus \{R\}$. By induction we have

$$D_k(P) = D(\mathcal{C}') \times \prod_{Q \in \mathcal{C}'} \operatorname{Inf}_{P/Q}^P(T_k(P/Q)).$$

Therefore, it suffices to show that

$$D(\mathcal{C}') = D(\mathcal{C}) \times \operatorname{Inf}_{P/R}^P(T_k(P/R)).$$

We have $D(\mathcal{C}) = D(\mathcal{C}') \cap \ker(\operatorname{Defres}_{P/R}^P)$. It follows from 7.5.10 applied to the group P/R that the image of $D(\mathcal{C}')$ under the map $\operatorname{Defres}_{P/R}^P$ is contained in the subgroup $T_k(P/R)$ because every nontrivial subgroup of P/R is of the form Q/R for a subgroup Q that properly contains R and hence belongs to $\mathcal{C}'$ by the minimality of R in $\mathcal{C}$. Moreover, it follows from 7.5.11 that $\operatorname{Inf}_{P/R}^P(T_k(P/R))$ is indeed a subgroup of $D(\mathcal{C}')$ that intersects $\ker(\operatorname{Defres}_{P/R}^P)$ trivially. This proves $D(\mathcal{C}') = D(\mathcal{C}) \times \operatorname{Inf}_{P/R}^P(T_k(P/R))$, hence $D_k(P) = D(\mathcal{C}) \times \prod_{Q \in \mathcal{C}} \operatorname{Inf}_{P/Q}^P(T_k(P/Q))$ for any upwardly closed set $\mathcal{C}$ of subgroups of P. By taking for $\mathcal{C}$ the set of all nontrivial subgroups of P, it follows again from 7.5.10 that $D_k(P) = \prod_Q \operatorname{Inf}_{P/Q}^P(T_k(P/Q))$, where now Q runs over all subgroups of P. Translating this back to endopermutation modules this means that if V is an indecomposable endopermutation kP-module with vertex P, then V is a direct summand of $\otimes_Q V_Q$, where $\otimes_Q$ is the tensor product taken over all subgroups Q of P, of indecomposable endotrivial kP/Q-modules V_Q inflated to P, which are uniquely determined up to isomorphism. If Q has index at most 2 in P, then V_Q is trivial, so we may ignore these factors, and if Q has index at least 3, then $V_Q \cong \Omega_{P/Q}^{n_Q}(k)$ for some integer n_Q thanks to the classification of endotrivial modules in 7.7.1. This completes the proof of 7.8.1. $\qquad\square$

The following result, due to Puig, shows that under some compatibility assumptions, families of endopermutation modules of quotients of a finite abelian p-group P can be 'glued together', in a way that is compatible with group actions on P. This should be viewed as a statement about the image of the map d in the exact sequence in 7.5.10. We prove this here as a consequence of 7.8.1; the original proof of 7.8.2 in [175, 3.6] circumvents 7.8.1, using a Möbius inversion argument instead.

Theorem 7.8.2 ([175, Proposition 3.6]) *Let P be a finite abelian p-group and E a subgroup of* $\mathrm{Aut}(P)$. *Let C be a nonempty upwardly closed E-stable set of subgroups of P; that is, if $Q \leq R \leq P$ with Q in C, then R and $y(Q)$ belong to C for all $y \in E$. For any subgroup Q in C let V_Q be an indecomposable endopermutation kP/Q-module with vertex P/Q. Suppose that if Q, R belong to C such that $Q \leq R$, then* $\mathrm{Defres}_{P/R}^{P/Q}([V_Q]) = [V_R]$, *where we have identified P/R and $(P/Q)/(R/Q)$. Then there is an indecomposable endopermutation kP-module V with vertex P such that* $\mathrm{Defres}_{P/Q}^{P}(V) = V_Q$ *for any subgroup Q in C. In addition, if* $^y(V_Q) \cong V_{y(Q)}$ *for all $y \in E$ and all Q in C, then $^yV \cong V$ for all $y \in E$.*

Proof Let Q be a subgroup of P contained in C. By 7.8.1 applied to P/Q, the module V_Q is isomorphic to a direct summand of the tensor product

$$\otimes_{Q \leq S \leq P} \mathrm{Inf}_{P/S}^{P/Q}(\Omega_{P/S}^{n(Q,S)}(k))$$

for some integers $n(Q, S)$, with S running over the subgroups of index at least 3 of P containing Q. If P/S is cyclic we chose $n(Q, S)$ to be 0 or 1 in order to ensure that these integers are uniquely determined. By 7.5.11, the group homomorphism $\mathrm{Defres}_{P/R}^{P}$ has $\mathrm{Inf}_{P/S}^{P}(T_k(P/S))$ in its kernel whenever R is not contained in S. The compatibility condition means therefore that if $Q \leq R \leq S$, then $n(Q, S) = n(R, S)$. In particular, $n(Q, S) = n(S, S)$. Define V as an indecomposable direct summand of $\otimes_{S \leq P} \mathrm{Inf}_{P/S}^{P}(\Omega_{P/S}^{n(S,S)}(k))$. Using 7.5.11 one verifies that V has the property $\mathrm{Defres}_{P/R}^{P}([V]) = [V_R]$ for all R in C. By 7.3.9, the kP-module $\mathrm{Defres}_{P/R}^{P}(V)$ is indecomposable, so that $\mathrm{Defres}_{P/R}^{P}(V) = V_R$. The E-stability of the family of modules V_Q is equivalent to $n(Q, S) = n(y(Q), y(S))$ for $Q \leq S \leq P$, with Q in C, and $y \in E$, hence to the E-stability of the isomorphism class of V. $\qquad\square$

The next result investigates under what circumstances an endopermutation kP-module can be extended to a finite group containing the abelian p-group P as a normal subgroup; this is the crucial ingredient for Dade's fusion splitting theorem 7.9.2 below.

Theorem 7.8.3 *Let P be an abelian normal p-subgroup of a finite group G and let V be an indecomposable endopermutation kP-module with vertex P that is G-stable; that is, that satisfies ${}^x V \cong V$ for all $x \in G$. Then there is an indecomposable kG-module W such that $\mathrm{Res}_P^G(W)$ is an endopermutation kP-module satisfying $\mathrm{Res}_P^G(W) \cong V \oplus V'$, where every indecomposable direct summand of V' has a proper subgroup of P as a vertex.*

Proof There is one case where this is immediate: if $V \cong \Omega_P^n(k)$ for some integer n, then $W = \Omega_G^n(k)$ satisfies $\mathrm{Res}_P^G(W) \cong V \oplus V'$ for some projective kP-module V'. The general case follows from this observation and Dade's classification in 7.8.1. Note first that if the hypotheses and conclusion hold for two endopermutation kP-modules V_1, V_2, then also for the unique summand of $V_1 \otimes_k V_2$ with vertex P. By 7.8.1, V is a direct summand of

$$U = \otimes_Q \mathrm{Inf}_Q^P(\Omega_{P/Q}^{n_Q}(k)),$$

where $\otimes_Q$ is the tensor product over k with Q running over the subgroups of P, and where $n_Q \in \mathbb{Z}$. We choose n_Q to be 0 or 1 if P/Q is cyclic. The G-stability of the isomorphism class of V is equivalent to $n_Q = n_R$ for any two G-conjugate subgroups Q, R of P. For a fixed subgroup Q of P, set

$$U_Q = \otimes_R \mathrm{Inf}_R^P(\Omega_{P/R}^{n_R}(k)),$$

where R runs over the subgroups of Q that are G-conjugate to Q. Note that $n_R = n_Q$ for any G-conjugate R of Q, hence U_Q is G-stable. Moreover, U is the tensor product of the U_Q, with Q running over a set of representatives of the G-conjugacy classes of subgroups of P. In order to prove the theorem, we may therefore assume that $U = U_Q$ for some fixed proper subgroup Q of P. We set $n = n_Q$ and

$$W = \mathrm{Ten}_{N_G(Q)}^G(\mathrm{Inf}_{N_G(Q)/Q}^{N_G(Q)}(\Omega_{N_G(Q)/Q}^n(k))).$$

If x runs over a set of representatives of $G/N_G(Q)$ then ${}^x Q$ runs over the set of different G-conjugates of Q. An easy verification shows that $\mathrm{Res}_P^G(W)$ is isomorphic to

$$\otimes_R \mathrm{Res}_P^{N_G(R)}(\mathrm{Inf}_{N_G(R)/R}^{N_G(R)}(\Omega_{N_G(R)/R}^n(k)))$$

which is clearly isomorphic to

$$\otimes_R \mathrm{Inf}_{P/R}^P (\mathrm{Res}_{P/R}^{N_G(R)/R}(\Omega_{N_G(R)/R}^n(k))),$$

where R runs over the different G-conjugates of Q. By our first observation this is isomorphic to $U \oplus U'$, where U' is a direct sum of endopermutation kP-modules with vertices strictly contained in P. By the construction of U, we also have $U \cong V \oplus V'$, with V' a direct sum of endopermutation kP-modules with vertices strictly contained in P. Thus $\mathrm{Res}_P^G(W)$ has V as a direct summand, and every other summand has a vertex strictly smaller than P. If W is not indecomposable, replace W by an indecomposable direct summand whose restriction to P has V as a summand. The result follows.   □

Corollary 7.8.4 *Let G be a finite group having a normal abelian Sylow p-subgroup P. Let V be an indecomposable G-stable endopermutation $\mathcal{O}P$-module with vertex P. Then there is an indecomposable $\mathcal{O}G$-module W such that $\mathrm{Res}_P^G(W) \cong V$. In particular, W has P as a vertex and V as a source.*

Proof Let W be an indecomposable $\mathcal{O}G$-module such that $\mathrm{Res}_P^G(W)$ has a direct summand isomorphic to V. Since P is a Sylow p-subgroup of G, it follows that P is a vertex of W, hence W is isomorphic to a direct summand of $\mathrm{Ind}_P^G(V)$. Since P is normal in G, the G-stability of V implies that $\mathrm{Res}_P^G(\mathrm{Ind}_P^G(V))$ is isomorphic to a direct sum of $|G : P|$ copies of V. Thus $\mathrm{Res}_P^G(W)$ is isomorphic to a direct sum of copies of V. By 7.3.17, $k \otimes_\mathcal{O} W$ remains an indecomposable kG-module whose restriction to kP is a direct sum of copies of $k \otimes_\mathcal{O} V$. By 7.8.3, the module W can be chosen in such a way that $\mathrm{Res}_P^G(W)$ has exactly one indecomposable direct summand with vertex P. The result follows.   □

Remark 7.8.5 Theorem 7.8.1 implies the following rationality property: if P is a finite abelian p-group, then every endopermutation kP-module is defined over the prime field $\mathbb{F}_p$ and lifts to an endopermutation $\mathcal{O}P$-module defined over the subring W of $\mathcal{O}$ isomorphic to the Witt vectors of $\mathbb{F}_p$.

7.9 Splitting of fusion in endopermutation modules

Dade's fusion splitting Theorem 7.9.2, announced in [56], is one of the cornerstones in block theory. The original proof in [57] has remained unpublished. Puig has given a proof in [168]. We state this theorem as in [168, (e)], which says, roughly speaking, that if S is an interior P-algebra over k with structural homomorphism $\sigma : P \to S^\times$ and a P-stable k-basis such that $S(P)$ is nonzero, then the Brauer homomorphism $\mathrm{Br}_P : S^P \to S(P)$ extends to a group homo-morphism $N_{S^\times}(P) \to S(P)^\times$ in a way that is compatible with the conjugation by elements in $N_{S^\times}(P)$, where $N_{S^\times}(P)$ denotes the normaliser in $S^\times$ of the sub-group $\sigma(P)$ of $S^\times$. We assume in this section that k is algebraically closed.

Theorem 7.9.1 *Let P be a finite p-group and V an endopermutation kP-module having an indecomposable direct summand with vertex P. Set $S = \mathrm{End}_k(V)$ and denote by $\sigma : P \to S^\times$ the structural map given by the action of P on V. Denote by $N_{S^\times}(P)$ the normaliser of $\sigma(P)$ in $S^\times$. There is a group homo-morphism $f : N_{S^\times}(P) \to S(P)^\times$ that satisfies $f(s) = \mathrm{Br}_P(s)$ for all $s \in (S^P)^\times$ and $f(y)\mathrm{Br}_P(s)f(y^{-1}) = \mathrm{Br}_P(ysy^{-1})$ for all $y \in N_{S^\times}(P)$ and all $s \in S^P$.*

Proof After replacing P by $P/\ker(\sigma)$ we may assume that σ is injective, or equivalently, that P acts faithfully on V. Set $N = N_{S^\times}(P)$ and $C = C_N(P) = (S^P)^\times$. Thus N/C is isomorphic to a subgroup of $\mathrm{Aut}(P)$; in particular, the group $E = N/C$ is finite and acts on P; denote by $P \rtimes E$ the corresponding semidirect product.

We consider first the case where P is abelian and V is indecomposable. By 7.8.3 there is an endopermutation kP-module V' all of whose indecomposable summands have vertices strictly smaller than P, such that $W = V \oplus V'$ is an endopermutation kP-module which extends to a $k(P \rtimes E)$-module. Any $k(P \rtimes E)$-module can be viewed as a $k(P \rtimes N)$-module via inflation along the canon-ical surjection $N \to E = N/C$. Set $T = \mathrm{End}_k(W)$. Denote by $\tau : P \rtimes N \to T^\times$ the group homomorphism determined by a $k(P \rtimes E)$-module structure of W extending that of P. Note that C is contained in the kernel of τ. We have $T(P) \cong S(P) \cong k$, where the first isomorphism uses that no summand of V' has vertex P, and the second isomorphism uses that V is indecomposable with vertex P. Denote by $i \in T$ the projection of W onto V with kernel V'. This is a primitive idempotent in T^P satisfying $\mathrm{Br}_P(i) = 1_k$ and $\mathrm{Br}_P(1 - i) = 0$. We identify S with iTi. Since V is, up to isomorphism, the unique summand of W with vertex P, there is a unique local point of P on T, and this unique local point contains i and all $T^\times$-conjugates of i. Thus, for $y \in N$, there is $c(y) \in (1 + \ker(Br_P)) \cap T^\times$ such that $^{\tau(y)}i = {}^{c(y)}i$. Then $c(y)^{-1}\tau(y)$ centralises i and acts on the image of P in $S = iTi$ exactly as y does. Thus

$$\rho(y) = c(y)^{-1}\tau(y)i$$

is an element in $S^\times$ that acts on the image of P in $S^\times$ as y. It follows that $y\rho(y)^{-1}$ acts trivially on the image of P in S, hence $y\rho(y)^{-1} \in (S^P)^\times$. We set

$$f(y) = \mathrm{Br}_P(y\rho(y)^{-1})$$

for all $y \in N$, and we will show that f determines a group homomorphism as claimed in the statement. The definition of f does not depend on the choices of the elements $c(y)$, since these were chosen in $1 + \ker(\mathrm{Br}_P)$. We show next that f extends the Brauer homomorphism. Suppose that $y \in C = (S^P)^{\times}$. Then $\tau(y) = 1_T$, by the construction of τ, and hence $\rho(y) \in i + \ker(\mathrm{Br}_P)$, implying that $f(y) = \mathrm{Br}_P(y)$ in that case. Thus f extends Br_P. Let $y \in N$ and $s \in S^P$. Using that $S(P) \cong k$ is commutative, we get that $f(y)\mathrm{Br}_P(s)f(y^{-1}) = \mathrm{Br}_P(s)$, and the induced action by N on $S(P)$ is trivial, so this is also equal to $\mathrm{Br}_P(ysy^{-1})$. Thus the compatibility of Br_P with conjugation by $f(y)$ and y is trivial in this case. We need to show that f is a group homomorphism. Let $y, z \in N$. Conjugation by $\rho(y)$ on S induces a conjugation action on S^P, which in turn induces the trivial action on $k \cong S(P)$. Thus $z\rho(z)^{-1}$ and its conjugate $\rho(y)z\rho(z)^{-1}\rho(y)^{-1}$ have the same image in $S(P)$. Multiplying both elements by $y\rho(y)^{-1}$ implies that the two elements $y\rho(y^{-1})z\rho(z)^{-1}$ and $yz\rho(z)^{-1}\rho(y)^{-1}$ have both the same image in $S(P)$. Since $i + \ker(\mathrm{Br}_P)$ is normalised by N, it follows that the second of these two elements is equal to $yz\rho(yz)^{-1}c$ for some invertible element $c \in 1 + \ker(\mathrm{Br}_P)$. This shows that these two elements have the same image in $S(P)$, and hence that f is a group homomorphism that extends Br_P as stated.

We consider next the case where P is still abelian, but V is no longer assumed to be indecomposable. Let $y \in N$. Choose $\theta(y) \in S(P)^{\times}$ such that the action induced by y on $S(P)$ via Br_P coincides with the conjugation action by $\theta(y)$; that is, such that

$$\mathrm{Br}_P(ysy^{-1}) = \theta(y)\mathrm{Br}_P(s)\theta(y)^{-1}$$

for all $s \in S^P$. The elements $\theta(y)$ are defined only up to a scalar, hence there is a 2-cocycle β with coefficients in $k^{\times}$ satisfying

$$\theta(y)\theta(z) = \beta(y, z)\theta(yz)$$

for all $y, z \in N$. We will show first that the $\theta(y)$ can be chosen such that β is constant 1, and then that $f(y) = \theta(y)$ satisfies the statement. Let i be a primitive idempotent in S^P belonging to the unique local point of P on S; that is, $i(V)$ is an indecomposable direct summand of V with vertex P. The image $j = \mathrm{Br}_P^S(i)$ is then a primitive idempotent in $S(P)$, and hence $(iSi)(P) \cong jS(P)j \cong k$. Since N stabilises the unique local point of P on S, for any $y \in N$ there is an element $d(y) \in (S^P)^{\times}$ such that $^{y}i = {}^{d(y)}i$. Then $d(y)^{-1}y$ centralises i, and $\bar{d}(y)^{-1}\theta(y)$ centralises the image j of i in $S(P)$, where $\bar{d}(y) = \mathrm{Br}_P(d(y))$. Since $jS(P)j \cong k$, we may modify $\theta(y)$ by a nonzero scalar such that $\bar{d}(y)^{-1}\theta(y)j = j$ for all $y \in N$. It follows from the above that

$$\sigma(y) = d(y)^{-1}yi$$

belongs to $N_{(iSi)^\times}(P)$ and acts on the image of P in iSi as the action of y on P. The first part implies that there is a group homomorphism $g:$ $N_{(iSi)^\times}(P) \to ((iSi)^P)^\times \cong k^\times$ extending Br_P^{iSi}. After possibly multiplying $d(y)$ by a nonzero scalar, we may assume that $\sigma(y) \in \ker(g)$. Let $y, z \in N$. Then $\sigma(y)\sigma(z)\sigma(yz)^{-1} \in \ker(g) \cap (iS^P i)^\times = i + \ker(\mathrm{Br}_P^{iSi})$. We have

$$\sigma(y)\sigma(z)\sigma(yz)^{-1} = d(y)^{-1}yd(z)^{-1}z(yz)^{-1}d(yz)i = d(y)^{-1}(^y(d(z)^{-1}))d(yz)i.$$

The left side of this equation is contained in $i + \ker(\mathrm{Br}_P^{iSi})$, hence applying Br_P yields the equation

$$j = \bar{d}(y)^{-1}(^y(\bar{d}(z)^{-1}))\bar{d}(yz)j.$$

A similar calculation, with the elements $\bar{d}(y)^{-1}\theta(y)j = j$ instead of the elements $\sigma(y)$ yields the equation

$$j = \beta(y, z)\bar{d}(y)^{-1}(^y(\bar{d}(z)^{-1}))\bar{d}(yz)j$$

for all $y, z \in N$, hence $\beta(y, z) = 1$. This shows that the map defined by $f(y) = \theta(y)$ is a group homomorphism from $N_{S^\times}(P)$ to $S(P)^\times$ that extends Br_P and that by the construction of the elements $\theta(y)$ above satisfies the compatibility property in the statement. This shows 7.9.1 in the case where P is abelian.

For general P we argue by induction over the order of P. Assume that P is not abelian. Set $Z = Z(P)$. Thus Z is a proper nontrivial characteristic subgroup of P. In particular, the group $N_{S^\times}(P)$ is contained in $N_{S^\times}(Z)$, hence acts on $S(Z)$. Since P centralises Z, it follows that the structural homomorphism $P \to S^\times$ induces a homomorphism $P/Z \to S(Z)^\times$, and we have $S(Z) \cong \mathrm{End}_k(W)$ for some endopermutation kP/Z-module W. Since $S(P) \cong S(Z)(P/Z)$, the induction hypothesis applied to P/Z implies that there is a group homomorphism $g : N_{S(Z)^\times}(P/Z) \to S(P)^\times$ that extends $\mathrm{Br}_{P/Z}$ in a way that is compatible with conjugation by elements in $N_{S(Z)^\times}(P/Z)$. Since Z is abelian, there is by the previous arguments also a group homomorphism $f : N_{S^\times}(Z) \to S(Z)^\times$ that extends Br_Z in a way that is compatible with conjugation by elements in $N_{S^\times}(Z)$. Thus the group homomorphism f sends $N_{S^\times}(P)$ to $N_{S(Z)^\times}(P/Z)$. It follows that the composition $g \circ f : N_{S^\times}(P) \to S(P)^\times$ satisfies the statement. $\square$

The group homomorphism f in Theorem 7.9.1 is not unique, as it can be 'modified' by any group homomorphism from $N_{S^\times}(P)$ to $k^\times$. The following slightly less precise reformulation is frequently useful.

Theorem 7.9.2 *Let P be a finite p-group and V an endopermutation kP-module having an indecomposable direct summand with vertex P. Set $S = \mathrm{End}_k(V)$ and denote by $\sigma : P \to S^\times$ the structural map given by the action of P on V. Let E be a group that acts on P and on S such that $\sigma(^yu) = {}^y\sigma(u)$*

for all $u \in P$ and $y \in E$. Then E acts on $S(P)$, and the central extensions of E by $k^{\times}$ determined by the actions of E on S and on $S(P)$ are isomorphic. In particular, if V is indecomposable, then the kP-module structure of V extends to a $k(P \rtimes E)$-module structure, or equivalently, the central extension of E by $k^{\times}$ determined by the action of E on S splits.

Proof We may assume that σ is injective, after possibly replacing P by $P/\ker(\sigma)$. By 7.9.1, there is a group homomorphism $f : N_{S^{\times}}(P) \to S(P)^{\times}$ which satisfies $f(s) = \mathrm{Br}_P(s)$ for all $s \in S^P$ and $f(x)\mathrm{Br}_P(s)f(x^{-1}) = \mathrm{Br}_P(xsx^{-1})$ for all $x \in N_{S^{\times}}(P)$ and all $s \in S^P$. For $y \in E$, choose $\hat{y} \in N_{S^{\times}}(P)$ acting on $S^{\times}$ as y; in particular, $\hat{y} \in N_{S^{\times}}(P)$. Let $\alpha \in Z^2(E; k^{\times})$ such that $\widehat{yz} = \alpha(y, z)\hat{y}\hat{z}$. Then $\tilde{y} = f(\hat{y})$ acts as y on $S(P)^{\times}$. Since f is a group homomorphism, it follows that $\tilde{y}\tilde{z} = \alpha(y, z)\widetilde{yz}$. Thus the central extensions of E determined by the actions of E on S and on $S(P)$ are isomorphic. If V is indecomposable, then $S(P) \cong k$, so the class determined by the action of E on $S(P)$ is trivial, and hence so is the class determined by the action of E on S, implying the last statement. □

The isomorphism between the central extensions of E in Theorem 7.9.2 is noncanonical, since the group homomorphism f in Theorem 7.9.1 is not unique.

7.10 Endopermutation modules and p-nilpotent groups

Blocks of p-nilpotent finite groups give rise to endopermutation modules, obtained as sources of simple modules. They are special cases of nilpotent blocks which we will consider in Section 8.11. Let G be a *p-nilpotent* finite group; that is, $G = MP$ for some normal p'-subgroup M of G and a Sylow p-subgroup P of G. We assume in this section that k and K are splitting fields for the finite groups and their subgroups which arise in the statements. Since the characteristic p of k does not divide the order of M, it follows that the group algebra $\mathcal{O}M$ is a direct product of matrix algebras indexed by the irreducible characters of M; that is,

$$\mathcal{O}M = \prod_{\eta \in \mathrm{Irr}_K(M)} S_{\eta},$$

where $S_{\eta} = \mathcal{O}Me(\eta) \cong \mathrm{End}_{\mathcal{O}}(Y_{\eta})$ for some $\mathcal{O}M$-module Y_{η} having η as its character, and where $e(\eta) = \frac{\eta(1)}{|M|} \sum_{y \in M} \eta(y^{-1})y$ is the central primitive idempotent in $Z(\mathcal{O}M)$ determined by η. Again since M is a p'-group, every simple kM-module is of the form $\bar{S}_{\eta} = k \otimes_{\mathcal{O}} S_{\eta}$, and the Brauer character of $\bar{S}_{\eta}$ is equal to η. The action of P on $\mathcal{O}M$ permutes the factors in the above decomposition of $\mathcal{O}M$ as a product of matrix algebras.

Theorem 7.10.1 ([55, Theorem 13.13]) *Let $G = MP$ be a finite p-nilpotent group with normal p'-subgroup M and Sylow p-subgroup P. Let $\eta \in \mathrm{Irr}_K(M)$ be the character of an $\mathcal{O}$-free $\mathcal{O}M$-module Y_η. Denote by Q the stabiliser in P of the matrix factor $S_\eta = \mathcal{O}Me(\eta) \cong \mathrm{End}_{\mathcal{O}}(Y_\eta)$ of $\mathcal{O}M$.*

   (i) *The $\mathcal{O}M$-module Y_η extends canonically to an $\mathcal{O}MQ$-module, still denoted Y_η, and $\mathrm{Res}_Q^{MQ}(Y_\eta)$ is an endopermutation $\mathcal{O}Q$-module having an indecomposable direct summand V_Q with vertex Q.*
  (ii) *We have $\mathcal{O}MQe(\eta) \cong \mathrm{End}_{\mathcal{O}}(Y_\eta) \otimes_{\mathcal{O}} \mathcal{O}Q$ as interior Q-algebras.*
 (iii) *The $\mathcal{O}G$-module $X_\eta = \mathrm{Ind}_{MQ}^G(Y_\eta)$ has an irreducible character χ_η and the kG-module $\bar{X}_\eta = k \otimes_{\mathcal{O}} X_\eta$ is simple.*
  (iv) *The modules X_η and $\bar{X}_\eta$ have Q as a vertex and endopermutation source V_Q and $\bar{V}_Q = k \otimes_{\mathcal{O}} V_Q$, respectively.*
   (v) *If b_η is the block of $\mathcal{O}G$ to which X_η belongs and $\bar{b}_\eta$ its image in kG, then $\bar{X}_\eta$ is, up to isomorphism, the unique simple $kG\bar{b}_\eta$-module.*
  (vi) *We have $b_\chi = \sum_{\eta'} e(\eta')$, where η' runs over the P-orbit of η in $\mathrm{Irr}_k(M)$.*
 (vii) *We have $\mathcal{O}Gb_\eta \cong M_n(\mathcal{O}Q)$ as $\mathcal{O}$-algebras, where $n = |P : Q|\eta(1)$.*
(viii) *The correspondence sending η to $\bar{X}_\eta$ and b_η induces bijections between the P-orbits in $\mathrm{Irr}_K(M)$, the isomorphism classes of simple kG-modules, and the blocks of $\mathcal{O}G$.*
  (ix) *Every simple kG-module has an endopermutation module as a source and every simple kG-module lifts to an $\mathcal{O}$-free $\mathcal{O}G$-module having an irreducible character.*

Proof We may assume that k is algebraically closed, since the statement involves only finitely many modules with irreducible characters. The point of this preamble is that this allows us to apply results from Section 5.3 without further comment; one could alternatively adapt the quoted results from that section along the lines of 5.3.11. Set $S = S_\eta$ and $Y = Y_\eta$. Since $\eta(1)$ divides $|M|$, hence is prime to p, it follows from 5.3.4 that the action of Q on the matrix algebra $S = \mathcal{O}Me(\eta)$ lifts uniquely to a group homomorphism $\sigma : Q \to S^\times$ satisfying $\det(\sigma(u)) = 1$ for all $u \in Q$, or equivalently, the $\mathcal{O}M$-module structure on Y extends canonically to $\mathcal{O}MQ$. Since S is a direct summand of $\mathcal{O}M$ with respect to the action of Q, it follows that S has a Q-stable basis. Thus the canonical isomorphism $Y \otimes_k Y^* \cong S$ implies that the restriction to Q of Y is an endopermutation $\mathcal{O}Q$-module. Its rank $\eta(1)$ is prime to p, and hence as an $\mathcal{O}Q$-module, Y has an indecomposable direct summand with vertex Q. This shows (i). We have $\mathcal{O}MQe(\eta) = SQ$ as interior Q-algebras. A straightforward verification shows that we have an $\mathcal{O}$-algebra isomorphism $SQ \cong S \otimes_{\mathcal{O}} \mathcal{O}Q$ sending su to $s\sigma(u) \otimes u$, for $u \in Q$ and $s \in S$. This is an isomorphism of interior Q-algebras, where the interior Q-structure on the right side is given by the diagonal map sending $u \in Q$ to $\sigma(u) \otimes u$, whence (ii). Note that this implies that $\mathcal{O}MQe(\eta)$ is Morita equivalent to $\mathcal{O}Q$, and hence that $\mathcal{O}MQe(\eta)$ is a block of $\mathcal{O}MQ$ with a unique isomorphism class of simple modules, namely that of $k \otimes_{\mathcal{O}} Y$. The statements (iii), (iv), (v), (vi), (vii) and (viii) spell out in detail the general Clifford theoretic results on blocks of normal subgroups in 6.8.9 and 6.8.3. Statement (ix) is clear from the previous statements. $\square$

A result of Berger [17], [18] and Feit [69] shows that simple modules of p-solvable finite groups are *algebraic*, a concept due to Alperin. Let G be a finite group. A finitely generated kG-module U is called *algebraic* if its image $[U]$ in the Green ring is the root of a nonzero polynomial with integer coefficients. Alperin proved that if U is an indecomposable kG-module with vertex P and kP-source V, then U is algebraic if and only if V is an algebraic kP-module. It is easy to see that an indecomposable endopermutation kP-module V with vertex P is algebraic if and only if its image in the Dade group is a torsion element. An immediate consequence of the result of Berger and Feit is that only torsion endopermutation modules can show up as sources of simple modules over a p-nilpotent group; this was noted independently by Puig [172] and by Boltje and Külshammer [21]. Our presentation follows Salminen [199]. This result requires the following fact which for odd p is presently known only thanks to the classification of finite simple groups: if M is a finite simple group of order prime to p, then the Sylow p-subgroups of Aut(M) are cyclic. In the case $p = 2$ the group M has odd order, hence is solvable by the Odd Order Theorem of Feit and Thompson [71]. Further basic group theoretic facts required include the following: if G is a finite group such that $G/Z(G)$ is cyclic, then $G = Z(G)$ is abelian, if $G = [G, G]$ has the property that $G/Z(G)$ is a direct product of isomorphic finite nonabelian simple groups, then the inverse images of these simple groups in G commute pairwise, and if $G = [G, G]$ such that $G/Z(G)$ is nonabelian simple, then any nontrivial automorphism of G induces a nontrivial automorphism of $G/Z(G)$.

Theorem 7.10.2 *Let G be a finite p-nilpotent group, X a simple kG-module, Q a vertex of X and V a kQ-source of X. Then V is an indecomposable endopermutation kQ-module with vertex Q whose image in the Dade group $D_k(Q)$ has finite order.*

Proof By 7.10.1 (ix), V is an endopermutation $\mathcal{O}P$-module. In order to show that the class of V is a torsion element of $D_k(Q)$, we proceed inductively. In view of 7.10.1 we may assume that $G = MQ$ for some normal p'-subgroup M of G, so that Q is the Sylow p-subgroup of G, and that V is an indecomposable direct summand of $\mathrm{Res}_Q^G(X)$ for some simple kG-module X whose restriction to kM remains simple with a character η such that $e(\eta)$ is the block idempotent of kG of the block to which X belongs. Suppose that N is a subgroup of M that is normal in G. Let $\theta \in \mathrm{Irr}_k(N)$ such that $e(\theta)e(\eta) \neq 0$. Then by 6.8.3, X is induced from the stabiliser of $e(\theta)$ in G. We may therefore assume that η is G-stable, hence uniquely determined by η. Repeating this, if necessary, we may assume that for any normal subgroup N of G contained in M there is a unique $\eta_N \in \mathrm{Irr}_k(N)$ satisfying $e(\eta_N)e(\eta) = e(\eta)$; in particular, G stabilises η_N. We may further assume that the action of Q on M is faithful; indeed, $C_Q(M)$ is a normal p-subgroup of G, hence acts trivially on any simple kG-module. The implication of this assumption is that Q acts by noninner automorphisms on M, because inner automorphisms of M have orders prime to p. If N is a normal subgroup of G contained in M such that $NZ(M) = M$, then $kNe(\eta) = kMe(\eta)$ because the image of $Z(M)$ in the matrix algebra $kMe(\eta)$ has dimension 1. Thus we may assume that any such normal subgroup is equal to M. We argue by induction over the order of $M/Z(M)$. If M is abelian, then $\dim_k(X) = 1$, hence V is the trivial kQ-module, and the conclusion holds trivially. Assume that M is not abelian. Note that a subgroup of M is normal in G if and only if it is normal in M and Q-stable. Let N be a normal subgroup of G such that $Z(M) \leq N \leq M$. We have $e(\eta_N)e(\eta) = e(\eta)$ and $kNe(\eta_N)$ is a matrix subalgebra of $kMe(\eta_N)$; multiplication by $e(\eta)$ yields that $T = kNe(\eta)$ is a matrix subalgebra of the matrix algebra $S = kMe(\eta)$. Thus $S = T \otimes_k T'$, where $T \cong k_\alpha(M/N)$ for some $\alpha \in Z^2(M/N; k^\times)$, and this tensor product is Q-stable. The algebra T' is isomorphic to a block algebra of a central p'-extension of M/N on which Q acts. This shows that $V \cong W \otimes_k W'$ with endopermutation kQ-modules W, W' satisfying $\mathrm{End}_k(W) \cong T$ and $\mathrm{End}_k(W') \cong T'$. If N is a proper subgroup of M strictly containing $Z(M)$, then by induction, the indecomposable summands of W, W' as kQ-modules having vertex Q yield elements of finite order in the Dade group $D_k(Q)$. Thus we may assume that any normal subgroup N of G such that $Z(M) \leq N \leq M$ satisfies either $Z(M) = N$ or $N = M$. Since M is nonabelian, this implies $M = Z(M)[M, M]$, and by the above assumptions therefore $M = [M, M]$. In particular, p is odd. Let Y be the inverse image in M of a minimal normal subgroup of $M/Z(M)$. If $Y = M$, then $M/Z(M)$ is simple nonabelian. By a standard argument, the action of Q on $M/Z(M)$ remains faithful. Since $M/Z(M)$ is a simple nonabelian finite group of order prime to p, this implies that Q is cyclic, and hence that the image of V in the Dade group has finite order. It remains to consider the case where Y is a proper subgroup of M.

Then Y is not Q-stable; denote by R the stabiliser of Y in Q. Then $N_G(Y) = MR$. If $u \in Q \setminus R$ then ${}^uY$ is isomorphic to Y and its image in $M/Z(M)$ is a simple normal subgroup different from the image of Y in $M/Z(M)$. Thus the different Q-conjugates of Y commute pairwise. It follows that M is equal to the product of the ${}^uY$, with u running over $[Q/R]$, because this product is a subgroup of M containing $Z(M)$ which is normal in G. Any two different factors of this product commute, and hence M is isomorphic to a quotient of the direct product $\prod_{u \in [Q/R]} {}^uY$ by a central subgroup of order prime to p. This means that $kMe(\eta)$ is isomorphic to a tensor product $\otimes_u k^u Y c_u$, with u running over $\lfloor Q/R \rfloor$ and blocks c_u of $k^u Y$, such that the factors in this tensor product are permuted by the action of Q. Writing $kY c_1 = \operatorname{End}_k(W)$ we get that W is an endopermutation kR-module such that $V \cong \operatorname{Ten}_R^Q(W)$. By induction, W has finite order in $D_k(R)$, hence V has finite order in $D_k(Q)$. This completes the proof. $\qquad \square$

The arguments in this proof can be extended to p-groups acting on defect zero blocks of finite groups which are not necessarily of order prime to p; see Salminen [200].

7.11 Endosplit p-permutation resolutions

Following work of Rickard in [187], most endopermutation modules have finite resolutions by permutation modules. This is a key ingredient for reinterpreting Morita equivalences between block algebras given by a bimodule with an endo-permutation source in terms of derived equivalences given by complexes of p-permutation bimodules – a theme that will be developed further in Chapter 9. A complex X over some additive category is said to have *homology concentrated in degree zero* if $H_n(X) = \{0\}$ for any nonzero integer n.

Definition 7.11.1 ([187, §7]) Let G be a finite group and M a finitely generated $\mathcal{O}G$-module. An *endosplit p-permutation resolution of* M is a bounded complex X of finitely generated p-permutation $\mathcal{O}G$-modules with homology concentrated in degree 0 such that $H_0(X) \cong M$ and such that the complex $X \otimes_{\mathcal{O}} X^*$ is a split complex of $\mathcal{O}G$-modules, with G acting diagonally on the tensor product.

The complex $X \otimes_{\mathcal{O}} X^*$ has homology concentrated in degree zero isomorphic to $M \otimes_{\mathcal{O}} M^*$; the condition that this complex is split means that $X \otimes_{\mathcal{O}} X^*$ is isomorphic to a direct sum of $M \otimes_{\mathcal{O}} M^*$ and a bounded contractible complex of $\mathcal{O}G$-modules, for the diagonal action of G on tensor products. Not every finitely generated $\mathcal{O}G$-module M can have an endosplit p-permutation resolution. If $\mathcal{O}$ has characteristic zero and M has an endosplit p-permutation resolution, then the character χ_M of M is equal to $\sum_{i \in \mathbb{Z}} (-1)^i \chi_i$, where χ_i is the character of the i-th term X_i of the complex X, for all integers i. In particular, χ_M is in the subgroup generated by the characters of p-permutation $\mathcal{O}G$-modules. For instance, if p is odd and P is a nontrivial finite p-group, then no nontrivial $\mathcal{O}P$-modules of $\mathcal{O}$-rank 1 has an endosplit p-permutation resolution. It follows from the classification of endopermutation modules, that for p odd, every endopermutation kP-module has an endosplit p-permutation resolution. For $p = 2$, there are examples, such as the 'exotic' 3-dimensional endotrivial module over the quaternion group Q_8, which do not have an endosplit p-permutation resolution; see [149, Théorème 6.2.4].

We will show that if P is a finite abelian p-group, then any endopermutation $\mathcal{O}P$-module has an endosplit p-permutation resolution. The proof of this result, due to Rickard [187], uses Dade's classification of endopermutation modules over abelian p-groups.

Proposition 7.11.2 *Let G be a finite group, and let M be a finitely generated $\mathcal{O}$-free $\mathcal{O}G$-module having an endosplit p-permutation resolution X. For any subgroup H of G, the complex $\mathrm{Hom}_{\mathcal{O}H}(X, X)$ has homology concentrated in degree zero, isomorphic $\mathrm{Hom}_{\mathcal{O}H}(M, M)$, and there is an algebra isomorphism*

$$\rho_H : \mathrm{Hom}_{K(\mathrm{mod}(\mathcal{O}H))}(X, X) \cong \mathrm{Hom}_{\mathcal{O}H}(M, M)$$

satisfying $\mathrm{Res}_H^G \circ \rho_G = \rho_H \circ \mathrm{Res}_H^G$ *and* $\mathrm{Tr}_H^G \circ \rho_H = \rho_G \circ \mathrm{Tr}_H^G$. *In particular, ρ_G induces a vertex and multiplicity preserving bijection between the sets of isomorphism classes of indecomposable direct summands of the module M and of noncontractible indecomposable direct summands of the complex X.*

Proof The complex $X \otimes_{\mathcal{O}} X^* \cong \mathrm{Hom}_{\mathcal{O}}(X, X)$ is a split complex of $\mathcal{O}G$-modules for the diagonal G-action; that is, it is a direct sum of $\mathrm{Hom}_{\mathcal{O}}(M, M)$ and a contractible complex. Thus, for any subgroup H of G, taking H-fixed points commutes with taking homology. The homology of $\mathrm{Hom}_{\mathcal{O}}(X, X)$ is $\mathrm{Hom}_{\mathcal{O}}(M, M)$, and taking H-fixed points yields $\mathrm{Hom}_{\mathcal{O}H}(M, M)$. This is also the homology of the complex $\mathrm{Hom}_{\mathcal{O}H}(X, X)$. By a standard fact, this is also isomorphic to $\mathrm{Hom}_{K(\mathrm{mod}(\mathcal{O}H))}(X, X)$, which yields an algebra homomorphism ρ_H satisfying the compatibility with restriction and relative traces as stated. Primitive idempotents in $\mathrm{End}_{K(\mathrm{mod}(\mathcal{O}G))}(X)$ lift to primitive idempotents in $\mathrm{End}_{\mathrm{Ch}(\mathrm{mod}(\mathcal{O}G))}(X)$. Any primitive idempotent in $\mathrm{End}_{\mathrm{Ch}(\mathrm{mod}(\mathcal{O}G))}(X)$ determines an indecomposable direct summand of X, and this summand is contractible if and only if the corresponding idempotent has zero image in $\mathrm{End}_{K(\mathrm{mod}(\mathcal{O}G))}(X)$. The last statement follows. $\square$

The next lemma would be an immediate consequence of the Universal Coefficient Theorem, but we include an elementary direct proof.

Lemma 7.11.3 *Let X be a complex of finitely generated free $\mathcal{O}$-modules such that $k \otimes_{\mathcal{O}} X$ has homology concentrated in degree zero. Then X is a split complex that has homology concentrated in degree zero, $H_0(X)$ is $\mathcal{O}$-free, and $k \otimes_{\mathcal{O}} H_0(X) \cong H_0(k \otimes_{\mathcal{O}} X)$.*

Proof Denote by $\delta = (\delta_i : X_i \to X_{i-1})_{i \in \mathbb{Z}}$ the differential of X. Denote by $\bar{\delta}$ the differential of $k \otimes_{\mathcal{O}} X$. Let n be a nonzero integer. Since $H_n(k \otimes_{\mathcal{O}} X) = \{0\}$, it follows that $\ker(\delta_n) \subseteq \operatorname{Im}(\delta_{n+1}) + J(\mathcal{O})X_n$. As $\ker(\delta_n)$ is a pure submodule of X_n containing $\operatorname{Im}(\delta_{n+1})$, this implies that $\ker(\delta_n) = \operatorname{Im}(\delta_{n+1}) + J(\mathcal{O})\ker(\delta_n)$, hence $\ker(\delta_n) = \operatorname{Im}(\delta_{n+1})$ by Nakayama's Lemma, whence $H_n(X) = \{0\}$. In order to show that $H_0(X)$ is free, it suffices to show that $\operatorname{Im}(\delta_1)$ is a pure submodule of X_0. If not, then the image of $\operatorname{Im}(\delta_1)$ in $k \otimes_{\mathcal{O}} X_0$ would have dimension strictly smaller than $\operatorname{rk}_{\mathcal{O}}(\operatorname{Im}(\delta_1))$, by 4.2.8. But then the dimension of $\ker(\bar{\delta}_1)$ would be bigger than the rank of $\ker(\delta_1)$, which in turn would imply that $H_1(k \otimes_{\mathcal{O}} X)$ would be nonzero, a contradiction. Thus $H_0(X)$ is $\mathcal{O}$-free. It follows that the submodules $\ker(\delta_i)$, $\operatorname{Im}(\delta_{i+1})$ are all direct summands of X_i, for all integers i, hence X is split and in particular, $k \otimes_{\mathcal{O}} H_0(X) \cong H_0(k \otimes_{\mathcal{O}} X)$. $\square$

The following result is more general than what we need in this section, but will be needed in the context of lifting splendid derived equivalences.

Proposition 7.11.4 ([187, 5.1]) *Let G be a finite group and X a complex of finitely generated p-permutation kG-modules such that the complex $\operatorname{Hom}_{kG}(X, X)$ has homology concentrated in degree 0. Then there is, up to isomorphism, a unique complex Y of p-permutation $\mathcal{O}G$-modules such that $k \otimes_{\mathcal{O}} Y \cong X$. In particular, if X is split, then Y is split.*

Proof Denote by π an element in $\mathcal{O}$ such that $J(\mathcal{O}) = \pi\mathcal{O}$. Denote by δ the differential of X. For any integer n we have $H_n(\operatorname{Hom}_{kG}(X, X)) \cong \operatorname{Hom}_{K(\operatorname{Mod}(kG))}(X, X[n])$, which by the assumptions, is zero for all nonzero n. By 5.11.2, the graded p-permutation kG-module X lifts uniquely, up to isomorphism, to a graded p-permutation module Y, and the differential δ lifts to a graded endomorphism ϵ_1 of degree -1 of Y. Since $\delta \circ \delta = 0$, the image of $\epsilon_1 \circ \epsilon$ is contained in $J(\mathcal{O})Y$. We construct inductively lifts ϵ_i of δ with the property that the images of $\epsilon_{i+1} - \epsilon_i$ and of $\epsilon_i \circ \epsilon_i$ are contained in $J(\mathcal{O})^i Y$. Using the completeness of $\mathcal{O}$, the limit ϵ of the ϵ_i is then a differential of Y lifting δ. Suppose we have already constructed $\epsilon_1, \epsilon_2, \ldots, \epsilon_i$ for some positive integer i with the properties described above. Then $\epsilon_i \circ \epsilon_i = \pi^i \alpha$ for some graded endomorphism α of degree -2 of Y. Clearly ϵ_i commutes with $\epsilon_i \circ \epsilon_i = \pi^i \alpha$, and hence with α, since the terms of Y are $\mathcal{O}$-free. Since ϵ_i lifts the differential δ of X, it follows that the graded endomorphism $\bar{\alpha}$ of degree -2 induced by α commutes with δ, hence is in fact a chain map $X \to X[2]$. But then $\bar{\alpha}$ is homotopic to zero, so there is a homotopy $\bar{h}$ such that $\bar{\alpha} = \delta \circ \bar{h} + \bar{h} \circ \delta$. Note that h is a map of degree 1 from X to $X[2]$, or equivalently, a map of degree -1 from X to X. Lift $\bar{h}$ to a graded endomorphism h of X of degree -1. Set $\epsilon_{i+1} = \epsilon_i - \pi^i h$. Then $\epsilon_{i+1} \circ \epsilon_{i+1} = \epsilon_i \circ \epsilon_i - \pi^i(\epsilon_i \circ h + h \circ \epsilon_i) + \pi^{2i} h \circ h$. This is zero modulo π^{i+1}, and hence ϵ_{i+1} has the required properties. This shows the existence of a complex Y such that $k \otimes_{\mathcal{O}} Y \cong X$. The uniqueness follows a similar pattern. Let (Y', ϵ') be another complex of p-permutation $\mathcal{O}G$-modules such that $k \otimes_{\mathcal{O}} Y' \cong X$. Since the terms of X lift uniquely, up to isomorphism, there is a graded isomorphism $\beta_1 : Y \cong Y'$ such that β induces the identity on X. Suppose that for some positive integer i we have already constructed graded isomorphisms $\beta_1, \beta_2, \ldots, \beta_i$ from Y to Y' with the property that $\beta_{i+1} - \beta_i$ and $\beta_i \circ \epsilon - \epsilon' \circ \beta_i$ have images contained in $J(\mathcal{O})^i Y$. Then $\beta_i \circ \epsilon - \epsilon' \circ \beta_i = \pi^i \psi$ for some graded map ψ of degree 1. Clearly $\pi^i \psi$, and hence ψ, commutes with the differentials of Y, Y'. Thus ψ is a chain map of degree -1, or equivalently, a chain map from Y to $Y'[1]$. By the assumptions, the induced chain map from X to $X[1]$ is homotopic to zero. Thus there is a graded map h from Y to Y' such that $\psi - (h \circ \epsilon + \epsilon \circ h)$ has image contained in $J(\mathcal{O})Y$. Define $\beta_{i+1} = \beta_i - \pi^i h$. Then $\beta_{i+1} \circ \epsilon - \epsilon' \beta_{i+1}$ has image contained in $J(\mathcal{O})^{i+1}Y$. Thus the sequence of the β_i converges to an isomorphism of complexes $Y \cong Y'$ as required. Since a split bounded complex is a finite direct sum of complexes having exactly two nonzero terms in consecutive degrees that are isomorphic, it follows that if X is split, then X has a split lift over Y. The uniqueness property of such lifts implies that any lift of X is then split. $\square$

Proposition 7.11.5 ([187, 7.1]) *Let G be a finite group, and let M be a finitely generated kG-module that has an endosplit p-permutation resolution X. Then there is up to isomorphism a unique $\mathcal{O}$-free $\mathcal{O}G$-module N having an endosplit p-permutation resolution Y such that $k \otimes_{\mathcal{O}} N \cong M$ and $k \otimes_{\mathcal{O}} Y \cong X$.*

Proof By 7.11.2 and 7.11.4, the complex X lifts uniquely, up to isomorphism, to a complex Y of p-permutation $\mathcal{O}G$-modules. It follows from 7.11.3 that Y has homology concentrated in degree zero, and that $N = H_0(Y)$ satisfies $k \otimes_{\mathcal{O}} N \cong M$. Since $X \otimes_k X^*$ is split, so is its lift $Y \otimes_{\mathcal{O}} Y^*$ by the last statement in 7.11.4, and hence Y is a p-permutation resolution of N. The uniqueness of Y, up to isomorphism, implies that of N. $\square$

Proposition 7.11.5 does not say that M lifts uniquely up to isomorphism to an endopermutation $\mathcal{O}P$-module having an endosplit p-permutation resolution –

the uniqueness statement in this proposition hinges on prescribing an endosplit p-permutation resolution of that lift. See the Exercise 7.11.11.

Proposition 7.11.6 ([187, §7]) *Let G be a finite group, H a normal subgroup of G, let M, N be finitely generated $\mathcal{O}$-free $\mathcal{O}G$-modules and W a finitely generated $\mathcal{O}$-free $\mathcal{O}G/N$-module.*

(i) *If Y is an endosplit p-permutation resolution of the $\mathcal{O}G/H$-module W, then the inflation to $\mathcal{O}G$ of Y is an endosplit p-permutation resolution of the inflation of W to $\mathcal{O}G$.*

(ii) *If Y_M and Y_N are endosplit p-permutation resolutions of M and N, respectively, then $Y_M \otimes_k Y_N$ is an endosplit p-permutation resolution of $M \otimes_{\mathcal{O}} N$.*

(iii) *If $M \oplus N$ has an endosplit p-permutation resolution Y, then $Y = Y_M \oplus Y_N$ for some endosplit p-permutation resolutions Y_M and Y_N of M and N, respectively.*

(iv) *If M has an endosplit p-permutation resolution, then any module that is a finite direct sum of summands of the modules $\mathrm{Ind}_H^G \mathrm{Res}_H^G(M)$, with H running over the subgroups of G, has an endosplit p-permutation resolution.*

(v) *If Y is an endosplit p-permutation resolution of M, then Y^* is an endosplit p-permutation resolution of M^*.*

Proof Statement (i) is obvious, because the properties of being a p-permutation module or a split complex are preserved under inflation. Note that $(Y_M \otimes_{\mathcal{O}} Y_N) \otimes_{\mathcal{O}} (Y_M \otimes_{\mathcal{O}} Y_N)^* \cong Y_M \otimes_{\mathcal{O}} Y_M^* \otimes_{\mathcal{O}} Y_N \otimes_{\mathcal{O}} Y_N^*$. This is a tensor product of two split complexes, hence split. Clearly all terms of $Y_M \otimes_{\mathcal{O}} Y_N$ are p-permutation $\mathcal{O}G$-modules, whence (ii). By 7.11.2, we have $\mathrm{End}_{K(\mathrm{mod}(\mathcal{O}G))}(Y) \cong \mathrm{End}_{\mathcal{O}G}(M \oplus N)$. The projections of $M \oplus N$ onto M and N yield orthogonal idempotents in $\mathrm{End}_{K(\mathrm{mod}(\mathcal{O}G))}(Y)$. Since Y is bounded, this algebra is finitely generated as an $\mathcal{O}$-module. Thus these idempotents lift, uniquely up to conjugacy, to orthogonal idempotents in $\mathrm{End}_{\mathrm{Ch}(\mathrm{mod}(\mathcal{O}G))}(Y)$, where they yield the required decomposition of Y as a direct sum of endosplit p-permutation resolutions of M and N. This proves (iii). By (iii), in order to prove (iv) we may assume that $N = \oplus_i \mathrm{Ind}_{H_i}^G \mathrm{Res}_{H_i}^G(M)$, where i runs over a finite indexing set of a family of subgroups $\{H_i\}$ of G. Set $Y = \oplus_i \mathrm{Ind}_{H_i}^G \mathrm{Res}_{H_i}^G(Y_M)$, where Y_M is an endosplit p-permutation resolution of M. Then $Y \otimes_{\mathcal{O}} Y^*$ is isomorphic to a direct sum of complexes of the form

$$\mathrm{Res}_{\Delta G}^{G \times G} \mathrm{Ind}_{H_i \times H_j}^{G \times G} \mathrm{Res}_{H_i \times H_j}^{G \times G} (Y_M \otimes_{\mathcal{O}} Y_M^*).$$

Mackey's formula implies that this is a direct sum of complexes of the form $\mathrm{Ind}_{\Delta(H_i \cap {}^x H_j)}^{\Delta G}(Y_M \otimes_{\mathcal{O}} Y_M^*)$. These are all split by the choice of Y_M. This shows (iv). Statement (v) is an obvious consequence of the fact that the dual of a p-permutation module is again a p-permutation module. $\square$

Proposition 7.11.7 *Let G be a finite group, n be a positive integer, and let*

$$0 \longrightarrow \Omega^n(\mathcal{O}) \longrightarrow Y_{n-1} \longrightarrow \cdots \longrightarrow Y_1 \longrightarrow Y_0 \longrightarrow \mathcal{O} \longrightarrow 0$$

be an exact sequence of $\mathcal{O}G$-modules such that Y_i is projective, for $0 \le i \le n - 1$. Then the complex

$$Y = \cdots \longrightarrow 0 \longrightarrow Y_{n-1} \longrightarrow \cdots \longrightarrow Y_1 \longrightarrow Y_0 \longrightarrow \mathcal{O} \longrightarrow 0 \longrightarrow \cdots$$

with Y_{n-1} in degree zero is a p-permutation resolution of $\Omega^n(\mathcal{O})$.

Proof Clearly the homology of the complex Y is concentrated in degree zero and isomorphic to $\Omega^n(\mathcal{O})$. The complex $Y \otimes_{\mathcal{O}} Y^*$ has homology concentrated in degree zero isomorphic to $\Omega^n(\mathcal{O}) \otimes_{\mathcal{O}} \Omega^n(\mathcal{O})^* \cong \mathcal{O} \oplus W$ for some projective $\mathcal{O}G$-module W. All terms of $Y \otimes_{\mathcal{O}} Y^*$ are projective except the degree zero term (which has a direct summand isomorphic to $\mathcal{O}$), and the complex is zero in positive degree. Thus the last nonzero map of $Y \otimes_{\mathcal{O}} Y^*$, if any, is split surjective, and similarly, the first nonzero map is split injective. An easy induction shows that $Y \otimes_{\mathcal{O}} Y^*$ is split. $\square$

Corollary 7.11.8 *Let G be a finite group and n an integer. Then $\Omega^n(\mathcal{O})$ has an endosplit p-permutation resolution.*

Proof For $n = 0$ this is trivial, and for n positive this follows from 7.11.7. Since $\Omega^n(\mathcal{O})^* \cong \Omega^{-n}(\mathcal{O})$, the case where n is negative follows from 7.11.7 and 7.11.6 (v). $\square$

Theorem 7.11.9 ([187, Theorem 7.2]) *Let P be a finite abelian p-group, and let E be a subgroup of Aut(P). Suppose that k is algebraically closed. Let M be a finitely generated indecomposable endopermutation kP-module with vertex P.*

 (i) *The kP-module M has an endosplit p-permutation resolution X, and if M is E-stable, then X can be chosen E-stable.*
 (ii) *There is an indecomposable endopermutation $\mathcal{O}P$-module N with vertex P having an endosplit p-permutation resolution Y such that $k \otimes_{\mathcal{O}} N \cong M$. If M is E-stable, then both N and Y can be chosen E-stable.*

Proof By Dade's classification 7.8.1, M is obtained from taking summands of inflations of tensor products of Heller translates of trivial modules for quotient groups of P. The existence of an endosplit p-permutation resolution X in (i) follows from 7.11.8, 7.11.6, and 7.3.7 (iv). By going through the construction, one sees that X depends canonically on M, and hence the E-stability passes from M to X. Statement (ii) follows from 7.11.5. $\square$

Proposition 7.11.10 ([187, §7.3]) *Let P be a finite p-group, and M a finitely generated endopermutation kP-module having an indecomposable direct summand with vertex P. Suppose that M has an endosplit p-permutation resolution Y. Let Q be a subgroup of P, and let W be an endopermutation $kN_P(Q)$-module such that $\mathrm{End}_k(W) \cong (\mathrm{End}_k(M))(Q)$ as $N_P(Q)$-algebras. Then there is an integer i such that the shifted complex $Y(Q)[i]$ is an endosplit p-permutation resolution of the $kN_P(Q)$-module W.*

Proof Clearly $Y(Q)$ is a bounded complex of permutation $kN_P(Q)$-modules (on which Q acts trivially). Since $Y \otimes_k Y^*$ is a split complex, taking homology commutes with applying the Brauer construction. Thus $(Y \otimes_k Y^*)(Q)$ has homology concentrated in degree zero, isomorphic to $(V \otimes_k V^*)(Q)$. By 5.8.10 we have $(Y \otimes_k Y^*)(Q) \cong Y(Q) \otimes_k Y^*(Q)$. It follows from the Künneth formula 2.21.7 that $Y(Q)$ has homology concentrated in a single degree i, isomorphic to a $kN_P(Q)$-module W. But then $W \otimes_k W^*$ is also the homology of $Y(Q) \otimes_k Y(Q)$ in degree 0, hence isomorphic to $(M \otimes_k M^*)(Q)$. The result follows from the isomorphisms $M \otimes_k M^* \cong \mathrm{End}_k(M)$ and the corresponding isomorphism for W. $\square$

Exercise 7.11.11 Suppose that $p = 2$ and that P is cyclic of order 2. Let X_0 be the complex equal to the trivial kP-module in degree 0. Let X_1 be a 2-term complex of the form $k \to kP$, where the map from k in degree 1 to kP in degree 0 is injective. Show that X_0, X_1 are both endosplit p-permutation resolutions. Show that X_0 lifts to an endosplit p-permutation resolution of the trivial $\mathcal{O}P$-module $\mathcal{O}$, while X_1 lifts to an endosplit p-permutation resolution of the unique nontrivial $\mathcal{O}P$-module of $\mathcal{O}$-rank 1 (on which the nontrivial element of P acts as -1).

Exercise 7.11.12 Let P be a nontrivial finite p-group. Consider the augmentation map $\mathcal{O}P \to \mathcal{O}$ as a two-term complex Y with $\mathcal{O}P$ in degree zero. Show that Y is a p-permutation resolution of $\Omega(\mathcal{O})$, and that for any nontrivial subgroup Q of P, the complex $Y(Q)[1]$ is a p-permutation resolution of the trivial kQ-module.

Exercise 7.11.13 Let P be a nontrivial finite p-group. Show that any endosplit p-permutation resolution of a finitely generated $\mathcal{O}$-free nonprojective $\mathcal{O}P$-module has at least one nonprojective term.

Exercise 7.11.14 Let G be a finite group, H a subgroup of G, and Y an endosplit p-permutation resolution of a finitely generated $\mathcal{O}$-free $\mathcal{O}H$-module W. Suppose that for any $x \in G$ there is an isomorphism of complexes of $\mathcal{O}(H \cap {}^xH)$-modules $\mathrm{Res}^H_{H \cap {}^xH}(Y) \cong \mathrm{Res}^{{}^xH}_{H \cap {}^xH}({}^xY)$. Show that $\mathrm{Ind}^G_H(Y)$ is an endosplit p-permutation resolution of $\mathrm{Ind}^G_H(W)$.

8

Local Structure

Sylow's Theorems attach to a finite group G a p-subgroup P, uniquely up to conjugation. Theorems of Burnside and Frobenius establish further connections between G and P, taking into account information on G-conjugacy in P. Alperin's Fusion Theorem implies that if two subgroups of P are G-conjugate, then any such conjugation can be realised by products of elements in normalisers of certain nontrivial subgroups of P. The underlying philosophy relating the structure of a finite G to its 'p-local structure' evolved into a sophisticated tool in finite group theory, with many applications in the context of the classification of finite simple groups. A category theoretic perspective, developed by Puig in [166], describes the p-local structure of a finite group G as a category with objects the p-subgroups of G and morphisms the group homomorphisms between p-subgroups induced by conjugation in G. As a consequence of work of Alperin and Broué [5] on Brauer pairs, the formal properties of the categories describing the p-local structure of finite groups carry over to categories associated with blocks, where p-subgroups are replaced by Brauer pairs. This led to Puig's definition of fusion systems in the early 1990s. Broto, Levi and Oliver laid in [33] the homotopy theoretic foundations of fusion systems. We will focus in this chapter on a crucial question in block theory: given a fusion system $\mathcal{F}$ on a finite p-group P, what are the structural invariants of a block with defect group P and fusion system $\mathcal{F}$? We will describe some cases in which a block algebra may be completely described in terms of invariants associated with its fusion system. This includes *nilpotent blocks* and their extensions.

8.1 Fusion systems

We give in this section a brief review on fusion systems; more detailed accounts can be found in [134], [50]. Throughout this section we denote by p a prime.

165

Given subgroups P, Q, R of a group G we denote by $\mathrm{Hom}_P(Q, R)$ the set of all group homomorphisms $\varphi : Q \to R$ for which there exists an element $x \in P$ such that $\varphi(u) = xux^{-1}$ for all $u \in Q$. We write $\mathrm{Aut}_P(Q) = \mathrm{Hom}_P(Q, Q)$. The group $\mathrm{Aut}_P(Q)$ is canonically isomorphic to $N_P(Q)/C_P(Q)$. The group $\mathrm{Aut}_Q(Q) \cong Q/Z(Q)$ is the group of inner automorphisms of Q.

Definition 8.1.1 A *category on a finite p-group P* is a category $\mathcal{F}$ having as objects the subgroups of P, and for any two subgroups Q, R, the set $\mathrm{Hom}_{\mathcal{F}}(Q, R)$ of morphisms in $\mathcal{F}$ from Q to R is a set of injective group homomorphisms with the following properties.

  (i) If Q is contained in R then the inclusion morphism $Q \leq R$ is a morphism in $\mathcal{F}$.
 (ii) If $\varphi : Q \to R$ is a morphism in $\mathcal{F}$ then the induced isomorphism $\varphi : Q \cong \varphi(Q)$ and its inverse are morphisms in $\mathcal{F}$.
(iii) Composition of morphisms in $\mathcal{F}$ is the usual composition of group homomorphisms.

Example 8.1.2 Let G be a group and P a finite p-subgroup of G. Denote by $\mathcal{F}_P(G)$ the category with objects the subgroups of P and morphism sets $\mathrm{Hom}_{\mathcal{F}_P(G)}(Q, R) = \mathrm{Hom}_G(Q, R)$ for any two subgroups Q, R of P. Clearly $\mathcal{F}_P(G)$ is a category on P in the sense of the above definition.

Definition 8.1.3 Let p be a prime, P a finite p-group and $\mathcal{F}$ a category on P. A subgroup Q of P is called *fully $\mathcal{F}$-centralised* if $|C_P(Q)| \geq |C_P(Q')|$ for any subgroup Q' of P such that there is an isomorphism $Q \cong Q'$ in $\mathcal{F}$, and Q of P is called *fully $\mathcal{F}$-normalised* if $|N_P(Q)| \geq |N_P(Q')|$ for any subgroup Q' of P such that there is an isomorphism $Q \cong Q'$ in $\mathcal{F}$.

Proposition 8.1.4 *Let G be a finite group, P a Sylow p-subgroup of G and Q a subgroup of P. Set $\mathcal{F} = \mathcal{F}_P(G)$. Then Q is fully $\mathcal{F}$-centralised (resp. fully $\mathcal{F}$-normalised) if and only if $C_P(Q)$ (resp. $N_P(Q)$) is a Sylow p-subgroup of $C_G(Q)$ (resp. $N_G(Q)$).*

Proof Let S be a Sylow p-subgroup of $C_G(Q)$ containing $C_P(Q)$. In particular, QS is a p-subgroup of G, and hence there is $x \in G$ such that ${}^x(QS) \leq P$. Then conjugation by x induces an isomorphism $\varphi : Q \cong {}^xQ$ belonging to the category $\mathcal{F}$, because both Q and ${}^xQ$ are contained in P. Moreover, ${}^xS \leq C_P({}^xQ)$. Thus $|C_P(Q)| \leq |S| \leq |C_P({}^xQ)|$, hence Q is fully $\mathcal{F}$-centralised if and only if $|C_P(Q)| = |S|$, hence if and only if $C_P(Q) = S$ since S was chosen to contain $C_P(Q)$. The same argument with normalisers instead of centralisers completes the proof. $\square$

While in Example 8.1.2 we did not require P to be a Sylow p-subgroup, this was essential in the proof of Proposition 8.1.4.

Definition 8.1.5 Let p be a prime, P a finite p-group and $\mathcal{F}$ a category on P. Let Q be a subgroup of P and $\varphi : Q \to P$ a morphism in $\mathcal{F}$. We denote by N_φ the subgroup of $N_P(Q)$ consisting of all elements $y \in N_P(Q)$ for which there exists an element $z \in N_P(\varphi(Q))$ such that $\varphi(yuy^{-1}) = z\varphi(u)z^{-1}$ for all $u \in Q$.

In other words, N_φ is the inverse image in $N_P(Q)$ of the subgroup $\mathrm{Aut}_P(Q) \cap (\varphi \circ \mathrm{Aut}_P(\varphi(Q)) \circ \varphi^{-1})$ of $\mathrm{Aut}_P(Q)$. Note that $QC_P(Q) \leq N_\varphi \leq N_P(Q)$.

Definition 8.1.6 (Puig) Let p be a prime and P a finite p-group. A *fusion system* on P is a category $\mathcal{F}$ on P such that $\mathrm{Hom}_P(Q, R) \subseteq \mathrm{Hom}_\mathcal{F}(Q, R)$ for any two subgroups Q, R of P, and such that the following hold:

(I) "Sylow axiom" The group $\mathrm{Aut}_P(P)$ is a Sylow p-subgroup of $\mathrm{Aut}_\mathcal{F}(P)$; and

(II) "Extension axiom" For every morphism $\varphi : Q \to P$ in $\mathcal{F}$ such that $\varphi(Q)$ is fully $\mathcal{F}$-normalised there exists a morphism $\psi : N_\varphi \to P$ in $\mathcal{F}$ such that $\psi|_Q = \varphi$.

Theorem 8.1.7 *Let G be a finite group and P be a Sylow p-subgroup of G. The category $\mathcal{F}_P(G)$ is a fusion system on P.*

Proof Set $\mathcal{F} = \mathcal{F}_P(G)$. Clearly $\mathcal{F}$ is a category on P and we have $\mathcal{F}_P(P) \subseteq \mathcal{F}$. Thus we only have to check the two axioms (I) and (II) from 8.1.6. We have $\mathrm{Aut}_\mathcal{F}(P) \cong N_G(P)/C_G(P)$, and the image of P in this quotient group is a Sylow p-subgroup, which implies that $\mathrm{Aut}_P(P) \cong P/Z(P)$ is a Sylow p-subgroup of $\mathrm{Aut}_\mathcal{F}(P)$. This proves (I). Let now Q be a subgroup of P and let $\varphi : Q \to P$ be a morphism in $\mathcal{F}$. Set $R = \varphi(Q)$ and suppose that R is fully $\mathcal{F}$-normalised. Let $x \in G$ such that $\varphi(u) = {}^x u$ for all $u \in Q$. We have

$$N_\varphi = \{y \in N_P(Q) \mid \exists z \in N_P(R) : \varphi({}^y u) = {}^z \varphi(u)(\forall u \in Q)\}.$$

Since φ is given by conjugation with x, this translates to

$$N_\varphi = \{y \in N_P(Q) \mid \exists z \in N_P(R) : {}^{xy} u = {}^{zx} u(\forall u \in Q)\}.$$

The equation ${}^{xy} u = {}^{zx} u$ for all $u \in Q$ means that $x^{-1}z^{-1}xy$ centralises Q, hence $z^{-1}xyx^{-1}$ centralises ${}^x Q = R$, and so we can write $xyx^{-1} = zc$ for some $c \in C_G(R)$. This shows that we have

$$ {}^x N_\varphi \subseteq N_P(R)C_G(R).$$

Since R is fully $\mathcal{F}$-normalised, by 8.1.4 the group $N_P(R)$ is a Sylow p-subgroup of $N_G(R)$, hence of $N_P(R)C_G(R)$. Thus there is an element $d \in C_G(R)$such that $^{dx}N_\varphi \subseteq N_P(R)$. Define $\psi : N_\varphi \to P$ by $\psi(y) = {}^{dx}y$ for all $y \in N_\varphi$. We claim that ψ extends φ. Indeed, if $u \in Q$, then $\psi(u) = {}^{dx}u = {}^d\varphi(u) = \varphi(u)$ because d centralises $R = \varphi(Q)$. This proves (II).   □

If P is properly contained in a Sylow p-subgroup S of G, then the category $\mathcal{F}_P(G)$ need not be a fusion system. For instance, a necessary condition for the Sylow axiom to hold would be $N_S(P) = PC_S(P)$. A fusion system that does not arise from a finite group having the underlying p-group as a Sylow subgroup is called *exotic*. Examples include the Solomon fusion systems [118], [119], and the Ruiz–Viruel fusion systems [198]. By a result of Park [162], every fusion system on a finite p-group is equal to $\mathcal{F}_P(G)$ for some finite group G containing P (but P need not be a Sylow p-subgroup of G). Dividing a finite group by a normal p'-subgroup leaves the fusion system on a Sylow p-subgroup unchanged:

Theorem 8.1.8 *Let G be a finite group, let P be a Sylow p-subgroup of G and let K be a normal p'-subgroup of G. Set $\bar{G} = G/K$ and denote by $\bar{P}$ the image of P in $\bar{G}$. The canonical group homomorphism $\alpha : G \to \bar{G}$ induces an isomorphism of fusion systems $\mathcal{F}_P(G) \cong \mathcal{F}_{\bar{P}}(\bar{G})$.*

Proof Since K is a p'-group the map α induces an isomorphism $P \cong \bar{P}$. Thus, for any two subgroups Q, R of P the map $\mathrm{Hom}_{\mathcal{F}}(Q, R) \to \mathrm{Hom}_{\bar{\mathcal{F}}}(\bar{Q}, \bar{R})$ induced by α is injective, where $\bar{Q}, \bar{R}$ are the canonical images of Q, R in $\bar{P}$. In order to show that it is surjective, let $\psi : \bar{Q} \to \bar{R}$ be a morphism in $\bar{\mathcal{F}}$. Let S be the inverse image in P of $\psi(\bar{Q})$. Then there is $x \in G$ such that conjugation by x induces ψ; in other words, $^xQ \subseteq KS$ and hence both ^xQ and S are Sylow p-subgroups of KS. Thus there is $y \in K$ such that $^{yx}Q = S$. Then $\varphi : Q \to R$ defined by $\varphi(u) = {}^{yx}u$ for $u \in Q$ induces the morphism ψ as required.   □

Corollary 8.1.9 *Let G be a finite group, let K be a normal p'-subgroup of G and let $\alpha : G \to G/K$ be the canonical surjection. Let T be a Sylow p-subgroup of G/K and let P be a Sylow p-subgroup of G such that $\alpha(P) = T$. Let $y \in G/K$. Then there is $x \in G$ such that $\alpha(x) = y$ and such that $\alpha(P \cap {}^xP) = T \cap {}^yT$.*

Proof Set $R = T \cap {}^yT$. Then $^{y^{-1}}R \subseteq T$. Thus conjugation by y^{-1} is a morphism $R \to {}^{y^{-1}}R$ in $\mathcal{F}_T(G/K)$. By 8.1.8 this lifts to a morphism $Q \to {}^{x^{-1}}Q$ for some $x \in G$ satisfying $\alpha(x) = y$, where Q is the inverse image of R in P. Then $Q \subseteq P \cap {}^xP$, hence $\alpha(Q) \subseteq \alpha(P \cap {}^xP) \subseteq \alpha(P) \cap \alpha({}^xP) = T \cap {}^yT = \alpha(Q)$ and hence equality.   □

It is sometimes convenient to consider the category obtained from a fusion system by "dividing out" by all inner automorphisms:

Definition 8.1.10 Let p be a prime, P a finite p-group and $\mathcal{F}$ a fusion system on P. The *orbit category of* $\mathcal{F}$ is the category $\bar{\mathcal{F}}$ having the subgroups of P as objects and, for any two subgroups Q, R of P we have $\mathrm{Hom}_{\bar{\mathcal{F}}}(Q, R) =$ $\mathrm{Aut}_R(R)\backslash\mathrm{Hom}_{\mathcal{F}}(Q, R)$, with composition of morphisms induced by that in $\mathcal{F}$.

This makes sense: with the notation above, the group $\mathrm{Aut}_R(R)$ acts on $\mathrm{Hom}_{\mathcal{F}}(Q, R)$ by composition of group homomorphisms, and if $\varphi, \varphi' \in$ $\mathrm{Hom}_{\mathcal{F}}(Q, R)$ are in the same $\mathrm{Aut}_R(R)$-orbit and $\psi, \psi' \in \mathrm{Hom}_{\mathcal{F}}(R, S)$ in the same $\mathrm{Aut}_S(S)$-orbit, a trivial verification shows that $\psi \circ \varphi, \psi' \circ \varphi' \in$ $\mathrm{Hom}_{\mathcal{F}}(Q, S)$ are in the same $\mathrm{Aut}_S(S)$-orbit as well, for any subgroups Q, R, S of P. The following propositions, due to Stancu (cf. [103]) show that the above definition of fusion systems is equivalent to that used in [33], where fusion systems are called *saturated fusion systems*.

Proposition 8.1.11 *Let p be a prime, P a finite p-group and $\mathcal{F}$ a fusion system on P. A subgroup Q of P is fully $\mathcal{F}$-normalised if and only if Q is fully $\mathcal{F}$-centralised and $\mathrm{Aut}_P(Q)$ is a Sylow p-subgroup of $\mathrm{Aut}_{\mathcal{F}}(Q)$.*

Proof Assume that Q is fully $\mathcal{F}$-normalised. We first show that then Q is also fully $\mathcal{F}$-centralised. Let $\varphi : R \to Q$ be an isomorphism in $\mathcal{F}$ such that R is fully $\mathcal{F}$-centralised. By the extension axiom 8.1.6 (II) there is a morphism $\psi : RC_P(R) \to P$ in $\mathcal{F}$ such that $\psi|_R = \varphi$. Hence ψ maps $C_P(R)$ to $C_P(Q)$, which implies that $|C_P(R)| \leq |C_P(Q)|$, hence equality since R is fully $\mathcal{F}$-centralised. Thus Q is fully $\mathcal{F}$-centralised. Choose Q to be of maximal order such that Q is fully $\mathcal{F}$-normalised but $\mathrm{Aut}_P(Q)$ is not a Sylow p-subgroup of $\mathrm{Aut}_{\mathcal{F}}(Q)$. Then Q is a proper subgroup of P by the Sylow axiom 8.1.6(I). Choose a p-subgroup S of $\mathrm{Aut}_{\mathcal{F}}(Q)$ such that $\mathrm{Aut}_P(Q)$ is a proper normal subgroup of S. Let $\varphi \in S \setminus \mathrm{Aut}_P(Q)$. Since φ normalises $\mathrm{Aut}_P(Q)$, it follows that for every $y \in N_P(Q)$ there is $z \in N_P(Q)$ such that $\varphi(^yu) = {}^z\varphi(u)$ for all $u \in Q$. In other words, $N_\varphi = N_P(Q)$. Since Q is fully $\mathcal{F}$-normalised, it follows from 8.1.6(II) that there is an automorphism ψ of $N_P(Q)$ in $\mathcal{F}$ such that $\psi|_Q = \varphi$. Since φ has p-power order, by decomposing ψ into its p-part and its p'-part we may in fact assume that ψ has p-power order. Let $\tau : N_P(Q) \to P$ be a morphism in $\mathcal{F}$ such that $\tau(N_P(Q))$ is fully $\mathcal{F}$-normalised. Now $\tau\psi\tau^{-1}$ is a p-element in $\mathrm{Aut}_{\mathcal{F}}(\tau(N_P(Q)))$, thus conjugate to an element in $\mathrm{Aut}_P(\tau(N_P(Q)))$. Therefore we may choose τ in such a way that there is $y \in N_P(\tau(N_P(Q)))$

satisfying $\tau \psi \tau^{-1}(v) = {}^y v$ for any $v \in \tau(N_P(Q))$. Since $\psi|_Q = \varphi$, the automorphism $\tau \psi \tau^{-1}$ of $\tau(N_P(Q))$ stabilises $\tau(Q)$. Thus $y \in N_P(\tau(Q))$. Since Q is fully $\mathcal{F}$-normalised we have $N_P(\tau(Q)) \subseteq \tau(N_P(Q))$, hence $\psi(u) = {}^{\tau^{-1}(y)} u$ for all $u \in N_P(Q)$. But then in particular $\varphi \in \mathrm{Aut}_P(Q)$, contradicting our initial choice of φ. Thus $\mathrm{Aut}_P(Q)$ is a Sylow p-subgroup in $\mathrm{Aut}_{\mathcal{F}}(Q)$. The converse is easy since $|N_P(Q)| = |\mathrm{Aut}_P(Q)| \cdot |C_P(Q)|$. This proves the result. $\qquad\square$

Lemma 8.1.12 *Let p be a prime, P a finite p-group and $\mathcal{F}$ a fusion system on P. Let Q, R be subgroups of P and let $\varphi : Q \to R$ be an isomorphism in $\mathcal{F}$ such that R is fully $\mathcal{F}$-normalised. Then there is an isomorphism $\psi : Q \to R$ in $\mathcal{F}$ such that $N_\psi = N_P(Q)$, or equivalently, such that ψ can be extended to a morphism from $N_P(Q)$ to P in $\mathcal{F}$.*

Proof The group $\varphi \circ \mathrm{Aut}_P(Q) \circ \varphi^{-1}$ is a p-subgroup of $\mathrm{Aut}_{\mathcal{F}}(R)$. Since R is fully $\mathcal{F}$-normalised, $\mathrm{Aut}_P(R)$ is a Sylow p-subgroup of $\mathrm{Aut}_{\mathcal{F}}(R)$ by 8.1.11. Thus there is $\beta \in \mathrm{Aut}_{\mathcal{F}}(R)$ such that $\beta \circ \varphi \circ \mathrm{Aut}_P(Q) \circ \varphi^{-1} \circ \beta^{-1} \subseteq \mathrm{Aut}_P(R)$. Set $\psi = \beta \circ \varphi$. The above inclusion means precisely that for any $y \in N_P(Q)$ there is $z \in N_P(R)$ such that $\psi \circ c_y \circ \psi^{-1} = c_z$, where c_y and c_z are the automorphisms of Q and R induced by conjugation with y and z, respectively. Equivalently, for any $y \in N_P(Q)$ there is $z \in N_P(R)$ such that $\psi \circ c_y = c_z \circ \psi$, which in turn means that $N_\psi = N_P(Q)$. The extension axiom (II-S) implies that ψ can be extended to a morphism from $N_P(Q)$ to P in $\mathcal{F}$. $\qquad\square$

Proposition 8.1.13 *Let p be a prime, P a finite p-group and $\mathcal{F}$ a fusion system on P. Let Q be a subgroup of P. Every morphism $\varphi : Q \to P$ such that $\varphi(Q)$ is fully $\mathcal{F}$-centralised extends to a morphism $\psi : N_\varphi \to P$ in $\mathcal{F}$ (that is, $\psi|_Q = \varphi$).*

Proof Let $\varphi : Q \to P$ be a morphism in $\mathcal{F}$ such that $\varphi(Q)$ is fully $\mathcal{F}$-centralised. Let $\rho : \varphi(Q) \to P$ be a morphism in $\mathcal{F}$ such that $R = \rho(\varphi(Q))$ is fully $\mathcal{F}$-normalised. By 8.1.12 we may choose ρ in such a way that $N_\rho = N_P(\varphi(Q))$. In particular, ρ extends to a morphism $\sigma : N_P(\varphi(Q)) \to P$. But then $N_\varphi \subseteq N_{\rho \circ \varphi}$, hence $\rho \circ \varphi$ extends to a morphism $\tau : N_\varphi \to P$. We show that this implies the inclusion $\tau(N_\varphi) \subseteq \sigma(N_P(\varphi(Q)))$. Let $y \in N_\varphi$. Thus there is $z \in N_P(\varphi(Q))$ such that $\varphi({}^y u) = {}^z \varphi(u)$ for all $u \in Q$. Since σ extends ρ it follows that $\rho(\varphi({}^y u)) = {}^{\sigma(z)} \rho(\varphi(u))$ for all $u \in Q$. Since τ extends $\rho \circ \varphi$ it follows that also $\rho(\varphi({}^y u)) = {}^{\tau(y)} \rho(\varphi(u))$ for all $u \in Q$. Thus conjugation by $\tau(y)$ and $\sigma(z)$ induce the same automorphism on R, and hence $\tau(y) = \sigma(z)c$ for some $c \in C_P(R)$. The hypothesis that $\varphi(Q)$ is fully $\mathcal{F}$-centralised implies that σ restricts to an isomorphism $C_P(\varphi(Q)) \cong C_P(R)$. Thus there is $d \in C_P(\varphi(Q))$ such that $\sigma(d) = c$. Then $\tau(y) = \sigma(zc)$. This shows the inclusion $\tau(N_\varphi) \subseteq \sigma(N_P(\varphi(Q)))$

as claimed. Composing τ with the inverse of σ restricted to $\tau(N_\varphi)$ yields a morphism $\psi = \sigma^{-1}|_{\tau(N_\varphi)} \circ \tau : N_\varphi \to P$. This morphism clearly extends φ, whence the result. $\qquad\square$

Proposition 8.1.14 *Let P be a finite abelian p-group and let $\mathcal{F}$ be a fusion system on P. Then $E = \mathrm{Aut}_\mathcal{F}(P)$ is a p'-subgroup of $\mathrm{Aut}(P)$ and we have $\mathcal{F} = \mathcal{F}_P(P \rtimes E)$.*

Proof Since P is abelian, the Sylow axiom implies that $E = \mathrm{Aut}_\mathcal{F}(P)$ is a p'-group. Again since P is abelian, every subgroup of P is fully $\mathcal{F}$-centralised with P as centraliser. Hence, by 8.1.13, every morphism in $\mathcal{F}$ extends to an automorphism in $\mathcal{F}$ of P. The result follows. $\qquad\square$

Corollary 8.1.15 (Burnside) *Let G be a finite group with an abelian Sylow p-subgroup P. Then $\mathcal{F}_P(G) = \mathcal{F}_P(N_G(P))$.*

Proof Since $\mathrm{Aut}_{\mathcal{F}_P(G)}(P) \cong N_G(P)/C_G(P) \cong \mathrm{Aut}_{\mathcal{F}_P(N_G(P))}(P)$, the result follows from 8.1.14. $\qquad\square$

Definition 8.1.16 Let $\mathcal{F}$ be a fusion system on a finite p-group P. The $\mathcal{F}$-*focal subgroup of P*, denoted $\mathfrak{foc}(\mathcal{F})$, is the subgroup generated by the commutators $[\mathrm{Aut}_\mathcal{F}(Q), Q]$, with Q running over the subgroups of P. The $\mathcal{F}$-*hyperfocal subgroup of P* is the subgroup generated by all commutators $[O^p(\mathrm{Aut}_\mathcal{F}(Q)), Q]$, with Q running over the subgroups of P.

By D. G. Higman's focal subgroup theorem, if $\mathcal{F}$ is the fusion system of a finite group G on one of its Sylow p-subgroups P, then $\mathfrak{foc}(\mathcal{F}) = [G, G] \cap P$, and by a theorem of Puig [178, §1.1], we have $\mathfrak{hyp}(\mathcal{F}) = O^p(G) \cap P$ in that case. In general, the focal subgroup of a fusion system $\mathcal{F}$ on a finite p-group P contains the derived subgroup $[P, P]$ of P, and we have $\mathfrak{foc}(P) = [P, P]\mathfrak{hyp}(\mathcal{F})$. See [10, §7] for more details and references.

8.2 Alperin's Fusion Theorem

Alperin showed in [1] that for P a Sylow p-subgroup of a finite group G, the category $\mathcal{F}_P(G)$ is completely determined by normalisers of certain subgroups of P. Goldschmidt refined this in [75], where he showed that it suffices to consider a class of subgroups, later called *essential subgroups* in work of Puig [166]. The underlying ideas carry over to arbitrary fusions systems, which is what we describe in this section. As before, p is a prime.

Definition 8.2.1 Let $\mathcal{F}$ be a category on a finite p-group P. A subgroup Q of P is called $\mathcal{F}$-*centric* if $C_P(R) = Z(R)$ for any subgroup R of P such that there is an isomorphism $R \cong Q$ in $\mathcal{F}$. We denote by $\mathcal{F}^c$ the full subcategory of $\mathcal{F}$ consisting of all $\mathcal{F}$-centric subgroups of P. If $\mathcal{F}$ is a fusion system, we denote by $\bar{\mathcal{F}}^c$ the full subcategory of the orbit category $\bar{\mathcal{F}}$ consisting of all $\mathcal{F}$-centric subgroups of P.

If Q is an $\mathcal{F}$-centric subgroup of P, then in particular Q, and every subgroup R of P isomorphic to Q in $\mathcal{F}$, is fully $\mathcal{F}$-centralised. Centric subgroups in fusion systems of finite groups can be characterised as follows.

Proposition 8.2.2 *Let G be a finite group, P a Sylow p-subgroup and set $\mathcal{F} = \mathcal{F}_P(G)$. Let Q be a subgroup of P. The following are equivalent.*

(i) Q is $\mathcal{F}$-centric.
(ii) $Z(Q)$ is a Sylow p-subgroup of $C_G(Q)$.
(iii) $C_G(Q) = Z(Q) \times O_{p'}(C_G(Q))$.

Proof The equivalence of (i) and (ii) follows from the characterisation of fully $\mathcal{F}$-centralised subgroups in 8.1.4. The equivalence between (ii) and (iii) follows from the Schur–Zassenhaus Theorem, implying that a finite group with a normal Sylow p-subgroup has a complement, necessarily of order prime to p, and if the Sylow p-subgroup is actually central, such a p'-complement is then unique, hence normal. $\square$

Proposition 8.2.3 *Let $\mathcal{F}$ be a fusion system on a finite p-group P and let Q be a fully $\mathcal{F}$-centralised subgroup of P. Then $QC_P(Q)$ is $\mathcal{F}$-centric.*

Proof Let $\varphi : QC_P(Q) \to R$ be an isomorphism in $\mathcal{F}$. Let $\psi : \varphi(Q) \to Q$ be the inverse of φ restricted to $\varphi(Q)$. Since Q is fully $\mathcal{F}$-centralised, ψ extends to a morphism $\tau : \varphi(Q)C_P(\varphi(Q)) \to P$ whose image must be contained in $QC_P(Q)$. Since $\varphi(Q) \subseteq R$ we have $RC_P(R) \subseteq \varphi(Q)C_P(\varphi(Q))$, hence $|RC_P(R)| \leq |QC_P(Q)| = |R|$, which forces $C_P(R) \subseteq R$ or $C_P(R) = Z(R)$. Thus $QC_P(Q)$ is $\mathcal{F}$-centric. $\square$

Proposition 8.2.4 *Let $\mathcal{F}$ be a category on a finite p-group P and let Q, R be subgroups of P such that $Q \subseteq R$. If Q is $\mathcal{F}$-centric then R is $\mathcal{F}$-centric and we have $Z(R) \subseteq Z(Q)$.*

Proof Let $\psi : R \to P$ be a morphism in $\mathcal{F}$. If Q is $\mathcal{F}$-centric, then $C_P(\psi(Q)) = Z(\psi(Q))$, hence $C_P(\psi(R)) \subseteq C_P(\psi(Q)) \subseteq \psi(R)$ and hence $C_P(\psi(R)) = Z(\psi(R))$. In particular, $Z(R) = C_P(R) \subseteq C_P(Q) = Z(Q)$. $\square$

For p a prime, the largest normal p-subgroup of a finite group G is denoted by $O_p(G)$. Given a fusion system $\mathcal{F}$ on a finite p-group P, for any subgroup Q of P we have $\mathrm{Aut}_Q(Q) \trianglelefteq \mathrm{Aut}_{\mathcal{F}}(Q)$, and hence $\mathrm{Aut}_Q(Q) \leq O_p(\mathrm{Aut}_{\mathcal{F}}(Q))$.

Definition 8.2.5 Let p be a prime, P a finite p-group and $\mathcal{F}$ a fusion system on P. A subgroup Q of P is called $\mathcal{F}$-*radical* if $O_p(\mathrm{Aut}_{\mathcal{F}}(Q)) = \mathrm{Aut}_Q(Q)$.

Definition 8.2.6 A proper subgroup H of a finite group G is called *strongly* p-*embedded* if p divides $|H|$ but p does not divide $|H \cap {}^xH|$ for any $x \in G \setminus H$.

We include in 8.2.13 below a list of equivalent reformulations of this notion; in particular, we will see that H is strongly p-embedded if and only if H contains a Sylow p-subgroup P of G and if $P \neq 1$ but $H \cap {}^xP = \{1\}$ for any $x \in G \setminus H$. This property does not depend on the choice of P in H. Denote by $S_p(G)$ the partially ordered set of all nontrivial p-subgroups of G. The $S_p(G)$ is disconnected if and only if G has a strongly p-embedded subgroup. Indeed, if $S_p(G)$ is disconnected, take for H the stabiliser in G of a connected component of $S_p(G)$; one easily checks that H is strongly p-embedded. Conversely, if H is a strongly p-embedded subgroup containing a Sylow p-subgroup P of G, then for any $x \in G \setminus H$ the conjugate group ${}^xP$ does not belong to the same connected component as P. If $O_p(G) \neq 1$ then $S_p(G)$ is connected, hence G has no strongly p-embedded subgroup in that case. For $\mathcal{F}$ a fusion system on a finite p-group P, we denote by $\bar{\mathcal{F}}$ its orbit category, as defined in 8.1.10.

Definition 8.2.7 Let p be a prime, P a finite p-group and $\mathcal{F}$ a fusion system on P. A subgroup Q of P is called $\mathcal{F}$-*essential* if Q is $\mathcal{F}$-centric and if $\mathrm{Aut}_{\bar{\mathcal{F}}}(Q)$ has a strongly p-embedded subgroup.

If Q is $\mathcal{F}$-essential then Q is $\mathcal{F}$-radical, but the converse need not be true. An $\mathcal{F}$-essential subgroup of P is always a proper subgroup of P because $\mathrm{Aut}_{\mathcal{F}}(P)/\mathrm{Aut}_P(P)$ is a p'-group by the Sylow axiom.

Theorem 8.2.8 (Alperin's Fusion Theorem) *Let $\mathcal{F}$ be a fusion system on a finite p-group P. Every isomorphism in $\mathcal{F}$ can be written as a composition of finitely many isomorphisms of the form $\varphi : R \to S$ in $\mathcal{F}$ for which there exists a subgroup Q containing both R, S and an automorphism $\alpha \in \mathrm{Aut}_{\mathcal{F}}(Q)$ such that $\alpha|_R = \varphi$ and such that either $Q = P$ or Q is fully $\mathcal{F}$-normalised essential and α has p-power order.*

Corollary 8.2.9 *Let $\mathcal{F}$ and $\mathcal{F}'$ be fusion systems on a finite p-group P. We have $\mathcal{F} = \mathcal{F}'$ if and only if $\mathrm{Aut}_{\mathcal{F}}(Q) = \mathrm{Aut}_{\mathcal{F}'}(Q)$ for any subgroup Q of P that is $\mathcal{F}$-essential or $\mathcal{F}'$-essential or equal to P.*

In many applications it is sufficient to know Alperin's Fusion Theorem with essential subgroups replaced by the potentially larger class of centric radical subgroups. We therefore organise the proof in such a way that this slightly weaker version is obtained in the first half of the proof, using only two of the following three lemmas which contain some of the technicalities needed in the proof.

Lemma 8.2.10 *Let $\mathcal{F}$ be a category on a finite p-group P and let Q be a fully $\mathcal{F}$-normalised subgroup of P. Let $\alpha \in \mathrm{Aut}_{\mathcal{F}}(Q)$. There is a unique subgroup R of $N_P(Q)$ containing $QC_P(Q)$ such that $\mathrm{Aut}_R(Q) = \mathrm{Aut}_P(Q) \cap (\alpha^{-1} \circ \mathrm{Aut}_P(Q) \circ \alpha)$, and we have $R = N_\alpha$.*

Proof The intersection $\mathrm{Aut}_P(Q) \cap (\alpha^{-1} \circ \mathrm{Aut}_P(Q) \circ \alpha)$ is a subgroup of $\mathrm{Aut}_P(Q) \cong N_P(Q)/C_P(Q)$, hence equal to $\mathrm{Aut}_R(Q)$ for a unique subgroup R of $N_P(Q)$ containing $QC_P(Q)$. By the definition of R, this group consists of all elements $y \in N_P(Q)$ for which there is $z \in N_P(Q)$ such that $c_y = \alpha^{-1} \circ c_z \circ \alpha$, or equivalently, such that $\alpha \circ c_y = c_z \circ \alpha$, where c_y, c_z are the automorphisms given by conjugation with y, z, respectively. Explicitly, the last equation means that $\alpha(^y u) = {}^z\alpha(u)$ for all $u \in Q$, which is equivalent to saying that y belongs to the group N_α. $\square$

Lemma 8.2.11 *Let $\mathcal{F}$ be a fusion system on a finite p-group P and let Q be a fully $\mathcal{F}$-normalised subgroup of P. There is a unique subgroup R of $N_P(Q)$ containing $QC_P(Q)$ such that $O_p(\mathrm{Aut}_{\mathcal{F}}(Q)) = \mathrm{Aut}_R(Q)$, and then every automorphism $\alpha \in \mathrm{Aut}_{\mathcal{F}}(Q)$ extends to an automorphism $\beta \in \mathrm{Aut}_{\mathcal{F}}(R)$.*

Proof Since Q is fully $\mathcal{F}$-normalised, $\mathrm{Aut}_P(Q) \cong N_P(Q)/C_P(Q)$ is a Sylow p-subgroup of $\mathrm{Aut}_{\mathcal{F}}(Q)$. Thus there is indeed a unique subgroup R of $N_P(Q)$ containing $C_P(Q)$ such that $\mathrm{Aut}_R(Q) = O_p(\mathrm{Aut}_{\mathcal{F}}(Q))$. Since $\mathrm{Aut}_Q(Q)$ is a normal p-subgroup of $\mathrm{Aut}_{\mathcal{F}}(Q)$ we also have $Q \subseteq R$. Let $\alpha \in \mathrm{Aut}_{\mathcal{F}}(Q)$. Since $\mathrm{Aut}_R(Q) = O_p(\mathrm{Aut}_{\mathcal{F}}(Q))$ we have $\alpha \circ \mathrm{Aut}_R(Q) \circ \alpha^{-1} = \mathrm{Aut}_R(Q)$. This means that $R \subseteq N_\alpha$ by 8.2.10, hence α extends to a morphism $\beta : R \to P$. For any $y \in R$ and any any $u \in Q$ we have $\alpha(^y u) = \beta(yuy^{-1}) = {}^{\beta(y)}\alpha(u)$, or equivalently, $\alpha \circ c_y \circ \alpha^{-1} = c_{\beta(y)}$, where c_y, $c_{\beta(y)}$ are the automorphisms of Q given by conjugation with y, $\beta(y)$. Since $\mathrm{Aut}_R(Q)$ is normal in $\mathrm{Aut}_{\mathcal{F}}(Q)$ this implies $c_{\beta(y)} \in \mathrm{Aut}_R(Q)$ and hence $\beta(y) \in R$ as $C_P(Q) \subseteq R$. Thus $\beta \in \mathrm{Aut}_{\mathcal{F}}(R)$. $\square$

Lemma 8.2.12 *Let $\mathcal{F}$ be a fusion system on a finite p-group P and let Q be a fully $\mathcal{F}$-normalised subgroup of P. Let $\alpha \in \mathrm{Aut}_{\mathcal{F}}(Q)$ such that $\mathrm{Aut}_Q(Q)$ is a proper subgroup of $\mathrm{Aut}_P(Q) \cap (\alpha \circ \mathrm{Aut}_P(Q) \circ \alpha^{-1})$. Then α extends to a morphism $\psi : R \to P$ in $\mathcal{F}$ for some subgroup R of $N_P(Q)$ that properly contains*

$QC_P(Q)$. *In particular, if α normalises $\mathrm{Aut}_P(Q)$, then α extends to an automorphism of $N_P(Q)$ in $\mathcal{F}$.*

Proof Let R be the unique subgroup of $N_P(Q)$ containing $QC_P(Q)$ such that $\mathrm{Aut}_R(Q) = \mathrm{Aut}_P(Q) \cap (\alpha \circ \mathrm{Aut}_P(Q) \circ \alpha^{-1})$. The hypotheses imply that $QC_P(Q)$ is a proper subgroup of R; in particular, Q is a proper subgroup of R. By 8.2.10 we have $R \subseteq N_\alpha$, and hence α extends to a morphism $\psi : R \to P$ as claimed. If α normalises $\mathrm{Aut}_P(Q)$, then $R = N_P(Q) = N_\alpha$, and hence α extends to a morphism $\psi : N_P(Q) \to P$. Since $\psi(Q) = \alpha(Q) = Q$, it follows that ψ preserves $N_P(Q)$, hence that ψ yields an automorphism of $N_P(Q)$. $\qquad\square$

Proof of Theorem 8.2.8 We denote by $\mathcal{A}$ the class of all isomorphisms in $\mathcal{F}$ which can be written as a composition of finitely many isomorphisms of the form $\psi : R \to R'$ for which there exists a subgroup Q of P containing R, R' and an automorphism $\alpha \in \mathrm{Aut}_\mathcal{F}(Q)$ such that $\alpha|_R = \psi$ and such that either $Q = P$ or Q is fully $\mathcal{F}$-normalised essential and α has p-power order. The class $\mathcal{A}$ is obviously closed under composition of isomorphisms, and if $\psi : R \to R'$ is an isomorphism in $\mathcal{A}$, then so is its inverse ψ^{-1} and any restriction of ψ to a subgroup V of R induces an isomorphism $\psi|_V : V \to V'$ in $\mathcal{A}$, where $V' = \psi(V)$. We have to show that in fact every isomorphism in $\mathcal{F}$ is in the class $\mathcal{A}$.

Let $\varphi : R \to S$ be an isomorphism in $\mathcal{F}$. We are going to show that φ can be written as composition of isomorphisms as claimed by induction over $|P : R|$. For $R = S = P$ there is nothing to prove, so we may assume that R is a proper subgroup of P. The basic argument is as follows: whenever N_φ contains R properly we can extend φ to N_φ, and then φ is in $\mathcal{A}$ by induction. We use this argument to first show that we can assume that S is fully $\mathcal{F}$-normalised. Indeed, let $\tau : S \to T$ be an isomorphism in $\mathcal{F}$ such that T is fully $\mathcal{F}$-normalised. By 8.1.12 we can choose τ such that τ can be extended to $N_P(S)$. Thus, by induction, τ belongs to the class $\mathcal{A}$, and hence, so does τ^{-1}. Therefore, in order to show that φ is in $\mathcal{A}$ it suffices to show that $\tau \circ \varphi$ is in $\mathcal{A}$. In other words, after replacing S by T and φ by $\tau \circ \varphi$, we may assume that S is fully $\mathcal{F}$-normalised. Then there is always some isomorphism $\psi : R \to S$ that extends to $N_P(R)$, and hence that is in $\mathcal{A}$ by induction. Thus it suffices to show that the automorphism $\varphi \circ \psi^{-1}$ of S is in $\mathcal{A}$. In other words, after replacing R by S and φ by $\varphi \circ \psi^{-1}$ we are down to assuming that $R = S$ is fully $\mathcal{F}$-normalised and $\varphi \in \mathrm{Aut}_\mathcal{F}(R)$. By induction and Lemma 8.2.11 we may assume in addition that R is $\mathcal{F}$-radical centric. Then in particular the group $U = \mathrm{Aut}_P(R)$ is a Sylow p-subgroup of $H = \mathrm{Aut}_\mathcal{F}(R)$. Note that the smallest normal subgroup $O^{p'}(H)$ of index prime to p is equal to the subgroup generated by all p-elements in H. By a Frattini argument, we have $H = N_H(U)O^{p'}(H)$; that is, φ is a product of

p-elements in $\text{Aut}_{\mathcal{F}}(R)$ and of an element $v \in U$. By 8.2.12, v extends to $N_P(R)$, hence belongs to $\mathcal{A}$.

This proves Alperin's Fusion Theorem for the class of $\mathcal{F}$-centric radical subgroups instead of essential subgroups. The last part of the proof shows that R may indeed be chosen to be essential.

Suppose that R is not essential. Set $X = \text{Aut}_P(R)$ and $Y = \varphi^{-1} \circ \text{Aut}_P(R) \circ \varphi$. Since R is fully $\mathcal{F}$-normalised, the groups X, Y are Sylow p-subgroups of $\text{Aut}_{\mathcal{F}}(R)$. Since R is not essential, the poset of nontrivial p-subgroups of $\text{Aut}_{\mathcal{F}}(R)/\text{Aut}_R(R)$ is connected. In other words, there is a finite family $\{X_0 | 1 \leq j \leq m\}$ of Sylow p-subgroups of $\text{Aut}_{\mathcal{F}}(R)$ such that $X = X_0$, $X_j \cap X_{j+1} \neq \text{Aut}_R(R)$ for $0 \leq j \leq m - 1$ and $X_m = Y$. For $0 \leq j \leq m$ let $\gamma_j \in \text{Aut}_{\mathcal{F}}(R)$ such that $\gamma_j \circ \text{Aut}_P(R) \circ \gamma_j^{-1} = X_j$; since $X_0 = \text{Aut}_P(R)$ we can take $\gamma_0 = \text{Id}_R$. We are going to prove, by induction over j, that the γ_j are in $\mathcal{A}$. For $j = 0$ this is trivial. Suppose $0 < j \leq m$. We have $X_{j-1} \cap X_j \neq \text{Aut}_R(R)$, hence

$$\text{Aut}_P(R) \cap (\gamma_{j-1}^{-1} \circ \gamma_j \circ \text{Aut}_P(R) \circ \gamma_j^{-1} \circ \gamma_{j-1}) \neq \text{Aut}_R(R).$$

Thus, by 8.2.12 and induction, the morphism $\gamma_j^{-1} \circ \gamma_{j-1}$ is in $\mathcal{A}$. By our induction over j the morphism γ_{j-1} is in $\mathcal{A}$, hence so is γ_j. In particular, γ_m is in $\mathcal{A}$. Since

$$Y = \gamma_m \circ \text{Aut}_P(R) \circ \gamma_m^{-1} = \varphi^{-1} \circ \text{Aut}_P(R) \circ \varphi$$

we get that $\varphi \circ \gamma_m$ normalises $\text{Aut}_P(R)$, hence extends to $N_P(R)$. Thus $\varphi \circ \gamma_m$ is in $\mathcal{A}$. But then φ is in $\mathcal{A}$ as well because γ_m and its inverse are so, which completes the proof. $\square$

Lemma 8.2.13 *Let G be a finite group, p a prime and H a subgroup of G. The following are equivalent.*

  (i) *H is strongly p-embedded in G.*
 (ii) *H has order divisible by p and contains $N_G(Q)$ for any nontrivial p-subgroup Q of H.*
(iii) *H contains a Sylow p-subgroup P of G that is nontrivial and H contains $N_G(Q)$ for any nontrivial subgroup Q of P.*
 (iv) *H contains a Sylow p-subgroup of G that is nontrivial, and we have $H \cap {}^x P = 1$ for any $x \in G \setminus H$ and any Sylow p-subgroup P of H.*
  (v) *H contains a Sylow p-subgroup of G that is nontrivial, and we have $P \cap {}^x P = 1$ for any $x \in G \setminus H$ and any Sylow p-subgroup P of H.*
 (vi) *H contains a Sylow p-subgroup P of G that is nontrivial, and that satisfies $P \cap {}^x P = 1$ for any $x \in G \setminus H$.*

Proof Suppose (i) holds. Then in particular $|H|$ is divisible by p. Let Q be a nontrivial p-subgroup of H and let $x \in N_G(Q)$. Then $Q \subseteq H \cap {}^xH$, hence $|H \cap {}^xH|$ is divisible by p and thus $x \in H$. This shows (i) $\Rightarrow$ (ii). Suppose (ii) holds. Then H has a nontrivial Sylow p-subgroup Q. Thus $N_G(Q) \subseteq H$. Thus, if P is a Sylow p-subgroup of G containing H we get $N_P(Q) \subseteq H$, which forces $N_P(Q) = Q$ and hence $P = Q$. This proves (ii) $\Rightarrow$ (iii). Suppose (iii) holds. By Alperin's Fusion Theorem, this implies that H controls fusion of G in any Sylow p-subgroup P of H. Let $x \in G - H$, set $Q = H \cap {}^xP$ and suppose, for contradiction, that $Q \neq 1$. There is $y \in H$ such that ${}^yQ \subseteq P$. Note that ${}^{x^{-1}}Q \subseteq P$. Thus conjugation by yx induces a group homomorphism from ${}^{x^{-1}}Q$ to ${}^yQ$ belonging to the fusion system $\mathcal{F}_P(G) = \mathcal{F}_P(H)$. Thus there is $h \in H$ such that $h^{-1}yx$ centralises the nontrivial p-subgroup ${}^{x^{-1}}Q$ of H. But then $h^{-1}yx \in H$, hence $x \in H$, a contradiction. This shows (iii) $\Rightarrow$ (iv). The implications (iv) $\Rightarrow$ (v) $\Rightarrow$ (vi) are trivial. Suppose that (vi) holds. Then p divides $|H|$. Let $x \in G \setminus H$ and let Q be a Sylow p-subgroup of $H \cap {}^xH$. Let R be a Sylow p-subgroup of H containing Q and let S be a Sylow p-subgroup of H such that ${}^xS$ contains Q. Let $r, s \in H$ such that ${}^rR = P = {}^sS$. Then ${}^rQ \subseteq P$ and also ${}^rQ \subseteq {}^{rx}S = {}^{rxs^{-1}}P$. Since $r, s \in H$ and $x \in G \setminus H$ we have $y = rxs^{-1} \in G \setminus H$ and ${}^rQ \subseteq P \cap {}^yP$, which forces $Q = 1$, whence the implication (vi) $\Rightarrow$ (i). $\qquad \square$

Under suitable hypotheses, the property of having a strongly p-embedded subgroup passes down to normal subgroups and quotients of a finite group:

Lemma 8.2.14 *Let G be a finite group, p a prime, N a normal subgroup of G and H a strongly p-embedded subgroup of G.*

(i) If p divides $|G/N|$ then HN/N is strongly p-embedded in G/N.
(ii) If p divides $|N|$ then $H \cap N$ is strongly p-embedded in N.

Proof If p divides $|G/N|$ and H contains a Sylow p-subgroup P of H, the quotient group HN/N has PN/N as nontrivial Sylow p-subgroup. Let $x \in G$ such that $xN \in G/N \setminus HN/N$. Then $x \in G \setminus HN$. Write $PN/N \cap {}^xPN/N = QN$ for some p-subgroup Q of G. Thus there are $m, n \in N$ such that ${}^mQ \subseteq P$ and ${}^nQ \subseteq {}^xP$, hence ${}^mQ \subseteq P \cap {}^{mn^{-1}x}P$. Since $x \in G \setminus NH$ we have $mn^{-1}x \in G \setminus NH$ and so this intersection is trivial. Thus HN/N is strongly p-embedded in G/N, which proves (i). Since N is normal in G and since H contains a Sylow p-subgroup of G, the intersection $H \cap N$ contains a Sylow p-subgroup of N. Thus if p divides $|N|$ then also p divides $|H \cap N|$. Let $y \in N \setminus (H \cap N)$. Then $y \notin H$, hence $H \cap {}^yH$ has order prime to p, and thus the same is true with H replaced by its subgroup $H \cap N$. This proves (ii). $\qquad \square$

8.3 Normalisers and centralisers in fusion systems

Just as in group theory, it is possible to define normalisers and centralisers of a subgroup Q of a finite p-group P in a fusion system $\mathcal{F}$ on P. We follow Puig's approach, defining the more general concept of K-normalisers, where K is a subgroup of $\mathrm{Aut}(Q)$. Normalisers and centralisers correspond to the particular choices $K = \mathrm{Aut}(Q)$ and $K = \{\mathrm{Id}_Q\}$, respectively.

Definition 8.3.1 Let P be a finite p-group, let Q be a subgroup of P and let K be a subgroup of $\mathrm{Aut}(Q)$. The *K-normaliser of Q in P* is the subgroup

$$N_P^K(Q) = \{y \in N_P(Q) \mid \exists \alpha \in K : \alpha(u) = yuy^{-1} \forall u \in Q\}.$$

We set $\mathrm{Aut}_P^K(Q) = K \cap \mathrm{Aut}_P(Q)$.

In other words, $N_P^K(Q)$ consists of all elements in $N_P(Q)$ that induce, by conjugation, an automorphism of Q belonging to the automorphism subgroup K. Note that $C_P(Q) \leq N_P^K(Q)$, and if K contains all inner automorphisms of Q then also $Q \leq N_P^K(Q)$. We have $\mathrm{Aut}_P^K(Q) \cong N_P^K(Q)/C_P(Q)$. There are various special cases of this construction we will encounter most frequently: if $K = \{\mathrm{Id}_Q\}$, then $N_P^K(Q) = C_P(Q)$ and $\mathrm{Aut}_P^K(Q) = \{\mathrm{Id}_Q\}$; if $K = \mathrm{Aut}_Q(Q)$, then $N_P^K(Q) = QC_P(Q)$ and $\mathrm{Aut}_P^K(Q) = \mathrm{Aut}_Q(Q)$; finally, if $K = \mathrm{Aut}(Q)$ then $N_P^K(Q) = N_P(Q)$ and $\mathrm{Aut}_P^K(Q) = \mathrm{Aut}_P(Q)$. For $\varphi : Q \to P$ an injective group homomorphism, the group $^\varphi K = \varphi \circ K \circ \varphi^{-1}$ is a subgroup of $\mathrm{Aut}(\varphi(Q))$, and it makes thus sense to consider $N_P^{\varphi K}(\varphi(Q))$.

Definition 8.3.2 Let $\mathcal{F}$ be a category on a finite p-group P, let Q be a subgroup of P and let K be a subgroup of $\mathrm{Aut}(Q)$. We say that Q is *fully K-normalised in $\mathcal{F}$* if $|N_P^K(Q)| \geq |N_P^{\varphi K}(\varphi(Q))|$ for any morphism $\varphi : Q \to P$ in $\mathcal{F}$.

Thus Q is fully $\mathrm{Aut}(Q)$-normalised in $\mathcal{F}$ if and only if Q is fully normalised, and Q is fully $\{\mathrm{Id}_Q\}$-normalised if and only if Q is fully $\mathcal{F}$-centralised. Note that being fully $\mathrm{Aut}_Q(Q)$-normalised is also equivalent to being fully centralised.

Definition 8.3.3 Let $\mathcal{F}$ be a fusion system on a finite p-group P, let Q be a subgroup of P and let K be a subgroup of $\mathrm{Aut}(Q)$. The *K-normaliser of Q in $\mathcal{F}$* is the subcategory $N_{\mathcal{F}}^K(Q)$ of $\mathcal{F}$ on $N_P^K(Q)$ having morphism sets

$$\mathrm{Hom}_{N_{\mathcal{F}}^K(Q)}(R, S)$$

$$= \{\varphi \in \mathrm{Hom}_{\mathcal{F}}(R, S) \mid \exists \psi \in \mathrm{Hom}_{\mathcal{F}}(QR, QS) : \psi|_R = \varphi, \psi|_Q \in K\}$$

for any two subgroups R, S in $N_P^K(Q)$. We set $\mathrm{Aut}_{\mathcal{F}}^K(Q) = K \cap \mathrm{Aut}_{\mathcal{F}}(Q)$.

If $K = \mathrm{Aut}(Q)$ we write $N_{\mathcal{F}}(Q)$ instead of $N_{\mathcal{F}}^K(Q)$ and call $N_{\mathcal{F}}(Q)$ the *normaliser of Q in $\mathcal{F}$*. Explicitly, the objects of $N_{\mathcal{F}}(Q)$ are the subgroups of $N_P(Q)$, and for any two subgroups R, S of $N_P(Q)$ the morphism set $\mathrm{Hom}_{N_{\mathcal{F}}(Q)}(R, S)$ consists of all morphisms $\varphi : R \to S$ in $\mathcal{F}$ that can be extended to a morphism $\psi : QR \to QS$ in $\mathcal{F}$ satisfying $\psi(Q) = Q$. In that case we have $\mathrm{Aut}_{\mathcal{F}}^K(Q) = \mathrm{Aut}_{\mathcal{F}}(Q)$. Similarly, if $K = \{\mathrm{Id}_Q\}$ we write $C_{\mathcal{F}}(Q)$ instead of $N_{\mathcal{F}}^K(Q)$ and call $C_{\mathcal{F}}(Q)$ the *centraliser of Q in $\mathcal{F}$*. Explicitly, the objects of $C_{\mathcal{F}}(Q)$ are the subgroups of $C_P(Q)$, and for any two subgroups R, S of $C_P(Q)$ the morphism set $\mathrm{Hom}_{C_{\mathcal{F}}(Q)}(R, S)$ consists of all morphisms $\varphi : R \to S$ in $\mathcal{F}$ that can be extended to a morphism $\psi : QR \to QS$ in $\mathcal{F}$ satisfying $\psi|_Q = \mathrm{Id}_Q$. In that case we have $\mathrm{Aut}_{\mathcal{F}}^K(Q) = \{\mathrm{Id}_Q\}$. The following result, due to Puig, shows that under suitable hypotheses on Q, K-normalisers are again fusion systems.

Theorem 8.3.4 *Let $\mathcal{F}$ be a fusion system on a finite p-group P, let Q be a subgroup of P and K a subgroup of $\mathrm{Aut}(Q)$. Suppose that Q is fully K-normalised in $\mathcal{F}$. Then $N_{\mathcal{F}}^K(Q)$ is a fusion system on $N_P^K(Q)$.*

Lemma 8.3.5 *Let $\mathcal{F}$ be a category on a finite p-group P such that $\mathcal{F}$ contains $\mathcal{F}_P(P)$, let Q be a subgroup of P and K a subgroup of $\mathrm{Aut}(Q)$.*

(i) *If Q is fully $\mathcal{F}$-centralised and if $\mathrm{Aut}_P^K(Q)$ is a Sylow p-subgroup of $\mathrm{Aut}_{\mathcal{F}}^K(Q)$, then Q is fully K-normalised in $\mathcal{F}$, and for any morphism $\varphi : Q \to P$ in $\mathcal{F}$ such that $\varphi(Q)$ is fully $\varphi \circ K \circ \varphi$-normalised in $\mathcal{F}$, the group $\varphi(Q)$ is fully $\mathcal{F}$-centralised and $\mathrm{Aut}_P^{\varphi \circ K \circ \varphi^{-1}}(\varphi(Q))$ is a Sylow p-subgroup of $\mathrm{Aut}_{\mathcal{F}}^{\varphi \circ K \circ \varphi^{-1}}(\varphi(Q))$.*

(ii) *If Q is fully $\mathcal{F}$-centralised and if $K \le \mathrm{Aut}_P(Q)$, then Q is fully K-normalised in $\mathcal{F}$.*

Proof Statement (i) follows from $|N_P^K(Q)| = |\mathrm{Aut}_P^K(Q)| \cdot |C_P(Q)|$. If $K \subseteq \mathrm{Aut}_P(Q)$ then $\mathrm{Aut}_P^K(Q) = \mathrm{Aut}_{\mathcal{F}}^K(Q)$, and hence (ii) is a special case of (i). $\square$

Proposition 8.3.6 *Let $\mathcal{F}$ be a fusion system on a finite p-group P, let Q be a subgroup of P and let K be a subgroup of $\mathrm{Aut}(Q)$.*

(i) *The group Q is fully K-normalised in $\mathcal{F}$ if and only if Q is fully $\mathcal{F}$-centralised and $\mathrm{Aut}_P^K(Q)$ is a Sylow p-subgroup of $\mathrm{Aut}_{\mathcal{F}}^K(Q)$.*

(ii) *Let $\varphi : Q \to R$ be an isomorphism in $\mathcal{F}$ and set $L = \varphi \circ K \circ \varphi^{-1}$. Suppose that R is fully L-normalised in $\mathcal{F}$. Then there are morphisms $\tau : Q \cdot N_P^K(Q) \to P$ in $\mathcal{F}$ and $\kappa \in K$ such that $\tau|_Q = \varphi \circ \kappa$.*

(iii) *Let $\psi : QN_P^K(Q) \to P$ be a morphism in $\mathcal{F}$. If Q is fully K-normalised in $\mathcal{F}$, then $\psi(Q)$ is fully $\psi \circ K \circ \psi^{-1}$-normalised in $\mathcal{F}$.*

(iv) *If H is a normal subgroup of K and if Q is fully K-normalised in $\mathcal{F}$, then Q is fully H-normalised in $\mathcal{F}$.*

Proof (i) Suppose that Q is fully K-normalised in $\mathcal{F}$. By 8.1.12 there is an isomorphism $\varphi : Q \to R$ in $\mathcal{F}$ such that R is fully $\mathcal{F}$-normalised and such that φ extends to a morphism $\psi : N_P(Q) \to P$ in $\mathcal{F}$. Set $L = \varphi \circ K \circ \varphi^{-1}$. Thus ψ maps $N_P^K(Q)$ to $N_P^L(R)$. Since Q is fully K-normalised we have $|N_P^K(Q)| \geq |N_P^L(R)|$ and hence ψ induces in fact an isomorphism $N_P^K(Q) \cong N_P^L(R)$. Any such isomorphism restricts to an isomorphism $C_P(Q) \cong C_R(Q)$. Since R is fully $\mathcal{F}$-normalised, R is fully $\mathcal{F}$-centralised by 8.1.11 and thus Q is fully $\mathcal{F}$-centralised. The group $\varphi \circ \text{Aut}_{\mathcal{F}}^K(Q) \circ \varphi^{-1}$ is a p-subgroup of the subgroup $\text{Aut}_{\mathcal{F}}^L(R)$ of $\text{Aut}_{\mathcal{F}}(R)$. By 8.1.11 the group $\text{Aut}_P(R)$ is a Sylow p-subgroup of $\text{Aut}_{\mathcal{F}}(R)$. Thus some conjugate of $\text{Aut}_P(R)$ intersected with $\text{Aut}_{\mathcal{F}}^L(R)$ is a Sylow p-subgroup of $\text{Aut}_{\mathcal{F}}^L(R)$, and this Sylow p-subgroup can, of course, be chosen to contain the p-subgroup $\varphi \circ \text{Aut}_{\mathcal{F}}^K(Q) \circ \varphi^{-1}$. Thus there is $\beta \in \text{Aut}_{\mathcal{F}}(R)$ such that

$$\varphi \circ \text{Aut}_{\mathcal{F}}^K(Q) \circ \varphi^{-1} \leq \beta \circ \text{Aut}_P(R) \circ \beta^{-1} \cap \text{Aut}_{\mathcal{F}}^L(R)$$

and such that the right side of this inclusion is a Sylow p-subgroup of $\text{Aut}_{\mathcal{F}}^L(R)$. Then $\text{Aut}_P(R) \cap \text{Aut}_{\mathcal{F}}^{\beta^{-1} \circ L \circ \beta}(R) = \text{Aut}_P^{\beta^{-1} \circ L \circ \beta}(R)$ is a Sylow p-subgroup of $\text{Aut}_{\mathcal{F}}^{\beta^{-1} \circ L \circ \beta}(R)$. Since Q is fully K-normalised we have $|\text{Aut}_P^K(Q)| \geq |\text{Aut}_P^{\beta^{-1} \circ L \circ \beta}(R)|$, which implies that $\text{Aut}_P^K(Q)$ is a Sylow p-subgroup of $\text{Aut}_{\mathcal{F}}^K(Q)$. The converse follows from 8.3.5 (i).

(ii) Since R is fully K-normalised in $\mathcal{F}$ it is in particular fully $\mathcal{F}$-centralised. Set $N = Q \cdot N_P^K(Q)$. Then $\varphi \circ N \circ \varphi^{-1}$ is a p-subgroup of $\text{Aut}_{\mathcal{F}}^L(R)$. By (i), $\text{Aut}_P^L(R)$ is a Sylow p-subgroup of $\text{Aut}_{\mathcal{F}}^L(R)$. Thus there is $\lambda \in \text{Aut}_{\mathcal{F}}^L(R)$ such that $\lambda \circ \varphi \circ N \circ \varphi^{-1} \circ \lambda^{-1} \leq \text{Aut}_P(R)$. This means that $N \leq N_{\lambda \circ \varphi}$, and hence $\lambda \circ \varphi$ extends to a morphism $\tau : Q \cdot N_P^K(Q) \to P$. Now $\lambda \circ \varphi = \varphi \circ (\varphi^{-1} \circ \lambda \circ \varphi)$ and $\kappa = \varphi^{-1} \circ \lambda \circ \varphi \in \text{Aut}_P^K(Q)$ is as required.

(iii) The map ψ sends $N_P^K(Q)$ to $N_P^{\psi \circ K \circ \psi^{-1}}(\psi(Q))$. In particular, $|N_P^K(Q)| \leq |N_P^{\psi \circ K \circ \psi^{-1}}(\psi(Q))|$, whence the result.

(iv) If Q is fully K-normalised in $\mathcal{F}$, then Q is fully $\mathcal{F}$-centralised and $\text{Aut}_P^K(Q)$ is a Sylow p-subgroup of $\text{Aut}_{\mathcal{F}}^K(Q)$, by (i). If also H is normal in K then $\text{Aut}_{\mathcal{F}}^H(Q)$ is normal in $\text{Aut}_{\mathcal{F}}^K(Q)$, and hence $\text{Aut}_P^H(Q) = \text{Aut}_P^K(Q) \cap \text{Aut}_{\mathcal{F}}^H(Q)$ is a Sylow p-subgroup in $\text{Aut}_{\mathcal{F}}^H(Q)$. Thus, again by (i), Q is fully H-normalised in $\mathcal{F}$. $\square$

Proof of Theorem 8.3.4 Clearly $N_{\mathcal{F}}^K(Q)$ is a category on $N_P^K(Q)$, and $N_{\mathcal{F}}^K(Q)$ contains the category $\mathcal{F}_{N_P^K(Q)}(N_P^K(Q))$. For any subgroup R of $N_P^K(Q)$ and any subgroup I of $\text{Aut}(R)$ we set

$$K * I = \{\alpha \in \text{Aut}(QR) \,|\, \alpha|_Q \in K, \alpha|_R \in I\}.$$

Then

(1) $N^I_{N^K_P(Q)}(R) = N^{K*I}_P(QR)$.

Indeed, both sides are equal to the subgroup of all $y \in N_P(Q) \cap N_P(R)$ such that conjugation by y induces on Q an automorphism in K and on R an automorphism in I. With this notation,

(2) the restriction map $\mathrm{Aut}^{K*I}_{\mathcal{F}}(QR) \to \mathrm{Aut}^I_{N^K_P(Q)}(R)$ is surjective.

Indeed, any $\beta \in \mathrm{Aut}_{N^K_{\mathcal{F}}(Q)}(R)$ extends to some $\alpha \in \mathrm{Aut}_{\mathcal{F}}(QR)$ with $\alpha|_Q \in K$, and since $\beta = \alpha|_R$, the surjectivity follows. We observe next that

(3) for any subgroup R of $N^K_P(Q)$ and any subgroup I of $\mathrm{Aut}(R)$ there is a morphism $\varphi : QR \to QN^K_P(Q)$ in $\mathcal{F}$ such that $\varphi|_Q \in K$ and such that $\varphi(QR) = Q\varphi(R)$ is fully $\varphi \circ (K * I) \circ \varphi^{-1}$-normalised.

To see this, let $\rho : QR \to P$ be a morphism in $\mathcal{F}$ such that $\rho(QR)$ is fully $\rho \circ (K * I) \circ \rho^{-1}$-normalised. Set $\sigma = \rho|_Q$. Since Q is fully K-normalised in $\mathcal{F}$, by 8.3.6 (ii) applied to σ^{-1} and $\sigma(Q)$, there is a morphism $\tau : \sigma(Q)N^{\sigma \circ K \circ \sigma^{-1}}_P(\sigma(Q)) \to P$ and $\kappa \in \sigma \circ K \circ \sigma^{-1}$ such that $\tau|_{\sigma(Q)} = \sigma^{-1} \circ \kappa$. Set $\varphi = \tau \circ \rho$. Then $\varphi|_Q = \tau \circ \rho|_Q = \tau \circ \sigma = \sigma^{-1} \circ \kappa \circ \sigma \in K$. Note that $\rho(QR)N^{\rho \circ (K*I) \circ \rho^{-1}}_P(\rho(QR)) \le \sigma(Q)N^{\sigma \circ K \circ \sigma^{-1}}_P(\sigma(Q))$ and that $\rho(QR)$ is fully $\rho \circ (K * I) \circ \rho^{-1}$-normalised in $\mathcal{F}$. It follows from 8.3.6 (iii) applied to the appropriate restriction of τ and $\rho(QR)$ that RQ is fully $K * I$-normalised in $\mathcal{F}$. This proves (3). We show next that

(4) if a subgroup R of $N^K_P(Q)$ is fully I-normalised in $N^K_{\mathcal{F}}(Q)$ for some subgroup I of $\mathrm{Aut}(R)$ then QR is fully $K * I$-normalised in $\mathcal{F}$.

Indeed, by (3) there exists a morphism $\varphi : QR \to QN^K_P(Q)$ in $\mathcal{F}$ such that $\varphi|_Q \in K$ and such that $Q\varphi(Q)$ is fully $\varphi \circ (K * I) \circ \varphi^{-1}$-normalised. We have $N^{K*I}_P(QR) = N^I_{N^K_P(Q)}(R)$ by (1) and $|N^I_{N^K_P(Q)}(R)| \ge |N^{\varphi \circ I \circ \varphi^{-1}}_{N^K_P(Q)}(\varphi(R))|$ because R is fully I-normalised. Now $N^{\varphi \circ I \circ \varphi^{-1}}_{N^K_P(Q)}(\varphi(R)) = N^{K*(\varphi \circ I \circ \varphi^{-1})}_P(\varphi(R))$ by (1) and since $\varphi|_Q \in K$ we have $K * (\varphi \circ I \circ \varphi^{-1}) = \varphi \circ (K * I) \circ \varphi^{-1}$, from which we get that the last group is equal to $N^{\varphi \circ (K*I) \circ \varphi^{-1}}_P(\varphi(R))$. In particular, $|N^{K*I}_P(QR)| \ge |N^{\varphi \circ (K*I) \circ \varphi^{-1}}_P(\varphi(R))|$, which proves the statement (4). We use this now to prove the stronger version of the Sylow axiom as in 8.1.11.

(5) if a subgroup R of $N^K_P(Q)$ is fully $N^K_{\mathcal{F}}(Q)$-normalised then R is fully $N^K_{\mathcal{F}}(Q)$-centralised and $\mathrm{Aut}_{N^K_P(Q)}(R)$ is a Sylow p-subgroup of $\mathrm{Aut}_{N^K_{\mathcal{F}}(Q)}(R)$.

In order to show (5) it suffices, by 8.3.5 (i), to show this for some subgroup R' of $N_P^K(Q)$ isomorphic to R in $N_{\mathcal{F}}^K(Q)$. Set $A = \mathrm{Aut}(R)$. By (3) there is a morphism $\varphi : QR \to QN_P^K(Q)$ such that $\varphi|_Q \in K$ and such that $Q\varphi(R)$ is fully $\varphi \circ (K * A) \circ \varphi^{-1}$-normalised in $\mathcal{F}$. By 8.3.6 (iv), $Q\varphi(R)$ is then also fully $\varphi \circ (K * \mathrm{Id}) \circ \varphi^{-1}$-normalised in $\mathcal{F}$, where Id is the trivial automorphism group of R. Since $\varphi|_Q \in K$ we have $\varphi \circ (K * \mathrm{Id}) \circ \varphi^{-1} = K * \mathrm{Id}$, where now Id denotes abusively also the trivial automorphism group of $\varphi(R)$. Note that $N_P^{K*\mathrm{Id}}(Q\varphi(R)) = C_{N_P^K(Q)}(\varphi(R))$. Thus, if $\psi : QR \to QN_P^K(Q)$ is any other morphism in $\mathcal{F}$ such that $\psi|_Q \in K$ then, using that $Q\varphi(R)$ is fully $K * \mathrm{Id}$-normalised in $\mathcal{F}$, we get

$$|C_{N_P^K(Q)}(\psi(R))| = |N_P^{K*\mathrm{Id}}(Q\psi(R))| \leq |N_P^{K*\mathrm{Id}}(Q\varphi(R))| = |C_{N_P^K(Q)}(\varphi(R))|.$$

This shows that $\varphi(R)$ is fully $N_P^K(Q)$-centralised. Set $B = \varphi \circ A \circ \varphi^{-1} = \mathrm{Aut}(\varphi(R))$. Since $Q\varphi(R)$ is fully $K * B$-normalised in $\mathcal{F}$ it follows from 8.3.6 (i) that $\mathrm{Aut}_P^{K*B}(Q\varphi(R))$ is a Sylow p-subgroup of $\mathrm{Aut}_{\mathcal{F}}^{K*B}(Q\varphi(R))$. The restriction map $\mathrm{Aut}_{\mathcal{F}}^{K*B}(Q\varphi(R)) \to \mathrm{Aut}_{N_{\mathcal{F}}^K(Q)}(\varphi(R))$ from (2) is surjective, and it maps $\mathrm{Aut}_P^{K*B}(Q\varphi(R))$ to $\mathrm{Aut}_{N_P^K(Q)}(\varphi(R))$, which is hence a Sylow p-subgroup of $\mathrm{Aut}_{N_{\mathcal{F}}^K(Q)}(\varphi(R))$. This proves (5).

For the proof of the extension axiom in $N_{\mathcal{F}}^K(Q)$, let R be a subgroup of $N_P^K(Q)$ and let $\varphi : R \to N_P^K(Q)$ be a morphism in $N_{\mathcal{F}}^K(Q)$ such that $\varphi(R)$ is fully $N_{\mathcal{F}}^K(Q)$-normalised. Note that then $\varphi(R)$ is in particular fully $N_{\mathcal{F}}^K(Q)$-centralised, by (5). We consider the group N_φ as defined in 8.1.5 for the morphism φ in the category $N_{\mathcal{F}}^K(Q)$; that is, N_φ is the subgroup of all $y \in N_{N_P^K(Q)}(R)$ for which there exists an element $z \in N_{N_P^K(Q)}(\varphi(R))$ satisfying $\varphi(yuy^{-1}) = z\varphi(u)z^{-1}$ for all $u \in R$. Set $I = \mathrm{Aut}_{N_\varphi}(R)$. Note that conjugation by any $y \in N_\varphi$ leaves Q invariant and induces an automorphism of Q belonging to K. Then $N_\varphi = N_{N_P^K(Q)}^I(R) = N_P^{K*I}(QR)$. We have $\varphi \circ I \circ \varphi^{-1} \leq \mathrm{Aut}_{N_P^K(Q)}(\varphi(Q)) \leq \mathrm{Aut}_{N_P^K(Q)}(\varphi(R))$, and hence $\varphi(R)$ is fully $\varphi \circ I \circ \varphi^{-1}$-normalised in $N_{\mathcal{F}}^K(Q)$ by 8.3.5 (ii). Thus $Q\varphi(R)$ is fully $K * (\varphi \circ I \circ \varphi^{-1})$-normalised by (4). Since φ is a morphism in $N_{\mathcal{F}}^K(Q)$ there exists a morphism $\psi : QR \to P$ in $\mathcal{F}$ such that $\psi|_Q \in K$ and $\psi|_R = \varphi$. So $\psi(QR) = Q\varphi(R)$ is fully $K * (\varphi \circ I \circ \varphi^{-1})$-normalised by the above. Now $\psi^{-1} \circ (K * (\varphi \circ I \circ \varphi^{-1})) \circ \psi = K * I$ because $\psi|_Q \in K$ and because $\psi^{-1}|_R \circ \varphi = \mathrm{Id}_R$. Applying 8.3.6 (ii) to ψ and QR yields the existence of a morphism $\tau : QR \cdot N_P^{K*I}(QR) \to P$ and $\kappa \in K * I$ such that $\tau|_{QR} = \psi \circ \kappa$. Since $\kappa|_R \in I = \mathrm{Aut}_R(R)$, there is $y \in R$ such that $\kappa|_R = c_y$, conjugation by y. Then $\tau|_{N_\varphi} \circ c_y^{-1} : N_\varphi \to P$ is a morphism in $N_{\mathcal{F}}^K(Q)$, because it extends to the morphism $\tau|_{QN_\varphi} \circ c_y : QN_\varphi \to P$ whose restriction to Q is $\tau|_Q \circ c_y^{-1}|_Q = \psi|_Q \circ \kappa|_Q \circ c_y^{-1}|_Q \in K$. Moreover, the morphism $\tau|_{N_\varphi} \circ c_y^{-1}$ restricted to R is equal to $\psi|_R \circ \kappa|_R \circ c_y^{-1}|_R = \psi|_R = \varphi$, and hence this morphism extends φ as required in the extension axiom. This completes the proof of 8.3.4. □

Proposition 8.3.7 *Let $\mathcal{F}$ be a fusion system on a finite p-group P, let Q be a fully $\mathcal{F}$-centralised subgroup of P and let S be a subgroup of $C_P(Q)$. Then S is fully $C_{\mathcal{F}}(Q)$-centralised if and only if QS is fully $\mathcal{F}$-centralised. In particular, if S is $C_{\mathcal{F}}(Q)$-centric, then QS is $\mathcal{F}$-centric.*

Proof Suppose that S is fully $C_{\mathcal{F}}(Q)$-centralised. Let $\psi : QS \to P$ be a morphism in $\mathcal{F}$. Denote by $\tau : \psi(Q) \to Q$ the inverse of the isomorphism $\psi|_Q : Q \cong \psi(Q)$ induced by ψ. Since Q is fully $\mathcal{F}$-centralised, τ extends to a morphism $\sigma : \psi(Q)C_P(\psi(Q)) \to QC_P(Q)$ in $\mathcal{F}$ (cf. 8.1.13). Note that $\psi(S) \leq C_P(\psi(Q))$. By construction, we have $\sigma \circ \psi|_Q = \mathrm{Id}_Q$. Thus $\sigma \circ \psi|_S$ is a morphism in $C_{\mathcal{F}}(Q)$. Since S is fully $C_{\mathcal{F}}(Q)$-centralised, we have

$$|C_P(QS)| = |C_{C_P(Q)}(S)| \geq |C_{C_P(Q)}(\sigma(\psi(S)))| = |C_P(Q\sigma(\psi(S)))|.$$

Now σ sends $C_P(\psi(QS))$ to $C_P(Q\sigma(\psi(S)))$, and hence $|C_P(Q\sigma(\psi(S)))| \geq |C_P(\psi(QS))|$, which proves that QS is fully $\mathcal{F}$-centralised. If S is $C_{\mathcal{F}}(Q)$-centric, then S is fully $C_{\mathcal{F}}(Q)$-centralised and $C_P(QS) = C_{C_P(Q)}(S) = Z(S) \leq QS$, hence $C_P(QS) = Z(QS)$. Since QS is fully $\mathcal{F}$-centralised by the first argument, it follows that QS is $\mathcal{F}$-centric. Conversely, suppose that QS is fully $\mathcal{F}$-centralised. Let $\tau : S \to C_P(Q)$ be a morphism in $C_{\mathcal{F}}(Q)$. That is, τ extends to a morphism $\sigma : QS \to P$ in $\mathcal{F}$ such that $\sigma|_Q = \mathrm{Id}_Q$. Since QS is fully $\mathcal{F}$-centralised, we have

$$|C_{C_P(Q)}(S)| = |C_P(QS)| \geq |C_P(\sigma(QS))| = |C_P(Q\tau(S))| = |C_{C_P(Q)}(\tau(S))|$$

which proves that S is fully $C_{\mathcal{F}}(Q)$-centralised. □

As in finite group theory, given a fusion system $\mathcal{F}$ on a finite p-group P, one of the questions one would like to be able to address is that of *control of fusion* by normalisers of p-subgroups, a question that we can now rephrase as follows: when do we have an equality $\mathcal{F} = N_{\mathcal{F}}(Q)$ for some normal subgroup Q of P? Burnside's Theorem 8.1.15 on fusion in finite groups with an abelian Sylow p-subgroup can be reformulated for arbitrary fusion systems as follows:

Proposition 8.3.8 *Let $\mathcal{F}$ be a fusion system on an abelian finite p-group P. Then $\mathcal{F} = N_{\mathcal{F}}(P)$.*

Proof This is a reformulation of 8.1.14. □

It has been shown by Stancu [209], [210] that the conclusion of 8.3.8 remains true if P is a metacyclic finite p-group for some odd prime p, or if P is a finite p-group of the form $P = E \times A$, where A is elementary abelian and where E

is extra-special of exponent p^2 if p is odd and not isomorphic to D_8 if $p = 2$. Finite p-groups P with the property that every fusion system $\mathcal{F}$ on P satisfies $\mathcal{F} = N_{\mathcal{F}}(P)$ are called *resistant*.

Theorem 8.3.9 *Let $\mathcal{F}$ be a fusion system on a finite p-group P. The following are equivalent:*

  (i) For any $Q \subseteq P$ the group $\mathrm{Aut}_{\mathcal{F}}(Q)$ is a p-group.
  (ii) $\mathcal{F} = \mathcal{F}_P(P)$.
  (iii) For any nontrivial fully $\mathcal{F}$-normalised subgroup Q of P we have $N_{\mathcal{F}}(Q) = \mathcal{F}_{N_P(Q)}(N_P(Q))$.

Proof If (i) holds and if Q is a fully $\mathcal{F}$-normalised subgroup of P, then $\mathrm{Aut}_{\mathcal{F}}(Q) = \mathrm{Aut}_P(Q)$ by the Sylow axiom. Thus (i) implies (ii) by Alperin's Fusion Theorem. The implications (ii) $\Rightarrow$ (iii) $\Rightarrow$ (i) are trivial. $\qquad\square$

Corollary 8.3.10 *Let $\mathcal{F}$ be a fusion system on a finite p-group P and $\mathcal{G}$ a fusion system on a subgroup Q of P such that $\mathcal{G} \subseteq \mathcal{F}$. If $\mathcal{F} = \mathcal{F}_P(P)$, then $\mathcal{G} = \mathcal{F}_Q(Q)$.*

Proof This follows from the equivalence of the statements (i) and (ii) in 8.3.9, applied to the fusion system $\mathcal{G}$ on Q. $\qquad\square$

When specialised to fusion systems of finite groups, normalisers and centralisers in the fusion system correspond to the fusion systems of the relevant normaliser and centralisers:

Proposition 8.3.11 *Let G be a finite group, let P be a Sylow p-subgroup and set $\mathcal{F} = \mathcal{F}_P(G)$. Let Q be a subgroup of P.*

  (i) If Q is fully $\mathcal{F}$-normalised, then $N_{\mathcal{F}_P(G)}(Q) = \mathcal{F}_{N_P(Q)}(N_G(Q))$.
  (ii) If Q is fully $\mathcal{F}$-centralised, then $C_{\mathcal{F}_P(G)}(Q) = \mathcal{F}_{C_P(Q)}(C_G(Q))$.
  (iii) If Q is normal in G, then $\mathcal{F} = N_{\mathcal{F}}(Q)$.

Proof The statements make sense: if Q is fully $\mathcal{F}$-normalised, then $N_P(Q)$ is a Sylow p-subgroup of $N_G(Q)$, and if Q is fully $\mathcal{F}$-centralised, then $C_P(Q)$ is a Sylow p-subgroup of $C_G(Q)$. The rest is a trivial verification. $\qquad\square$

8.4 Quotients and products of fusions systems

If $\mathcal{F}$ is a fusion system on a finite p-group P such that $\mathcal{F} = N_{\mathcal{F}}(Q)$ for some normal subgroup Q of P, we can define a quotient category $\mathcal{F}/Q$ as follows:

Definition 8.4.1 Let $\mathcal{F}$ be a fusion system on a finite p-group P such that $\mathcal{F} = N_{\mathcal{F}}(Q)$ for some normal subgroup Q of P. We define the category $\mathcal{F}/Q$ on P/Q as follows: for any two subgroups R, S of P containing Q, a group homomorphism $\psi : R/Q \to S/Q$ is a morphism in the category $\mathcal{F}/Q$ if there exists a morphism $\varphi : R \to S$ in $\mathcal{F}$ satisfying $\varphi(u)Q = \psi(uQ)$ for all $u \in R$.

Theorem 8.4.2 (Puig) *Let $\mathcal{F}$ be a fusion system on a finite p-group P such that $\mathcal{F} = N_{\mathcal{F}}(Q)$ for some normal subgroup Q of P. The category $\mathcal{F}/Q$ is a fusion system on P/Q.*

Proof Clearly $\mathcal{F}/Q$ is a category on P/Q. Since $\mathrm{Aut}_{\mathcal{F}/Q}(P/Q)$ is a quotient of $\mathrm{Aut}_{\mathcal{F}}(P)$, the group $\mathrm{Aut}_{P/Q}(P/Q)$ is a Sylow p-subgroup of $\mathrm{Aut}_{\mathcal{F}/Q}(P/Q)$; thus the Sylow axiom I-S holds. It remains to show that the extension axiom holds as well. Let R, S be subgroups of P containing Q, and let $\psi : R/Q \to S/Q$ be an isomorphism in $\mathcal{F}/Q$ such that S/Q is fully $\mathcal{F}$-normalised. Since $N_{P/Q}(S/Q) = N_P(S)/Q$ this implies that S is fully $\mathcal{F}$-normalised. Thus, by 8.1.12, there is a morphism $\rho : N_P(R) \to P$ such that $\rho(R) = S$. Denote by $\sigma : N_P(R)/Q \to P/Q$ the morphism induced by ρ. In order to show that ψ extends to N_ψ it suffices to show that $\psi \circ \sigma^{-1}|_{S/Q}$ extends to $N_{\psi \circ \sigma^{-1}|_{S/Q}}$, because the inverse image of N_ψ is contained in $N_P(R)$. In other words, we may assume that $R = S$ is fully $\mathcal{F}$-normalised and that $\psi \in \mathrm{Aut}_{\mathcal{F}/Q}(R/Q)$. Let K be the kernel of the canonical surjective map $\mathrm{Aut}_{\mathcal{F}}(R) \to \mathrm{Aut}_{\mathcal{F}/Q}(R/Q)$. Then $X = K \cap \mathrm{Aut}_P(R)$ is a Sylow p-subgroup of K. The Frattini argument implies that $\mathrm{Aut}_{\mathcal{F}}(R) = K N_{\mathrm{Aut}_{\mathcal{F}}(R)}(X)$. Note that $\mathrm{Aut}_P(R)$ normalises X, so $\mathrm{Aut}_P(R)$ remains a Sylow p-subgroup of $N_{\mathrm{Aut}_{\mathcal{F}}(R)}(X)$. Thus we have a short exact sequence of finite groups

$$1 \longrightarrow K \cap N_{\mathrm{Aut}_{\mathcal{F}}(R)}(X) \longrightarrow N_{\mathrm{Aut}_{\mathcal{F}}(R)}(X) \overset{\alpha}{\longrightarrow} \mathrm{Aut}_{\mathcal{F}/Q}(R/Q) \longrightarrow 1.$$

The group $L = (K \cap N_{\mathrm{Aut}_{\mathcal{F}}(R)}(X))/X$ is a p'-group, and we have a short exact sequence

$$1 \longrightarrow L \longrightarrow N_{\mathrm{Aut}_{\mathcal{F}}(R)}(X)/X \longrightarrow \mathrm{Aut}_{\mathcal{F}/Q}(R/Q) \longrightarrow 1.$$

By 8.2.10, the image of N_ψ in $\mathrm{Aut}_{\mathcal{F}/Q}(R/Q)$ is the intersection of two Sylow p-subgroups $\mathrm{Aut}_{P/Q}(R/Q) \cap (\psi^{-1} \circ \mathrm{Aut}_{P/Q}(R/Q) \circ \psi)$. By 8.1.9 there is $\tau \in N_{\mathrm{Aut}_{\mathcal{F}}(R)}(X)$ which lifts ψ such that the canonical map α sends $\mathrm{Aut}_P(R) \cap (\tau^{-1} \circ \mathrm{Aut}_P(R) \circ \tau)$ onto this intersection, or equivalently, the canonical map $N_\tau \to N_\psi$ is surjective. This means that if $\delta : N_\tau \to P$ is a morphism in $\mathcal{F}$ that extends τ, then its image γ modulo Q is a morphism in $\mathcal{F}/Q$ that extends ψ. This proves that the extension axiom holds for the category $\mathcal{F}/Q$. $\square$

Theorem 8.4.2 applies in particular when $\mathcal{F} = C_{\mathcal{F}}(Z)$ for some subgroup Z of $Z(P)$. In that case one can be more precise about connections between $\mathcal{F}$ and $\mathcal{F}/Z$. The following two theorems are well-known.

Theorem 8.4.3 *Let $\mathcal{F}$ be a fusion system on a finite p-group P and let Z be a subgroup of $Z(P)$ such that $\mathcal{F} = C_{\mathcal{F}}(Z)$. Let Q be a subgroup of P containing Z. Then the following hold.*

(i) *The canonical group homomorphism $\mathrm{Aut}_{\mathcal{F}}(Q) \to \mathrm{Aut}_{\mathcal{F}/Z}(Q/Z)$ is surjective, having as kernel an abelian p-group.*

(ii) *If Q/Z is $\mathcal{F}/Z$-centric, then Q is $\mathcal{F}$-centric.*

(iii) *If Q is $\mathcal{F}$-radical, then Q/Z is $\mathcal{F}/Z$-radical.*

(iv) *If Q is $\mathcal{F}$-centric radical, then Q/Z is $\mathcal{F}/Z$-centric radical.*

Proof Let $\varphi \in \mathrm{Aut}_{\mathcal{F}}(Q)$. Suppose that φ induces the identity on Q/Z. That is, $\varphi(u) = u\zeta(u)$ for $u \in Q$ and suitable $\zeta(u) \in Z$. One checks that ζ is in fact a group homomorphism $\zeta : Q \to Z$. The map sending such a φ to ζ in turn is easily seen to be an injective group homomorphism from $\ker(\mathrm{Aut}_{\mathcal{F}}(Q) \to \mathrm{Aut}_{\mathcal{F}/Z}(Q/Z))$ to the abelian p-group $\mathrm{Hom}(Q, Z)$, with group structure induced by that of Z. This proves (i). If Q/Z is $\mathcal{F}/Z$-centric then $C_{P/Z}(R/Z) \leq R/Z$ for every subgroup R isomorphic to Q in $\mathcal{F}$. Thus $C_P(R) \leq R$ for any such R, hence Q is $\mathcal{F}$-centric. This proves (ii). If Q is $\mathcal{F}$-radical, the kernel of the canonical map $\mathrm{Aut}_{\mathcal{F}}(Q) \to \mathrm{Aut}_{\mathcal{F}/Z}(Q/Z)$, being a p-group, must be contained in $\mathrm{Aut}_Q(Q)$, hence in that case we get an isomorphism $\mathrm{Aut}_{\mathcal{F}}(Q)/\mathrm{Aut}_Q(Q) \cong \mathrm{Aut}_{\mathcal{F}/Z}(Q/Z)/\mathrm{Aut}_{Q/Z}(Q/Z)$, and hence Q/Z is $\mathcal{F}/Z$-radical. This proves (iii). Suppose now that Q is $\mathcal{F}$-centric radical. We may assume that Q/Z is fully $\mathcal{F}/Z$-centralised. The group Q/Z is $\mathcal{F}/Z$-radical by (iii). The kernel K of the canonical map $\mathrm{Aut}_{\mathcal{F}}(Q) \to \mathrm{Aut}_{\mathcal{F}/Z}(Q/Z)$ is a p-group, by (i). Since Q is $\mathcal{F}$-radical this implies $K \leq \mathrm{Aut}_Q(Q)$. Let C be the inverse image in P of $C_{P/Z}(Q/Z)$. That is, the image in P/Z of any element in C centralises Q/Z, and hence $\mathrm{Aut}_C(Q) \leq K$. Thus $\mathrm{Aut}_C(Q) \leq \mathrm{Aut}_Q(Q)$, which implies $C \leq QC_P(Q)$. Since Q is $\mathcal{F}$-centric this implies in fact that $C \leq Q$. Taking images in P/Z yields $C_{P/Z}(Q/Z) \leq Q/Z$, or equivalently, $C_{P/Z}(Q/Z) = Z(Q/Z)$. Since Q/Z was chosen to be fully $\mathcal{F}/Z$-centralised, this implies that Q/Z is $\mathcal{F}/Z$-centric, whence (iv). $\qquad\square$

Theorem 8.4.4 *Let $\mathcal{F}$, $\mathcal{F}'$ be fusion systems on a finite p-group P such that $\mathcal{F} = C_{\mathcal{F}}(Z)$ and $\mathcal{F}' = C_{\mathcal{F}'}(Z)$ for some subgroup Z of $Z(P)$. Suppose that $\mathcal{F} \subseteq \mathcal{F}'$. Then $\mathcal{F} = \mathcal{F}'$ if and only if $\mathcal{F}/Z = \mathcal{F}'/Z$.*

Proof Suppose that $\mathcal{F}/Z = \mathcal{F}'/Z$. Let Q be an $\mathcal{F}'$-centric radical subgroup of P. Then, by 8.4.3, the kernel K of the canonical map $\mathrm{Aut}_{\mathcal{F}'}(Q) \to$

$\text{Aut}_{\mathcal{F}/Z}(Q/Z)$ is contained in $\text{Aut}_Q(Q)$. Thus K is also the kernel of the canonical map $\text{Aut}_{\mathcal{F}}(Q) \to \text{Aut}_{\mathcal{F}/Q}(Q/Z)$. Since $\text{Aut}_{\mathcal{F}/Z}(Q/Z) = \text{Aut}_{\mathcal{F}'/Q}(Q/Z)$ it follows that $\text{Aut}_{\mathcal{F}}(Q)$ and $\text{Aut}_{\mathcal{F}'}(Q)$ have the same order. The assumption $\mathcal{F} \subseteq \mathcal{F}'$ implies $\text{Aut}_{\mathcal{F}}(Q) = \text{Aut}_{\mathcal{F}'}(Q)$. The equality $\mathcal{F} = \mathcal{F}'$ follows now from Alperin's Fusion Theorem. The converse is trivial. $\qquad\Box$

The next theorem is a slight generalisation of 8.4.4:

Theorem 8.4.5 ([98, 3.4]) *Let P be a finite p-group, let Q be a normal subgroup of P and let $\mathcal{F}, \mathcal{G}$ be fusion systems on P such that $\mathcal{F} = PC_{\mathcal{F}}(Q)$ and such that $\mathcal{G} \subseteq \mathcal{F}$. Let R be a normal subgroup of P containing Q. We have $\mathcal{G} = N_{\mathcal{F}}(R)$ if and only if $\mathcal{G}/Q = N_{\mathcal{F}/Q}(R/Q)$.*

Proof Suppose that $\mathcal{G}/Q = N_{\mathcal{F}/Q}(R/Q)$. In order to show the equality $\mathcal{G} = N_{\mathcal{F}}(R)$ we proceed by induction over the order of Q. If $Q = 1$ there is nothing to prove. If $Q \neq 1$ then $Z = Q \cap Z(P) \neq 1$ because Q is normal in P. Since $\mathcal{F} = PC_{\mathcal{F}}(Q)$ we have in fact $\mathcal{F} = C_{\mathcal{F}}(Z)$. Set $\bar{\mathcal{F}} = \mathcal{F}/Z$ and $\bar{\mathcal{G}} = \mathcal{G}/Z$. Similarly, set $\bar{P} = P/Z$, $\bar{Q} = Q/Z$ and $\bar{R} = R/Z$. Then $\bar{\mathcal{F}}$, $\bar{\mathcal{G}}$ are fusion systems on $\bar{P}$ satisfying $\bar{\mathcal{F}} = \bar{P}C_{\bar{\mathcal{F}}}(\bar{Q})$ and $\bar{\mathcal{G}} \subseteq \bar{\mathcal{F}}$. The canonical isomorphism $\bar{P}/\bar{Q} \cong P/Q$ induces obviously isomorphisms of fusion systems $\bar{\mathcal{G}}/\bar{Q} \cong \mathcal{G}/Q$ and $N_{\bar{\mathcal{F}}/\bar{Q}}(\bar{R}/\bar{Q}) \cong N_{\mathcal{F}/Q}(R/Q)$. Thus, by induction, we get that $\bar{\mathcal{G}} = N_{\bar{\mathcal{F}}}(\bar{R})$. Note that $N_{\bar{\mathcal{F}}}(\bar{R}) = N_{\mathcal{F}}(R)/Z$. Any $\bar{\mathcal{G}}$-centric radical subgroup of $\bar{P}$ contains $\bar{R}$. Thus, by 8.4.3, any $\mathcal{G}$-centric radical subgroup of P contains R. Since $\bar{\mathcal{G}} = N_{\bar{\mathcal{F}}}(\bar{R})$ this implies that $\mathcal{G} = N_{\mathcal{G}}(R)$. Since also $\mathcal{G} \subseteq \mathcal{F}$ it follows that $\mathcal{G} \subseteq N_{\mathcal{F}}(R)$. Thus, by 8.4.4 applied to $\mathcal{G}$ and $N_{\mathcal{F}}(R)$ we get the equality $\mathcal{G} = N_{\mathcal{F}}(R)$. The converse is trivial. $\qquad\Box$

When specialised to fusion systems of finite groups, the notion of quotients of fusion systems coincides with what one expects:

Proposition 8.4.6 *Let G be a finite group, let P be a Sylow p-subgroup and let Q be a subgroup of P that is normal in G. Set $\mathcal{F} = \mathcal{F}_P(G)$. Then $\mathcal{F} = N_{\mathcal{F}}(Q)$ and we have $\mathcal{F}/Q = \mathcal{F}_{P/Q}(G/Q)$.*

Proof Trivial verification. $\qquad\Box$

Products of fusion systems are defined as follows:

Definition 8.4.7 Let P, Q be finite p-groups, $\mathcal{F}$ a category on P and $\mathcal{G}$ a category on Q. We define a category $\mathcal{F} \times \mathcal{G}$ on $P \times Q$ as follows. Let R, S be subgroups of $P \times Q$. Denote by R_1, R_2 the images in P, Q, under the canonical projections from $P \times Q$ to P, Q, respectively. A group homomorphism $\varphi : R \to S$ is a morphism in $\mathcal{F} \times \mathcal{G}$ if and only if there are morphisms $\varphi_1 : R_1 \to P$, $\varphi_2 : R_2 \to Q$, such that φ is equal to the restriction to R followed by the inclusion of S in $P \times Q$ of the homomorphism $\varphi_1 \times \varphi_2$ from $R_1 \times R_2$ to $P \times Q$ sending (u_1, u_2) to $\varphi_1(u), \varphi_2(u_2))$ for all $u_1 \in R_1, u_2 \in R_2$.

Lemma 8.4.8 *Let P, Q be finite p-groups, $\mathcal{F}$ a category on P and $\mathcal{G}$ a category on Q. Let R be a subgroup of $P \times Q$. Denote by R_1 and R_2 the images in P and Q under the canonical projections from $P \times Q$ to P and Q, respectively. Then R is fully $\mathcal{F} \times \mathcal{G}$-centralised if and only if R_1 is fully $\mathcal{F}$-centralised and R_2 is fully $\mathcal{G}$-centralised.*

Proof This follows from the obvious equality $C_{P \times Q}(R) = C_P(R_1) \times C_Q(R_2)$. □

Proposition 8.4.9 (cf. [33, Lemma 1.5]) *Let P, Q be finite p-groups, $\mathcal{F}$ a fusion system on P and $\mathcal{G}$ a fusion system on Q. Then the category $\mathcal{F} \times \mathcal{G}$ is a fusion system on $P \times Q$.*

Proof Since $\mathcal{F}, \mathcal{G}$ are fusion systems, the Sylow axiom implies that the group $\mathrm{Aut}_{P \times Q}(P \times Q) \cong \mathrm{Aut}_P(P) \times \mathrm{Aut}_Q(Q)$ is a Sylow p-subgroup of $\mathrm{Aut}_{\mathcal{F} \times \mathcal{G}}(P \times Q) \cong \mathrm{Aut}_{\mathcal{F}}(P) \times \mathrm{Aut}_{\mathcal{G}}(Q)$. Thus the Sylow axiom holds for $\mathcal{F} \times \mathcal{G}$. Let $\varphi : R \to S$ be an isomorphism in $\mathcal{F} \times \mathcal{G}$ such that S is fully $\mathcal{F} \times \mathcal{G}$-centralised. Denote by R_1, S_1 the images in P of R, S, respectively, under the canonical projection $P \times Q \to P$, and denote by R_2, S_2 the images in Q of R, S, respectively, under the canonical projection $P \times Q \to Q$. Denote by $\varphi_1 : R_1 \to S_1$ and $\varphi_2 : R_2 \to S_2$ the (necessarily unique) morphisms in $\mathcal{F}, \mathcal{G}$, respectively, such that φ is the restriction of $\varphi_2 \times \varphi_2$. By 8.4.8, S_1 is fully $\mathcal{F}$-centralised and S_2 is fully $\mathcal{G}$-centralised. Thus φ_1 and φ_2 extend, in $\mathcal{F}$ and $\mathcal{G}$, to the groups N_{φ_1} and N_{φ_2}, respectively. One checks that N_φ is contained in $N_{\varphi_1} \times N_{\varphi_2}$, and hence φ extends to N_φ, completing the proof of the proposition. □

For fusion systems of finite groups, their product is the fusion system of the direct product of the finite groups:

Proposition 8.4.10 *Let G, H be finite groups with Sylow p-subgroups P, Q, respectively. Then $P \times Q$ is a Sylow p-subgroup of $G \times H$ and we have $\mathcal{F}_{P \times Q}(G \times H) = \mathcal{F}_P(G) \times \mathcal{F}_Q(H)$.*

Proof Trivial verification. □

Proposition 8.4.11 *Let $\mathcal{F}$ be a fusion system on a finite p-group P and let Q be a fully $\mathcal{F}$-centralised subgroup of P. Set $\Delta Q = \{(u, u) | u \in Q\}$. Then the fusion systems $C_{\mathcal{F} \times \mathcal{F}}(\Delta Q)$ and $C_{\mathcal{F}}(Q) \times C_{\mathcal{F}}(Q)$ on $C_{P \times P}(\Delta Q) = C_P(Q) \times C_P(Q)$ are equal.*

Proof Since Q is fully $\mathcal{F}$-centralised it follows from 8.4.8 that ΔQ is fully $\mathcal{F} \times \mathcal{F}$-centralised. Thus both $C_{\mathcal{F} \times \mathcal{F}}(\Delta Q)$ and $C_{\mathcal{F}}(Q) \times C_{\mathcal{F}}(Q)$ are fusion systems on $C_{P \times P}(\Delta Q) = C_P(Q) \times C_P(Q)$. Let R be a subgroup of $C_P(Q) \times C_P(Q)$. Denote by R_1, R_2 the images of R in $C_P(Q)$ under the two projections of the direct product $C_P(Q) \times C_P(Q)$ onto its first and second factor, respectively. Let $\varphi : R \to C_P(Q) \times C_P(Q)$ be a morphism in $C_{\mathcal{F} \times \mathcal{F}}(\Delta Q)$. That is, φ extends to a morphism $\psi : R \Delta Q \to P \times P$ in $\mathcal{F} \times \mathcal{F}$ such that $\psi|_{\Delta Q} = \mathrm{Id}_{\Delta Q}$. The images in P of $R \Delta Q$ under the two projections of $P \times P$ onto its factors are equal to $R_1 Q$ and $R_2 Q$. Thus ψ being a morphism in $\mathcal{F} \times \mathcal{F}$ is equivalent to the existence of morphisms $\psi_1 : R_1 Q \to P$ and $\psi_2 : R_2 Q \to P$ which are the identity on Q and which induce ψ. The result follows. $\qquad\qquad\square$

8.5 Fusion systems of blocks

Every block of a finite group gives rise to a fusion system on any of its defect groups, defined in a way that is similar to the fusion system of a finite group as in 8.1.2 above, replacing p-subgroups by Brauer pairs. Let p be a prime and $\mathcal{O}$ a complete local Noetherian ring with residue field k of characteristic p.

Definition 8.5.1 Let G be a finite group, let b be a block of $\mathcal{O}G$ and let (P, e) be a maximal (G, b)-Brauer pair. For any subgroup Q of P denote by e_Q the unique block of $kC_G(Q)$ such that $(Q, e_Q) \leq (P, e)$. We define a category $\mathcal{F} = \mathcal{F}_{(P,e)}(G, b)$ as follows: the objects of $\mathcal{F}$ are the subgroups of P, and for any two subgroups Q, R of P, the morphism set $\mathrm{Hom}_{\mathcal{F}}(Q, R)$ is the set of all group homomorphisms $\psi : Q \to R$ for which there exists an element $x \in G$ satisfying $\varphi(u) = {}^x u$ for all $u \in Q$ and such that ${}^x(Q, e_Q) \leq (R, e_R)$.

If S is a Sylow p-subgroup of G containing P then clearly $\mathcal{F}_{(P,e)}(G, b)$ is a subcategory of $\mathcal{F}_S(G)$, but it need not be a full subcategory, because in the definition of morphisms in the fusion system of a block there is an additional compatibility condition involving (G, b)-Brauer pairs. Since Brauer pairs are blocks of centralisers over the residue field k, the fusion systems of a block of $\mathcal{O}G$ coincide with those of the corresponding block of kG. The definition of the category $\mathcal{F}_{(P,e)}(G, b)$ does not require the field k to be large enough. For the next result, showing that this category is a fusion system, we will need a hypothesis on k being large enough for the block e of $kC_G(P)$ in order to ensure that the Sylow axiom holds.

Theorem 8.5.2 *Let G be a finite group, b a block of $\mathcal{O}G$ and (P, e) a maximal (G, b)-Brauer pair. Suppose that k is a splitting field for the block e of $kC_G(P)$. The category $\mathcal{F}_{(P,e)}(G, b)$ is a fusion system on P.*

The proof of this theorem follows the lines of the proof of 8.1.2 for fusion systems of finite groups. We start with characterisations of $\mathcal{F}_{(P,e)}(G, b)$-centric, fully centralised or normalised subgroups, similar to those for fusion systems of finite groups in 8.2.2 and 8.1.4.

Proposition 8.5.3 *Let G be a finite group, let b be a block of kG, and let (P, e) be a maximal (G, b)-Brauer pair. Let (Q, f) be a (G, b)-Brauer pair such that $(Q, f) \leq (P, e)$. Set $\mathcal{F} = \mathcal{F}_{(P,e)}(G, b)$.*

  (i) *The group $C_P(Q)$ is contained in a defect group of $kC_G(Q)f$, and Q is fully $\mathcal{F}$-centralised if and only if $C_P(Q)$ is a defect group of $kC_G(Q)f$.*
  (ii) *The group $N_P(Q)$ is contained in a defect group of $kN_G(Q, f)f$, and Q is fully $\mathcal{F}$-normalised if and only $N_P(Q)$ is a defect group of $kN_G(Q, f)f$.*
  (iii) *The group Q is $\mathcal{F}$-centric if and only if $Z(Q)$ is a defect group of $kC_G(Q)f$. In that case, Q has a unique local point δ on $\mathcal{O}Gb$ satisfying $\mathrm{Br}_Q(\delta)f \neq 0$.*

Proof By 6.3.9, the group $C_P(Q)$ is contained in a defect group of f as a block of $kC_G(Q)$, and there is $x \in G$ such that $^x(Q, f) \leq (P, e)$ and such that $C_P(^xQ)$ is a defect group of ^xf as block of $kC_G(^xQ)$. Since the defect groups of the blocks f and ^xf of $kC_G(Q)$ and $kC_G(^xQ)$, respectively, have the same order, this implies that $C_P(Q)$ has maximal possible order if and only if $C_P(Q)$ is a defect group of f, whence (i). By 6.2.6, f remains a block of $kN_G(Q, f)$. Again by 6.3.9, $N_P(Q)$ is contained in a defect group of f as block of $kN_G(Q, f)$, and there is $x \in G$ such that $^x(Q, f) \leq (P, e)$ and such that $N_P(^xQ)$ is a defect group of ^xf as block of $kN_G(^x(Q, f))$. This proves (ii). The first statement in (iii) is a special case of (i). Thus if Q is $\mathcal{F}$-centric, then $kC_G(Q)f$ has the central subgroup $Z(Q)$ as a defect group. By 6.6.5, $kC_G(Q)f$ has a unique isomorphism class of simple modules, hence a unique conjugacy class of primitive idempotents. This lifts to a unique conjugacy class δ of primitive idempotents in $(\mathcal{O}Gb)^Q$ satisfying $\mathrm{Br}_Q(\delta)f \neq 0$, implying the second statement in (iii). $\square$

Proof of Theorem 8.5.2 Clearly $\mathcal{F} = \mathcal{F}_{(P,e)}(G, b)$ is a category on P. By Brauer's First Main Theorem 6.7.6, the group $N_G(P, e)/PC_G(P)$ is a p'-group (the inertial quotient of b), and hence the group $\mathrm{Aut}_{\mathcal{F}}(P) \cong N_G(P, e)/C_G(P)$ has $\mathrm{Aut}_P(P)$ as Sylow p-subgroup. In particular, the Sylow axiom holds. It remains to verify that $\mathcal{F}$ satisfies the extension axiom. Let Q, R be subgroups of P such that R is fully $\mathcal{F}$-normalised and such that there is an isomorphism $\varphi : Q \to R$ in $\mathcal{F}$. By 8.5.3, $N_P(R)$ is a defect group of e_R as a block of $kN_G(R, e_R)$. Let $x \in G$ such that $^x(Q, e_Q) = (R, e_R)$ and such that $\varphi(u) = {}^xu$ for all $u \in Q$. Then

$$N_\varphi = \{y \in N_P(Q) \mid \exists z \in N_P(R) : {}^{xy}u = {}^{zx}u \ (\forall u \in Q)\}.$$

Thus $^x N_\varphi \leq N_P(R) C_G(R)$. Since R is fully $\mathcal{F}$-normalised, $N_P(R)$ is a defect group of e_R viewed as block of $k N_G(R, e_R)$, and hence $N_P(R)$ is still a defect group of e_R viewed as block of $N_P(R) C_G(R)$. Therefore $(N_P(R), e_{N_P(R)})$ is a maximal $(N_P(R) C_G(R), e_R)$-Brauer pair and contains hence a $C_G(R)$-conjugate of every other $(N_P(R) C_G(R), e_R)$-Brauer pair, by 6.3.7. Thus there is $c \in C_G(R)$ such that $^{cx}(N_\varphi, e_{N_\varphi}) \leq (N_P(R), e_{N_P(R)})$. Hence $\psi : N_\varphi \to P$ defined by $\psi(n) = {}^{cx}n$ for all $n \in N_\varphi$ is a morphism in $\mathcal{F}$ that extends φ. $\qquad\square$

The following result generalises 8.3.11 to fusion systems of blocks.

Proposition 8.5.4 *Let G be a finite group, let b be a block of kG, and let (P, e) be a maximal b-Brauer pair. For every subgroup Q of P, denote by e_Q the unique block of $k C_G(Q)$ such that $(Q, e_Q) \subseteq (P, e)$. Set $\mathcal{F} = \mathcal{F}_{(P,e)}(G, b)$.*

(i) *If Q is a fully $\mathcal{F}$-centralised subgroup of P then $(C_P(Q), e_{Q C_P(Q)})$ is a maximal $(C_G(Q), e_Q)$-Brauer pair and we have $\mathcal{F}_{(C_P(Q), e_{Q C_P(Q)})}(C_G(Q), e_Q) = C_\mathcal{F}(Q)$.*

(ii) *If Q is a fully $\mathcal{F}$-normalised subgroup of P then $(N_P(Q), e_{N_P(Q)})$ is a maximal $(N_G(Q, e_Q), e_Q)$-Brauer pair and therefore we have $\mathcal{F}_{(N_P(Q), e_{N_P(Q)})}(N_G(Q, e_Q), e_Q) = N_\mathcal{F}(Q)$.*

Proof Suppose that Q is fully $\mathcal{F}$-centralised. By 8.5.3 (i), $C_P(Q)$ is a defect group of e_Q as a block of $C_G(Q)$. We have $C_{C_G(Q)}(C_P(Q)) = C_G(Q C_P(Q))$, hence $(C_P(Q), e_{Q C_P(Q)})$ is a maximal $(C_G(Q), e_Q)$-Brauer pair. Similarly, for any subgroup R of $C_P(Q)$, the pair (R, e_{QR}) is a $(C_G(Q), e_Q)$-Brauer pair contained in $(C_P(Q), e_{Q C_P(Q)})$. If R, S are subgroups of $C_P(Q)$ and $x \in C_G(Q)$ such that $^x(R, e_{QR}) \subseteq (S, e_{QS})$, then the group homomorphism from R to S induced by conjugation with x extends to a group homomorphism from QR to QS which is the identity on Q. Statement (i) follows. For (ii), suppose that Q is fully $\mathcal{F}$-normalised. By 8.5.3 (ii), $N_P(Q)$ is a defect group of e_Q as block of $N_G(Q, e_Q)$. We have $C_{N_G(Q)}(C_P(Q)) = C_G(N_P(Q))$, hence $(N_P(Q), e_{N_P(Q)})$ is a maximal $(N_G(Q, e_Q), e_Q)$-Brauer pair. Similarly, for any subgroup R of $N_P(Q)$, the pair (R, e_{QR}) is a $(N_G(Q, e_Q), e_Q)$-Brauer pair contained in $(N_P(Q), e_{N_P(Q)})$. If R, S are subgroups of $N_P(Q)$ and $x \in N_G(Q, e_Q)$ such that $^x(R, e_{QR}) \leq (S, e_{QS})$, then the group homomorphism from R to S induced by conjugation with x extends to a group homomorphism from QR to QS which restricts to an automorphism of Q in $\mathrm{Aut}_\mathcal{F}(Q)$. The result follows. $\qquad\square$

The fusion system of a principal block is just the fusion system of the underlying finite group itself; this is an immediate consequence of Brauer's Third Main Theorem.

Proposition 8.5.5 *Let G be a finite group, b a block of principal type of $\mathcal{O}G$, and (P, e) a maximal (G, b)-Brauer pair. Suppose that k is a splitting field for the block e of $kC_G(P)$. We have $\mathcal{F}_{(P,e)}(G, b) = \mathcal{F}_P(G)$.*

Proof Since b is of principal type, we have $e = \mathrm{Br}_{\Delta P}(b)$, and more generally, for any subgroup Q of P, the element $e_Q = \mathrm{Br}_{\Delta Q}(b)$ is the unique block of $kC_G(Q)$ satisfying $(Q, e_Q) \leq (P, e)$. Since $\mathrm{Br}_{\Delta Q}$ maps $Z(\mathcal{O}G)$ to $kC_G(Q)^{\Delta N_G(Q)}$, it follows that $N_G(Q, e_Q) = N_G(Q)$. The equality of fusion systems as stated follows from Alperin's Fusion Theorem 8.2.8 or its corollary 8.2.9.   □

Note that this puts some restrictions on which p-subgroups can arise as defect groups of blocks of principal type, since for an arbitrary p-subgroup P of G, the category $\mathcal{F}_P(G)$ need not be a fusion system.

Corollary 8.5.6 *Let G be a finite group, b the principal block of $\mathcal{O}G$, P a Sylow p-subgroup of G and e the principal block of $kC_G(P)$. Then (P, e) is a maximal (G, b)-Brauer pair and we have $\mathcal{F}_{(P,e)}(G, b) = \mathcal{F}_P(G)$.*

Proof By Brauer's Third Main Theorem 6.3.14, the principal block of $\mathcal{O}G$ is a block of principal type having (P, e) as maximal (G, b)-Brauer pair. Thus 8.5.6 is a special case of 8.5.5.   □

Proposition 8.5.7 *Let G be a finite group, b a block of $\mathcal{O}G$, (P, e) a maximal (G, b)-Brauer pair and (Q, f) a (G, b)-Brauer pair contained in (P, e). Suppose that P is abelian and that k is a splitting field for $kC_G(P)e$.*

(i) *The pair (P, e) is a maximal $(C_G(Q), f)$-Brauer pair and a maximal $(N_G(Q, f), f)$-Brauer pair. In particular, P is a defect group of f both as a block of $kC_G(Q)$ and as a block of $kN_G(Q, f)$.*

(ii) *We have $N_G(Q, f) \subseteq N_G(P, e)C_G(Q)$.*

(iii) *Set $E = N_G(P, e)/C_G(P)$. We have $N_G(Q, f)/C_G(Q) \cong E/C_E(Q)$.*

Proof Set $\mathcal{F} = \mathcal{F}_{(P,e)}(G, b)$. Since P is abelian, we have $C_P(Q) = N_P(Q) = P$, and hence Q is a fully $\mathcal{F}$-normalised subgroup of P by 8.5.3. It follows from 8.5.4 that (P, e) is a maximal Brauer pair for both $(C_G(Q), f)$ and $(N_G(Q, f), f)$, whence (i). We have $\mathcal{F} = N_{\mathcal{F}}(P)$ by 8.3.8. Thus every automorphism of Q induced by an element in $N_G(Q, f)$ extends to an automorphism of P induced by an element in $N_G(P, e)$, whence (ii). Statement (iii) follows from (ii).   □

We will see in 8.7.8 that fusion systems are invariant under the reduction in 6.8.3.

8.6 Fusion systems of blocks and Clifford Theory

Let p be a prime and $\mathcal{O}$ a complete local Noetherian ring with residue field k of characteristic p. If N is a normal subgroup of a finite group G and c a G-stable block of $\mathcal{O}N$, then $\mathcal{O}Nc$ is a primitive G-algebra. Since G permutes the elements of N, it follows that $\mathcal{O}Nc$ is a p-permutation G-algebra. Thus the notion of $(G, \mathcal{O}Nc)$-Brauer pairs, the inclusion of Brauer pairs, and the main properties of this inclusion from 5.9.6 apply to this situation. In particular, the maximal $(G, \mathcal{O}Nc)$-Brauer pairs are conjugate. The defect groups of $\mathcal{O}Nc$ as a G-algebra are the maximal p-subgroups P of G satisfying $\mathrm{Br}_P(c) \neq 0$. If Q is a defect group in N of the block $\mathcal{O}Nc$, then Q is contained in a defect group P of the G-algebra $\mathcal{O}Nc$, and then necessarily $Q = P \cap N$, because $\mathrm{Br}_{P \cap N}(c) \neq 0$. Since c is G-stable and since N permutes the maximal (N, c)-Brauer pairs transitively, it follows that if P is any defect group of the G-algebra $\mathcal{O}Nc$, then $P \cap N$ is a defect group of the block algebra $\mathcal{O}Nc$. We slightly modify the notation as follows.

Definition 8.6.1 Let G be a finite group, N a normal subgroup of G, and let c be a G-stable block of $\mathcal{O}N$. A (G, N, c)-*Brauer pair* is a pair (Q, f) consisting of a p-subgroup Q of G and a block f of $kC_N(Q)$ satisfying $\mathrm{Br}_Q(c)f \neq 0$.

Thus a (G, N, c)-Brauer pair is a $(G, \mathcal{O}Nc)$-Brauer pair, in the notation of 5.9.1. If $G = N$, then a (G, N, c)-Brauer pair is a (G, c)-Brauer pair, in the notation of 6.3.1. We extend the definition of fusion systems of blocks to this situation.

Definition 8.6.2 ([103, 3.3]) Let G be a finite group, N a normal subgroup of G, and let c be a G-stable block of $\mathcal{O}N$. Let (P, f) be a maximal (G, N, c)-Brauer pair. For any subgroup Q of P denote by f_Q the unique block of $kC_N(Q)$ such that we have an inclusion of (G, N, c)-Brauer pairs $(Q, f_Q) \leq (P, f)$. Denote $\mathcal{F}_{(P,f)}(G, N, c)$ the category on P having as morphisms, for any two subgroups Q and R of P, the set of group homomorphisms $\varphi : Q \to R$ for which there exists an element $x \subset G$ satisfying $\varphi(u) = xux^{-1}$ for all $u \in Q$ and $^x(Q, f_Q) \leq (R, f_R)$.

The block f_Q of $C_N(Q)$ need not be $C_G(Q)$-stable; we write $C_G(Q, f_Q) = C_G(Q) \cap N_G(Q, f_Q)$. Setting $\mathcal{F} = \mathcal{F}_{(P,e)}(G, N, c)$, we have $\mathrm{Aut}_{\mathcal{F}}(Q) \cong N_G(Q, f_Q)/C_G(Q, f_Q)$. For $G = N$ we have $\mathcal{F}_{(P,f)}(G, N, c) = \mathcal{F}_{(P,f)}(G, c)$. Setting $R = P \cap N$, the category $\mathcal{F}_{(P,f)}(G, N, c)$ contains the fusion system $\mathcal{F}_{(R,f_R)}(N, c)$; note that then (R, f_R) is automatically a maximal (N, c)-Brauer pair.

Theorem 8.6.3 ([103, 3.4]) *Let G be a finite group, N a normal subgroup of G, and let c be a G-stable block of $\mathcal{O}N$. Let (P, f) be a maximal (G, N, c)-Brauer pair. Suppose that k is a splitting field for the subgroups of N. Then the category $\mathcal{F}_{(P,f)}(G, N, c)$ is a fusion system on P. Moreover, the image of P in G/N is a Sylow p-subgroup of G/N.*

The proof of this theorem – which has been generalised to a larger class of p-permutation G-algebras in [10, IV, Theorem 3.2] – follows that for block fusion systems. We need the following lemma.

Lemma 8.6.4 ([103, 3.2]) *Let G be a finite group, N a normal subgroup of G, and let c be a G-stable block of $\mathcal{O}N$. Let (Q, f) be a (G, N, c)-Brauer pair. Let H be a subgroup of $N_G(Q, f)$ containing $QC_N(Q)$. Let S be a p-subgroup of H containing Q, and let e be a block of $kC_N(S)$. The following are equivalent.*

(i) (S, e) is a (G, N, c)-Brauer pair satisfying $(Q, f) \leq (S, e)$.
(ii) (S, e) is an $(H, C_N(Q), f)$-Brauer pair.

Proof Note that $C_N(Q)$ is a normal subgroup of H and that f is an H-stable block of $kC_N(Q) = kC_H(Q)$. Suppose that (i) holds. Since S normalises Q, it follows from 5.9.9 that f is the unique S-stable block of $kC_H(Q)$ satisfying $\mathrm{Br}_S(f)e = e$. This shows that (S, e) is an $(H, C_N(Q), f)$-Brauer pair. Conversely, suppose that (ii) holds; that is, $\mathrm{Br}_S(f)e = e$. This implies that $\mathrm{Br}_S(c)e = \mathrm{Br}_S(c)\mathrm{Br}_S(f)e = \mathrm{Br}_S(\mathrm{Br}_Q(c))\mathrm{Br}_S(f)e = \mathrm{Br}_S(\mathrm{Br}_Q(c)f)e = \mathrm{Br}_S(f)e = e$. Thus (S, e) is a (G, N, c)-Brauer pair. By 5.9.6, applied to the G-algebra kNc, there is a unique (G, N, c)-Brauer pair (Q, f') such that $(Q, f') \leq (S, e)$. Then f' is S-stable, and $\mathrm{Br}_S(f')e = e$. By the assumptions, we also have $\mathrm{Br}_S(f)e = e$. We will show that $f' = f$. If not, then $ff' = 0$. Since Br_S is an algebra homomorphism from $(kC_N(Q))^S$ to onto $kC_N(S)$, this would imply that $\mathrm{Br}_S(f)\mathrm{Br}_S(f') = 0$. Hence, using that e is a central idempotent, we get that $e = \mathrm{Br}_S(f)e\mathrm{Br}_S(f')e = 0$, a contradiction. Thus $f' = f$, and hence $(Q, f) \leq (S, e)$. □

Proof of Theorem 8.6.3 Set $\mathcal{F} = \mathcal{F}_{(P,f)}(G, N, c)$. For any subgroup Q of P denote by f_Q the unique block of $kC_N(Q)$ such that $(Q, f_Q) \leq (P, f)$. Since conjugation by an element in P stabilises f, hence conjugates the pairs contained in (P, f), it follows that $\mathcal{F}$ contains $\mathcal{F}_P(P)$. We verify the Sylow axiom. Since (P, f) is a maximal (G, N, c)-Brauer pair, we have $c \in (kNc)_P^G$. Thus $\mathrm{Br}_P(c) = \mathrm{Tr}_1^{N_G(P)/P}(d)$ for some $d \in kN_N(P)$. Since the maximal (G, N, c)-Brauer pairs are all conjugate, we have $\mathrm{Br}_P(c) = \mathrm{Tr}_{N_G(P,f)}^{N_G(P)}(f)$, and the different $N_G(P)$-conjugates of f are pairwise orthogonal. It follows that the map sending $u \in (kC_N(P)f)^{N_G(P,f)}$ to $\mathrm{Tr}_{N_G(P,f)}^{N_G(P)}(u) \in (kC_N(P)\mathrm{Br}_P(c))^{N_G(P)}$ is an algebra isomorphism, with inverse given by multiplication with f. Let $x \in N_G(P)$ and $u \in kC_N(P)$. Then

$$\mathrm{Tr}_P^{N_G(P)}(u(^x f)) = \mathrm{Tr}_{N_G(P,^x f)}^{N_G(P)}(\mathrm{Tr}_P^{N_G(P,^x f)}(u(^x f))) = \mathrm{Tr}_{N_G(P,^x f)}^{N_G(P)}(\mathrm{Tr}_P^{N_G(P,^x f)}(u)^x f).$$

The trace from $N_G(P, f)$ to $N_G(P)$ is a sum indexed by $N_G(P)/N_G(P, f)$. Multiplying this expression by f annihilates all summands except the one indexed by the class of x^{-1}. Thus

$$\mathrm{Tr}_P^{N_G(P)}(u(^x f))f = {}^{x^{-1}}\mathrm{Tr}_P^{N_G(P,^x f)}(u(^x f)) = \mathrm{Tr}_P^{N_G(P,f)}((^x u)f).$$

This shows that the image of the ideal $\mathrm{Tr}_P^{N_G(P)}(kC_N(P)\mathrm{Br}_P(c))$ under the map given by multiplication with f is equal to the ideal $\mathrm{Tr}_P^{N_G(P,f)}(kC_N(P)f)$. In particular, since $\mathrm{Br}_P(c) = \mathrm{Tr}_P^{N_G(P)}(d)$, it follows that $f = \mathrm{Tr}_P^{N_G(P,f)}(d')$ for some $d' \in kC_N(P)f$. The transitivity of relative trace maps implies that $f = \mathrm{Tr}_{PC_G(P,f)}^{N_G(P,f)}(\mathrm{Tr}_P^{PC_G(P,f)}(d'))$. Then $\mathrm{Tr}_P^{PC_G(P,f)}(d') \in Z(kC_N(P)f)$, and this is a local algebra. As k is large enough, this is a split local algebra, and so

$$\mathrm{Tr}_P^{PC_G(P,f)}(d') = \alpha f + v$$

for some $\alpha \in k$ and some element $v \in J(Z(kC_N(P)f))$ which is fixed by $PC_G(Q, f)$. Applying the trace from $PC_G(P, f)$ to $N_G(P, f)$ implies that

$$f = \alpha|N_G(P, f) : PC_G(P, f)| + r$$

for some $r \in J(Z(kC_N(P)f)$. As f is an idempotent, it is not contained in the radical, and hence p does not divide the order of the group $N_G(P, f)/PC_G(P, f)$. This group is isomorphic to $\mathrm{Aut}_{\mathcal{F}}(P)/\mathrm{Aut}_P(P)$, which shows that the Sylow axiom holds. For the extension axiom, let Q be a subgroup of P that is fully $\mathcal{F}$-normalised. Set $R = N_P(Q)$. Set $H = RC_G(Q, f_Q)$. We show first Q being fully $\mathcal{F}$-normalised implies that (R, f_R) is a maximal $(H, C_N(Q), f_Q)$-Brauer pair. By 8.6.4, the pair (R, f_R) is an $(H, C_N(Q), f_Q)$-Brauer pair. Suppose that (S, e) is an $(H, C_N(Q), f_Q)$-Brauer pair such that $(Q, f_Q) \leq (S, e)$. Then by 8.6.4, the pair (S, e) is a (G, N, c)-Brauer pair satisfying $(Q, f_Q) \leq (S, e)$. Since all maximal (G, N, c)-Brauer pairs are G-conjugate, there is an element $x \in G$ such that $^x(S, e) \leq (P, f)$. Thus we also have $^x(Q, f_Q) \leq (P, e)$; that is, Q and $^x Q$ are isomorphic in $\mathcal{F}$. We have $^x S \leq N_P(^x Q)$. The maximality of R implies that $S = R$. This shows that (R, f_R) is a maximal $(H, C_N(Q), f_Q)$-Brauer pair as claimed. Let $y \in G$ such that $^y(Q, f_Q) \leq (P, f)$. Conjugation by y^{-1} yields a morphism $\varphi : {}^y Q \to Q$ in $\mathcal{F}$. We need to show that this morphism extends to the group N_φ.

To this end, we first show that $N_\varphi = \{v \in N_P(^y Q)|y^{-1}vy \in H\}$. We have $v \in N_\varphi$ if and only if there exists $w \in N_P(Q) = R$ such that $\varphi(u^v) = \varphi(u)^w$ for all

$u \in {}^y Q$. Since φ is given by conjugation with y^{-1}, this is equivalent to $u^{vy} = u^{yw}$ for all $u \in {}^y Q$, hence to $ywy^{-1}v^{-1} \in C_G({}^y Q)$, and thus to $wy^{-1}v^{-1}y \in C_G(Q)$. Since ${}^y(Q, f_Q) \leq (P, f)$, it follows that $N_P({}^y Q)$ stabilises ${}^y f_Q$. Thus $N_P({}^y Q)^y$ stabilises f_Q, and clearly $R = N_P(Q)$ stabislies t_Q. Therefore, the previous inclusion is equivalent to $wy^{-1}v^{-1}y \in C_G(Q, f_Q)$. This, in turn, is equivalent to $y^{-1}v^{-1}y \in RC_G(Q, f_Q)$, as claimed. In particular, we have $y^{-1}N_\varphi y \leq H$. For the extension axiom, it suffices to show that there is an element $z \in C_G(Q, f_Q)$ such that $z^{-1}y^{-1}N_\varphi yz \leq R = N_P(Q)$; indeed, conjugation by $(yz)^{-1}$ yields then a morphism in $\mathcal{F}$ from N_φ to $N_P(Q)$ which extends φ, because conjugation by z is trivial on Q. Since ${}^y(Q, f_Q) \leq (P, f_P)$ and $N_\varphi \leq N_P({}^y Q)$, it follows that we have an inclusion of (G, N, c)-Brauer pairs ${}^y(Q, f_Q) \leq (N_\varphi, f_{e_\varphi})$, or equivalently, $(Q, f_Q) \leq {}^{y^{-1}}(N_\varphi, e_\varphi)$. By 8.6.4, applied with $S = {}^{y^{-1}}N_\varphi$ and $e = {}^{y^{-1}}e_{N_\varphi}$, it follows that ${}^{y^{-1}}(N_\varphi, e_\varphi)$ is an $(H, C_N(Q), f_Q)$-Brauer pair. Since (R, f_R) is a maximal $(H, C_N(Q), F_Q)$-Brauer pair, there exists an element $z \in H$ such that ${}^{z^{-1}y^{-1}}N_\varphi \leq R = N_P(Q)$. Since $H = RC_G(Q, f_Q)$, we may choose z in $C_G(Q, f_Q)$. This shows that $\mathcal{F}_{(P,f)}(G, N, c)$ is a fusion system on P. If b is a block of $\mathcal{O}G$ covering c and R a defect group of b, then $\mathrm{Br}_R(b) = \mathrm{Br}_R(bc) \neq 0$, hence $\mathrm{Br}_R(c) \neq 0$. Thus S contains a conjugate of R. It follows from Theorem 6.8.9 (iii) that the image of P in G/N is a Sylow p-subgroup of G/N. $\qquad\square$

Theorem 8.6.5 ([103, 3.5]) *Let G be a finite group, N a normal subgroup of G, c a G-stable block of kN, and b a block of kG such that $bc = b$. Let (P, e) be a maximal (G, b)-Brauer pair. Suppose that k is a splitting field for the subgroups of G. Then there exists a maximal (G, N, c)-Brauer pair (S, f) such that $\mathcal{F}_{(P,e)}(G, b)$ is a subfusion system of $\mathcal{F}_{(S,f)}(G, N, c)$. Moreover the following hold.*

(i) We have $P \cap N = S \cap N$.

(ii) Set $R = S \cap N$ and denote by f_R the unique block of $kC_N(R)$ such that $(R, f_R) \leq (S, f)$. Then (R, f_R) is a maximal (N, c)-Brauer pair.

(iii) The fusion system $\mathcal{F}_{(R, f_R)}(N, c)$ is an invariant subfusion system in $\mathcal{F}_{(S,f)}(G, N, c)$.

Proof Let e' be a central primitive idempotent in $kC_N(P)$ such that $\mathrm{Br}_P(c)e' = e'$ and such that $e'e \neq 0$ Thus (P, e') is a (G, N, c)-Brauer pair. Let (S, f) be a maximal (G, N, c)-Brauer pair such that $(P, e') \leq (S, f)$. Let Q be a subgroup of P. Let e_Q, e'_Q be the blocks of $kC_G(Q)$ and $kC_N(Q)$, respectively, such that $(Q, e_Q) \leq (P, e)$, and such that $(Q, e'_Q) \leq (S, f)$. We will show that $e_Q e'_Q \neq 0$. Let J be a primitive decomposition of 1 in $(kN)^P$. Then $1 = \sum_{j \in J} \mathrm{Br}_P(j) \in kC_N(P)$. Multiplying by $e_P e'_P$ yields $e_P e'_P = \sum_{j \in J} \mathrm{Br}_P(j)e_P e'_P$.

Since $e_P e'_P \neq 0$, it follows that there is a primitive idempotent $j \in (kN)^P$ satisfying $\mathrm{Br}_P(j) e_P e'_P \neq 0$. As Br_P maps $(kN)^P$ onto $kC_N(P)$, it follows that $\mathrm{Br}_P(j)$ is a primitive idempotent in $kC_N(P)$, hence $\mathrm{Br}_P(j) e'_P = \mathrm{Br}_P(j)$. Let now I be a primitive decomposition of j in $(kG)^P$. Then $\mathrm{Br}_P(j) = \sum_{i \in I} \mathrm{Br}_P(i)$. Thus $0 \neq \mathrm{Br}_P(j) e_P e'_P = \sum_{i \in I} \mathrm{Br}_P(i) e_P e'_P$. Thus there is a primitive idempotent $i \in (kG)^P$ satisfying $ij = i = ji$ and $\mathrm{Br}_P(i) e_P e'_P \neq 0$. As $\mathrm{Br}_P(i)$ is primitive in $kC_G(P)$ this implies that $\mathrm{Br}_P(i) e_P = \mathrm{Br}_P(i)$. But then also $\mathrm{Br}_Q(i) e_Q \neq 0$ and $\mathrm{Br}_Q(j) e'_Q \neq 0$, by the definition of inclusion of Brauer pairs. More precisely, we have $\mathrm{Br}_Q(i) \mathrm{Br}_Q(j) e_Q e'_Q = \mathrm{Br}_Q(i) \mathrm{Br}_Q(j) = \mathrm{Br}_Q(i) \neq 0$. This shows that $e_Q e'_Q \neq 0$. Consider now the set $\mathcal{B}$ of all $N_G(Q, e_Q)$-conjugates of the block e'_Q of $kC_N(Q)$. As $e_Q e'_Q \neq 0$, we have $e_Q({}^x e'_Q) \neq 0$ for all $x \in N_G(Q, e_Q)$. The group $C_G(Q)$ acts transitively on the set of all blocks f' of $kC_N(Q)$ satisfying $e_Q f' \neq 0$. Thus $C_G(Q)$ acts transitively on the set $\mathcal{B}$. A Frattini argument implies that

$$N_G(Q, e_Q) = C_G(Q)(N_G(Q, e_Q) \cap N_G(Q, e'_Q)).$$

Thus $N_G(Q, e_Q)/C_G(Q) \cong (N_G(Q, e_Q) \cap N_G(Q, e'_Q))/(C_G(Q) \cap N_G(Q, e'_Q))$. This shows that the automorphism group of Q in the fusion system $\mathcal{F}_{(P,e)}(G, b)$ is contained in the automorphism group of Q in $\mathcal{F}_{(S,f)}(G, N, c)$. Alperin's Fusion Theorem implies that $\mathcal{F}_{(P,e)}(G, b)$ is a subfusion system of $\mathcal{F}_{(S,f)}(G, N, c)$. By 6.8.9, the group $R = P \cap N$ is a defect group of c. Since $\mathrm{Br}_{S \cap N}(c) \neq 0$, the group $S \cap N$ is contained in a defect group of kNc, whence $S \cap N = P \cap N = R$. Thus (R, e'_R) is a maximal (N, c)-Brauer pair, and $\mathcal{F}_{(R, e'_R)}(N, c)$ is the subcategory of $\mathcal{F}_{(S,f)}(G, N, c)$ of all subgroups of R and morphisms induced by conjugation in N. Since N is normal in G, it follows easily that this is an invariant subcategory. $\square$

With the notation of 8.6.5, the fusion systems $\mathcal{F}_{(R, f_R)}(N, c)$ and $\mathcal{F}_{(P,e)}(G, b)$ are both subfusion systems of $\mathcal{F}_{(S,f)}(G, N, c)$, but $\mathcal{F}_{(R, f_R)}(N, c)$ need not be a subsystem of $\mathcal{F}_{(P,e)}(G, b)$. There is one case where this inclusion does hold.

Theorem 8.6.6 *Let G be a finite group, N a normal subgroup of G, and c a G-stable block of kN. Suppose that G/N is a p-group and that k is a splitting field for the subgroups of G. Let (Q, f) be a a maximal (N, c)-Brauer pair. There exists a maximal (G, c)-Brauer pair (P, e) such that $\mathcal{F}_{(Q,f)}(N, c)$ is a subfusion system of $\mathcal{F}_{(P,e)}(G, b)$.*

Proof By 6.8.11, c remains a block of $\mathcal{O}G$. Let R be a subgroup of Q and denote by f_R the unique block of $kC_N(R)$ such that $(R, f_R) \leq (Q, f)$. Since $C_N(R)$ is normal of p-power index in $C_G(R)$, it follows from 6.8.11 that f_R determines a unique block e_R that covers f_R, and we have $f_R = e_R f_R$. Thus $e_R \mathrm{Br}_R(c) \neq 0$, and hence (R, e_R) is a (G, c)-Brauer pair. The same reasoning applied to Q instead of R yields a unique block e_Q of $kC_G(Q)$ covering f, and then (Q, e_Q) is a (G, c)-Brauer pair. Choose (P, e) to be a maximal (G, c)-Brauer pair such that $(Q, e_Q) \leq (P, e)$. The uniqueness of e_R implies that $N_N(R, f_R) \leq N_G(R, e_R)$ for any subgroup R of Q, and hence Alperin's Fusion Theorem implies the inclusion of fusion systems as stated.   □

The theme of p-permutation G-algebras and fusion systems is developed further in work of Kessar [10, Part IV]. In particular, it is shown in [10, IV.3.9] that $\mathcal{F}_{(R,f_R)}(N, c)$ is *normal* in $\mathcal{F}_{(S,f)}(G, N, c)$ in the stronger sense of the notion of normal subsystems due to Aschbacher (cf.[10, I.6.1]).

8.7 Fusion systems of almost source algebras

The main result of this section, due to Puig [171], states that the fusion system of a block with respect to a chosen maximal Brauer pair can be read off the associated source algebras of the block. We present this result in the slightly more general context of almost source algebras, following [136]. Let p be a prime and $\mathcal{O}$ a complete local Noetherian ring with residue field k of characteristic p. We assume in this section that k is large enough for all blocks, in order to ensure that the associated categories on defect groups are fusion systems.

Recall from §6.4 that given a finite group G and a block b of $\mathcal{O}G$ with defect group P, an *almost source idempotent* is an idempotent i in $(\mathcal{O}Gb)^P$ satisfying $\mathrm{Br}_P(i) \neq 0$ such that for any subgroup Q of P there is a unique block e_Q with the property $\mathrm{Br}_Q(i)e_Q \neq 0$, or equivalently, with the property $\mathrm{Br}_Q(i) \in kC_G(Q)e_Q$. The interior P-algebra $A = i\mathcal{O}Gi$ is called an *almost source algebra* of the block b. The pair (P, e_P) is then in particular a maximal (G, b)-Brauer pair, and by 6.3.3, we have $(Q, e_Q) \leq (P, e_P)$ for all $Q \leq P$. By the results of the preceding section, any almost source idempotent i determines thus a fusion system $\mathcal{F} = \mathcal{F}_{(P,e_P)}(G, b)$ on P, and we will call $\mathcal{F}$ the *fusion system of the almost source algebra A* or the *fusion system on P determined by the almost source idempotent i*. We will show that the $\mathcal{O}P$-$\mathcal{O}P$-bimodule structure of A determines $\mathcal{F}$. Fixed points and Brauer homomorphism are in this section considered consistently with respect to the conjugation action, which is why we use the simplified notation $(\mathcal{O}Gb)^P$ instead of $(\mathcal{O}Gb)^{\Delta P}$.

Theorem 8.7.1 *Let G be a finite group, b be a block of $\mathcal{O}G$ with defect group P, let i be an almost source idempotent in $(\mathcal{O}Gb)^P$ and set $A = i\mathcal{O}Gi$. Denote by $\mathcal{F}$ the fusion system of A on P. Let Q, R be subgroups of P. Then A determines $\mathcal{F}$ as follows.*

(i) *Every indecomposable direct summand of A as an $\mathcal{O}Q$-$\mathcal{O}R$-bimodule is isomorphic to $\mathcal{O}Q \otimes_{\mathcal{O}S} {}_\varphi \mathcal{O}R$ for some subgroup S of Q and some morphism $\varphi : S \to R$ belonging to $\mathcal{F}$.*

(ii) *If $\varphi : Q \to R$ is an isomorphism in $\mathcal{F}$ such that R is fully $\mathcal{F}$-centralised, then ${}_\varphi \mathcal{O}R$ is isomorphic to a direct summand of A as an $\mathcal{O}Q$-$\mathcal{O}R$-bimodule.*

(iii) *Suppose that Q is fully $\mathcal{F}$-centralised and that $\varphi \in \mathrm{Aut}(Q)$. Then $\varphi \in \mathrm{Aut}_{\mathcal{F}}(Q)$ if and only if ${}_\varphi \mathcal{O}Q$ is isomorphic to a direct summand of A as an $\mathcal{O}Q$-$\mathcal{O}Q$-bimodule.*

In particular, $\mathcal{F}$ is determined by the $\mathcal{O}P$-$\mathcal{O}P$-bimodule structure of A.

We break up the proof in several intermediate results. The first observation shows that even though the Brauer homomorphism Br_Q need not send source idempotents to source idempotents, it does send almost source idempotents to almost source idempotents if Q is fully $\mathcal{F}$-centralised.

Proposition 8.7.2 *Let G be a finite group, let b be a block of $\mathcal{O}G$, let P be a defect group of b and let i be an almost source idempotent in $(\mathcal{O}Gb)^P$. Let $\mathcal{F}$ be the fusion system determined by i on P, let Q be a fully $\mathcal{F}$-centralised subgroup of P and let e be the unique block of $kC_G(Q)$ such that $\mathrm{Br}_Q(i)e \neq 0$. The idempotent $\mathrm{Br}_Q(i)$ in $(kC_G(Q)e)^{C_P(Q)}$ is an almost source idempotent of the block e of $kC_G(Q)$.*

Proof Since Q is fully $\mathcal{F}$-centralised, the group $C_P(Q)$ is a defect group of e. The idempotent $\mathrm{Br}_Q(i)$ belongs to $kC_G(Q)^{C_P(Q)}$ and satisfies $\mathrm{Br}_{C_P(Q)}(\mathrm{Br}_Q(i)) = \mathrm{Br}_{QC_P(Q)}(i)$ which is nonzero because even $\mathrm{Br}_P(i)$ is nonzero. Moreover, for every subgroup S of $C_P(Q)$ we have $C_{C_P(Q)}(S) = C_P(QS)$ and $C_{C_G(Q)}(S) = C_G(QS)$. Thus the unique block f of $C_G(QS)$ satisfying $\mathrm{Br}_{QS}(i)f \neq 0$ is also the unique block of $kC_{C_G(Q)}(S)$ satisfying $\mathrm{Br}_S(\mathrm{Br}_Q(i))f \neq 0$, and so $\mathrm{Br}_Q(i)$ is an almost source idempotent. $\square$

One of the technical issues with nonprincipal blocks is that $(\mathcal{O}Gb)(Q) = kC_G(Q)\mathrm{Br}_Q(b)$ may decompose as a product of more than one block algebra. The next result shows that at the level of an almost source algebra A this problem does not arise: the algebra $A(Q)$ remains indecomposable and Morita equivalent to the unique block of $kC_G(Q)$ determined by A so long as one chooses Q to be fully centralised with respect to the fusion system of A on P.

Proposition 8.7.3 *Let G be a finite group, b a block of $\mathcal{O}G$, P a defect group of b and let i be an almost source idempotent in $(\mathcal{O}Gb)^P$. Set $A = i\mathcal{O}Gi$ and let $\mathcal{F}$ be the fusion system of A on P. Let Q be a fully $\mathcal{F}$-centralised subgroup of P and let e be the unique block of $kC_G(Q)$ such that $\mathrm{Br}_Q(i)e \neq 0$.*

(i) *The $kC_G(Q)e$-$A(Q)$-bimodule $kC_G(Q)\mathrm{Br}_Q(i)$ and its dual induce a Morita equivalence between the block algebra $kC_G(Q)e$ and its almost source algebra $A(Q)$.*

(ii) *For any local point δ of Q on $\mathcal{O}Gb$ satisfying $\mathrm{Br}_Q(\delta)e \neq \{0\}$ we have $\delta \cap A \neq \emptyset$.*

(iii) *The canonical map $A(Q) \otimes_{kC_P(Q)} A(Q) \to A(Q)$ induced by multiplication in $A(Q)$ splits as a homomorphism of $A(Q)$-$A(Q)$-bimodules.*

(iv) *Suppose that Q is $\mathcal{F}$-centric. Then Q has a unique local point δ on $\mathcal{O}Gb$ satisfying $\mathrm{Br}_Q(\delta)e \neq 0$. Moreover, $\delta \cap A$ is the unique local point of Q on A, and we have $N_G(Q_\delta) = N_G(Q, e)$.*

Proof By 8.7.2, the idempotent $\mathrm{Br}_Q(i)$ is an almost source idempotent, and thus 6.4.6 applies to $kC_G(Q)e$ and $\mathrm{Br}_Q(i)$. This implies statement (i). If δ is a local point of Q on $\mathcal{O}Gb$ satisfying $\mathrm{Br}_Q(\delta)e \neq \{0\}$, then $\mathrm{Br}_Q(\delta)$ is a conjugacy class of primitive idempotents in $kC_G(Q)e$. Since $kC_G(Q)e$ is Morita equivalent to $A(Q) = \mathrm{Br}_Q(i)kC_G(Q)\mathrm{Br}_Q(i)$ it follows that there is $j \in \delta$ such that $\mathrm{Br}_Q(j) \in \mathrm{Br}_Q(i)kC_G(Q)\mathrm{Br}_Q(i)$. The lifting theorems for idempotents imply that j can be chosen in $A^Q = i(kG)^Qi$, whence statement (ii). Statement (iii) follows from 6.4.7 applied to $kC_G(Q)$, e, $\mathrm{Br}_Q(i)$ and $C_P(Q)$ instead of $\mathcal{O}G$, b, i and P, respectively. Suppose that Q is $\mathcal{F}$-centric. We noted in 8.5.3 (iii) that Q has a unique local point δ on $\mathcal{O}Gb$ satisfying $\mathrm{Br}_Q(\delta)e \neq 0$. It follows from (ii) that $\delta \cap A$ is the unique local point of Q on A. The uniqueness of δ implies that $N_G(Q_\delta) = N_G(Q, e)$. $\qquad\square$

Proof of Theorem 8.7.1 Let Y be an indecomposable direct summand of A as an $\mathcal{O}Q$-$\mathcal{O}R$-bimodule. Then Y has a $Q \times R$-stable $\mathcal{O}$-basis on which Q and R act freely on the left and on the right, respectively. Thus

$$Y \cong \mathcal{O}Q \otimes_{\mathcal{O}S} {}_\varphi\mathcal{O}R$$

for some subgroup S of Q and some injective group homomorphism $\varphi : S \to R$. Set $T = \varphi(S)$. Restricting Y to $S \times T$ shows that $_\varphi\mathcal{O}T$ is isomorphic to a direct summand of Y, hence of A, as an $\mathcal{O}S$-$\mathcal{O}T$-bimodule. Now A is a direct summand of $\mathcal{O}G$ as an $\mathcal{O}P$-$\mathcal{O}P$-bimodule. In particular, $_\varphi\mathcal{O}T$ is isomorphic to a direct summand of $\mathcal{O}G$ as an $\mathcal{O}S$-$\mathcal{O}T$-bimodule, hence isomorphic to $\mathcal{O}Sy^{-1} = y^{-1}\mathcal{O}T$ for some element $y \in G$ such that ${}^yS = T$ and such that ${}^ys = \varphi(s)$ for all $s \in S$. Then $\mathcal{O}S$ is isomorphic to a direct summand of $i\mathcal{O}Giy = i\mathcal{O}Gy^{-1}iy$ as an $\mathcal{O}S$-$\mathcal{O}S$-bimodule. Thus $\mathrm{Br}_S(i\mathcal{O}Gy^{-1}iy) \neq 0$ by 5.8.8. Since $\mathrm{Br}_S(i) \in kC_G(S)e_S$ this forces also that $\mathrm{Br}_S(y^{-1}iy)e_S \neq 0$. Conjugating by y yields that $\mathrm{Br}_T(i)^ye_S \neq 0$. But then necessarily ${}^ye_S = e_T$ because e_T is the unique block of $kC_G(T)$ with the property $\mathrm{Br}_T(i)e_T \neq 0$. This shows that φ is a morphism in the fusion system

$\mathcal{F}$, whence (i). Let $\varphi : Q \to R$ be an isomorphism in $\mathcal{F}$ such that R is fully $\mathcal{F}$-centralised. Thus there is an element $x \in G$ such that $\varphi(u) = {}^x u$ for all $u \in Q$ and such that ${}^x e_Q = e_R$. Let μ be a local point of Q on $\mathcal{O}Gb$ such that $\mu \cap A \neq \emptyset$. Set $\nu = {}^x \mu$; that is, ν is the local point of R on $\mathcal{O}Gb$ such that ${}^x (Q_\delta) = R_\nu$. Since $\mu \cap A \neq \emptyset$ we have $\mathrm{Br}_Q(\mu) e_Q \neq 0$. Conjugating by x implies that $\mathrm{Br}_R(\nu) e_R \neq 0$. Since R is fully $\mathcal{F}$-centralised, it follows from 8.7.3 (ii) that also $\nu \cap A \neq \emptyset$. Let $m \in \mu \cap A$ and let $n \in \nu \cap A$. Note that n and ${}^x m$ belong both to ν, hence are conjugate in $(A^R)^\times$. Since $\mathrm{Br}_Q(m) \neq 0$ we get $(m\mathcal{O}Gm)(Q) \neq \{0\}$, hence $m\mathcal{O}Gm$ has a direct summand isomorphic to $\mathcal{O}Q$ as an $\mathcal{O}Q$-$\mathcal{O}Q$-bimodule. Therefore, $m\mathcal{O}Gmx^{-1} = m\mathcal{O}Gxmx^{-1} \cong m\mathcal{O}Gn = mAn$ has a direct summand isomorphic to $\mathcal{O}Q_{\varphi^{-1}} \cong {}_\varphi \mathcal{O}R$ as an $\mathcal{O}Q$-$\mathcal{O}R$-bimodule. This proves (ii), and (iii) follows from (i) and (ii). $\qquad\square$

Detecting fusion in terms of the bimodule structure of an almost source algebra of a block can be rephrased in a number of ways as follows.

Theorem 8.7.4 *Let G be a finite group, let b be a block of $\mathcal{O}G$ with a defect group P, let $i \in (\mathcal{O}Gb)^P$ be an almost source idempotent and set $A = i\mathcal{O}Gi$. Denote by $\mathcal{F}$ the fusion system of A on P. Let Q be a fully $\mathcal{F}$-centralised subgroup of P and let $\varphi : Q \to P$ be an injective group homomorphism. Set $R = \varphi(Q)$. Denote by e_Q, e_R the unique blocks of $kC_G(Q)$, $kC_G(R)$ satisfying $\mathrm{Br}_Q(i)e_Q \neq 0$ and $\mathrm{Br}_R(i)e_R \neq 0$. The following are equivalent.*

(i) *The group homomorphism φ is a morphism in the fusion system $\mathcal{F}$.*

(ii) *For any primitive idempotent n in $(\mathcal{O}Gb)^R$ satisfying $\mathrm{Br}_R(n)e_R \neq 0$ there is a primitive idempotent m in A^Q satisfying $\mathrm{Br}_Q(m) \neq 0$ such that $m\mathcal{O}G \cong {}_\varphi(n\mathcal{O}G)$ as $\mathcal{O}Q$-$\mathcal{O}Gb$-bimodules and such that $\mathcal{O}Gm \cong (\mathcal{O}Gn)_\varphi$ as $\mathcal{O}Gb$-$\mathcal{O}Q$-bimodules.*

(iii) *For any primitive idempotent n in A^R satisfying $\mathrm{Br}_R(n) \neq 0$ there is a primitive idempotent m in A^Q satisfying $\mathrm{Br}_Q(m) \neq 0$ such that $mA \cong {}_\varphi(nA)$ as $\mathcal{O}Q$-A-bimodules and such that $Am \cong (An)_\varphi$ as A-$\mathcal{O}Q$-bimodules.*

(iv) *For any primitive idempotent n in A^R satisfying $\mathrm{Br}_R(n) \neq 0$ there is a primitive idempotent m in A^Q satisfying $\mathrm{Br}_Q(m) \neq 0$ and an element $c \in A^\times$ such that we have $cumc^{-1} = \varphi(u)n$ for all $u \in Q$.*

(v) *There exist a primitive idempotent n in A^R satisfying $\mathrm{Br}_R(n) \neq 0$, a primitive idempotent m in A^Q satisfying and $\mathrm{Br}_Q(m) \neq 0$, and an element $c \in A^\times$ such that $cumc^{-1} = \varphi(u)n$ for all $u \in Q$.*

Proof Suppose that (i) holds. Then there is an element $x \in G$ such that $\varphi(u) = xux^{-1}$ for all $u \in Q$ and such that ${}^x(e_Q) = e_R$. Let n be a primitive idempotent in $(\mathcal{O}Gb)^R$ satisfying $\mathrm{Br}_R(n)e_R \neq 0$. Denote by ν be the local point of R on $\mathcal{O}Gb$ containing n. Since ${}^x Q = R$ it follows that there is a local point μ of Q

on $\mathcal{O}Gb$ such that $v = {}^x\mu$. Since $\mathrm{Br}_R(v)e_R \neq 0$ we have $\mathrm{Br}_Q(\mu)e_Q \neq 0$. This implies that $\mathrm{Br}_Q(\mu)$ is in fact a conjugacy class of primitive idempotents in $kC_G(Q)e_Q$. Now Q is fully $\mathcal{F}$-centralised, and hence by 8.7.3 (ii), μ contains an element m such that $m \in A$. Moreover, n^x and m belong both to the same point μ of Q on $\mathcal{O}Gb$, and hence there is an element $c \in ((\mathcal{O}Gb)^Q)^\times$ such that ${}^{xc}m = n$. The map sending $ma \in m\mathcal{O}G$ to $xcma = nxca$ is the required isomorphism $m\mathcal{O}G \cong {}_\varphi(n\mathcal{O}G)$. The map sending am to $amc^{-1}x^{-1} = ac^{-1}x^{-1}n$ is the required isomorphism $\mathcal{O}Gm \cong (\mathcal{O}Gn)_\varphi$. This proves that (i) implies (ii). We will show that (ii) implies (iii) via the Morita equivalence from 6.4.6 between $\mathcal{O}Gb$ and A. Let n be a primitive idempotent in A^R satisfying $\mathrm{Br}_R(n) \neq 0$. The unit element i of A satisfies $\mathrm{Br}_R(i)e_R = \mathrm{Br}_R(i)$ by the uniqueness of the inclusion of Brauer pairs. Thus $\mathrm{Br}_R(n)e_R \neq 0$. Statement (ii) yields a primitive idempotent $m \in A^Q$ satisfying $\mathrm{Br}_Q(m) \neq 0$ such that there are isomorphisms $m\mathcal{O}G \cong {}_\varphi(n\mathcal{O}G)$ and $\mathcal{O}Gm \cong (\mathcal{O}Gn)_\varphi$. Multiplying these isomorphisms by i on the right and on the left, respectively, yields the isomorphisms as stated in (ii). Suppose that (iii) holds. An $\mathcal{O}Q$-A-bimodule isomorphism $mA \cong {}_\varphi(nA)$ is in particular an isomorphism as right A-modules, hence induced by left multiplication with an element $c \in A^\times$ satisfying $cm = nc$. Since this is also an isomorphism as left $\mathcal{O}Q$-modules, it follows that for all $u \in Q$ we have $cum = \varphi(u)cm = \varphi(u)nc$. Multiplying this by c^{-1} on the right side by c^{-1} yields (iv). Statement (iv) implies (v) trivially. Left multiplication by any element c as in (v) is an isomorphism of $\mathcal{O}Q$-A-bimodules $mA \cong {}_\varphi(nA)$. In order to prove (i), it suffices to show that the inverse ψ of the isomorphism $\varphi : Q \cong R$ is a morphism in $\mathcal{F}$. The above $\mathcal{O}Q$-A-bimodule isomorphism $mA \cong {}_\varphi(nA)$ is also an $\mathcal{O}R$-A-bimodule isomorphism ${}_\psi(mA) \cong nA$. Since $\mathrm{Br}_Q(m) \neq 0$, it follows from 5.8.8 that $\mathcal{O}Q$ is isomorphic to a direct summand of mA as an $\mathcal{O}Q$-$\mathcal{O}Q$-bimodule. Thus ${}_\psi(\mathcal{O}Q)$ is isomorphic to a direct summand of ${}_\psi(mA) \cong nA$ as an $\mathcal{O}R$-$\mathcal{O}Q$-bimodule. In particular, ${}_\psi(\mathcal{O}Q)$ is isomorphic to a direct summand of A as an $\mathcal{O}R$-$\mathcal{O}Q$-bimodule. It follows from 8.7.1 that ψ, hence φ, is a morphism in $\mathcal{F}$. $\qquad\square$

In view of Alperin's Fusion Theorem, we specialise statement (v) of the above result to automorphisms of centric subgroups.

Proposition 8.7.5 *Let G be a finite group, let b be a block of $\mathcal{O}G$ with a defect group P, let $i \in (\mathcal{O}Gb)^P$ be an almost source idempotent and set $A = i\mathcal{O}Gi$. Denote by $\mathcal{F}$ the fusion system of A on P. Let Q be an $\mathcal{F}$-centric subgroup of P, let δ be the unique local point of Q on A and let $j \in \delta$. We have canonical group isomorphisms*

$$N_G(Q_\delta)/C_G(Q) \cong \mathrm{Aut}_\mathcal{F}(Q) \cong N_{(jAj)^\times}(Qj)/(jA^Qj)^\times.$$

Proof Let e be the block of $kC_G(Q)$ such that $\mathrm{Br}_Q(i)e \neq 0$. Thus $\mathrm{Aut}_{\mathcal{F}}(Q)$ is canonically isomorphic to $N_G(Q, e)/C_G(Q)$. By 8.7.3, there is a unique local point δ of Q on $\mathcal{O}Gb$ satisfying $\mathrm{Br}_Q(\delta)e \neq 0$, and we have $N_G(Q_\delta) = N_G(Q, e_Q)$. This implies the first isomorphism. Note that $Q \cong Qj$ because jAj is projective as a left (and right) $\mathcal{O}Q$-module. Let $\varphi \in \mathrm{Aut}_{\mathcal{F}}(Q)$. By 8.7.4 (iv) there is a primitive idempotent j' in A^Q satisfying $\mathrm{Br}_Q(j') \neq 0$ and an element $c \in A^\times$ such that $cujc^{-1} = \varphi(u)j'$ for all $u \in Q$. Since δ is the unique local point of Q on A, the idempotents j and j' both belong to δ, and hence they are conjugate by an element in $(A^Q)^\times$. Therefore, after modifying c if necessary, we may assume that $j' = j$. Then c centralises j. Thus $cj \in (jAj)^\times$, and conjugation by cj induces the automorphism φ on the image Qj of Q in jAj. Conjugation by cj induces the identity on Qj if an only if $cj \in (jA^Qj)^\times$. Conversely, if ψ is an automorphism of Q such that there exists $d \in (jAj)^\times$ with the property that conjugation by d induces the automorphism of Qj determined by ψ, then the element $c = i - j + d$ belongs to $A^\times$, we have $c^{-1} = i - j + d^{-1}$, where d^{-1} is the inverse of d in jAj. It follows that c satisfies $cujc^{-1} = dujd^{-1} = \psi(u)j$ for all $u \in Q$. By 8.7.4 this forces $\psi \in \mathrm{Aut}_{\mathcal{F}}(Q)$, whence the result. $\qquad\square$

Proposition 8.7.6 *Let G, H be finite groups, let b, c be blocks of $\mathcal{O}G$, $\mathcal{O}H$ with defect group P, Q, respectively. Let $i \in (\mathcal{O}Gb)^P$ and $j \in (\mathcal{O}Hc)^Q$ be almost source idempotents, let $\mathcal{F}$ be the fusion system on P determined by i and let $\mathcal{G}$ be the fusion system on Q determined by j. Identify $\mathcal{O}G \otimes_{\mathcal{O}} \mathcal{O}H$ and $\mathcal{O}(G \times H)$ through the canonical isomorphism.*

(i) $b \otimes c$ is a block of $\mathcal{O}(G \times H)$ and $P \times Q$ is a defect group of $b \otimes c$.
(ii) $i \otimes j$ is an almost source idempotent of $b \otimes c$.
(iii) The fusion system determined by $i \otimes j$ on $P \times Q$ is equal to $\mathcal{F} \times \mathcal{G}$.

Proof Statement (i) is an easy consequence of the characterisation 6.2.1 (ii) of defect groups of blocks. For R a subgroup of $P \times Q$, denote by R_1 the image of R under the projection $P \times Q \to P$ and by R_2 the image of R under the projection $P \times Q \to Q$. Then $R \subseteq R_1 \times R_2$ and $C_{G \times H}(R) = C_G(R_1) \times C_H(R_2)$. Thus $\mathrm{Br}_R(i \otimes j) = \mathrm{Br}_{R_1}(i) \otimes \mathrm{Br}_{R_2}(j)$. Therefore, if e_1, f_2 are the unique blocks of $kC_G(R_1)$, $kC_H(R_2)$, respectively, satisfying $\mathrm{Br}_{R_1}(i)e_1 \neq 0$ and $\mathrm{Br}_{R_2}(j)f_2 \neq 0$, then $e = e_1 \otimes f_2$ is the unique block of $C_{G \times H}(R)$ satisfying $\mathrm{Br}_R(i \otimes j)e \neq 0$. This shows (ii). Statement (iii) follows from the characterisation 8.7.1 of fusion in almost source algebras in terms of the bimodule structure of the almost source algebras. $\qquad\square$

Proposition 8.7.7 *Let G be a finite group, let b be a block of $\mathcal{O}G$, let P be a defect group of b and let $i \in (\mathcal{O}Gb)^P$ be an almost source idempotent. Let α be the anti-automorphism of $\mathcal{O}G$ sending $x \in G$ to x^{-1}. Set $b^0 = \alpha(b)$ and*

$i^0 = \alpha(i)$. Then b^0 is a block of $\mathcal{O}G$ having P as defect group and $i^0 \in (\mathcal{O}Gb^0)^P$ is an almost source idempotent of b^0. Moreover, the fusion systems on P determined by i and by i^0 are equal.

Proof Using the fact that α sends $(\mathcal{O}G)_P^G$ to itself one sees that b, b^0 have both the same defect groups, and that i^0 is an almost source idempotent of b^0. The anti-automorphisms induced by α on centralisers of subgroups of P send (G, b)-Brauer pairs to (G, b^0)-Brauer pairs in such a way that if two (G, b)-Brauer pairs are conjugate by an element $x \in G$ then the corresponding (G, b^0) Brauer pairs are conjugate by x^{-1}. The result follows. $\square$

Combining the fact that source algebras determine fusion systems and the fact that the Clifford theoretic reduction 6.8.3 preserves source algebras yields the invariance of fusion systems under the reduction 6.8.3 as an immediate consequence.

Proposition 8.7.8 *Let G be a finite group, N a normal subgroup, c a block of $\mathcal{O}N$. Let H be the stabiliser of c in G and let b be a block of $\mathcal{O}G$ satisfying $bc \neq 0$. Let d be the unique block of $\mathcal{O}Hc$ such that $\mathrm{Tr}_H^G(d) = b$. Let (P, f) be a maximal (H, d)-Brauer pair. There is a unique maximal (G, b)-Brauer pair (P, e) such that $ef \neq 0$, and then $\mathcal{F}_{(P,e)}(G, b) = \mathcal{F}_{(P,f)}(H, d)$.*

Proof Let $i \in (\mathcal{O}Hd)^P$ be a source idempotent associated with f; that is, $\mathrm{Br}_P(i)f \neq 0$. By 6.8.3, i remains a source idempotent of $\mathcal{O}Gb$, and $i\mathcal{O}Hi = i\mathcal{O}Gi$. If e is a block of $kC_G(P)$ satisfying $ef \neq 0$, then $\mathrm{Br}_P(ef) = ef \neq 0$, and hence i satisfies $\mathrm{Br}_P(i)e \neq 0$. Thus the condition $ef \neq 0$ determines the unique block e of $kC_G(P)$ such that (P, e) is a maximal (G, b)-Brauer pair associated with i. By 8.7.1, source algebras of blocks determine fusion systems, whence the result. $\square$

The fact that the bimodule structure of an almost source algebra determines a fusion system can be reformulated in terms of permutation bases. For P a finite group, Q a subgroup of P and $\varphi : Q \to P$ an injective group homomorphism, we define a P-P-biset $P \times_{(Q,\varphi)} P = P \times P/ \sim$, where $\sim$ is the equivalence relation on $P \times P$ given by $(uw, v) \sim (u, \varphi(w)v)$ for all $u, v \in P$ and $w \in Q$. With our earlier notation, we have an isomorphism of $\mathcal{O}P$-$\mathcal{O}P$-bimodules $\mathcal{O}[P \times_{(Q,\varphi)} P] \cong \mathcal{O}P \otimes_{\mathcal{O}Q} (_\varphi\mathcal{O}P)$. For X a P-P-biset and Q, φ as before, we denote by $_\varphi X$ the Q-P-biset obtained from letting $u \in Q$ act on the left as $\varphi(u)$, and with the right action of P unchanged. We denote by $_Q X$ the Q-P-biset obtained from restricting the left action to Q; that is, $_Q X = {}_\iota X$, where $\iota : Q \to P$ is the inclusion homomorphism. We use the analogous notation for restrictions to Q on the right side.

Definition 8.7.9 Let P be a finite p-group and $\mathcal{F}$ a fusion system on P. A *characteristic biset of* $\mathcal{F}$ is a finite P-P-biset X with the following properties.

(a) Every transitive P-P-sub-biset of X is isomorphic to $P \times_{(Q,\varphi)} P$ for some subgroup Q of P and some $\varphi \in \mathrm{Hom}_{\mathcal{F}}(Q, P)$.
(b) $\frac{|X|}{|P|}$ is prime to p.
(c) For any subgroup Q of P and any $\varphi \in \mathrm{Hom}_{\mathcal{F}}(Q, P)$, we have an isomorphism of Q-P-bisets $_QX \cong {_\varphi}X$ and an isomorphism of P-Q-bisets $X_Q \cong X_\varphi$.

Property (c) is also called the *left and right $\mathcal{F}$-stability of X*. By a result of Broto, Levi, Oliver [33, 5.5], every fusion system has a characteristic biset. It had been previously noted by the author and P. J. Webb, that the existence of such a biset implies the existence of a direct factor of BP regarded as a spectrum, with cohomology the $\mathcal{F}$-stable elements in $H^*(P; k)$. See Ragnarsson [183] for further details and references on spectra associated with fusion systems.

Since an almost source algebra A of a block with defect group P has a P-P-stable basis X that is determined, up to isomorphism of P-P-bisets, by the $\mathcal{O}P$-$\mathcal{O}P$-bimodule structure, we can reformulate previous results as follows.

Proposition 8.7.10 *Let A be an almost source algebra of a block of a finite group algebra over k with defect group P and fusion system $\mathcal{F}$. Let X be a P-P-stable k-basis of A. The following hold.*

(i) *Every transitive P-P-sub-biset of X is isomorphic to $P \times_{(Q,\varphi)} P$ for some subgroup Q of P and some $\varphi \in \mathrm{Hom}_{\mathcal{F}}(Q, P)$.*
(ii) *Let Q be a fully $\mathcal{F}$-centralised subgroup of P and let $\varphi \in \mathrm{Aut}(Q)$. We have $\varphi \in \mathrm{Aut}_{\mathcal{F}}(Q)$ if and only if $_\varphi Q$ is isomorphic to a Q-Q-orbit in X.*

Proof This is a reformulation in terms of bisets of the properties 8.7.1 (i) and (iii). □

The biset X in 8.7.10 comes close to being a characteristic biset for $\mathcal{F}$, but it is not known whether this holds in general. A sufficient criterion for X to be a characteristic biset is as follows.

Proposition 8.7.11 *Let A be a source algebra of a block of a finite group algebra over k, assumed to be large enough, with defect group P and fusion system $\mathcal{F}$. Let X be a P-P-stable k-basis of A. Suppose that X is contained in $A^\times$. Then X is a characteristic biset of $\mathcal{F}$.*

Proof By 8.7.10, the biset X satisfies property (a) of 8.7.9. Property (b) follows from 6.15.1 By Alperin's Fusion Theorem, it suffices to verify the stability property (c) for Q an $\mathcal{F}$-centric subgroup and $\varphi \in \mathrm{Aut}_{\mathcal{F}}(Q)$. By 8.7.10, there is $x \in X$ such that $Qx = xQ \cong {_\varphi}Q$ as Q-Q-bisets. If w is the image of x under such a biset isomorphism $xQ \cong {}_\varphi Q$, then this isomorphism sends xw^{-1} to 1. After replacing x by xw^{-1}, we may assume that there is a biset isomorphism $xQ \cong {}_\varphi Q$ sending x to 1, hence uxv to $\varphi(u)v$ for all $u, v \in Q$. But $\varphi(u)v$ is also the image of $x\varphi(u)v$ under that map. This implies that $ux = x\varphi(u)$ for all $u \in Q$. Since x is invertible in A, it follows that left multiplication by x on A induces an isomorphism of $\mathcal{O}Q$-$\mathcal{O}P$-bimodules ${}_\varphi A \cong {}_Q A$. Since the $\mathcal{O}Q$-$\mathcal{O}P$-bimodule structure of ${}_Q A$ and ${}_\varphi A$ determines the biset structure, up to isomorphism, of the permutation bases ${}_Q X$ and ${}_\varphi X$ of these bimodules, it follows that ${}_Q X \cong {}_\varphi X$ as Q-P-bisets. This shows the required $\mathcal{F}$-stability of X in the left side. A similar argument for the $\mathcal{F}$-stability on the right side concludes the proof. □

Exercise 8.7.12 Let G be a finite group and P a Sylow p-subgroup of G.

(a) Show that G, regarded as a P-P-biset, is a characteristic biset of the fusion system $\mathcal{F} = \mathcal{F}_P(G)$.
(b) Let X be a P-P-stable basis of the principal block algebra B of kG. Using that B is an almost source algebra, show that X is a characteristic biset of $\mathcal{F}$.

8.8 Almost source algebras and p-permutation modules

Let p be a prime and $\mathcal{O}$ a complete local Noetherian ring with residue field k of characteristic p. Assume that k is large enough for all block algebras in this section. The rather technical results in this section on the interplay between fusion and vertices of p-permutation modules for almost source algebras will be needed in the context of splendid equivalences and p-permutation equivalences in §9.5. Since an almost source algebra A of a block with defect group P is in particular an A-$\mathcal{O}P$-bimodule, we can apply, for any subgroup R of P, the Brauer construction with respect to any other subgroup Q of P to the left A-module $A \otimes_{\mathcal{O}R} \mathcal{O}$. Since A has a $P \times P$-stable $\mathcal{O}$-basis, its canonical image in the A-module $A \otimes_{\mathcal{O}R} \mathcal{O}$ is a P-stable $\mathcal{O}$-basis with respect to the left action by P. We will need to consider fixed points with respect to actions that may not necessarily be induced by conjugation, and hence, whenever appropriate, we use the notation $A(\Delta P)$ rather than $A(P)$ to avoid confusion.

Theorem 8.8.1 ([136, Theorem 6.2]) *Let A be an almost source algebra of a block with defect group P and fusion system $\mathcal{F}$. Let Q, R be subgroups of P such that Q is fully $\mathcal{F}$-centralised. Let Y be an indecomposable direct summand of the $A(\Delta Q)$-module $(A \otimes_{\mathcal{O}R} \mathcal{O})(Q)$. Then Y is isomorphic to a direct summand of $A(\Delta Q)m \otimes_{kS} k$ for some subgroup S of $C_P(Q)$ containing $Z(Q)$ and some* *primitive idempotent* $m \in A(\Delta Q)^{\Delta S}$ *satisfying* $\mathrm{Br}_{\Delta S}(m) \neq 0$ *such that S is fully* $C_{\mathcal{F}}(Q)$*-centralised and QS is isomorphic in* $\mathcal{F}$ *to a subgroup of R.*

For the proof of this theorem we will need the following fact:

Lemma 8.8.2 *Let A be an almost source algebra of a block with defect group P and fusion system* $\mathcal{F}$*. Let R be a subgroup of P and let X be an indecomposable direct summand of the A-module* $A \otimes_{\mathcal{O}R} \mathcal{O}$*. Then X is isomorphic to a direct summand of* $Am \otimes_{\mathcal{O}Q} \mathcal{O}$ *for some fully* $\mathcal{F}$*-centralised subgroup Q of P that is isomorphic, in the fusion system* $\mathcal{F}$*, to a subgroup of R, and some primitive idempotent* $m \in A^{\Delta Q}$ *satisfying* $\mathrm{Br}_{\Delta Q}(m) \neq 0$ *and* $X(\Delta Q) \neq \{0\}$.

Proof Replace R by a minimal subgroup with the property that X is isomorphic to a direct summand of $A \otimes_{\mathcal{O}R} \mathcal{O}$. In particular, X is a permutation $\mathcal{O}P$-module. By 2.6.12, X is a direct summand of $X \otimes_{\mathcal{O}R} V$ for some indecomposable direct summand V of X as an $\mathcal{O}R$-module. Thus $V \cong \mathrm{Ind}_S^R(\mathcal{O})$ for some subgroup S of R. The minimality of R implies $R = S$, and hence X has a trivial direct summand as an $\mathcal{O}R$-module. (This would also follow from the more general statement 6.4.10.) Since X is indecomposable, X is isomorphic to a direct summand of $An \otimes_{\mathcal{O}R} \mathcal{O}$ for some primitive idempotent $n \in A^{\Delta R}$. The minimality of R, together with 5.12.8, implies that $\mathrm{Br}_{\Delta R}(n) \neq 0$. Let Q be a fully $\mathcal{F}$-centralised subgroup of P such that there is an isomorphism $\varphi : Q \cong R$ in $\mathcal{F}$. By an appropriate version of 8.7.4 (ii) there is a primitive idempotent $m \in A^{\Delta Q}$ such that $\mathrm{Br}_{\Delta Q}(m) \neq 0$ and such that $Am \cong (An)_\varphi$ as A-$\mathcal{O}Q$-bimodules. Then $An \otimes_{\mathcal{O}R} \mathcal{O} \cong Am \otimes_{\mathcal{O}Q} \mathcal{O}$, and hence in particular, X is isomorphic to a direct summand of $A \otimes_{\mathcal{O}Q} \mathcal{O}$ as stated, and Q is minimal with this property (otherwise $A \otimes_{\mathcal{O}Q} \mathcal{O}$ would not have a trivial direct summand as an $\mathcal{O}R$-module). But then the same argument as above implies that X has a trivial direct summand as an $\mathcal{O}Q$-module, and hence $X(\Delta Q) \neq \{0\}$, which concludes the proof. $\square$

Proof of Theorem 8.8.1 By 8.7.3, $A(\Delta Q)$ is isomorphic to a direct summand of $A(\Delta Q) \otimes_{kC_P(Q)} A(\Delta Q)$. Tensoring with $- \otimes_{A(\Delta Q)} Y$ implies that Y is isomorphic to a direct summand of $A(\Delta Q) \otimes_{kC_P(Q)} Y$. Since Y is indecomposable, Y is in fact isomorphic to a direct summand of $A(\Delta Q) \otimes_{kC_P(Q)} W$ for some indecomposable direct summand of Y as $kC_P(Q)$-module. Thus we need to determine the $kC_P(Q)$-module structure of Y. Let X be a P-P-stable $\mathcal{O}$-basis of A. We use the notation $X \otimes 1$ for the image of the set X in $A \otimes_{\mathcal{O}R} \mathcal{O}$. The set $X \otimes 1$ is a P-stable $\mathcal{O}$-basis of $A \otimes_{\mathcal{O}R} \mathcal{O}$. Thus, in order to compute $(A \otimes_{\mathcal{O}R} \mathcal{O})(Q)$ we need to determine the Q-fixed points in the set $X \otimes 1$. For $x, y \in X$ the images $x \otimes 1$ and $y \otimes 1$ in $X \otimes 1$ are equal if and only if there is an element $r \in R$ such that $y = xr$. Therefore, $x \otimes 1 \in (X \otimes 1)^Q$ if and only if for every $u \in Q$ there is $r_u \in R$ such that $ux = xr_u$. In that case, r_u is then uniquely determined by u because P acts freely on the right of X. Since P acts also freely on the left of X, the map sending $u \in Q$ to $r_u \in R$ is an injective group homomorphism $\varphi : Q \to R$. Set

$$U_\varphi = \{x \in X \mid ux = x\varphi(u)\, (\forall u \in Q)\}.$$

Note that by 8.7.1 (i) any φ arising in this way belongs to the fusion system $\mathcal{F}$ because $\mathcal{O}Qx = x\mathcal{O}\varphi(Q) \cong {}_\varphi\mathcal{O}\varphi(Q)$ is a direct summand of A as $\mathcal{O}Q$-$\mathcal{O}\varphi(Q)$-bimodule. It follows from the above that

$$(X \otimes 1)^Q = \bigcup_\varphi U_\varphi \otimes 1$$

where φ runs over the set $\mathrm{Hom}_\mathcal{F}(Q, R)$. Let φ, $\psi \in \mathrm{Hom}_\mathcal{F}(Q, R)$ and denote by $\tilde{\varphi}$, $\tilde{\psi}$ their images in the orbit space $\mathrm{Hom}_{\tilde{\mathcal{F}}}(Q, R) = \mathrm{Inn}(R)\backslash\mathrm{Hom}_\mathcal{F}(Q, R)$, where the group of inner automorphisms $\mathrm{Inn}(R)$ of R acts by composition of group homomorphisms on the set $\mathrm{Hom}_\mathcal{F}(Q, R)$. Suppose there is an element $x \in U_\varphi$ and $y \in U_\psi$ such that $x \otimes 1 = y \otimes 1$. Then there is an element $r \in R$ such that $y = xr$, and for all $u \in U$, we have $ux = x\varphi(u)$ and $uy = y\psi(u)$. Thus $uxr = x\varphi(u)r = xr\psi(u)$, hence $\varphi(u)r = \psi(u)r$, or equivalently, $\varphi(u) = {}^r\psi(u)$ for all $u \in Q$. This means that $\tilde{\varphi} = \tilde{\psi}$. It follows that $U_\varphi \otimes 1 = U_\psi \otimes 1$ if $\tilde{\varphi} = \tilde{\psi}$ and that $U_\varphi \otimes 1 \cap U_\psi \otimes 1 = \emptyset$ if $\tilde{\varphi} \neq \tilde{\psi}$. In other words, $(X \otimes 1)^Q$ is in fact the disjoint union

$$(X \otimes 1)^Q = \bigsqcup_\varphi U_\varphi \otimes 1$$

with φ running over a set of representatives in $\mathrm{Hom}_\mathcal{F}(Q, R)$ of $\mathrm{Hom}_{\tilde{\mathcal{F}}}(Q, R) = \mathrm{Inn}(R)\backslash\mathrm{Hom}_\mathcal{F}(Q, R)$. Therefore, by 5.8.1, we have

$$(A \otimes_{OR} \mathcal{O})(Q) \cong k((X \otimes 1)^Q) = \oplus_\varphi k(U_\varphi \otimes 1)$$

with φ running as before over a set of representatives in $\mathrm{Hom}_\mathcal{F}(Q, R)$ of $\mathrm{Inn}(R)\backslash\mathrm{Hom}_\mathcal{F}(Q, R)$. This is a decomposition of $(A \otimes_{OR} \mathcal{O})(Q)$ as a $kC_P(Q)$-module because the subsets U_φ of X are invariant under the left action of $C_P(Q)$. Thus the direct summand W of Y as a $kC_P(Q)$-module is isomorphic to a direct summand of $k(U_\varphi \otimes 1)$ for some $\varphi \in \mathrm{Hom}_\mathcal{F}(Q, R)$. Now W is a permutation $kC_P(Q)$-module, and hence, in order to determine its structure, it suffices to determine the stabilisers in $C_P(Q)$ of basis elements $x \otimes 1 \in U_\varphi \otimes 1$. Let x, $y \in U_\varphi$ such that $x \otimes 1 = y \otimes 1$. Then, on one hand, $y = xr$ for some $r \in R$, and on the other hand, $ux = x\varphi(u)$ and $uy = y\varphi(u)$ for all $u \in Q$. Thus $uxr = x\varphi(u)r = xr\varphi(u)$, which implies $r \in C_P(\varphi(Q))$. It follows that the stabiliser S in $C_P(Q)$ of an element $x \otimes 1 \in U_\varphi \otimes 1$ consists of all elements $z \in C_P(Q)$

such that $zx = xr_z$ for some $r_z \in C_P(\varphi(Q))$. In particular $Z(Q) \subseteq S$ because $zx = x\varphi(z)$ for $z \in Z(Q)$. It follows that there is a well-defined group homomorphism $\psi : QS \to R$ given by $\psi(uz) = \varphi(u)r_z$ for all $u \in Q$ and $z \in S$. Note that ψ has the property $uzx = x\psi(uz)$, so $\mathcal{O}(QS)x = x\mathcal{O}(\psi(QS)) \cong {}_\psi \mathcal{O}\psi(QS)$ is a direct summand of A as an $\mathcal{O}(QS)$-$\mathcal{O}\psi(QS)$-bimodule. Hence ψ belongs to the fusion system $\mathcal{F}$ by 8.7.1 (i). This shows that Y is isomorphic to a direct summand of $A(\Delta Q) \otimes_{kS} k$ for some subgroup S of $C_P(Q)$ containing $Z(Q)$ for which there exists a morphism $\psi : QS \to R$ in $\mathcal{F}$. We need to show that S can furthermore be chosen to be fully $C_{\mathcal{F}}(Q)$-centralised. Applying 8.8.2 to $A(\Delta Q)$ and its fusion system $C_{\mathcal{F}}(Q)$ shows that Y is isomorphic to a direct summand of $A(\Delta Q)m \otimes_{kT} k$ for some fully $C_{\mathcal{F}}(Q)$-centralised subgroup T of $C_P(Q)$ for which there exists a morphism $\tau : T \to S$ in $C_{\mathcal{F}}(Q)$ and some primitive idempotent $m \in A(\Delta Q)^{\Delta T}$ satisfying $\mathrm{Br}_{\Delta T}(m) \neq 0$. By the definition of $C_{\mathcal{F}}(Q)$, the morphism τ extends to a morphism $\sigma : QT \to QS$ in $\mathcal{F}$ satisfying $\sigma|_Q = \mathrm{Id}_Q$. Thus replacing S by T and ψ by $\psi \circ \sigma$ concludes the proof. $\qquad\square$

Corollary 8.8.3 *Let A be an almost source algebra of a block with defect group P and fusion system $\mathcal{F}$. Let Q, R be subgroups of P such that Q is fully $\mathcal{F}$-centralised. If $\mathrm{Hom}_{\mathcal{F}}(Q, R) = \emptyset$, then $(A \otimes_{\mathcal{O}R} \mathcal{O})(Q) = \{0\}$.*

Proof The hypotheses imply that Q is not isomorphic, in $\mathcal{F}$, to a subgroup of R, and hence the statement follows from 8.8.1. $\qquad\square$

Corollary 8.8.4 *Let A, B be almost source algebras of blocks of finite groups having a common defect group P and the same fusion system $\mathcal{F}$. Let Q, R be subgroups of P such that Q is fully $\mathcal{F}$-centralised. If $\mathrm{Hom}_{\mathcal{F}}(Q, R) = \emptyset$, then $(A \otimes_{\mathcal{O}R} B)(\Delta Q) = \{0\}$.*

Proof Note that $\mathcal{F} \times \mathcal{F}$ is the fusion system of the almost source algebra $A \otimes_{\mathcal{O}} B^{\mathrm{op}}$, by 8.7.6, and that ΔQ is fully $\mathcal{F} \times \mathcal{F}$-centralised in $P \times P$, by 8.4.8. The hypotheses imply that ΔQ is not isomorphic, in the fusion system $\mathcal{F} \times \mathcal{F}$ on $P \times P$, to a subgroup of ΔR. Since $A \otimes_{\mathcal{O}R} B$ is isomorphic, as an $A \otimes_{\mathcal{O}} B^{\mathrm{op}}$-module, to $(A \otimes B^{\mathrm{op}}) \otimes_{\mathcal{O}\Delta R} \mathcal{O}$, the statement follows from 8.8.3 applied to $A \otimes_{\mathcal{O}} B^{\mathrm{op}}$ instead of A. $\qquad\square$

Corollary 8.8.5 *Let A be an almost source algebra of a block with defect group P and fusion system $\mathcal{F}$. Let Q be a fully $\mathcal{F}$-centralised subgroup of P. Every indecomposable direct summand of $(A \otimes_{\mathcal{O}Q} \mathcal{O})(Q)$ is isomorphic to $A(\Delta Q)m \otimes_{kZ(Q)} k$ for some primitive idempotent $m \in A(\Delta Q)$.*

Proof By 8.8.1, an indecomposable direct summand Y of $(A \otimes_{\mathcal{O}Q} \mathcal{O})(Q)$ is isomorphic to a direct summand of $A(\Delta Q)m \otimes_{kS} k$ for some subgroup S of

$C_P(Q)$ containing $Z(Q)$ and some primitive idempotent $m \in A(\Delta Q)^{\Delta S}$ satisfying $\mathrm{Br}_{\Delta S}(m) \neq 0$ such that QS is isomorphic to a subgroup of Q in $\mathcal{F}$. This forces $S \subseteq Q$, hence $S = Z(Q)$. Since the image of $Z(Q)$ in $A(\Delta Q)^\times$ belongs to $Z(A(\Delta Q))$, a primitive idempotent in $A(\Delta Q)^{Z(Q)}$ is primitive in $A(\Delta Q)$ and hence the module $A(\Delta Q)m \otimes_{kZ(Q)} k$ is indecomposable since it is an indecomposable projective module for the algebra $A(\Delta Q) \otimes_{kZ(Q)} k$. □

We conclude this section with translations of earlier results - notably from Section 5.8 – to almost source algebras, via the canonical Morita equivalences from 6.4.6. If P is a finite p-group and $\varphi : M \to N$ a homomorphism of $\mathcal{O}P$-modules, we denote by $\varphi(Q) : M(Q) \to N(Q)$ the homomorphism induced by φ, for any subgroup Q of P.

Proposition 8.8.6 *Let A be an almost source algebra of a block with defect group P and fusion system $\mathcal{F}$ and let M be an indecomposable direct summand of $A \otimes_{\mathcal{O}R} \mathcal{O}$ for some subgroup R of P. Then M is projective if and only if $M(Q) = \{0\}$ for any nontrivial fully $\mathcal{F}$-centralised subgroup Q of P satisfying $\mathrm{Hom}_{\mathcal{F}}(Q, R) \neq \emptyset$.*

Proof If M is projective, then M remains projective upon restriction to $\mathcal{O}P$, and hence $M(Q) = \{0\}$ for any nontrivial subgroup of P. Suppose conversely that $M(Q) = \{0\}$ for any fully $\mathcal{F}$-centralised subgroup Q of P satisfying $\mathrm{Hom}_{\mathcal{F}}(Q, R) \neq \emptyset$. It follows from 8.8.2 that M is isomorphic to a direct summand of $A \otimes_{\mathcal{O}S} \mathcal{O}$ for some fully $\mathcal{F}$-centralised subgroup S of P such that $X(\Delta S) \neq \{0\}$. By 8.8.3 we have $\mathrm{Hom}_{\mathcal{F}}(S, R) \neq \emptyset$. By the assumptions this forces $S = \{1\}$, and hence M is projective. □

Proposition 8.8.7 *Let A be an almost source algebra of a block with defect group P and fusion system $\mathcal{F}$. Let M, N be finitely generated A-modules that are direct sums of summands of the A-modules $A \otimes_{\mathcal{O}Q} \mathcal{O}$, with Q running over the set of subgroups of P. An A-homomorphism $\varphi : M \to N$ induces an isomorphism in the relatively $\mathcal{O}$-stable category $\underline{\mathrm{Mod}}(A)$ if and only if $\varphi(Q) : M(Q) \to N(Q)$ is an isomorphism of $A(\Delta Q)$-modules for any fully $\mathcal{F}$-centralised nontrivial subgroup Q of P. In particular, if φ is surjective and $\varphi(Q)$ is an isomorphism for any fully $\mathcal{F}$-centralised subgroup Q of P, then φ is split surjective and $\ker(\varphi)$ is a projective A-module.*

Proof Let b be a block of a finite group algebra $\mathcal{O}G$ having a defect group P and i an almost source idempotent such that $A = i\mathcal{O}Gi$. Set $U = \mathcal{O}Gi \otimes_A M$, $V = \mathcal{O}Gi \otimes_A N$, and $\psi = \mathrm{Id}_{\mathcal{O}Gi} \otimes \varphi$. If φ induces an isomorphism in $\underline{\mathrm{Mod}}(A)$, then ψ induces an isomorphism in $\underline{\mathrm{Mod}}(\mathcal{O}Gb)$, hence $\psi(Q)$ is an isomorphism for all nontrivial subgroups by 5.8.12. Multiplying ψ by the idempotent i yields φ, and hence $\varphi(Q)$ is an isomorphism for all subgroups Q of P. Suppose conversely that $\varphi(Q)$ is an isomorphism for any full $\mathcal{F}$-centralised subgroups Q of P. By 5.8.12 it suffices to show that $\psi(Q)$ is an isomorphism for all nontrivial subgroups Q of P. Clearly this is the case if and only if $\psi(Q)$ induces an isomorphism $(jU)(Q) \cong (jV)(Q)$ for any primitive idempotent j in $(\mathcal{O}Gb)^\Delta$. If $\mathrm{Br}_{\Delta Q}(j) = 0$, both sides are zero, so we only need to consider the case where Q is a nontrivial subgroup of P and j a primitive idempotent in $(\mathcal{O}Gb)^{\Delta Q}$ such that $\mathrm{Br}_{\Delta Q}(j) \neq \{0\}$. We need to show that the induced map $(jU)(Q) \to (jV)(Q)$ is an isomorphism. Since this property is invariant under conjugation in G, we may assume that j belongs to a local point of Q on A such that Q is a fully $\mathcal{F}$-centralised subgroup of P. Then $jU = jM$ and $jV = jN$. Since $\varphi(Q) : M(Q) \cong N(Q)$ is an isomorphism it follows that $\varphi(Q)$ induces an isomorphism $(jU)(Q) \cong (jV)(Q)$. The result follows as in 5.8.12. □

Proposition 8.8.8 *Let A be an almost source algebra of a block with defect group P and fusion system $\mathcal{F}$. Let X be a bounded complex of finitely generated A-modules that are direct sums of summands of the bimodules $A \otimes_{\mathcal{O}Q} \mathcal{O}$, with Q running over the subgroups of P. Then X is homotopy equivalent to a bounded complex of finitely generated projective A-modules if and only if the complex of $A(\Delta Q)$-modules $X(Q)$ is acyclic for every nontrivial fully $\mathcal{F}$-centralised subgroup Q of P.*

Proof This is played back to Bouc's Theorem 5.11.10, following the same pattern as in the two proofs before. Writing $A = i\mathcal{O}Gi$ for some almost source idempotent in a block algebra $\mathcal{O}Gb$ having P as a defect group, consider the complex $Y = \mathcal{O}Gi \otimes_A X$. Then $iY = X$, and X is homotopy equivalent to a bounded complex of finitely generated projective A-modules if and only if Y is homotopy equivalent to a bounded complex of finitely generated projective $\mathcal{O}Gb$-modules. By 5.11.10, this is equivalent to $Y(Q)$ being acyclic for every nontrivial subgroup Q of P, which in turn is the case if and only if $(jY)(Q)$ is acyclic for every nontrivial subgroup Q of P and any primitive idempotent j in $(\mathcal{O}Gb)^{\Delta Q}$ satisfying $\mathrm{Br}_{\Delta Q}(j) \neq 0$. As in the two previous proofs, this property is invariant under conjugation in G, and we therefore need to consider only nontrivial fully $\mathcal{F}$-centralised subgroups of P. The result follows. □

8.9 Subsections

Let p be a prime, and let $\mathcal{O}$ be a complete discrete valuation ring having a residue field $k = \mathcal{O}/J(\mathcal{O})$ of characteristic p and a field of fractions K of characteristic 0. We assume that k and K are splitting fields for all finite groups and their subgroups in this section.

Definition 8.9.1 (Brauer [27])   Let G be a finite group, b a block of $\mathcal{O}G$ and (u, e) a (G, b)-Brauer element. Denote by $\hat{e}$ the unique block of $\mathcal{O}C_G(u)$ whose image in $kC_G(u)$ is equal to e. For any class function $\chi : G \to K$ we denote by $\chi^{(u,e)} : G \to K$ the unique class function satisfying

$$\chi^{(u,e)}(x) = \chi(\hat{e}x)$$

for any $x \in G$ such that the p-part of x is equal to u, and $\chi^{(u,e)}(x) = 0$ for any $x \in G$ such that the p-part of x is not conjugate to u.

With the notation from 5.15.13 and 6.13.3 we have

$$\chi^{(u,e)} = t_G^u \circ d_G^{(u,e)}(\chi)$$

which shows that $\chi^{(u,e)}$ is a well-defined class function on G. Note that if $u = 1$, then $\hat{e} = b$. Recall that $C_G(u) = C_G(u^{-1})$, hence if (u, e) is a Brauer element, then so is (u^{-1}, e), and if (u, e) over a set of representatives of the G-conjugacy classes of (G, b)-Brauer elements, then so does (u^{-1}, e). We collect the basic properties of the functions $\chi^{(u,e)}$; this is the theme of *subsections* in Brauer's terminology, developed in [27]. For simplicity, in what follows, we write $d^{(u,e)}$ instead of $d_G^{(u,e)}$.

Proposition 8.9.2   *Let G be a finite group and b a block of $\mathcal{O}G$. Let $\chi, \psi : G \to K$ be class functions associated with b. The following hold.*

(i) *For any (G, b)-Brauer element (u, e) the class function $\chi^{(u,e)}$ is associated with b.*

(ii) *For any two (G, b)-Brauer elements (u, e), (v, f) that are G-conjugate we have $\chi^{(u,e)} = \chi^{(v,f)}$.*

(iii) *For any two (G, b)-Brauer elements (u, e), (v, f) that are not G-conjugate we have*

$$\langle \chi^{(u,e)}, \psi^{(v^{-1},f)} \rangle = 0.$$

(iv) *Let (u, e) be a (G, b)-Brauer element. We have*

$$\langle \chi^{(u,e)}, \psi^{(u^{-1},e)} \rangle = \langle d^{(u,e)}(\chi), d^{(u^{-1},e)}(\psi) \rangle'_{C_G(u)}.$$

(v) *We have $\chi = \sum_{(u,e)} \chi^{(u,e)}$, where (u, e) runs over a set of representatives of the G-conjugacy classes of (G, b)-Brauer elements.*

(vi) *We have $\langle \chi, \psi \rangle = \sum_{(u,e)} \langle \chi^{(u,e)}, \psi^{(u^{-1},e)} \rangle$, where (u, e) runs over a set of representatives of the G-conjugacy classes of (G, b)-Brauer elements.*

Proof Since $\chi^{(u,e)} = (t_G^u \circ d_G^{(u,e)})(\chi)$ it follows from 6.13.2 that $\chi^{(u,e)}$ is associated with b. This proves (i). Statement (ii) is trivial. Statement (iii) is trivial if u and v are not conjugate. If u, v are conjugate, but (u, e), (v, f) are not, we may assume (after possibly replacing (v, f) by a conjugate) that $u = v$ and $e \neq f$. Using 5.15.14 and the fact that $d^u \circ t^u$ is the identity on $\mathrm{Cl}_K(C_G(u)_{p'})$ we get that

$$\langle \chi^{(u,e)}, \psi^{(u,f)} \rangle = \langle t^u(d^{(u,e)}(\chi)), t^u(d^{(u,f)}(\psi)) \rangle = \langle d^{(u,e)}(\chi), d^{(u,f)}(\psi) \rangle' = 0$$

where the last equation is from 6.5.2. The same argument with $f = e$ yields (iv). For statement (v), observe first that if e, e' are different blocks of $kC_G(u)$ such that (u, e), (u, e') are (G, b)-Brauer elements, then (u, e) and (u, e') are not G-conjugate. Indeed, if $(u, e') = {}^x(u, e)$ for some $x \in G$ then $x \in C_G(u)$. But then x fixes any block of $kC_G(u)$, hence $e' = e$. Let $\hat{b}$ be the central idempotent in $\mathcal{O}C_G(u)$ that lifts $\mathrm{Br}_Q(b)$. Thus $b - \hat{b} \in \ker(\mathrm{Br}_Q)$, and hence $\chi(x) = \chi(xb) = \chi(x\hat{b})$ by 5.12.16. Now $\hat{b}$ lifts the sum of all blocks e of $kC_G(u)$ such that (u, e) is a (G, b)-Brauer element. Thus $\chi(x) = \sum_e \chi^{(u,e)}(x)$, where e runs over the blocks of $kC_G(u)$ such that (u, e) is a (G, b)-Brauer pair. Since $\chi^{(v,f)}(x) = 0$ for any (G, b)-Brauer pair (v, f) such that v is not conjugate to u we get that (v) holds. Combining (iii) and (iv) yields (vi). □

Proposition 8.9.3 *Let G be a finite group and b a block of $\mathcal{O}G$. Let $\chi : G \to K$ be a class function associated with b. For any (G, b)-Brauer elements (u, e) and (v, f) we have $(\chi^{(u,e)})^{(u,e)} = \chi^{(u,e)}$, and if (u, e), (v, f) are not G-conjugate, then $(\chi^{(u,e)})^{(v,f)} = 0$.*

Proof The class function $(\chi^{(u,e)})^{(v,f)}$ vanishes on group elements whose p-part is not conjugate to v. Let $x \in G$ such that the p-part of x is equal to v; that is, $x = vs$ for some $s \in C_G(v)_{p'}$. We have $(\chi^{(u,e)})^{(v,f)}(x) = \chi^{(u,e)}(vs\hat{f})$. By 5.15.17 we have $s\hat{f} = a + c$, where a is a linear combination of p'-elements in $C_G(v)$ and where c is an additive commutator. Note that vc is still an additive commutator, as v is in the centre of $C_G(v)$. Since central functions vanish on additive commutators, it follows that the previous expression is equal to $\chi^{(u,e)}(va)$. If v is not conjugate to u, this is zero. If $v = u$, this expression in equal to $\chi(ua\hat{e})$. By reversing the above argument, replacing a by $a + c$, this expression is also equal to $\chi(us\hat{f}\hat{e})$, so zero unless $\hat{f} = \hat{e}$, in which case it is equal to $\chi(us\hat{e}) = \chi^{(u,e)}(us)$, whence the result. □

Proposition 8.9.4 ([27, (5C)]) *Let G be a finite group, b a block of $\mathcal{O}G$ and (u, e) a (G, b)-Brauer element. We have*

$$\sum_{\chi \in \mathrm{Irr}_K(G,b)} \langle \chi^{(u,e)}, \chi^{(u^{-1},e)} \rangle = \ell(kC_G(u)e).$$

Proof By 8.9.2 (iv), the sum in the statement is equal to the sum

$$\sum_{\chi \in \mathrm{Irr}_K(G,b)} \langle d^{(u,e)}(\chi), d^{(u^{-1},e)}(\chi) \rangle'$$

where this is a scalar product of class functions on $C_G(u)_{p'}$. By 6.13.6, we have

$$d^{(u,e)}(\chi) = \sum_{\varphi \in \mathrm{IBr}_k(C_G(u),e)} d^{(u,e)}_{\chi,\varphi} \varphi.$$

Together with the corresponding expression for $d^{(u^{-1},e)}(\chi)$ it follows that the previous sum is equal to the sum

$$\sum_{\chi,\varphi,\psi} d^{(u,e)}_{\chi,\varphi} \langle \varphi, \psi \rangle' d^{(u^{-1},e)}_{\chi,\psi},$$

where χ runs over $\mathrm{Irr}_K(G,b)$, and φ, ψ run over $\mathrm{IBr}_k(C_G(u),e)$. Denote by $c(\varphi,\psi)$ the entries of the Cartan matrix C_u of $kC_G(u)e$ and by $c'(\varphi,\psi)$ the entries of its inverse matrix. By 5.14.11 we have

$$\langle \varphi, \psi \rangle' = c'(\varphi,\psi)$$

and by 6.13.8 we have

$$\sum_{\chi} d^{(u,e)}_{\chi,\varphi} d^{(u^{-1},e)}_{\chi,\psi} = c(\varphi,\psi).$$

Together we get that

$$\sum_{\chi} \langle \chi^{(u,e)}, \chi^{(u^{-1},e)} \rangle = \sum_{\varphi,\psi} c(\varphi,\psi) c'(\varphi,\psi).$$

This is the trace of the matrix $C_u \cdot C_u^{-1} = \mathrm{Id}$, hence this trace is equal to $\ell(kC_G(u)e)$. □

8.10 Characters and local structure

The $*$-construction, due to Broué and Puig [40], associates with a character in a block and a fusion stable character of a defect group another generalised character of the block. The main ingredients used for this construction are the functions $\chi^{(u,e)}$ from the previous section and Brauer's characterisation of characters. Let p be a prime, and let $\mathcal{O}$ be a complete discrete valuation ring having a residue field $k = \mathcal{O}/J(\mathcal{O})$ of characteristic p and a field of fractions K of characteristic 0. We assume that k and K are splitting fields for all finite groups and their subgroups in this section.

Definition 8.10.1 Let $\mathcal{F}$ be a fusion system on a finite p-group P. A class function $\eta : P \to K$ is called $\mathcal{F}$-*stable* if $\eta(u) = \eta(\varphi(u))$ for any $u \in P$ and any $\varphi \in \mathrm{Hom}_{\mathcal{F}}(\langle u \rangle, P)$.

Definition 8.10.2 ([40, 2.4]) Let G be a finite group, b a block of $\mathcal{O}G$ and (P, e_P) a maximal (G, b)-Brauer pair. Denote by $\mathcal{F} = \mathcal{F}_{(P,e_P)}(G, b)$ the associated fusion system on P. For any class function $\chi : G \to K$ and any $\mathcal{F}$-stable class function $\eta : P \to K$ we define a class function $\eta * \chi : G \to K$ by

$$\eta * \chi = \sum_{(u,e)} \eta(u) \chi^{(u,e)}$$

where (u, e) runs over a set of representatives of the G-conjugacy classes of (G, b)-Brauer elements contained in (P, e_P), and where $\chi^{(u,e)}$ is the function defined in 8.9.1.

The $\mathcal{F}$-stability assumption is necessary for $\eta * \chi$ to be independent of the choice of (u, e) in its conjugacy class. Indeed, if (u, e) and (v, f) are (G, b)-Brauer elements contained in (P, e_P) and if there is an element $x \in G$ such that $(v, f) = {}^x(u, e)$, then conjugation by x induces an isomorphism $\langle u \rangle \cong \langle v \rangle$ which belongs to the fusion system $\mathcal{F}$ and which sends u to v. Thus $\eta(v) = \eta(u)$, and hence $\eta(v) \chi^{(v,f)} = \eta(u) \chi^{(u,e)}$. Using this definition to calculate $\eta * \chi$ on group elements requires some care. We have $(\eta * \chi)(x) = 0$ if the p-part of x is not G-conjugate to an element in P. If the p-part of x is equal to an element $u \in P$, then

$$(\eta * \chi)(x) = \sum_{(v,f)} \eta(v) \chi({}^{y(v,f)}x\hat{f})$$

where (v, f) runs over a set of representatives of the G-conjugacy classes of (G, b)-Brauer elements contained in (P, e_P) such that there exists an element $y(v, f) \in G$ satisfying ${}^{y(v,f)}u = v$. Here $\hat{f} \in Z(\mathcal{O}C_G(v))$ is the block idempotent that lifts f. Note that the values $\eta(u)$, $\eta(v)$ may be different since even though u, v are G-conjugate, there need not be an isomorphism $\langle u \rangle \cong \langle v \rangle$ in the fusion system $\mathcal{F}$ sending u to v. The property (iv) in 8.9.2 means that $1 * \chi = \chi$, where 1 is here the trivial character of P. The group basis does not sit well with the block decomposition of $\mathcal{O}G$. The above formula for $\eta * \chi$ is better suited for calculating its values at local pointed elements, because these are partitioned according to the block decomposition of $\mathcal{O}G$; this is the point of view taken in [169, §5].

Proposition 8.10.3 *Let G be a finite group, b a block of $\mathcal{O}G$ and (P, e_P) a maximal (G, b)-Brauer pair. Denote by $\mathcal{F} = \mathcal{F}_{(P,e_P)}(G, b)$ the associated fusion* *system on P. Let* $\chi : G \to K$ *be a class function, and let* $\eta : P \to K$ *be an* $\mathcal{F}$-*stable class function. Let* (u, e) *be a* (G, b)-*Brauer element contained in* (P, e_P). *We have*

$$(\eta * \chi)^{(u,e)} = \eta(u)\chi^{(u,e)} = \eta * \chi^{(u,e)}.$$

Proof We have $\eta * \chi^{(u,e)} = \sum_{(v,f)} \eta(v)(\chi^{(u,e)})^{(v,f)}$, where (v, f) runs over a set of representatives of the conjugacy classes of (G, b)-Brauer elements. By 8.9.3, all summands are zero except the summand indexed by (u, e), which is equal to $\eta(u)\chi^{(u,e)}$. Similarly, we have $(\eta * \chi)^{(u,e)} = \sum_{(v,f)} \eta(v)(\chi^{(v,f)})^{(u,e)}$. Using 8.9.3 again, this expression is also equal to $\eta(u)\chi^{(u,e)}$. □

Corollary 8.10.4 *With the notation of 8.10.3, let* j *be an idempotent in* $(\mathcal{O}G)^{(u)}$ *such that* $\mathrm{Br}_{(u)}(j) \in kC_G(u)e$. *Then*

$$(\eta * \chi)(uj) = \eta(u)\chi(uj).$$

In particular, for any idempotent i *in* $\mathcal{O}Gb$ *we have*

$$(\eta * \chi)(i) = \eta(1)\chi(i)$$

and for any p'-*element* x *in* G *we have*

$$(\eta * \chi)(xb) = \eta(1)\chi(xb).$$

Proof By 5.12.16 we may replace j by an idempotent $j' \in \mathcal{O}C_G(u)\hat{e}$ whose image in $kC_G(u)e$ is equal to $\mathrm{Br}_{(u)}(j)$. Here $\hat{e}$ is the unique block of $\mathcal{O}C_G(u)$ that lifts e. By 5.15.16 we can write $j' = a + c$ with a a linear combination of p'-elements in $C_G(u)$ and c an additive commutator. Note that the space of additive commutators is a module over the centre of an algebra. We have $uj' = uj'\hat{e} = ua\hat{e} + uc\hat{e}$, and $uc\hat{e}$ belongs still to the additive commutator space of $\mathcal{O}C_G(u)\hat{e}$. Thus we may replace uj' on both sides by $ua\hat{e}$. Then the left side in the statement is equal to $(\eta * \chi)^{(u,e)}$ evaluated at ua, and the right side is $\eta * \chi^{(u,e)}$ evaluated at ua. By 8.10.3, these are equal. If $u = 1$, then $e = \bar{b}$, the image of b in kG, whence the second equation. The last equation is immediate from the definition of $\eta * \chi$, and is in fact equivalent to the second equation using 5.15.16. □

Theorem 8.10.5 ([40, 2.6]) *Let* G *be a finite group,* b *a block of* $\mathcal{O}G$ *and* (P, e_P) *a maximal* (G, b)-*Brauer pair. Denote by* $\mathcal{F} = \mathcal{F}_{(P,e_P)}(G, b)$ *the fusion system determined by* (P, e_P) *on* P. *For any generalised character* χ *of* G *with values in* K *associated with* b *and any* $\mathcal{F}$-*stable generalised character* η *of* P *with values in* K *the class function* $\eta * \chi$ *is a generalised character of* G *with values in* K *associated with* b.

Both the original proof in [40] as well as the approach in [169, §5] use Brauer's characterisation of characters, and so will the proof we present here. An elementary subgroup, of a finite group G, as defined in 3.7.1, and regardless for which prime, can always be written in the form QH for some p-group Q and some p'-subgroup of $C_G(Q)$. The generalised characters of groups of this form can be described as follows.

Lemma 8.10.6 *Let G be a finite group, Q a p-subgroup of G and H a p'-subgroup of $C_G(Q)$. A class function $\lambda : QH \to K$ is a generalised character of QH if and only if for any primitive idempotent $j \in OH$ the map sending $u \in Q$ to $\lambda(uj)$ is a generalised character of Q.*

Proof The blocks of OH correspond to the characters of H; that is, if $\mu \in \mathrm{Irr}_K(H)$, then $OHe(\mu)$ is a matrix algebra of rank $\mu(1)^2$, and $OQHe(\mu) \cong OQ \otimes_O OHe(\mu)$ is a matrix algebra over OQ. In order to check whether λ is a generalised character, we may proceed blockwise; that is, we may assume that λ is associated with a block $OQHe(\mu)$ of OQH. Let j be a primitive idempotent in the matrix algebra $OHe(\mu)$. Any character of QH associated with $e(\mu)$ is of the form $\theta\mu$ for some $\theta \in \mathrm{Irr}(Q)$. Multiplication by j yields a Morita equivalence between $OQHe(\mu)$ and OQ through which $\theta\mu$ corresponds to θ. This Morita equivalence yields hence an isomorphism between the groups of generalised characters of Q and of QH associated with $e(\mu)$. The character μ of H is the trace map on the matrix algebra $OHe(\mu)$, hence sends j to 1. The result follows. $\qquad\square$

Proof of Theorem 8.10.5 By Brauer's characterisation of characters, it suffices to show that the restriction of $\eta * \chi$ to any subgroup of the form QH is a generalised character, where Q is a p-subgroup of G and H a p'-subgroup of $C_G(Q)$. By the previous lemma, it suffices to show that the map $u \mapsto (\eta * \chi)(uj)$, with $u \in Q$, is a generalised character of Q, for any primitive idempotent $j \in OH$. By decomposing j in $OC_G(Q)$, we may replace j by a primitive idempotent in $OC_G(Q)$. After possibly conjugating Q by a suitable element in G we may assume that j belongs to the unique block $\hat{e}_Q$ which lifts the block e_Q of $kC_G(Q)$ satisfying $(Q, e_Q) \le (P, e_P)$. By 5.12.16, we may replace j by a primitive idempotent $\hat{j}$ belonging to a local point ϵ of Q on OG. The uniqueness property of the inclusion of Brauer pairs as in 5.9.6 implies that if $u \in Q$ and if e is the block of $kC_G(u)$ such that $(u, e) \le (P, e_P)$, then $(u, e) \le (Q, e_Q)$, and we have $\mathrm{Br}_{\langle u \rangle}(\hat{j}) \in kC_G(u)e$. Therefore, by 8.10.4, for all $u \in Q$, we have $(\eta * \chi)(u\hat{j}) = \eta(u)\chi(u\hat{j})$. The right side is the product of two generalised characters of Q, and hence the left side is a generalised character of Q, as required. $\qquad\square$

Corollary 8.10.7 ([40, Corollary]) *Let G be a finite group, b a block of $\mathcal{O}G$ and (P, e_P) a maximal (G, b)-Brauer pair. Denote by $\mathcal{F} = \mathcal{F}_{(P,e_P)}(G, b)$ the fusion system determined by (P, e_P) on P. Let $\chi \in \mathrm{Irr}_K(Gb)$ and let η be an $\mathcal{F}$-stable character of P of degree 1. We have $\eta * \chi \in \mathrm{Irr}_K(G, b)$ and $\eta * \chi - \chi \in L^0(G, b)$; in particular, χ and $\eta * \chi$ have the same degree.*

Proof By 8.10.5, $\eta * \chi$ is a generalised character associated with b. It follows from 8.9.2 that $1 = \langle \chi, \chi \rangle = \sum_{(u,e)} \langle \chi^{(u,e)}, \chi^{(u,e)} \rangle$, where (u, e) runs over a set of representatives of the (G, b)-Brauer elements. Similarly, we have $\langle \eta * \chi, \eta * \chi \rangle = \sum_{(u,e)} \langle \eta(u)\chi^{(u,e)}, \eta(u)\chi^{(u,e)} \rangle$, with (u, e) as before. This sum is equal to $\sum_{(u,e)} \eta(u)\eta(u^{-1})\langle \eta(u)\chi^{(u,e)}, \eta(u)\chi^{(u,e)} \rangle$. Since η has degree 1, we have $\eta(u^{-1}) = \eta(u)^{-1}$, and hence this sum is equal to $\langle \chi, \chi \rangle = 1$. This shows that $\eta * \chi$ or its negative is an irreducible character. By 8.10.4, the class functions χ and $\eta * \chi$ coincide on p'-elements, an hence their difference is in $L^0(G, b)$. In particular, χ, $\eta * \chi$ have the same degree, showing that $\eta * \chi$ is irreducible. $\qquad\square$

With the notation of the previous corollary, a linear character $\eta : P \to \mathcal{O}^\times$ is $\mathcal{F}$-stable if and only if $\ker(\eta)$ contains the $\mathcal{F}$-focal subgroup $\mathfrak{foc}(\mathcal{F})$. Thus the set of $\mathcal{F}$-stable linear characters of P forms a group that can be identified with the group of linear characters $\mathrm{Hom}(P/\mathfrak{foc}(\mathcal{F}), \mathcal{O}^\times)$. Since $P/\mathfrak{foc}(\mathcal{F})$ is abelian and since K is large enough, the group $\mathrm{Hom}(P/\mathfrak{foc}(\mathcal{F}), \mathcal{O}^\times)$ is the dual group of $P/\mathfrak{foc}(\mathcal{F})$, hence (noncanonically) isomorphic to $P/\mathfrak{foc}(\mathcal{F})$. Corollary 8.10.7 implies that the group $\mathrm{Hom}(P/\mathfrak{foc}(\mathcal{F}), \mathcal{O}^\times)$ acts on $\mathrm{Irr}_K(G, b)$ via $(\eta, \chi) \mapsto \eta * \chi$. By 8.10.4, this action preserves character degrees, and hence in particular, permutes the sets of characters with a fixed height. This action is faithful; it is in fact free on the subset of height zero characters.

Theorem 8.10.8 *Let G be a finite group, b a block of $\mathcal{O}G$ and (P, e_P) a maximal (G, b)-Brauer pair. Denote by $\mathcal{F} = \mathcal{F}_{(P,e_P)}(G, b)$ the fusion system determined by (P, e_P) on P. The map sending an $\mathcal{F}$-stable linear character η of P and an irreducible character $\chi \in \mathrm{Irr}(G, b)$ to $\eta * \chi$ induces a free action of the group $\mathrm{Hom}(P/\mathfrak{foc}(\mathcal{F}), \mathcal{O}^\times)$ on the set of characters of height zero in $\mathrm{Irr}_K(G, b)$. In particular, the number of height zero characters in $\mathrm{Irr}(G, b)$ is divisible by $|P/\mathfrak{foc}(\mathcal{F})|$.*

Proof Let $i \in (\mathcal{O}G)^P$ be a source idempotent of b associated with e_P; that is, $\mathrm{Br}_P(i) \in kC_G(P)e_P$. Let η be a nontrivial $\mathcal{F}$-stable character of P, and let $\chi \in \mathrm{Irr}_K(G, b)$ such that $\mathrm{ht}(\chi) = 0$. We need to show that $\eta * \chi \neq \chi$. Let $u \in P$ such that $\eta(u) \neq 1$. Let e be the unique block of $kC_G(u)$ such that (u, e) is a (G, b)-Brauer element contained in (P, e_P). By 5.9.6 we have $\mathrm{Br}_{\langle u \rangle}(i) \in kC_G(u)e$.

Thus 8.10.4 applies, showing that $(\eta * \chi)(ui) = \eta(u)\chi(ui)$. By 6.11.12 we have $\chi(ui) \neq 0$. Since $\eta(u) \neq 1$ this shows that $\eta * \chi \neq \chi$. $\qquad\square$

For characters with arbitrary heights we have the following result, due to Robinson.

Theorem 8.10.9 ([192, Theorem 2]) *Let G be a finite group, b a block of $\mathcal{O}G$ and (P, e_P) a maximal (G, b)-Brauer pair. Denote by $\mathcal{F} = \mathcal{F}_{(P,e_P)}(G, b)$ the fusion system determined by (P, e_P) on P. The length of any orbit of the action of $\mathrm{Hom}(P/\mathfrak{foc}(\mathcal{F}), \mathcal{O}^\times)$ on the set $\mathrm{Irr}_K(G, b)$ via the $*$-construction is divisible by $|Z(P) : Z(P) \cap \mathfrak{foc}(\mathcal{F})|$. In particular, for any nonnegative integer h, the number of characters in $\mathrm{Irr}(G, b)$ of height h is divisible by $|Z(P) : Z(P) \cap \mathfrak{foc}(\mathcal{F})|$.*

Proof Let i be a source idempotent in $(\mathcal{O}Gb)^\Gamma$ such that $\mathrm{Br}_P(i) \in kC_G(P)e_P$. Let $\chi \in \mathrm{Irr}_K(B)$ and $\lambda \in \mathrm{Hom}(P/\mathfrak{foc}(\mathcal{F}), \mathcal{O}^\times)$ such that $\lambda * \chi = \chi$. Let $u \in Z(P)$. Using 8.10.4 as in the previous proof, we have $\chi(ui) = (\lambda * \chi)(ui) = \lambda(u)\chi(ui)$. Since $\chi(ui) \neq 0$ by 6.11.10, it follows that $\lambda(u) = 1$. Thus the stabiliser of χ in $\mathrm{Hom}(P/\mathfrak{foc}(\mathcal{F}), \mathcal{O}^\times)$ is contained in the subgroup $\mathrm{Hom}(P/Z(P)\mathfrak{foc}(\mathcal{F}), \mathcal{O}^\times)$ of index $\frac{|P:\mathfrak{foc}(\mathcal{F})|}{|P:Z(P)\mathfrak{foc}(\mathcal{F})|} = |Z(P) : Z(P) \cap \mathfrak{foc}(\mathcal{F})|$. It follows that every orbit of the action of $\mathrm{Hom}(P/\mathfrak{foc}(\mathcal{F}), \mathcal{O}^\times)$ on $\mathrm{Irr}_K(G, b)$ has length at least $|Z(P) : Z(P) \cap \mathfrak{foc}(\mathcal{F})|$. Since all orbit lengths are powers of p, they are actually divisible by $|Z(P) : Z(P) \cap \mathfrak{foc}(\mathcal{F})|$. $\qquad\square$

Corollary 8.10.10 *With the notation of 8.10.9, if P is abelian, then $\mathrm{Hom}(P/\mathfrak{foc}(\mathcal{F}), \mathcal{O}^\times)$ acts freely on $\mathrm{Irr}_K(G, b)$. In particular, $|\mathrm{Irr}_K(G, b)|$ is divisible by $|P : \mathfrak{foc}(\mathcal{F})|$.*

Proof This is an immediate consequence of 8.10.9. $\qquad\square$

Modulo Brauer's height zero conjecture, 8.10.10 would also follow from 8.10.8.

8.11 Nilpotent blocks

Let p be a prime, and let $\mathcal{O}$ be a complete discrete valuation ring having a residue field $k = \mathcal{O}/J(\mathcal{O})$ of characteristic p and a field of fractions K. We allow the case $\mathcal{O} = k$ unless stated otherwise. Nilpotent blocks were introduced by Broué and Puig, generalising a description, due to Frobenius, of p-nilpotent finite groups to blocks of finite groups.

Definition 8.11.1 ([41]) Let G be a finite group, b a block of $\mathcal{O}G$ with maximal (G, b)-Brauer pair (P, e) and associated fusion system $\mathcal{F} = \mathcal{F}_{(P,e)}(G, b)$. The block b is called *nilpotent* if $\mathcal{F} = \mathcal{F}_P(P)$.

Any block with a central defect group is trivially nilpotent. Since the definition of a nilpotent block refers only to its fusion system, a block b of $\mathcal{O}G$ is nilpotent if and only if its image $\bar{b}$ in kG is a nilpotent block. The following criterion for a block to be nilpotent is the original definition of nilpotent blocks in [41].

Proposition 8.11.2 *Let G be a finite group and b a block of $\mathcal{O}G$ with maximal (G, b)-Brauer pair (P, e). Suppose that k is a splitting field for the block e of $kC_G(P)$. Then the block b is nilpotent if and only if $N_G(Q, f)/C_G(Q)$ is a p-group for any (G, b)-Brauer pair (Q, f).*

Proof The category $\mathcal{F} = \mathcal{F}_{(P,e)}(G, b)$ is a fusion system by 8.5.2; we use here the assumption that k is large enough. By Alperin's Fusion Theorem, $\mathcal{F}$ is completely determined by the automorphism groups $\mathrm{Aut}_{\mathcal{F}}(Q)$ of fully $\mathcal{F}$-normalised subgroups Q of P. For any such Q, the subgroup $\mathrm{Aut}_P(Q)$ is a Sylow p-subgroup of $\mathrm{Aut}_{\mathcal{F}}(Q)$. Thus, for any fully $\mathcal{F}$-normalised subgroup Q of P, the group $\mathrm{Aut}_{\mathcal{F}}(Q) \cong N_G(Q, f)/C_G(Q)$ is a p-group if and only if $\mathrm{Aut}_{\mathcal{F}}(Q) = \mathrm{Aut}_P(Q)$. Since every subgroup of P is isomorphic, in $\mathcal{F}$, to a fully $\mathcal{F}$-normalised subgroup, the result follows. $\square$

The fusion system of the principal block of $\mathcal{O}G$ is equal to that of the finite group G on a Sylow p-subgroup of G. Thus the principal block of $\mathcal{O}G$ is nilpotent if and only if $N_G(Q)/C_G(Q)$ is a p-group for all p-subgroups Q of G.

Proposition 8.11.3 *Let G be a finite group and b a block of $\mathcal{O}G$ with maximal (G, b)-Brauer pair (P, e). Suppose that k is a splitting field for the block e of $kC_G(P)$. Assume that P is abelian. Then the block b is nilpotent if and only if the inertial quotient $E = N_G(P, e)/C_G(P)$ of b is trivial.*

Proof It follows from 8.1.14 or 8.5.7 that the fusion system of b on P determined by (P, e) is equal to $\mathcal{F}_P(P \rtimes E)$. Thus b is nilpotent if and only if E is trivial. $\square$

The property of being nilpotent passes down to Brauer pairs:

Proposition 8.11.4 *Let G be a finite group and b a nilpotent block of $\mathcal{O}G$ with maximal (G, b)-Brauer pair (P, e). Suppose that k is a splitting field for all subgroups of G. For any (G, b)-Brauer pair (Q, f), the block f of $kC_G(Q)$ is nilpotent. Moreover, if $(Q, f) \leq (P, e)$ then $C_P(Q)$ is a defect group of $kC_G(Q)f$.*

Proof The property of being nilpotent is invariant under conjugation with elements in G, and every (G, b)-Brauer pair is conjugate to a Brauer pair contained in (P, e). Thus we may assume $(Q, f) \leq (P, e)$. By the assumptions, the fusions system $\mathcal{F} = \mathcal{F}_{(P,e)}(G, b)$ is equal to the trivial fusion system $\mathcal{F}_P(P)$. In particular, every subgroup of P is fully $\mathcal{F}$-centralised. It follows from 8.5.4 (ii) that $C_P(Q)$ is a defect group of f and that the fusion system of f on $C_P(Q)$ is equal to $\mathcal{F}_{C_P(Q)}(C_P(Q))$. Equivalently, the block f of $kC_G(Q)$ is nilpotent. □

The main result of this section, due to Puig [170], describes the structure of the source algebras of nilpotent blocks:

Theorem 8.11.5 ([170]) *Let G be a finite group, b a nilpotent block of $\mathcal{O}G$ with defect group P, and denote by $\bar{b}$ the image of b in kG. Suppose that k is a splitting field for all subgroups of G. Then the following hold.*

(i) *Any source algebra of b is of the form $S \otimes_{\mathcal{O}} \mathcal{O}P$, where $S = \mathrm{End}_{\mathcal{O}}(V)$ for some indecomposable endopermutation $\mathcal{O}P$-module V with vertex P and determinant 1.*

(ii) *As an $\mathcal{O}$-algebra, $\mathcal{O}Gb$ is isomorphic to a matrix algebra over $\mathcal{O}P$; in particular, $\mathcal{O}Gb$ and $\mathcal{O}P$ are Morita equivalent.*

(iii) *The k-algebra $kG\bar{b}$ has a unique isomorphism class of simple modules, or equivalently, a unique irreducible Brauer character φ, and the Cartan matrix of $kG\bar{b}$ consists of the single entry $|P|$.*

(iv) *Any simple $kG\bar{b}$-module has P as a vertex and $k \otimes_{\mathcal{O}} V$ as a source.*

(v) *If $\mathrm{char}(K) = 0$, then b has an ordinary irreducible character χ such that $d_G(\chi) = \varphi$; equivalently, there is an $\mathcal{O}$-free $\mathcal{O}Gb$-module X such that $k \otimes_{\mathcal{O}} X$ is a simple $kG\bar{b}$-module.*

The statements (ii), (iii), (iv), (v) are immediate consequences of statement (i) in this theorem, because block algebras are Morita equivalent to their source algebras, in a way that preserves vertices and sources of modules. The interior P-algebra structure of the source algebra $S \otimes_{\mathcal{O}} \mathcal{O}P$ is given by the group homomorphism $P \to (S \otimes_{\mathcal{O}} \mathcal{O}P)^{\times}$ which sends $y \in P$ to $\sigma(y) \otimes y$, where $\sigma(y)$ is the endomorphism of V given by the action of y on V; that is, $\sigma(y)(v) = yv$ for all $y \in P$ and $v \in V$. Note that by 7.4.5, the unit element of $S \otimes_{\mathcal{O}} \mathcal{O}P$ remains primitive in $(S \otimes_{\mathcal{O}} \mathcal{O}P)^{\Delta P}$, where the ΔP-fixed points are understood as fixed points with respect to the conjugation action of the image of P under the group homomorphism $P \to (S \otimes_{\mathcal{O}} \mathcal{O}P)^{\times}$ described above.

Corollary 8.11.6 *Let G be a finite group and P a Sylow p-subgroup of G. The principal block b of $\mathcal{O}G$ is nilpotent if and only if $\mathcal{O}Gb \cong \mathcal{O}P$ as interior P-algebras.*

Proof If the principal block is nilpotent, then its block algebra and any of its source algebras are isomorphic to matrix algebras over $\mathcal{O}P$. In particular, $\mathcal{O}Gb \cong M_n(\mathcal{O}P)$ for some positive integer n. But the principal block algebra $\mathcal{O}Gb$ also has a character of degree 1, which forces $n = 1$. Thus b remains primitive in $(\mathcal{O}Gb)^{\Delta P}$, and hence $\mathcal{O}Gb$ is isomorphic to its source algebra $\mathcal{O}P$ as an interior P-algebra. Conversely, if $\mathcal{O}Gb \cong \mathcal{O}P$ is an isomorphism of interior P-algebras, then $\mathcal{O}P$ is a source algebra of b, and hence the fusion system of G on P is the trivial fusion system $\mathcal{F}_P(P)$ by 8.7.1. The result follows. $\square$

A finite group G is called *p-nilpotent* if $G = N \rtimes P$ for some normal p'-subgroup N of G and a Sylow p-subgroup P of G, or equivalently, if the largest normal p'-subgroup $O_{p'}(G)$ of G is equal to the smallest normal subgroup $O^p(G)$ of G having p-power index in G. Frobenius' characterisation of p-nilpotent finite groups, can be proved as a special case of the structure theorem on nilpotent blocks (which motivates the use of the term 'nilpotent' for these blocks).

Corollary 8.11.7 *(Frobenius) Let G be a finite group. The following are equivalent.*

(i) G is p-nilpotent.
(ii) For any p-subgroup Q of G, the group $N_G(Q)/C_G(Q)$ is a p-group.

Proof The principal block of $\mathcal{O}G$ is nilpotent if and only if $N_G(Q)/C_G(Q)$ is a p-group for all p-subgroups Q of G. Suppose that (ii) holds. Then 8.11.6 implies that $\mathcal{O}Gb \cong \mathcal{O}P$, where P is a Sylow p-subgroup of G. Let N be the kernel of the canonical group homomorphism $G \to (\mathcal{O}Gb)^{\times} \cong (\mathcal{O}P)^{\times}$. The isomorphism $\mathcal{O}Gb \cong \mathcal{O}P$ induces a bijection $\mathrm{Irr}_K(G, b) \cong \mathrm{Irr}_K(P)$ which sends the trivial character of G to the trivial character of P. Thus the induced homomorphism $\mathcal{O}G \to \mathcal{O}P$ sends the augmentation ideal $I(\mathcal{O}G)$ to the augmentation ideal $I(\mathcal{O}P)$ of $\mathcal{O}P$. It follows that if $x \in G$, then the image of xb in $\mathcal{O}P$ belongs to $1 + I(\mathcal{O}P)$. Thus if x has order prime to p, then 4.1.8 (v) implies that the image of xb in $\mathcal{O}P$ is equal to 1. This shows that N contains all p'-elements of G but no nontrivial element of P, and hence $N = O_{p'}(G)$, which implies that (i) holds. The converse implication is an easy group theoretic exercise. $\square$

Corollary 8.11.8 *Let G be a finite group. The following are equivalent.*

(i) G is p-nilpotent.
(ii) The principal block of kG is nilpotent.
(iii) The algebra kG has a nilpotent block of principal type with a Sylow p-subgroup of G as a defect group.

Proof The statements (ii) and (iii) are both equivalent to the statement that $N_G(Q)/C_G(Q)$ is a p-group for any p-subgroup Q of G, hence to (i) by the previous corollary. □

Before embarking on the proof of 8.11.5, we state the following characterisation of nilpotent blocks (which is a consequence of 8.11.5 and later results on Morita equivalences given by a bimodule with endopermutation source):

Theorem 8.11.9 *Let G be a finite group and b a nilpotent block of $\mathcal{O}G$ with defect group P. Suppose that k is a splitting field for all subgroups of G. The block b is nilpotent if and only if there is a Morita equivalence between $\mathcal{O}Gb$ and $\mathcal{O}P$ given by an $\mathcal{O}Gb$-$\mathcal{O}P$-bimodule M with the property that if M is viewed as an $\mathcal{O}(G \times P)$-module, then M has ΔP as a vertex and an endopermutation $\mathcal{O}\Delta P$-module V with determinant 1 as a source.*

Proof Combined with the structure theorem 8.11.5, this is a special case of Theorem 9.11.5, which describes the connections between the algebra structure of Morita equivalent block algebra, where the Morita equivalence is given by a bimodule with 'diagonal' vertex and endopermutation source. □

It has been shown by Puig in [177, Theorem 8.2] that if the characteristic of $\mathcal{O}$ is zero, then a block is nilpotent if and only if it is Morita equivalent to $\mathcal{O}P$. A key step in the proof of 8.11.5 is the following statement, which implies that 8.11.5 holds if $\mathcal{O} = k$. The proof we present here is based on the lifting theorem 4.8.2 due to Külshammer, Okuyama, and Watanabe.

Proposition 8.11.10 *Let G be a finite group, b a nilpotent block of $\mathcal{O}G$ with a defect group P, and let $i \in (\mathcal{O}Gb)^{\Delta P}$ be a source idempotent of b. Set $A = i\mathcal{O}Gi$ and $B = \mathcal{O}P \otimes_{\mathcal{O}} A$, viewed as an interior P-algebra via the diagonal map from P to $B^{\times}$. Denote by $\pi : B \to A$ the homomorphism of interior P-algebras mapping $u \otimes a$ to a, for any $u \in P$ and any $a \in A$. Suppose that there is an $\mathcal{O}$-free A-module V such that $k \otimes_{\mathcal{O}} V$ is a simple $k \otimes_{\mathcal{O}} A$-module. Set $S = \mathrm{End}_{\mathcal{O}}(V)$ and denote by $\sigma : A \to S$ the structural map.*

(i) There is a homomorphism of interior P-algebras $\tau : A \to B$ such that $\pi \circ \tau = \mathrm{Id}_A$.

(ii) The map $(\sigma \otimes \mathrm{Id}_{\mathcal{O}P}) \circ \tau$ is an isomorphism of interior P-algebras $A \cong S \otimes_{\mathcal{O}} \mathcal{O}P$.

(iii) The restriction to $\mathcal{O}P$ of V is an indecomposable endopermutation module with vertex P.

Proof The crucial point is that because b is nilpotent, every indecomposable direct summand of B as an $\mathcal{O}P$-$\mathcal{O}P$-bimodule is of the form $kP \otimes_{kQ} kP$ for some subgroup Q of P; we use here that by 8.7.1 the bimodule structure of a source algebra encodes its fusion system. Let X be a $P \times P$-stable $\mathcal{O}$-basis of A. Then the set $P \times X$ is a $P \times P$-stable basis of $B = kP \otimes_{\mathcal{O}} A$. The indecomposable direct summands of A as an $\mathcal{O}P$-$\mathcal{O}P$-bimodule correspond to the $P \times P$-orbits in X. Since any such summand is isomorphic to $\mathcal{O}P \otimes_{kQ} \mathcal{O}P$, the corresponding orbit in X contains an element x satisfying $ux = xu$ if and only if $u \in Q$. But then $1 \otimes x$ has the same property, hence generates a direct summand of B as an $\mathcal{O}P$-$\mathcal{O}P$-bimodule isomorphic to $\mathcal{O}P \otimes_{kQ} \mathcal{O}P$. This shows that π is split surjective as a homomorphism of $\mathcal{O}P$-$\mathcal{O}P$-bimodules. Since by theorem 6.4.7 the algebra A is relatively $\mathcal{O}P$-separable, theorem 4.8.2 applies and shows that actually π is split surjective as a homomorphism of interior P-algebras. This proves (i). It follows from Nakayama's Lemma that in order to prove (ii) we may assume that $\mathcal{O} = k$. Let V be a simple A-module. Then $kP \otimes_k V$ becomes a B-module. This can be viewed as an A-module via σ. Since the involved algebras and algebra homomorphisms are interior P-algebra homomorphisms, restricting this further to kP via the structural homomorphism $kP \to A$, we get a 'diagonal' action of $u \in P$ on $kP \otimes_k V$ by $uy \otimes iuv$, where $y \in kP$ and $v \in V$. That implies that $kP \otimes_k V$ is projective as module over kP, for this action. But then $kP \otimes V$ is projective as A-module by 6.4.8. A composition series of kP yields a composition series of $kP \otimes_k V$ as B-modules, hence as A-modules. Thus $kP \otimes_k V$ is a projective A-module all of whose composition factors are isomorphic to the simple A-module V. This forces V to be the only simple A-module, up to isomorphism, by 4.10.5, and thus $kP \otimes_k V$ is its projective cover, as as A-module. This is then also a progenerator, so the structural map $A \to \mathrm{End}_k(kP \otimes_k V)$ is injective. The structural map $B \to \mathrm{End}_k(kP \otimes_k V)$ has $kP \otimes J(A)$ in its kernel because V is a simple A-module. Thus the composition of algebra homomorphisms

$$A \to B = kP \otimes_k A \to kP \otimes_k A/J(A)$$

is injective. The quotient $A/J(A)$ is a matrix algebra $M_n(k)$, where $n = \dim_k(V)$, so comparing dimensions shows that this map is an isomorphism of algebras. This is, in fact, an isomorphism of interior P-algebras $A \cong kP \otimes_k \mathrm{End}_k(V)$. In order to prove (iii), note first that V is indecomposable of vertex P when viewed as an $\mathcal{O}P$-module because otherwise the idempotent i would either not be primitive or not local. The isomorphism in (ii) implies that $S = \mathrm{End}_{\mathcal{O}}(V)$ is a direct summand of A with respect to the conjugation action of P on A. Thus S has a stable basis for the conjugation action, because this is true for A. This shows that V is an endopermutation $\mathcal{O}P$-module. $\square$

Corollary 8.11.11   *Theorem 8.11.5 holds if $\mathcal{O} = k$.*

Proof Statement 8.11.10 (ii) implies 8.11.5 (i); in particular, the source algebra A and the defect group algebra kP are Morita equivalent. By 6.4.6 the algebra kGb is Morita equivalent to its source algebra A, hence to the local algebra kP. Thus statement 8.11.5 (ii) follows from 2.8.9, and 8.11.10 (iii) implies that P is a vertex and V a source of any simple kGb-module. $\qquad\square$

It follows from 8.11.10 that in order to prove 8.11.5 over a more general coefficient ring $\mathcal{O}$, it suffices to lift a simple module over k to an $\mathcal{O}$-free module. We note first that this is equivalent to lifting the endopermutation module arising over k in 8.11.5 to an endopermutation module over $\mathcal{O}$.

Lemma 8.11.12 *Let G be a finite group and b a nilpotent block of $\mathcal{O}G$ with a defect group P. Denote by $\bar{b}$ the image of b in kG. Let $\bar{V}$ be an indecomposable endopermutation kP-module such that $\bar{S} \otimes_k kP$ is a source algebra of $kG\bar{b}$, where $\bar{S} = \mathrm{End}_k(V)$. The following are equivalent.*

(i) There is an endopermutation $\mathcal{O}P$-module V such that $k \otimes_{\mathcal{O}} V \cong \bar{V}$.

(ii) There is an $\mathcal{O}$-free $\mathcal{O}Gb$-module X such that $k \otimes_{\mathcal{O}} X$ is a simple kGb-module.

Proof Suppose that (i) holds. Then S has a ΔP-stable basis X. The set $X \otimes P$ is then a $P \times P$-stable basis of $S \otimes_{\mathcal{O}} \mathcal{O}P$. It follows from the uniqueness property 6.4.9, that $S \otimes_{\mathcal{O}} \mathcal{O}P$ is a source algebra of $\mathcal{O}Gb$. Thus V, with the trivial action of $1 \otimes P$, yields an $\mathcal{O}$-free $S \otimes_{\mathcal{O}} \mathcal{O}P$ module such that $k \otimes_{\mathcal{O}} V$ is the unique simple $\bar{S} \otimes_k kP$-module, up to isomorphism. The canonical Morita equivalence between $S \otimes_{\mathcal{O}} \mathcal{O}P$ and $\mathcal{O}Gb$ implies that (ii) holds. The converse implication follows from 8.11.10. $\qquad\square$

One way to conclude the proof of Theorem 8.11.5 would be to observe that thanks to the classification of endopermutation kP-modules, every endopermutation kP-module lifts to an endopermutation $\mathcal{O}P$-module. For P abelian, this follows from Dade's classification in 7.8.1. Theorem 8.11.5 would follow then from 8.11.12 and 8.11.10.

The original proof in [41] for the existence of an irreducible character lifting the unique irreducible Brauer character circumvents the use of any such classification of endopermutation modules. It is based on calculations of generalised decomposition numbers in case K is a splitting field, with the punchline being the inequality between the arithmetic and geometric means, followed by a Galois descent argument, making use of a simplification due to Thévenaz in [215, (51.10)]. We start with the case where K is a splitting field of characteristic zero. The fact that 8.11.5 holds over k implies in particular that if b is a nilpotent block of a finite group algebra then b has a unique irreducible Brauer character, or equivalently, $\ell(b) = 1$. Since the property of being nilpotent passes down to Brauer pairs, it follows that we also have $\ell(f) = 1$ for any (G, b)-Brauer pair (Q, f). This is used in the next technical observation. For φ, ψ two K-valued class functions defined on a subset containing the set of p'-elements in a finite group G, we use the notation $\langle \varphi, \psi \rangle'_G = \frac{1}{|G|} \sum_{x \in G_{p'}} \varphi(x)\psi(x^{-1})$ introduced in 5.14.7.

Lemma 8.11.13 *Let G be a finite group and b a nilpotent block of $\mathcal{O}G$ with defect group P. Suppose that k is a splitting field for all subgroups of G and that $\text{char}(K) = 0$. Let (u, e) be a (G, b)-Brauer element. Then e has a unique irreducible Brauer character φ, and we have*

$$\langle \varphi, \varphi \rangle'_{C_G(u)} = \frac{1}{|C_P(u)|}.$$

If φ' is an irreducible Brauer character of $kC_G(u)$ different from φ, then $\langle \varphi, \varphi' \rangle'_{C_G(u)} = 0$.

Proof By 8.11.4, the block $kC_G(u)e$ is nilpotent, hence has a unique isomorphism class of simple modules by 8.11.11. Denote by Φ the character of a projective indecomposable $\mathcal{O}C_G(u)\hat{e}$-module, where $\hat{e}$ is the block lifting e. Since $C_P(u)$ is a defect group of e and of $\hat{e}$, and since φ is the unique irreducible Brauer character of e, it follows from 6.11.8 that the unique Cartan invariant of e is equal to $|C_P(u)|$ and that $\text{Res}^G_{G_{p'}}(\Phi) = |C_P(u)| \cdot \varphi$. Brauer's reciprocity 5.14.8, implies that $\langle \Phi, \varphi \rangle'_{C_G(u)} = 1$ and that $\langle \Phi, \varphi' \rangle'_{C_G(u)} = 0$. Dividing these two equalities by $|C_P(u)|$ yields the result. $\square$

Lemma 8.11.14 *Let G be a finite group and b a nilpotent block of $\mathcal{O}G$ with a defect group P. Let γ be a local point of P on $\mathcal{O}Gb$, and let $i \in \gamma$ be a source idempotent of b. Suppose that $\text{char}(K) = 0$ and that K is a splitting field for all subgroups of G.*

(i) *For any $u \in P$ there is a unique local point $\epsilon(u)$ on $\mathcal{O}Gb$ such that $u_{\epsilon(u)} \in P_\gamma$.*

(ii) *If $\chi \in \text{Irr}_K(G, b)$ such that $\chi(i)$ is not divisible by p, then $\chi(u_{\epsilon(u)}) \neq 0$ for all $u \in P$.*

(iii) *Let U be a cyclic subgroup of P and ϵ a local point of U on $\mathcal{O}Gb$. If $\chi \in \text{Irr}_K(G, b)$ such that $\chi(i)$ is not divisible by p, then $\prod_u \chi(u_\epsilon)$ is a nonzero rational integer, where u runs over the set of elements in U satisfying $\langle u \rangle = U$.*

Proof If u_ϵ is a local pointed element contained in P_γ, then u_ϵ corresponds to an irreducible Brauer character φ_ϵ of the unique block f of $kC_G(u)$ determined by γ. By 8.11.4, $kC_G(u)f$ is nilpotent, hence has a unique irreducible Brauer character by 8.11.11. This shows (i). Let J be a primitive decomposition of i in $(i\mathcal{O}Gi)^P$. If $j \in J$ does not belong to $\epsilon(u)$, then j belongs to a nonlocal point of $\langle u \rangle$ on $\mathcal{O}Gb$, and hence $\chi(uj) = 0$, by 5.12.16. Thus $\chi(ui) = m\chi(uj)$, where $m = |J \cap \epsilon(u)|$ is the multiplicity of $\epsilon(u)$ on $i\mathcal{O}Gi$. Now $\chi(ui)$ and $\chi(i)$ have the same image in k, which is nonzero as p does not divide $\chi(i)$, and hence $\chi(u_{\epsilon(u)})$ is nonzero as claimed. The product in (iii) is a rational integer by 5.15.4, and nonzero by (ii). $\qquad\square$

Proposition 8.11.15 *Let G be a finite group, b a nilpotent block of $\mathcal{O}G$ with a defect group P, and let $i \in (\mathcal{O}Gb)^{\Delta P}$ be a source idempotent of b. Suppose that $\mathrm{char}(K) = 0$ and that K is a splitting field for all subgroups of G. There is $\chi \in \mathrm{Irr}_K(G, b)$ such that $\chi(i)$ is not divisible by p. If X is an $\mathcal{O}$-free $\mathcal{O}Gb$-module with χ as character then $k \otimes_{\mathcal{O}} X$ is a simple kGb-module.*

Proof Denote by φ the unique irreducible Brauer character associated with b. By 6.15.1 there is $\chi \in \mathrm{Irr}_K(G, b)$ such that $\chi(i)$ is coprime to p. We need to show that $d_\varphi^\chi = 1$. By the first orthogonality relations we have

$$1 = \langle \chi, \chi \rangle = \frac{1}{|G|} \sum_{x \in G} \chi(x)\chi(x^{-1}).$$

By writing each $x \in G$ as a commuting product $x = us = su$ of a p-element u and a p'-element s, we obtain

$$1 = \frac{1}{|G|} \sum_u \sum_s \chi(us)\chi(s^{-1}u^{-1}),$$

where u runs over the set G_p of p-elements in G and s runs over the set of p'-elements $C_G(u)_{p'}$ in $C_G(u)$. We use the notation from 5.15.2 and 5.15.3. For each summand indexed by u and s, write $\chi(us) = \sum_\epsilon \chi(u_\epsilon)\varphi_\epsilon(s)$, where ϵ runs over the local points of $\langle u \rangle$ on $\mathcal{O}G$ and where φ_ϵ is the irreducible Brauer character of $C_G(u)$ associated with ϵ. By Brauer's Second Main Theorem 6.13.1, it suffices to sum over the local points ϵ of $\langle u \rangle$ on $\mathcal{O}Gb$. Note that $\langle u \rangle = \langle u^{-1} \rangle$, hence any local point ϵ of $\langle u \rangle$ is also a local point of $\langle u^{-1} \rangle$. Therefore, together with the analogous expression for $\chi(s^{-1}u^{-1})$, we obtain the equality

$$1 = \frac{1}{|G|} = \sum_u \sum_{\epsilon,\delta} \chi(u_\epsilon)\chi(u_\delta^{-1})\left(\sum_s \varphi_\epsilon(s)\varphi_\delta(s)\right).$$

As before, u runs over G_p, ϵ, δ run over the local points of $\langle u \rangle$ on $\mathcal{O}Gb$, and s runs over $C_G(u)_{p'}$. By 8.11.13, the third sum vanishes if $\epsilon \neq \delta$, and if $\epsilon = \delta$, this sum is equal to $\frac{|C_G(u)|}{|C_P(u)|}$. Thus we have

$$1 = \frac{1}{|G|} \sum_u \sum_\epsilon \frac{|C_G(u)|}{|C_P(u)|} \chi(u_\epsilon) \chi(u_\epsilon^{-1}).$$

In view of Brauer's Theorem on splitting fields 3.8.1 we may assume that K is a subfield of the complex number field $\mathbb{C}$. Then the number $\chi(u_\epsilon)$ and $\chi(u_\epsilon^{-1})$ are complex conjugates, since they are the evaluations of the character of the $\langle u \rangle$-module jX at u and at u^{-1}, where $j \in \epsilon$ and X is an $\mathcal{O}$-free $\mathcal{O}Gb$-module with character χ. Thus $\chi(u_\epsilon) \chi(u_\epsilon^{-1}) = |\chi(u_\epsilon)|^2$. It follows that

$$1 = \frac{1}{|G|} \sum_{u_\epsilon} |\chi(u_\epsilon)|^2,$$

where u_ϵ runs over the set of local pointed elements on $\mathcal{O}Gb$. The value $\chi(u_\epsilon)$ depends only on the G-conjugacy class of u_ϵ. The stabiliser in G of u_ϵ contains $C_G(u)$, and is actually equal to $C_G(u)$ because the block b is nilpotent. Thus the G-conjugacy class of u_δ has $|G : C_G(u)|$ elements. It follows that

$$1 = \sum_{u_\epsilon} \frac{1}{|C_P(u)|} |\chi(u_\epsilon)|^2,$$

where now u_ϵ runs over a set of representatives of the G-conjugacy classes of local pointed elements on $\mathcal{O}Gb$. For any $u \in P$ there is a unique local point $\epsilon(u)$ on $\mathcal{O}Gb$ such that $u_{\epsilon(u)} \in P_\gamma$, where γ is the local point of P on $\mathcal{O}Gb$ containing the source idempotent i. By 5.5.16, any local pointed element on $\mathcal{O}Gb$ has conjugate that is contained in P_γ. Thus the previous equation, rewritten as a sum over all elements in P instead of conjugacy class representatives, takes the form

$$1 = \frac{1}{|P|} \sum_{u \in P} |\chi(u_{\epsilon(u)})|^2.$$

In other words, the arithmetic mean of the numbers $|\chi(u_{\epsilon(u)})|^2$ is 1. Consider next the product $\prod_{u \in P} |\chi(u_{\epsilon(u)})|^2$. It follows from 8.11.14 that this product is a positive integer, hence at least equal to 1. Taking the $|P|$-th root of this product yields still a number greater or equal to 1, and this is the geometric mean of the numbers $|\chi(u_{\epsilon(u)})|^2$. Since the geometric mean is bounded above by the arithmetic mean, it follows that this product is equal to 1. An equality of the geometric and arithmetic means forces however that all numbers $|\chi(u_{\epsilon(u)})|^2$ are equal, hence equal to 1. Thus all generalised decomposition numbers $\chi(u_{\epsilon(u)})$ have norm 1. In particular, applied to $u = 1$ we get that the ordinary decomposition number d_φ^χ is equal to 1. This shows that φ is the Brauer character of

$k \otimes_{\mathcal{O}} X$ for some $\mathcal{O}$-free $\mathcal{O}Gb$-module with χ as its character. This completes the proof. $\qquad\square$

Corollary 8.11.16 *Theorem 8.11.5 holds if* $\mathrm{char}(K) = 0$ *and if* K *is a splitting field for all subgroups of* G.

Proof This follows from combining 8.11.15 and 8.11.10 (ii). $\qquad\square$

It remains to prove 8.11.5 if $\mathrm{char}(K) = 0$, without necessarily K being a splitting field. We modify the original argument, avoiding Schur multipliers, by considering commutative quotients of basic algebras. We will make use of the elementary exercises 4.2.10, 4.2.11, and 4.2.12.

Proof of 8.11.5 for general $\mathcal{O}$ The ring $\mathcal{O}$ contains the unique unramified complete discrete valuation ring with residue field k. Therefore, in view of 6.4.16, we may assume that $\mathcal{O}$ is unramified. By the assumptions in 8.11.5, the residue field k is a splitting field for all subgroups of G, hence contains enough p'-roots of unity. Since these lift to $\mathcal{O}^{\times}$, it follows that the quotient field K contains enough p'-roots of unity, and that it suffices to add a suitable p-power root of unity ζ to p for the field $K' = K(\zeta)$ to be a splitting field for all subgroups of G. In particular, by 8.11.16, the conclusions of 8.11.5 hold for $\mathcal{O}'$ instead of $\mathcal{O}$. Let Γ be the Galois group of the Galois extension K'/K. The action of Γ on $\mathcal{O}'$ induces an action of Γ on $\mathcal{O}'Gb$ such that taking Γ-fixed points yields $\mathcal{O}Gb$. See 1.14.14 and 3.1.26 for background. If P is a defect group of $\mathcal{O}Gb$ and $i \in (\mathcal{O}Gb)^P$ a source idempotent, then i remains a source idempotent in $(\mathcal{O}'Gb)^P$ that is Γ-stable. Thus the action of Γ on $\mathcal{O}'Gb$ induces an action on the source algebra $A' = i\mathcal{O}'Gi$, such that, setting $A = i\mathcal{O}Gi$, we can identify $A' = \mathcal{O}' \otimes_{\mathcal{O}} A$. The action of Γ on A' fixes P, because P is contained in A. By 8.11.5 applied to $\mathcal{O}'$, we have an interior P-algebra isomorphism $A' \cong S \otimes_{\mathcal{O}'} \mathcal{O}'P$, where $S = \mathrm{End}_{\mathcal{O}'}(V)$ for some indecomposable endopermutation $\mathcal{O}'P$-module V with vertex P and determinant 1. Consider $\bar{V} = V \otimes_{\mathcal{O}'} \mathcal{O}'$ as an A'-module, with the natural action of S on V and with $1 \otimes u$ acting as identity, where $u \in P$. For any γ in Γ, the module $_{\gamma}\bar{V}$ obtained via restriction along γ, has again determinant 1 on the image of P in A' (cf. 3.1.26). The standard Morita equivalence between $A' = S \otimes_{\mathcal{O}'} \mathcal{O}'P$ and $\mathcal{O}'P$ implies that every A'-module is of the form $\bar{V} \otimes_{\mathcal{O}'} W$ for a uniquely determined $\mathcal{O}'P$-module W. In particular, we have $_{\gamma}\bar{V} \cong \bar{V} \otimes_{\mathcal{O}'} W$ for some rank one $\mathcal{O}'P$-module W. But then the determinant of P on $_{\gamma}\bar{V}$ is equal to $\det_W^n$, where n is the rank of V. Since n is prime to p, this forces $\det_W = 1$, hence W is trivial. Thus the A'-module $\bar{V}$ is Γ-stable. Since this module extends to a simple $K' \otimes_{\mathcal{O}'} A'$-module X, it follows that X is Γ-stable of dimension n. We follow this action through to commutative quotients of basic algebras, in order to play this back to 1.14.14. Let j be a primitive idempotent in A. By 4.7.3, the idempotent j remains primitive in A', and hence we have an $\mathcal{O}'$-algebra isomorphism $jA'j \cong \mathcal{O}'P$. Note that jX is then a one-dimensional simple $K'P$-module through this isomorphism. The basic algebra jAj of is identified through this isomorphism to an $\mathcal{O}$-subalgebra C of $\mathcal{O}'P$ such that $\mathcal{O}' \otimes_\mathcal{O} C = \mathcal{O}'P$. In this way, the action of Γ on $\mathcal{O}' \otimes_\mathcal{O} C$ induces an action of Γ on $\mathcal{O}'P$. We do not know whether this is the canonical action (which would have $\mathcal{O}P$ as fixed points, and which would allow us immediately to conclude that C has a rank one module, hence a lift of the unique simple module). We argue instead as follows. The largest $\mathcal{O}$-free commutative quotient of $\mathcal{O}'P$ is $\mathcal{O}'P/P'$, where P' is the derived subgroup of P. Note that jX remains a one-dimensional simple $K'P/P'$-module. One checks (cf. 4.2.12) that if $\bar{C}$ is the largest $\mathcal{O}$-free commutative quotient of C, then $\mathcal{O}' \otimes_\mathcal{O} \bar{C} \cong \mathcal{O}'P/P'$ as $\mathcal{O}'$-algebras. The action of Γ on $\mathcal{O}'P$ induces an action on the quotient $\mathcal{O}'P/P'$, with fixed points $\bar{C}$. Since this action was obtained from the action on A', it follows that Γ stabilises the isomorphism class of the one-dimensional simple $K'P/P'$-module corresponding to jX. It follows from 1.14.14 that the K-algebra $K \otimes_\mathcal{O} \bar{C}$, and hence $K \otimes_\mathcal{O} C$, has a one-dimensional module. Thus C has a module of rank one. This implies that A has an $\mathcal{O}$-free module of rank n that lifts the (up to isomorphism) unique simple $k \otimes_\mathcal{O} A$-module. The result follows from 8.11.12. This concludes the proof of 8.11.5. $\qquad\square$

Theorem 8.11.17 *Let G be a finite group, b a nilpotent block of $\mathcal{O}G$ with a defect group P. Let $i \in (\mathcal{O}Gb)^P$ be a source idempotent, and let V be an indecomposable endopermutation $\mathcal{O}P$-module with determinant 1 such that $i\mathcal{O}Gi \cong S \otimes_\mathcal{O} \mathcal{O}P$, where $S = \mathrm{End}_\mathcal{O}(V)$. Suppose that K is a splitting field for KGb and that $\mathrm{char}(K) = 0$. There is a height preserving bijection $\mathrm{Irr}_K(G, b) \cong \mathrm{Irr}_K(P)$ sending χ to λ_χ, and the endopermutation $\mathcal{O}P$-module V has a character η with nonzero integer values. Denote by $\omega(u)$ the sign of the nonzero integer $\eta(u)$ for all $u \in P$. The generalised decomposition matrix of b is equal to the matrix $(\omega(u)\lambda(u))_{\lambda,u}$, where λ runs over $\mathrm{Irr}_K(P)$ and u over a set of representatives of the conjugacy classes of P.*

Proof For $\chi \in \mathrm{Irr}_K(G, b)$, it follows from 6.11.11 that $p^{\mathrm{ht}(\chi)}$ is the highest power of p dividing the degree $\chi(i)$ of the irreducible character of $i\mathcal{O}Gi$ corresponding to χ through the standard Morita equivalence between $\mathcal{O}Gb$ and the source algebra $i\mathcal{O}Gi$. Since $i\mathcal{O}Gi \cong S \otimes_\mathcal{O} \mathcal{O}P$ and the $\mathcal{O}$-rank of S is prime to p, it follows that $p^{\mathrm{ht}(\chi)}$ is also the highest power of p dividing the degree of the character λ of $\mathcal{O}P$ corresponding to χ through these Morita equivalences. Since P is a p-group, it follows that the highest power of p dividing $\lambda(1)$ is $p^{\mathrm{ht}(\lambda)}$, and hence $\mathrm{ht}(\chi) = \mathrm{ht}(\lambda)$. Since V is defined over the unramified subring of $\mathcal{O}$ with residue field k, it follows from 7.3.13 that the character η of V has nonzero integer values. The description of the generalised decomposition numbers is a special case of 7.4.3, applied to the algebras $A = \mathcal{O}P$ and $A' = S \otimes_{\mathcal{O}} \mathcal{O}P$. □

In the above theorem, the block algebra $\mathcal{O}Gb$ is Morita equivalent to the source algebra $i\mathcal{O}Gi \cong S \otimes_{\mathcal{O}} \mathcal{O}P$, hence to $\mathcal{O}P$. We will see later that any Morita equivalence between block algebras induces a *perfect isometry* (cf. 9.3.4) and that any perfect isometry preserves heights (cf. 9.2.5).

8.12 Extensions of nilpotent blocks

We consider in this section the situation where N is a normal subgroup of a finite group G and c a nilpotent block of $\mathcal{O}N$. By 6.8.3 we may assume without loss of generality that c is G-stable. The structure theory of the blocks of $\mathcal{O}G$ covering c is due to Külshammer and Puig [115]. This is one of the corner stones of block theory, involving substantial technical difficulties. Let p be a prime, and let $\mathcal{O}$ be a complete discrete valuation ring having a residue field $k = \mathcal{O}/J(\mathcal{O})$ of characteristic p and a field of fractions K. We allow the case $\mathcal{O} = k$ unless stated otherwise. For convenience, we assume in the remaining sections of this Chapter that k is algebraically closed. One can replace this condition by requiring that k is a perfect field that is large enough – although one needs to take some care as to what 'large enough' means in this context, since we will be making use of some earlier results formulated for algebraically closed fields.

We start with two significantly easier special cases, namely p-extensions and p'-extensions of nilpotent blocks. By a result of Cabanes, if G/N is a p-group, then any block b of $\mathcal{O}Gc$ is again nilpotent. We prove this as a consequence of a slightly more general result.

Theorem 8.12.1 *Let G be a finite group, N a normal subgroup of G, and c a G-stable block of kN. Suppose that G/N is a p-group. Let (S, e) be a maximal (G, N, c)-Brauer pair, and let (Q, f) be a maximal (N, c)-Brauer pair. We have $\mathcal{F}_{(Q,f)}(N, c) = \mathcal{F}_Q(Q)$ if and only if $\mathcal{F}_{(S,e)}(G, N, c) = \mathcal{F}_S(S)$.*

Proof By 8.6.5 we may choose notation that $Q = S \cap N$, and such that $(Q, f) \leq (S, e)$, where here (Q, f) is regarded as a (G, N, c)-Brauer pair. It follows from 8.6.5 that $\mathcal{F}_{(Q,f)}(N, c)$ is a subfusion system of $\mathcal{F}_{(S,e)}(G, N, c)$. Thus if $\mathcal{F}_{(S,e)}(G, N, c) = \mathcal{F}_S(S)$, then $\mathcal{F}_{(Q,f)}(N, c) = \mathcal{F}_Q(Q)$ by 8.3.10. Conversely, suppose that $\mathcal{F}_{(Q,f)}(N, c) = \mathcal{F}_Q(Q)$. Let P be a subgroup of S, and let f_P be the block of $kC_N(P)$ such that $(P, f_P) \leq (S, e)$. Set $R = P \cap N$, and denote by f_R the unique block of $kC_N(R)$ such that $(R, f_R) \leq (Q, f)$. We need to show that $N_G(P, f_P)/C_G(P, f_P)$ is a p-group. Since $C_G(P, f_P) \cap N = C_N(P)$ has p-power index in $C_G(P, f_P)$, this is equivalent to showing that $N_G(P, f_P)/C_N(P)$ is a p-group. Consider the chain of subgroups

$$N_G(P, f_P) \geq N_N(P, f_P) \geq N_N(P, f_P) \cap C_G(R) \geq C_N(P).$$

It suffices to show that every subgroup in this chain has p-power index in the preceding subgroup. Since G/N is a p-group, so is $N_G(P, f_P)/N_N(P, f_P)$. Note that $N_G(P)$ normalises $R = P \cap N$, as N is normal in G. The uniqueness of inclusion of Brauer pairs implies that $N_G(P, f_P) \leq N_G(R, f_R)$. Thus $N_G(P, f_P)$ normalises $C_G(R)$, and $N_G(P, f_P) \cap C_G(R) = N_G(P, f_P) \cap C_G(R, f_R)$. The group $N_G(P, f_P)/(N_G(P, f_P) \cap C_G(R)) \cong N_G(P, f_P)C_G(R)/C_G(R)$ is therefore isomorphic to a subgroup of $N_G(R, f_R)/C_G(R, f_R)$. This group, in turn has a subgroup of p-power index $N_N(R, e_R)/C_N(R)$, which is a p-group by the assumptions on the fusion system of kNc. This shows that the second inclusion in the above chain has p-power index. For the last inclusion, let $x \in N_N(P, f_P) \cap C_G(R)$. Conjugation by x induces an automorphism of P that restricts to the identity on R. Since $x \in N$ normalises P, we have $xux^{-1}u^{-1} \in P \cap N = R$ for all $u \in P$. Thus we may write ${}^x u = u\rho(u)$ for every $u \in P$ and some $\rho(u) \in R$. Since x centralises $\rho(u)$ it follows that for any positive integer n we have ${}^{x^n} u = u\rho(u)^n$. Thus, if $n = |R|$, then conjugation by x^n is the identity on P. This shows that $C_N(P)$ has p-power index in $N_N(P, f_P) \cap C_G(R)$, whence the result. $\square$

Theorem 8.12.2 ([43, Theorem 2]) *Let G be a finite group, N a normal subgroup of G, and c a G-stable block of $\mathcal{O}N$. Suppose that G/N is a p-group. Let Q be a defect group of $\mathcal{O}Nc$. The idempotent c remains a block of $\mathcal{O}G$, with a defect group P satisfying $G = NP$ and $Q = N \cap P$. Moreover, c is a nilpotent block of $\mathcal{O}N$ if and only if c is a nilpotent block of $\mathcal{O}G$.*

Proof Since G/N is a p-group, it follows from 6.8.11 that c remains a block of $\mathcal{O}G$. Let P be a defect group of $\mathcal{O}Gc$. Again by 6.8.11 we have $G = NP$ by and $Q = N \cap P$. Suppose that $\mathcal{O}Nc$ is nilpotent. It follows from 8.6.5 that if e_P is a block of $kC_G(P)$ such that (P, e_P) is a maximal (G, b)-Brauer pair, then $\mathcal{F}_{(P, e_P)}(G, c)$ is a subfusion system of $\mathcal{F}_{(S, e_S)}(G, N, c)$ for some maximal (G, N, c)-Brauer pair (S, e_S). By 8.12.1 we have $\mathcal{F}_{(S, e_S)}(G, N, c) = \mathcal{F}_S(S)$ and hence $\mathcal{F}_{(P, e_P)}(G, c) = \mathcal{F}_P(P)$; that is, $\mathcal{O}Gc$ is nilpotent. Suppose conversely that $\mathcal{O}Gc$ is nilpotent. It follows from 8.6.6 and 8.3.10 that $\mathcal{O}Nc$ is nilpotent. $\square$

In the situation of the preceding theorem, $\mathcal{O}Gc$ is isomorphic, as an $\mathcal{O}$-algebra, to a matrix algebra over $\mathcal{O}P$ and $\mathcal{O}Nc$ is isomorphic to a matrix algebra over $\mathcal{O}Q$. We will need a more precise statement on the source algebras of a p-extension of a nilpotent block.

Theorem 8.12.3 *Let G be a finite group, N a normal subgroup of G, and c a G-stable block of $\mathcal{O}N$. Suppose that c is a nilpotent block of $\mathcal{O}N$ and that G/N is a p-group. Let Q be a defect group of the block c of $\mathcal{O}N$ and P a defect group of c as a block of $\mathcal{O}G$ such that $Q \leq P$. There is a source idempotent $i \in (\mathcal{O}Gc)^P$ of c as a block of $\mathcal{O}G$ such that i is contained in $(\mathcal{O}Nc)^P$ and an indecomposable endopermutation $\mathcal{O}P$-module V with vertex P, such that, setting $S = \mathrm{End}_\mathcal{O}(V)$, we have an isomorphism of interior P-algebras*

$$i\mathcal{O}Gi \cong S \otimes_\mathcal{O} \mathcal{O}P$$

which restricts to a Q-interior P-algebra isomorphism

$$i\mathcal{O}Ni \cong S \otimes_\mathcal{O} \mathcal{O}Q.$$

Proof Let (P, f_P) be a maximal (G, N, c)-Brauer pair. For $R \leq Q$ let f_R be the unique block of $kC_N(R)$ such that $(R, f_R) \leq (P, f_P)$. By 6.8.11 there is a unique block e_R of $kC_G(R)$ covering f_R, and this block satisfies $e_R f_R = f_R$. Thus $\mathrm{Br}_R(c)e_R \neq 0$ since $\mathrm{Br}_R(c)f_R \neq 0$. In other words, (R, e_R) is a (G, c)-Brauer pair. In particular, (P, e_P) is a maximal (G, c)-Brauer pair, and (Q, f_Q) is a maximal (N, c)-Brauer pair.

Let i be a primitive idempotent in $(\mathcal{O}Nc)^P$. Then $\mathrm{Br}_P(i)$ is a primitive idempotent in $kC_N(P)$ belonging to a block that appears in the block decomposition of $kC_N(P)\mathrm{Br}_P(c)$. Since the maximal (G, N, c)-Brauer pairs are conjugate, we may assume that $\mathrm{Br}_P(i) \in kC_G(P)f_P$, and since $e_P f_P = f_P$, we also have $\mathrm{Br}_P(i) \in kC_G(P)e_P$. By 5.12.10, i remains primitive in $(\mathcal{O}Gc)^P$. Set $A = i\mathcal{O}Ni$ and $B = i\mathcal{O}Gi$. Since the block $\mathcal{O}Gc$ is nilpotent with defect group P, it follows from 8.11.5 that $B \cong S \otimes_\mathcal{O} \mathcal{O}P$ as interior P-algebras, where $S = \mathrm{End}_\mathcal{O}(V)$ for some indecomposable endopermutation $\mathcal{O}P$-module with vertex P and determinant 1. Note that the interior P-algebra structure of S is determined by its P-algebra structure and the condition that P has determinant 1 on V; this follows from the fact that V has rank prime to p and 5.3.4. Denote by J the ideal in B corresponding to $S \otimes_\mathcal{O} I(\mathcal{O}P)$ under this isomorphism. Then J is a P-stable ideal in B contained in $J(B)$ such that $B/J \cong S$ as interior P-algebras. In particular, the composition of the obvious maps $A \subseteq B \to S$ is a Q-interior P-algebra homomorphism. We will show that this homomorphism has a P-stable $\mathcal{O}$-algebra section. To do this, we first observe that it suffices to show that this map splits as an $\mathcal{O}P$-homomorphism, with respect to the conjugation action of P.

Indeed, by 2.16.17 and 4.8.5, S is P-stably $\mathcal{O}$-separable; that is, the canonical map $S \otimes_\mathcal{O} S \to S$ has a section as an S-S-bimodule homomorphism that commutes with the action of P on S and the diagonal action of P on $S \otimes_\mathcal{O} S$. The equivariant version 4.8.4 of the Lifting Theorem of Külshammer, Okuyama, and Watanabe, applied with $\mathcal{O}$, P instead of C, G, respectively, implies that if the map $A \to S$ splits as an $\mathcal{O}P$-module homomorphism, then it splits in fact P-equivariantly as a homomorphism of $\mathcal{O}$-algebras.

Note that A, B and S are permutation $\mathcal{O}P$-modules with respect to the conjugation action of P. Therefore, by 5.8.11, in order to show that $A \to S$ is split surjective as an $\mathcal{O}P$-module homomorphism, it suffices to show that for any subgroup R of P, the induced map $A(R) \to B(R) \to S(R)$ is surjective. Since V is an endopermutation module with vertex P, it follows that $S(R)$ is a matrix algebra over k. The blocks f_R of $kC_N(R)$ and e_R of $kC_G(R)$ containing $\mathrm{Br}_R(i)$ are both nilpotent, by 6.8.11, we have $e_R = \mathrm{Tr}_{C_G(R, f_R)}^{C_G(R)}(f_R)$, and the algebras $kC_G(R)e_R$ and $kC_G(R, f_R)f_R$ are Morita equivalent. Thus the algebras $kC_N(R)f_R$ and $kC_G(R, f_R)f_R$ have each a unique conjugacy class of primitive idempotents, and any primitive idempotent in $kC_N(R)f_R$ remains primitive in $kC_G(R, f_R)f_R$, by 5.12.5. It follows from 4.7.10 that $kC_G(R, f_R)f_R = kC_N(R)f_R + J(kC_G(R, f_R)f_R)$. Multiplying this on the left and right by $\mathrm{Br}_R(i)$ yields $B(R) = A(R) + J(A(R))$. Thus the unique simple quotient $S(R)$ of $B(R)$ is also the unique simple quotient of $A(R)$, which shows that the induced map $A(R) \to S(R)$ is surjective.

This shows that the map $A \to S$ splits P-equivariantly as an algebra homomorphism (that is, it splits as a P-algebra homomorphism). Any P-equivariant section $S \to A$ of this homomorphism yields a P-stable subalgebra T of A that is isomorphic to S as a P-algebra. By 4.8.6, T is conjugate to S by an element in $1 + J^P$. Thus there is an interior P-algebra isomorphism $\alpha : S \otimes_{\mathcal{O}} \mathcal{O}P \cong B$ such that $\alpha(S)$ is contained in B. Since this homomorphism maps the diagonal image of Q in $S \otimes_{\mathcal{O}} \mathcal{O}P$ to Qi, which is also contained in A, it follows that α sends $S \otimes_{\mathcal{O}} \mathcal{O}Q$ to A. Now $S \otimes_{\mathcal{O}} \mathcal{O}Q$ is $\mathcal{O}$-pure in $S \otimes_{\mathcal{O}} \mathcal{O}P$, so in order to show that $\alpha(S \otimes_{\mathcal{O}} \mathcal{O}Q) = A$, it suffices to show that $S \otimes \mathcal{O}Q$ and A have the same $\mathcal{O}$-rank. We have $G/N \cong P/Q$. Since $i \in \mathcal{O}N$, it follows that $|P| \cdot \mathrm{rk}_{\mathcal{O}}(S) = \mathrm{rk}_{\mathcal{O}}(B) = |P : Q| \cdot \mathrm{rk}_{\mathcal{O}}(A)$, hence $\mathrm{rk}_{\mathcal{O}}(A) = |Q| \cdot \mathrm{rk}_{\mathcal{O}}(S) = \mathrm{rk}_{\mathcal{O}}(S \otimes_{\mathcal{O}} \mathcal{O}Q)$ as claimed. $\square$

We consider next the case of p'-extensions of nilpotent blocks.

Theorem 8.12.4 ([143]) *Let G be a finite group, N a normal subgroup of G, c a G-stable block of $\mathcal{O}N$, and b a block of $\mathcal{O}G$ such that $bc = b$. Suppose that c is a nilpotent block of $\mathcal{O}N$ and that the quotient group $E = G/N$ has order prime to p. Let P_γ be a defect pointed group of $\mathcal{O}Nc$. Let $i \in \gamma$; that is, i is a source idempotent of $\mathcal{O}Nc$.*

  (i) *We have $G = NN_G(P_\gamma)$ and $N_N(P_\gamma) = PC_N(P)$.*
  (ii) *The canonical homomorphism $E \cong N_G(P_\gamma)/PC_N(P) \to \mathrm{Out}(P)$ lifts uniquely, up to conjugation in $\mathrm{Aut}(P)$, to a group homomorphism $\rho : E \to \mathrm{Aut}(P)$. We denote by $P \rtimes E$ the corresponding semidirect product.*

(iii) *Let $\alpha \in H^2(E; k^\times)$ be the canonical class whose restriction along $G \to$*
$G/N \cong E$ is opposite to the class determined as in 5.3.12 by the action
of G on the unique simple quotient of ONc. Let V be an indecomposable
endopermutation OP-module such that $iONi \cong S \otimes_O OP$ as interior P-
algebras, where $S = \mathrm{End}_O(V)$. Then

$$iOGi \cong S \otimes_O O_\alpha(P \rtimes E)$$

as interior P-algebras.

(iv) *If $\rho : E \to \mathrm{Aut}(P)$ is injective, then $b = c$, the idempotent i remains prim-*
itive in $(OGb)^{\Delta P}$, and the algebra $S \otimes_O O_\alpha(P \rtimes E)$ is a source algebra of
OGb.

Proof Since c is G-stable, any G-conjugate of P_γ is a defect pointed group
of ONc. The group N acts transitively on the set of defect pointed groups of
ONc, and hence the equality $G = NN_G(P_\gamma)$ follows from a Frattini argument.
The equality $N_N(P_\gamma) = PC_N(P)$ holds because ONc is nilpotent. This shows
(i). By the assumptions, E is a p'-group, and hence its image in $\mathrm{Out}(P)$ has
an inverse image in $\mathrm{Aut}(P)$ with $\mathrm{Inn}(P)$ as a normal Sylow p-subgroup, and
therefore $\mathrm{Inn}(P)$ has a complement in this inverse image that lifts E and is
unique up to conjugacy, by the Schur–Zassenhaus Theorem. This shows (ii).
In view of the Lifting Theorem 6.4.9, in order to prove (iii), we may assume
that $O = k$. Set $A = kGc$ and $B = kNc$. Since N is normal in G of index prime
to p, it follows from 1.11.10 that $J(A) = J(B)A = AJ(B)$. Since B is nilpotent, the quotient algebra $T = B/J(B)$ is a matrix algebra over k. Let $x \in G$.
Then x acts on B and on the matrix algebra $T = B/J(B)$. By 5.3.12, there is a
choice of elements $t_x \in T^\times$, for any $x \in G$, such that x and t_x act in the same
way on T, and such that $t_{xy} = t_x\bar{y}$ for all $x \in G$, $y \in N$, where $\bar{y}$ is the canonical image of yc in T. Choose in addition $t_1 = \mathrm{Id}_T$. Then $\alpha(x, x') \in k^\times$ defined
by $t_x t_{x'} = \alpha^{-1}(x, x')t_{xx'}$ for $x, x' \in G$ is a normalised 2-cocycle representing the
class opposite to that determined by the action of G on T, and again by 5.3.12,
this class is the restriction to G of a canonical class in $H^2(E; k^\times)$, which we will
abusively again denote by α. The standard argument from the Fong–Reynolds
reduction 6.8.13 implies that there is an algebra isomorphism

$$A/J(A) \cong T \otimes_k k_\alpha E$$

induced by the map which sends $x \in G$ to $t_x \otimes e_x$, where $e_x = xN$ in $E = G/N$.
Since $iBi = S \otimes_k kP$, we have $iBi/J(iBi) = iTi \cong S$ as interior P-algebras, and
hence this extends to an isomorphism of interior P-algebras $iAi/J(iAi) \cong S \otimes_k$
$k_\alpha E$. For $e \in E$, let x_e be a representative of e in $N_G(P_\gamma)$. Then conjugation by x_e
sends i to an element $x_e i(x_e)^{-1}$ which belongs still to γ, hence which is equal to
$a_e i(a_e)^{-1}$ for some $a_e \in B^{\Delta P}$. Then the element $u_e = (a_e)^{-1}x_e i = i(a_e)^{-1}x_e$ is

invertible in iBi and acts as e on the image of P in iBi. We have $iAi = \bigoplus_{e \in E} iBie_u$. This means that the kP-kP-bimodule structure of iAi is obtained by taking the direct sum of $|E|$ copies of iBi, with the right kP-module structure on the copy labelled by $e \in E$ twisted by the automorphism of P induced by e via ρ. But as $iBi = S \otimes_k kP$, this is the same as the kP-kP-bimodule structure of $S \otimes_k k_\alpha(P \rtimes E)$. The algebra iAi is relatively kP-separable. Thus the lifting theorem 4.8.2 applies, and yields a homomorphism of interior P-algebras $iAi \to S \otimes_k k_\alpha(P \rtimes E)$ which lifts the map $iAi \to iAi/J(iAi) \cong S \otimes_k k_\alpha E$. This proves (iii). If ρ is injective, hence E a subgroup of $\mathrm{Aut}(P)$, then $C_G(P) \subseteq N$, and Brauer's First Main Theorem implies that $b = c$ is the unique block lying above c. It follows from 7.4.5 that the interior P-algebra $S \otimes_\mathcal{O} \mathcal{O}_\alpha(P \rtimes E)$ is primitive, hence a source algebra of $\mathcal{O}Gb$. This completes the proof. $\qquad \square$

We state the general case of extensions of nilpotent blocks, due to Külshammer and Puig.

Theorem 8.12.5 ([115]) *Let G be a finite group. Let N be a normal subgroup of G and let c be a G-stable nilpotent block of $\mathcal{O}N$. Let (P, f_P) be a maximal (G, N, c)-Brauer pair. Set $Q = P \cap N$. Let f_Q be the unique block of $kC_N(Q)$ such that $(Q, f_Q) \leq (P, f_P)$. Then (Q, f_Q) is a maximal (N, c)-Brauer pair. Let γ be the local point of P on $\mathcal{O}Nc$ associated with f_P, and let $i \in \gamma$. Then multiplication by i induces a Morita equivalence between $\mathcal{O}Gc$ and $i\mathcal{O}Gi$, and we have an isomorphism of interior P-algebras*

$$i\mathcal{O}Gi \cong S \otimes_\mathcal{O} \mathcal{O}_\alpha L$$

where $S = \mathrm{End}_\mathcal{O}(V)$ for some indecomposable endopermutation $\mathcal{O}P$-module V with vertex P, where L is a finite group having P as a Sylow p-subgroup such that Q is normal in L, $L/Q \cong G/N$, and such that $\mathcal{F}_P(L) = \mathcal{F}_P(G, N, c)$, and where $\alpha \in H^2(L; k^\times)$ is inverse to the class induced by the action of $N_G(Q, f_Q)$ on the unique simple quotient of $\mathcal{O}C_N(Q)f_Q$. Moreover, multiplication by i induces an isomorphism $Z(\mathcal{O}Gc) \cong Z(\mathcal{O}_\alpha L)$; in particular, multiplication by i induces a bijection between the blocks of $\mathcal{O}G$ covering c and the blocks of $\mathcal{O}_\alpha L$.

One can be slightly more precise regarding the class of α in the two previous theorems, using Dade's fusion splitting theorem; see the Remark 8.12.8 below. The algebra $i\mathcal{O}Gi$ is not itself a source algebra of any particular block of $\mathcal{O}G$ covering c. To determine a source algebra for a block b covering c, one needs to multiply $\mathcal{O}_\alpha L$ with the block that corresponds to b, determine a defect group of that block (which may be a proper subgroup of P) and then multiply on both sides with a primitive local idempotent with respect to this defect group.

Lemma 8.12.6 *With the notation from Theorem 8.12.5, the following hold.*

(i) *The group $Q = P \cap N$ is a defect group of $\mathcal{O}Nc$, and (Q, f_Q) is a maximal (N, c)-Brauer pair.*

(ii) *The image of P in G/N is a Sylow p-subgroup in G/N.*

(iii) *We have $G = N_G(Q, f_Q)N$ and $N_N(Q, f_Q) = QC_N(Q)$.*

Proof Statement (i) follows from 6.8.9, and (ii) follows from 8.6.3. Since c is G-stable, for any $x \in G$, the pair $^x(Q, f_Q)$ is also a maximal (N, c)-Brauer pair, and hence $^x(Q, f_Q) = {}^y(Q, f_Q)$ for some $y \in N$. Then $y^{-1}x \in N_G(Q, f_Q)$, whence the equality $G = N_G(Q, f_Q)N$. The equality $N_N(Q, f_Q) = QC_N(Q)$ holds because c is nilpotent. $\square$

The construction of the group L is described in the following lemma.

Lemma 8.12.7 *With the notation from Theorem 8.12.5, set $H = N_G(Q, f_Q)$, $\bar{L} = H/C_N(Q)$, and $\bar{P} = P/Z(Q)$. We have canonical isomorphisms $G/N \cong \bar{L}$ and $PN/P \cong P/Q$. In particular, $\bar{P}$ is isomorphic to a Sylow p-subgroup of $\bar{L}$. Identify $\bar{P}$ with its image in $\bar{L}$. The following hold.*

(i) *We have a commutative diagram of finite groups with exact rows and injective vertical maps:*

$$
\begin{array}{ccccccccc}
1 & \longrightarrow & C_N(Q) & \longrightarrow & H & \longrightarrow & \bar{L} & \longrightarrow & 1 \\
& & \uparrow & & \uparrow & & \uparrow & & \\
1 & \longrightarrow & Z(Q) & \overset{\iota}{\longrightarrow} & P & \overset{\tau}{\longrightarrow} & \bar{P} & \longrightarrow & 1
\end{array}
$$

(ii) *Let R be a subgroup of P containing Q, and set $\bar{R} = R/Z(Q)$. The canonical map $N_H(R) \to N_{\bar{L}}(\bar{R})$ is surjective.*

(iii) *Denote by ζ_P the class in $H^2(\bar{P}; Z(Q))$ represented by the second row in (i). Denote by $\bar{\mathcal{F}}$ the fusion system of $\bar{L}$ on $\bar{P}$. Then ζ_P is $\bar{\mathcal{F}}$-stable.*

(iv) *There is a unique class $\zeta \in H^2(\bar{L}; Z(Q))$ whose restriction to $\bar{P}$ is ζ_P; equivalently, there is a commutative diagram of finite groups with exact rows and injective vertical maps of the form*

$$
\begin{array}{ccccccccc}
1 & \longrightarrow & Z(Q) & \longrightarrow & L & \longrightarrow & \bar{L} & \longrightarrow & 1 \\
& & \| & & \uparrow & & \uparrow & & \\
1 & \longrightarrow & Z(Q) & \longrightarrow & P & \longrightarrow & \bar{P} & \longrightarrow & 1
\end{array}
$$

where the second row and the right vertical map are as in (i).

Proof By 8.12.6 (iii) we have $G = HN$ and $H \cap N = QC_N(Q)$, hence $G/N \cong H/QC_Q(N) = \bar{L}$ as stated. We have $P \cap N = Q$, hence $PN/N \cong P/Q$. The exactness of the first row in (i) is trivial. Since $P \cap C_N(Q) = Z(Q)$, the exactness of the second row and the commutativity of this diagram follow immediately. For (ii), let $\bar{x} \in N_{\bar{L}}(\bar{R})$. Let $x \in H$ be an inverse image of $\bar{x}$. Then x normalises the inverse image $RC_N(Q)$ of $\bar{R}$ in H. Note that f_Q remains a block of $kRC_N(Q)$, and that R is a defect group of f_Q as a block of $kRC_N(Q)$. Since x also normalises f_Q of $C_N(Q)$, it follows that ^xR is again a defect group of f_Q as a block of $RC_N(Q)$, hence conjugate to R by an element in $C_N(Q)$. Thus there is $c \in C_N(Q)$ such that $c^{-1}x$ normalises R. The canonical image of $c^{-1}x$ in $\bar{L}$ is unchanged equal to $\bar{x}$. This implies (ii). In order to show (iii), by Alperin's Fusion Theorem, it suffices to show that the restriction ζ_R of ζ_P to a class in $H^2(\bar{R}; Z(Q))$ is stable under $N_{\bar{L}}(\bar{R})$, for R a subgroup of P containing Q. The class ζ_R is represented by the obvious short exact sequence

$$1 \xrightarrow{} Z(Q) \xrightarrow{\;\iota\;} R \xrightarrow{\;\tau\;} \bar{R} \xrightarrow{} 1.$$

Let $x \in N_G(R)$. Denote by $\bar{x}$ its image in $N_{\bar{L}}(\bar{R})$ and by φ and ψ the automorphisms on R and $\bar{R}$ induced by conjugation with x and $\bar{x}$, respectively. Note that ψ restricts to an automorphism of $Z(Q)$, and the induced automorphism does not depend on the choice of the lift x of $\bar{x}$, so this yields an action of $N_{\bar{L}}(\bar{R})$ on $Z(Q)$. An easy verification shows that the class $^{\bar{x}}\zeta_R$ is represented by the short exact sequence

$$1 \xrightarrow{} Z(Q) \xrightarrow{\;\psi^{-1}\circ\iota\;} R \xrightarrow{\;\varphi\circ\tau\;} \bar{R} \xrightarrow{} 1.$$

These two extensions are equivalent, since the diagram

$$
\begin{array}{ccccccccc}
1 & \longrightarrow & Z(Q) & \xrightarrow{\;\iota\;} & R & \xrightarrow{\;\tau\;} & \bar{R} & \longrightarrow & 1 \\
 & & \| & & \big\uparrow{\psi} & & \| & & \\
1 & \longrightarrow & Z(Q) & \xrightarrow{\;\psi^{-1}\circ\iota\;} & R & \xrightarrow{\;\varphi\circ\tau\;} & \bar{R} & \longrightarrow & 1
\end{array}
$$

is clearly commutative. Since every element in $N_{\bar{L}}(\bar{R})$ lifts to an element in $N_G(R)$ by (ii), the stability of ζ_P follows. This proves (iii). By a result of Cartan–Eilenberg, the $\bar{\mathcal{F}}$-stable elements in $H^2(\bar{P}; Z(Q))$ are exactly those in the image of the restriction map $H^2(\bar{L}; Z(Q)) \to H^2(\bar{P}; Z(Q))$, whence (iv). $\qquad\square$

Proof of Theorem 8.12.5 By 8.12.3, applied with NP instead of G, there is an indecomposable endopermutation $\mathcal{O}P$-module V of determinant 1, such that setting $S = \text{End}_{\mathcal{O}}(V)$, we have a commutative diagram

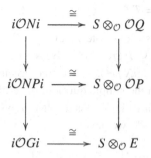

where the vertical arrows are inclusions and where E is the centraliser of the matrix algebra S in $i\mathcal{O}Gi$; we make here use of 4.6.6. The first row is an isomorphism of Q-interior P-algebras, and the remaining two rows are isomorphisms of interior P-algebras. Note that the image of P in $i\mathcal{O}Gi$ maps to the diagonal image $\{\sigma(u) \otimes u\}_{u \in P}$ on the right side under the isomorphisms in the second and third row, where $\sigma : P \to S^{\times}$ is the structural homomorphism. Let δ be the unique local point of Q on $\mathcal{O}Nc$ associated with f_Q. Then $Q_\delta \leq P_\gamma$. We choose $j \in \delta$ such that $j \in S^Q$; that is, such that j belongs to the unique local point of Q on S. This is possible because a primitive idempotent in S^Q remains primitive in $S \otimes_{\mathcal{O}} \mathcal{O}Q$, hence in $(i\mathcal{O}Ni)^Q$. Set $T = jSj$. Multiplying the above diagram on both sides by j yields the diagram:

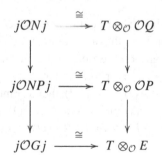

Let $x \in N_G(Q, f_Q) = N_G(Q_\delta)$. Then x stabilises the unique simple quotient of $\mathcal{O}Nc$, and x stabilises δ. Thus there is $d(x) \in ((\mathcal{O}Nc)^Q)^{\times}$ such that $d(x)$ acts as x on the simple quotient of $\mathcal{O}Nc$ and such that $xjx^{-1} = d(x)jd(x)^{-1}$. Thus $d(x)^{-1}x$ centralises j. It follows that $s(x) = d(x)^{-1}xj = jd(x)^{-1}x$ is an invertible element in $(j\mathcal{O}Gj)^{\times}$ that acts as x on the image Qj of Q in $j\mathcal{O}Gj$; that is, we have

$$(uj)^{s(x)} = (u^x)j$$

for all $u \in Q$. The rank of T is prime to p, and we have

$$jONj = T \otimes_{\mathcal{O}} \mathcal{O}Q = T \oplus T \otimes_{\mathcal{O}} I(\mathcal{O}Q).$$

Moreover, $T \otimes_{\mathcal{O}} I(\mathcal{O}Q)$ is a Q-stable ideal of $jONj$ contained in $J(jONj)$, and we have $(J(jONj))^Q \subseteq J((jONj)^Q)$. By 4.8.6, $T^{s(x)}$ and T are conjugate via an element in $1 + J((jONj)^Q)$, so we may assume that $T^{s(x)} = T$, without affecting the action on the simple quotient of $iONi$ (this will be important when it comes to identifying the 2-class α). It follows from 4.1.6 that modulo identifying $jOGj = T \otimes_{\mathcal{O}} E$, we have

$$s(x) = t(x) \otimes c(x)$$

for some $t(x) \in T^{\times}$ and $c(x) \in E$. Both $t(x)$ and $c(x)$ act as x on the respective images of Q in T and in E. Let $\mathcal{R}$ be a set of representatives in $N_G(Q_\delta)$ of $N_G(Q_\delta)/QC_N(Q) \cong G/N$. Multiplying the decomposition $\mathcal{O}G = \oplus_{x \in \mathcal{R}} \mathcal{O}Nx$ on both sides by j and using $xj = xjx^{-1}x = d(x)jd(x)^{-1}x = d(x)js(x)$ shows that

$$jOGj = \oplus_{x \in \mathcal{R}} jONjs(x) = \oplus_{x \in \mathcal{R}} T \otimes \mathcal{O}Qc(x)$$

and since $j \in S \subseteq ON$ we have

$$\mathcal{O}Qc(x) = E \cap ONx$$

$$E = \oplus_{x \in \mathcal{R}} \mathcal{O}Qc(x).$$

This decomposition implies in particular that the $\mathcal{O}$-rank of E is equal to $|Q| \cdot |G/N| = |Z(Q)| \cdot |\bar{L}|$. Note that any summand $\mathcal{O}Qc(x)$ in this direct sum decomposition depends only on the image of x in $G/N \cong N_G(Q_\delta)/QC_N(Q)$. That is, if $yQC_N(Q) = xQC_N(Q)$ for some $x, y \in N_G(Q_\delta)$, then $c(y) = vc(x)$ for some $v \in (\mathcal{O}Q)^{\times}$, and if $yC_N(Q) = xC_N(Q)$, then x, y induce the same automorphism on Q, hence $c(y) = vc(x)$ for some $v \in Z(\mathcal{O}Q)^{\times}$. Choose a set of representatives S of the classes of $\bar{L} = N_G(Q_\delta)/C_N(Q)$ in $N_G(Q_\delta)$. By the above, we have

$$c(x)c(y) = \lambda(x, y)c(x, y)$$

for $x, y \in S$, and some $\lambda(x, y) \in Z(\mathcal{O}Q)^{\times}$. Thus $\lambda \in Z^2(\bar{L}; Z(\mathcal{O}Q)^{\times})$. Denote by X the subgroup of $E^{\times}$ generated by $Z(\mathcal{O}Q)^{\times}$ and the set $\{c(x)|x \in S\}$. The map sending $c(x)$ to $xC_N(Q)$ induces a surjective group homomorphism $X \to \bar{L}$ with kernel $Z(\mathcal{O}Q)^{\times}$, hence a group extension

$$1 \longrightarrow Z(\mathcal{O}Q)^{\times} \longrightarrow X \longrightarrow \bar{L} \longrightarrow 1$$

representing the class of λ in $H^2(\bar{L}; Z(\mathcal{O}Q)^\times)$. Now $Z(\mathcal{O}Q)^\times = k^\times \times (1 + J(Z(\mathcal{O}Q)))$, and hence we have

$$\lambda = \alpha\beta$$

for some $\alpha \in Z^2(\bar{L}; k^\times)$ and some $\beta \in Z^2(\bar{L}; 1 + J(Z(\mathcal{O}Q)))$. In order to finish the proof of 8.12.5, we need to show that α represents the class inverse to that obtained from the action of $N_G(Q_\delta)$ on the simple quotient of $\mathcal{O}Nc$, and that β can be chosen in $H^2(\bar{L}; Z(Q))$ corresponding to the extension L of $\bar{L}$ by $Z(Q)$ described in 8.12.7.

In order to show that β can be chosen in $H^2(\bar{L}; Z(Q))$ we need to show that the group extension $Y = X/k^\times$ of $\bar{L}$ by $1 + J(Z(\mathcal{O}Q))$ contains the group extension L of $\bar{L}$ by $Z(Q)$ described in 8.12.7. Observe first that the copy of P in $E^\times$ is contained in the subgroup X, and we have $P \cap Z(\mathcal{O}Q)^\times = Z(Q)$. Set $J = J(Z(\mathcal{O}Q))$. The map $X \to L$ from the above extension induces a map $Y \to \bar{L}$. Thus we have a commutative diagram of extensions

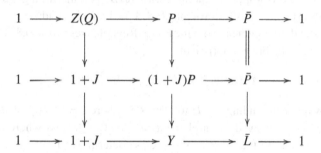

where the vertical arrows are the canonical injections. The third row represents the class β. The second row represents the class β_P obtained via the restriction map $H^2(\bar{L}; 1 + J) \to H^2(\bar{P}; 1 + J)$. The first row represents the class ζ_P as in 8.12.7. This class lifts to a unique class ζ in $H^2(\bar{L}; Z(Q))$ by 8.12.7, corresponding to the extension L as described in 8.12.7. Consider the commutative diagram:

$$
\begin{array}{ccc}
H^2(\bar{L}; Z(Q)) & \longrightarrow & H^2(\bar{L}; 1 + J) \\
\downarrow & & \downarrow \\
H^2(\bar{P}; Z(Q)) & \longrightarrow & H^2(\bar{P}; 1 + J)
\end{array}
$$

The horizontal maps are induced by the inclusions $Z(Q) \to 1 + J$. The vertical maps are induced by the restriction from $\bar{L}$ to $\bar{P}$, and these maps are injective by 1.2.16 and 4.7.24. It follows that the unique class $\zeta \in H^2(\bar{L}; Z(Q))$ lifting

ζ_P maps to $\beta \in H^2(\bar{L}; 1 + J)$ under the map induced by the inclusion $Z(Q) \to 1 + J$. That is, we have a commutative diagram of extensions

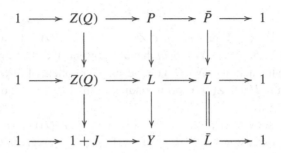

where the three rows represent ζ_P, ζ, β. Thus β is the image of ζ under the canonical map $H^2(\bar{L}; Z(Q)) \to H^2(\bar{L}; 1 + J)$ as required.

In order to describe α, we may assume $\mathcal{O} = k$. Set $A = kGc$ and $B = kNc$. Note that since c is nilpotent, the algebra $B/J(B)$ is a matrix algebra over k. Since N is normal in G, it follows that $J(B)A = AJ(B)$ is an ideal contained in $J(A)$. The standard argument from the Fong–Reynolds reduction 6.8.13 implies that there is an algebra isomorphism

$$A/J(B)A \cong B/J(B) \otimes_k k_\alpha H/C_N(Q)$$

induced by the map sending $x \in H$ to $d(x) \otimes \bar{x}$, where $\bar{x} = xC_N(Q)$ is the image of x in $H/C_N(Q)$, viewed as an element in $k_\alpha H/C_N(Q)$, and where α is a 2-cocycle representing the inverse of the class determined by the action of H on $B/J(B)$. With the identification $jBj = jkNj = T \otimes_k kQ$, we have $J(jBj) = T \otimes_k J(kQ)$. Thus multiplying the above isomorphism by j on both sides yields

$$jAj/J(jBj)jAj = T \otimes_k E/J(kQ)E$$

and since multiplication by j is a Morita equivalence, this yields an isomorphism $E/J(kQ)E \cong k_\alpha L/Q$ mapping $c(x)$ to the image of x in $L/Q \cong H/QC_N(Q)$. This shows that α is as stated.

Since $Y = X/k^\times$, this shows that we may choose the elements $c(x) \in X$ such that $c(x)c(y) = \alpha(x, y)\zeta(x, y)c(xy)$ for $x, y \in S$, and hence $E \cong \mathcal{O}_\alpha L$ as claimed, where α is inflated to L via the canonical surjection $L \to \bar{L}$. $\quad\square$

Remark 8.12.8 Consider the situation as in the statement of Theorem 8.12.5.

(a) A block b of $\mathcal{O}Gc$ corresponds to a unique block d of $\mathcal{O}_\alpha L$, such that

$$i\mathcal{O}Gbi \cong S \otimes_\mathcal{O} \mathcal{O}_\alpha Ld.$$

By Proposition 1.2.18, the algebra $\mathcal{O}_\alpha Ld$ is a block algebra of $\mathcal{O}L'$ for some finite central p'-extension L' of L. Since $\mathcal{O}Gb$ and $i\mathcal{O}Gbi$ are Morita equivalent via the p-permutation bimodule $\mathcal{O}Gbi$, it follows that $\mathcal{O}Gb$ and $\mathcal{O}_\alpha Ld$ are Morita equivalent via a bimodule with a diagonal vertex ΔR, where R is a defect group of b contained in P, and with an endopermutation $\mathcal{O}R$-module as a source, obtained as a direct summand of $\mathrm{Res}_R^P(V)$. Morita equivalences of this form will be investigated further in Theorem 9.11.9, where it is shown that such a Morita equivalence preserves fusion systems.

(b) The class α is also the inverse of the class determined by the action of $N_G(Q, f_Q)$ on the simple quotient of $kC_N(Q)f_Q$. This follows from Dade's Fusion Splitting Theorem 7.9.2. As a consequence, for any block b of $\mathcal{O}Gc$, there is a Morita equivalence between $\mathcal{O}Gb$ and its Harris–Knörr correspondent block of $\mathcal{O}N_G(Q, f_Q)$ covering the block $\mathcal{O}C_N(Q, f_Q)f_Q$. This Morita equivalence is given by a bimodule with endopermutation source. This follows from the fact that $\mathcal{O}_\alpha L$ is determined by $N_G(Q, f_Q)$ and $\mathcal{O}C_N(Q)f_Q$. This Morita equivalence is not canonical, since the splitting in Dade's Fusion Splitting Theorem is not.

Centric Brauer pairs give rise to extensions of nilpotent blocks.

Corollary 8.12.9 *Let G be a finite group, b a block of $\mathcal{O}G$, and (P, e) a maximal (G, b)-Brauer pair. Denote by $\mathcal{F}$ the fusion system of b on P determined by the choice of (P, e). Let Q be an $\mathcal{F}$-centric subgroup of P and let f be the block of $kC_G(Q)$ such that $(Q, f) \leq (P, e)$. Suppose that $G = N_G(Q, f)$ and set $C = QC_G(Q)$. The following hold.*

(i) *The image of b in kG is equal to f, and as a block of C, f is nilpotent.*
(ii) *The algebra $\mathcal{O}Gb$ is Morita equivalent to $\mathcal{O}_\alpha L$, where L is a finite group having P as a Sylow p-subgroup such that Q is normal in L, $L/Q \cong G/C$, and $\mathcal{F}_P(L) = \mathcal{F}$, and where $\alpha \in H^2(L; k^\times)$ is inverse to the class induced by the action of G on the unique simple quotient of kCf.*
(iii) *The algebra $k(G/Q)\bar{f}$ is Morita equivalent to the algebra $k_\alpha \mathrm{Aut}_{\bar{\mathcal{F}}}(Q)$, where $\bar{f}$ is the image of f in kG/Q.*

Proof By the assumptions, Q is normal in G, and hence b is contained in $\mathcal{O}C_G(Q)$. Since f is G-stable, it remains a block of kG, and hence f is the image of b in kG. Thus b is also the unique block of $\mathcal{O}C$ that lifts f. As a block of $kC_G(Q)$, f has the central defect group $Z(Q)$ because (Q, f) is assumed to be a centric Brauer pair. In particular, f remains a nilpotent block of kC with defect group Q. This shows (i). Let i be a primitive local idempotent in $(\mathcal{O}Cb)^P$. Since P is a defect group of b as a block of $\mathcal{O}G$, it follows that $\mathcal{O}Gb$ and $i\mathcal{O}Gi$ are

Morita equivalent. Thus 8.12.5 implies that $\mathcal{O}Gb$ and $\mathcal{O}_\alpha L$ are Morita equivalent, with α and L in the statement, except that we need to verify that $\mathcal{F}$ is indeed the fusion system $\mathcal{F}_P(G, C, f)$. Since $\mathcal{F} = N_{\mathcal{F}}(Q)$ by the assumptions, we only need to show that $\operatorname{Aut}_{\mathcal{F}}(R) = \operatorname{Aut}_{\mathcal{F}_{(G,C,f)}}(R)$ for $Q \le R \le P$. But for any such R we have $C_G(R) \le C_G(Q) \le C$, so this is clear. This shows (ii). For (iii) we need to show that the Morita equivalence in (ii) is compatible with dividing by Q. Since Q acts trivially on the unique simple quotient of kCf, it follows that Q acts trivially on the algebra S in the algebra isomorphism in 8.12.5, and hence dividing both sides in this algebra isomorphism by Q yields (iii). $\quad\square$

Remark 8.12.10 The existence of a finite group L as in 8.12.9 is part of the much broader theme of *centric linking systems* associated with fusion systems; cf. [33]. It is shown in [34, Prop. C] that if $\mathcal{F}$ is a fusion system on a finite p-group P such that $\mathcal{F} = N_{\mathcal{F}}(Q)$ for some $\mathcal{F}$-centric subgroup Q of P, then $\mathcal{F}$ is the fusion system of a finite group L having P as a Sylow p-subgroup, such that Q is normal in L and satisfies $C_L(Q) = Z(L)$. Fusion systems of the form $\mathcal{F} = N_{\mathcal{F}}(Q)$ for some $\mathcal{F}$-centric subgroup Q are called *constrained*.

8.13 Subdivisions of fusion systems

Let k be an algebraically closed field of prime characteristic p. Let $\mathcal{F}$ be a fusion system on a finite p-group P. The map sending a subgroup Q of P to its automorphism group $\operatorname{Aut}_{\mathcal{F}}(Q)$ is not functorial in general, since a morphism $Q \to R$ in $\mathcal{F}$ need not induce a group homomorphism between $\operatorname{Aut}_{\mathcal{F}}(Q)$ and $\operatorname{Aut}_{\mathcal{F}}(R)$ in either direction. In order to achieve functoriality of taking automorphism groups of objects, we will need to replace $\mathcal{F}$ by its subdivision. This is relevant for applications, for instance in the context of Külshammer–Puig classes and Alperin's weight conjecture. The passage to subdivisions amounts essentially to considering chains

$$Q_0 < Q_1 < \cdots < Q_m$$

of subgroups Q_i of P, where m is a nonnegative integer. Such a chain can be regarded as a faithful functor

$$\sigma : [m] \to \mathcal{F},$$

where $[m]$ is the finite category with object set $\{1, 2, \ldots, m\}$ and exactly one morphism $i \to j$ whenever $i \le j$. The functor σ sends i to Q_i and the morphism $i \to i+1$ in $[m]$ to the inclusion map $Q_i < Q_{i+1}$. The faithfulness of σ is equivalent to requiring that Q_i is a proper subgroup of Q_{i+1}, for $0 \le i \le m-1$. The set of chains becomes a category with morphisms induced by those in $\mathcal{F}$.

There is a lot of flexibility at this point. Instead of just considering chains given by proper inclusions, we could consider the larger set of chains of nonisomorphisms in $\mathcal{F}$. As it turns out, this leads to an equivalent category of chains. We could also restrict attention to chains with additional conditions, such as requiring that the Q_i are elementary abelian, or that the Q_i are normal in Q_m, or that the Q_i are all $\mathcal{F}$-centric, amongst many other possibilites. To capture some of this generality, we develop subdivisions for the slightly larger class of finite EI-categories, a notion due to Lück [144]. A small category C is called an EI-category if any endomorphism of any object in C is an automorphism of that object. A fusion system $\mathcal{F}$ on a finite p-group P is clearly a finite EI-category, with the additional property that all morphisms in $\mathcal{F}$ are monomorphisms.

Let C be a finite EI-category. Denote by $[C]$ the set of isomorphism classes of objects in C. For any object X in C denote by $[X]$ its isomorphism class in $[C]$. An elementary argument shows that the EI-property implies that the set $[C]$ is partially ordered, with partial order $[X] \le [Y]$ if the morphism set $\mathrm{Hom}_C(X, Y)$ is nonempty, where X, Y are objects in C. Subdivisions of EI-categories go back to work of Słomińska [206]. We follow here the notation used in [138].

Definition 8.13.1 Let C be an EI-category. The *subdivision category* $S(C)$ is the category that has as objects the faithful functors $\sigma : [m] \to C$, where m is a nonnegative integer and $[m]$ is the totally ordered set $\{0, 1, \dots, m\}$. The integer m is called the *length of* σ, denoted by $|\sigma|$. A morphism in $S(C)$ from σ to another object $\tau : [n] \to C$ is a pair (β, φ) consisting of an order preserving map $\beta : [m] \to [n]$ and an isomorphism of functors $\varphi : \sigma \cong \tau \circ \beta$. Composition of morphisms in $S(C)$ is induced by that in C.

The objects of $S(C)$ can be represented as sequences of nonisomorphisms

$$Q_0 \to Q_1 \to \cdots \to Q_m$$

in C, where the Q_i are objects in C. The map β in the above definition of morphisms in $S(C)$ is necessarily injective, as the functors σ, τ are faithful. If we describe τ as before as a chain of nonisomorphisms $R_0 \to R_1 \to \cdots \to R_n$, then a morphism from τ to σ is a family of isomorphisms $\varphi_i : Q_i \cong R_{\beta(i)}$ in C making the obvious diagram commutative. The fact that a morphism in $S(C)$ is a family of certain isomorphisms in C implies that every morphism in $S(\mathbb{C})$ is a monomorphism and that $S(C)$ is again an EI-category. The objects of length 0 in $S(C)$ are the objects of C, but C is not a subcategory of $S(C)$, because the morphisms in $S(C)$ between two chains of length zero are isomorphisms in C.

There is, however, a canonical functor

$$S(\mathcal{C}) \to \mathcal{C}$$

sending a chain $\sigma : [m] \to \mathcal{C}$ to the maximal term $Q_m = \sigma(m)$ of that chain. This functor sends a morphism $(\beta, \varphi) : \sigma \to \tau$ as in Definition 8.13.1 to the morphism

$$Q_m \cong R_{\beta(m)} \to R_n$$

where the first isomorphism is $\varphi(m)$ and the second morphism is $\tau(\beta(m) \to n)$; that is, the composition of all morphisms in the chain τ starting at $R_{\beta(m)}$ and ending at the maximal term R_n. There is an analogous contravariant functor sending σ to its minimal term $Q_0 = \sigma(0)$. If in addition all morphisms in $\mathcal{C}$ are monomorphisms, then an endomorphism of σ in $S(\mathcal{C})$ is completely determined by the induced automorphism of the maximal term $Q_m = \sigma(m)$; that is, $\mathrm{Aut}_{S(\mathcal{C})}(\sigma)$ can be identified with a subgroup of $\mathrm{Aut}_{\mathcal{C}}(Q_m)$. This is in particular the case if $\mathcal{C}$ is a full subcategory of a fusion system $\mathcal{F}$ on a finite p-group P. In that case, the subgroups $Q_i = \sigma(i)$ of a chain σ of length m in $S(\mathcal{F})$ satisfy $|Q_{i+1}| > |Q_i|$ for $0 \le i \le m - 1$.

Definition 8.13.2 Let $\mathcal{F}$ be a fusion system on a finite p-group P. We define by $S_<(\mathcal{F})$ the full subcategory of $S(\mathcal{F})$ consisting of all chains $\sigma : [m] \to \mathcal{F}$ in $S(\mathcal{F})$ with the property that the nonisomorphisms $\sigma(i \to i+1) : Q_i \to Q_{i+1}$ are (necessarily proper) inclusion maps $Q_i < Q_{i+1}$ for $0 \le i \le m - 1$, where $Q_i = \sigma(i)$ for $0 \le i \le m$. We denote by $S_\lhd(\mathcal{F})$ the full subcategory of $S_<(\mathcal{F})$ consisting of those chains of proper inclusions $Q_0 < Q_1 < \cdots < Q_m$ that have the additional property that the Q_i are normal in the maximal term Q_m, for $0 \le i \le m$.

By the above, the category $S(\mathcal{F})$ and its subcategories $S_<(\mathcal{F})$ and $S_\lhd(\mathcal{F})$ are again *EI*-categories. Using that any homomorphism in $\mathcal{F}$ is a composition of an isomorphism and an inclusion of subgroups of P, one obtains the following observation.

Proposition 8.13.3 *Let $\mathcal{F}$ be a fusion system on a finite p-group P. The inclusion of categories $S_<(\mathcal{F}) \to S(\mathcal{F})$ is an equivalence of categories. In particular, this inclusion induces an isomorphism of partially ordered sets $[S_<(\mathcal{F})] \cong [S(\mathcal{F})]$.*

Proof We need to show that any chain of nonisomorphisms in $\mathcal{F}$ is isomorphic, in $S(\mathcal{F})$, to a chain of proper inclusions. For chains of length zero – consisting of a single subgroup of P – this is trivial. We check that chains of length 1 are isomorphic to chains consisting of an inclusion. Let $\varphi : Q_0 \to Q_1$ be a nonisomorphism in $\mathcal{F}$. Then $\varphi(Q_0) < Q_1$. The obvious commutative diagram

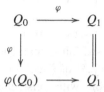

is an isomorphism, in $S(\mathcal{F})$, from the chain $\quad Q_0 \overset{\varphi}{\longrightarrow} Q_1 \quad$ to the chain $\varphi(Q_0) < Q_1$. This argument extends easily to chains of arbitrary positive lengths, whence the result. $\qquad \square$

Definition 8.13.4 ([133, 2.1]) An *EI*-category $\mathcal{C}$ is called *regular* if for any two objects X, Y in $\mathcal{C}$ and any two morphisms φ, $\varphi' \in \mathrm{Hom}_{\mathcal{C}}(X, Y)$ there is a unique automorphism ψ of X such that $\varphi' = \varphi \circ \psi$, or equivalently, if for any two objects X, Y in $\mathcal{C}$, the group $\mathrm{Aut}_{\mathcal{C}}(X)$ acts regularly on $\mathrm{Hom}_{\mathcal{C}}(X, Y)$ whenever this set is nonempty.

If $\mathcal{C}$ is a regular *EI*-category, then any morphism $\varphi : X \to Y$ in $\mathcal{C}$ induces a group homomorphism $\mathrm{Aut}_{\mathcal{C}}(Y) \to \mathrm{Aut}_{\mathcal{C}}(X)$ as follows. If ϵ is an automorphism of Y, then φ and $\epsilon \circ \varphi$ are two morphisms from X to Y. By the regularity property, there is a unique automorphism ψ of X such that $\varphi \circ \epsilon = \varphi \circ \psi$. An easy verification shows that the map $\epsilon \mapsto \psi$ defined in this way is a group homomorphism $\mathrm{Aut}_{\mathcal{C}}(Y) \to \mathrm{Aut}_{\mathcal{C}}(X)$. Moreover, this construction is functorial; that is, the assignment $X \mapsto \mathrm{Aut}_{\mathcal{C}}(X)$ becomes a contravariant functor from $\mathcal{C}$ to the category of groups. The passage to subdivision ensures the regularity property.

Proposition 8.13.5 (cf. [133, 1.3]) *Let $\mathcal{C}$ be an EI-category. The category $S(\mathcal{C})$ is regular. In particular, for $\mathcal{F}$ a fusion system, the category $S(\mathcal{F})$ and its subcategories $S_<(\mathcal{F})$ and $S_\lhd(\mathcal{F})$ are regular EI-categories.*

Proof Let (β, φ), (β', φ') be morphisms in $S(\mathcal{C})$ from the object $\sigma : [m] \to \mathcal{C}$ in $S(\mathcal{C})$ to the object $\tau \cdot [n] \to \mathcal{C}$. Then β, β' are two injective order preserving maps from $[m]$ to $[n]$ such that we have isomorphisms $\varphi(i) : \sigma(i) \cong \tau(\beta(i))$ and $\varphi'(i) : \sigma(i) \cong \tau(\beta'(i))$, for $0 \le i \le m$. In particular, $\tau(\beta(i))$ and $\tau(\beta'(i))$ are isomorphic objects in $\mathcal{C}$. Since – by the definition of $S(\mathcal{C})$ - the functor τ is faithfull, this forces $\beta(i) = \beta'(i)$, and hence $\beta = \beta'$. Thus, setting $\psi(i) = \varphi(i)^{-1} \circ \varphi'(i)$, it follows that the family of automorphisms $(\psi(i))_{0 \le i \le m}$ determines an automorphism $(\mathrm{Id}_{[m]}, \psi)$ of σ satisfying $(\beta, \varphi') = (\beta, \varphi) \circ (\mathrm{Id}_{[m]}, \psi)$. The uniqueness of this automorphism is obvious, as is the passage of the regularity to the subcategories $S_<(\mathcal{F})$ and $S_\lhd(\mathcal{F})$ of $S(\mathcal{F})$. $\qquad \square$

This proposition highlights the fact that because φ is a family of isomorphisms in $\mathcal{F}$, this family determines β. For notational convenience, if no confusion arises, we suppress β from the notation of a morphism in $S(\mathcal{C})$, and denote a morphism (β, φ) simply by φ. In the above considerations we can replace $\mathcal{F}$ by any upwardly closed full subcategory of $\mathcal{F}$. In particular, we can replace $\mathcal{F}$ by the full subcategory $\mathcal{F}^c$ of $\mathcal{F}$-centric subgroups of P. By a result of Słomińska [206, 1.5], cohomology with constant coefficients is invariant under the passage to subdivisions via the canonical functor. We mention here without proof, that combining [206, 1.5] with [138, 6.2] yields the following result on cohomology with coefficients in the constant contravariant functors sending every object to the abelian group $k^{\times}$.

Proposition 8.13.6 *Let $\mathcal{F}$ be a fusion system on a finite p-group P. The canonical functors induce isomorphisms*

$$H^*(S(\mathcal{F}^c); k^{\times}) \cong H^*(\mathcal{F}^c; k^{\times}) \cong H^*(\bar{\mathcal{F}}^c; k^{\times}) \cong H^*(S(\bar{\mathcal{F}}^c), k^{\times}).$$

If $\mathcal{F}$ is the fusion system of a block of a finite group G on one of its defect groups, then the partially ordered set $[S(\mathcal{F})]$ is closely related to certain simplicial complexes of p-subgroups of G. This theme, of which we merely scratch the surface here, goes back a long way, relating functor cohomology, equivariant homotopy theory, and simplicial complexes. See for instance [23], [42], [78], [106], [133], [182], [211], [212], [225], [226].

Definition 8.13.7 Let G be a finite group. We denote by $\mathcal{P}_G$ the set of all chains of the form $Q_0 < Q_1 < \cdots < Q_m$, where m runs over the nonnegative integers and the Q_i are nontrivial p-subgroups of G. We denote by $\mathcal{N}_G$ the subset of $\mathcal{P}_G$ consisting of those chains for which Q_i is normal in Q_m for $0 \leq i \leq m$. We denote by $\mathcal{E}_G$ the subset of $\mathcal{N}_G$ of those chains for which the Q_i are elementary abelian.

The set $\mathcal{P}_G$ is a G-set, with G acting by conjugation, and $\mathcal{P}_G$ is partially ordered by passing to subchains. In particular, $\mathcal{P}_G$ is a simplicial complex, called the *Brown complex*. The sets $\mathcal{N}_G$ and $\mathcal{E}_G$ are partially ordered G-subsets of $\mathcal{P}_G$. In particular, $\mathcal{E}_G$ is a simplicial complex, called the *Quillen complex*. We denote by $\mathcal{P}_G/G$ a set of representatives of the G-conjugacy classes of chains in $\mathcal{P}_G$, and use the analogous notation $\mathcal{N}_G/G$ and $\mathcal{E}_G/G$. We use similar notation for partially ordered G-sets of chains of Brauer pairs. The independence of choices of representatives is understood as an implicit statement in any context where we use this notation.

Definition 8.13.8 Let G be a finite group and b a block of kG. We denote by $\mathcal{P}_{G,b}$ the set of all chains of the form $(Q_0, e_0) < (Q_1, e_1) < \cdots < (Q_m, e_m)$, where m runs over the nonnegative integers and the (Q_i, e_i) are (G, b)-Brauer pairs with Q_i nontrivial. We denote by $\mathcal{N}_{G,b}$ the subset of $\mathcal{P}_{G,b}$ consisting of those chains for which Q_i is normal in Q_m for $0 \leq i \leq m$. We denote by $\mathcal{E}_{G,b}$ the subset of $\mathcal{N}_{G,b}$ of those chains for which the Q_i are elementary abelian.

As before, the sets $\mathcal{P}_{G,b}$, $\mathcal{N}_{G,b}$, $\mathcal{E}_{G,b}$ are partially ordered G-sets, and we denote by $\mathcal{P}_{G,b}/G$, $\mathcal{N}_{G,b}/G$, $\mathcal{E}_{G,b}/G$ sets of representatives of the G-conjugacy classes in $\mathcal{P}_{G,b}$, $\mathcal{N}_{G,b}$, $\mathcal{E}_{G,b}$, respectively. Note that in the above definitions we have excluded the trivial subgroup of G and the trivial Brauer pair $(1, b)$; some of the above mentioned papers have different conventions. If b is the principal block of kG, then Brauer's Third Main Theorem implies that we have canonical isomorphism of G-posets $\mathcal{P}_{G,b} \cong \mathcal{P}_G$, $\mathcal{N}_{G,b} \cong \mathcal{N}_G$, and $\mathcal{E}_{G,b} \cong \mathcal{E}_G$, induced by the correspondence sending a (G, b)-Brauer pair (Q, e) to its first component Q.

If we fix a Sylow p-subgroup P of G, then any chain $Q_0 < Q_1 < \cdots < Q_m$ is G-conjugace to chain of subgroups of P. More generally, for b a block of kG, if we fix a maximal (G, b)-Brauer pair (P, e), then any chain $(Q_0, e_0) < (Q_1, e_1) < \cdots < (Q_m, e_m)$ in $\mathcal{P}_{G,b}$ is G-conjugate to a chain of (G, b)-Brauer pairs contained in (P, e). From this we get immediately the following fact.

Proposition 8.13.9 ([133, 4.6]) *Let G be a finite group, b a block of kG and (P, e) a maximal (G, b)-Brauer pair. Set $\mathcal{F} = \mathcal{F}_{(P,e)}(G, b)$ and denote by $\mathcal{C}$ the full subcategory of $\mathcal{F}$ of all nontrivial subgroups of P. We have a canonical isomorphism of partially ordered sets $\mathcal{P}_{G,b}/G \cong [S(\mathcal{C})]$, induced by the map sending a G-conjugacy class in $\mathcal{P}_{G,b}$ to a representative consisting of Brauer pairs contained in (P, e).*

Proof Note that by 8.13.3, we have $[S(\mathcal{C})] \cong [S_<(\mathcal{C})]$. Let $(Q_0, e_0) < (Q_1, e_1) < \cdots < (Q_m, e_m)$ and $(R_0, f_0) < (R_1, f_1) < \cdots < (R_m, f_m)$ be two chains in $\mathcal{P}_{G,b}$ such that (Q_m, e_m) and (R_m, f_m) are contained in (P, e). We need to show that these two chains are G-conjugate if and only if they are isomorphic in $S(\mathcal{C})$, or equivalently, in $S(\mathcal{F})$. But this is obvious, since G-conjugation between Brauer pairs contained in (P, e) is exactly what defines morphisms in $\mathcal{F}$. $\square$

Specialising this Proposition to the case where b is the principal block yields the following immediate consequence.

Corollary 8.13.10 ([133, 3.4]) *Let G be a finite group, P a Sylow p-subgroup, and let $\mathcal{C}$ be the full subcategory of $\mathcal{F}_P(G)$ of all nontrivial subgroups of P. We have a canonical isomorphism of partially ordered sets $\mathcal{P}_G/G \cong [S(\mathcal{C})]$.*

The next results are technical tools to compare alternating sums indexed by chains of p-subgroups to sums indexed by chains of Brauer pairs. For σ a chain of p-subgroups or of Brauer pairs of a finite group G, we denote by $N_G(\sigma)$ its stabiliser with respect to the conjugation action by G.

Lemma 8.13.11 *Let m be a nonnegative integer and let $\sigma = (Q_0, e_0) < (Q_1, e_1) < \cdots < (Q_m, e_m)$ be a chain of Brauer pairs on kG. Denote by $\tau = Q_0 < Q_1 < \cdots < Q_m$ the underlying chain of p-subgroups of G. Denote by $N_G(\tau, e_m)$ the stabiliser in $N_G(\tau)$ of the block e_m of $kC_G(Q_m)$. We have*

$$N_G(\tau) = \cap_{i=0}^{m} N_G(Q_i),$$

$$N_G(\sigma) = \cap_{i=0}^{m} N_G(Q_i, e_i) = N_G(\tau, e_m).$$

Moreover, the block e_m of $kC_G(Q_m)$ remains a block of $kN_G(\sigma)$.

Proof Conjugation by an element $x \in G$ stabilises the chain τ if and only if it stabilises all subgroups in that chain, whence the equality $N_G(\tau) = \cap_{i=0}^{m} N_G(Q_i)$. A similar argument yields the equality $N_G(\sigma) = \cap_{i=0}^{m} N_G(Q_i, e_i)$. The third equality follows from this together with the uniqueness of the inclusion of Brauer pairs. The last statement follows from 6.5.17. $\qquad\square$

In view of this Lemma, we use the following notation. For G a finite group and b a central idempotent in kG, we set $\mathbf{k}(b) = \dim_k(Z(kGb))$; this is the number of ordinary irreducible characters associated with the sum of blocks b of kG. As before, we denote be $\ell(b)$ the number of isomorphism classes of simple kGb-modules. This notation is slightly abusive in that b could be contained in a subgroup algebra kH for which the corresponding numbers $\ell(kHb)$ and $\dim_k(Z(kHb))$ are different from those for G, so when using this notation, we will need to specify the ambient group G.

For any chain $\tau = Q_0 < Q_1 < \cdots < Q_m$ of p-subgroups in G, we set $b_\tau = \mathrm{Br}_{Q_m}(b)$, regarded as a sum of blocks of $kN_G(\tau)$, possibly zero. For any chain $\sigma = (Q_0, e_0) < (Q_1, e_1) \cdots < (Q_n, e_n)$ of (G, b)-Brauer pairs, we set $e_\sigma = e_n$, and we regard e_σ as a block of $kN_G(\sigma)$, using the last statement of the previous Lemma.

Proposition 8.13.12 ([133, 4.5]) *Let G be a finite group and b a block of kG. With the notation above, we have*

$$\sum_{\tau \in \mathcal{P}_G/G} (-1)^{|\tau|} \mathbf{k}(b_\tau) = \sum_{\sigma \in \mathcal{P}_{G,b}/G} (-1)^{|\sigma|} \mathbf{k}(e_\sigma),$$

$$\sum_{\tau \in \mathcal{P}_G/G} (-1)^{|\tau|} \ell(b_\tau) = \sum_{\sigma \in \mathcal{P}_{G,b}/G} (-1)^{|\sigma|} \ell(e_\sigma).$$

*These equalities hold with $\mathcal{P}_G$, $\mathcal{P}_{G,b}$ replaced by $\mathcal{N}_G$, $\mathcal{N}_{G,b}$ or $\mathcal{E}_G$, $\mathcal{E}_{G,b}$, respectively. They also hold with **k** replaced by any function **f** on the set of pairs (H, c) consisting of a subgroup H of G and a central idempotent c of kH to an abelian group $\mathcal{A}$ having the property that source algebra equivalent blocks of subgroups of G are sent to the same element in $\mathcal{A}$ and having the additivity property $\mathbf{f}(c + c') = \mathbf{f}(c) + \mathbf{f}(c')$ for any two central idempotents c, c' of kH satisfying $cc' = 0$, for any subgroup H of G.*

Proof Let $\tau = Q_0 < Q_1 < \cdots < Q_m$ be a chain in $\mathcal{P}_G$. By the above lemma or by 6.5.17, all block idempotents of $kN_G(\tau)$ are contained in $kC_G(Q_m)$. If Q_m is not conjugate to a subgroup of a defect group of b, then $b_\tau = 0$. If Q_m is contained in a defect group of b, then b_τ is a sum of blocks of $kN_G(\tau)$. Let c be such a block. Note that $C_G(Q_m)$ is a normal subgroup of $N_G(\tau)$ and that c is contained in $kC_G(Q_m)$. Thus we are in a standard Clifford theoretic situation: by 6.8.3, we have $c = \mathrm{Tr}_{N_G(\tau,e)}^{N_G(\tau)}(e)$ for some block e of $kC_G(Q_m)$. The block e is unique up to conjugation by an element in $N_G(\tau)$. Moreover, e remains a block of $N_G(\tau, e)$, and the block algebras $kN_G(\tau)c$ and $kN_G(\tau, e)e$ are source algebra equivalent. Thus in particular, we have $\mathbf{k}(c) = \mathbf{k}(e)$. Set $e_m = e$. For any of the subgroups Q_i in the chain τ, denote by e_i the unique block of $kC_G(Q_i)$ satisfying $(Q_i, e_i) \leq (Q_m, e_m)$. Set $\sigma = (Q_0, e_0) < (Q_1, e_1) < \cdots < (Q_m, e_m)$. This is now a chain in $\mathcal{P}_{G,b}$, and $e_\sigma = e_m$ is a block of its stabiliser $N_G(\sigma) = N_G(\tau, e)$, where in the last equality we used 8.13.11. In order to finish the proof, we need to check that the correspondence sending (τ, c) as above to σ becomes bijective upon taking G-conjugacy classes. Every chain σ shows up as the image of some (τ, c); indeed, given a chain of nontrivial (G, b)-Brauer pairs σ, this follows from taking for τ the underlying chain of p-subgroups in σ, and for c the block $\mathrm{Tr}_{N_G(\tau,e_m)}^{N_G(\tau)}(e_m)$ as in the above construction. All we need to observe is that in the construction of the correspondence going from the pair (τ, c) to the chain σ, the one point where we made a choice – namely when choosing e – a different choice yields a chain conjugate to σ. This, however, is an easy consequence of the fact that e is unique up to conjugation by an element in $N_G(\tau)$, and hence a different choice for e yields a chain of Brauer pairs that is conjugate to σ via an element in $N_G(\tau)$. A chain τ of nontrivial p-subgroups satisfying $b_\tau \neq 0$ is in $\mathcal{N}_G$ if and only if for any block c occurring in b_τ the corresponding chain σ is in $\mathcal{N}_{G,b}$, and similarly for $\mathcal{E}_G$ and $\mathcal{E}_{G,b}$, which implies that we may replace $\mathcal{P}_G$ and $\mathcal{P}_{G,b}$ as stated. The only properties of the function $\mathbf{k}$ required for this proof are the properties as stated at the end. The result follows. $\square$

The functions sending a chain τ to $\mathbf{k}(b_\tau)$ or to $\ell(b_\tau)$ as in the above proposition have the property that they only depend on $N_G(\tau)$, rather than τ. In order to calculate the alternating sums as in the above proposition for functions with this property, we may replace $\mathcal{P}_G$ by any of $\mathcal{N}_G$ or $\mathcal{E}_G$.

Proposition 8.13.13 ([106, 3.3]) *Let G be a finite group and b a block of kG. Let $\mathbf{f}$ be a function from the set of subgroups of G to an abelian group $\mathcal{A}$ such that $\mathbf{f}$ is constant on conjugacy classes of subgroups of G. We have*

$$\sum_{\tau \in \mathcal{P}_G/G} (-1)^{|\tau|}\, \mathbf{f}(N_G(\tau)) = \sum_{\tau \in \mathcal{N}_G/G} (-1)^{|\tau|}\, \mathbf{f}(N_G(\tau)) = \sum_{\tau \in \mathcal{E}_G/G} (-1)^{|\tau|}\, \mathbf{f}(N_G(\tau))$$

$$\sum_{\sigma \in \mathcal{P}_{G,b}/G} (-1)^{|\sigma|}\, \mathbf{f}(N_G(\sigma)) = \sum_{\sigma \in \mathcal{N}_{G,b}/G} (-1)^{|\sigma|}\, \mathbf{f}(N_G(\sigma)) = \sum_{\sigma \in \mathcal{E}_G/G} (-1)^{|\sigma|}\, \mathbf{f}(N_G(\sigma)).$$

Proof Let $\sigma = Q_0 < Q_1 < \cdots < Q_m$ be a chain in $\mathcal{P}_G$. Define a chain σ' as follows. If σ belongs to $\mathcal{E}_G$, then set $\sigma' = \sigma$. If σ does not belong to $\mathcal{E}_G$, then Q_m is not elementary abelian, so has a nontrivial proper Frattini subgroup $\Phi(Q_m)$. Let j be the smallest nonnegative integer with the property that $\Phi(Q_m) \leq Q_j$. Thus Q_{j-1} is does not contain $\Phi(Q_m)$ (if $j = 0$, we set $Q_{-1} = 1$ for notational convenience). Then $Q_{j-1} < \Phi(Q_m)Q_{j-1} \leq Q_j$. If $\Phi(Q_m)Q_{j-1} < Q_j$, form σ' by inserting $\Phi(Q_m)Q_{j-1}$. Note that then $|\sigma'| = |\sigma| + 1$. If $\Phi(Q_m)Q_{j-1} = Q_j$, then $j < m$, so we form σ' by removing Q_j from σ. Note that then $|\sigma'| = |\sigma| - 1$. One easily verifies that the corresponndence $\sigma \mapsto \sigma'$ is an involution; that is, $(\sigma')' = \sigma$. By construction, this involution fixes exactly the chains in $\mathcal{E}_G$. Note that σ and σ' have the same maximal term Q_m. Since $\Phi(Q_m)$ is a characteristic subgroup of Q_m, it follows that $N_G(\sigma) = N_G(\sigma')$. Since $(-1)^{|\sigma|} = -(-1)^{|\sigma'|}$ whenever $\sigma' \neq \sigma$, it follows that the terms in the first sum indexed by σ and σ' cancel each other whenever σ does not belong to $\mathcal{E}_G$. This shows the equality between the first and the third sum. Again since $\Phi(Q_m)$ is characteristic in Q_m, it follows that σ belongs to $\mathcal{N}_G$ if and only if σ' belongs to $\mathcal{N}_G$. Thus the same reasoning with chains in $\mathcal{N}_G$ yields the equality between the second and the third sum. A similar argument shows the corresponding two equalities for chains of Brauer pairs. □

Similar statements hold for other simplicial complexes, such as the complex of chains $Q_0 < Q_1 < \cdots < Q_m$ in $\mathcal{N}_G$ with the additional property that Q_m/Q_0 is elementary abelian, or the *Bouc complex* of chains $Q_0 < Q_1 < \cdots < Q_m$ in $\mathcal{P}_G$ with the additional property that $Q_i = O_p(N_G(Q_i))$ for $0 \leq i \leq m$. We mention two more results, without proofs, which are based on extending some of the above methods. The involution on chains in the proof of Proposition 8.13.13 and variations thereof can be used to construct homotopies on certain chain complexes, leading to the following result.

Theorem 8.13.14 ([137, Theorem 1.1]) *Let $\mathcal{F}$ be a fusion system on a finite p-group P and let $\mathcal{C}$ be an upwardly closed full subcategory in $\mathcal{F}$. Then the partially ordered sets $[S(\mathcal{C})]$ and $[S_{\lhd}(\mathcal{C})]$ are contractible, when viewed as topological spaces.*

For fusion systems of finite groups, this was first conjectured by Webb [225], [226] and proved by Symonds [211]. For fusion systems of blocks, this is due to Barker [12]. The contractibility of $[S(\mathcal{C})]$ as above has in turn implications for the nature of the base change spectral sequences associated with the canonical functors from $S(\mathcal{C})$ to $\mathcal{C}$ and $\mathcal{C}^{op}$, and yields in particular the following finiteness result.

Theorem 8.13.15 (cf. [138, Theorem 1.1]) *Let $\mathcal{F}$ be a fusion system on a finite p-group P and let $\mathcal{C}$ be an upwardly closed full subcategory in $\mathcal{F}$. Then the abelian group $H^2(\mathcal{C}; k^{\times})$ is finite of order prime to p. In particular, $H^2(\mathcal{F}^c; k^{\times})$ is a finite abelian p'-group.*

8.14 Külshammer–Puig classes

In addition to defect groups and fusion systems, the third fundamental local invariant of a block is the family of *Külshammer–Puig classes* of a block, so named since these classes are special cases of the 2-classes arising in the context of extensions of nilpotent blocks. Their construction is as in 8.12.9 above; we review this for convenience. We assume in this section that k is an algebraically closed field of prime characteristic p.

Let G be a finite group, b a block of kG, and let (P, e_P) be a maximal (G, b)-Brauer pair. Denote by $\mathcal{F}$ the fusion system of the block b on its defect group P determined by the choice of (P, e_P). For any subgroup Q of P denote by e_Q the unique block of $kC_G(Q)$ satisfying $(Q, e_Q) \leq (P, e_P)$. Any block with a central defect group is trivially nilpotent, and hence any extension of a block with a central defect group gives rise to such a class. This situation arises for any $\mathcal{F}$-centric subgroup Q of P. Note that then $Z(Q)$ is a defect group of $kC_G(Q)e_Q$ and that $Z(Q)$ is contained in the centre of $C_G(Q)$. Denote by $\bar{e}_Q$ the image of e_Q in $kC_G(Q)/Z(Q)$. Thus $\bar{e}_Q$ is a block having defect zero of $kC_G(Q)/Z(Q)$, and therefore the algebra

$$S = kC_G(Q)/Z(Q)\bar{e}_Q$$

is a matrix algebra over k. The group $N_G(Q, e_Q)$ acts on $C_G(Q)$ and stabilises e_Q. This induces an action of $N_G(Q, e_Q)$ on the algebra S, with Q acting trivially on S. Moreover, the action of the normal subgroup $C_G(Q)$ of $N_G(Q, e_Q)$ lifts in a canonical way to a group homomorphism $C_G(Q) \to S^\times$ sending $c \in C_G(Q)$ to $\bar{c}e_Q$, where $\bar{c}$ is the image of c in $C_G(Q)/Z(Q)$. This is exactly the situation considered in 5.3.12, with $N_G(Q, e_Q)$, $C_G(Q)$, and $\mathrm{Aut}_{\mathcal{F}}(Q)$ instead of G, N, and G/N, respectively. Thus 5.3.12 gives rise to a canonical class α_Q in $H^2(\mathrm{Aut}_{\mathcal{F}}(Q); k^\times)$ such that restricting α_Q along the canonical surjection $N_G(Q, e_Q) \to \mathrm{Aut}_{\mathcal{F}}(Q)$ yields the class opposite to the class determined by the action of $N_G(Q, e_Q)$ on S. The class α_Q in $H^2(\mathrm{Aut}_{\mathcal{F}}(Q); k^\times)$ is called the *Külshammer–Puig class of Q*.

We show that Külshammer–Puig classes of a block are invariants of its source algebras. More precisely, we show that the central extension of $\mathrm{Aut}_{\mathcal{F}}(Q)$ by $k^\times$ determined by the class α_Q can be described in terms of a source algebra.

Theorem 8.14.1 *Let G be a finite group, b a block of kG, and $i \in (kGb)^P$ a source idempotent. Set $A = ikGi$. Denote by $\mathcal{F}$ the fusion system on P determined by A. Let Q be an $\mathcal{F}$-centric subgroup of P and let j be a primitive idempotent in A^Q belonging to the unique local point of Q on A. Denote by M the normaliser in $(jAj)^\times$ of Qj. The canonical map $M \to \mathrm{Aut}(Q)$ induces a central extension*

$$1 \longrightarrow k^\times \longrightarrow M/1 + J(jA^Q j) \longrightarrow \mathrm{Aut}_{\mathcal{F}}(Q) \longrightarrow 1$$

which represents the Külshammer–Puig class α in $H^2(\mathrm{Aut}_{\mathcal{F}}(Q); k^\times)$. In particular, the family of Külshammer-Puig classes of the $\mathcal{F}$-centric subgroups of P is determined in terms of the source algebra A.

Proof By 8.7.5 we have a canonical short exact sequence

$$1 \longrightarrow (jA^Q j)^\times \longrightarrow M \longrightarrow \mathrm{Aut}_{\mathcal{F}}(Q) \longrightarrow 1.$$

Since $jA^Q j$ is local, it follows that $(jA^Q j)^\times = k^\times \times (1 + J(jA^Q j))$. Thus dividing the first two nonzero terms of the above sequence by $1 + J(jA^Q j)$ yields a canonical short exact sequence

$$1 \longrightarrow k^\times \longrightarrow M/1 + J(jA^Q j) \longrightarrow \mathrm{Aut}_{\mathcal{F}}(Q) \longrightarrow 1.$$

In order to compare this to the construction of α, we will need to specialise the proof of 8.12.5 to the situation of the normaliser of a centric subgroup at hand.

Let e be the block of $kC_G(Q)$ such that $\mathrm{Br}_Q(i)e \neq 0$. The class α is, by the above, obtained from the special case of 8.12.5 applied to $N_G(Q, e)$, $C_G(Q)$, $N_P(Q)$ instead of G, N, P, respectively. The block $kC_G(Q)e$ has the central defect group $Z(Q)$, and hence if j' is a primitive idempotent in $kC_G(Q)e =$

$(kC_G(Q)e)^{Z(Q)}$, then we have an isomorphism of interior $Z(Q)$-algebras

$$j'kC_G(Q)j' \cong kZ(Q).$$

In other words, the matrix algebra T in the proof of 8.12.5 is trivial. By the construction of the classes α and ζ in the proof of 8.12.5, the algebra

$$E = j'kN_G(Q, e)j'$$

has the property that $E^\times$ contains a subgroup $\hat{L}$, generated by $k^\times$ and a set of elements invertible elements $c(x)$, with x running over a set of representatives $\mathcal{S}$ of $N_G(Q, e)/C_G(Q)$, such that conjugation by $c(x)$ induces the same automorphism as x on Q via the isomorphism $Qj' \cong Q$, and such that

$$c(x)c(y) = \alpha(x, y)\zeta(x, y)c(xy)$$

for all $x, y \in \mathcal{S}$. Thus $\hat{L}/Z(Q)$ is the central extension of $\bar{L} = \mathrm{Aut}_{\mathcal{F}}(Q)$ by $k^\times$ which represents α. We need to show that there is an isomorphism

$$\hat{L}/Z(Q) \cong M/1 + J(jA^Q j)$$

that induces an isomorphism of extensions of $\mathrm{Aut}_{\mathcal{F}}(Q)$ by $k^\times$, or equivalently, that induces the identity on the images of $k^\times$ in these two groups and that commutes with the canonical maps from these groups to $\mathrm{Aut}_{\mathcal{F}}(Q)$. It suffices to show that there is a surjective group homomorphism

$$\hat{L} \to M/1 + J(jA^Q j)$$

with the compatibility with $k^\times$ and the canonical maps to $\mathrm{Aut}_{\mathcal{F}}(Q)$. Consider the inclusion

$$E = j'kN_G(Q, e)j' \subseteq j'kGj'.$$

Since j' is primitive in $kC_G(Q)e$, it lifts via the Brauer homomorphism Br_Q to a primitive idempotent in the unique local point of Q on kGb associated with e. We may therefore choose j such that

$$j' = j + j_0$$

where j_0 is an idempotent in $(j'kGj')^Q$ that commutes with j and that belongs to $\ker(\mathrm{Br}_Q)$. We show that the map sending $c(x)$ to $jc(x)j$ and preserving $k^\times$ induces a surjective group homomorphism $\hat{L} \to M/1 + J(jA^Q j)$ as claimed. Since the elements $c(x)$ normalise Qj', they normalise $(j'kGj')^Q$ and $\ker(\mathrm{Br}_Q)$. We have

$$j = jc(x)c(x^{-1})j = jc(x)jc(x^{-1})j + jc(x)j_0 c(x^{-1})j.$$

The second term belongs to $\ker(\mathrm{Br}_Q)$, because j_0 does, hence also its conjugate $c(x)j_0c(x)^{-1}$. This shows that

$$jc(x)jc(x)^{-1} \in j + \ker(\mathrm{Br}_Q).$$

Since j is primitive local, we have $\ker(\mathrm{Br}_Q) \cap jA^Q j \subseteq J(jA^Q j)$. It follows that $jc(x)j$ is invertible in jAj and normalises Qj, hence belongs to the subgroup M of $(jAj)^{\times}$. It follows further that the inverse of the image of $c(x)$ in $M/1 + J(jA^Q j)$ is the image of $jc(x)^{-1}j$. Clearly $c(x)$ and $jc(x)j$ induce the same automorphism of Q via the canonical isomorphisms $Qj \cong Q \cong Qj'$. It suffices therefore to show that the assignment sending $c(x)$ to the image of $jc(x)j$ in $M/1 + J(jA^Q j)$ is multiplicative. Let x, $y \in S$. As before, since $\ker(\mathrm{Br}_Q) \cap jA^Q j \subseteq J(jA^Q j)$, it suffices to show that

$$jc(x)jc(y) = jc(x)c(y)j(j + r)$$

for some $r \in \ker(\mathrm{Br}_Q)$. The argument showing that $jc(x)j$ is invertible shows that $jc(x)c(y)j$ is invertible, belongs to M, and that its image in $M/1 + J(jA^Q j)$ has as inverse the image of $jc(y)^{-1}c(x)^{-1}j$. Thus it suffices to show that

$$jc(x)jc(y)jc(y)^{-1}c(x)^{-1}j$$

belongs to $j + \ker(\mathrm{Br}_Q)$. The following calculation takes place within the interior Q-algebra $j'kGj'$. Using that $j' = j + j_0$, we have

$$j = jc(x)c(y)c(y)^{-1}c(x)^{-1}j$$
$$= jc(x)jc(y)jc(y)^{-1}c(x)^{-1}j + jc(x)j_0c(y)jc(y)^{-1}c(x)^{-1}j$$
$$+ jc(x)c(y)j_0c(y)^{-1}c(x)^{-1}j.$$

We need to show that the last two terms are in $\ker(\mathrm{Br}_Q)$. We rewrite these terms as products of Q-stable elements by inserting brackets as appropriate. We have

$$jc(x)j_0c(y)jc(y)^{-1}c(x)^{-1}j = j(c(x)(j_0c(y)jc(y)^{-1})c(x)^{-1})j.$$

Since j_0 belongs to $\ker(\mathrm{Br}_Q)$, so does $j_0(c(y)jc(y)^{-1})$, hence also $c(x)(j_0c(y) jc(y)^{-1})c(x)^{-1}$ and $jc(x)j_0c(y)jc(y)^{-1}c(x)^{-1}j$. Similarly, since j_0 belongs to $\ker(\mathrm{Br}_Q)$, so do $c(x)c(y)j_0c(y)^{-1}c(x)^{-1}$ and $jc(x)c(y)j_0c(y)^{-1}c(x)^{-1}j$. This shows that the map sending $c(x)$ to $jc(x)j$ induces the required group homomorphism, whence the result. $\qquad\square$

With the notation of Theorem 8.14.1, denote by $\mathcal{F}^c$ is the full subcategory of $\mathcal{F}$ consisting of all $\mathcal{F}$-centric subgroups of P. If α is a class in $H^2(\mathcal{F}^c; k^{\times})$, then the restriction of α to automorphism groups yields a family of classes α_Q in $H^2(\mathrm{Aut}_{\mathcal{F}}(Q); k^{\times})$, with Q running over the $\mathcal{F}$-centric subgroups of P. The

2-class gluing conjecture predicts that the family of Külshammer–Puig classes arise in this way.

Conjecture 8.14.2 (2-class gluing conjecture [132, 4.2]) *With the notation of 8.14.1, there is a class* $\alpha \in H^2(\mathcal{F}^c; k^\times)$ *such that for any $\mathcal{F}$-centric subgroup Q of P, the restriction of α to $\mathrm{Aut}_{\mathcal{F}}(Q)$ is the Külshammer-Puig class α_Q of Q.*

By work of Park [163], the class α, if it exists, need not be uniquely determined by the family of Külshammer-Puig classes $(\alpha_Q)_{Q \in \mathcal{F}^c}$. As mentioned in Theorem 8.13.15, the group $H^2(\mathcal{F}^c; k^\times)$ is a finite abelian p'-group.

Remark 8.14.3 The gluing conjecture 8.14.2 is stated in [132, 4.2] using the orbit category $\bar{\mathcal{F}}$, obtained from $\mathcal{F}$ by taking quotients modulo inner automorphisms, but this makes no difference. For Q an $\mathcal{F}$-centric subgroup, we have $\mathrm{Aut}_{\bar{\mathcal{F}}}(Q) = \mathrm{Aut}_{\mathcal{F}}(Q)/\mathrm{Inn}(Q)$. Since $\mathrm{Inn}(Q)$ is a p-group, we have $H^i(\mathrm{Inn}(Q); k^\times) = 0$ for $i \geq 1$, and hence

$$H^*(\mathrm{Aut}_{\mathcal{F}}(Q); k^\times) \cong H^*(\mathrm{Aut}_{\bar{\mathcal{F}}}(Q); k^\times).$$

We use the same notation α_Q for the image of the Külshammer–Puig 2-class α_Q under the isomorphism

$$H^2(\mathrm{Aut}_{\mathcal{F}}(Q); k^\times) \cong H^2(\mathrm{Aut}_{\bar{\mathcal{F}}}(Q); k^\times).$$

A similar isomorphism hold for classes over the subcategories $\mathcal{F}^c$ and $\bar{\mathcal{F}}^c$ of $\mathcal{F}$-centric subgroups of $\mathcal{F}$ and $\bar{\mathcal{F}}$, respectively. More precisely, by Proposition 8.13.6, we have an isomorphism

$$H^*(\mathcal{F}^c; k^\times) \cong H^*(\bar{\mathcal{F}}^c; k^\times)$$

where $k^\times$ denotes the constant contravariant functor with value $k^\times$ on the categories $\mathcal{F}^c$ and $\bar{\mathcal{F}}^c$.

Using the three local invariants of a block, we can reformulate Alperin's weight conjecture 6.10.2 for blocks as follows. As earlier, for any finite-dimensional k-algebra A, we denote by $\ell(A)$ the number of isomorphism classes of simple A-modules and by $w(A)$ the number of isomorphism classes of simple projective A-modules.

Theorem 8.14.4 *Alperin's weight conjecture 6.10.2 is equivalent to the following statement. For any finite group G and any block algebra B of kG with defect group P and fusion system $\mathcal{F}$ on P, we have*

$$\ell(B) = \sum_Q w(k_{\alpha_Q} \mathrm{Aut}_{\bar{\mathcal{F}}}(Q)).$$

where Q runs over a set of representatives of the $\mathcal{F}$-isomorphism classes of $\mathcal{F}$-centric subgroups of P.

Proof By 6.10.8, if (Q, f) is a B-Brauer pair such that $w(k(N_G(Q, f)/Q)\bar{f}) \neq 0$, where $\bar{f}$ is the image of f in $kQC_G(Q)/Q \cong kC_G(Q)/Z(Q)$, then $Z(Q)$ is a defect group of f as a block of $kC_G(Q)$. Thus Q is $\mathcal{F}$-centric in the fusion system $\mathcal{F}$ of B defined by a maximal B-Brauer pair (P, e) containing (Q, f). Therefore the sum on the right side of the blockwise version 6.10.2 of Alperin's weight conjecture involves only $\mathcal{F}$-centric subgroups of P. It suffices to show that $w(k(N_G(Q, f)/Q)\bar{f}) = w(k_{\alpha_Q}\mathrm{Aut}_{\mathcal{F}}(Q))$. This is an immediate consequence of the fact that the algebras $k(N_G(Q, f)/Q)\bar{f}$ and $k_{\alpha_Q}\mathrm{Aut}_{\mathcal{F}}(Q)$ are Morita equivalent by 8.12.9.   □

The 2-class gluing conjecture suggests that the family of Külshammer–Puig classes should be a limit, taken over the category $\mathcal{F}^c$. The assignment $Q \mapsto H^2(\mathrm{Aut}_{\mathcal{F}}(Q); k^{\times})$ is, however, not functorial on $\mathcal{F}^c$. In order to interpret Külshammer–Puig classes as a limit, we need to replace the category $\mathcal{F}^c$ by its subdivision, as reviewed in the last section. The interpretation of the family of Külshammer–Puig classes as a limit requires some basic functor cohomology; since these techniques are not used elsewhere in this book, we refer to the literature as we go.

Via the canonical functor $S(\mathcal{F}) \to \mathcal{F}$ the notion of Külshammer–Puig classes extends to any chain

$$\sigma = (Q_0 < Q_1 < \cdots < Q_m)$$

of $\mathcal{F}$-centric subgroups Q_i of P as follows. The group $\mathrm{Aut}_{S(\mathcal{F})}(\sigma)$ can be canonically identified with the subgroup of $\mathrm{Aut}_{\mathcal{F}}(Q_m)$ consisting of all automorphisms $\varphi \in \mathrm{Aut}_{\mathcal{F}}(Q_m)$ which satisfy $\varphi(Q_i) = Q_i$ for $0 \leq i \leq m$. We denote by α_{σ} the class in $H^2(\mathrm{Aut}_{S(\mathcal{F})}(\sigma); k^{\times})$ obtained from restricting the Külshammer–Puig class $\alpha_{Q_m} \in H^2(\mathrm{Aut}_{\mathcal{F}}(Q_m); k^{\times})$ to this subgroup. Extending the terminology in the obvious way, we call α_{σ} the *Külshammer–Puig class of the chain* σ. The following fact, which interprets the family of Külshammer–Puig classes as a limit, is stated without proof in [138, §1].

Theorem 8.14.5 *Let G be a finite group, b a block of kG, and (P, e) a maximal (G, b)-Brauer pair. Denote by $\mathcal{F}$ the fusion system of b on P determined by the choice of (P, e), and denote by $\mathcal{C}$ the full subcategory of $\mathcal{F}$-centric subgroups of $\mathcal{F}$. For any nonempty strictly increasing chain σ of $\mathcal{F}$-centric subgroups of P denote by $\alpha_{\sigma} \in H^2(\mathrm{Aut}_{S(\mathcal{C})}(\sigma); k^{\times})$ the Külshammer–Puig class. There is a functor $\mathcal{A}^2$ from the partially ordered set $[S(\mathcal{C})]$ of isomorphism classes of chains σ as above to the category of abelian groups sending σ to*

$H^2(\mathrm{Aut}_{S(C)}(\sigma); k^\times)$, *and the family* (α_σ) *of the Külshammer–Puig classes determines an element in* $\lim_{[S(C)]}(\mathcal{A}^2)$.

Theorem 8.14.5 is used in work of Libman [120], [121] and Park [163] on structural aspects and special cases of the 2-class gluing conjecture.

Proof of Theorem 8.14.5 For any subgroup Q of P denote by e_Q the unique block of $kC_G(Q)$ such that (Q, e_Q) is a (G, b)-Brauer pair satisfying $(Q, e_Q) \leq (P, e)$. Denote by $\mathcal{F}$ the fusion system on P of the block b determined by the choice of (P, e). Denote by $\mathcal{C}$ the full subcategory of $\mathcal{F}$ consisting of all $\mathcal{F}$-centric subgroups of P. Denote by $S(\mathcal{C})$, $S_<(\mathcal{C})$, $S_\lhd(\mathcal{C})$ the corresponding full subcategories of $S(\mathcal{F})$, $S_<(\mathcal{F})$, $S_\lhd(\mathcal{F})$, respectively, consisting of chains of $\mathcal{F}$-centric subgroups. In what follows we write Aut instead of $\mathrm{Aut}_{S(C)}$. Any automorphism of a chain restricts to an automorphism on any subchain, and hence the assignment $\sigma \mapsto \mathrm{Aut}(\sigma)$ is contravariant functorial. Thus there is a covariant functor on $S(\mathcal{C})$ sending a chain σ to $H^2(\mathrm{Aut}(\sigma); k^\times)$. Since inner automorphisms of $\mathrm{Aut}(\sigma)$ act trivially on $H^2(\mathrm{Aut}(\sigma); k^\times)$, it follows from the regularity of the EI-category $S(\mathcal{C})$ that this functor factors through the canonical functor $S(\mathcal{C}) \to [S(\mathcal{C})]$, hence induces a covariant functor

$$\mathcal{A}^2 : [S(\mathcal{C})] \to \mathbf{Ab}.$$

Since $S(\mathcal{C})$ is equivalent to $S_<(\mathcal{C})$ it follows that $[S(\mathcal{C})] = [S_<(\mathcal{C})]$. The functor $\mathcal{A}^2$ has the property that its value at the isomorphism class $[\sigma]$ of a chain σ depends only on $\mathrm{Aut}(\sigma)$. It is shown in [137, 4.7, 4.11] that this implies that the restriction of $\mathcal{A}^2$ to the subposet $[S_\lhd(\mathcal{C})]$ induces an isomorphism on higher limits (including in degree zero). It follows that it suffices to show that the family of (α_σ) determines an element in $\lim_{[S_\lhd(\mathcal{C})]}(\mathcal{A}^2)$. Thus it suffices to show that if σ is a subchain of τ belonging to $S_\lhd(\mathcal{C})$, then α_τ is the restriction of α_σ along the group homomorphism $\mathrm{Aut}(\tau) \to \mathrm{Aut}(\sigma)$ induced by the inclusion of chains $\sigma \to \tau$. If the maximal terms Q_m of σ and R_n of τ are equal, then α_σ and α_τ are both the restriction of the same class $\alpha_{Q_m} = \alpha_{R_n}$, so there is nothing to prove.

The one nontrivial case is where the maximal term R_n of τ is strictly bigger than the maximal term Q_m of σ. The classes α_σ and α_τ are the restrictions of the classes α_{Q_m} and α_{R_n}, respectively. Thus, setting $Q = Q_m$ and $R = R_n$, it suffices to consider the case where $\sigma = Q$ and $\tau = (Q \lhd R)$. Set

$$M = N_G(Q \lhd R, e_R) = N_G(R, e_R) \cap N_G(Q, e_Q);$$

the second equality uses the uniqueness of the inclusion of Brauer pairs. We have canonical identifications $\mathrm{Aut}(\sigma) = N_G(Q, e_Q)/C_G(Q)$ and

$\mathrm{Aut}(\tau) = M/C_G(R)$. Note that M contains $C_G(R)$ but not necessarily $C_G(Q)$. We consider $M/C_M(Q)$ as a subgroup of $\mathrm{Aut}(\sigma)$. Set

$$S = kC_G(Q)/Z(Q)\bar{e}_Q \quad \text{and} \quad T = kC_G(R)/Z(R)\bar{e}_R,$$

where as before $\bar{e}_Q$ and $\bar{e}_R$ are the canonical images of e_Q in $kC_G(Q)/Z(Q)$ and of e_R in $kC_G(R)/Z(R)$, respectively. The algebras S and T are matrix algebras, and the group M acts on both S and T. Since Q is centric, we have $C_G(R) \cap Z(Q) = Z(R)$. Thus we may consider $C_G(R)/Z(R)$ as a subgroup of $C_G(Q)/Z(Q)$, and hence we may consider $kC_G(R)/Z(R)$ as a subalgebra of $kC_G(Q)/Z(Q)$. In particular, $\bar{e}_R$ becomes, with this identification, an idempotent in $kC_G(Q)/Z(Q)$ which is fixed by the action of M. Thus the idempotent $e = \bar{e}_R\bar{e}_Q$ belongs to S^M. By 5.3.5, for the purpose of calculating Külshammer–Puig classes, we may replace S by eSe. The equality $\mathrm{Br}_R(e_Q)e_R = e_R$ from 6.3.4 implies that

$$(eSe)(R) = T.$$

Dade's Fusion Splitting Theorem 7.9.1 implies in turn that the Brauer homomorphism $\mathrm{Br}_R^\times : (eS^Re)^\times \to T^\times$ extends to a group homomorphism $f : N_{(eSe)^\times}(R) \to T^\times$ satisfying

$$f(u)\mathrm{Br}_R(s)f(u^{-1}) = \mathrm{Br}_R(usu^{-1})$$

for all $u \in N_{(eSe)^\times}(R)$ and $s \in eS^Re$.

The class $\alpha_Q = \alpha_\sigma$ in $H^2(\mathrm{Aut}(\sigma); k^\times)$ restricts to a class of $H^2(M/C_M(Q); k^\times)$ that is, by the construction of Külshammer–Puig classes, obtained as follows: there is a choice of elements $s_x \in (eSe)^\times$ for every $x \in M$ such that s_x acts as x on eSe, such that s_c is the canonical image in eSe of c for any $c \in C_M(Q)$, and such that $s_{xc} = s_xs_c$ for all $x \in M$ and all $c \in C_M(Q)$. The corresponding 2-cocycle α satisfying $s_xs_y = \alpha(x, y)s_{xy}$ depends only on the images of x, y in $M/C_M(Q)$. Thus α induces a 2-cocycle on $M/C_M(Q)$ which represents the class $\mathrm{Res}_{M/C_M(Q)}^{\mathrm{Aut}_F(Q)}(\alpha_Q)$. Since M normalises R, it follows that $s_x \in N_{(eSe)^\times}(R)$ for all $x \in M$. Set $t_x = f(s_x)$ for all $x \in M$. Then, since f is a group homomorphism, it follows that t_x acts as x on T. Since Br_R is the identity on $C_G(R)$, it follows that Br_R sends the canonical image of $c \in C_G(R) \subseteq C_M(Q)$ in eS^Re to the canonical image of c in T. Using again that f is a group homomorphism, the equality $s_xs_y = \alpha(x, y)s_{xy}$ yields the equality $t_xt_y = \alpha(x, y)t_{xy}$ for all x, $y \in M$. Together, this shows that the elements t_x satisfy the properties that are required for the definition of the Külshammer–Puig class α_τ. This concludes the proof of Theorem 8.14.5. $\square$

8.15 Weights for fusion systems

We assume in this section that k is an algebraically closed field of prime characteristic p. The formulation of Alperin's weight conjecture in 8.14.4 suggests the following generalisation of weights to fusion systems.

Definition 8.15.1 Let $\mathcal{F}$ be a fusion system on a finite p-group P and let $\alpha \in H^2(\bar{\mathcal{F}}^c; k^\times)$. A *weight of* $(\mathcal{F}, \alpha)$ is a pair (Q, V) consisting of an $\mathcal{F}$-centric subgroup Q of P and a projective simple $k_\alpha \mathrm{Aut}_{\bar{\mathcal{F}}}(Q)$-module V. Denote by $w(Q, \alpha)$ the number of isomorphism classes of projective simple $k_\alpha \mathrm{Aut}_{\bar{\mathcal{F}}}(Q)$-modules for any $\mathcal{F}$-centric subgroup Q of P. Set

$$w(\mathcal{F}, \alpha) = \sum_Q{}' w(Q, \alpha),$$

where Q runs over a set of representatives of the $\mathcal{F}$-isomorphism classes of $\mathcal{F}$-centric subgroups in P. If α represents the trivial class, we write $w(\mathcal{F})$ instead of $w(\mathcal{F}, \alpha)$.

The above definition of $w(\mathcal{F}, \alpha)$ involves only the family $(\alpha_Q)_{Q \in \mathcal{F}^c}$ of restrictions of α to automorphism groups of centric subgroups. Therefore, modulo replacing α by the family of Külshammer–Puig 2-classes, the number $w(\mathcal{F}, \alpha)$ makes sense for blocks even if the 2-class gluing conjecture is not known to hold for that block. One can view $w(\mathcal{F}, \alpha)$ as the number of isomorphism classes of weights of $(\mathcal{F}, \alpha)$, where two weights (Q, V), (R, W) are called isomorphic if there is an isomorphism $\varphi : Q \to R$ in $\mathcal{F}$ such that V and W correspond to each other via the algebra isomorphism $k_\alpha \mathrm{Aut}_{\bar{\mathcal{F}}}(Q) \cong k_\alpha \mathrm{Aut}_{\bar{\mathcal{F}}}(R)$ induced by φ. The Sylow axiom for fusion systems implies that (P, V) is a weight for any simple $k_\alpha \mathrm{Aut}_{\bar{\mathcal{F}}}(P)$-module V, because $\mathrm{Aut}_{\bar{\mathcal{F}}}(P)$ is a p'-group. With the above terminology, Alperin's Weight Conjecture is equivalent to the equality

$$w(\mathcal{F}, \alpha) = \ell(B),$$

where B is a block with associated fusion system $\mathcal{F}$ and 2-class (or family of Külshammer–Puig 2-classes) α, and where $\ell(B)$ is the number of isomorphism classes of simple B-modules.

Alperin's Weight Conjecture can also be used to predict the number of ordinary irreducible characters associated with B in terms of $(\mathcal{F}, \alpha)$. For any fusion system $\mathcal{F}$ on a finite p-group P and any fully $\mathcal{F}$-centralised subgroup Q of P we have a canonical functor

$$C_{\mathcal{F}}(Q) \longrightarrow \mathcal{F}$$

sending a subgroup R of $C_P(Q)$ to QR and a morphism $\varphi : R \to S$ in $C_{\mathcal{F}}(Q)$ to the unique morphism $\psi : QR \to QS$ in $\mathcal{F}$ satisfying $\psi|_R = \varphi$ and $\psi|_Q = \mathrm{Id}_Q$. By 8.3.7, this functor maps $C_{\mathcal{F}}(Q)^c$ to $\mathcal{F}^c$. Thus, restriction induces a map

$$H^2(\mathcal{F}^c; k^\times) \longrightarrow H^2(C_{\mathcal{F}}(Q)^c; k^\times)$$

and for $\alpha \in H^2(\mathcal{F}^c; k^\times)$ we denote by $\alpha(Q)$ the image of α in $H^2(C_{\mathcal{F}}(Q)^c; k^\times)$. If $Q = \langle u \rangle$, we write $\alpha(u)$ instead of $\alpha(\langle u \rangle)$. By 8.5.4, if $\mathcal{F}$ is the fusion system of a block B of a finite group having P as a defect group, then $C_{\mathcal{F}}(Q)$ is the fusion system on $C_P(Q)$ of a block e_Q of $kC_G(Q)$ such that (Q, e_Q) is a B-Brauer pair. Moreover, since $C_{C_G(Q)}(S) = C_G(QS)$ for any subgroup S of $C_P(Q)$, it follows that $\mathrm{Aut}_{C_{\mathcal{F}}(Q)}(S)$ can be identified with a subgroup of $\mathrm{Aut}_{\mathcal{F}}(QS)$. Note that if S is $C_{\mathcal{F}}(Q)$-centric, then $Z(Q) \le Z(C_P(Q)) \le S$, and hence $Z(QS) = Z(S)$. One easily checks that restricting the Külshammer–Puig class of QS in $\mathcal{F}$ yields the corresponding Külshammer–Puig class of S in $C_{\mathcal{F}}(S)$ for any $C_{\mathcal{F}}(Q)$-centric subgroup S of Q. In other words, with the notation above, if α glues the Külshammer–Puig classes of B together, then $\alpha(Q)$ glues the Külshammer–Puig classes of e_Q together.

Definition 8.15.2 Let $\mathcal{F}$ be a fusion system on a finite p-group P and $\alpha \in H^2(\mathcal{F}^c; k^\times)$. Set

$$\mathbf{k}(\mathcal{F}, \alpha) = \sum_u w(C_{\mathcal{F}}(u); \alpha(u))$$

where the sum runs over a set of representatives of elements in P, up to isomorphism in $\mathcal{F}$, such that $\langle u \rangle$ is fully $\mathcal{F}$-centralised. We write $\mathbf{k}(\mathcal{F})$ instead of $\mathbf{k}(\mathcal{F}, \alpha)$ if α is trivial.

Alperin's Weight Conjecture, in conjunction with Brauer's Theorem 6.13.12, implies that if B is a block with associated fusion system $\mathcal{F}$ and 2-class α, then

$$\mathbf{k}(\mathcal{F}, \alpha) = \mathbf{k}(B),$$

where $\mathbf{k}(B)$ is the number of ordinary irreducible characters associated with B.

Remark 8.15.3 The numbers $\ell(\mathcal{F}, \alpha)$ and $\mathbf{k}(\mathcal{F}, \alpha)$ make perfectly sense even if $\mathcal{F}$ is an exotic fusion system, and one may wonder whether these numbers have an interpretation as numerical invariants of some algebra.

The following considerations regarding fusion systems with one weight are from [97]. Modulo Alperin's Weight Conjecture, every block B with a unique isomorphism class of simple modules should have a fusion system $\mathcal{F}$ and associated 2-class α satisfying $w(\mathcal{F}, \alpha) = 1$.

Theorem 8.15.4 ([97, Theorem 1.1]) *Let $\mathcal{F}$ be a fusion system on a finite p-group P and let $\alpha \in H^2(\mathcal{F}^c; k^\times)$. The following are equivalent:*

(i) $\mathcal{F} = \mathcal{F}_P(P)$.
(ii) $w(\mathcal{F}) = 1$ and $\alpha = 1$.
(iii) $w(C_{\mathcal{F}}(Q), \alpha(Q)) = 1$ for any fully $\mathcal{F}$-centralised subgroup Q of P.

Corollary 8.15.5 *Let $\mathcal{F}$ be a fusion system on a finite p-group P. We have $w(\mathcal{F}) = 1$ if and only if $\mathcal{F} = \mathcal{F}_P(P)$.*

The proof of Theorem 8.15.4 adapts arguments due to A. Watanabe [224] in the context of blocks with a single isomorphism class of simple modules to general fusion systems.

Lemma 8.15.6 *Let $\mathcal{F}$ be a fusion system on a finite p-group P. Suppose that for any fully $\mathcal{F}$-centralised nontrivial subgroup Q of P we have $C_{\mathcal{F}}(Q) = \mathcal{F}_{C_P(Q)}(C_P(Q))$. Then for any nontrivial subgroup Q of P the following hold.*

(i) Any p'-subgroup A of $\mathrm{Aut}_{\mathcal{F}}(Q)$ acts freely on $Z(Q) \setminus \{1\}$.
(ii) We have $H^2(\mathrm{Aut}_{\mathcal{F}}(Q); k^\times) = \{0\}$.

Proof Suppose there is a p'-subgroup of $\mathrm{Aut}_{\mathcal{F}}(Q)$ that does not act freely on $Z(Q) \setminus \{1\}$. Then there is a nontrivial p'-subgroup A of $\mathrm{Aut}_{\mathcal{F}}(Q)$ such that $R = C_{Z(Q)}(A)$ is nontrivial. Since $R \subseteq Z(Q)$ we have $Q \subseteq C_P(R)$. Thus any morphism $\varphi : R \to P$ extends to a morphism $\psi : Q \to P$, and hence we may assume that R is fully $\mathcal{F}$-centralised. Now A is a subgroup of $\mathrm{Aut}_{C_{\mathcal{F}}(R)}(Q)$, contradicting the fact that $C_{\mathcal{F}}(R) = \mathcal{F}_{C_P(R)}(C_P(R))$. This contradiction proves (i). It follows from the structure theorems on groups with a free action in [76, Ch. 5, Thm. 4.11], that any Sylow subgroup S of $\mathrm{Aut}_{\mathcal{F}}(Q)$ is either a p-group, cyclic of prime order $\ell \neq p$ or generalised quaternion; in all three cases we have $H^2(S; k^\times) = \{0\}$ whence (ii). □

Lemma 8.15.7 *Let $\mathcal{F}$ be a fusion system on a finite p-group P such that $\mathrm{Aut}_{\mathcal{F}}(P) = \mathrm{Aut}_P(P)$ but such that $\mathcal{F} \neq \mathcal{F}_P(P)$. Let Q be a subgroup of P of maximal order such that $\mathrm{Aut}_{\mathcal{F}}(Q)$ is not a p-group. Set $\mathrm{Aut}_{\bar{\mathcal{F}}}(Q) = \mathrm{Aut}_{\mathcal{F}}(Q)/\mathrm{Aut}_Q(Q)$. Then*

$$\mathrm{Aut}_{\bar{\mathcal{F}}}(Q) = K \rtimes S$$

for some nontrivial p'-subgroup K of $\mathrm{Aut}_{\bar{\mathcal{F}}}(Q)$ and some p-subgroup S of $\mathrm{Aut}_{\bar{\mathcal{F}}}(Q)$ such that S acts freely on $K \setminus \{1\}$ and on the set of nontrivial irreducible characters of K. In particular, $\mathrm{Aut}_{\bar{\mathcal{F}}}(Q)$ has a p-block of defect zero.

Proof The maximality of Q implies that Q is centric radical; we may choose Q to be fully $\mathcal{F}$-normalised. Then, by [34, Prop. C] there is a finite group L with $N_P(Q)$ as Sylow p-subgroup such that Q is normal in L, $C_L(Q) = Z(Q)$ and such that $\mathcal{F}_L(N_P(Q)) = N_{\mathcal{F}}(Q)$. Note that $L/Q \cong \mathrm{Aut}_{\bar{\mathcal{F}}}(Q)$. The maximality of Q with $\mathrm{Aut}_{\mathcal{F}}(Q) \neq \mathrm{Aut}_P(Q)$ implies that $N_{\mathcal{F}}(Q)/Q = \mathcal{F}_{N_P(Q)/Q}(N_P(Q)/Q)$ and hence L/Q is p-nilpotent. Thus $L/Q = K \rtimes S$ for some nontrivial p'-subgroup K of L and some p-subgroup S of L/Q. Since Q is fully normalised we may choose S in such a way that $S = N_P(Q)/Q$. Let s be a nontrivial element in S, denote by $t \in N_P(Q)$ an inverse image of s, let $x \in C_K(s)$ and denote by $y \in L$ an inverse image of y in L. Then y normalises the group $Q\langle t \rangle$. Since $\mathrm{Aut}_{\mathcal{F}}(Q\langle t \rangle)$ is a p-group and since $C_L(Q\langle t \rangle) \subseteq Z(Q)$ it follows that y is a p-element. But then $x = 1$ because K is a p'-group. This shows that S acts freely on $K \setminus \{1\}$. But then S acts also freely on the set of nontrivial conjugacy classes in K. Indeed, if a nontrivial element in S stabilises a conjugacy class in K, then it fixes at least one element in that conjugacy class because the conjugacy classes of K have length coprime to p. Since $K \neq \{1\}$ the group S has at least one regular orbit on the set of conjugacy classes of K, and hence, by Brauer's Permutation Lemma, S acts freely on the set of nontrivial irreducible characters of K; any regular orbit of S gives rise to a p-block of defect zero of $\mathrm{Aut}_{\bar{\mathcal{F}}}(Q)$ as claimed. $\qquad\square$

Lemma 8.15.8 *Let P be a finite p-group and set $\mathcal{F} = \mathcal{F}_P(P)$. Then $H^n(\mathcal{F}^c; k^\times) = \{0\}$ for any positive integer n.*

Proof Denote by $\bar{\mathcal{F}}$ the orbit category of $\mathcal{F}$. Then P is a final object in $\bar{\mathcal{F}}$, hence $H^n(\bar{\mathcal{F}}^c; k^\times) = \{0\}$ for n positive. By [138, Theorem 11.2] we have $H^*(\mathcal{F}^c; k^\times) \cong H^*(\bar{\mathcal{F}}^c; k^\times)$, whence the result. $\qquad\square$

Proof of Theorem 8.15.4 If (i) holds, then $\alpha = 1$ by 8.15.8, and $w(\mathcal{F}) = 1$, hence (i) implies (ii). Similarly, if (i) holds then $C_{\mathcal{F}}(Q) = \mathcal{F}_{C_P(Q)}(C_P(Q))$ for any fully $\mathcal{F}$-centralised subgroup Q of P, which implies $w(C_{\mathcal{F}}(Q)) = 1$. Moreover, in that case we have $\alpha(Q) = 1$ by 8.15.8, and thus (i) implies (iii). Suppose now that (ii) holds. We argue by contradiction; that is, $w(\mathcal{F}) = 1$ but $\mathcal{F} \neq \mathcal{F}_P(P)$. Since $w(\mathcal{F}) = 1$ we have $\mathrm{Aut}_{\mathcal{F}}(P) = \mathrm{Aut}_P(P)$ because every simple $k\mathrm{Aut}_{\bar{\mathcal{F}}}(P)$-module is projective, as this is a p'-group. Let Q be a fully $\mathcal{F}$-normalised subgroup of P of maximal order such that $\mathrm{Aut}_{\mathcal{F}}(Q)$ is not a p-group. Then 8.15.7 shows that $k\mathrm{Aut}_{\bar{\mathcal{F}}}(Q)$ has a projective simple module, contradicting the fact that $w(\mathcal{F}) = 1$. This shows that (ii) implies (i). It remains to show that (iii) implies (i). Suppose that (iii) holds. If $\mathcal{F} = C_{\mathcal{F}}(Z)$ for some nontrivial subgroup Z of $Z(P)$ then the hypothesis (iii) passes down to $\mathcal{F}/Z$, and so (i) holds for $\mathcal{F}/Z$ by induction, but then (i) holds for $\mathcal{F}$ by. We may therefore suppose that $C_{\mathcal{F}}(Q)$ is a proper subsystem of $\mathcal{F}$ for any nontrivial fully $\mathcal{F}$-centralised subgroup Q of P. By 8.3.7, hypothesis (iii) passes down to centralisers. Thus, arguing by induction, we get $C_{\mathcal{F}}(Q) = \mathcal{F}_{C_P(Q)}(C_P(Q))$ for any nontrivial fully $\mathcal{F}$-centralised subgroup Q of P. Then $H^2(\mathrm{Aut}_{\mathcal{F}}(Q); k^\times) = \{0\}$ for any nontrivial subgroup Q of P, by 8.15.6. Thus in fact $w(\mathcal{F}) = w(\mathcal{F}, \alpha) = 1$, by hypothesis (iii) applied to $Q = 1$. The implication (ii) $\Rightarrow$ (i) implies that $\mathcal{F} = \mathcal{F}_P(P)$. Thus (iii) implies (i) and the proof is complete. $\square$

Conjecture 8.15.9 *If $w(\mathcal{F}, \alpha) = 1$ then $\mathcal{F}$ is the fusion system of a finite p-solvable group; in particular, $\mathcal{F}$ is constrained.*

9

Isometries and Bimodules

Extending partial isometries between groups of virtual characters has long been one of the standard techniques in finite group theory. The notions of perfect isometries and isotypies, due to Broué [38], provide a theoretical framework for this technique. A perfect isometry between two p-blocks of finite groups is a character bijection 'with signs' satisfying certain arithmetic conditions. An isotypy is a 'compatible' family of perfect isometries between corresponding Brauer pairs of two blocks with the same local structure. Isotypies can be refined to character correspondences which are induced by virtual characters of p-permutation bimodules, a concept due to Boltje and Xu [22]. It was shown by Broué that a derived equivalence between two blocks induces a perfect isometry, and that a 'splendid' derived equivalence induces an isotypy. The character theoretic version of Broué's Abelian Defect Conjecture predicts that there should be a perfect isometry between a block with an abelian defect group and its Brauer correspondent.

Throughout this chapter we denote by $\mathcal{O}$ a complete discrete valuation ring with quotient field K of characteristic zero and residue field $k = \mathcal{O}/J(\mathcal{O})$ of prime characteristic p.

9.1 Perfect bimodules and Grothendieck groups

Let A and B be $\mathcal{O}$-algebras. As in 2.17.10 we denote by perf(A, B) the full $\mathcal{O}$-linear subcategory of mod$(A \otimes_{\mathcal{O}} B^{\mathrm{op}})$ consisting of all A-B-bimodules M with the property that M is finitely generated projective as a left A-module and as a right B-module. The associated Grothendieck groups inherit ring and bimodule structures; this is based on the following observation (used already in the proof of 2.17.11):

Proposition 9.1.1 *Let A, B, C be $\mathcal{O}$-algebras, M be an A-B-bimodule in* $\mathrm{perf}(A, B)$ *and N be a B-C-bimodule in* $\mathrm{perf}(B, C)$. *Then the A-C-bimodule* $M \otimes_B N$ *is in* $\mathrm{perf}(A, C)$.

Proof The hypotheses on M imply that as a right B-module, M is isomorphic to a direct summand of B^n for some positive integer n. Thus, as a right C-module, $M \otimes_B N$ is isomorphic to a direct summand of $B^n \otimes_B N \cong N^n$. This is a finitely generated projective right C-module by the assumptions on N. A similar argument shows that $M \otimes_B N$ is finitely generated projective as a left A-module. $\qquad\square$

Since an A-B-bimodule is, by convention, an $A \otimes_{\mathcal{O}} B^{\mathrm{op}}$-module, a projective A-B-bimodule is a projective $A \otimes_{\mathcal{O}} B^{\mathrm{op}}$-module. If A and B are finitely generated projective as $\mathcal{O}$-modules then the projective A-B-bimodule $A \otimes_{\mathcal{O}} B$ belongs to $\mathrm{perf}(A, B)$.

Proposition 9.1.2 *Let A, B, C be $\mathcal{O}$-algebras, M an A-B-bimodule in* $\mathrm{perf}(A, B)$ *and N a B-C-bimodule in* $\mathrm{perf}(B, C)$. *Suppose that A, B, C are finitely generated projective as $\mathcal{O}$-modules. If M is projective as an $A \otimes_{\mathcal{O}} B^{\mathrm{op}}$-module or if N is projective as a $B \otimes_{\mathcal{O}} C^{\mathrm{op}}$-module, then $M \otimes_B N$ is projective as an $A \otimes_{\mathcal{O}} C^{\mathrm{op}}$-module.*

Proof Suppose that M is projective as an $A \otimes B^{\mathrm{op}}$-module. Note that M is automatically finitely generated as an $A \otimes_{\mathcal{O}} B^{\mathrm{op}}$-module since it is even finitely generated as a left A-module. Thus M, viewed as an A-B-bimodule, is isomorphic to a direct summand of $(A \otimes_{\mathcal{O}} B)^n$ for some positive integer n. Therefore $M \otimes_B N$ is isomorphic to a direct summand of $(A \otimes_{\mathcal{O}} N)^n$, and since N is isomorphic to a direct summand of B^m, as a right B-module, for some positive integer m, the statement follows. $\qquad\square$

Definition 9.1.3 Let A, B be $\mathcal{O}$-algebras. We denote by $\mathcal{P}(A, B)$ the Grothendieck group of $\mathrm{perf}(A, B)$ with respect to split exact sequences. We set $\mathcal{P}(A) = \mathcal{P}(A, A)$.

If A and B are finitely generated as $\mathcal{O}$-modules, then so is $A \otimes_{\mathcal{O}} B^{\mathrm{op}}$, and hence the Krull–Schmidt Theorem holds for finitely generated $A \otimes_{\mathcal{O}} B^{\mathrm{op}}$-modules. In that case, the abelian group $\mathcal{P}(A, B)$ is free, having as a basis the set of isomorphism classes of indecomposable bimodules in $\mathrm{perf}(A, B)$. We denote by $[M]$ the isomorphism class of an A-B-bimodule M. If A, B, C are $\mathcal{O}$-algebras that are finitely generated projective as $\mathcal{O}$-modules, then by 9.1.1, the tensor product over B induces a bilinear map

$$- \cdot_B - : \mathcal{P}(A, B) \times \mathcal{P}(B, C) \to \mathcal{P}(A, C)$$

sending $([M], [N])$ to $[M] \cdot_B [N] = [M \otimes_B N]$. Applied to $A = B = C$ this induces an associative multiplication $- \cdot_A -$ on $\mathcal{P}(A)$; in this way, $\mathcal{P}(A)$ becomes a ring with unit element $[A]$. Similarly, the tensor products over A and over B induce a $\mathcal{P}(A)$-$\mathcal{P}(B)$-bimodule structure on the abelian group $\mathcal{P}(A, B)$.

Definition 9.1.4 Let A, B be $\mathcal{O}$-algebras. We say that elements X in $\mathcal{P}(A, B)$ and Y in $\mathcal{P}(B, A)$ *induce a virtual Morita equivalence between A and B* if $X \cdot_B Y = [A]$ and $Y \cdot_A X = [B]$.

A Morita equivalence induces a virtual Morita equivalence, but the converse need not be true. Any element X in $\mathcal{P}(A, B)$ is a finite $\mathbb{Z}$-linear combination of isomorphism classes of A-B-bimodules in perf(A, B). By collecting all summands with a positive coefficient and all summands with a negative coefficient we can write X in the form $X = [M] - [M']$ for some bimodules M, M' in perf(A, B), and after cancelling isomorphic summands in M, M', this way of writing X is unique thanks to the Krull–Schmidt Theorem. Similarly, we can write $Y = [N] - [N']$ for some N, N' in perf(B, A). The equality $X \cdot_B Y = [A]$ is thus equivalent to

$$[M \otimes_B N] - [M \otimes_B N'] - [M' \otimes_B N] + [M' \otimes_B N'] = [A].$$

By bringing negative summands to the other side, this becomes equivalent to an isomorphism of A-B-bimodules

$$M \otimes_B N \oplus M' \otimes_B N' \cong A \oplus M \otimes_B N' \oplus M' \otimes_B N.$$

The equality $Y \cdot_A X = [B]$ can be rewritten in a similar fashion. Any Rickard equivalence induces a virtual Morita equivalence:

Proposition 9.1.5 *Let A, B be $\mathcal{O}$-algebras, let X be a bounded complex of A-B-bimodules in* perf(A, B) *and Y a bounded complex of B-A-bimodules in* perf(B, A). *Suppose that $X \otimes_B Y \simeq A$ and that $Y \otimes_A X \simeq B$, where A and B are considered as complexes concentrated in degree zero. Then $\sum_{i \in \mathbb{Z}} (-1)^i [X_i]$ and $\sum_{i \in \mathbb{Z}} (-1)^i [Y_i]$ induce a virtual Morita equivalence between A and B.*

Proof For any bounded complex W of modules W_i in perf(A, B) write $[W] = \sum_{i \in \mathbb{Z}} (-1)^i [W_i]$; this is an element in $\mathcal{P}(A, B)$. If W is contractible then W is a direct sum of finitely many complexes which have exactly two nonzero identical terms in consecutive degrees, and hence $[W] = 0$. Thus $[X \otimes_B Y] = [A]$ and $[Y \otimes_A X] = [B]$. The definition of the tensor product of two complexes implies that $[X \otimes_B Y] = \sum_{i,j \in \mathbb{Z}} (-1)^{i+j} [X_i \otimes_B Y_j]$, and this is the product (induced by the tensor product over B) of the two elements $[X] = \sum_{i \in \mathbb{Z}} (-1)^i [X_i]$ and $[Y] = \sum_{i \in \mathbb{Z}} (-1)^i [Y_i]$. Similarly for $[Y \otimes_A X]$, whence the result. □

Proposition 9.1.6 *Let A, B be symmetric $\mathcal{O}$-algebras. The map sending an A-B-bimodule M to its $\mathcal{O}$-dual $M^* = \mathrm{Hom}_{\mathcal{O}}(M, \mathcal{O})$ induces a group isomorphism $\mathcal{P}(A, B) \cong \mathcal{P}(B, A)$. For $A = B$ this is an anti-automorphism of $\mathcal{P}(A)$ whose square is the identity on $\mathcal{P}(A)$.*

Proof Duality commutes with finite direct sums, hence induces a group homomorphism $\mathcal{P}(A, B) \to \mathcal{P}(B, A)$. Applying duality again yields the inverse of this group homomorphism. If M, N are in $\mathrm{perf}(A, A)$ then $(M \otimes_A N)^* \cong M^* \otimes_A N^*$, and hence this yields an anti-automorphism of order two of $\mathcal{P}(A)$ as claimed. □

9.2 Perfect isometries

Perfect isometries are certain character bijections 'with signs' between blocks of finite groups which preserve many numerical and structural invariants. This concept and all results in this section are due to Broué [38, §1]. We assume that K is a splitting fields for the finite groups and their subgroups in this section.

Let G, H be finite groups and b, c be blocks of $\mathcal{O}G$, $\mathcal{O}H$, respectively. Then $b \otimes c$ corresponds to a unique block of $\mathcal{O}(G \times H)$ via the canonical isomorphism $\mathcal{O}(G \times H) \cong \mathcal{O}G \otimes_{\mathcal{O}} \mathcal{O}H$, and we denote this block abusively again by $b \otimes c$. Let μ be a K-valued class function on $G \times H$. Then, for any $x \in G$, the map $\mu(x, -)$ sending $y \in H$ to $\mu(x, y)$ is a class function on H, and for any $y \in H$ the map $\mu(-, y)$ sending $x \in G$ to $\mu(x, y)$ is a class function on G. For any K-valued class function η on H the map χ defined by

$$\chi(x) = \langle \mu(x, -), \eta \rangle_H = \frac{1}{|H|} \sum_{y \in H} \mu(x, y^{-1}) \eta(y)$$

for all $x \in G$ is a K-valued class function on G. If $\mu \in \mathbb{Z}\mathrm{Irr}_K(G \times H)$ and $\eta \in \mathbb{Z}\mathrm{Irr}_K(H)$, then $\chi \in \mathbb{Z}\mathrm{Irr}_K(G)$; in other words, if μ, η are generalised characters of $G \times H$, H, respectively, then χ, defined as above, is a generalised character of G. Thus any $\mu \in \mathbb{Z}\mathrm{Irr}_K(G \times H)$ induces a group homomorphism $\Phi_\mu : \mathbb{Z}\mathrm{Irr}_K(H) \to \mathbb{Z}\mathrm{Irr}_K(G)$ defined by the formula

$$\Phi_\mu(\eta)(x) = \frac{1}{|H|} \sum_{y \in H} \mu(x, y^{-1}) \eta(y)$$

for all $\eta \in \mathbb{Z}\mathrm{Irr}_K(H)$ and $x \in G$. Conversely, any group homomorphism $\Phi : \mathbb{Z}\mathrm{Irr}_K(H) \to \mathbb{Z}\mathrm{Irr}_K(G)$ determines a generalised character $\mu_\Phi \in \mathbb{Z}\mathrm{Irr}_K(G \times H)$

by the formula

$$\mu_\Phi(x, y) = \sum_{\eta \in \mathrm{Irr}_K(H)} \Phi(\eta)(x)\eta(y)$$

for all $(x, y) \in G \times H$. It follows from the first orthogonality relations that the assignments $\mu \mapsto \Phi_\mu$ and $\Phi \mapsto \mu_\Phi$ are inverse to each other. This correspondence is compatible with blocks in the following sense. If μ is associated with the block $b \otimes c$ of $G \times H$, then the group homomorphism Φ_μ sends $\mathbb{Z}\mathrm{Irr}_K(H, c)$ to $\mathbb{Z}\mathrm{Irr}_K(G, b)$ and is zero outside of $\mathbb{Z}\mathrm{Irr}_K(H, c)$. Conversely, given a group homomorphism $\Phi : \mathbb{Z}\mathrm{Irr}_K(H, c) \to \mathbb{Z}\mathrm{Irr}_K(G, b)$, extending this by zero on $\mathrm{Irr}_K(H, c')$ for all blocks c' of $\mathcal{O}H$ different from c, the corresponding character μ_Φ is associated with $b \otimes c$.

Definition 9.2.1 ([38, 1.1]) Let G, H be finite groups. A generalised K-valued character $\mu \in \mathbb{Z}\mathrm{Irr}_K(G \times H)$ is called *perfect* if the following two conditions hold for any $x \in G$ and any $y \in H$:

(i) we have $\frac{\mu(x,y)}{|C_G(x)|} \in \mathcal{O}$ and $\frac{\mu(x,y)}{|C_H(y)|} \in \mathcal{O}$; in other words, $\mu(x, y)$ is divisible in $\mathcal{O}$ by the orders of $C_G(x)$ and $C_H(y)$; and
(ii) if exactly one of x, y has order prime to p but the other has order divisible by p then $\mu(x, y) = 0$.

The groups $\mathbb{Z}\mathrm{Irr}_K(G, b)$, $\mathbb{Z}\mathrm{Irr}_K(H, c)$ are endowed with scalar products; an *isometry* between b and c is an isomorphism $\mathbb{Z}\mathrm{Irr}_K(H, c) \cong \mathbb{Z}\mathrm{Irr}_K(G, b)$ preserving scalar products. Any such isometry preserves in particular elements of norm 1, thus sends $\eta \in \mathrm{Irr}_K(H, c)$ to $\delta_\eta \chi_\eta$ for some $\delta_\eta \in \{\pm 1\}$ and some $\chi_\eta \in \mathrm{Irr}_K(G, b)$. In other words, an isometry between b and c is a character bijection 'with signs'.

Definition 9.2.2 ([38, 1.4]) Let G, H be finite groups and let b, c be blocks of $\mathcal{O}G, \mathcal{O}H$, respectively. An isometry $\Phi : \mathbb{Z}\mathrm{Irr}_K(H, c) \cong \mathbb{Z}\mathrm{Irr}_K(G, b)$ is called *perfect* if the associated generalised character $\mu \in \mathbb{Z}\mathrm{Irr}_K(G \times H)$ defined by $\mu(x, y) = \sum_{\eta \in \mathrm{Irr}_K(H,c)} \Phi(\eta)(x)\eta(y)$ for all $(x, y) \in G \times H$ is perfect.

Any character bijection $\eta \mapsto \chi_\eta$ between $\mathrm{Irr}_K(H, c)$ and $\mathrm{Irr}_K(G, b)$ induces an isomorphism of centres $Z(KHc) \cong Z(KGb)$ over K, sending the primitive idempotent $e(\eta) = \frac{\eta(1)}{|H|} \sum_{y \in H} \eta(y^{-1})y$ in $Z(KHc)$ to the primitive idempotent $e(\chi_\eta)$ in $Z(KGb)$. Not any such isomorphism will restrict to an isomorphism $Z(\mathcal{O}Hc) \cong Z(\mathcal{O}Gb)$, however. One of the important properties of perfect isometries is that they always do induce isomorphisms of the centres over $\mathcal{O}$, and this is where the arithmetic conditions in the definition of perfect characters come into play:

Theorem 9.2.3 ([38, 1.5 (1)]) *Let G, H be finite groups and let b, c be blocks of $\mathcal{O}G$, $\mathcal{O}H$, respectively. Let $\Phi : \mathbb{Z}\mathrm{Irr}_K(H, c) \cong \mathbb{Z}\mathrm{Irr}_K(G, b)$ be a perfect isometry. For $\eta \in \mathrm{Irr}_K(H, c)$ let $\delta_\eta \in \{\pm 1\}$ and $\chi_\eta \in \mathrm{Irr}_K(G, b)$ such that $\Phi(\eta) = \delta_\eta \chi_\eta$. The map sending $e(\eta)$ to $e(\chi_\eta)$ induces an isomorphism $Z(\mathcal{O}Hc) \cong Z(\mathcal{O}Gb)$ as $\mathcal{O}$-algebras.*

Proof Define $\mu \in \mathbb{Z}\mathrm{Irr}_K(G \times H)$ as above by $\mu(x, y) = \sum_{\eta \in \mathrm{Irr}_K(H,c)} \delta_\eta \chi_\eta(x)\eta(y)$, for all $(x, y) \in G \times H$. By the assumptions, μ is perfect. Define a map $\rho : Z(KHc) \rightarrow Z(KGb)$ by

$$\rho(u) = \sum_{x \in G} \left(\frac{1}{|H|} \sum_{y \in H} \mu(x, y^{-1})u(y) \right) x$$

for all $u = \sum_{y \in H} u(y)y$ in $Z(KHc)$. Similarly, define a map $\sigma : Z(KGb) \rightarrow Z(KHc)$ by

$$\sigma(v) = \sum_{y \in H} \left(\frac{1}{|G|} \sum_{x \in G} \mu(x^{-1}, y)v(x) \right) y$$

for all $v = \sum_{x \in G} v(x)x$ in $Z(KGb)$. The first orthogonality relations imply that

$$\rho(e(\eta)) = \frac{|H| \cdot \delta_\eta \chi_\eta(1)}{|G| \cdot \eta(1)} e(\chi_\eta)$$

$$\sigma(e(\chi_\eta)) = \frac{|G| \cdot \eta(1)}{|H| \cdot \delta_\eta \chi_\eta(1)} e(\eta)$$

for all $\eta \in \mathrm{Irr}_K(H, c)$. This shows that ρ and σ are K-linear maps that are inverse to each other. We show next that ρ maps $Z(\mathcal{O}Hc)$ to $Z(\mathcal{O}Gb)$. Let u be the conjugacy class sum of a group element $y \in H$. Then $uc \in Z(\mathcal{O}Hc)$ and $\eta(uc) = \eta(u) = |H : C_H(y)|\eta(y)$ for all $\eta \in \mathrm{Irr}_K(H, c)$. Thus $\rho(uc) = \sum_{x \in G} (\frac{1}{|H|} \sum_{\eta \in \mathrm{Irr}_K(H,c)} |H : C_H(y)|\eta(y)\chi_\eta(x^{-1}))x = \sum_{x \in G} \frac{\mu(x,y^{-1})}{|C_H(y)|}x$, and this belongs to $Z(\mathcal{O}Gb)$ thanks to the divisibility property of perfect characters. Since elements of the form uc, with u an H-conjugacy class sum, generate $Z(\mathcal{O}Hc)$ as an $\mathcal{O}$-module it follows that ρ maps $Z(\mathcal{O}Hc)$ to $Z(\mathcal{O}Gb)$. A similar argument shows that σ maps $Z(\mathcal{O}Gb)$ to $Z(\mathcal{O}Hc)$. Define K-linear maps $\alpha : Z(KHc) \rightarrow Z(KGb)$ and $\beta : Z(KGb) \rightarrow Z(KHc)$ by setting

$$\alpha(u) = \rho(u\sigma(b))$$

$$\beta(v) = \sigma(v\rho(c))$$

for all $u \in Z(KHc)$ and $v \in Z(KGb)$. Since ρ maps $Z(\mathcal{O}Hc)$ to $Z(\mathcal{O}Gb)$, so does α. Similarly for β. Using the fact that $b = \sum_{\chi \in \mathrm{Irr}_K(G,b)} e(\chi)$ we get that

$$\alpha(e(\eta)) = \rho\left(\sum_{\eta' \in \mathrm{Irr}_K(H,c)} e(\chi_\eta)\sigma(e(\chi_{\eta'}))\right) = e(\chi_\eta).$$

Similarly, $\beta(e(\chi_\eta)) = e(\eta)$. This shows that α and β are in fact algebra isomorphisms between $Z(KHc)$ and $Z(KGb)$ that are inverse to each other. By the above α and β induce inverse isomorphisms $Z(\mathcal{O}Hc) \cong Z(\mathcal{O}Gb)$, whence the result. $\qquad\square$

Corollary 9.2.4 ([38, 1.6]) *Let G, H be finite groups and let b, c be blocks of $\mathcal{O}G$, $\mathcal{O}H$, respectively. Let $\Phi : \mathbb{Z}\mathrm{Irr}_K(H, c) \cong \mathbb{Z}\mathrm{Irr}_K(G, b)$ be a perfect isometry. For $\eta \in \mathrm{Irr}_K(H, c)$ let $\delta_\eta \in \{\pm 1\}$ and $\chi_\eta \in \mathrm{Irr}_K(G, b)$ such that $\Phi(\chi) = \delta_\eta \chi_\eta$. Then for any $\eta \in \mathrm{Irr}_K(H, c)$ we have $\frac{|H| \cdot \delta_\eta \chi_\eta(1)}{|G| \cdot \eta(1)} \in \mathcal{O}^\times$, and the image in $k^\times$ of this number does not depend on η.*

Proof With the notation of the proof of 9.2.3, we have

$$\rho(c) = \sum_{\eta \in \mathrm{Irr}_K(H,c)} \left(\frac{|H| \cdot \delta_\eta \chi_\eta(1)}{|G| \cdot \eta(1)}\right) e(\eta).$$

By 6.5.8 it suffices to show that $\rho(c)$ is invertible in $Z(\mathcal{O}Gb)$. The algebra isomorphism β defined in the proof of 9.2.3 is the composition of multiplication by $\rho(c)$ followed by the linear isomorphism σ; thus multiplication by $\rho(c)$ is a linear automorphism of $Z(\mathcal{O}Gb)$, and hence $\rho(c)$ is invertible in $Z(\mathcal{O}Gb)$. $\qquad\square$

Corollary 9.2.5 ([38, 1.5 (2)]) *Let G, H be finite groups and let b, c be blocks of $\mathcal{O}G$, $\mathcal{O}H$ with defect groups P, Q, respectively. Let $\Phi : \mathbb{Z}\mathrm{Irr}_K(H, c) \cong \mathbb{Z}\mathrm{Irr}_K(G, b)$ be a perfect isometry.*

(i) *We have $|P| = |Q|$.*
(ii) *For $\eta \in \mathrm{Irr}_K(H, c)$, $\chi \in \mathrm{Irr}_K(G, b)$, and $\delta \in \{\pm 1\}$ such that $\Phi(\eta) = \delta\chi$, we have $\mathrm{ht}(\chi) = \mathrm{ht}(\eta)$.*

Proof Let η, χ be as in (ii). By the definition of character heights, the p-part of $\chi(1)$ is $p^{a-d+\mathrm{ht}(\chi)}$, where p^a is the p-part of $|G|$ and where $p^d = |P|$. Thus the p-part of $\frac{|G|}{\chi(1)}$ is equal to $p^{d-\mathrm{ht}(\chi)}$. Similarly, the p-part of $\frac{|H|}{\eta(1)}$ is equal to $p^{d'-\mathrm{ht}(\eta)}$, where $p^{d'} = |Q|$. It follows from 9.2.4 that $p^{d-\mathrm{ht}(\chi)} = p^{d'-\mathrm{ht}(\eta)}$. Applied to a character χ of height zero (which exists in any block by 6.11.7) it follows that $d \leq d'$, and applied to a character η of height zero, it follows that $d \geq d'$. This proves (i). But then $p^{d-\mathrm{ht}(\chi)} = p^{d-\mathrm{ht}(\eta)}$ for any pair of characters χ, η as in (ii), and hence $\mathrm{ht}(\chi) = \mathrm{ht}(\eta)$, which proves (ii). $\qquad\square$

The subgroup $\mathrm{Pr}_{\mathcal{O}}(G, b)$ of $\mathbb{Z}\mathrm{Irr}_K(G, b)$ generated by the characters of finitely generated projective $\mathcal{O}Gb$-modules consists precisely of all generalised characters in $\mathbb{Z}\mathrm{Irr}_K(G, b)$ that vanish on all p-singular elements. Its orthogonal subgroup $L^0(G, b)$ consists of all generalised characters in $\mathbb{Z}\mathrm{Irr}_K(G, b)$ that vanish on all p'-elements. These subgroups are preserved under perfect isometries:

Proposition 9.2.6 *Let G, H be finite groups and let b, c be blocks of $\mathcal{O}G$, $\mathcal{O}H$, respectively. Let $\Phi : \mathbb{Z}\mathrm{Irr}_K(H, c) \cong \mathbb{Z}\mathrm{Irr}_K(G, b)$ be a perfect isometry. Then Φ maps $\mathrm{Pr}_{\mathcal{O}}(H, c)$ onto $\mathrm{Pr}_{\mathcal{O}}(G, b)$, and Φ maps $L^0(H, c)$ onto $L^0(G, b)$.*

Proof Let μ be the generalised character of $\mathcal{O}(G \times H)$ such that $\Phi = \Phi_\mu$. Let $\eta \in \mathrm{Pr}_{\mathcal{O}}(H, c)$; that is, $\eta(y) = 0$ for all p-singular $y \in H$. The formula for Φ_μ implies that $\Phi(\eta)(x) = 0$ for all p-singular elements $x \in G$. Thus Φ maps $\mathrm{Pr}_{\mathcal{O}}(H, c)$ to $\mathrm{Pr}_{\mathcal{O}}(G, b)$. Its inverse maps $\mathrm{Pr}_{\mathcal{O}}(G, b)$ to $\mathrm{Pr}_{\mathcal{O}}(H, c)$. Similarly for $L^0(H, c)$ and $L^0(G, b)$. □

The subgroup $L^0(G, b)$ of $\mathbb{Z}\mathrm{Irr}_K(G, b)$ is the kernel of the decomposition map $d_G : \mathbb{Z}\mathrm{Irr}_K(G, b) \to \mathbb{Z}\mathrm{IBr}_k(G, b)$, which by 5.14.1 is surjective. By the preceding result, this subgroup is preserved under perfect isometries.

Corollary 9.2.7 (cf. [38, 1.3]) *Let G, H be finite groups and let b, c be blocks of $\mathcal{O}G$, $\mathcal{O}H$, respectively. Let $\Phi : \mathbb{Z}\mathrm{Irr}_K(H, c) \cong \mathbb{Z}\mathrm{Irr}_K(G, b)$ be a perfect isometry. There is a unique group isomorphism $\bar{\Phi} : \mathbb{Z}\mathrm{IBr}_k(G, b) \to \mathbb{Z}\mathrm{IBr}_k(G, b)$ such that $d_G \circ \Phi = \bar{\Phi} \circ d_H$. Moreover, $\bar{\Phi}$ induces an isomorphism $\mathrm{Pr}_k(H, c) \cong \mathrm{Pr}_k(G, b)$.*

The order of the finite abelian group $\mathbb{Z}\mathrm{IBr}_k(G, b)/\mathrm{Pr}_k(G, b)$ is the determinant of the Cartan matrix of b, and the orders of the cyclic direct factors of this group are the elementary divisors of the Cartan matrix. Thus we obtain:

Corollary 9.2.8 *Let G, H be finite groups and let b, c be blocks of $\mathcal{O}G$, $\mathcal{O}H$, respectively. Let $\Phi : \mathbb{Z}\mathrm{Irr}_K(H, c) \cong \mathbb{Z}\mathrm{Irr}_K(G, b)$ be a perfect isometry. The Cartan matrices of b and c have the same determinant and elementary divisors (with the same multiplicities).*

Exercise 9.2.9 Let G be a finite group. Show that the identity map on $\mathrm{Irr}_K(G)$ is a perfect isometry. (We will see in the next section that this can be deduced from more general results on Morita and derived equivalences between block algebras.)

Exercise 9.2.10 Show that there is a perfect isometry between the character groups of the dihedral group D_8 and the quaternion group Q_8. (*Hint:* use that D_8 and Q_8 have the same character table, and then use the previous exercise.)

Exercise 9.2.11 Using the notation from §3.4, show that there is a perfect isometry between the character groups of A_4 and of the principal 2-block of A_5 given by the assignment $\eta_1 \mapsto \chi_1$, $\eta_2 \mapsto -\chi_2$, $\eta_3 \mapsto -\chi_3$, $\eta_4 \mapsto -\chi_5$.

Exercise 9.2.12 Using the notation from §3.4, show that there is a perfect isometry between the character group of the principal 5-block of A_5 and the character group of the normaliser H in A_5 of the 5-cycle $(1, 2, 3, 4, 5)$ (which is a dihedral group of order 10, hence a Frobenius group), given by an assignment $\chi_1 \mapsto \psi_1$, $\chi_2 \mapsto -\psi_3$, $\chi_3 \mapsto -\psi_4$, $\chi_4 \mapsto -\psi_2$, where χ_1, ψ_1 are the trivial characters of A_5, H, respectively, ψ_2 is the nontrivial degree 1 character of H and ψ_3, ψ_4 have degree 2.

9.3 Perfect isometries, derived and stable equivalences

Following Broué [38, §3], a Morita equivalence and, more generally, a derived equivalence between two blocks induces a perfect isometry. To show this, we will need the following criterion for when a character of $G \times H$ is perfect:

Proposition 9.3.1 ([38, 1.2]) *Let G, H be finite groups and M an $\mathcal{O}G$-$\mathcal{O}H$-bimodule such that the restrictions of M on the left and right are finitely generated projective. Then the character μ of M, viewed as an $\mathcal{O}(G \times H)$-module, is perfect.*

Proof Since M is projective as a left $\mathcal{O}G$-module, by Higman's criterion 2.6.2, there is an element $\varphi \in \mathrm{End}_{\mathcal{O}}(M)$ such that $\mathrm{Tr}_1^G(\varphi) = \mathrm{Id}_M$, where $\mathrm{Tr}_1^G(\varphi)(m) = \sum_{x \in G} x\varphi(x^{-1}m)$ for all $m \in M$. Consider $\mathrm{End}_{\mathcal{O}}(M)$ as $\mathcal{O}G$-$\mathcal{O}H$-bimodule via $(x \cdot \varphi \cdot y)(m) = x\varphi(m)y$, where $x \in G$, $y \in H$, $m \in M$ and $\varphi \in \mathrm{End}_{\mathcal{O}}(M)$. Let $z \in Z(\mathcal{O}G)$ and $y \in H$. The character value $\mu(z, y)$ is the trace of the endomorphism $z \cdot \mathrm{Id}_M \cdot y^{-1}$; thus

$$\mu(z, y) = \mathrm{tr}_M(z \cdot \mathrm{Tr}_1^G(\varphi) \cdot y^{-1}) = \mathrm{tr}_M(\mathrm{Tr}_1^G(z \cdot \varphi \cdot y^{-1})) = |G| \cdot \mathrm{tr}_M(z \cdot \varphi \cdot y^{-1}).$$

Thus $\frac{\mu(z,y)}{|G|} \in \mathcal{O}$. By taking for z the conjugacy class sum of an element $x \in G$ we get that $\frac{\mu(x,y)}{|C_G(x)|} \in \mathcal{O}$. Suppose now that y is a p'-element in H. The restriction of M, viewed as an $\mathcal{O}(G \times H)$-module, to the subgroup $G \times \langle y \rangle$ is then a projective module for this group. Thus its character vanishes at all p-singular elements in $G \times \langle y \rangle$, and hence if $\mu(x, y) \neq 0$ for some $x \in G$, then x is necessarily a p'-element in G as well. $\square$

We show next that if a perfect character μ of $G \times H$ is the character of an $\mathcal{O}G$-$\mathcal{O}H$-bimodule M that is finitely generated projective as a left and right module, then the group homomorphism $\Phi_M : \mathbb{Z}\mathrm{Irr}_K(H) \to \mathbb{Z}\mathrm{Irr}_K(G)$ is equal to the group homomorphism induced by the functor $M \otimes_{\mathcal{O}H} -$ from $\mathrm{mod}(\mathcal{O}H)$ to $\mathrm{mod}(\mathcal{O}G)$. We show more generally that this statement can be extended to bounded complexes. The character χ_U of a bounded chain complex U of $\mathcal{O}$-free $\mathcal{O}G$-modules is defined as the alternating sum $\chi_U = \sum_{i \in \mathbb{Z}}(-1)^i \chi_i$, where χ_i is the character of the i-th component U_i of U. In general, χ_U is a virtual character. Two homotopy equivalent bounded complexes of $\mathcal{O}$-free $\mathcal{O}G$-modules have the same character because a bounded contractible complex has the zero character as it is a direct sum of finitely many complexes of the form $0 \to U \to U \to 0$, with U an $\mathcal{O}G$-module in two consecutive degrees.

Proposition 9.3.2 *Let G, H be finite groups and X a bounded complex of $\mathcal{O}G$-$\mathcal{O}H$-bimodules such that the components of X are finitely generated projective as right $\mathcal{O}Hc$-modules. Denote by μ the character of X as a complex of $\mathcal{O}(G \times H)$-modules. Let V be a bounded complex of finitely generated $\mathcal{O}$-free $\mathcal{O}H$-modules and denote by η the character of V. Then the components of the bounded complex $X \otimes_{\mathcal{O}H} V$ are finitely generated $\mathcal{O}$-free, and its character χ satisfies*

$$\chi(x) = \frac{1}{|H|} \sum_{y \in H} \mu(x, y^{-1})\eta(y)$$

for all $x \in G$; in other words, we have $\chi = \Phi_\mu(\eta)$.

Proof We consider first the case where X, V are modules, viewed as complexes concentrated in degree zero. The module $X \otimes_{\mathcal{O}H} V$ is finitely generated $\mathcal{O}$-free because V is and X is finitely generated projective as a right $\mathcal{O}H$-module. The statement on the characters of these modules depends only on their scalar extensions over K. Thus it suffices to show the character formula for the character, abusively again denoted by η, of a simple KH-module W instead of V and a simple KG-KH-bimodule instead of X. Moreover, after extending K, if necessary, we may assume that K is a splitting field for $G \times H$. Then any simple KG-KH-bimodule is of the form $S \otimes_K T^*$, where S is a simple KG-module and T a simple KH-module. The character μ of this module, as a $K(G \times H)$-module, is given by $\mu(x, y) = \chi(x)\psi(y^{-1})$, where χ is the character of S and ψ is the character of T. By 2.12.8 we have $T^* \otimes_{KH} W \cong K$ if $W \cong T$, and zero otherwise. Thus $(S \otimes_K T^*) \otimes_{KH} W \cong X$ if $W \cong T$, and zero otherwise. By the orthogonality relations, the character of this module is in both cases equal to $\frac{1}{|H|}\chi(x) \sum_{y \in H} \psi(y^{-1})\eta(y)$, which proves the result in the case where X, V are concentrated in degree zero. Back to the general case, the component of $X \otimes_{\mathcal{O}H} V$ in degree $n \in \mathbb{Z}$ is equal to $\oplus_{i+j=n} X_i \otimes_{\mathcal{O}H} V_j$. The character μ of X is equal to $\sum_{n \in \mathbb{Z}} \mu_n$, where μ_n is the character of X_n; similarly, the character η

of V is equal to $\sum_{j\in\mathbb{Z}} \eta_j$, where η_j is the character of V_j. Thus the character of $X \otimes_{\mathcal{O}H} V$ is equal to $\sum_{n\in\mathbb{Z}} \sum_{i+j=n} (1-)^i (-1)^j \Phi_{\mu_i}(\eta_j) = \sum_i (-1)^i \Phi_{\mu_i}(\eta)) = \Phi_\mu(\eta)$ as claimed. $\qquad\square$

This proposition implies that a bimodule that induces a Morita equivalence between two blocks has a perfect character which induces a perfect isometry between the two blocks. This is true, more generally, for derived equivalences. If two block algebras are derived equivalent, then some derived equivalence is realised by tensoring with a Rickard complex. In conjunction with the previous proposition, this shows that derived equivalences imply the existence of perfect isometries:

Corollary 9.3.3 ([38, 3.1]) *Let G, H be finite groups and let b, c be blocks of $\mathcal{O}G$, $\mathcal{O}H$, respectively. Suppose that K is a splitting field for $G \times H$. Let X be a bounded complex of $\mathcal{O}Gb$-$\mathcal{O}Hc$-bimodules that are finitely generated projective as left and right modules such that $X \otimes_{\mathcal{O}H} X^* \simeq \mathcal{O}Gb$ and $X^* \otimes_{\mathcal{O}G} X \simeq \mathcal{O}Hc$. Denote by μ the character of X. Then $\Phi_\mu : \mathbb{Z}\mathrm{Irr}_K(H, c) \cong \mathbb{Z}\mathrm{Irr}_K(G, b)$ is a perfect isometry.*

Proof By the assumptions on X and by 9.3.2, the group homomorphisms Φ_μ and Φ_{μ^*} are inverse to each other, where μ^* is the character of X^*. Since the components of X are finitely generated projective as left and right modules, the character μ is perfect, by 9.3.1, whence the result. $\qquad\square$

The same argument works for virtual Morita equivalences. If we write an element X in $\mathcal{P}(\mathcal{O}Gb, \mathcal{O}Hc)$ in the form $X = [M] - [M']$ for some M, M' in $\mathrm{perf}(\mathcal{O}Gb, \mathcal{O}Hc)$ then the character of X is defined as the difference $\mu_M - \mu_{M'}$ of the characters of M and M'. By 9.3.1, the character μ of X is perfect, and induces a group homomorphism $\Phi_\mu : \mathbb{Z}\mathrm{Irr}_K(H, c) \to \mathbb{Z}\mathrm{Irr}_K(G, b)$.

Corollary 9.3.4 ([38, 3.1]) *Let G, H be finite groups and let b, c be blocks of $\mathcal{O}G$, $\mathcal{O}H$, respectively. Suppose that K is a splitting field for $G \times H$. Let X be an element in $\mathcal{P}(\mathcal{O}Gb, \mathcal{O}Hc)$ such that X and X^* induce a virtual Morita equivalence between $\mathcal{O}Gb$ and $\mathcal{O}Hc$. Denote by μ the character of X. Then $\Phi_\mu : \mathbb{Z}\mathrm{Irr}_K(H, c) \to \mathbb{Z}\mathrm{Irr}_K(G, b)$ is a perfect isometry.*

Proof The proof is an obvious variation of the argument in the proof of 9.3.3. $\qquad\square$

Not every perfect isometry is induced by a virtual Morita equivalence; see the Remark 9.3.8 below. Since a Rickard equivalence induces a virtual Morita equivalence, it follows that 9.3.4 implies 9.3.3. By contrast, a stable equivalence of Morita type between two block algebras $\mathcal{O}Gb$ and $\mathcal{O}Hc$ need not induce an isometry. We will show that it induces a partial isometry between the groups $L^0(H, c)$ and $L^0(G, b)$ of generalised characters which are perpendicular to all projective characters. The first obstacle to extending a partial isometry to an isometry between $\mathbb{Z}\mathrm{Irr}_K(H, c)$ and $\mathbb{Z}\mathrm{Irr}_K(G, b)$ is that we do not know that these two groups have the same rank. What we can show is that whenever this partial isometry does indeed extend, then the resulting isometry is perfect. If $A = \mathcal{O}Gb$, where G is a finite group and b a block of $\mathcal{O}G$, the group $\mathbb{Z}\mathrm{Irr}_K(G, b)$ is canonically isomorphic to the Grothendieck group $R_K(A)$ of finite dimensional $K \otimes_{\mathcal{O}} A$-modules. Using this identification, the existence of partial isometries induced by stable equivalences of Morita type holds for arbitrary symmetric $\mathcal{O}$-algebras whose Cartan matrices over k are invertible. We use the results and notation from Section 4.17, notably the notation $R_K(A)$, $R_k(A)$ for the Grothendieck groups of finite-dimensional modules over $K \otimes_{\mathcal{O}} A$, $k \otimes_{\mathcal{O}} A$, respectively, as well as $L^0(A)$ for the kernel of the decomposition map $d_A :$ $R_K(A) \to R_k(A)$. If M is in perf(A, B) we denote by $\Phi_M : R_K(B) \to R_K(A)$ the group homomorphism induced by the exact functor $M \otimes_B -$. If $A = \mathcal{O}Gb$ and $B = \mathcal{O}Hc$ for some blocks b, c of finite groups G, H, then the map Φ_M is equal to the map Φ_μ induced by the character μ of M, thanks to 9.3.2. Recall that if A, B are symmetric $\mathcal{O}$-algebras, an A-B-bimodule M is said to induce a stable equivalence of Morita type if M is in perf(A, B) and if $M \otimes_B M^* \cong A \oplus V$ for some projective A-A-bimodule V and $M^* \otimes_A M \cong B \oplus W$ for some projective B-B-bimodule W.

Proposition 9.3.5 *Let A, B be symmetric $\mathcal{O}$-algebras such that $K \otimes_{\mathcal{O}} A$ and $K \otimes_{\mathcal{O}} B$ are split semisimple and such that $k \otimes_{\mathcal{O}} A$ and $k \otimes_{\mathcal{O}} B$ are split. Assume that the Cartan matrices C_A and C_B of $k \otimes_{\mathcal{O}} A$ and $k \otimes_{\mathcal{O}} B$ are non singular. Let M be a B-A-bimodule in* perf(A, B). *Suppose that M induces a stable equivalence of Morita type between A and B.*

(i) Φ_M and Φ_{M^*} induce inverse isomorphisms $R_K(A)/\mathrm{Pr}_{\mathcal{O}}(A) \cong R_K(B)/$ $\mathrm{Pr}_{\mathcal{O}}(B)$.

(ii) φ_M and φ_{M^*} induce inverse isomorphisms $R_k(A)/\mathrm{Pr}_k(A) \cong R_k(B)/\mathrm{Pr}_k(B)$; in particular, $|\det(C_B)| = |\det(C_A)|$.

(iii) Φ_M and Φ_{M^*} induce inverse isometries $L^0(A) \cong L^0(B)$.

Proof The Cartan matrix of $k \otimes_{\mathcal{O}} A$ is non singular, and hence we have $\mathrm{Pr}(A) \cap$ $L^0(A) = \{0\}$, by 4.17.1. Similarly for B. Note that Φ_M sends $\mathrm{Pr}_{\mathcal{O}}(A)$ to $\mathrm{Pr}_{\mathcal{O}}(B)$ and $L^0(A) = \ker(d_A)$ to $L^0(B) = \ker(d_B)$, thus induces group homomorphisms $R_K(A)/\mathrm{Pr}_{\mathcal{O}}(A) \to R_K(B)/\mathrm{Pr}_{\mathcal{O}}(B)$ and $R_k(A)/\mathrm{Pr}_k(A) \to R_k(B)/\mathrm{Pr}_k(B)$. The map $\Phi_{M^*} \circ \Phi_M$ is the map induced by tensoring with the bimodule $M^* \otimes_B M \cong A \oplus W$, where W is a projective A-A-bimodule. Thus Φ_W maps

$R_K(A)$ to $\mathrm{Pr}_{\mathcal{O}}(A)$, and hence for any $\chi \in R_K(A)$ we have $\Phi_{M^*}(\Phi_M(\chi)) = \chi + \zeta$ for some $\zeta \in \mathrm{Pr}_{\mathcal{O}}(A)$. Since $\Phi_{M^*} \circ \Phi_M$ maps $L^0(A)$ to itself and since $\mathrm{Pr}(A) \cap L^0(A) = \{0\}$ this implies that if $\chi \in L^0(A)$ then $\zeta = 0$. Thus Φ_M and Φ_{M^*} induce inverse isomorphisms $R_K(A)/\mathrm{Pr}_{\mathcal{O}}(A) \cong R_K(B)/\mathrm{Pr}_{\mathcal{O}}(B)$ and $L^0(A) \cong L^0(B)$. They also induce inverse isomorphisms $R_k(A)/\mathrm{Pr}_k(A) \cong R_k(B)/\mathrm{Pr}_k(B)$, as observed already much earlier in 4.14.13. Since the functors $M \otimes_A -$ and $M^* \otimes_B -$ are adjoint we get in particular for $\chi, \chi' \in L^0(A)$ that

$$\langle \Phi_M(\chi), \Phi_M(\chi') \rangle_B = \langle \chi, \Phi_{M^*}(\Phi_M(\chi')) \rangle_A = \langle \chi, \chi' \rangle_A$$

which shows that the isomorphisms between $L^0(A)$ and $L^0(B)$ are isometries. $\square$

The surjectivity of the decomposition map is invariant under stable equivalences of Morita type:

Proposition 9.3.6 ([99, 3.2]) *Let A, B be symmetric $\mathcal{O}$-algebras such that $K \otimes_{\mathcal{O}} A$, $K \otimes_{\mathcal{O}} B$ are split semisimple and such that $k \otimes_{\mathcal{O}} A$, $k \otimes_{\mathcal{O}} B$ are split. Suppose there is a stable equivalence of Morita type between A and B. Then the decomposition map $d_A : R_K(A) \to R_k(A)$ is surjective if and only if the decomposition map $d_B : R_K(B) \to R_k(B)$ is surjective.*

Proof Let M be a B-A-bimodule that is finitely generated projective as a left B-module and as a right A-module and that induces a stable equivalence of Morita type between A and B. Suppose that d_A is surjective. Then the map $\bar{d}_A : R_K(A)/\mathrm{Pr}_{\mathcal{O}}(A) \to R_k(A)/\mathrm{Pr}_k(A)$ induced by d_A is surjective. Thus, using the fact that d_A and d_B commute with the maps Φ_M and $\bar{\Phi}_M$ as in 4.17.4, the map $\bar{d}_B : R_K(B)/\mathrm{Pr}_{\mathcal{O}}(B) \to R_k(B)/\mathrm{Pr}_k(B)$ induced by d_B is surjective. Since $\mathrm{Im}(d_B)$ contains $\mathrm{Pr}_k(B) = d_B(\mathrm{Pr}_{\mathcal{O}}(B))$ it follows that d_B is surjective. $\square$

We show next that if an isometry $L^0(A) \cong L^0(B)$ between block algebras A, B is induced by a stable equivalence of Morita type given by a bimodule M and extends to an isometry $\Phi : R_K(A) \cong R_K(B)$, then this is a perfect isometry. In fact, the next proposition is slightly more precise than that: it also shows that the 'difference' between the character of M and that associated with Φ is a projective character.

Theorem 9.3.7 ([99, 3.3]) *Let G, H be finite groups, A a block algebra of $\mathcal{O}G$ and B a block algebra of $\mathcal{O}H$. Suppose that K, k are splitting fields for G and H. Let M be a B-A-bimodule that is finitely generated projective as left B-module and as right A-module such that M induces a stable equivalence of Morita type between A and B. Assume that the isometry $L^0(A) \cong L^0(B)$ induced by Φ_M*

extends to an isometry $\Phi : R_K(A) \cong R_K(B)$. Then $\chi_\Phi - \chi_M \in \mathrm{Pr}_\mathcal{O}(A \otimes_\mathcal{O} B^{op})$; in particular, Φ is a perfect isometry.

Proof For $x \in G$ denote by $c(x)$ the conjugacy class of x in G and by τ_x the restriction to A of the $\mathcal{O}$-linear map $\mathcal{O}G \to \mathcal{O}$ sending $x' \in c(x)$ to 1 and every other group element to 0. If x is p-singular then $\tau_x \in K \otimes_\mathbb{Z} L^0(A)$, and if x runs over the p-singular elements of G then τ_x runs over a spanning set of the K-space $K \otimes_\mathbb{Z} L^0(A)$ of K-valued central functions on $K \otimes_\mathcal{O} A$ that are orthogonal to $\mathrm{Pr}_\mathcal{O}(A)$. Extend Φ, Φ_M in the obvious way to $K \otimes_\mathbb{Z} R_K(A)$. Since Φ and Φ_M coincide on $L^0(A)$ we have $\Phi_M(\tau_x) = \Phi(\tau_x)$. Since Φ_M is induced by tensoring with M we also have

$$\Phi_M(\tau_x)(y) - \frac{1}{|G|} \sum_{s \in G} \chi_M(y, s) \tau_x(s) = \frac{|c(x)|}{|G|} \chi_M(y, x)$$

and a similar reasoning yields

$$\Phi(\tau_x)(y) = \frac{|c(x)|}{|G|} \chi_\Phi(y, x).$$

This shows that $\chi_\Phi(y, x) = \chi_M(y, x)$ for any p-singular $x \in G$ and any $y \in H$. Exchanging the roles of A, B shows that also $\chi_\Phi(y, x) = \chi_M(y, x)$ for any p-singular $y \in H$ and any $x \in G$. Thus $\chi_\Phi - \chi_M$ vanishes outside the p-regular elements of $H \times G$, hence belongs to $\mathrm{Pr}_\mathcal{O}(A \otimes_\mathcal{O} B^{op})$. Since χ_M is perfect by the assumptions on M this implies that χ_Φ is perfect. $\square$

Remark 9.3.8 Not every perfect isometry is induced by a virtual Morita equivalence. Suppose that $p = 2$. Since D_8 and Q_8 have the same character tables, there is a perfect isometry mapping $\mathrm{Irr}_K(D_8)$ to $\mathrm{Irr}_K(Q_8)$ (cf. Exercise 9.2.10). One can show that there is no virtual Morita equivalence between $\mathcal{O}D_8$ and $\mathcal{O}Q_8$. In fact, there is not even a virtual Morita equivalence between kD_8 and kQ_8. Using the transfer maps in Hochschild cohomology from [130], it follows that if there were a virtual Morita equivalence between kD_8 and kQ_8, then the Hilbert series of $HH^*(kD_8)$ and $HH^*(kQ_8)$ would have to be equal. This is, however, not the case, since kQ_8 is periodic as a bimodule, while kD_8 is not.

9.4 Trivial source bimodules and Grothendieck groups

We have seen so far that a derived equivalence between block algebras induces a virtual Morita equivalence, and that a virtual Morita equivalence induces a perfect isometry. The bimodules involved in these constructions were assumed to be finitely generated projective as left and right modules. If one assumes in addition that the involved bimodules are *p*-permutation modules and that the two block algebras have the same fusion system, then one gets families of equivalences and perfect isometries between corresponding Brauer pairs, compatible with the Brauer construction. Prompted by Broué's conjectures in [38], Rickard developed in [187] the structural background of certain derived equivalences, called *splendid equivalences*, which would yield those families. This is done explicitly for blocks of principal type in [187]; the adaptation to arbitrary blocks (which essentially requires replacing block algebras by almost source algebras) is in [129], [131]. Further generalisations are in [177].

We assume in this section that *k* is a splitting fields for all finite groups, blocks, and almost source algebras in this section; this hypothesis ensures that the categories on defect groups defined using Brauer pairs are fusion systems.

Definition 9.4.1 Let A, B be almost source algebras of blocks of finite groups with a common defect group P. We denote by $\mathcal{T}(A, B)$ the subgroup of $\mathcal{P}(A, B)$ generated by the finite direct sums of summands of the A-B-bimodules $A \otimes_{\mathcal{O}Q} B$, where Q runs over the subgroups of P.

Since A, B are projective as left and right $\mathcal{O}P$-modules, the same is true for the bimodules $A \otimes_{\mathcal{O}Q} B$, where Q is a subgroup of P, and hence the group $\mathcal{T}(A, B)$ is indeed a subgroup of $\mathcal{P}(A, B)$ defined in 9.1.3.

Theorem 9.4.2 ([131, 2.3], [136, 8.1]) *Let A, B, C be almost source algebras of p-blocks of finite groups with a common defect group P all having the same fusion system $\mathcal{F}$ on P. Let M be an indecomposable direct summand of the A-B-bimodule $A \otimes_{\mathcal{O}Q} B$, and let N be an indecomposable direct summand of the B-C-bimodule $B \otimes_{\mathcal{O}R} C$, where Q, R are subgroups of P. Let X be an indecomposable direct summand of the A-C-bimodule $M \otimes_B N$. Then X is isomorphic to a direct summand of $A \otimes_{\mathcal{O}S} C$ for some fully $\mathcal{F}$-centralised subgroup S of P such that $X(\Delta S) \neq \{0\}$. In particular, the tensor product over B induces a bilinear map*

$$- \cdot_B - : \mathcal{T}(A, B) \times \mathcal{T}(B, C) \longrightarrow \mathcal{T}(A, C).$$

Proof By the assumptions, the bimodule X is isomorphic to a direct summand of

$$A \otimes_{\mathcal{O}Q} B \otimes_{\mathcal{O}R} C.$$

Since X is indecomposable, there is an indecomposable direct summand W of B as $\mathcal{O}Q$-$\mathcal{O}R$-bimodule such that X is isomorphic to a direct summand of

$$A \otimes_{\mathcal{O}Q} W \otimes_{\mathcal{O}R} C.$$

By 8.7.1 (i), the $\mathcal{O}Q$-$\mathcal{O}R$-bimodule W is isomorphic to $\mathcal{O}Q \otimes_{\mathcal{O}S} {}_{\varphi}\mathcal{O}R$ for some subgroup S of Q and some morphism $\varphi : S \to R$ belonging to the fusion system $\mathcal{F}$. Thus X is isomorphic to a direct summand of

$$A \otimes_{\mathcal{O}S} {}_{\varphi}(C).$$

Since X is indecomposable, X is in fact isomorphic to a direct summand of $An \otimes_{\mathcal{O}S} {}_{\varphi}(sC)$ for some primitive idempotent $n \in A^{\Delta S}$ and some primitive idempotent $s \in C^{\Delta \varphi(S)}$. By 5.12.8 we may assume that n, s belong to local points of S, $\varphi(S)$ on A, C, respectively. Let T be a fully $\mathcal{F}$-centralised subgroup of P such that there is an isomorphism $\psi : T \to S$ in $\mathcal{F}$. By 8.7.4 there are primitive idempotents $m \in A^{\Delta T}$ and $r \in C^{\Delta T}$ such that $Am \cong An_{\psi}$ as A-$\mathcal{O}T$-bimodules, and such that $rC \cong {}_{\varphi \psi}sC$ as $\mathcal{O}T$-C-bimodules. Thus X is a direct summand of

$$Am \otimes_{\mathcal{O}T} rC.$$

This shows that the isomorphism class of X belongs to $\mathcal{T}(A, C)$. The fact that the subgroup S in the statement can be chosen to satisfy $X(\Delta S) \neq \{0\}$ follows from 8.8.2. $\qquad\square$

Corollary 9.4.3 *Let A be an almost source algebra of a p-block of a finite group with defect group P. The abelian group $\mathcal{T}(A)$ is a unital subring of $\mathcal{P}(A)$.*

Proof Applying 9.4.2 to $A = B = C$ shows that $\mathcal{T}(A)$ is ring. Since A is isomorphic to a direct summand of $A \otimes_{\mathcal{O}P} A$ by 6.4.7, this subring contains the unit element $[A]$ of $\mathcal{P}(A)$. $\qquad\square$

Corollary 9.4.4 *Let A, B be almost source algebras of p-blocks of finite groups with a common defect group P and the same fusion system. Then $\mathcal{T}(A, B)$ is a $\mathcal{T}(A)$-$\mathcal{T}(B)$-bimodule, with left and right module structure induced by the tensor products over A and B, respectively.*

Proof This follows from 9.4.2 applied to the cases $A = B$ and $B = C$, respectively. $\qquad\square$

Proposition 9.4.5 *Let A, B be almost source algebras of blocks of finite groups with a common defect group P. Taking $\mathcal{O}$-duals of bimodules induces a group isomorphism $\mathcal{T}(A, B) \cong \mathcal{T}(B, A)$ and an anti-automorphisms of the ring $\mathcal{T}(A)$.*

Proof Since A, B, $\mathcal{O}P$ are symmetric, the $\mathcal{O}$-dual of an A-B-bimodule of the form $A \otimes_{\mathcal{O}Q} B$ for some subgroup Q of P is isomorphic to the B-A-bimodule $B \otimes_{\mathcal{O}Q} A \cong B \otimes_{\mathcal{O}Q} A$. The first statement follows. The second statement follows from 9.1.6. $\qquad\square$

Theorem 9.4.6 ([136, 9.1]) *Let A, B be almost source algebras of p-blocks b, c of finite groups G, H, respectively, with a common defect group P having the same fusion system $\mathcal{F}$ on P. Let Q be a fully $\mathcal{F}$-centralised subgroup of P. For any $X \in \mathcal{T}(A, B)$ we have $X(\Delta Q) \in \mathcal{T}(A(\Delta Q), B(\Delta Q))$.*

Proof Let $i \in (\mathcal{O}Gb)^{\Delta P}$ and $j \in (\mathcal{O}Hc)^{\Delta P}$ be almost source idempotents such that $A = i\mathcal{O}Gi$ and $B = j\mathcal{O}Hj$. Denote by c^0 and j^0 the images in $\mathcal{O}H$ of c and j, respectively, of the anti-automorphism of $\mathcal{O}H$ sending $y \in H$ to y^{-1}. Then $B^{\mathrm{op}} \cong j^0\mathcal{O}Hj^0$ is an almost source algebra of the block c^0 of $\mathcal{O}H$. Let R be a subgroup of P. We need to show that $(A \otimes_{\mathcal{O}R} B)(\Delta Q)$ is a direct sum of summands of the $A(\Delta Q)$-$B(\Delta Q)$-bimodules $A(\Delta Q) \otimes_{kS} B(\Delta Q)$, with S running over the subgroups of $C_P(Q)$. Let Y be an indecomposable direct summand of $(A \otimes_{\mathcal{O}R} B)(\Delta Q)$. As $A \otimes_{\mathcal{O}} B^{\mathrm{op}}$-modules we have an isomorphism

$$A \otimes_{\mathcal{O}R} B \cong (A \otimes_{\mathcal{O}} B^{\mathrm{op}}) \otimes_{\mathcal{O}\Delta R} \mathcal{O}$$

sending $x \otimes y$ to $(x \otimes y) \otimes 1_{\mathcal{O}}$, for $x \in A$ and $y \in B$. The algebra $A \otimes_{\mathcal{O}} B^{\mathrm{op}}$ is an almost source algebra of the block $b \otimes c^0$ of $G \times H$. By the assumptions, A, B and hence B^{op} determine the same fusion system $\mathcal{F}$ on P. Since Q is fully $\mathcal{F}$-centralised, ΔQ is fully $\mathcal{F} \times \mathcal{F}$-centralised by 8.4.8. We apply now 8.8.1 to $A \otimes_{\mathcal{O}} B^{\mathrm{op}}$, $P \times P$, $\mathcal{F} \times \mathcal{F}$, ΔQ, ΔR instead of A, P, $\mathcal{F}Q$, R, respectively. The conclusion of 8.8.1 yields a fully $C_{\mathcal{F}\times\mathcal{F}}(\Delta Q)$-centralised subgroup S of $C_P(Q) \times C_P(Q)$ and a morphism $\psi : \Delta Q \cdot S \to \Delta R$ in $\mathcal{F} \times \mathcal{F}$ such that Y is isomorphic to a direct summand of $(A \otimes_{\mathcal{O}} B^{\mathrm{op}})(\Delta Q) \otimes_{kS} k$ as $(A \otimes_{\mathcal{O}} B^{\mathrm{op}})(\Delta Q)$-modules. Note that we have obvious isomorphisms

$$(A \otimes_{\mathcal{O}} B^{\mathrm{op}})(\Delta Q) \cong A(\Delta Q) \otimes_k B^{\mathrm{op}}(\Delta Q) \cong A(\Delta Q) \otimes_k B(\Delta Q)^{\mathrm{op}}.$$

Since $\psi : \Delta Q \cdot S \to \Delta R$ is an injective group homomorphism, the group S must be of the form

$$S = \{(t, \tau(t)) \mid t \in T\}$$

for some subgroup T of $C_P(Q)$ and some injective group homomorphism $\tau : T \to C_P(Q)$. Since ψ is a morphism in $\mathcal{F} \times \mathcal{F}$ there are morphisms $\psi_1 : QT \to P$ and $\psi_2 : Q\tau(T) \to P$ such that

$$\psi(ut, v\tau(s)) = (\psi_1(ut), \psi_2(v\tau(t)))$$

for all u, $v \in Q$ and t, $s \in T$. Since $\psi(\Delta Q) \subseteq \Delta R$ we have $\psi_1(u) = \psi_2(u)$ for all $u \in Q$. Since also $\psi(S) \subseteq \Delta R$ we have $\psi_1(t) = \psi_2(\tau(t))$ for all $t \in T$. Thus $\psi_2^{-1} \circ \psi_1|_Q = \mathrm{Id}_Q$ and $\psi_2^{-1} \circ \psi_1|_T = \tau$. This proves that τ is a morphism in $C_{\mathcal{F}}(Q)$. Thus we have an isomorphism

$$\Delta(QT) \cong \Delta Q \cdot S$$

in $\mathcal{F} \times \mathcal{F}$ mapping (ut, ut) to $(u, u) \cdot (t, \tau(t))$ for all $u \in Q$ and $t \in T$. Now S is fully $C_{\mathcal{F} \times \mathcal{F}}(\Delta Q)$-centralised and $C_{\mathcal{F} \times \mathcal{F}}(\Delta Q) = C_{\mathcal{F}}(Q) \times C_{\mathcal{F}}(Q)$ by 8.4.11. Thus, by 8.4.8, in particular T is fully $C_{\mathcal{F}}(Q)$-centralised. But then, by 8.4.8 again, ΔT is fully $C_{\mathcal{F}}(Q) \times C_{\mathcal{F}}(Q)$-centralised. Thus, by combining 8.7.4 and 8.8.1 we may replace S by ΔT. This, however, shows that Y is isomorphic to a direct summand of

$$(A(\Delta Q) \otimes_k B(\Delta Q)^{\mathrm{op}}) \otimes_{k\Delta T} k \cong A(\Delta Q) \otimes_{kT} B(\Delta Q).$$

Thus the isomorphism class of Y belongs to $\mathcal{T}(A(\Delta Q), B(\Delta Q))$. The result follows. □

It was shown by Rickard [187] that for blocks of principal type of finite groups with the same fusion system the Brauer construction 'commutes' with tensor products of certain p-permutation bimodules. This has been extended to almost source algebras of arbitrary blocks with the same fusion system in [131] and slightly further to linear source modules in [136]; we do not state this in the most general form but restrict attention to p-permutation bimodules.

Theorem 9.4.7 ([131, 2.3], [136, 9.2]) *Let A, B, C be almost source algebras of p-blocks of finite groups with a common defect group P having the same fusion system $\mathcal{F}$ on P and let Q be a fully $\mathcal{F}$-centralised subgroup of P. For any A-B-bimodule M and any B-C-bimodule N, there is a canonical homomorphism of $A(\Delta Q)$-$C(\Delta Q)$-bimodules*

$$M(\Delta Q) \otimes_{B(\Delta Q)} N(\Delta Q) \to (M \otimes_B N)(\Delta Q).$$

If M is a finite direct sum of summands of the bimodules $A \otimes_{OR} B$, with R running over the subgroups of P, then this map is an isomorphism

$$M(\Delta Q) \otimes_{B(\Delta Q)} N(\Delta Q) \cong (M \otimes_B N)(\Delta Q).$$

In particular, for any $X \in \mathcal{T}(A, B)$ and any $Y \in \mathcal{P}(B, C)$ we have

$$(X \cdot_B Y)(\Delta Q) = X(\Delta Q) \cdot_{B(\Delta Q)} Y(\Delta Q),$$

and, the map sending $X \in \mathcal{T}(A)$ to $X(\Delta Q)$ is a ring homomorphism

$$\rho_Q : \mathcal{T}(A) \to \mathcal{T}(A(\Delta Q)).$$

Proof Let M be an A-B-bimodule and let N be a B-C-bimodule. Consider the obvious maps

$$M^{\Delta Q} \otimes_{\mathcal{O}} N^{(\Delta Q)} \to (M \otimes_{\mathcal{O}} N)^{(\Delta Q)} \to (M \otimes_B N)^{(\Delta Q)} \to (M \otimes_B N)(\Delta Q).$$

The composition of these maps induces a homomorphism of $A(\Delta Q)$-$C(\Delta Q)$-bimodules

$$M(\Delta Q) \otimes_{B(\Delta Q)} N(\Delta Q) \to (M \otimes_B N)(\Delta Q).$$

If we fix N and let M vary, this map is additive functorial in M. Thus, in order to show that this is an isomorphism if M is a finite direct sum of summands of bimodules of the form $A \otimes_{OR} B$, we may assume that M is a direct summand of $A \otimes_{OR} B$ as OQ-B-bimodule for some subgroup R of P. It suffices to show that this map is a k-linear isomorphism, and so we may ignore the left $A(\Delta Q)$-module structure. Thus we may in fact assume that $M = W \otimes_{OR} B$ for some indecomposable direct summand W of A as an OQ-OR-bimodule, for some subgroup R of P. By 8.7.1, we have $W \cong OQ \otimes_{OT} {}_\varphi OR$ for some subgroup T of Q and some group homomorphism $\varphi : T \to R$ belonging to the common fusion system of A and B on P. If T is a proper subgroup of Q then both sides in the above map are zero. Assume that $Q = T$, hence $M \cong {}_\varphi B$ as an OQ-B-bimodule. By decomposing M further we may in fact assume that $M = {}_\varphi n B$ for some primitive idempotent $n \in B^{\Delta \varphi(Q)}$. Then again, if $\mathrm{Br}_{\Delta \varphi(Q)}(n) = 0$, both sides in the above map are easily seen to be zero. Thus we may assume that $\mathrm{Br}_{\Delta \varphi(Q)}(n) \neq 0$. As Q is fully $\mathcal{F}$-centralised, we can apply 8.7.4, which shows that there is a primitive idempotent $m \in B^{\Delta Q}$ such that ${}_\varphi n B \cong m B$ as OQ-B-bimodules. Thus we may assume $M = m B$. But then M is a direct summand of B itself, so we may assume that $M = B$ as an OQ-B-bimodule. But in that case, both sides in the above map are canonically isomorphic to $N(\Delta Q)$. The last statement follows from applying this to $A = B = C$. $\qquad\square$

Theorem 9.4.8 *Let A, B be almost source algebras of p-blocks of finite groups with a common defect group P having the same fusion system $\mathcal{F}$ on P and let Q be a fully $\mathcal{F}$-centralised subgroup of P. If $X \in \mathcal{T}(A, B)$ induces a virtual Morita equivalence between A and B, then $X(\Delta Q)$ induces a virtual Morita equivalence between $A(\Delta Q)$ and $B(\Delta Q)$.*

Proof If X induces a virtual Morita equivalence, then $X \cdot_B X^* = [A]$ and $X^* \cdot_A X = [B]$. By 9.4.5 we have $X^* \in \mathcal{T}(B, A)$. It follows from 9.4.7 that $X(\Delta Q) \cdot_{B(\Delta Q)} X^*(\Delta Q) = (X \cdot_B X^*)(\Delta Q) = [A(\Delta Q)]$, and similarly we get $X^*(\Delta Q) \cdot_{A(\Delta Q)} X(\Delta Q) = [B(\Delta Q)]$, whence the result. $\qquad\square$

The following two results are tools for switching back and forth between block algebras and almost source algebras.

Proposition 9.4.9 (cf. [136, 4.7]) *Let G be a finite group, let b be a block of OG, let P be a defect group of b and let i be an almost source idempotent in $(OGb)^{\Delta P}$. Set $A = iOGi$ and let $\mathcal{F}$ be the fusion system of A on P. Let Q be* a fully $\mathcal{F}$-centralised subgroup of P and let e be the unique block of $kC_G(Q)$ such that $\mathrm{Br}_{\Delta Q}(i)e \neq 0$. For any A-module M there is a natural isomorphism of $kC_G(Q)e$-modules

$$e((\mathcal{O}Gi \otimes_A M)(Q)) \cong kC_G(Q)\mathrm{Br}_{\Delta Q}(i) \otimes_{A(\Delta Q)} M(Q).$$

Proof Since $\mathrm{Br}_{\Delta Q}(i) \in kC_G(Q)e$, both sides in the statement are indeed $kC_G(Q)e$-modules. Multiplication by $\mathrm{Br}_{\Delta Q}(i)$ induces a Morita equivalence between $kC_G(Q)e$ and $A(\Delta Q)$. Thus, in order to show that there is a natural isomorphism as stated, it suffices to show this after multiplying both sides with $\mathrm{Br}_{\Delta Q}(i)$. The left side becomes

$$\mathrm{Br}_{\Delta Q}(i)(kGi \otimes_A M)(Q)) = (ikGi \otimes_A M)(Q) = (A \otimes_A M)(Q) \cong M(Q)$$

and the right side becomes

$$\mathrm{Br}_{\Delta Q}(i)kC_G(Q)\mathrm{Br}_{\Delta Q}(i) \otimes_{A(\Delta Q)} M(Q) \cong M(Q);$$

both sides are naturally isomorphic as $A(\Delta Q)$-modules.   □

Proposition 9.4.10 *Let G be a finite group, b a block of $\mathcal{O}G$, P a defect group of b and $i \in (\mathcal{O}Gb)^{\Delta P}$ an almost source idempotent. Let Q be a subgroup of P and let e, e' be two different blocks of $kC_G(Q)$. For any subgroup R of P and any direct summand Y of $\mathcal{O}Gi \otimes_{\mathcal{O}R} i\mathcal{O}G$ we have $eY(\Delta Q)e' = \{0\}$.*

Proof We may assume $\mathcal{O} = k$. Since i is an almost source idempotent, for any subgroup S of P there is a unique block e_S of $kC_G(S)$ satisfying $\mathrm{Br}_{\Delta S}(i)e_S \neq 0$. The image of the set $G \times G$ in $kG \otimes_{kR} kG$ is a k-basis that is stable under the action of ΔQ. In order to compute $Y(\Delta Q)$ we have to determine the ΔQ-fixed points in this basis. For $x, y \in G$ the image $x \otimes y$ in $kG \otimes_{kR} kG$ is fixed under the action of ΔQ if for every $u \in Q$ there is $r_u \in R$ such that $ux = xr_u$ and $yu^{-1} = (r_u)^{-1}y$, or equivalently, if and only if $xy \in C_G(Q)$ and $Q^x = {}^yQ \subseteq R$. Thus we have

$$(kG \otimes_{kR} kG)^{\Delta Q} = \sum_{(x,y)} k(x \otimes y) \oplus \ker(\mathrm{Br}_{\Delta Q})$$

where (x, y) runs over the set of pairs in $G \times G$ satisfying $xy \in C_G(Q)$ and $Q^x \subseteq R$. Consequently, we get that

$$(ekGi \otimes_{kR} ikGe')^{\Delta Q} = \sum_{(x,y)} k(exi \otimes iye') + \ker(\mathrm{Br}_{\Delta Q})$$

with (x, y) as before. Write $exi = e({}^xi)x$. Since i is an almost source idempotent we have $e({}^xi) \in \ker(\mathrm{Br}_{\Delta Q})$ unless $e = {}^x(e_{Q^x})$. Similarly, we have $({}^yi)e' \in \ker(\mathrm{Br}_{\Delta Q})$ unless $e' = (e_{{}^yQ})^y$. But these two equalities would imply the contradiction $e = e'$ since $xy \in C_G(Q)$. Thus at least one of $e({}^x i)$, $({}^y i) e'$ is contained in $\ker(\mathrm{Br}_{\Delta Q})$, whence the result. $\qquad\square$

9.5 Isotypies and p-permutation equivalences

Perfect isometries between block algebras tend to come in families that are compatible with the local structure and the decomposition maps of the considered blocks. This line of thought is at the heart of work of Broué [38] and has subsequently been extended to other concepts, most notably to derived equivalences in work of Rickard [187]. We assume that k and K are splitting fields for the blocks, their Brauer pairs and their almost source algebras in this section.

For G a finite group and u a p-element in G, the decomposition map d_G^u sends a class function χ on G to the class function $d_G^u(\chi)$ on $C_G(u)_{p'}$ defined by $d_G^u(\chi)(s) = \chi(us)$ for all p'-elements s in $C_G(u)$. Each decomposition map d_G^u can be written as a sum of maps $d_G^{(u,e)}$, with e running over the blocks of $kC_G(u)$, defined as follows. Given a block e of $kC_G(u)$, denote by $\hat{e}$ the unique block of $\mathcal{O}C_G(u)$ that lifts e. For any class function χ in $\mathrm{Cl}_K(G)$ define a class function $d_G^{(u,e)}(\chi)$ in $\mathrm{Cl}_K(C_G(u)_{p'})$ by setting

$$d_G^{(u,e)}(\chi)(s) = \chi(\hat{e}us)$$

for all p'-elements s in $C_G(u)$. Note that this is equal to $d_{C_G(u)}^u(\chi^{(u,e)})$, where $\chi^{(u,e)}$ is as in 8.9.1. Since the sum of the blocks of $kC_G(u)$ is the unit element of $kC_G(u)$, it follows that the sum of the maps $d_G^{(u,e)}$, with e running over the blocks of $kC_G(u)$, is equal to d_G^u. In particular, if χ is the character of a finitely generated $\mathcal{O}$-free $\mathcal{O}G$-module M, and if we denote by ψ the character of the $\mathcal{O}C_G(u)$-module $\hat{e}M$, then

$$d_G^{(u,e)}(\chi) = d_{C_G(u)}^u(\psi).$$

The map $d_G^{(u,e)}$ from $\mathrm{Cl}_K(G)$ to $\mathrm{Cl}_K(C_G(u)_{p'})$ sends $\mathbb{Z}\mathrm{Irr}_K(G)$ to $\mathbb{Z}[\zeta]\mathrm{IBr}_k(C_G(u), e)$ for some p-power root of unity ζ in K.

Definition 9.5.1 ([38, §4]) Let G, H be finite groups, b a block of $\mathcal{O}G$ and c a block of $\mathcal{O}H$. Suppose that b and c have a common defect group P, let $i \in (\mathcal{O}Gb)^{\Delta P}$ and $j \in (\mathcal{O}Hc)^{\Delta P}$ be source idempotents. Suppose further that the fusion systems of the source algebras $i\mathcal{O}Gi$ and $j\mathcal{O}Hj$ on P are equal. For any subgroup Q of P denote by e_Q the unique block of $kC_G(Q)$ satisfying $\mathrm{Br}_{\Delta Q}(i)e_Q \neq 0$ and by f_Q the unique block of $kC_H(Q)$ satisfying $\mathrm{Br}_{\Delta Q}(j)f_Q \neq 0$. Denote by $\hat{e}_Q$ and $\hat{f}_Q$ the blocks of $\mathcal{O}C_G(Q)$ and $\mathcal{O}C_H(Q)$ lift e_Q and

f_Q, respectively. An *isotypy between b and c* is a family of perfect isometries

$$\Psi_Q : \mathbb{Z}\mathrm{Irr}_K(C_H(Q), \hat{f}_Q) \cong \mathbb{Z}\mathrm{Irr}_K(C_G(Q), \hat{e}_Q)$$

for every subgroup Q of P, with the following properties.

(1) For any isomorphism $\varphi : Q \cong R$ in the common fusion system $\mathcal{F}$ we have
$^{\varphi}\Psi_Q = \Psi_R$, where $^{\varphi}\Psi_Q$ is obtained from composing Ψ_Q with the isomor-
phisms $\mathbb{Z}\mathrm{Irr}_K(C_G(Q), \hat{e}_Q) \cong \mathbb{Z}\mathrm{Irr}_K(C_G(R), \hat{e}_R)$ and $\mathbb{Z}\mathrm{Irr}_K(C_H(Q), \hat{f}_Q) \cong$
$\mathbb{Z}\mathrm{Irr}_K(C_H(R), \hat{f}_R)$ given by conjugation with elements $x \in G$, $y \in H$ sat-
isfying $\varphi(u) = xux^{-1} = yuy^{-1}$ for all $u \in Q$.
(2) For any subgroup Q of P, any element $u \in C_P(Q)$, setting $R = Q\langle u \rangle$, we
have an equality of maps

$$d_{(C_G(Q),e_u)}^{(u,e_R)} \circ \Psi_Q = \bar{\Psi}_R \circ d_{(C_H(Q),f_Q)}^{(u,e_R)}$$

from $\mathbb{Z}\mathrm{Irr}_K(C_H(Q), \hat{f}_Q)$ to $K \otimes_{\mathbb{Z}} \mathbb{Z}\mathrm{IBr}_K(C_G(R), \hat{e}_R)$, where $d_{(C_G(Q),e_Q)}^{(u,e_R)}$ is
the generalised decomposition map from $\mathbb{Z}\mathrm{Irr}_K(C_G(Q), \hat{e}_Q)$ to $K \otimes_{\mathbb{Z}}$
$\mathbb{Z}\mathrm{IBr}_K(C_G(R), \hat{e}_R)$, and $d_{(C_H(Q),f_Q)}^{(u,f_R)}$ is defined analogously.

A virtual Morita equivalence between two block algebras induces a perfect
isometry by 9.3.4, and a Rickard equivalence induces a virtual Morita equivalence by 9.1.5. A more precise picture arises if we require the involved bimodules to be *p*-permutation bimodules. The following definition is due to Boltje
and Xu [22] for block algebras, adapted to almost source algebras in [136] (and
this definition does not require the hypothesis on k and K being large enough).

Definition 9.5.2 Let A, B be almost source algebras of blocks b, c of finite
groups G, H, respectively, with a common defect group P. A *p-permutation
equivalence between A and B* is an element $X \in \mathcal{T}(A, B)$ inducing a virtual
Morita equivalence between A and B; that is, X satisfies $X \cdot_B X^* = [A]$ in the
Grothendieck group of A-A-bimodules and $X^* \cdot_A X = [B]$ in the Grothendieck
group of B-B-bimodules.

One can express this in terms of bimodules for block algebras $\mathcal{O}Gb$ and
$\mathcal{O}Hc$ instead of the almost source algebras $A = i\mathcal{O}Gi$ and $B = j\mathcal{O}Hj$, where
$i \in (\mathcal{O}Gb)^{\Delta P}$ and $j \in (\mathcal{O}Hc)^{\Delta P}$ are almost source idempotents. The Morita
equivalences between A and $\mathcal{O}Gb$ and between B and $\mathcal{O}Hc$ imply that if X is
a virtual Morita equivalence between A and B, then $Y = [\mathcal{O}Gi] \cdot_A X \cdot_B [j\mathcal{O}H]$
induces a virtual Morita equivalence between $\mathcal{O}Gb$ and $\mathcal{O}Hc$. We clearly have
$iYj = X$, where iYj means the virtual A-B-bimodule obtained by multiplying
the underlying actual bimodules with i on the left and j on the right. If in addition X is in $\mathcal{T}(A, B)$, then Y is a virtual sum of bimodules that are summands
of the bimodules $\mathcal{O}Gi \otimes_{\mathcal{O}Q} j\mathcal{O}H$, where Q runs over the subgroups of P. The

notion of 'splendid' in [187] would allow summands of $\mathcal{O}Gb \otimes_{\mathcal{O}Q} \mathcal{O}Hc$, with Q running over the subgroups of P. For principal blocks (and more generally, blocks of principal type) this makes no difference because principal block idempotents are almost source idempotents by Brauer's Third Main Theorem. For nonprincipal blocks, it is slightly more restrictive to only allow summands of $\mathcal{O}Gi \otimes_{\mathcal{O}Q} j\mathcal{O}H$ for some fixed choice of almost source idempotents i, j, and this slightly more restricted use of the term 'splendid' is needed to generalise results such as in [187] or [22] to arbitrary blocks.

Remark 9.5.3 One could define a p-permutation equivalence between $\mathcal{O}Gb$ and $\mathcal{O}Hc$ without reference to defect groups and source idempotents as a virtual $\mathcal{O}Gb$-$\mathcal{O}Hc$-bimodule Y that is a linear combination of p-permutation bimodules that are finitely generated projective as left and as right modules, satisfying $Y \otimes_{\mathcal{O}Hc} Y^* = [\mathcal{O}Gb]$ and $Y^* \otimes_{\mathcal{O}Gb} Y = [\mathcal{O}Hc]$ in the appropriate Grothendieck groups of bimodules. Boltje and Perepelitsky showed that this gives rise to a p-permutation equivalence as in 9.5.2. More precisely, they showed that given Y as above there is an isomorphism between defect groups of b and c such that, after identifying defect groups, the blocks b and c have the same fusion system $\mathcal{F}$ on a common defect group P, the vertices of the indecomposable summands of Y are diagonal subgroups of the form ΔQ, where Q runs over the subgroups of P, and that there are source idempotents i and j of b and c, respectively, such that iYj yields a p-permutation equivalence in the sense of the above definition 9.5.2. They also showed that for $\mathcal{F}$-centric subgroups Q of P the Külshammer-Puig classes in $H^2(\mathrm{Aut}_{\mathcal{F}}(Q); k^\times)$ are preserved under p-permutation equivalences.

The following result shows that p-permutation equivalences induce isotypies. For blocks with abelian defect groups this was first proved in [22, 1.11].

Theorem 9.5.4 ([136, Theorem 1.4]) *Let A, B be almost source algebras of blocks b, c of finite groups G, H, respectively, having a common defect group P and the same fusion system $\mathcal{F}$ on P. If there is a p-permutation equivalence between A and B, then the blocks b and c are isotypic.*

Proof Write $A = i\mathcal{O}Gi$ and $B = j\mathcal{O}Hj$ for some source idempotents $i \in (\mathcal{O}Gb)^{\Delta P}$ and $j \in (\mathcal{O}Hc)^{\Delta P}$. For any subgroup Q of P denote by e_Q and f_Q the unique blocks of $kC_G(Q)$ and $kC_H(Q)$ satisfying $\mathrm{Br}_{\Delta Q}(i)e_Q = \mathrm{Br}_{\Delta Q}(i)$ and $\mathrm{Br}_{\Delta Q}(j)f_Q = \mathrm{Br}_{\Delta Q}(j)$, respectively. Denote by $\hat{e}_Q$ and $\hat{f}_Q$ the blocks of $\mathcal{O}C_G(Q)$ and $\mathcal{O}C_H(Q)$ that lift e_Q and f_Q, respectively. Suppose there is $X \in \mathcal{T}(A, B)$ such that X induces a virtual Morita equivalence. Set $Y = [\mathcal{O}Gi] \cdot_A X \cdot_B [j\mathcal{O}H]$. Since A and B are Morita equivalent to $\mathcal{O}Gb$ and $\mathcal{O}Hc$ via the bimodules $\mathcal{O}Gi$, $\mathcal{O}Hj$ and their duals $i\mathcal{O}G$, $j\mathcal{O}H$, respectively, it follows that Y

induces a virtual Morita equivalence between $\mathcal{O}Gb$ and $\mathcal{O}Hc$. With some obvious abuse of notation, we have $iYj = X$. By 9.4.8, if Q is a fully $\mathcal{F}$-centric subgroup Q of P, then $X(\Delta)$ induces a virtual Morita equivalence between $A(\Delta Q)$ and $B(\Delta Q)$. The algebras $A(\Delta Q)$ and $B(\Delta Q)$ are almost source algebras of $kC_G(Q)e_Q$ and $kC_H(Q)f_Q$, hence Morita equivalent to $kC_G(e_Q)$ and $kC_H(Q)f_Q$. Thus we get a virtual Morita equivalence between $kC_G(Q)e_Q$ and $kC_H(Q)f_Q$ given by the virtual bimodule

$$[kC_G(Q)i] \cdot_{A(\Delta Q)} X(\Delta Q) \cdot_{B(\Delta Q)} [ikC_H(Q)].$$

This virtual bimodule is isomorphic to $Y_Q = e_Q Y(\Delta Q)f_Q$. Since p-permutation modules lift uniquely, up to isomorphism, we obtain in fact a virtual Morita equivalence between $\mathcal{O}C_G(Q)e_Q$ and $\mathcal{O}C_H(Q)f_Q$ given by the unique lift, denoted $\hat{Y}_Q$ of the virtual bimodule Y_Q. If Q is not fully centralised, there are $x \in G$ and $y \in H$ such that $^x(Q, e_Q) \subseteq (P, e_P)$ and $^y(Q, f_Q) \subseteq (P, f_P)$ are fully $\mathcal{F}$-centralised and such that $^xu = {}^yu$ for all $u \in Q$. Since the Brauer construction commutes with conjugation, it follows that $\hat{Y}_Q$ induces a virtual Morita equivalence for any Q of P, and hence we obtain a perfect isometry

$$\Psi_Q : \mathbb{Z}\mathrm{Irr}_K(C_H(Q), \hat{f}_Q) \cong \mathbb{Z}\mathrm{Irr}_K(C_G(Q), \hat{e}_Q)$$

induced by the virtual Morita equivalence $\hat{Y}_Q$ for any subgroup Q of P. It follows from 5.15.12 that this family of perfect isometries is compatible with the decomposition maps. $\qquad\square$

9.6 Galois conjugate blocks are isotypic

We encountered the notion of Galois conjugate blocks in the context of Kessar's rationality conjecture 6.12.11. We show in this section that Galois conjugate blocks of a finite group are isotypic; in particular, Galois conjugate blocks have the same fusion systems. The material of this section is due to Kessar [95]. We assume that k is an algebraic closure of $\mathbb{F}_p$, and denote by W the absolutely unramified discrete valuation ring contained in $\mathcal{O}$ with k as residue field. We denote by

$$\sigma : k \to k$$

the Frobenius automorphism of k, defined by $\sigma(\lambda) = \lambda^p$ for all $\lambda \in k$. Given a finite group G, we use the same letter σ for the extension of σ to the ring automorphism of kG sending $\sum_{x \in G} \lambda_x x$ to $\sum_{x \in G} \lambda_x^p x$. Note that σ is a ring automorphism of kG, but not a k-algebra automorphism, since we have $\sigma(\lambda a) = \lambda^p \sigma(a)$ for all $\lambda \in k$ and $a \in kG$. This ring automorphism permutes the blocks of kG.

We say that two blocks c, c' of kG are *Galois conjugate* if $c' = \sigma^n(c)$ for some integer $n \geq 0$. Through the canonical bijection between the blocks of $\mathcal{O}G$ and of kG, we extend this notion in the obvious way: we say that two blocks b, b' of $\mathcal{O}G$ are *Galois conjugate* if their canonical images in kG are Galois conjugate. The uniqueness of W implies that the ring automorphism σ of k lifts to a unique ring automorphism σ_W of W, with the property that for any p'-root of unity η in $\mathcal{O}^\times$ we have $\sigma_W(\eta) = \eta^p$. As before, we use the same letter σ_W for the ring automorphism of WG sending $\sum_{x \in G} \beta_x x$ in WG to $\sum_{x \in G} \sigma_W(\beta_x)x$. By 6.5.5, the blocks of $\mathcal{O}G$ are contained in WG, and hence σ_W permutes the blocks of WG in a way that is compatible with the permutation induced by σ on the blocks of kG. In other words, two blocks b, b' of $\mathcal{O}G$ are Galois conjugate if and only if $b' = \sigma_W^n(b)$ for some integer $n \geq 0$. For future reference, we point out the elementary fact that the ring automorphisms σ, σ_0, σ_W of kG, K_0G, WG, respectively, commute with all algebra automorphisms that are induced by group automorphisms because they fix the group bases in these algebras. In particular, they commute with conjugation by elements in G.

Theorem 9.6.1 ([95, Theorem 1.1]) *Let G be a finite group. Suppose that K contains a primitive $|G|$-th root of unity. Then any two Galois conjugate blocks of $\mathcal{O}G$ are isotypic.*

We keep the notation and hypothesis introduced above. In addition, we denote by K_0 the algebraic closure of $\mathbb{Q}$ in K, and we choose a field automorphism σ_0 of K_0 that for any $|G|$-th root of unity satisfies $\sigma_0(\zeta) = \zeta^p$ if the order of ζ is prime to p, and $\sigma_0(\zeta) = \zeta$ if the order of η is a power of p. Note that $K_0 \cap W$ contains a primitive $|G|_{p'}$-root η of unity, and that σ_W and σ_0 coincide on $\mathbb{Q}[\eta] \cap W$, because σ_W and σ_0 both send η to η^p. Moreover, K_0G contains all block idempotents of $\mathcal{O}G$. Thus two blocks b, b' of $\mathcal{O}G$ are Galois conjugate if and only if $b' = \sigma_0^n(b)$ for some integer $n \geq 0$. The proof of Theorem 9.6.1 constructs explicitly an isotypy between a block b and its Galois conjugate $\sigma_W(b)$.

Definition 9.6.2 Let H be a subgroup of G. We denote by I^H the K-linear endomorphism on the space of K-valued class functions $\mathrm{Cl}_K(H)$ on H defined by

$$I^H(\psi)(x) = \psi(x_p x_{p'}^p)$$

for all $\psi \in \mathrm{Cl}_K(H)$ and $x \in H$, where x_p and $x_{p'}$ denote the p-part and p'-part of x.

If a class function $\psi \in \mathrm{Cl}_K(H)$ takes values in K_0, we denote by $\sigma_0(\psi)$ the class function sending $x \in H$ to $\sigma_0(\psi(x))$. Similarly, if a class function

$\psi \in \mathrm{Cl}_K(H)$ takes values in W, we denote by $\sigma_W(\psi)$ the class function sending $x \in H$ to $\sigma_W(\psi(x))$. Note that the K-valued irreducible characters of H take values in K_0. We use the same convention for class functions that are defined only on the set $H_{p'}$ of p'-elements in H. In particular, the Brauer characters of H take values in K_0. Thus the statements in the following lemma make sense.

Lemma 9.6.3 *With the notation and hypotheses of 9.6.1, let H be a subgroup of G, and let c be a block of $\mathcal{O}H$. Set $c' = \sigma_0(c)$.*

(i) For any $\chi \in \mathrm{Irr}_K(H)$ we have $I^H(\chi) = \sigma_0(\chi) \in \mathrm{Irr}_K(H)$.

(ii) The map $\chi \mapsto I^H(\chi)$ is a bijection from $\mathrm{Irr}_K(H,c)$ to $\mathrm{Irr}_K(H,c')$.

(iii) The restriction of I^H to $\mathrm{Irr}_K(H,c)$ induces a perfect isometry between $\mathcal{O}Hc$ and $\mathcal{O}Hc'$.

(iv) For any $\varphi \in \mathrm{IBr}_{\mathcal{O}}(H)$ we have $I^H(\varphi) = \sigma_0(\varphi)$.

Proof Let $\chi \in \mathrm{Irr}_K(H)$ and $n = \chi(1)$. By 3.8.1, there is a representation $\rho : H \to \mathrm{GL}_n(K_0)$ affording χ; that is, $\chi(x)$ is equal to the trace of the matrix $\rho(x)$ for all $x \in H$. The automorphism σ_0 of K_0 induces a group automorphism of $\mathrm{GL}_n(K_0)$, which we will again denote by σ_0. Then $\sigma_0 \circ \rho$ is a representation of H affording the character $\sigma_0(\chi)$; clearly this is again an irreducible character of H. In order to show (i), we need to show that the character afforded by $\sigma_0 \circ \rho$ is equal to the class function $I^H(\chi)$. Let $x \in H$, write $x = us$, where u is the p-part of x and s is the p'-part of x. Since K_0 contains all eigenvalues of $\rho(x)$, we may assume that $\rho(x)$ is a diagonal matrix. Since u and s are powers of x, it follows that $\rho(u)$ is a diagonal matrix with entries $\zeta_1, \zeta_2, \ldots, \zeta_n$, where the ζ_i are p-power roots of unity, and $\rho(s)$ is a diagonal matrix with entries $\eta_1, \eta_2, \ldots, \eta_n$, where the η_i are roots of unity of order prime to p. Thus $\rho(x) = \rho(u)\rho(s)$ is the diagonal matrix with entries $\zeta_1\eta_1, \zeta_2\eta_2, \ldots, \zeta_n\eta_n$. Similarly, $\rho(us^p)$ is the diagonal matrix with entries $\zeta_1\eta_1^p, \zeta_2\eta_2^p, \ldots, \zeta_n\eta_n^p$. Since $\zeta_i\eta_i^p = \sigma_0(\zeta_i\eta_i)$ it follows that $I^H(\chi)(x) = \sigma_0(\chi(x)) = \sigma_0(\chi)(x)$. This proves (i). The ring automorphism σ_0 of K_0G sends $e(\chi)$ to $e(\sigma_0(\chi))$, and clearly $c' = \sigma_0(c) = \sum_{\chi \in \mathrm{Irr}_K(H,c)} e(\sigma_0(\chi))$. Statement (ii) follows easily. For (iii) we need to show that the $H \times H$-character μ defined for $x, y \in H$ by

$$\mu(x,y) = \sum_{\chi \in \mathrm{Irr}_K(H,c)} \chi(x) I^H(\chi)(y^{-1})$$

satisfies the divisibility conditions in 9.2.1. We use the fact that the character ι of the $\mathcal{O}(H \times H)$-module $\mathcal{O}H$ is perfect, by 9.3.1. By 6.5.3, we have

$$\iota(x,y) = \sum_{\chi \in \mathrm{Irr}_K(H,c)} \chi(x)\chi(y^{-1}).$$

Thus

$$\mu(x, y) = \iota(x, y_p y_{p'}^p).$$

Since $C_G(y) = C_G(y_p y_{p'}^p)$, the divisibility conditions for μ follow from those of ι. This proves (iii). For statement (iv), note that this involves functions defined on p'-elements in H, hence functions with values that are sums of p'-order roots of unity, hence in $\mathbb{Q}[\eta] \cap W$. Since σ_W and σ_0 coincide on this intersection and both lift σ, the statement follows easily. Alternatively, (iv) follows from (i) and the surjectivity of the decomposition map.   □

Let P be a p-subgroup of G. The canonical maps $(\mathcal{O}G)^P \to (kG)^P$ and $(WG)^P \to (kG)^P$ are both surjective, with kernels in the radical. Thus they both induce bijections between the sets of points of P on $\mathcal{O}G$, WG, and kG, and these bijections restrict to bijections between local points. In particular, every point γ of P on $\mathcal{O}G$ has a representative i in WG. Then $\sigma_W(i)$ belongs to a point denoted $\sigma_W(\gamma)$ of P on $\mathcal{O}G$; clearly this is independent of the choice of i in γ. If P_γ belongs to the block b of $\mathcal{O}G$, then $P_{\sigma_W(\gamma)}$ belongs to the Galois conjugate block $\sigma_W(b)$. Generalised decomposition numbers are invariant under Galois conjugation.

Lemma 9.6.4 *With the notation and hypotheses of 9.6.1, let H be a subgroup of G, let $\chi \in \mathrm{Irr}_K(H)$ and u_ϵ a local pointed element on $\mathcal{O}H$. Set $\chi' = \sigma_0(\chi)$ and $\epsilon' = \sigma_W(\epsilon)$. We have $\chi'(u_{\epsilon'}) = \chi(u_\epsilon)$. In particular, for any $\varphi \in \mathrm{IBr}_\mathcal{O}(H)$, setting $\varphi' = \sigma_0(\varphi)$, we have*

$$d_{\varphi'}^{\chi'} = d_\varphi^\chi.$$

Proof If X is an $\mathcal{O}H$-module with character χ, then $\chi(u_\epsilon)$ is the trace of u on the $\mathcal{O}\langle u \rangle$-module jX, for some (hence any) choice $j \in \epsilon$. This is a linear combination of p-power roots of unity, hence fixed by σ_0. The first statement follows. The last statement is the special case $u = 1$.   □

Lemma 9.6.5 *With the notation and hypotheses of 9.6.1, let b be a block of $\mathcal{O}G$ and let (P, e), (Q, f) be Brauer pairs on $\mathcal{O}G$. Set $b' = \sigma_0(b)$.*

(i) *(P, e) is a (G, b)-Brauer pair if and only if $(P, \sigma(e))$ is a (G, b')-Brauer pair.*

(ii) *We have $(Q, f) \leq (P, e)$ if and only if $(R, \sigma(f)) \leq (Q, \sigma(e))$.*

(iii) *(P, e) is a maximal (G, b)-Brauer pair if and only if $(P, \sigma(e))$ is a maximal (G, b')-Brauer pair, and in that case we have an equality of fusion systems $\mathcal{F}_{(P,e)}(G, b) = \mathcal{F}_{(P,\sigma(e))}(G, b').$*

Proof Since σ_0, restricted to $K_0 \cap W$, lifts σ, it follows that $\sigma(\mathrm{Br}_P(b)e) = \mathrm{Br}_P(\sigma_0(b))\sigma(e)$. The left side is nonzero if and only if (P, e) is a (G, b)-Brauer pair, and the right side is nonzero if and only if $(P, \sigma(e))$ is a (G, b')-Brauer pair. This shows (i). Suppose that $(Q, f) \leq (P, e)$. By definition, there is a primitive idempotent $i \in (kG)^P$ such that $\mathrm{Br}_P(i)e$ and $\mathrm{Br}_Q(i)f$ are both nonzero. Note that $\sigma(i)$ is again a primitive idempotent in $(kG)^P$. Thus, applying σ implies that $(Q, \sigma(f)) \leq (P\sigma(e))$. This shows (ii). Conjugation of Brauer pairs by elements in G clearly commutes with the action of σ on the set of Brauer pairs, and hence (iii) follows from (i) and (ii). $\qquad\square$

Proof of Theorem 9.6.1 Let b be a block of $\mathcal{O}G$. It suffices to show that b and $b' = \sigma_0(b)$ are isotypic. Let (P, e) be a maximal (G, b)-Brauer pair. Set $e' = \sigma(e)$. By 9.6.5, (P, e') is a maximal (G, b')-Brauer pair, and if $(Q, f) \leq (P, e)$, then $(Q, f') \leq (P, e')$, where $f' = \sigma(f)$. Moreover, the fusion systems $\mathcal{F}_{(P,e)}(G, b)$ and $\mathcal{F}_{(P,e')}(G, b')$ are equal.

Let $(Q, f) \leq (P, e)$, and set $f' = \sigma(f)$. Denote by

$$I^Q : \mathcal{F}_K(C_G(Q), \hat{f}) \to \mathcal{F}_K(C_G(Q), \hat{f}')$$

the linear map obtained from restricting $I^{C_G(Q)}$ to the blocks $\hat{f}$ and $\hat{f}'$ of $\mathcal{O}C_G(Q)$ lifting f and f', respectively. By 9.6.3 the map I^Q is a perfect isometry. As mentioned earlier, the maps I^Q are compatible with conjugation by elements in G. It remains to check the compatibility condition (2) in Definition 9.5.1. This follows from the invariance of the generalised decomposition numbers in 9.6.4. $\qquad\square$

Remark 9.6.6 One can use Theorem 9.6.1 to prove an earlier rationality result on the centre of a block. Let b be a block of $\mathcal{O}G$ and let $b' = \sigma_0(b)$. Denote by $\bar{b}$ and $\bar{b}'$ the canonical images of b and b' in kG. Since $\mathcal{O}Gb$ and $\mathcal{O}Gb'$ are isotypic, there is an isomorphism of $\mathcal{O}$-algebras $Z(\mathcal{O}Gb') \cong Z(\mathcal{O}Gb)$. This induces an isomorphism of k-algebras $\tau : Z(kG\bar{b}') \cong Z(kG\bar{b})$. Precomposing this with the ring isomorphism $\sigma : Z(kG\bar{b}) \cong Z(kG\bar{b}')$ yields a ring automorphism $\Phi = \tau \circ \sigma$ of $Z(kG\bar{b})$ satisfying $\Phi(\lambda_z) = \lambda^r \Psi(z)$ for all $\lambda \in k$ and all $z \in Z(kG\bar{b})$. It follows from 1.14.17 that the Φ-fixed points in $Z(kG\bar{b})$ yield an $\mathbb{F}_p$-form of $Z(kG\bar{b})$, proving thus again 6.17.10.

Remark 9.6.7 The Galois conjugation on blocks considered in this section extends in the obvious way to an action on the Külshammer–Puig classes of the blocks. Since the Külshammer–Puig classes of a block need not be invariant under the action on $k^\times$ of the Frobenius automorphism σ, it follows that the isotypy from 9.6.1 between Galois conjugate blocks need not be induced by a p-permutation equivalence, since by a result of Boltje and Perepelitsky, a p-permutation equivalence preserves the Külshammer–Puig classes (see 9.5.3).

9.7 Splendid Morita and derived equivalences

Since Rickard's paper [187], it has become customary to attach the adjective 'splendid' to any of the derived, stable, or Morita equivalences and their virtual analogues if these are induced by trivial source bimodules. A p-permutation equivalence is, for instance, a splendid virtual Morita equivalence. Block algebras related by a splendid stable, derived, or Morita equivalence have isomorphic defect groups. More generally, slightly extending the arguments of [177, 7.4], block algebras that are separably equivalent via a bimodule with a source of rank prime to p have isomorphic defect groups.

Proposition 9.7.1 *Let G, H be finite groups, b a block of $\mathcal{O}G$ with a defect group P and c a block of $\mathcal{O}H$ with a defect group Q. Let M be an indecomposable $\mathcal{O}Gb$-$\mathcal{O}Hc$-bimodule that is finitely generated projective as a left and right module. Suppose that $\mathcal{O}Gb$ is isomorphic to a direct summand of $M \otimes_{\mathcal{O}Hc} M^*$ as an $\mathcal{O}Gb$-$\mathcal{O}Gb$-bimodule, and that $\mathcal{O}Hc$ is isomorphic to a direct summand of $M^* \otimes_{\mathcal{O}Gb} M$ as an $\mathcal{O}Hc$-$\mathcal{O}Hc$-bimodule. Let R be a subgroup of $P \times Q$ that is a vertex of M as an $\mathcal{O}(G \times H)$-module and let W be an $\mathcal{O}R$-source of M. If the $\mathcal{O}$-rank of W is prime to p, then there is an isomorphism $\varphi : P \cong Q$ such that $R = \{(u, \varphi(u)) | u \in P\}$.*

Proof By 5.11.8, R is of the form $R = \{(u, \varphi(u)) | u \in T\}$ for some subgroup T of P and some injective group homomorphism $\varphi : T \to Q$. By 5.1.16, every indecomposable direct summand of $M \otimes_{\mathcal{O}Hc} M^*$ has a vertex of order at most $|R| = |T|$. Since $M \otimes_{\mathcal{O}Hc} M^*$ has a direct summand isomorphic to $\mathcal{O}Gb$, which has ΔP as a vertex, it follows that $T = P$. A similar argument applied to $M^* \otimes_{\mathcal{O}Gb} M$ shows that $\varphi(T) = Q$. $\square$

Proposition 9.7.1 applies in particular if M has an endopermutation module as a source. In fact, it is shown in [177, 7.4] that the hypotheses of 9.7.1 imply already that W is an endopermutation module.

Remark 9.7.2 If M is a p-permutation $\mathcal{O}Gb$-$\mathcal{O}Hc$-bimodule inducing a Morita equivalence, then M is indecomposable and has a trivial source. Thus 9.7.1 applies and yields an isomorphism of defect groups $P \cong Q$. If we identify P and Q through this isomorphism, then M is a direct summand of

$\mathrm{Ind}_{\Delta P}^{G \times H}(\mathcal{O}) \cong \mathcal{O}G \otimes_{\mathcal{O}P} \mathcal{O}H$, hence of $\mathcal{O}Gb \otimes_{\mathcal{O}P} \mathcal{O}Hc$. The indecomposability of M implies that M is a direct summand of $\mathcal{O}Gi \otimes_{\mathcal{O}P} j\mathcal{O}H$ for some primitive idempotents $i \in \mathcal{O}Gb^{\Delta P}$ and $j \in \mathcal{O}Hc^{\Delta P}$. Moreover, 9.7.1 implies that ΔP is a vertex of M. This in turn implies that the idempotents i and j are source idempotents; indeed, if one of $\mathrm{Br}_{\Delta P}(i)$ or $\mathrm{Br}_{\Delta P}(j)$ is zero, then every indecomposable summand of $\mathcal{O}Gi \otimes_{\mathcal{O}P} j\mathcal{O}H$ has a vertex strictly contained in ΔP by 5.12.8.

Definition 9.7.3 Let G, H be finite groups and b, c blocks of $\mathcal{O}G$, $\mathcal{O}H$, respectively, having a common defect group P. Let $i \in \mathcal{O}Gb^{\Delta P}$ and $j \in \mathcal{O}Hc^{\Delta P}$ be almost source idempotents. Set $A = i\mathcal{O}Gi$ and $B = j\mathcal{O}Hj$. A Morita equivalence between $\mathcal{O}Gb$ and $\mathcal{O}Hc$ is called *splendid with respect to the choice of the almost source idempotents i and j* if it is given by an $\mathcal{O}Gb$-$\mathcal{O}Hc$-bimodule M that is isomorphic to a direct summand of $\mathcal{O}Gi \otimes_{\mathcal{O}P} j\mathcal{O}H$. A Morita equivalence between A and B is called *splendid* if it is given by an A-B-bimodule N that is isomorphic to a direct summand of $A \otimes_{\mathcal{O}P} B$.

The canonical Morita equivalences 6.4.6 between block algebras and almost source algebras imply that if M induces a splendid Morita equivalence between $\mathcal{O}Gb$ and $\mathcal{O}Hc$, then iMj induces a splendid Morita equivalence between A and B. Similarly, if N induces a splendid Morita equivalence between A and B, then $\mathcal{O}Gi \otimes_A N \otimes_B j\mathcal{O}H$ induces a splendid Morita equivalence between $\mathcal{O}Gb$ and $\mathcal{O}Hc$ with respect to the choice of i and j. A splendid Morita equivalence between two block algebras is equivalent to a source algebra isomorphism; this is due independently to Scott [205] and Puig [177]. The presentation follows [131, §4].

Theorem 9.7.4 *Let G, H be finite groups, let b, c be blocks of $\mathcal{O}G$, $\mathcal{O}H$, respectively, having a common defect group P, and let $i \in (\mathcal{O}Gb)^{\Delta P}$ and $j \in (\mathcal{O}Hc)^{\Delta P}$ be source idempotents. The following are equivalent.*

(i) There is an indecomposable direct summand M of the $\mathcal{O}Gb$-$\mathcal{O}Hc$-bimodule $\mathcal{O}Gi \otimes_{\mathcal{O}P} j\mathcal{O}H$ such that M and its dual M^* induce a Morita equivalence between $\mathcal{O}Gb$ and $\mathcal{O}Hc$.

(ii) There is an isomorphism of interior P-algebras $\varphi : i\mathcal{O}Gi \to j\mathcal{O}Hj$.

The correspondence sending φ as in (ii) to the $\mathcal{O}Gb$-$\mathcal{O}Hc$-bimodule $\mathcal{O}Gi \otimes_{i\mathcal{O}Gi} {}_\varphi(j\mathcal{O}H)$ induces a bijection between $((j\mathcal{O}Hj)^{\Delta P})^\times$-conjugacy classes of isomorphisms of interior P-algebras $i\mathcal{O}Gi \cong j\mathcal{O}Hj$ and the set of isomorphism classes of direct summands of $\mathcal{O}Gi \otimes_{\mathcal{O}P} j\mathcal{O}H$ inducing a Morita equivalence.

Proof We will use without further comment the fact proved in 6.2.1, 6.4.7, that block algebras and source algebras are relatively separable over their defect groups. If (ii) holds, then the canonical Morita equivalences between $i\mathcal{O}Gi$ and $\mathcal{O}Gb$ and between $j\mathcal{O}Hj$ and $\mathcal{O}Hc$ imply that $M = \mathcal{O}Gi \otimes_{i\mathcal{O}Gi \ \varphi}(j\mathcal{O}H)$ and its dual induce a Morita equivalence between $\mathcal{O}Gb$ and $\mathcal{O}Hc$. Moreover, since $i\mathcal{O}Gi$ is isomorphic a direct summand of the $i\mathcal{O}Gi$-$i\mathcal{O}Gi$-bimodule $i\mathcal{O}Gi \otimes_{\mathcal{O}P} i\mathcal{O}Gi$ it follows that M is isomorphic to a direct summand of $\mathcal{O}Gi \otimes_{\mathcal{O}P} j\mathcal{O}H$. Thus (ii) implies (i). Suppose conversely that (i) holds. Let M be an indecomposable direct summand of $\mathcal{O}Gi \otimes_{\mathcal{O}P} j\mathcal{O}H$ inducing a Morita equivalence between $\mathcal{O}Gb$ and $\mathcal{O}Hc$. As an $\mathcal{O}Hc$-$\mathcal{O}P$-bimodule, $\mathcal{O}Hj$ is indecomposable because j is primitive in $(\mathcal{O}Hc)^{\Delta P}$. The functor $M \otimes_{\mathcal{O}Hc} -$ induces a Morita equivalence between $\mathcal{O}Gb \otimes_{\mathcal{O}} \mathcal{O}P$ and $\mathcal{O}Hc \otimes_{\mathcal{O}} \mathcal{O}P$, and hence $Mj \cong M \otimes_{\mathcal{O}Hc} \mathcal{O}Hj$ is indecomposable as an $\mathcal{O}Gb$-$\mathcal{O}P$-bimodule. Thus Mj is isomorphic to an indecomposable direct summand of $\mathcal{O}Gi \otimes_{\mathcal{O}P} j\mathcal{O}Hj$, hence of $\mathcal{O}Gi \otimes_{\mathcal{O}P} W$ for some indecomposable direct summand W of $j\mathcal{O}Hj$ as an $\mathcal{O}P$-$\mathcal{O}P$-bimodule. Then $M^* \otimes_{\mathcal{O}Gb} Mj \cong \mathcal{O}Hj$ is isomorphic to a direct summand of $M^*i \otimes_{\mathcal{O}P} W$. Since $\mathrm{Br}_{\Delta P}(j) \neq 0$ we have $(M^*i \otimes_{\mathcal{O}P} W)(\Delta P) \neq \{0\}$, which in conjunction with 8.7.1 forces $W \cong k[yP]$ for some $y \in N_H(P, f)$, where f is the block of $kN_H(P)$ satisfying $\mathrm{Br}_{\Delta P}(j)f \neq 0$. Thus ^yj and j are conjugate in $((\mathcal{O}Hc)^{\Delta P})^\times$, which implies that $Mj \cong Mjy^{-1}$ as $\mathcal{O}Gb$-$\mathcal{O}P$-bimodules. Thus multiplication by y^{-1} on $\mathcal{O}Gi \otimes_{\mathcal{O}P} W$ shows that Mj is isomorphic to a direct summand of $\mathcal{O}Gi$ as an $\mathcal{O}Gb$-$\mathcal{O}P$-bimodule. Since $\mathcal{O}Gi$ is indecomposable as an $\mathcal{O}Gb$-$\mathcal{O}P$-bimodule it follows that we have an isomorphism $\alpha : Mj \cong \mathcal{O}Gi$ as $\mathcal{O}Gb$-$\mathcal{O}P$-bimodules. This is in particular an isomorphism of left $\mathcal{O}Gb$-modules, hence induces an interior P-algebra isomorphism $\mathrm{End}_{\mathcal{O}Gb}(Mj) \cong \mathrm{End}_{\mathcal{O}Gb}(i\mathcal{O}Gi)$. The left side in this isomorphism is $(j\mathcal{O}Hj)^{\mathrm{op}}$, and the right side is $(i\mathcal{O}Gi)^{\mathrm{op}}$. Passing to opposite algebras shows that (i) implies (ii). The last statement in the theorem follows from the explicit constructions relating φ and M in the above proof of the equivalence of (i) and (ii). $\square$

Applied to $G = H$ this shows that $\mathrm{Out}_P(i\mathcal{O}Gi)$ is canonically isomorphic to the subgroup of $\mathrm{Pic}(B)$ of splendid Morita equivalences given by bimodule summands of $\mathcal{O}Gi \otimes_{\mathcal{O}P} i\mathcal{O}G$ (cf. Exercise 6.16.5). The group $\mathrm{Out}_P(i\mathcal{O}Gi)$ is not necessarily isomorphic to the group of all splendid Morita equivalences of B given by a bimodule with ΔP as a vertex. The notion of being splendid extends to Rickard complexes.

Definition 9.7.5 Let A, B be almost source algebras of blocks b, c of finite groups G, H, respectively, with a common defect group P. Let $i \in (\mathcal{O}Gb)^{\Delta P}$ and $j \in (\mathcal{O}Hc)^{\Delta P}$ be almost source idempotents. Set $A = i\mathcal{O}Gi$ and $B = j\mathcal{O}Hj$. A complex X of $\mathcal{O}Gb$-$\mathcal{O}Hc$-bimodules is called *splendid* with respect to the choice of i and j if the components of X are finite direct sums of summands of the $\mathcal{O}Gb$-$\mathcal{O}Hc$-bimodules $\mathcal{O}Gi \otimes_{\mathcal{O}Q} j\mathcal{O}H$, with Q running over the subgroups of P. A *splendid derived equivalence between $\mathcal{O}Gb$ and $\mathcal{O}Hc$* is a derived equivalence induced by a splendid Rickard complex with respect to some choice of almost source idempotents. A complex Y of A-B-bimodules is called *splendid* if the components of Y are finite sum of direct summands of the bimodules $A \otimes_{\mathcal{O}Q} B$, with Q running over the subgroups of P. A *splendid derived equivalence between A and B* is a derived equivalence induced by a splendid Rickard complex.

With this notation, if X is a splendid (Rickard) complex of $\mathcal{O}Gb$-$\mathcal{O}Hc$-bimodules with respect to the almost source idempotents i and j, then iXj is a splendid (Rickard) complex of A-B-bimodules. Similarly, if Y is a splendid (Rickard) complex of A-B-bimodules, then $\mathcal{O}Gi \otimes_A Y \otimes_B j\mathcal{O}H$ is a splendid (Rickard) complex of $\mathcal{O}Gb$-$\mathcal{O}Hc$-bimodules with respect to i and j. Broué's abelian defect conjecture 6.7.10 admits the following refinement.

Conjecture 9.7.6 (Broué's abelian defect group conjecture) *Let G be a finite group, b a block of $\mathcal{O}G$, P a defect group of b, and denote by c the block of $\mathcal{O}N_G(P)$ corresponding to b under the Brauer correspondence. If P is abelian, then there is a splendid derived equivalence between $\mathcal{O}Gb$ and $\mathcal{O}N_G(P)c$.*

As in the case of splendid Morita equivalences, results regarding splendid Rickard complexes can be formulated for block algebras or for almost source algebras. The following theorem, formulated for almost source algebras, is an amalgam of [129, Theorem 1.1], [136, Theorem 1.5], extending [187, Theorem 1.4] and [22, Theorem 1.5]. In view of Remark 9.5.3 we could have defined splendid Rickard equivalences between block algebras more generally without the hypothesis on the defect groups being isomorphic and without reference to almost source idempotents.

Theorem 9.7.7 *Let A, B be almost source algebras of blocks b, c of finite groups G, H, respectively, having a common defect group P and the same fusion system $\mathcal{F}$. Let X be a splendid Rickard complex of A-B-bimodules. Assume that k is a splitting field for the subgroups of G and H.*

(i) *For any fully $\mathcal{F}$-centralised subgroup Q of P, the complex $X(\Delta Q)$ is a splendid Rickard complex of $A(\Delta Q)$-$B(\Delta Q)$-modules.*
(ii) *The image of X in $\mathcal{T}(A, B)$ is a p-permutation equivalence. In particular, if K is a splitting field for all subgroups of G and H, then the blocks b and c are isotypic.*

Proof The Brauer construction is functorial and, when applied to complexes, maps a homotopy to a homotopy. Thus, applying the Brauer construction to a homotopy equivalence $X \otimes_B X^* \simeq A$ yields a homotopy equivalence $(X \otimes_B X^*)(\Delta Q) \simeq A(\Delta Q)$. Therefore, if Q is fully $\mathcal{F}$-centralised, it follows from 9.4.7 that $X(\Delta Q) \otimes_{B(\Delta Q)} X(\Delta Q)^* \cong (X \otimes_B X^*)(\Delta Q) \simeq A(\Delta Q)$ and similarly that $X(\Delta Q)^* \otimes_{A(\Delta Q)} X(\Delta Q) \cong (X^* \otimes_A X)(\Delta Q) \simeq B(\Delta Q)$. This shows (i). By 9.1.5, the image of X in $\mathcal{T}(A, B)$ is a virtual Morita equivalence, hence a p-permutation equivalence. The last statement follows from 9.5.4. $\quad\square$

The version of this theorem for block algebras reads as follows:

Theorem 9.7.8 *Let b, c be blocks of finite groups G, H having a common defect group P and the same fusion system with respect to a choice of source idempotents i and j of b and c, respectively. Assume that k is a splitting field for all subgroups of G and H. If X is a splendid Rickard complex of $\mathcal{O}Gb$-$\mathcal{O}Hc$-bimodules with respect to i and j, then for any subgroup Q of P, the complex $e_Q X(\Delta Q) f_Q$ is a splendid Rickard complex of $kC_G(Q)e_Q$-$kC_G(Q)f_Q$-modules with respect to the almost source idempotents $\mathrm{Br}_{\Delta Q}(i)$ and $\mathrm{Br}_{\Delta Q}(j)$, where e_Q and f_Q are the unique blocks of the centralisers of Q in G and H determined by i and j, respectively.*

Proof This follows from using the canonical Morita equivalences 6.4.6 between the block algebra $\mathcal{O}Gb$ and its almost source algebra $A = i\mathcal{O}Gi$, together with the analogous Morita equivalences between $kC_G(Q)e_Q$ and $A(\Delta Q)$, where Q is a subgroup of P. $\quad\square$

The conclusion in 9.7.8 holds for all subgroups Q of P, not just the fully $\mathcal{F}$-centralised subgroups of P. The reason is that when working with complexes of $\mathcal{O}Gb$-$\mathcal{O}Hc$-bimodules, we can replace Q by a fully $\mathcal{F}$-centralised subgroup via simultaneously conjugating Q by elements $x \in G$ and $y \in H$ such that $^xQ = {}^yQ$ and such that $^xu = {}^yu$ for all $u \in Q$. This is possible because the fusion systems of the blocks b and c determined by i and j coincide.

Theorem 9.7.9 *Suppose that k is large enough. Let A, B, C be source algebras of blocks of finite groups, a common defect group P and a common fusion system $\mathcal{F}$ on P. Let X be a splendid Rickard complex of A-B-bimodules and let Y be a splendid Rickard complex of B-C-bimodules. Then $X \otimes_B Y$ is a splendid Rickard complex of A-C-bimodules, and for any fully $\mathcal{F}$-centralised subgroup Q of P we have $(X \otimes_B Y)(\Delta Q) \cong X(\Delta Q) \otimes_{B(\Delta Q)} Y(\Delta Q)$.*

Proof By 2.12.5 we have $(A \otimes_B Y)^* \cong Y^* \otimes_B X^*$. Since $X \otimes_B X^* \simeq A$ and $Y \otimes_C Y^* \simeq B$, it follows that $(X \otimes_B Y) \otimes_C (X \otimes_B Y)^* \cong X \otimes_B Y \otimes_C Y^* \otimes_B X^* \simeq X \otimes_B X^* \simeq A$ in the category of complexes of A-A-bimodules. Thus

$X \otimes_B Y$ is a Rickard complex. The fact that $X \otimes_B Y$ is splendid follows from 9.4.2. The last isomorphism follows from 9.4.7. □

As before, this can be formulated at the level of splendid Rickard complexes over three block algebras with a common defect group and a common fusion system for some choice of almost source idempotents.

9.8 Splendid stable equivalences

Let $\mathcal{O}Gb$ and $\mathcal{O}Hc$ be two block algebras, and let M be an indecomposable p-permutation $\mathcal{O}Gb$-$\mathcal{O}Hc$-bimodule inducing a stable equivalence of Morita type. As in the case of splendid Morita equivalences 9.7.2, it follows from 9.7.1 that a defect group P of b can be identified with a defect group of c in such a way that M is a summand of $\mathcal{O}Gi \otimes_{\mathcal{O}P} j\mathcal{O}H$ for some source idempotents $i \in (\mathcal{O}Gb)^{\Delta P}$ and $j \in (\mathcal{O}Hc)^{\Delta P}$.

Definition 9.8.1 Let G, H be finite groups and b, c blocks of $\mathcal{O}G, \mathcal{O}H$, respectively, having a common defect group P. Let $i \in \mathcal{O}Gb^{\Delta P}$ and $j \in \mathcal{O}Hc^{\Delta P}$ be almost source idempotents. Set $A = i\mathcal{O}Gi$ and $B = j\mathcal{O}Hj$. A stable equivalence of Morita type between $\mathcal{O}Gb$ and $\mathcal{O}Hc$ is called *splendid with respect to the choice of the almost source idempotents i and j* if it is given by an $\mathcal{O}Gb$-$\mathcal{O}Hc$-bimodule M that is isomorphic to a direct summand of $\mathcal{O}Gi \otimes_{\mathcal{O}P} j\mathcal{O}H$. A stable equivalence of Morita type between A and B is called *splendid* if it is given by an A-B-bimodule N that is isomorphic to a direct summand of $A \otimes_{\mathcal{O}P} B$.

Splendid stable equivalences of Morita type preserve fusion systems and induce splendid Morita equivalences between nontrivial Brauer pairs. For principal blocks this has first been noted by Alperin; the general case appears in [131] and [177]. A version of this result where G and H are assumed to have the same local structure is stated also in Broué [39, 6.3]. We formulate this in parallel for block algebras and their source algebras.

Theorem 9.8.2 (cf. [131, Theorem 3.1]) *Let G, H be finite groups, let b, c be blocks of $\mathcal{O}G, \mathcal{O}H$, respectively, having a common defect group P, and let $i \in (\mathcal{O}Gb)^{\Delta P}$ and $j \in (\mathcal{O}Hc)^{\Delta P}$ be source idempotents. Set $A = i\mathcal{O}Gi$ and $B = j\mathcal{O}Hj$. For any subgroup Q of P denote by e_Q and f_Q the unique block of $kC_G(Q)$ and of $kC_H(Q)$, respectively, such that $\mathrm{Br}_Q(i)e_Q \neq 0$ and $\mathrm{Br}_Q(j)f_Q \neq 0$. Let M be an indecomposable direct summand of the $\mathcal{O}Gb$-$\mathcal{O}Hc$-bimodule $\mathcal{O}Gi \otimes_{\mathcal{O}P} j\mathcal{O}H$, and set $N = iMj$. The following are equivalent.*

(i) *The $\mathcal{O}Gb$-$\mathcal{O}Hc$-bimodule M and its dual induce a stable equivalence of Morita type between $\mathcal{O}Gb$ and $\mathcal{O}Hc$.*

(ii) *The A-B-bimodule N and its dual induce a stable equivalence of Morita type between A and B.*

(iii) *For any nontrivial subgroup Q of P the bimodule $e_Q M(\Delta Q) f_Q$ and its dual induce a Morita equivalence between $kC_G(Q)e_Q$ and $kC_H(Q)f_Q$, and the fusion systems of b and c on P determined by the choices of the source idempotents i and j are equal.*

(iv) *For any nontrivial fully $\mathcal{F}$-centralised subgroup Q of P the bimodule $(iMj)(\Delta Q)$ and its dual induce a Morita equivalence between $A(\Delta Q)$ and $B(\Delta Q)$, and the fusion systems of b and c on P determined by A and B are equal.*

Proof We may assume that P is nontrivial. The equivalence of (i) and (ii) follows from the standard Morita equivalences between block algebras and their source algebras. Similarly, the equivalence between (iii) and (iv) follows from the standard Morita equivalences between $kC_G(Q)e_Q$ and $A(\Delta Q)$ for any fully $\mathcal{F}$-centralised subgroup of P (and the analogous Morita equivalences between $kC_H(Q)f_Q$ and $B(\Delta Q)$). Suppose that (i) holds. Then $M \otimes_{\mathcal{O}Hc} M^* \cong \mathcal{O}Gb \oplus X$ for some projective $\mathcal{O}Gb$-$\mathcal{O}Hc$-bimodule X and $M^* \otimes_{\mathcal{O}Gb} M \cong \mathcal{O}Hc \oplus Y$ for some projective $\mathcal{O}Hc$-$\mathcal{O}Hc$-bimodule Y. Thus $X(\Delta Q) = \{0\}$ for every nontrivial subgroup Q of P, similarly for Y. We show that the fusion systems on P determined by the source algebras A and B coincide. The $\mathcal{O}Hc$-$\mathcal{O}P$-bimodule $\mathcal{O}Hj$ is indecomposable nonprojective. By 2.17.11, the $\mathcal{O}Gb$-$\mathcal{O}P$-bimodule $Mj \cong M \otimes_{\mathcal{O}Hc} \mathcal{O}Hj$ is the direct sum of an indecomposable nonprojective $\mathcal{O}Gb$-$\mathcal{O}P$-bimodule and a projective $\mathcal{O}Gb$-$\mathcal{O}P$-bimodule. By the hypotheses on M, the bimodule is also a direct summand of $\mathcal{O}Gi \otimes_{\mathcal{O}P} B$, and hence the nonprojective part of Mj is isomorphic to a direct summand of $\mathcal{O}Gi \otimes_{\mathcal{O}P} W$ for some indecomposable direct summand W of B as an $\mathcal{O}P$-$\mathcal{O}P$-bimodule. Thus on one hand we have $M^* \otimes_{\mathcal{O}Gb} Mj \cong \mathcal{O}Hj \oplus Yj$, and on the other hand, $M^* \otimes_{\mathcal{O}Gb} Mj$ is isomorphic to a direct summand of $M^* \otimes_{\mathcal{O}P} W$. Since $\mathrm{Br}_{\Delta P}(j) \neq 0$, this forces $W \cong k[yP]$ for some $y \in N_H(P, f_P)$, by 8.7.1. Since conjugation by y stabilises f_P, hence the local point of P on $\mathcal{O}Gb$ to which j belongs, we have $Mj \cong Mjy^{-1}$, and so the nonprojective part of Mj is isomorphic to a direct summand of $\mathcal{O}Gi$. This shows that

$$Mj \cong \mathcal{O}Gi \oplus U$$

for some projective $\mathcal{O}Gb$-$\mathcal{O}P$-module U. Tensoring this over $\mathcal{O}Gb$ with its dual $jM^* \cong i\mathcal{O}G \oplus U^*$ and using $jM^* \otimes_{\mathcal{O}Gb} Mj \cong B \oplus jYj$ yields an isomorphism of $\mathcal{O}P$-$\mathcal{O}P$-bimodules

$$B \oplus V \cong A \oplus W$$

for some projective $\mathcal{O}P$-$\mathcal{O}P$-bimodules V, W. But then A and B determine the same fusion system $\mathcal{F}$ on P thanks to 8.7.1. In order to show that $e_Q M(\Delta Q) f_Q$ induces a Morita equivalence, we may therefore assume that Q is fully $\mathcal{F}$-centralised. As mentioned before, since block algebras are Morita equivalent to their almost source algebras, it suffices to show that $N(\Delta Q)$ induces a Morita equivalence between $A(\Delta Q)$ and $B(\Delta Q)$. Since A and B determine the same fusion systems, this follows from applying 9.4.7: we have $N(\Delta Q) \otimes_{B(\Delta Q)} N(\Delta Q)^* \cong (N \otimes_B N^*)(\Delta Q) \cong A(\Delta Q) \oplus (iXi)(\Delta Q) \cong A(\Delta Q)$. Similarly we have $N(\Delta Q)^* \otimes_A N(\Delta Q) \cong B(\Delta Q)$, hence (i) implies (iii) and (iv). Suppose that (iii) holds. The hypotheses imply that $N = iMj$ is a direct summand of $A \otimes_{\mathcal{O}P} B$ and that $N(\Delta Q)$ induces a Morita equivalence for all fully $\mathcal{F}$-centralised subgroups Q of P, where $\mathcal{F}$ is the common fusion system on P determined by i and j. Consider the adjunction map $N \otimes_B N^* \to A$. This map induces an isomorphism $N(\Delta Q) \otimes_{B(\Delta Q)} N(\Delta Q)^* \cong (N \otimes_B N^*)(\Delta Q) \cong A(\Delta Q)$ for any fully $\mathcal{F}$-centralised subgroup Q of P. The fully $\mathcal{F} \times \mathcal{F}$-centralised subgroups of the subgroup ΔP of $P \times P$ are all of the form ΔQ for some fully $\mathcal{F}$-centralised subgroup Q of P, by 8.4.8. Since A is indecomposable as an A-A-bimodule, it follows from 8.8.7 that $N \otimes_B N^* \cong A \oplus X$ for some projective A-A-bimodule X. Similarly, $N^* \otimes_A N \cong B \oplus Y$ for some projective B-B-bimodule Y. This implies that M induces a stable equivalence of Morita type and shows therefore that (iii) implies (i) and (ii). $\qquad\square$

In order to verify whether induction and restriction functors, truncated by block idempotents, induce a stable equivalence of Morita type between a block algebra and its Brauer correspondent, it suffices to determine the bimodule structure of a block restricted to its Brauer correspondent.

Proposition 9.8.3 *Let G be a finite group, b a block of $\mathcal{O}G$, P a defect group of b and H a subgroup of G containing $N_G(P)$. Let c be the block of $\mathcal{O}H$ with defect group P corresponding to b. Then $\mathcal{O}Hc$ is isomorphic to a direct summand of $c\mathcal{O}Gbc$ as an $\mathcal{O}Hc$-$\mathcal{O}Hc$-bimodule. If $c\mathcal{O}Gbc \cong \mathcal{O}Hc \oplus Y$ for some projective $\mathcal{O}Hc$-$\mathcal{O}Hc$-bimodule Y, then the $\mathcal{O}Gb$-$\mathcal{O}Hc$-bimodule $M = b\mathcal{O}Gc$ and its dual $M^* \cong c\mathcal{O}Gb$ induce a stable equivalence of Morita type between $\mathcal{O}Gb$ and $\mathcal{O}Hc$.*

Proof By 6.7.2 (ii), $\mathcal{O}Hc$ is isomorphic to a direct summand of $c\mathcal{O}Gbc$. Suppose that $c\mathcal{O}Gbc \cong \mathcal{O}Hc \oplus Y$ for some projective $\mathcal{O}Hc$-$\mathcal{O}Hc$-bimodule Y. We have $\mathrm{Br}_P(b) = \mathrm{Br}_P(c) \neq 0$, hence $\mathrm{Br}_P(bc) \neq 0$. It follows from 6.4.6 that multiplication by c induces a Morita equivalence between $\mathcal{O}Gb$ and $c\mathcal{O}Gbc$, and hence it suffices to show that induction and restriction induce a stable equivalence of Morita type between $c\mathcal{O}Gbc$ and $\mathcal{O}Hc$. By 4.14.12, it suffices to show that $c\mathcal{O}Gbc$ is relatively $\mathcal{O}Hc$-separable. It follows from 6.4.7 that the canonical bimodule homomorphism $\mathcal{O}Gbc \otimes_{\mathcal{O}P} c\mathcal{O}Gb \to \mathcal{O}Gb$ is split surjective, hence the homomorphism $\mathcal{O}Gbc \otimes_{\mathcal{O}Hc} c\mathcal{O}Gb \to \mathcal{O}Gb$, is split surjective. Multiplying by c on the left and right implies that $c\mathcal{O}Gbc$ is relatively $\mathcal{O}Hc$-separable, as required. The result follows. $\qquad\square$

Proposition 9.8.3 applied to the case $H = N_G(P)$ yields in conjunction with earlier results the following useful version at the source algebra level.

Proposition 9.8.4 *Let G be a finite group, b a block of $\mathcal{O}G$, P a defect group of b and $i \in (\mathcal{O}Gb)^{\Delta P}$ a source idempotent of b. Set $A = i\mathcal{O}Gi$ and $B = \mathcal{O}_\alpha(P \rtimes E)$ a source algebra of the block c of $\mathcal{O}N_G(P)$ corresponding to b via the Brauer correspondence. Identify B to its image in A via the homomorphism from 6.15.1, and let Y be a complement of B in A as a B-B-bimodule. Then the A-B-bimodule A_B and the B-A-bimodule ${}_BA$ induce a stable equivalence of Morita type between A and B if and only if the B-B-bimodule Y is projective.*

Proof This follows from 9.8.3 applied to the case $H = N_G(P)$, combined with the standard Morita equivalences between A and $\mathcal{O}Gb$ as well as B and $\mathcal{O}N_G(P)$, and the split injective algebra homomorphism $B \to A$ from 6.15.1. Alternatively, this follows also directly from 4.14.15. $\qquad\square$

By 5.8.2, the bimodule Y in 9.8.4 is projective if and only if $Y(S) = \{0\}$ for any subgroup S of $P \times P$. If the fusion system of A is equal to that of $P \rtimes E$, then it suffices to test this on diagonal subgroups (this applies in particular when P is abelian).

Corollary 9.8.5 *With the notation of 9.8.4, suppose that A and B have the same fusion system on P. The following are equivalent.*

  (i) *The A-B-bimodule A_B and the B-A-bimodule ${}_BA$ induce a stable equivalence of Morita type between A and B.*
 (ii) *The B-B-bimodule Y is projective.*
(iii) *We have $Y(\Delta Q) = \{0\}$ for any nontrivial subgroup Q of P.*

Proof The equivalence of (i) and (ii) is from 9.8.4, or also from 4.14.15, and clearly (ii) implies (iii). Suppose that (iii) holds. Then $A(\Delta Q) \cong B(\Delta Q)$ for any nontrivial subgroup Q of P. Since A and B have the same fusion system on P, statement (i) follows from 9.8.2. $\qquad\square$

Theorem 9.8.6 *Let G be a finite group, b a block of $\mathcal{O}G$ and P a defect group of b. Let H be a subgroup of G containing P such that $P \cap {}^x P = \{1\}$ for every $x \in G \setminus H$. Then H contains $N_G(Q)$ for any nontrivial subgroup Q of P. Denote by c the block of $\mathcal{O}H$ with P as a defect group corresponding to b through the Brauer correspondence. The following hold.*

(i) *For any nontrivial subgroup Q of P we have $\mathrm{Br}_Q(b) = \mathrm{Br}_Q(c)$.*
(ii) *The $\mathcal{O}Gb$-$\mathcal{O}Hc$-bimodule $M = b\mathcal{O}Gc$ and its dual $M^* \cong c\mathcal{O}Gb$ induce a stable equivalence of Morita type between $\mathcal{O}Gb$ and $\mathcal{O}Hc$.*

Proof For $a \in (\mathcal{O}G)^P$ and Q a nontrivial subgroup of P we have $\mathrm{Tr}_P^G(a) = \sum_{x \in [Q \setminus G/P]} \mathrm{Tr}_{Q \cap {}^x P}^Q({}^x a)$. If $x \in G \setminus H$, then $Q \cap {}^x P = \{1\}$, and hence Br_Q annihilates the summands in the previous sum corresponding to terms indexed by elements in $G \setminus H$. Thus, applying Br_Q yields $\mathrm{Br}_Q(\mathrm{Tr}_P^G(a)) = \mathrm{Br}_Q(\mathrm{Tr}_P^H(a))$. In particular, $\mathrm{Br}_Q((\mathcal{O}G)_P^G) = \mathrm{Br}_Q((\mathcal{O}H)_P^H)$, and this is an ideal in the subalgebra $\mathrm{Br}_Q(Z(\mathcal{O}G))$ of $Z(kC_G(Q)) = Z(kC_H(Q))$. Now b is a primitive idempotent in $(\mathcal{O}G)_P^G$ satisfying $\mathrm{Br}_P(b) \neq 0$, hence $\mathrm{Br}_Q(b) \neq 0$. The lifting theorems for idempotents imply that there is a unique primitive idempotent c' in $(\mathcal{O}H)_P^H$ satisfying $\mathrm{Br}_Q(b) = \mathrm{Br}_Q(c')$. But then, if $R = N_P(Q)$, we also have $\mathrm{Br}_R(b) = \mathrm{Br}_R(c')$. Inductively we get $\mathrm{Br}_P(b) = \mathrm{Br}_P(c')$. Since c is the unique block of $\mathcal{O}H$ with P as a defect group and satisfying $\mathrm{Br}_P(b) = \mathrm{Br}_P(c)$ we get that $c' = c$. It follows that $(\mathcal{O}Gb)(Q) = kC_G(Q)\mathrm{Br}_Q(b) = kC_H(Q)\mathrm{Br}_Q(c) = (\mathcal{O}Hc)(Q)$, and that $M(\Delta Q) = kC_G(Q)\mathrm{Br}_Q(b)$ induces the trivial Morita equivalence on this algebra. One way to conclude the proof would be to play this back to a special case of 9.8.2, but one can deduce this also more directly from 9.8.3. We have $\mathrm{Br}_Q(c - bc) = 0$ for any nontrivial subgroup Q of P. Since $c - bc \in (\mathcal{O}G)_P^H$ it follows that $c - bc \in (\mathcal{O}G)_1^H$. Since $P \cap {}^x P = \{1\}$ for $x \in G \setminus H$, it follows that $P \times P$ acts freely on $G \setminus H$ via $(u, v) \cdot x = uxv^{-1}$, where u, $v \in P$ and $x \in G - H$, and therefore $\mathcal{O}[G \setminus H]$ is projective as an $\mathcal{O}P$-$\mathcal{O}P$-bimodule. Thus $c\mathcal{O}Gc = \mathcal{O}Hc \oplus c\mathcal{O}[G \setminus H]c$ as $\mathcal{O}Hc$-$\mathcal{O}Hc$-bimodules, and the second summand is projective. Since $\mathcal{O}Hc$ is a direct summand of $c\mathcal{O}Gbc$, which is a direct summand of $c\mathcal{O}Gc$, it follows that $c\mathcal{O}Gbc \simeq \mathcal{O}Hc \oplus Y$ for some direct summand Y of $c\mathcal{O}[G \setminus H]c$, and then Y is projective by the preceding argument. The result follows from 9.8.3. $\square$

Definition 9.8.7 Let A, B be symmetric $\mathcal{O}$-algebras. A bounded complex X of A-B-bimodules in $\mathrm{perf}(A, B)$ is said to *induce a stable equivalence* if there are isomorphisms of complexes of bimodules $X \otimes_B X^* \cong A \oplus Y$ and $X^* \otimes_A X \cong B \oplus Z$ with Y and Z homotopy equivalent to bounded complexes of projective A-A-bimodules and B-B-bimodules, respectively.

If Y and Z are homotopic to zero in the above definition, then X is a Rickard complex. The following result, due to Rouquier, is a criterion for when a splendid complex between two blocks with the same fusion system induces a stable equivalence.

Theorem 9.8.8 ([195, 5.6]) *Let G, H be finite groups, b, c blocks of $\mathcal{O}G$, $\mathcal{O}H$, respectively, having a common defect group P, let $i \in (\mathcal{O}Gb)^{\Delta P}$ and $j \in (\mathcal{O}Hc)^{\Delta P}$ be source idempotents. Suppose that i and j determine the same fusion system $\mathcal{F}$ on P. Set $A = i\mathcal{O}Gi$ and $B = j\mathcal{O}Hj$. For any subgroup Q of P denote by e_Q and f_Q the unique blocks of $kC_G(Q)$ and $kC_H(Q)$ satisfying $\mathrm{Br}_{\Delta Q}(i)e_Q \neq 0$ and $\mathrm{Br}_{\Delta Q}(j)f_Q \neq 0$. Let X be a bounded complex of $\mathcal{O}Gb$-$\mathcal{O}Hc$ bimodules whose terms are finite direct sums of summands of the $\mathcal{O}Gb$-$\mathcal{O}Hc$-bimodules $\mathcal{O}Gi \otimes_{\mathcal{O}Q} j\mathcal{O}H$, where Q runs over the subgroups of P. The following are equivalent:*

(i) *The complex X induces a stable equivalence.*
(ii) *For every nontrivial subgroup Q of P the complex $e_Q X(\Delta Q)f_Q$ is a Rickard complex of $kC_G(Q)e_Q$-$kC_H(Q)f_Q$-bimodules.*
(iii) *For every nontrivial fully $\mathcal{F}$-centralised subgroup Q of P the complex $(iXj)(\Delta Q)$ is a Rickard complex of $A(\Delta Q)$-$B(\Delta Q)$-bimodules.*

Proof The proof we present here is from [136, Appendix]; it is an adaptation of Rouquier's proof for principal blocks in [195, 5.6]. The equivalence of (ii) and (iii) follows from the Morita equivalences 8.7.3. Suppose that (i) holds. Write $X \otimes_{\mathcal{O}Hc} X^* \cong \mathcal{O}Gb \oplus Y$ for some bounded complex Y of $\mathcal{O}Gb$-$\mathcal{O}Gb$-bimodules that is homotopy equivalent to a complex of projective $\mathcal{O}Gb$-$\mathcal{O}Gb$-bimodules. Since the Brauer construction with respect to a nontrivial p-subgroup sends any projective module to zero and preserves homotopies, we get that $Y(\Delta Q) \simeq \{0\}$ for any nontrivial subgroup Q of P. Thus, by 9.4.7, translated back to block algebras using 9.4.9, we get that

$$(e_Q X(\Delta Q)f_Q) \otimes_{kC_H(Q)f_Q} (f_Q X^*(\Delta Q)e_Q) \cong e_Q(X \otimes_{\mathcal{O}Hc} X^*)(\Delta Q)e_Q$$

$$\simeq e_Q(\mathcal{O}Gb)(\Delta Q)e_Q = kC_G(Q)e_Q$$

for any nontrivial fully $\mathcal{F}$-centralised subgroup Q of P. Since any subgroup of P is isomorphic, in the fusion system $\mathcal{F}$, to a fully $\mathcal{F}$-centralised subgroup we get this isomorphism for any nontrivial subgroup Q of P. In particular, $e_Q X(\Delta Q)f_Q$ is a Rickard complex for any nontrivial subgroup Q of P. Suppose conversely that (ii) holds. Let Y be the mapping cone of the adjunction unit $\mathcal{O}Gb \to X \otimes_{\mathcal{O}Hc} X^*$. By the assumptions, the induced map

$$kC_G(Q)e_Q \to (e_Q X(\Delta Q)f_Q) \otimes_{kC_H(Q)f_Q} (f_Q X^*(\Delta Q)e_Q)$$

is a homotopy equivalence, and so its mapping cone is contractible, for every nontrivial subgroup Q of P. But as before, the right side is canonically isomorphic to $e_Q(X \otimes_{\mathcal{O}Hc} X^*)(\Delta Q)e_Q$, and so the corresponding mapping cone is $e_Q Y(\Delta Q) f_Q$. This is contractible, for every nontrivial Q. But then in fact $Y(\Delta Q) \simeq 0$ for every nontrivial subgroup Q of P, thanks to 9.4.10. Bouc's Theorem 5.11.10 implies that Y is homotopy equivalent to a bounded complex of projective bimodules, whence (i). $\qquad\square$

Remark 9.8.9 Let P be a finite p-group. Let A and B be relatively $\mathcal{O}P$-separable symmetric primitive interior P-algebras with P-stable $\mathcal{O}$-bases such that A, B are projective as left and right $\mathcal{O}P$-modules. The terminology of splendid Morita, derived, and stable equivalences extends to this situation in the obvious way. An A-B-bimodule is called *splendid* if it is isomorphic to a finite direct summand of direct summands of the A-B-bimodules $A \otimes_{\mathcal{O}Q} B$, where Q runs over the subgroups of P. The assumptions on A and B imply that the dual of a splendid A-B-bimodule is a splendid B-A-bimodule. A Morita equivalence or a stable equivalence of Morita type between A and B is called *splendid* if it is given by a splendid bimodule. Similarly, a derived equivalence between A and B is called *splendid* if it is given by a Rickard complex consisting of splendid bimodules.

9.9 Fusion-stable endopermutation modules

The Morita equivalence 8.11.9 between a nilpotent block of a finite group and one of its defect groups need not be splendid but has an endopermutation module as a source. We collect in this section some technical results which will be used to show in the next section that a number of results on splendid stable and Morita equivalences carry over to stable and Morita equivalences induced by bimodules with endopermutation sources. The starting point for what follows is the fact from 7.3.7 which states that for P a finite p-group, an endopermutation $\mathcal{O}P$-module V has at most one isomorphism class of indecomposable direct summands with vertex Q, for any subgroup Q of P. Equivalently, Q has at most one local point on the P-algebra $S = \mathrm{End}_\mathcal{O}(V)$.

Definition 9.9.1 Let P be a finite p-group and $\mathcal{F}$ a fusion system on P. Let Q be a subgroup of P and V be an endopermutation $\mathcal{O}Q$-module. We say that V is $\mathcal{F}$-*stable* if for any subgroup R of Q and any morphism $\varphi : R \to Q$ in $\mathcal{F}$ the sets isomorphism classes of indecomposable direct summands with vertex R of the $\mathcal{O}Q$-modules $\mathrm{Res}^Q_R(V)$ and $_\varphi V$ are equal (including the possibility that both sets may be empty).

By 7.11.2 there is a vertex preserving bijection between the indecomposable direct summands of an endopermutation module V and the noncontractible indecomposable direct summands of an endosplit p-permutation resolution Y of V. We use this to extend the notion of fusion stability to Y in the obvious way.

Definition 9.9.2 Let P be a finite p-group and $\mathcal{F}$ a fusion system on P. Let V be an endopermutation $\mathcal{O}P$-module having an indecomposable direct summand with vertex P, and let Y be an endosplit p-permutation resolution of V. We say that Y is $\mathcal{F}$-stable, if for any subgroup Q of P and any morphism $\varphi : Q \to P$, the sets of isomorphism classes of indecomposable direct summands with vertex Q of the complexes of permutation $\mathcal{O}Q$-modules $\mathrm{Res}_Q^P(Y)$ and $_\varphi Y$ are equal.

With the notation of 9.9.1, the property of V being $\mathcal{F}$-stable does not necessarily imply that $\mathrm{Res}_R^Q(V)$ and $_\varphi V$ have to be *isomorphic* as $\mathcal{O}R$-modules, where $\varphi : R \to Q$ is a morphism in $\mathcal{F}$. What the $\mathcal{F}$-stability of V means is that the indecomposable direct factors of $\mathrm{Res}_R^Q(V)$ and $_\varphi V$ with vertex R, if any, are isomorphic, but they may occur with different multiplicities in direct sum decompositions. By [141, 3.7], every class in $D_\mathcal{O}(P)$ having an $\mathcal{F}$-stable representative has a representative W satisfying the stronger stability condition $\mathrm{Res}_R^P(W) \cong {}_\varphi W$ for any morphism $\varphi : R \to P$ in $\mathcal{F}$.

Example 9.9.3 Let P be a finite p-group, Q a subgroup of P and n an integer. Let $\varphi : Q \to P$ be an injective group homomorphism. Then $_\varphi \Omega_P^n(\mathcal{O}) \cong \mathrm{Res}_Q^P(\Omega_P^n(\mathcal{O}))$. In particular, $\Omega_P^n(\mathcal{O})$ is $\mathcal{F}$-stable for any fusion system $\mathcal{F}$ on P. Indeed, $_\varphi \Omega_P^n(\mathcal{O})$ and $\mathrm{Res}_Q^P(\Omega_P^n(\mathcal{O}))$ both have $\Omega_Q^n(\mathcal{O})$ as their unique indecomposable nonprojective direct summand, and since they have the same $\mathcal{O}$-rank, they are isomorphic.

It follows from Alperin's Fusion Theorem 8.2.8 that in order to check whether an endopermutation $\mathcal{O}P$-module V with an indecomposable direct summand of vertex P is $\mathcal{F}$-stable, it suffices to verify that $\mathrm{Res}_R^P(V)$ and $_\varphi V$ have isomorphic summands with vertex R for any $\mathcal{F}$-essential subgroup R of P and any p'-automorphism φ of R in $\mathrm{Aut}_\mathcal{F}(R)$. In particular, if P is abelian, then an indecomposable endopermutation $\mathcal{O}P$-module V with vertex P is $\mathcal{F}$-stable if and only if $V \cong {}_\varphi V$ for any $\varphi \in \mathrm{Aut}_\mathcal{F}(P)$. In many cases where Definition 9.9.1 is used we will have $Q = P$. One notable exception arises in the context of bimodules, where we consider the fusion systems of the form $\mathcal{F} \times \mathcal{F}'$ on $P \times P$, with the diagonal subgroup ΔP playing the role of Q. Note that if Q is a subgroup of P and $\tau : \Delta Q \to \Delta P$ a morphism in $\mathcal{F} \times \mathcal{F}'$, then $\tau(u, u) = (\varphi(u), \varphi(u))$ for all $u \in Q$, where $\varphi : Q \to P$ is a morphism that belongs to both $\mathcal{F}$ and to $\mathcal{F}'$. If, for instance $\mathcal{F}'$ is a subfusion system of

$\mathcal{F}$, then being $\mathcal{F} \times \mathcal{F}'$-stable for an endopermutation $\mathcal{O}\Delta P$-module V means that V is $\mathcal{F}'$-stable when viewed as an $\mathcal{O}P$-module. Thus if one of $\mathcal{F}$, $\mathcal{F}'$ is the trivial fusion system $\mathcal{F}_P(P)$, then every $\mathcal{P}\Delta P$-endopermutation module is $\mathcal{F} \times \mathcal{F}'$-stable. The key argument exploiting the $\mathcal{F}$-stability of an endopermutation $\mathcal{O}P$-module V having an indecomposable direct summand with vertex P goes as follows: if Q is a subgroup of P and $\varphi : Q \to P$ a morphism in $\mathcal{F}$, then the restriction to ΔQ of $V \otimes_{\mathcal{O}} {}_\varphi V^*$ is again a permutation module, or equivalently, $V \otimes_{\mathcal{O}} V^*$ remains a permutation module for the twisted diagonal subgroup $\Delta_\varphi Q = \{(u, \varphi(u)) | u \in Q\}$ of $P \times P$.

Proposition 9.9.4 *Let A be an almost source algebra of a block of a finite group algebra over $\mathcal{O}$ with a defect group P and fusion system $\mathcal{F}$ on P. Let Q be a subgroup of P and let V be an $\mathcal{F}$-stable endopermutation $\mathcal{O}Q$-module having an indecomposable direct summand with vertex Q. Set $U = A \otimes_{\mathcal{O}Q} V$. The following hold:*

 (i) *As an $\mathcal{O}Q$-module, U is an endopermutation module, and has a direct summand isomorphic to V.*
 (ii) *Let R be a subgroup of Q. The A-module structure on U induces an $A(\Delta R)$-module structure on $U' = \mathrm{Defres}^Q_{RC_Q(R)/R}(U)$ extending that of $kC_Q(R)$ and such that $(\mathrm{End}_{\mathcal{O}}(U))(\Delta R) \cong \mathrm{End}_k(U')$ as $A(\Delta R)$-$A(\Delta R)$-bimodules.*

Proof By 8.7.1, every indecomposable direct summand of A as an $\mathcal{O}Q$-$\mathcal{O}Q$-bimodule is isomorphic to $\mathcal{O}Q \otimes_{\mathcal{O}R} {}_\varphi \mathcal{O}Q$ for some subgroup R of Q and some morphism $\varphi \in \mathrm{Hom}_{\mathcal{F}}(R, Q)$, and at least one summand is isomorphic to $\mathcal{O}Q$. Thus every indecomposable direct summand of U is isomorphic to $\mathrm{Ind}^Q_R({}_\varphi V_{\varphi(R)})$ for some subgroup R of Q and some $\varphi \in \mathrm{Hom}_{\mathcal{F}}(R, Q)$, where $V_{\varphi(R)}$ is an indecomposable direct summand of vertex $\varphi(R)$ of $\mathrm{Res}^Q_{\varphi(R)}(V)$, and V is a summand of U as an $\mathcal{O}Q$-module. Since V is $\mathcal{F}$-stable, we have ${}_\varphi(V_{\varphi(R)}) \cong V_R$, which implies that the restriction to $\mathcal{O}Q$ of U is an endopermutation $\mathcal{O}Q$-module. Statement (i) follows. For statement (ii) we consider the structural algebra homomorphism $A \to \mathrm{End}_{\mathcal{O}}(U)$ given by the action of A on U. This is a homomorphism of interior Q-algebras. Applying the Brauer construction with respect to ΔR, where R is a subgroup of Q, yields a homomorphism of interior $C_Q(R)$-algebras $A(\Delta R) \to (\mathrm{End}_{\mathcal{O}}(U))(\Delta R)$. Since U is an endopermutation $\mathcal{O}Q$-module, we have $(\mathrm{End}_{\mathcal{O}}(U))(\Delta R) \cong \mathrm{End}_k(U')$ as interior $C_Q(R)$-algebras. This yields a homomorphism $A(\Delta R) \to \mathrm{End}_k(U')$, hence a canonical $A(\Delta R)$-module structure on U' with the properties as stated. $\square$

Statement (ii) in 9.9.4 is particularly useful when Q is fully $\mathcal{F}$-centralised, since in that case $C_P(Q)$ is a defect group of the unique block e_Q of $kC_G(Q)$

satisfying $\mathrm{Br}_{\Delta Q}(i)e_Q \neq 0$, and the algebras $A(\Delta Q)$ and $kC_G(Q)e_Q$ are Morita equivalent. The bimodule version of 9.9.4 is as follows.

Proposition 9.9.5 *Let A and B be almost source algebras of blocks of finite group algebras over $\mathcal{O}$ with a common defect group P, and with fusion systems $\mathcal{F}$ and $\mathcal{F}'$ on P, respectively. Let V be an $\mathcal{F}$-stable indecomposable endopermutation $\mathcal{O}P$-module with vertex P, viewed as an $\mathcal{O}\Delta P$-module through the canonical isomorphism $\Delta P \cong P$. Set $U = A \otimes_{\mathcal{O}P} \mathrm{Ind}_{\Delta P}^{P \times P}(V) \otimes_{\mathcal{O}P} B$. As an $\mathcal{O}\Delta P$-module, U is an endopermutation module having V as a direct summand, and for any subgroup Q of P, the A-B-bimodule structure on U induces an $A(\Delta Q)$-$B(\Delta Q)$-bimodule structure on $U' = \mathrm{Defres}_{\Delta QC_P(Q)/Q}^{\Delta P}(U)$ such that we have an isomorphism of $A(\Delta Q) \otimes_k B(\Delta Q)^{\mathrm{op}}$-bimodules $\mathrm{End}_{\mathcal{O}}(U)(\Delta Q) \cong \mathrm{End}_k(U')$.*

Proof Using 8.7.6 and 8.7.7 one sees that this is the special case of 9.9.4 with $P \times P$, $\mathcal{F} \times \mathcal{F}'$, ΔP, ΔQ, $A \otimes_{\mathcal{O}} B^{\mathrm{op}}$, instead of P, $\mathcal{F}$, Q, R, A, respectively. $\qquad\qquad\square$

Remark 9.9.6 We will need technical statements that involve the tensor product of two bimodules. This yields a priori four module structures, and keeping track of those is essential. If the algebras under consideration are group algebras, we play this back to two actions via the usual 'diagonal' convention: given two finite groups G, H and two $\mathcal{O}G$-$\mathcal{O}H$-bimodules S, S', we consider $S \otimes_{\mathcal{O}} S'$ as an $\mathcal{O}G$-$\mathcal{O}H$-bimodule via the diagonal left action by G and the diagonal right action by H; explicitly, $x \cdot (s \otimes s') \cdot y = xsy \otimes xs'y$, where $x \in G, y \in H, s \in S$, and $s' \in S'$. This is equivalent to the diagonal $G \times H$-action if we interpret the $\mathcal{O}G$-$\mathcal{O}H$-bimodules as $\mathcal{O}(G \times H)$-modules in the usual way.

Proposition 9.9.7 *Let A and B be almost source algebras of blocks of finite group algebras over $\mathcal{O}$ with a common defect group P. Suppose that the fusion system $\mathcal{F}$ of A is contained in the fusion system $\mathcal{F}'$ of B on P. Let V be an $\mathcal{F}$-stable indecomposable endopermutation $\mathcal{O}P$-module with vertex P. Consider V as an $\mathcal{O}\Delta P$-module through the canonical isomorphism $\Delta P \cong P$. Set $U = A \otimes_{\mathcal{O}P} \mathrm{Ind}_{\Delta P}^{P \times P}(V) \otimes_{\mathcal{O}P} B$. Let Q be a fully $\mathcal{F}'$-centralised subgroup of P, and let U' be the $A(\Delta Q)$-$B(\Delta Q)$-bimodule from 9.9.5 such that $(\mathrm{End}_{\mathcal{O}}(U))(\Delta Q) \cong \mathrm{End}_k(U')$. Then $\mathrm{End}_{B^{\mathrm{op}}}(U)$ is a ΔQ-subalgebra of $\mathrm{End}_{\mathcal{O}}(U)$, the algebra homomorphism*

$$\beta : \mathrm{End}_{B^{\mathrm{op}}}(U)(\Delta Q) \to \mathrm{End}_{\mathcal{O}}(U)(\Delta Q)$$

induced by the inclusion $\text{End}_{B^{\text{op}}}(U) \subseteq \text{End}_{\mathcal{O}}(U)$ *is injective, and there is a commutative diagram of algebra homomorphisms*

$$
\begin{array}{ccc}
\text{End}_{\mathcal{O}}(U)(\Delta Q) & \xrightarrow{\cong} & \text{End}_k(U') \\
{\scriptstyle\beta}\big\uparrow & & \big\uparrow \\
\text{End}_{B^{\text{op}}}(U)(\Delta Q) & \xrightarrow[\gamma]{} & \text{End}_{B(\Delta Q)^{\text{op}}}(U')
\end{array}
$$

where the right vertical arrow is the obvious inclusion map. In particular, the algebra homomorphism γ is injective. There is a similar commutative diagram using the inclusion $\text{End}_A(U) \to \text{End}_{\mathcal{O}}(U)$ instead, provided that the fusion system of B is contained in that of A.

Proof For $\epsilon \in \text{End}_{\mathcal{O}}(U)$, $y \in Q$, and $u \in U$ we have

$$
{}^{\Delta y}\epsilon(u) = \Delta y \cdot \epsilon(\Delta y^{-1} \cdot u) = y\epsilon(y^{-1}uy)y^{-1},
$$

where $\Delta y = (y, y)$. If $\epsilon \in \text{End}_{B^{\text{op}}}(U)$, then in particular ϵ commutes with the right action by Q, and hence we have ${}^{\Delta y}\epsilon(u) = y\epsilon(y^{-1}u) = {}^{(y,1)}\epsilon(u)$, which shows that ${}^{\Delta y}\epsilon$ is again a B^{op}-homomorphism. The algebra of ΔQ-fixed points in $\text{End}_{B^{\text{op}}}(U)$ is equal to $\text{End}_{\mathcal{O}Q \otimes_{\mathcal{O}} B^{\text{op}}}(U)$. The existence of a commutative diagram as in the statement is formal: if $\varphi \in \text{End}_{\mathcal{O}Q \otimes_{\mathcal{O}} B^{\text{op}}}(U)$, then in particular $b \cdot \varphi = \varphi \cdot b$ for all $b \in B$, hence for all $b \in B^{\Delta Q}$, and applying $\text{Br}_{\Delta Q}$ yields that the image of φ in $\text{End}_{\mathcal{O}}(U)(\Delta Q)$ commutes with the elements in $B(\Delta Q)$. Since the upper horizontal map is a bimodule isomorphism, it follows that the image of φ in $\text{End}_k(U')$ commutes with the elements in $B(\Delta Q)$, hence lies in the subalgebra $\text{End}_{B(\Delta Q)^{\text{op}}}(U')$. In order to show that β is injective, we first note that this injectivity does not make use of the left A-module structure of U but only of the left $\mathcal{O}Q$-module structure. Thus we may decompose U by decomposing A as an $\mathcal{O}Q$-$\mathcal{O}P$-bimodule. By 8.7.1, every summand of A as an $\mathcal{O}Q$-$\mathcal{O}P$-bimodule is of the form $\mathcal{O}Q \otimes_{\mathcal{O}R} {}_{\varphi}\mathcal{O}P$ for some subgroup R of Q and some homomorphism $\varphi : R \to P$ belonging to the fusion system $\mathcal{F}$ of A, hence to $\mathcal{F}'$ by the assumptions. Using the isomorphism $\mathcal{O}P$-B-isomorphism $\text{Ind}_{\Delta P}^{P \times P}(V) \otimes_{\mathcal{O}P} B \cong V \otimes_{\mathcal{O}} B$ from 2.4.13 it suffices therefore to show that applying $\text{Br}_{\Delta Q}$ to the inclusion map

$$
\text{Hom}_{B^{\text{op}}}(\mathcal{O}Q \otimes_{\mathcal{O}R} {}_{\varphi}(V \otimes_{\mathcal{O}} B), \mathcal{O}Q \otimes_{\mathcal{O}S} {}_{\psi}(V \otimes_{\mathcal{O}} B))
$$

$$
\subseteq \text{Hom}_{\mathcal{O}}(\mathcal{O}Q \otimes_{\mathcal{O}R} {}_{\varphi}(V \otimes_{\mathcal{O}} B), \mathcal{O}Q \otimes_{\mathcal{O}S} {}_{\psi}(V \otimes_{\mathcal{O}} B))
$$

remains injective, where R, S are subgroups of Q and where $\varphi \in \text{Hom}_{\mathcal{F}}(R, P)$, $\psi \in \text{Hom}_{\mathcal{F}}(S, P)$. If one of R, S is a proper subgroup of Q, then both sides are zero. Thus it suffices to show that the map

$$\mathrm{Hom}_{B^{\mathrm{op}}}({}_\varphi(V \otimes_{\mathcal{O}} B), {}_\psi(V \otimes_{\mathcal{O}} B))(\Delta Q)$$

$$\to \mathrm{Hom}_{\mathcal{O}}({}_\varphi(V \otimes_{\mathcal{O}} B), {}_\psi(V \otimes_{\mathcal{O}} B))(\Delta Q)$$

is injective, where φ, $\psi \in \mathrm{Hom}_{\mathcal{F}}(Q, P)$. The summands of ${}_\varphi V$, ${}_\psi V$ with vertices smaller than Q yield summands of $V \otimes_{\mathcal{O}} B$ that vanish on both sides upon applying $\mathrm{Br}_{\Delta Q}$. The fusion stability of V implies that indecomposable summands with vertex Q of ${}_\varphi V$, ${}_\psi V$ are all isomorphic to an indecomposable direct summand W with vertex Q of $\mathrm{Res}^P_Q(V)$. Thus it suffices to show that the map

$$\mathrm{End}_{B^{\mathrm{op}}}(W \otimes_{\mathcal{O}} {}_\varphi B)(\Delta Q) \to \mathrm{End}_{\mathcal{O}}(W \otimes_{\mathcal{O}} {}_\psi B)(\Delta Q)$$

is injective, where W is an indecomposable direct summand of $\mathrm{Res}^P_Q(V)$ with vertex Q. An idecomposable direct summand of ${}_\varphi B$ as an $\mathcal{O}Q$-B-bimodule is of the form ${}_\varphi nB$ for some primitive idempotent n in $B^{\Delta \varphi(Q)}$. If n does not belong to a local point, then applying $\mathrm{Br}_{\Delta Q}$ to this summand yields zero. Thus we only need to consider summands with n belonging to a local point of $\varphi(Q)$. Since Q is fully $\mathcal{F}$-centralised, it follows from 8.7.4 that we have an isomorphism of $\mathcal{O}Q$-B-bimodules ${}_\varphi nB \cong mB$ for some primitive local idempotent m in $B^{\Delta Q}$. Thus it suffices to show that the map

$$\mathrm{End}_{B^{\mathrm{op}}}(W \otimes_{\mathcal{O}} B)(\Delta Q) \to \mathrm{End}_{\mathcal{O}}(W \otimes_{\mathcal{O}} B)(\Delta Q)$$

is injective. Using the natural isomorphism

$$\mathrm{End}_{\mathcal{O}}(W \otimes_{\mathcal{O}} B) \cong \mathrm{Hom}_{\mathcal{O}}(B, W^* \otimes_{\mathcal{O}} W \otimes_{\mathcal{O}} B)$$

from 2.9.5 it suffices to show that the map

$$\mathrm{Hom}_B(B, W^* \otimes_{\mathcal{O}} W \otimes_{\mathcal{O}} B)(\Delta Q) \to \mathrm{Hom}_{\mathcal{O}}(B, W^* \otimes_{\mathcal{O}} W \otimes_{\mathcal{O}} B)(\Delta Q)$$

is injective. Now $W^* \otimes_{\mathcal{O}} W$ is a direct sum of a trivial $\mathcal{O}Q$-module $\mathcal{O}$ and indecomposable permutation $\mathcal{O}Q$-modules with vertices strictly smaller than Q. Thus it suffices to show that the map

$$\mathrm{End}_{B^{\mathrm{op}}}(B)(\Delta Q) \to \mathrm{End}_{\mathcal{O}}(B)(\Delta Q)$$

is injective. The canonical isomorphism $\mathrm{End}_{B^{\mathrm{op}}}(B) \cong B$ yields an isomorphism $\mathrm{End}_{B^{\mathrm{op}}}(B)(\Delta Q) \cong B(\Delta Q)$. Since B has a ΔQ-stable $\mathcal{O}$-basis, it follows from 5.8.6 that $\mathrm{End}_{\mathcal{O}}(B)(\Delta Q) \cong \mathrm{End}_k(B(\Delta Q))$. Using these isomorphisms, the last map is identified with the structural homomorphism

$$B(\Delta Q) \to \mathrm{End}_k(B(\Delta Q)),$$

which is clearly injective. Exchanging the roles of A and B yields a similar diagram. $\qquad\square$

We will need the special case where $B = \mathcal{O}P$. In that case the A-$\mathcal{O}P$-bimodule U in Proposition 9.9.7 is isomorphic to $A \otimes_{\mathcal{O}P} \text{Ind}_{\Delta P}^{P \times P}(V) \cong A \otimes_{\mathcal{O}} V$, viewed as an A-$\mathcal{O}P$-bimodule with A acting on the left on A, and P acting on the right diagonally by $(a \otimes v) \cdot y = ay \otimes y^{-1}v$, where $a \in A$, $v \in V$ and $y \in P$. The diagonal action of P on $A \otimes_{\mathcal{O}} V$ is given by $\Delta y \cdot (a \otimes v) = yay^{-1} \otimes yv$, where $\Delta y = (y, y)$. In this way, $A \otimes_{\mathcal{O}} V$ is an endopermutation $\mathcal{O}P$-module having V as a direct summand; this is a special case of 9.9.5, but can be seen in this case directly because A is a permutation $\mathcal{O}\Delta P$-module having a trivial direct summand. In order to identify the bimodule U' as in 9.9.7 in this situation, we start with a trivial technical observation.

Lemma 9.9.8 *Let A be an almost source algebra of a block of a finite group with a defect group P and fusion system $\mathcal{F}$. Let V be a finitely generated $\mathcal{O}$-free $\mathcal{O}P$-module and set $S = \text{End}_{\mathcal{O}}(V)$. For $a \in A$ denote by ρ_a the A-endomorphism of A given by right multiplication with a. We have a commutative diagram of $\mathcal{O}$-algebras*

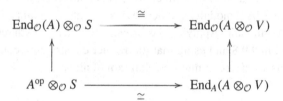

where the top horizontal map is the canonical isomorphism sending $\alpha \otimes s$ to the map $a \otimes v \mapsto \alpha(a) \otimes s(v)$, the bottom horizontal map is the canonical isomorphism from 1.12.10 sending $a \otimes s$ to $\rho_a \otimes s$, the left vertical map sends $a \otimes s$ to $\rho_a \otimes s$, and where the right vertical map is the inclusion.

Proof This is a straightforward verification. $\qquad\square$

This lemma is used to refine 9.9.7; in particular, if $B = \mathcal{O}P$, then the algebra homomorphism γ in 9.9.7 is an isomorphism.

Proposition 9.9.9 *Let A be an almost source algebra of a block of a finite group with a defect group P and fusion system $\mathcal{F}$. Let V be an indecomposable endopermutation $\mathcal{O}P$-module with vertex P, and let Q be a subgroup of $Z(P)$. Denote by V_Q the endopermutation kP-module satisfying $\text{End}_{\mathcal{O}}(V)(\Delta Q) \cong \text{End}_k(V_Q)$. Then $A \otimes_{\mathcal{O}} V$ is an endopermutation $\mathcal{O}\Delta P$-module having a summand isomorphic to V, we have $\text{Defres}_{\Delta P/Q}^{\Delta P}(A \otimes_{\mathcal{O}} V) \cong A(\Delta Q) \otimes V_Q$, and we* *have a commutative diagram of canonical algebra homomorphisms*

$$
\begin{array}{ccc}
\mathrm{End}_{\mathcal{O}}(A \otimes_{\mathcal{O}} V)(\Delta Q) & \xrightarrow{\ \cong\ } & \mathrm{End}_k(A(\Delta Q) \otimes_k V_Q) \\
\big\uparrow & & \big\uparrow \\
\mathrm{End}_A(A \otimes_{\mathcal{O}} V)(\Delta Q) & \xrightarrow[\ \cong\ ]{} & \mathrm{End}_{A(\Delta Q)}(A(\Delta Q) \otimes_k V_Q)
\end{array}
$$

where the top isomorphism is from 9.9.5, the bottom isomorphism is the map γ from 9.9.7 with A instead of B^{op}, where the right vertical map is the inclusion, and where the left vertical map is the homomorphism β from 9.9.7 with A instead of B^{op}.

Proof This diagram is a special case of the analogue of the diagram in 9.9.7, with A instead of B^{op}, and with $\mathcal{O}P$ instead of B, so we would only need to show that the bottom horizontal algebra homomorphism is an isomorphism, but it takes little extra effort to give a proof that is independent of 9.9.7. As mentioned before, A is a permutation $\mathcal{O}\Delta P$-module, and hence $A \otimes_{\mathcal{O}} V$ is an endopermutation $\mathcal{O}\Delta P$-module. Since $A(\Delta P)$ is nonzero, A has a trivial direct summand as an $\mathcal{O}\Delta P$-module, and hence $A \otimes_{\mathcal{O}} V$ has a summand isomorphic to V as an $\mathcal{O}\Delta P$-module. Applying the Brauer construction with respect to ΔQ to the diagram in 9.9.8 and using that the Brauer construction commutes with tensor products yields a commutative diagram of algebras

$$
\begin{array}{ccc}
(\mathrm{End}_{\mathcal{O}}(A))(\Delta Q) \otimes_{\mathcal{O}} S(\Delta Q) & \xrightarrow{\ \cong\ } & \mathrm{End}_{\mathcal{O}}(A \otimes_{\mathcal{O}} V)(\Delta Q) \\
\big\uparrow & & \big\uparrow \\
A^{\mathrm{op}}(\Delta Q) \otimes_{\mathcal{O}} S(\Delta Q) & \xrightarrow[\ \cong\ ]{} & \mathrm{End}_A(A \otimes_{\mathcal{O}} V)(\Delta Q)
\end{array}
$$

By 5.8.6 we have a canonical isomorphism $(\mathrm{End}_{\mathcal{O}}(A))(\Delta Q) \cong \mathrm{End}_k(A(\Delta Q))$. Applying 9.9.8 to $A(\Delta Q)$ and V_Q instead of A and V, respectively, yields a commutative diagram of algebras

$$
\begin{array}{ccc}
\mathrm{End}_k(A(\Delta Q)) \otimes_k S(\Delta Q) & \xrightarrow{\ \cong\ } & \mathrm{End}_k(A(\Delta Q) \otimes_k V_Q) \\
\big\uparrow & & \big\uparrow \\
A(\Delta Q)^{\mathrm{op}} \otimes_k S(\Delta Q) & \xrightarrow[\ \cong\ ]{} & \mathrm{End}_{A(\Delta Q)}(A(\Delta Q) \otimes_k V_Q)
\end{array}
$$

Combining the two commutative diagrams above yields the diagram as stated. The fact that the top horizontal map is an isomorphism implies that $\mathrm{Defres}^{\Delta P}_{\Delta P/Q}(A \otimes_{\mathcal{O}} V) \cong A(\Delta Q) \otimes_k V_Q$. □

9.10 Bimodules with endopermutation source and separable equivalence

The purpose of this section is to identify those bimodules with endopermutation source of two block algebras with a common defect group that yield a separable equivalence between the two block algebras. If P is a finite group and V an $\mathcal{O}P$-module, then we abusively consider V without any further comment as an $\mathcal{O}\Delta P$-module via the canonical isomorphism $P \cong \Delta P$ whenever this is useful.

Proposition 9.10.1 *Let A, B be almost source algebras of blocks of finite groups with a common defect group and the same fusion system $\mathcal{F}$ on P. Let V be an indecomposable $\mathcal{F}$-stable endopermutation $\mathcal{O}P$-module with vertex P. Set $U = A \otimes_{\mathcal{O}P} \mathrm{Ind}^{P \times P}_{\Delta P}(V) \otimes_{\mathcal{O}P} B$. Every indecomposable direct summand of $U \otimes_B U^*$ and of $U \otimes_{\mathcal{O}P} U^*$ is isomorphic to a direct summand of $A \otimes_{\mathcal{O}Q} A$ for some fully $\mathcal{F}$-centralised subgroup Q of P. In particular, $U \otimes_B U^*$ and $U \otimes_{\mathcal{O}P} U^*$ are p-permutation bimodules.*

Proof Since B is isomorphic to a direct summand of $B \otimes_{\mathcal{O}P} B$, it follows that $U \otimes_B U^*$ is isomorphic to a direct summand of $U \otimes_{\mathcal{O}P} U^*$. Thus it suffices to prove the statement for $U \otimes_{\mathcal{O}P} U^*$. Using the isomorphisms from 2.4.12, we get isomorphisms as $\mathcal{O}P$-$\mathcal{O}P$-bimodules

$$\mathrm{Ind}^{P \times P}_{\Delta P}(V^*) \otimes_{\mathcal{O}P} B \otimes_{\mathcal{O}P} B \otimes_{\mathcal{O}P} \mathrm{Ind}^{P \times P}_{\Delta P}(V)$$

$$\cong V^* \otimes_{\mathcal{O}} B \otimes_{\mathcal{O}P} B \otimes_{\mathcal{O}} V \cong (V \otimes_{\mathcal{O}} V^*) \otimes_{\mathcal{O}} (B \otimes_{\mathcal{O}P} B),$$

where the right side is to be understood as a tensor product of two $\mathcal{O}P$-$\mathcal{O}P$-bimodules with the conventions from 9.9.6. Every indecomposable summand of $B \otimes_{\mathcal{O}P} B$ as an $\mathcal{O}P$-$\mathcal{O}P$-bimodule is isomorphic to $\mathcal{O}P \otimes_{\mathcal{O}Q} {}_{\varphi}\mathcal{O}P \cong \mathrm{Ind}^{P \times P}_{\Delta_{\varphi}Q}(\mathcal{O})$ for some subgroup Q of P and some morphism $\varphi : Q \to P$ in $\mathcal{F}$, where $\Delta_{\varphi}Q = \{(u, \varphi(u)) | u \in Q\}$. Thus any indecomposable direct summand of $(V \otimes_{\mathcal{O}} V^*) \otimes_{\mathcal{O}} (B \otimes_{\mathcal{O}P} B)$ is isomorphic to a direct summand of an $\mathcal{O}P$-$\mathcal{O}P$-bimodule of the form $\mathrm{Ind}^{P \times P}_{\Delta_{\varphi}Q}(V \otimes_{\mathcal{O}} V^*)$. The restriction to $\Delta_{\varphi}Q$ of $V \otimes_{\mathcal{O}} V^*$ is a permutation module thanks to the stability of V, and hence the indecomposable direct summands of $\mathrm{Ind}^{P \times P}_{\Delta_{\varphi}Q}(V \otimes_{\mathcal{O}} V^*)$ are of the form $\mathcal{O}P \otimes_{\mathcal{O}R} {}_{\varphi}\mathcal{O}P$, where R is a subgroup of Q and where we use abusively the same letter φ for the restriction of φ to any such subgroup. Thus any indecomposable direct summand Y of $U \otimes_{\mathcal{O}P} U^*$ is isomorphic to a direct summand of $A \otimes_{\mathcal{O}R} {}_\varphi A$, with R and φ as before. Exactly as in the proof of Theorem 9.4.2 one concludes that Y is a summand of $A \otimes_{\mathcal{O}T} A$ for some fully $\mathcal{F}$-centralised subgroup T of P. $\square$

For a bimodule M over two almost source algebras A, B, it may or may not be the case that A is isomorphic to a direct summand of $M \otimes_B M^*$ as an A-A-bimodule. The following proposition provides a sufficient criterion in the case where the bimodule has an endopermutation source.

Proposition 9.10.2 *Let A, B be almost source algebras of blocks of finite groups with a common defect group P and the same fusion system $\mathcal{F}$ on P. Let V be an indecomposable $\mathcal{F}$-stable endopermutation $\mathcal{O}P$-module with vertex P. Let M be an indecomposable direct summand of the A-B-bimodule*

$$A \otimes_{\mathcal{O}P} \mathrm{Ind}_{\Delta P}^{P \times P}(V) \otimes_{\mathcal{O}P} B.$$

The following statements are equivalent.

 (i) A is isomorphic to a direct summand of the A-A-bimodule $M \otimes_B M^*$.
 (ii) A is isomorphic to a direct summand of the A-A-bimodule $M \otimes_{\mathcal{O}P} M^*$.
 (iii) $(M \otimes_B M^*)(\Delta P) \neq \{0\}$.
 (iv) $(M \otimes_{\mathcal{O}P} M^*)(\Delta P) \neq \{0\}$.

Proof Since B is isomorphic to a direct summand of the B-B-bimodule $B \otimes_{\mathcal{O}P} B$, it follows that $M \otimes_B M^*$ is isomorphic to a direct summand of $M \otimes_{\mathcal{O}P} M^*$. This yields the implications (i) $\Rightarrow$ (ii) and (iii) $\Rightarrow$ (iv). Since $A(\Delta P) \neq \{0\}$, we trivially have the implications (i) $\Rightarrow$ (iii) and (ii) $\Rightarrow$ (iv). Since M is finitely generated projective as a left A-module and as a right B-module (hence also as a right $\mathcal{O}P$-module), we have $M \otimes_B M^* \cong \mathrm{End}_{B^{op}}(M)$, and $M \otimes_{\mathcal{O}P} M^* \cong \mathrm{End}_{(\mathcal{O}P)^{op}}(M)$. It follows from 2.6.18 that if A is isomorphic to a direct summand of $M \otimes_{\mathcal{O}P} M^*$, then the canonical algebra homomorphism $A \to \mathrm{End}_{(\mathcal{O}P)^{op}}(M)$ is split injective as a bimodule homomorphism. This algebra homomorphism factors through the inclusion $\mathrm{End}_{B^{op}}(M) \subseteq \mathrm{End}_{(\mathcal{O}P)^{op}}(M)$, which implies that the canonical algebra homomorphism $A \to \mathrm{End}_{B^{op}}(M)$ is also split injective as a bimodule homomorphism. This shows the implication (ii) $\Rightarrow$ (i). Suppose that (iv) holds. By 9.10.1, $M \otimes_{\mathcal{O}P} M^*$ is a permutation $\mathcal{O}(P \times P)$-module such that $\mathcal{O}P$ acts freely on the left and on the right. Thus 6.15.2 applies, showing that the map $A \to \mathrm{End}_{\mathcal{O}P^{op}}(M)$ is split injective as an A-A-bimodule homomorphism. This shows the implication (iv) $\Rightarrow$ (ii), completing the proof. $\square$

Proposition 9.10.3 *Let A, B be almost source algebras of blocks of finite groups with a common defect group and the same fusion system $\mathcal{F}$ on P. Let*

V be an indecomposable $\mathcal{F}$-stable endopermutation $\mathcal{O}P$-module with vertex P. Let M be an indecomposable direct summand of the A-B-bimodule

$$A \otimes_{\mathcal{O}P} \operatorname{Ind}_{\Delta P}^{P \times P}(V) \otimes_{\mathcal{O}P} B.$$

The following are equivalent.

 (i) *The A-A-bimodule A is isomorphic to a direct summand of $M \otimes_B M^*$.*
 (ii) *The B-B-bimodule B is isomorphic to a direct summand of $M^* \otimes_A M$.*
 (iii) *We have $\operatorname{End}_{B^{op}}(M)(\Delta P) \neq \{0\}$.*
 (iv) *We have $\operatorname{End}_A(M)(\Delta P) \neq \{0\}$.*

In particular, if any one of these statements holds, then the algebras A, B are separably equivalent.

Proof The equivalence of (i) and (iii) follows from 9.10.2, using the isomorphism $M \otimes_B M^* \cong \operatorname{End}_{B^{op}}(M)$. Similarly, the equivalence of (ii) and (iv) follows from 9.10.2 with reversed roles of A and B, using the isomorphism $M^* \otimes_A M \cong \operatorname{End}_A(M)$. Suppose that A is isomorphic to a direct summand of $M \otimes_B M^*$, but that B is not isomorphic to a direct summand of $M^* \otimes_A M$. Then $(M^* \otimes_A M)(\Delta P)$ is zero, by the equivalence of (ii) and (iv). By 9.10.1, applied to B, A, V^* instead of A, B, V, respectively, that $M^* \otimes_B M$ is a direct sum of summands of bimodules of the form $B \otimes_{\mathcal{O}Q} B$, with Q running over a family of subgroups of P. By 8.8.2, if X is an indecomposable direct summand of $M^* \otimes_B M$, then X is isomorphic to a direct summand of $B \otimes_{\mathcal{O}Q} B$ for some subgroup Q of P satisfying $X(\Delta Q) \neq \{0\}$. However, since $(M^* \otimes_B M)(\Delta P)$ is zero, any such summand X is a direct summand of $B \otimes_{\mathcal{O}Q} B$ for some proper subgroup Q of P. Thus $M \otimes_B M^* \otimes_A M \otimes_B M^*$ is a direct sum of summands of bimodules of the form $M \otimes_{\mathcal{O}Q} M^*$, with Q running over a family of proper subgroups of P. In particular, we have $(M \otimes_B M^* \otimes_A M \otimes_B M^*)(\Delta P) = \{0\}$. But $A \cong A \otimes_A A$ is a summand of $M \otimes_B M^* \otimes_A M \otimes_B M^*$, hence $(M \otimes_B M^* \otimes_A M \otimes_B M^*)(\Delta P) \neq \{0\}$. This contradiction shows that B is isomorphic to a direct summand of $M^* \otimes_A M$. This shows the implication (i) $\Rightarrow$ (ii). Exchanging the roles of A and B yields the converse implication (ii) $\Rightarrow$ (i). $\square$

At the level of block algebras rather than almost source algebras it is possible to add a further criterion to the two propositions above, in terms of vertices and sources.

Proposition 9.10.4 *Let G, H be finite groups, b a block of $\mathcal{O}G$ and c a block of $\mathcal{O}H$. Suppose that b and c have a common defect group P. Let $i \in (\mathcal{O}Gb)^{\Delta P}$ and $j \in (\mathcal{O}Hc)^{\Delta P}$ be almost source idempotents. Suppose that i and j determine the* same fusion system $\mathcal{F}$ on P. Let V be an $\mathcal{F}$-stable indecomposable endopermutation $\mathcal{O}P$-module with vertex P. Let M be an indecomposable direct summand of the $\mathcal{O}Gb$-$\mathcal{O}Hc$-bimodule

$$\mathcal{O}Gi \otimes_{\mathcal{O}P} \operatorname{Ind}_{\Delta P}^{P \times P}(V) \otimes_{\mathcal{O}P} j\mathcal{O}H.$$

The following are equivalent.

(i) $\mathcal{O}Gb$ *is isomorphic to a direct summand of the* $\mathcal{O}Gb$-$\mathcal{O}Gb$-*bimodule* $M \otimes_{\mathcal{O}Gb} M^*$.

(ii) $\mathcal{O}Hc$ *is isomorphic to a direct summand of the* $\mathcal{O}Hc$-$\mathcal{O}Hc$-*bimodule* $M^* \otimes_{\mathcal{O}Hc} M$.

(iii) M *has vertex* ΔP *and source* V.

Proof Set $A = i\mathcal{O}Gi$ and $B = j\mathcal{O}Hj$. The equivalence of (i) and (ii) is a reformulation of 9.10.3 at the level of block algebras, via the standard Morita equivalences between block algebras and almost source algebras. The bimodule M has a vertex ΔQ contained in ΔP, for some subgroup Q of P. If this vertex is smaller than ΔP, then 5.1.16 implies that $(M \otimes_{\mathcal{O}Hc} M^*)(\Delta P) = \{0\}$, so also $(iMj \otimes_B jM^*i)(\Delta P) = \{0\}$. Thus 9.10.2 implies that $iMj \otimes_B jM^*i$ has no summand isomorphic to B, hence $M \otimes_{\mathcal{O}Hc} M^*$ has no summand isomorphic to $\mathcal{O}Hc$. Conversely, if the vertex of M is ΔP, then V is a source of M. By 6.4.10, M has a vertex source pair (P', V') such that $P' \subseteq P \times P$ and such that V' is a direct summand of iMj as an $\mathcal{O}P'$-module. It follows from 5.1.6 that iMj has an indecomposable direct summand with vertex P' and source V'. Green's Indecomposability Theorem implies that $\operatorname{Ind}_{P'}^{P \times P}(V')$ is a summand of iMj as an $\mathcal{O}P$-$\mathcal{O}P$-bimodule, hence of $A \otimes_{\mathcal{O}P} \operatorname{Ind}_{\Delta P}^{P \times P}(V) \otimes_{\mathcal{O}P} B$. Using the bimodule structure of A and B, it follows that P' is a 'twisted' diagonal subgroup of the form $\{(\varphi(u), \psi(u)) | u \in P\}$ for some φ, $\psi \in \operatorname{Aut}_{\mathcal{F}}(P)$. Since $A \cong {}_\varphi A$ as $\mathcal{O}P$-A-bimodules and $B \cong B_\psi$ as B-$\mathcal{O}P$-bimodules, it follows that iMj has a direct summand isomorphic to to $\operatorname{Ind}_{\Delta P}^{P \times P}(V')$, and then $V' \cong V$ by the stability of V. But then $iMj \otimes_{\mathcal{O}P} jM^*i$ has a summand isomorphic to $\operatorname{Ind}_{\Delta P}^{P \times P}(V) \otimes_{\mathcal{O}P} \operatorname{Ind}_{\Delta P}^{P \times P}(V^*) \cong \operatorname{Ind}_{\Delta P}^{P \times P}(V \otimes V^*)$. Since $V \otimes_{\mathcal{O}} V^*$ has a trivial summand, it follows that $iMj \otimes_{\mathcal{O}P} jM^*i$ has a summand isomorphic to $\mathcal{O}P$, which implies that $(iMj \otimes_{\mathcal{O}P} jM^*i)(\Delta P) \neq \{0\}$. Proposition 9.10.2 implies that A is isomorphic to a direct summand of $iMj \otimes_B jM^*i$, and hence $\mathcal{O}Gb$ is isomorphic to a direct summand of $M \otimes_{\mathcal{O}Hc} M^*$, completing the proof. $\square$

Combining this with results from the previous sections, we obtain the following indecomposability result which is a generalisation of 6.4.15.

Theorem 9.10.5 *Let A be a source algebra of a block b of a finite group G with a defect group P and fusion system $\mathcal{F}$. Let V be an $\mathcal{F}$-stable indecomposable endopermutation $\mathcal{O}P$-module with vertex P. Let Q be a subgroup of $Z(P)$ and let*

V_Q be an endopermutation kP-module satisfying $(\mathrm{End}_\mathcal{O}(V))(\Delta Q) \cong \mathrm{End}_k(V_Q)$ as interior P-algebras. The following hold.

  (i) The A-$\mathcal{O}P$-bimodule $A \otimes_\mathcal{O} V$ has, up to isomorphism, a unique indecomposable direct summand M satisfying $(\mathrm{End}_A(M))(\Delta P) \neq \{0\}$.
 (ii) The $A(\Delta Q)$-kP-bimodule $A(\Delta Q) \otimes_k V_Q$ has up to isomorphism a unique indecomposable direct summand M_Q satisfying $\mathrm{End}_{A(\Delta Q)}(M_Q)(\Delta P) \neq \{0\}$.
(iii) We have an isomorphism of interior P-algebras $(\mathrm{End}_A(M))(\Delta Q) \cong \mathrm{End}_{A(\Delta Q)}(M_Q)$.

Proof Write $A = i\mathcal{O}Gi$ for some source idempotent $i \in (\mathcal{O}Gb)^{\Delta P}$. Then the $\mathcal{O}Gb$-$\mathcal{O}P$-bimodule $\mathcal{O}Gi$ is indecomposable. By 7.3.18, the $\mathcal{O}Gb$-$\mathcal{O}P$-bimodule $\mathcal{O}Gi \otimes_\mathcal{O} V$, viewed as an $\mathcal{O}(G \times P)$-module, has up to isomorphism a unique indecomposable direct summand X with ΔP as a vertex, and any other direct summand has a proper subgroup of ΔP as a vertex. Note that by 2.4.12 we have $A \otimes_\mathcal{O} V \cong A \otimes_{\mathcal{O}P} \mathrm{Ind}_{\Delta P}^{P \times P}(V)$. It follows from 9.10.4 and the standard Morita equivalence between the categories of $\mathcal{O}Gb$-$\mathcal{O}P$-bimodules and of A-$\mathcal{O}P$-bimodules that $M = iX$ is the unique direct summand of $A \otimes_\mathcal{O} V$ satisfying $(\mathrm{End}_A(M))(\Delta P) \neq \{0\}$. This shows (i). Since Q is a subgroup of $Z(P)$, the algebra $A(\Delta Q)$ is a source algebra of $kC_G(Q)e_Q$, where e_Q is the unique block of $kC_G(Q)$ satisfying $(Q, e_Q) \leq (P, e)$. Thus (ii) is the special case of (i) applied to $A(\Delta Q)$ and V_Q instead of A and V, respectively. Consider now the algebra isomorphism $\mathrm{End}_A(A \otimes_\mathcal{O} V)(\Delta Q) \cong \mathrm{End}_{A(\Delta Q)}(A(\Delta Q) \otimes_k V_Q)$ from 9.9.9. This is an isomorphism of interior P-algebras through the right diagonal action of P on $A \otimes_\mathcal{O} V$ and on $A(\Delta Q) \otimes_k V_Q$, and both algebras have stable bases for the induced conjugation action of P. Note that the algebra of P-fixed points in $\mathrm{End}_A(A \otimes_\mathcal{O} V)$ is $\mathrm{End}_{A \otimes_\mathcal{O} \mathcal{O}P^{\mathrm{op}}}(A \otimes_\mathcal{O} V)$, and hence M corresponds to a primitive idempotent in this algebra. The above isomorphism sends $\mathrm{End}_{A \otimes_\mathcal{O} \mathcal{O}P^{\mathrm{op}}}(A \otimes_\mathcal{O} V)$ onto the P-fixed points in $(\mathrm{End}_A(A \otimes_\mathcal{O} V))(\Delta Q)$, and hence onto $\mathrm{End}_{A(\Delta Q) \otimes_k kP^{\mathrm{op}}}(A(\Delta Q) \otimes_k V_Q)$. Thus the image of the primitive idempotent corresponding to M is an idempotent in

$$\mathrm{End}_{A(\Delta Q) \otimes_k kP^{\mathrm{op}}}(A(\Delta Q) \otimes_k V_Q)$$

that is still primitive, hence that corresponds to an indecomposable direct summand M_Q of $A(\Delta Q) \otimes_k V_Q$ satisfying (iii). $\qquad \square$

9.11 Stable and Morita equivalences with endopermutation source

If M is a bimodule for two block algebras such that M induces a separable equivalence and such that M has an endopermutation module as a source, then

9.7.1 applies and provides an identification of defect groups of the two blocks. We will see that under such an identification, if M and its dual induce a stable equivalence of Morita type, then the fusion systems of the two blocks coincide, and the endopermutation source is stable with respect to this fusion system. Let G, H be finite groups, b a block of $\mathcal{O}G$ and c a block of $\mathcal{O}H$. Suppose that b and c have a common defect group P. Let M be an indecomposable $\mathcal{O}Gb$-$\mathcal{O}Hc$-bimodule having ΔP as a vertex, when considered as an $\mathcal{O}(G \times H)$-module. Denote by W an $\mathcal{O}\Delta P$-source of M. Then M is isomorphic to a direct summand of

$$\operatorname{Ind}_{\Delta P}^{G \times H}(W) \cong \mathcal{O}G \otimes_{\mathcal{O}P} \operatorname{Ind}_{\Delta P}^{P \times P}(W) \otimes_{\mathcal{O}P} \mathcal{O}H.$$

The indecomposability of M implies that M is isomorphic to a direct summand of

$$\mathcal{O}Gi \otimes_{\mathcal{O}P} \operatorname{Ind}_{\Delta P}^{P \times P}(W) \otimes_{\mathcal{O}P} j\mathcal{O}H$$

for some primitive idempotents $i \in (\mathcal{O}G)^{\Delta P}$ and $j \in (\mathcal{O}H)^{\Delta P}$. Since M is an $\mathcal{O}Gb$-$\mathcal{O}Hc$-bimodule, we have $i \in (\mathcal{O}Gb)^{\Delta P}$ and $j \in (\mathcal{O}Hc)^{\Delta P}$. Moreover, i and j are then source idempotents – that is, they satisfy $\operatorname{Br}_{\Delta P}(i) \neq 0$ and $\operatorname{Br}_{\Delta P}(j) \neq 0$. Indeed, otherwise a vertex of M would have order smaller than $|P|$ as a consequence of 5.12.8. In order to show that a stable equivalences of Morita type given by a bimodule with endopermutation source preserves the fusion systems, we need a technical statement in which we consider as earlier the tensor product of two bimodules, with the conventions described in 9.9.6.

Lemma 9.11.1 *Let P be a finite p-group, let φ, $\psi \in \operatorname{Aut}(P)$, and let S be an $\mathcal{O}P$-$\mathcal{O}P$-bimodule such that for any subgroup R of $P \times P$, every indecomposable direct summand of $\operatorname{Res}_R^{P \times P}(S)$ is absolutely indecomposable. Suppose that the $\mathcal{O}P$-$\mathcal{O}P$-bimodule $S \otimes_{\mathcal{O}} {}_\psi\mathcal{O}P$ has a direct summand isomorphic to ${}_\varphi\mathcal{O}P$. Then the images of φ and ψ in $\operatorname{Out}(P) = \operatorname{Aut}(P)/\operatorname{Inn}(P)$ are equal, and $\operatorname{Res}_{\Delta P}^{P \times P}({}_{\varphi^{-1}}S)$ has a trivial direct summand. In particular, φ is an inner automorphism if and only if ψ is an inner automorphism.*

Proof The hypotheses imply that $\mathcal{O}P$ is isomorphic to a direct summand of ${}_{\varphi^{-1}}(S \otimes_{\mathcal{O}} {}_\psi\mathcal{O}P)$. We may therefore assume that $\varphi = \operatorname{Id}_P$. We need to show that ψ is an inner automorphism of P and that $\operatorname{Res}_{\Delta P}^{P \times P}(S)$ has a trivial direct summand. We have

$$S \otimes_{\mathcal{O}} {}_\psi\mathcal{O}P \cong {}_\psi({}_{\psi^{-1}}S \otimes_{\mathcal{O}} \mathcal{O}P) \cong {}_\psi({}_{\psi^{-1}}S \otimes_{\mathcal{O}} \operatorname{Ind}_{\Delta P}^{P \times P}(\mathcal{O}))$$

$$\cong {}_\psi\operatorname{Ind}_{\Delta P}^{P \times P}(\operatorname{Res}_{\Delta P}^{P \times P}({}_{\psi^{-1}}S)).$$

Green's Indecomposability Theorem 5.12.3 implies that $\mathrm{Res}_{\Delta P}^{P \times P}({}_{\psi^{-1}}S)$ has a direct summand U of $\mathcal{O}$-rank 1 such that $\mathcal{O}P \cong {}_{\psi}\mathrm{Ind}_{\Delta P}^{P \times P}(U)$. The left side is a permutation bimodule, and hence U must be a trivial $\mathcal{O}\Delta P$-module. In particular, ${}_{\psi^{-1}}S$, restricted to ΔP, has a trivial direct summand. Thus $\mathcal{O}P \cong {}_{\psi}\mathrm{Ind}_{\Delta P}^{P \times P}(\mathcal{O}) \cong {}_{\psi}\mathcal{O}P$, which forces ψ to be an inner automorphism. Thus ψ^{-1} is an inner automorphism of P. It follows that ${}_{\psi^{-1}}S$ is isomorphic to S as an $\mathcal{O}\Delta P$-module, and hence $\mathrm{Res}_{\Delta P}^{P \times P}(S)$ has a trivial direct summand. This concludes the proof. $\qquad\square$

As remarked at the beginning of this section, a stable equivalence between blocks given by a bimodule with endopermutation source provides an identification between the defect groups of the two blocks. The next result shows that this identification preserves fusion systems. We use the following notation. Let P, Q be finite p-groups and $\varphi : P \to Q$ an isomorphism. If $\mathcal{F}$ is a fusion system on P, we denote by ${}^{\varphi}\mathcal{F}$ the fusion system on Q obtained from $\mathcal{F}$ and the isomorphism φ. We set

$$\Delta\varphi = \{(u, \varphi(u)) \mid u \in P\}$$

and whenever useful, we consider an $\mathcal{O}\Delta\varphi$-module V as an $\mathcal{O}P$-module and vice versa via the isomorphism $P \cong \Delta\varphi$ sending $u \in P$ to $(u, \varphi(u))$.

Theorem 9.11.2 (cf. [177, 7.6]) *Let G, H be finite groups, b a block of $\mathcal{O}G$ and c a block of $\mathcal{O}H$. Let P be a defect group of b and Q a defect group of c. Let $i \in (\mathcal{O}Gb)^{\Delta P}$ and $j \in (\mathcal{O}Hc)^{\Delta Q}$ be source idempotents. Denote by $\mathcal{F}$ the fusion system on P determined by i and by $\mathcal{G}$ the fusion system on Q determined by j. Let M be an indecomposable $\mathcal{O}Gb$-$\mathcal{O}Hc$-module inducing a stable equivalence of Morita type between $\mathcal{O}Gb$ and $\mathcal{O}Hc$. Suppose that as an $\mathcal{O}(G \times H)$-module, M has an endopermutation module as a source, for some vertex. Then there is an isomorphism $\varphi : P \to Q$ and an indecomposable endopermutation $\mathcal{O}P$-module V with vertex P such that M is isomorphic to a direct summand of the $\mathcal{O}Gb$-$\mathcal{O}Hc$-bimodule*

$$\mathcal{O}Gi \otimes_{\mathcal{O}P} \mathrm{Ind}_{\Delta\varphi}^{P \times Q}(V) \otimes_{\mathcal{O}Q} j\mathcal{O}H.$$

Moreover, the following hold.

(i) *The group $\Delta\varphi$ is a vertex of M, the group $\Delta\varphi^{-1}$ is a vertex of its dual M^*, the module V is an $\mathcal{O}\Delta\varphi$-source of M, and its dual V^* is an $\mathcal{O}\Delta\varphi^{-1}$-source of M^*.*

(ii) *We have ${}^{\varphi}\mathcal{F} = \mathcal{G}$.*

(iii) *The endopermutation $\mathcal{O}P$-module V is $\mathcal{F}$-stable.*

Proof The existence of an isomorphism $\varphi : P \to Q$ and an indecomposable endopermutation $\mathcal{O}P$-module V with vertex P such that M has vertex $\Delta\varphi$ and source V follows from 9.7.1 and the fact that V has $\mathcal{O}$-rank prime to p. Thus M is isomorphic to a direct summand of

$$ b \cdot \operatorname{Ind}_{\Delta\varphi}^{G\times H}(V) \cdot c = \mathcal{O}Gb \otimes_{\mathcal{O}P} \operatorname{Ind}_{\Delta\varphi}^{P\times Q}(V) \otimes_{\mathcal{O}Q} \mathcal{O}Hc. $$

Statement (i) follows in conjunction with 5.1.9.

Since M is indecomposable, we may replace b and c by primitive idempotents in $(\mathcal{O}Gb)^{\Delta P}$ and $(\mathcal{O}Hc)^{\Delta Q}$. Since M has a vertex of order $|P|$, it follows that M is a direct summand of the $\mathcal{O}Gb$-$\mathcal{O}Hc$-bimodule

$$ \mathcal{O}Gi \otimes_{\mathcal{O}P} \operatorname{Ind}_{\Delta\varphi}^{P\times Q}(V) \otimes_{\mathcal{O}Q} j\mathcal{O}H $$

for some source idempotents i, j. Local points of P on $\mathcal{O}Gb$ are $N_G(P)$-conjugate. Thus, up to modifying φ, if necessary, by an automorphism of P, we may choose for i the initially chosen source idempotent in the statement. Similarly, we may choose j to be the source idempotent chosen in the statement. This shows the statement preceding (i). It follows that in order to prove (ii) and (iii), without loss of generality, we may identify P and Q via φ.

The proofs of (ii) and (iii) follow in part the arguments of the proof of 9.8.2. In order to show (ii), we need to show that i and j determine the same fusion system $\mathcal{F}$ on P. The $\mathcal{O}Hc$-$\mathcal{O}P$-bimodule $\mathcal{O}Hj$ is indecomposable nonprojective. By 2.17.11, the $\mathcal{O}Gb$-$\mathcal{O}P$-bimodule $Mj \cong M \otimes_{\mathcal{O}Hc} \mathcal{O}Hj$ is the direct sum of an indecomposable nonprojective $\mathcal{O}Gb$-$\mathcal{O}P$-bimodule and a projective $\mathcal{O}Gb$-$\mathcal{O}P$-bimodule. By the hypotheses on M, the bimodule Mj is also a direct summand of $\mathcal{O}Gi \otimes_{\mathcal{O}P} \operatorname{Ind}_{\Delta P}^{P\times P}(V) \otimes_{\mathcal{O}P} j\mathcal{O}Hj$, and hence the nonprojective part of Mj is isomorphic to a direct summand of $\mathcal{O}Gi \otimes_{\mathcal{O}P} \operatorname{Ind}_{\Delta P}^{P\times P}(V) \otimes_{\mathcal{O}P} W$, for some indecomposable direct summand W of $j\mathcal{O}Hj$ as an $\mathcal{O}P$-$\mathcal{O}P$-bimodule. Thus on one hand we have $M^* \otimes_{\mathcal{O}Gb} Mj \cong \mathcal{O}Hj \oplus Yj$, and on the other hand we have that $M^* \otimes_{\mathcal{O}Gb} Mj$ is isomorphic to a direct summand of $M^* \otimes_{\mathcal{O}P} \operatorname{Ind}_{\Delta P}^{P\times P}(V) \otimes_{\mathcal{O}P} W$. Since $\operatorname{Br}_{\Delta P}(j) \neq 0$, this forces $W \cong k[yP]$ for some $y \in N_H(P, f_P)$, by 8.7.1. Since conjugation by y stabilises f_P, hence the local point of P on $\mathcal{O}Gb$ to which j belongs, we have $Mj \cong Mjy^{-1}$, and so the nonprojective part of Mj is isomorphic to a direct summand of $\mathcal{O}Gi \otimes_{\mathcal{O}P} \operatorname{Ind}_{\Delta P}^{P\times P}(V)$. Thus, as an $\mathcal{O}P$-$\mathcal{O}P$-bimodule, $j\mathcal{O}Hj$ is isomorphic to a direct summand of $jM^* \otimes_{\mathcal{O}Gb} Mj$, and the nonprojective part of this $\mathcal{O}P$-$\mathcal{O}P$-bimodule is isomorphic to a direct summand of

$$ \operatorname{Ind}_{\Delta P}^{P\times P}(V^*) \otimes_{\mathcal{O}P} i\mathcal{O}Gi \otimes_{\mathcal{O}P} \operatorname{Ind}_{\Delta P}^{P\times P}(V). $$

As in the proof of 9.10.1, using the isomorphisms from 2.4.12, this bimodule is isomorphic to

$$V^* \otimes_{\mathcal{O}} i\mathcal{O}Gi \otimes_{\mathcal{O}} V \cong (V \otimes_{\mathcal{O}} V^*) \otimes_{\mathcal{O}} i\mathcal{O}Gi,$$

where the right side is to be understood as a tensor product of two $\mathcal{O}P$-$\mathcal{O}P$-bimodules, with the 'diagonal action convention' from 9.9.6. Let R be a subgroup of P and $\psi \in \mathrm{Aut}(R)$. It follows from 9.11.1 that if $_\psi \mathcal{O}R$ is isomorphic to a direct summand of $j\mathcal{O}Hj$ as an $\mathcal{O}R$-$\mathcal{O}R$-bimodule, then $_\psi \mathcal{O}R$ is isomorphic to a direct summand of $i\mathcal{O}Gi$ as an $\mathcal{O}R$-$\mathcal{O}R$-bimodule. Thus, again by 8.7.1 and Alperin's Fusion Theorem 8.2.8, the fusion system of c determined by j is contained in that of b determined by i. Exchanging the roles of G and H implies the equality of the fusion systems, whence (ii). It follows further from 9.11.1 that as an $\mathcal{O}\Delta R$-module, $_{\psi^{-1}}V \otimes V^*$ has a trivial summand. But then by 5.1.13, the indecomposable summands with vertex R of the restrictions to R of $_{\psi^{-1}}V$ and V^* are dual to each other, and hence $_{\psi^{-1}}V_R \cong V_R$, where V_R is an indecomposable direct summand of $\mathrm{Res}^P_R(V)$ with vertex R. In view of Alperin's Fusion Theorem 8.2.8 this shows that V is $\mathcal{F}$-stable, where $\mathcal{F}$ is the common fusion system of b and c determined by i and j. This shows (iii). □

Corollary 9.11.3 *With the notation and hypotheses of the previous theorem, let n be an integer. The $\mathcal{O}Gb$-$\mathcal{O}Hc$-bimodule $\Omega^n_{G \times H}(M)$ is isomorphic to a direct summand of*

$$\mathcal{O}Gi \otimes_{\mathcal{O}P} \mathrm{Ind}^{P \times Q}_{\Delta\varphi}(\Omega^n_{\Delta\varphi}(V)) \otimes_{\mathcal{O}Q} j\mathcal{O}H$$

and induces, together with its dual, a stable equivalence of Morita type between $\mathcal{O}Gb$ and $\mathcal{O}Hc$. In particular, $\Omega^n_{G \times H}(M)$ is indecomposable with vertex $\Delta\varphi$ and source $\Omega^n_{\Delta\varphi}(V)$.

Proof We may assume $P = Q$ and $\varphi = \mathrm{Id}_P$. The functor on stable categories induced by the bimodule $\Omega^n_{G \times H}(M)$ and its dual can be viewed as the composition of the two stable equivalences of Morita type given by the bimodule M and by the functor Ω^n_G, hence yields a stable equivalence of Morita type by 2.17.13. Applying the exact functor $\mathcal{O}Gi \otimes_{\mathcal{O}P} \mathrm{Ind}^{P \times P}_{\Delta P}(-) \otimes_{\mathcal{O}P} j\mathcal{O}H$ to a projective or relatively $\mathcal{O}$-injective resolution of V yields a projective or relatively $\mathcal{O}$-injective resolution of M, respectively, and hence $\Omega^n_{G \times H}(M)$ is isomorphic to a direct summand of $\mathcal{O}Gi \otimes_{\mathcal{O}P} \mathrm{Ind}^{P \times P}_{\Delta P}(\Omega^n_P(V)) \otimes_{\mathcal{O}P} j\mathcal{O}H$. The statements on vertices and sources follow from 9.11.2. □

The proof of the next result is based on the methods, due to Rickard, developed for the proof of [187, Theorem 7.8].

Proposition 9.11.4 *Let* G, H *be finite groups, let* b, c *be blocks of* $\mathcal{O}G$, $\mathcal{O}H$, *respectively, having a common defect group* P, *and let* $i \in (\mathcal{O}Gb)^{\Delta P}$ *and* $j \in (\mathcal{O}Hc)^{\Delta P}$ *be source idempotents. Suppose that* i *and* j *determine the same fusion system* $\mathcal{F}$ *on* P. *Let* V *be an indecomposable endopermutation* $\mathcal{O}P$-*module with vertex* P. *Suppose that* V *has an* $\mathcal{F}$-*stable endosplit* p-*permutation resolution* Y_V. *Set*

$$U = \mathcal{O}Gi \otimes_{\mathcal{O}P} \mathrm{Ind}_{\Delta P}^{P \times P}(V) \otimes_{\mathcal{O}P} j\mathcal{O}H,$$

$$Y = \mathcal{O}Gi \otimes_{\mathcal{O}P} \mathrm{Ind}_{\Delta P}^{P \times P}(Y_V) \otimes_{\mathcal{O}P} j\mathcal{O}H.$$

Then the following hold.

(i) *The complex* Y *is a bounded splendid complex of* $\mathcal{O}Gb$-$\mathcal{O}Hc$-*bimodules, with respect to the choice of* i *and* j, *and* Y *has homology concentrated in degree zero, isomorphic to* U.

(ii) *The complex of* $\mathcal{O}Gb$-$\mathcal{O}Gb$-*bimodules* $Y \otimes_{\mathcal{O}Hc} Y^*$ *is split.*

(iii) *For any direct summand* M *of* U *there is a direct summand* X *of* Y *such that* X *has homology concentrated in degree zero, isomorphic to* M.

(iv) *Let* Q *be a subgroup of* P, *and let* e_Q *and* f_Q *be the unique blocks of* $kC_G(Q)$ *and of* $kC_H(Q)$ *satisfying* $\mathrm{Br}_Q(i)e_Q \neq 0$ *and* $\mathrm{Br}_Q(j)f_Q \neq 0$, *respectively. Then* $e_Q Y(\Delta Q)f_Q$ *is a splendid complex of* $kC_G(Q)e_Q$-$kC_H(Q)f_Q$-*bimodules, and has homology concentrated in a single degree.*

Proof Since Y_V is bounded, so is Y. The terms of Y_V are permutation $\mathcal{O}P$-modules, and hence Y is splendid. The homology of Y_V is concentrated in degree zero, isomorphic to V, and thus the homology of Y is concentrated in degree zero, isomorphic to U. This shows (i). In order to show that $Y \otimes_{\mathcal{O}Hc} Y^*$ is split, it suffices to show that every indecomposable direct summand of $Y \otimes_{\mathcal{O}Hc} Y^*$ is split. Any such summand is isomorphic to a direct summand of

$$\mathcal{O}Gi \otimes_{\mathcal{O}P} \mathrm{Ind}_{\Delta P}^{P \times P}(Y_V) \otimes_{\mathcal{O}P} W \otimes_{\mathcal{O}P} \mathrm{Ind}_{\Delta P}^{P \times P}(Y_V^*) \otimes_{\mathcal{O}P} i\mathcal{O}G$$

for some indecomposable direct summand W of $j\mathcal{O}Hj$. By 8.7.1, we have $W \cong \mathcal{O}P \otimes_{\mathcal{O}Q} {}_\varphi \mathcal{O}P$ for some subgroup Q of P and some morphism $\varphi : Q \to P$ in $\mathcal{F}$. Since any functor sends a split complex to a split complex, it suffices to show that the complex of $\mathcal{O}P$-$\mathcal{O}P$-bimodules

$$\mathrm{Ind}_{\Delta P}^{P \times P}(Y_V) \otimes_{\mathcal{O}Q} {}_\varphi \mathrm{Ind}_{\Delta P}^{P \times P}(Y_V^*)$$

is split. The indecomposable summands of this complex are isomorphic to summands of complexes of the form

$$\mathrm{Ind}_{\Delta Q}^{P \times P}(Y_V \otimes_{\mathcal{O}} ({}_\varphi Y_V^*))$$

and hence it suffices to observe that the complex

$$Y_V \otimes_{\mathcal{O}} (_\varphi Y_V^*)$$

is split as a complex of $\mathcal{O}Q$-modules. This, however, is an immediate conse-
quence of the fact that $Y_V \otimes_{\mathcal{O}} Y_V^*$ is split as a complex of $\mathcal{O}\Delta P$-modules by
virtue of being an endosplit p-permutation resolution, together with the fact
that Y_V is assumed to be $\mathcal{F}$-stable. This proves (ii). Since the terms of Y are
finitely generated projective as right $\mathcal{O}Hc$-modules, we have an isomorphism

$$Y \otimes_{\mathcal{O}Hc} Y^* \cong \mathrm{End}_{\mathcal{O}Hc^{\mathrm{op}}}(Y).$$

By (ii), this complex is split. Thus taking G-fixed points commutes with
taking homology. Taking homology on the left side yields $U \otimes_{\mathcal{O}Hc} U^* \cong$
$\mathrm{End}_{\mathcal{O}Hc^{\mathrm{op}}}(U)$, thus taking G-fixed points yields $\mathrm{End}_{\mathcal{O}Gb \otimes_{\mathcal{O}} \mathcal{O}Hc^{\mathrm{op}}}(U)$. On the
right side, taking G-fixed points yields the complex $\mathrm{End}_{\mathcal{O}Gb \otimes_{\mathcal{O}} \mathcal{O}Hc^{\mathrm{op}}}(Y)$, and the
homology of this complex is $\mathrm{End}_{K(\mathcal{O}Gb \otimes_{\mathcal{O}} \mathcal{O}Hc^{\mathrm{op}})}(Y)$. Thus we have an algebra
isomorphism

$$\mathrm{End}_{K(\mathcal{O}Gb \otimes_{\mathcal{O}} \mathcal{O}Hc^{\mathrm{op}})}(Y) \cong \mathrm{End}_{\mathcal{O}Gb \otimes \mathcal{O}Hc^{\mathrm{op}}}(U).$$

Now a direct summand M of U corresponds to an idempotent on the right side,
hence on the left side. This idempotent lifts uniquely, up to conjugacy, to an
idempotent in $\mathrm{End}_{\mathrm{Ch}(\mathcal{O}Gb \otimes_{\mathcal{O}} \mathcal{O}Hc^{\mathrm{op}})}(Y)$, hence yields a direct summand X of Y
satisfying

$$\mathrm{End}_{K(\mathcal{O}Gb \otimes_{\mathcal{O}} \mathcal{O}Hc^{\mathrm{op}})}(X) \cong \mathrm{End}_{\mathcal{O}Gb \otimes \mathcal{O}Hc^{\mathrm{op}}}(M).$$

The left side is the homology in degree zero of $X \otimes_{\mathcal{O}Hc} X^*$, and being a direct
summand of Y implies that the homology of X is concentrated in degree zero,
isomorphic to M by construction. This completes the proof of (iii). It follows
from 9.7.7 that $e_Q X(\Delta Q) f_Q$ is a splendid complex of $kC_G(Q)e_Q\text{-}kC_H(Q)f_Q$-
bimodules. In order to show that its cohomology is concentrated in a single
degree, it suffices to show this for the complex $(iXj)(\Delta Q)$, by the standard
Morita equivalence between blocks and their source algebras. It follows from
the $\mathcal{O}Q$-$\mathcal{O}Q$-bimodule structure of $i\mathcal{O}Gi$ and $j\mathcal{O}Hj$ that the restriction to $\mathcal{O}\Delta Q$
of iXj is a direct sum of complexes of the form $\mathrm{Ind}_{\Delta R}^{\Delta Q}(_{\Delta\psi}Y_V)$, where R is a subgroup of Q and $\psi : R \to P$ a morphism in $\mathcal{F}$. Applying the Brauer construction
with respect to ΔQ to such a summand yields 0 unless $R = Q$, and if $R = Q$,
then the indecomposable summands of $(_{\Delta\psi}Y_V)(\Delta Q)$ are isomorphic to those
of $Y_V(\Delta Q)$ by the stability of Y_V. It follows from 7.11.10 that this complex has
homology concentrated in a single degree, whence (iv). $\qquad\square$

Specialised to Rickard complexes, the previous Proposition yields the following result (of which statement (i) is from [129, 1.3]).

Theorem 9.11.5 *Let G, H be finite groups, let b, c be blocks of $\mathcal{O}G$, $\mathcal{O}H$, respectively, having a common defect group P, and let $i \in (\mathcal{O}Gb)^{\Delta P}$ and $j \in (\mathcal{O}Hc)^{\Delta P}$ be source idempotents. For any subgroup Q of P denote by e_Q and f_Q the unique block of $kC_G(Q)$ and of $kC_H(Q)$, respectively, such that $\mathrm{Br}_Q(i)e_Q \neq 0$ and $\mathrm{Br}_Q(j)f_Q \neq 0$. Let V be an indecomposable endopermutation $\mathcal{O}P$-module with vertex P. Let M be an indecomposable direct summand of the $\mathcal{O}Gb$-$\mathcal{O}Hc$-bimodule*

$$U = \mathcal{O}Gi \otimes_{\mathcal{O}P} \mathrm{Ind}_{\Delta P}^{P \times P}(V) \otimes_{\mathcal{O}P} j\mathcal{O}H.$$

Suppose that M and M^* induce a Morita equivalence between $\mathcal{O}Gb$ and $\mathcal{O}Hc$. Then i and j determine the same fusion system $\mathcal{F}$ on P. Suppose that V has an $\mathcal{F}$-stable endosplit p-permutation resolution Y_V. Then the complex

$$Y = \mathcal{O}Gi \otimes_{\mathcal{O}P} \mathrm{Ind}_{\Delta P}^{P \times P}(Y_V) \otimes_{\mathcal{O}P} j\mathcal{O}H$$

has a direct summand X with the following properties.

(i) *The complex X is a splendid Rickard complex of $\mathcal{O}Gb$-$\mathcal{O}Hc$-bimodules having homology concentrated in degree zero isomorphic to M.*

(ii) *For any subgroup Q of P, the complex $e_Q X(\Delta Q)f_Q$ is a splendid Rickard complex of $kC_G(Q)e_Q$-$kC_H(Q)f_Q$)-bimodules with homology M_Q concentrated in a single degree, such that M_Q induces a Morita equivalence between $kC_G(Q)e_Q$ and $kC_H(Q)f_Q$.*

(iii) *If Q is a fully $\mathcal{F}$-centralised subgroup of P, then M_Q has $\Delta C_P(Q)$ as a vertex and a source W_Q that is isomorphic to $\mathrm{Defres}_{QC_P(Q)/Q}^{P}(V)$ as a module over $C_P(Q)/Z(Q) \cong QC_P(Q)/Q$.*

(iv) *If $\mathrm{char}(K) = 0$ and K is large enough, then the character of X is a p-permutation equivalence between b and c; in particular, the blocks b and c are isotypic.*

Proof We may assume that P is nontrivial. The fact that i and j determine the same fusion system on P follows from 9.11.2. By 9.11.4, the complex $Y \otimes_{\mathcal{O}Hc} Y^*$ is a split complex of $\mathcal{O}Gb$-$\mathcal{O}Gb$-bimodules, and Y has a direct summand X with homology concentrated in degree zero, isomorphic to M. Then $X \otimes_{\mathcal{O}Hc} X^*$ is split, hence homotopy equivalent to its homology $M \otimes_{\mathcal{O}Hc} M^* \cong \mathcal{O}Gb$. This shows that X is a Rickard complex with the properties as stated in (i). Let Q be a subgroup of P. By 9.7.7, $e_Q X(\Delta Q)f_Q$ is a splendid Rickard complex of $kC_G(Q)e_Q$-$kC_H(Q)f_Q$)-bimodules. with homology M_Q concentrated in a single degree. Thus $M_Q \otimes_{kC_H(Q)f_Q} M_Q^*$ is the homology of $e_Q(X \otimes_{\mathcal{O}Hc} X^*)(\Delta Q)e_Q \simeq kC_G(Q)e_Q$, and hence M_Q induces a Morita equivalence, which shows (ii). Using 7.11.10 and the preceding arguments shows (iii). Statement (iv) follows from 9.5.4. □

Formulated at the source algebra level, the statements (i) and (ii) in the previous theorem take a slightly simpler form.

Theorem 9.11.6 *Let A, B be source algebras of blocks of finite group algebras over $\mathcal{O}$ having a common defect group P and the same fusion system $\mathcal{F}$ on P. Let V be an $\mathcal{F}$-stable indecomposable endopermutation $\mathcal{O}P$-module with vertex P, having an $\mathcal{F}$-stable endosplit p-permutation resolution Y_V. Let M be an indecomposable direct summand of the A-B-bimodule*

$$A \otimes_{\mathcal{O}P} \operatorname{Ind}_{\Delta P}^{P \times P}(V) \otimes_{\mathcal{O}P} B.$$

Suppose that M and M* induce a Morita equivalence between A and B. Then the complex of A-B-bimodules

$$Y = A \otimes_{\mathcal{O}P} \operatorname{Ind}_{\Delta P}^{P \times P}(Y_V) \otimes_{\mathcal{O}P} B$$

has a direct summand X with the following properties.

(i) *The complex X is a splendid Rickard complex of A-B-bimodules having homology concentrated in degree zero isomorphic to M.*
(ii) *For any fully $\mathcal{F}$-centralised subgroup Q of P, the complex $X(\Delta Q)$ is a splendid Rickard complex of $A(\Delta Q)$-$B(\Delta Q)$-bimodules with homology M_Q concentrated in a single degree, such that M_Q induces a Morita equivalence between $A(\Delta Q)$ and $B(\Delta Q)$.*

Proof This follows from the previous theorem, together with the standard Morita equivalences from 8.7.3. □

Remark 9.11.7 One obtains an obvious variation of Theorem 9.11.5 with the weaker assumption that M and M^* induce a stable equivalence of Morita type. In that case, the conclusion of (i) is that X induces a stable equivalence in the sense of 9.8.7, the statements (ii) and (iii) remain unchanged for Q nontrivial, and (iv) would lead to a notion of 'stable p-permutation equivalence' (but it would no longer be true in general that b and c are isotypic).

If we drop the assumption that V has an endosplit p-permutation resolution, then we obtain the following criterion for stable equivalences. We formulate this at the source algebra level, since this is slightly less technical.

Theorem 9.11.8 *Let A, B be source algebras of blocks of finite group algebras over $\mathcal{O}$ having a common defect group P and the same fusion system $\mathcal{F}$ on P. Let V be an $\mathcal{F}$-stable indecomposable endopermutation $\mathcal{O}P$-module with vertex P.*

Let M be an indecomposable direct summand of the A-B-bimodule

$$A \otimes_{\mathcal{O}P} \mathrm{Ind}_{\Delta P}^{P \times P}(V) \otimes_{\mathcal{O}P} B.$$

Suppose that $(M \otimes_B M^*)(\Delta P) \neq \{0\}$ *and that for any nontrivial fully* $\mathcal{F}$-*centralised subgroup* Q *of* P, *the canonical* $A(\Delta Q)$-$B(\Delta Q)$-*bimodule* M_Q *satisfying* $\mathrm{End}_k(M_Q) \cong (\mathrm{End}_{\mathcal{O}}(M))(\Delta Q)$ *induces a Morita equivalence between* $A(\Delta Q)$ *and* $B(\Delta Q)$. *Then M and its dual* M^* *induce a stable equivalence of Morita type between A and B.*

Proof Since $(M \otimes_B M^*)(\Delta P) \neq \{0\}$, it follows from 9.10.2 that $M \otimes_B M^* \cong A \oplus X$ for some A-A-bimodule X with the property that every indecomposable direct summand of X is isomorphic to a direct summand of $A \otimes_{\mathcal{O}Q} A$ for some fully $\mathcal{F}$-centralised subgroup Q of P. In what follows we use the canonical isomorphism $M \otimes_B M^* \cong \mathrm{End}_{B^{\mathrm{op}}}(M)$ and analogous versions. By 9.9.7, for any fully $\mathcal{F}$-centralised subgroup Q of P we have an injective algebra homomorphism

$$\mathrm{End}_{B^{\mathrm{op}}}(M)(\Delta Q) \to \mathrm{End}_{B(\Delta Q)^{\mathrm{op}}}(M_Q).$$

The left term is isomorphic to $A(\Delta Q) \oplus X(\Delta Q)$. If Q is nontrivial and fully $\mathcal{F}$-centralised, then the right term is isomorphic to $A(\Delta Q)$ by the assumptions on M_Q. This forces $X(\Delta Q) = \{0\}$ for any nontrivial fully $\mathcal{F}$-centralised subgroup Q of P. In view of 9.10.1 and 8.8.2, or alternatively using 8.8.6, this implies that X is projective as an A-A-bimodule. Similarly, 9.10.3 implies that $M^* \otimes_A M \cong B \oplus Y$ for some B-B-bimodule Y, and the same argument with the roles of A and B exchanged shows that Y is projective. $\qquad\square$

Morita equivalences between block algebras given by a bimodule with endopermutation source are, at the source algebra level, the Morita equivalences obtained from the action of the Dade group described in 7.4.1.

Theorem 9.11.9 *Let* G, H *be finite groups, let* b, c *be blocks of* $\mathcal{O}G$, $\mathcal{O}H$, *respectively, having a common defect group* P, *and let* $i \in (\mathcal{O}Gb)^{\Delta P}$ *and* $j \in (\mathcal{O}Hc)^{\Delta P}$ *be source idempotents. Set* $A = i\mathcal{O}Gi$ *and* $B = j\mathcal{O}Hj$. *Let* V *be an indecomposable endopermutation* $\mathcal{O}P$-*module with vertex* P. *The following are equivalent.*

  (i) *There is an indecomposable direct summand* M *of the* $\mathcal{O}Gb$-$\mathcal{O}Hc$-*bimodule*

$$\mathcal{O}Gi \otimes_{\mathcal{O}P} \mathrm{Ind}_{\Delta P}^{P \times P}(V) \otimes_{\mathcal{O}P} j\mathcal{O}H$$

  inducing a Morita equivalence between $\mathcal{O}Gb$ *and* $\mathcal{O}Hc$.
  (ii) *There is an isomorphism of interior* P-*algebras* $A \cong e(S \otimes_{\mathcal{O}} B)e$ *for some primitive idempotent* $e \in (S \otimes_{\mathcal{O}} B)^{\Delta P}$ *satisfying* $\mathrm{Br}_{\Delta P}(e) \neq 0$.

(iii) There is an isomorphism of interior P-algebras $B \cong f(S^{op} \otimes_{\mathcal{O}} A)f$ for some primitive idempotent $f \in (S^{op} \otimes_{\mathcal{O}} B)^{\Delta P}$ satisfying $\mathrm{Br}_{\Delta P}(f) \neq 0$.

Moreover, if these two equivalent statements hold, then A and B determine the same fusion system $\mathcal{F}$ on P, and the endopermutation module V is $\mathcal{F}$-stable.

Proof The equivalence of (ii) and (iii) is a reformulation of 7.4.1. Suppose that (i) holds. It follows from 9.11.2 that A and B determine the same fusion system $\mathcal{F}$ on P and that V is $\mathcal{F}$-stable. Since M induces a Morita equivalence, we have $\mathcal{O}Gb \cong \mathrm{End}_{\mathcal{O}Hc^{op}}(M)$. The Morita equivalences induced by multiplication with i and j imply that $A \cong \mathrm{End}_{B^{op}}(iMj)$. Let $f \in \mathrm{End}_{A\otimes_{\mathcal{O}}B^{op}}(A \otimes_{\mathcal{O}P} \mathrm{Ind}_{\Delta P}^{P \times P}(V) \otimes_{\mathcal{O}P} B)$ be a projection $A \otimes_{\mathcal{O}P} \mathrm{Ind}_{\Delta P}^{P \times P}(V) \otimes_{\mathcal{O}P} B$ onto iMj. Thus

$$A \cong e(\mathrm{End}_{B^{op}}(A \otimes_{\mathcal{O}P} \mathrm{Ind}_{\Delta P}^{P \times P}(V) \otimes_{\mathcal{O}P} B))e.$$

Since A is a primitive interior P-algebra, it follows that the idempotent e remains primitive in $\mathrm{End}_{\mathcal{O}P \otimes_{\mathcal{O}} B^{op}}(A \otimes_{\mathcal{O}P} \mathrm{Ind}_{\Delta P}^{P \times P}(V) \otimes_{\mathcal{O}P} B)$. Thus, after possibly replacing e by a suitable conjugate, there is an indecomposable direct summand W of A as an $\mathcal{O}P$-$\mathcal{O}P$-bimodule such that e is a primitive idempotent in $\mathrm{End}_{\mathcal{O}P \otimes_{\mathcal{O}} B^{op}}(W \otimes_{\mathcal{O}P} \mathrm{Ind}_{\Delta P}^{P \times P}(V) \otimes_{\mathcal{O}P} B)$ satisfying

$$A \cong e(\mathrm{End}_{B^{op}}(W \otimes_{\mathcal{O}P} \mathrm{Ind}_{\Delta P}^{P \times P}(V) \otimes_{\mathcal{O}P} B))e.$$

By our standard arguments, the property $A(\Delta P) \neq \{0\}$ forces $\mathrm{Br}_{\Delta P}(e) \neq 0$ as well as $W \cong \mathcal{O}P_{\varphi}$ for some $\varphi \in \mathrm{Aut}_{\mathcal{F}}(P)$, where we use as in previous proofs the bimodule structure of source algebras from 8.7.1. The $\mathcal{F}$-stability of V implies that we may assume $\varphi = \mathrm{Id}_P$, and hence $A \cong e(\mathrm{End}_{B^{op}}(V \otimes_{\mathcal{O}} B))e \cong \mathrm{End}_{B^{op}}(e(V \otimes_{\mathcal{O}} B))$. Thus A and B are Morita equivalent via the bimodule $e(V \otimes_{\mathcal{O}} B)$. Statement (ii) follows from the description of this Morita equivalence in 7.4.1. Suppose conversely that (ii) holds. Consider $e(V \otimes_{\mathcal{O}} B)$ as an A-B-bimodule via the isomorphism $A \cong e(S \otimes_{\mathcal{O}} B)e$. By 6.4.7, the interior P-algebra A is relatively $\mathcal{O}P$-separable. Thus $e(V \otimes_{\mathcal{O}} B)$ is isomorphic to a direct summand of the A-B-bimodule $A \otimes_{\mathcal{O}P} e(V \otimes_{\mathcal{O}} B)$, hence of

$$A \otimes_{\mathcal{O}P} (V \otimes_{\mathcal{O}} B) \cong A \otimes_{\mathcal{O}P} \mathrm{Ind}_{\Delta P}^{P \times P}(V) \otimes_{\mathcal{O}P} B,$$

where this isomorphism is from 2.4.12. Thus the $\mathcal{O}Gb$-$\mathcal{O}Hc$-bimodule M corresponding to the A-B-bimodule $e(V \otimes_{\mathcal{O}} B)$ through the canonical Morita equivalences between source algebras and block algebras is isomorphic to a direct summand of $\mathcal{O}Gi \otimes_{\mathcal{O}P} \mathrm{Ind}_{\Delta P}^{P \times P}(V) \otimes_{\mathcal{O}P} j\mathcal{O}H$. This shows that (ii) implies (i), completing the proof. □

If two block algebras are Morita equivalent via some bimodule with an endopermutation source, then that source need not be unique.

Proposition 9.11.10 *Let P be a finite p-group, E a p'-subgroup of* $\mathrm{Aut}(P)$, *and* $\alpha \in H^2(E; \mathcal{O}^\times)$. *Let V, V' be two E-stable indecomposable endopermutation $\mathcal{O}P$-modules such that* $k \otimes_{\mathcal{O}} V \cong k \otimes_{\mathcal{O}} V'$. *Set $S = \mathrm{End}_{\mathcal{O}}(V)$ and $S' = \mathrm{End}_{\mathcal{O}}(V')$. There is an isomorphism of interior P-algebras*

$$S \otimes_{\mathcal{O}} \mathcal{O}_\alpha(P \rtimes E) \cong S' \otimes_{\mathcal{O}} \mathcal{O}_\alpha(P \rtimes E).$$

Proof By 7.3.12, there is an $\mathcal{O}P$-module W of $\mathcal{O}$-rank one such that $V' \cong W \otimes_{\mathcal{O}} V$. We may assume $V' = W \otimes_{\mathcal{O}} V$. Denote by ζ the character of W. Since V, V' are E-stable, so is W, hence ζ, wich amounts to $\zeta(^e u) = \zeta(u)$ for all $u \in P$ and $e \in E$. If $\sigma : P \to S^\times$ is the structural homomorphism of the interior P-algebra S, then the map σ' sending $u \in P$ to $\zeta(u)\mathrm{Id}_W \otimes \sigma(u) \in \mathrm{End}_{\mathcal{O}}(W \otimes_{\mathcal{O}} V) = S'$ is the structural map of the interior P-algebra S'. Using the E-stability of ζ one verifies that the map sending $s \otimes ue$ to $(\zeta(u)\mathrm{Id}_W \otimes s) \otimes ue$ yields an isomorphism as required. $\square$

Corollary 9.11.11 *Let P be an abelian finite p-group, E a p'-subgroup of* $\mathrm{Aut}(P)$, *and* $\alpha \in H^2(E; \mathcal{O}^\times)$. *Let V be an E-stable indecomposable endopermutation $\mathcal{O}P$-modules, and set $S = \mathrm{End}_{\mathcal{O}}(V)$. Let $\mathcal{O}'$ be the unramified subring of $\mathcal{O}$ with residue field k. There is an indecomposable endopermutation $\mathcal{O}'P$-module V' with vertex P such that $k \otimes_{\mathcal{O}'} V' \cong k \otimes_{\mathcal{O}} V$. In particular, setting $S' = \mathrm{End}_{\mathcal{O}}(\mathcal{O} \otimes_{\mathcal{O}'} V')$, there is an isomorphism of interior P-algebras*

$$S \otimes_{\mathcal{O}} \mathcal{O}_\alpha(P \rtimes E) \cong S' \otimes_{\mathcal{O}} \mathcal{O}_\alpha(P \rtimes E).$$

Proof This follows from the classification of endopermutation $\mathcal{O}P$-modules in 7.8.1 together with 9.11.10. $\square$

Remark 9.11.12 For many of the results in this section, we have chosen to make use of the identification of defect groups P, Q along an isomorphism $\varphi : P \cong Q$, determined by a stable equivalence of Morita type with endopermutation source between two block algebras $\mathcal{O}Gb$, $\mathcal{O}Hc$ thanks to Proposition 9.7.1. This is largely unproblematic – except in the situation where $G = H$, $b = c$, and $P = Q$, because in that case we have already an identification of defect groups, while φ may be a nontrivial automorphism of P. Thus when it comes to calculating (stable) Picard groups, we need the more precise version of Theorem 9.11.2 with $P = Q$ but φ possibly nontrivial. Theorem 9.11.2 implies that $^\varphi \mathcal{F} = \mathcal{F}$ in that case; in other words, φ is an automorphism of the fusion system $\mathcal{F}$.

10

Structural Results in Block Theory

Throughout this chapter we denote by p a prime, k a field of characteristic p and $\mathcal{O}$ a complete local principal ideal domain having k as a residue field. Denote by K the field of fractions of $\mathcal{O}$. Unless stated otherwise, we allow the case $K = \mathcal{O} = k$.

10.1 On dimensions of block algebras

We assume that K and k are splitting fields for all finite groups and their subgroups arising in this section. We determined in Theorem 6.7.13 the p-part of the dimension of a block algebra of kG in terms of the orders of G and of a defect group of the block. The following result, due to Brauer, is more precise.

Theorem 10.1.1 ([30]) *Let G be a finite group and B a block algebra of kG. Let a and d be the nonnegative integers such that p^a is the order of a Sylow p-subgroup of G and such that p^d is the order of a defect group of b. Let u_B be the positive integer such that $p^{a-d}u_B$ is the greatest common divisor of the dimensions of the simple B-modules. Then u_B is prime to p, and there is a positive integer $v_B \geq \ell(B)$ that is prime to p, such that*

$$\dim_k(B) = p^{2a-d} \cdot u_B^2 \cdot v_B.$$

In particular, we have $\dim_k(B) \geq p^{2a-d} \cdot u_B^2 \cdot \ell(B)$.

Proof By 4.8.1, B contains a semisimple subalgebra $S \cong B/J(B)$. By 6.5.12, the dimension of any simple B-module is divisible by $p^{a-d}u_B$, and by 6.11.7, the integer u_B is prime to p. Each simple factor of S is therefore a matrix algebra of dimension divisible by $m = p^{2a-2d}u_B^2$. Setting $T = M_m(k)$, we have $S = T \otimes_k S'$ for some split semisimple k-algebra S'. By 4.6.6 we have $B \cong T \otimes_k B'$, where B' is the centraliser of T in B, and B' is Morita equivalent to B. In particular, we have $\ell(B') = \ell(B)$. The dimension of a projective indecomposable

B-module is divisible by p^a, hence the dimension of a projective indecomposable B'-module is divisible by p^d. Since B' has at least $\ell(B)$ summands, it follows that $\dim_k(B') = p^d \cdot v_B$ for some integer $v_B \geq \ell(B)$. Thus $\dim_k(B) = p^{2a-d} u_B^2 v_B$. By 6.7.13, the integer p^{2a-d} is the largest power of p dividing $\dim_k(B)$, and hence v_B is prime to p, implying the result. $\qquad\square$

The surjectivity of the decomposition map implies that u_B is also equal to the greatest common divisor of the degrees of the ordinary irreducible characters associated with B. Theorem 10.1.1 implies that the minimal possible dimension of a block algebra is p^{2a-d}. We will characterise blocks of minimal dimension in 10.1.6 below. This will require the following source algebra version of Theorem 10.1.1.

Theorem 10.1.2 (cf. [139, Theorem 1]) *Let G be a finite group, B a block algebra of kG, and A a source algebra of B. Let d be the nonnegative integers such that p^d is the order of a defect group of b. Let u_A be the greatest common divisor of the dimension of the simple A-modules. Then u_A is prime to p, and there is an integer $v_A \geq \ell(B)$ that is prime to p such that*

$$\dim_k(A) = p^d \cdot u_A^2 \cdot v_A.$$

In particular, we have $\dim_k(A) \geq p^d \cdot \ell(B)$.

Proof Arguing as in the proof of 10.1.1, we have $A \cong T \otimes_k A'$, where T is a matrix algebra of dimension u_A^2. The dimension of any projective indecomposable A-module is divisible by p^d. By 6.11.11, the integer u_A is prime to p. Thus the dimension of any projective indecomposable A'-module is still divisible by p^d. Since A' has at least $\ell(B)$ summands, it follows that $\dim_k(A') = p^d \cdot v_A$ for some integer $v_A \geq \ell(B)$. By 6.15.1, p^d is the highest power of p dividing $\dim_k(A)$, hence also the highest power of p dividing $\dim_k(A')$. This shows that v_A is prime to p. The result follows. $\qquad\square$

The blocks for which the inequality in 10.1.2 is an equality can be characterised in terms of their source algebras.

Theorem 10.1.3 ([139, Theorem 2]) *Let G be a finite group. Let A be a source algebra of a block B of kG with defect group P. Let d be the positive integer such that $p^d = |P|$. The following are equivalent.*

(i) *We have $\dim_k(A) = p^d \cdot \ell(B)$.*
(ii) *We have $A \cong k(P \rtimes E)$ as interior P-algebras, for some abelian p'-subgroup E of $\mathrm{Aut}(P)$.*

Proof Suppose that (i) holds. Let J be a primitive decomposition of 1_A in A. Then $A = \oplus_{j \in J} A j$ as left A-module, and Aj is projective as a kP-module for each $j \in J$. Thus $\dim_k(A) = p^d \sum_{j \in J} \frac{\dim_k(Aj)}{p^d}$. The equality in (i) is therefore equivalent to $|J| = \ell(B)$ and $\dim_k(Aj) = p^d$ for all $j \in J$. Thus each point of A has multiplicity 1, or equivalently, A is a basic k-algebra, hence each simple A-module has dimension 1, and each module Aj restricted to kP is isomorphic to the regular module kP. Thus the radical of Aj as an A-module is equal to the radical of Aj as kP-module, hence $J(kP)A = J(A)$. The same argument yields $J(A) = AJ(kP)$. By 8.7.1 and 6.15.1, as a kP-kP-bimodule, A is isomorphic to a direct sum of $k(P \rtimes E)$ for some p'-subgroup E of $\mathrm{Aut}(P)$ and indecomposable direct summands of the form $kP \otimes_{kQ} {}_\varphi kP$, where Q is a proper subgroup of P and $\varphi : Q \to P$ is an injective group homomorphism. The equality $J(kP)A = AJ(kP)$ forces that there is no summand of that form. To see this, note first that $J(kP)A$ is a kP-kP-submodule of A, thus so is the quotient $A/J(kP)A$. The elements of P act as identity on the left of $A/J(kP)A$, hence also on the right, but the elements in P outside of $\varphi(Q)$ do not act as identity on the right side of the kP-kP-bimodule $(kP \otimes_{kQ} {}_\varphi kP)/J(kP)(kP \otimes_{kQ} {}_\varphi kP)) \cong k \otimes_{kQ} {}_\varphi kP$. This implies that $A = k(P \rtimes E)$ as a kP-kP-bimodule. Again by 6.15.1, A is isomorphic, as an interior P-algebra to a twisted group algebra $k_\alpha(P \rtimes E)$, for some $\alpha \in H^2(E; k^\times)$, inflated to $P \rtimes E$. Since A has a simple module of dimension 1, it follows from 1.2.8 that the class α is trivial. Since every simple A-module has dimension 1 this forces E to be abelian. Thus (i) implies (ii). The converse is easy. $\qquad\square$

A block B satisfying the equivalent conditions of Theorem 10.1.3 is splendidly Morita equivalent to its Brauer correspondent, hence satisfies Alperin's weight conjecture, its refinements by Dade and Robinson, and Broué's abelian defect conjecture if P is abelian.

Theorem 10.1.4 ([139, Theorem 3]) *Let G be a finite group and B a block algebra of kG. Let a and d be the nonnegative integers such that p^a is the order of a Sylow p-subgroup of G and such that p^d is the order of a defect group of b. Let A be a source algebra of B. We have*

$$\dim_k(B) \geq p^{2a-2d}\dim_k(A),$$

with equality if and only if the block idempotent 1_B remains primitive in $(kG)^S$ for any Sylow p-subgroup S of G.

Proof Choose a Sylow p-subgroup S of G and a defect group P of B such that $P \leq S$. Since $\mathrm{Br}_P(1_B) \neq 0$ there is a primitive idempotent $e \in B^S$ such that $\mathrm{Br}_P(e) \neq 0$. Thus there is a primitive idempotent $i \in (eBe)^P$ such that

$\text{Br}_P(i) \neq 0$. Then e belongs to a point σ of S on B and i belongs to a local point γ of P on B such that $P_\gamma \leq S_\sigma$. Since P is maximal with the property $\text{Br}_P(1_B) \neq 0$ it follows that P_γ is a defect pointed group of S_σ, hence $\sigma \subseteq B_P^S$. Theorem 5.12.8 implies that $kGe \cong kGi \otimes_{kP} kS$ as $k(G \times S)$-modules. The k-dual of $kGi \otimes_{kP} kS$ is isomorphic to the $k(S \times G)$-module $kS \otimes_{kP} ikG$. Tensoring these two modules over kG yields an isomorphism of kS-kS-bimodules $eBe \cong kS \otimes_{kP} ikGi \otimes_{kP} kS$. Since $\text{Br}_P(i) \neq 0$, the algebra $ikGi = iBi$ is a source algebra of B; in particular, $\dim_k(ikGi) = \dim_k(A)$. Clearly $\dim_k(kS \otimes_{kP} ikGi \otimes_{kP} kS) = p^{2a-2d}\dim_k(A)$ and $\dim_k(B) \geq \dim_k(eBe)$. This shows the inequality, and also shows that the equality holds if and only if $1_B = e$ is primitive in $(kG)^S$. Since 1_B is a central idempotent, 1_B is therefore primitive in $(kG)^{S'}$ for any Sylow p-subgroup S' of G, whence the second statement. $\qquad\square$

Remark 10.1.5 The inequality in Theorem 10.1.4 is rather coarse in that it does not take into account the multiplicity of a local point of P on B. For this reason there is no obvious connection between u_A and u_B. Examples where $u_A = 1$ and u_B is an arbitrary p'-integer arise from blocks of $H \times P$, where H is a finite p'-group having an ordinary irreducible character of degree u_B; in such a situation, u_B is precisely the multiplicity of the unique point of P on that block algebra.

Theorem 10.1.6 ([135, Theorem]) *Let G be a finite group and B a block algebra of kG. Let a and d be the nonnegative integers such that p^a is the order of a Sylow p-subgroup of G and such that p^d is the order of a defect group of b. If $\dim_k(B) = p^{2a-d}$, then B is a nilpotent block with source algebra kP, where P is a defect group of B.*

Proof Combining 10.1.4 and 10.1.2 yields

$$\dim_k(B) \geq p^{2a-2d}\dim_k(A) \geq p^{2a-d}\ell(B).$$

Thus the equality $\dim_k(B) = p^{2a-d}$ forces $\ell(B) = 1$ and $\dim_k(A) = p^d$. Since by 10.1.3 also $A \cong k(P \rtimes E)$ for some abelian p'-group E, this implies that $E = 1$, and hence kP is a source algebra of B. In particular, B is nilpotent. $\qquad\square$

10.2 Characterisations of nilpotent blocks

We assume that K and k are splitting fields for the finite groups in this section. Blocks with a single isomorphism class of simple modules need not be nilpotent. The following example illustrates this.

Example 10.2.1 Suppose that p is odd. Denote by C_p a cyclic group of order p. The cyclic group C_2 order 2 acts on C_p in such a way that the nontrivial element of C_2 inverts the elements of C_p. Set $P = C_p \times C_p$ and $H = (C_p \rtimes C_2) \times (C_p \rtimes C_2) = P \rtimes V_4$, where V_4 is a Klein four group. Through the canonical surjection $Q_8 \to V_4$ we consider Q_8 acting on P such that $Z = Z(Q_8)$ acts trivially on P. Set $G = P \rtimes Q_8$. Then $Z(G) = Z$ and $C_G(P) = P \times Z$. Moreover, we have $G/Z = H$. Since P is normal in G, every block of $\mathcal{O}G$ has P as a defect group, and every block idempotent is contained in $\mathcal{O}C_G(P)$. Thus $\mathcal{O}G$ has two blocks, corresponding to the two irreducible characters of Z, namely the principal block $b_0 = \frac{1}{2}(1 + t)$, and a nonprincipal block $b_1 = \frac{1}{2}(1 - t)$. We have $\mathcal{O}Gb_0 \cong \mathcal{O}H$, and we have $\mathcal{O}Gb_1 \cong \mathcal{O}_\alpha H$, where α is the class of the nonsplit extension of V_4 by Z corresponding to Q_8. Since $\mathcal{O}_\alpha V_4 \cong M_2(\mathcal{O})$, it follows that $\mathcal{O}Gb_1$ has a unique isomorphism class of simple modules. The block b_1 is, however, not nilpotent, since it has the nontrivial inertial quotient V_4.

The block b_1 is the nonprincipal block of $\mathcal{O}G$ in the above example, and this is no coincidence.

Theorem 10.2.2 *Let G be a finite group. The principal block b_0 of $\mathcal{O}G$ is nilpotent if and only if $\mathcal{O}Gb_0$ has a unique isomorphism class of simple modules.*

Proof Suppose that $\mathcal{O}Gb_0$ has a unique isomorphism class of simple modules. Then the trivial module k is, up to isomorphism, the unique simple module. Moreover, $\mathcal{O}Gb_0$ is a matrix algebra over its basic algebra. But then $\mathcal{O}Gb_0$ is in fact isomorphic to its basic algebra precisely because it has a one-dimensional module. Thus $\mathcal{O}Gb_0$ has $\mathcal{O}$-rank $|P|$ by 6.11.8. Since $\mathcal{O}P$ is a subalgebra of $\mathcal{O}Gb_0$, it follows that $\mathcal{O}Gb_0 \cong \mathcal{O}P$ as interior P-algebra. Therefore b_0 is nilpotent by 8.11.6. The converse holds by 8.11.5. $\qquad\square$

Nilpotent blocks with an abelian defect group are characterised by the following result due to Okuyama and Tsushima. What is remarkable about this result is that it is a characterisation purely in terms of the algebra structure over k of the block, and it completely answers the question when a block is Morita equivalent to a commutative algebra.

Theorem 10.2.3 ([160]) *Let G be a finite group, b a block of kG, and P a defect group of b. Set $B = kGb$. The following are equivalent.*

(i) *The block b is nilpotent and its defect group P is abelian.*
(ii) *The block algebra B is Morita equivalent to a commutative k-algebra.*
(iii) *The algebra $Z(B)$ is symmetric.*
(iv) *We have $J(B) = J(Z(B))B$.*

Proof The statements (ii), (iii) and (iv) are equivalent by 4.12.14. If b is nilpotent, then B is Morita equivalent to kP by 8.11.5. Thus (i) implies (ii). Suppose that (iii) holds. That is, $Z(B)$ is symmetric, and hence B is Morita equivalent to $Z(B)$, by 4.12.14. In other words, a basic algebra C of B is split local commutative; in particular, $Z(B) \cong C$. By 6.11.8, we have $\dim_k(C) = |P|$. A source algebra A of B is Morita equivalent to C, hence isomorphic to $M_n(C)$ for some positive integer n. By 6.15.1, the integer $n^2 = \frac{\dim_k(A)}{|P|}$ is prime to p; thus n is prime to p. Therefore the space $[M_n(k), M_n(k)]$ of trace zero matrices intersects trivially $k \cdot 1$, and hence $[A, A] \cap Z(A) = \{0\}$. By 5.4.12 we have $\ker(\mathrm{Br}_P) \cap Z(A) \subseteq [A, A] \cap Z(A) = \{0\}$. In conjunction with 6.17.3 this implies that $Z(A) \cong \mathrm{Br}_P(Z(A)) \cong kZ(P)^E$, where E is the inertial quotient of b. Since $\dim_k(Z(A)) = |P|$, this forces $P = Z(P)$ and $E = \{1\}$. Thus P is abelian, and since any fusion system on an abelian p-group is determined by its automorphism group of P, it follows that b is nilpotent. Thus (iii) implies (i). $\square$

By Theorem 6.5.7, it is possible to determine from the character table of a finite group G whether two irreducible characters belong to the same block. It seems, however, unknown, whether the property of a block being nilpotent can be read off the character table. A conjecture of Malle and Navarro in [145] states that a block is nilpotent if and only if all of its height zero characters have the same degree – if true, this would in particular imply that nilpotent blocks are indeed determined by the character table. Navarro and Robinson proved in [155], using the classification of finite simple groups, that if all characters in a block B have p-power degree, then B is nilpotent. We show here a slightly weaker result, without using the classification of finite simple groups. We write $\mathrm{Irr}(B)$ instead of $\mathrm{Irr}_K(B)$ for the set of irreducible K-valued characters of B and $\mathrm{IBr}(B)$ instead of $\mathrm{IBr}_k(B)$ for the set of irreducible Brauer characters of B. We denote as before by $\mathrm{ht}(\chi)$ the height of χ in $\mathrm{Irr}(B)$, and $\ell(B) = |\mathrm{IBr}(B)|$ is the number of isomorphism classes of simple $k \otimes_\mathcal{O} B$-modules.

Theorem 10.2.4 *Let G be a finite group, B a block algebra of $\mathcal{O}G$, and let P be a defect group of B. Suppose that $\mathrm{char}(K) = 0$, that $\chi(1)$ is a power of p for all $\chi \in \mathrm{Irr}(B)$, and that $\sum_{\chi \in \mathrm{Irr}(B)} p^{2\mathrm{ht}(\chi)}$ is a power of p. Then B is nilpotent, and $\mathcal{O}P$ is a source algebra of B.*

The technicalities of the proof are collected in the following lemma.

Lemma 10.2.5 *Let G be a finite group, and let B be a block of $\mathcal{O}G$. Suppose that $\mathrm{char}(K) = 0$ and that there is a positive integer m such that $\chi(1) = mp^{\mathrm{ht}(\chi)}$ for all $\chi \in \mathrm{Irr}(B)$. Suppose further that the integer $\sum_{\chi \in \mathrm{Irr}(B)} p^{2\mathrm{ht}(\chi)}$ is a power of p. Then the following hold.*

(i) *We have $\ell(B) = 1$.*

(ii) *The character of the projective indecomposable B-modules is $\sum_{\chi \in \mathrm{Irr}(B)} p^{\mathrm{ht}(\chi)} \chi$. In particular, all ordinary decomposition numbers of B are powers of p.*

(iii) *The integer $\sum_{\chi \in \mathrm{Irr}(B)} p^{2\mathrm{ht}(\chi)}$ is equal to the order of the defect groups of B.*

(iv) *All irreducible characters of height zero in $\mathrm{Irr}(B)$ lift the unique irreducible Brauer character of B.*

Proof The statements (iii) and (iv) are immediate consequences of (ii), but we will first prove (iii) and then deduce (i) and (ii). Let a, d be the nonnegative integers such that p^a is the p-part of $|G|$ and such that p^d is the order of a defect group of B. Thus the p-part of $\chi(1)$ is $p^{a-d+\mathrm{ht}(\chi)}$ for any $\chi \in \mathrm{Irr}(B)$. By the assumptions, all height zero characters in $\mathrm{Irr}(B)$ have degree m, and the p-part of m is equal to p^{a-d}. Set

$$\rho = \sum_{\chi \in \mathrm{Irr}(B)} \chi(1)\chi;$$

that is, ρ is the regular character of B. The $\mathcal{O}$-rank of B is equal to

$$\rho(1) = \sum_{\chi \in \mathrm{Irr}(B)} \chi(1)^2 = m^2 \sum_{\chi \in \mathrm{Irr}(B)} p^{2\mathrm{ht}(\chi)}.$$

By Theorem 6.7.13, the p-part of $\rho(1)$ is equal to p^{2a-d}. Since the p-part of m^2 is equal to p^{2a-2d}, it follows that the p-part of the sum $\sum_{\chi \in \mathrm{Irr}(B)} p^{2\mathrm{ht}(\chi)}$ is equal to p^d. But this sum is assumed to be a power of p, so this sum is equal to p^d. This shows (iii), and it follows that $\rho(1) = m^2 p^d$. Set

$$\theta = \sum_{\chi \in \mathrm{Irr}(B)} p^{\mathrm{ht}(\chi)} \chi.$$

By the assumptions on character degrees, we have $\rho = m\theta$. In particular, θ is a generalised character that vanishes on p'-elements, and hence $\theta = \psi_1 - \psi_2$ for some characters ψ_1, ψ_2 of projective B-modules U_1, U_2, respectively. After cancelling common summands, we may assume that U_1, U_2 have no nonzero isomorphic summands. Since $\rho - m\theta - m\psi_1 - m\psi_2$ is an actual character of a projective B-module, it follows that $\psi_2 = 0$, and hence θ is the character of the projective B-module $U = U_1$. Again, since $\rho = m\theta$, it follows that $B \cong U^m$, the direct sum of m copies of U, as a left B-module. The equations $m^2 p^d = \rho(1) = m\theta(1)$ imply that $\theta(1) = mp^d$. Any projective B-module has $\mathcal{O}$-rank divisible by p^a. All characters of B have degree divisible by m, and therefore the surjectivity of the decomposition map implies that all $k \otimes_{\mathcal{O}} B$-modules have dimension divisible by m. Thus m divides the $\mathcal{O}$-rank of any projective B-module. It follows that the least common multiple of m

and p^a divides the $\mathcal{O}$-rank of any projective B-module. Since the p-part of m is p^{a-d}, this least common multiple is mp^d. Since we showed already that this is also the $\mathcal{O}$-rank of U, it follows that U is indecomposable, and hence U is up to isomorphism the unique projective indecomposable B-module, implying (i) and (ii).   □

The integer m in the above lemma is the degree of the height zero characters in $\mathrm{Irr}_K(B)$, and hence the Malle–Navarro conjecture in [145] would imply that B is nilpotent. If B is nilpotent, then B has all the properties stated in the above lemma, since in that case B is a matrix algebra over $\mathcal{O}P$, by Theorem 8.11.5.

Proof of Theorem 10.2.4 Let a, d be the integers such that p^a is the p-part of $|G|$ and such that $p^d = |P|$. The assumptions imply that $\chi(1) = p^{a-d}p^{\mathrm{ht}(\chi)}$ for all $\chi \in \mathrm{Irr}(B)$. It follows from 10.2.5, that $\sum_{\chi \in \mathrm{Irr}(B)} p^{2\mathrm{ht}(\chi)} = p^d$, and hence the $\mathcal{O}$-rank of B is equal to $\sum_{\chi \in \mathrm{Irr}(B)} \chi(1)^2 = p^{2a-2d}p^d = p^{2a-d}$. In other words, B is a block of minimal $\mathcal{O}$-rank. The conclusion follows from 10.1.6.   □

For B a block algebra of $\mathcal{O}G$ we denote by $\mathrm{Irr}_0(B)$ the set of irreducible characters of B that have height zero, and if $\mathcal{F}$ is a fusion system on a defect group P of B, we denote by $\mathfrak{foc}(\mathcal{F})$ its focal subgroup. For n a positive integer, we denote by n_p the highest power of p that divides n. We mention without proof the following results.

Theorem 10.2.6 ([101, Theorem 1.1]) *Let G be a finite group, B a block algebra of $\mathcal{O}G$ with a defect group P, source idempotent i, and associated fusion system $\mathcal{F}$ on P. Suppose that $\mathrm{char}(K) = 0$. The following are equivalent.*

(i) The block B is nilpotent.
(ii) We have $|\sum_{\chi \in \mathrm{Irr}_0(B)} \chi(1)^2|_p = |G : P|_p^2 \cdot |P : \mathfrak{foc}(\mathcal{F})|$.
(iii) We have $|\sum_{\chi \in \mathrm{Irr}_0(B)} \chi(i)^2|_p = |P : \mathfrak{foc}(\mathcal{F})|$.
(iv) We have $|\mathrm{Irr}_0(B)| = |P : \mathfrak{foc}(\mathcal{F})|$.

By 8.10.8, the group $P/\mathfrak{foc}(\mathcal{F})$ acts freely on $\mathrm{Irr}_0(B)$ via the $*$-construction, and so statement (iv) in the above theorem is equivalent to the statement that $P/\mathfrak{foc}(\mathcal{F})$ acts transitively on $\mathrm{Irr}_0(B)$. In particular, (iv) implies that all height zero characters have the same degree. Thus Theorem 10.2.6 would also follow from the aforementioned conjecture of Malle and Navarro in [145].

Theorem 10.2.7 ([111, A. Theorem]) *Let G be a finite group and B be a block algebra of $\mathcal{O}G$ with defect group P. Suppose that $\mathrm{char}(K) = 0$. If $|\mathrm{IBr}(B)| = 1$ and $|\mathrm{Irr}(B)| \leq 4$, then B is nilpotent and $|P| = |\mathrm{Irr}_K(B)|$; in particular, p is either 2 or 3.*

10.3 On vertices of simple modules

Any vertex of a simple module of a finite group algebra is centric in a fusion system of the block to which the simple module belongs. The proof of this result, due to Knörr [104], involves multiplicity modules. We state this in a slightly more general form. We assume that k is a splitting field for all finite groups and algebras arising in this section.

Theorem 10.3.1 (cf. [104, Theorem 3.3]) *Let G be a finite group and let A be a primitive interior G-algebra over $\mathcal{O}$ with defect pointed group Q_δ. Let b be the unique block of $\mathcal{O}G$ that is mapped to 1_A under the structural homomorphism $\mathcal{O}G \to A$. Suppose that a multiplicity module M_δ of Q_δ on A is simple. Then there is a block f of $kC_G(Q)$ such that the image $\bar{f}$ of f in $kC_G(Q)/Z(Q)$ does not annihilate M_δ. If f is such a block, then (Q, f) is a (G, b)-Brauer pair, and $Z(Q)$ is a defect group of $kC_G(Q)f$. In particular, if (P, e) is a maximal (G, b)-Brauer pair such that $(Q, f) \leq (P, e)$, then Q is an $\mathcal{F}$-centric subgroup of P, where $\mathcal{F} = \mathcal{F}_{(P,e)}(G, b)$ is the fusion system of the block b on P determined by the choice of (P, e).*

Proof By the assumptions, the structural homomorphism $\mathcal{O}G \to A$ induces an algebra homomorphism $\mathcal{O}Gb \to A$. Applying the Brauer construction with respect to the diagonal action of Q on both algebras yields an algebra homomorphism $kC_G(Q)\mathrm{Br}_{\Delta Q}(b) \to A(\Delta Q)$. Denote by $S(\delta)$ the simple quotient of $A(\Delta Q)$ determined by δ, and write $S(\delta) \cong \mathrm{End}_k(M_\delta)$. Thus the image of $\mathrm{Br}_{\Delta Q}(b)$ in $kC_G(Q)/Z(Q)$ acts as identity on M_δ. It follows that any block f of $kC_G(Q)$ whose image $\bar{f}$ in $kC_G(Q)/Z(Q)$ does not annihilate M_δ occurs in $\mathrm{Br}_{\Delta Q}(b)$, or equivalently, has the property that (Q, f) is a (G, b)-Brauer pair. As a module over a suitable twisted group algebra $k_\alpha N_G(Q_\delta)/Q$, M_δ is projective (by 5.7.5) and simple (by the assumptions). Thus the restriction to $kC_G(Q)/Z(Q)$ of M_δ is projective and semisimple. Therefore, if f is a block of $kC_G(Q)$ that does not annihilate M_δ, then $\bar{f}M_\delta$ is nonzero projective and semisimple as a $kC_G(Q)/Z(Q)\bar{f}$-module. This forces $\bar{f}$ to be a block of defect zero of $kC_G(Q)/Z(Q)$. It follows from 6.6.6 that f is a block of $kC_G(Q)$ having $Z(Q)$ as its defect group. By 8.5.3 this is equivalent to the last statement on being centric in a block fusion system. $\square$

Corollary 10.3.2 (cf. [104, Corollary 3.7]) *Let G be a finite group, b a block of $\mathcal{O}G$, and U a finitely generated indecomposable $\mathcal{O}Gb$-module with vertex Q. Suppose that $\underline{\mathrm{End}}_{\mathcal{O}Gb}(U) \cong \mathcal{O}/J(\mathcal{O})^m$ for some positive integer m. Then there is a block f of $kC_G(Q)$ such that (Q, f) is a (G, b)-Brauer pair and $Z(Q)$ is a defect group of $kC_G(Q)f$.*

Proof Since U is an indecomposable $\mathcal{O}Gb$-module, the algebra $A = \mathrm{End}_{\mathcal{O}}(U)$ is a primitive interior G-algebra with a structural homomorphism sending b to $\mathrm{Id}_U = 1_A$. By 5.7.8, a multiplicity module associated with a vertex and source of U is simple. Thus 10.3.2 is a special case of 10.3.1. $\qquad \square$

Corollary 10.3.3 *Let G be a finite group, b a block of $\mathcal{O}G$ with an abelian defect group P, and let U be a finitely generated indecomposable $\mathcal{O}Gb$-module such that $\underline{\mathrm{End}}_{\mathcal{O}Gb}(U) \cong \mathcal{O}/J(\mathcal{O})^m$ for some positive integer m. Then P is a vertex of U.*

Proof Since P is abelian, no proper subgroup of P is centric with respect to a fusion system on P. Equivalently, if (Q, f) is a (G, b)-Brauer pair with $Q \leq P$, then P is a defect group of $kC_G(Q)f$. Thus 10.3.3 follows from 10.3.2. $\qquad \square$

Corollary 10.3.4 *Let G be a finite group and b a block of $\mathcal{O}G$ with an abelian defect group P. Denote by $\bar{b}$ the image of b in kG. Suppose that $\mathrm{char}(K) = 0$.*

(i) Let S be a simple $kG\bar{b}$-module. Then P is a vertex of S.
(ii) Let $\chi \in \mathrm{Irr}_K(G, b)$ and let U be a finitely generated $\mathcal{O}$-free $\mathcal{O}Gb$-module with character χ. Then P is a vertex of U.

Proof The module S in (i) and the module U in (ii) satisfy the assumptions of the module denoted U in 10.3.3, whence the result. $\qquad \square$

Combining the above with earlier results yields the following theorem, due to Knörr.

Theorem 10.3.5 ([105, Theorem]) *Suppose that $\mathrm{char}(K) = 0$. Let G be a finite group, N a normal subgroup of G such that G/N is a p-group, and let b be a G-stable block of $\mathcal{O}N$. Denote by $\bar{b}$ the image of b in kN. Then b remains a block of $\mathcal{O}G$. Suppose that $\mathcal{O}Gb$ has an abelian defect group P. Then the following hold.*

(i) We have $G = NP$, and $P \cap N$ is a defect group of $\mathcal{O}N b$.
(ii) Every simple $kG\bar{b}$-module restricts to a simple $kN\bar{b}$-module, and this induces a bijection between the isomorphism classes of simple $kG\bar{b}$-modules and $kN\bar{b}$-modules, respectively. In particular, we have $\ell(kG\bar{b}) = \ell(kN\bar{b})$.
(iii) Every irreducible character in $\mathrm{Irr}_K(G, b)$ restricts to an irreducible character in $\mathrm{Irr}_K(N, b)$, and every irreducible character in $\mathrm{Irr}_K(N, b)$ extends to exactly $|G/N|$ different characters in $\mathrm{Irr}_K(G, b)$. In particular, we have $|\mathrm{Irr}_K(G, b)| = |G/N| \cdot |\mathrm{Irr}_K(N, b)|$.
(iv) The decomposition matrix of $\mathcal{O}Gb$ is obtained from repeating $|G/N|$-times each row of the decomposition matrix of $\mathcal{O}N b$.

(v) *The Cartan matrix $C_{G,b}$ of $kG\bar{b}$ is equal to $|G/N| \cdot C_{N,b}$, where $C_{N,b}$ is the Cartan matrix of $kN\bar{b}$.*

Proof The fact that b remains a block of $\mathcal{O}G$ and statement (i) follow from 6.8.11. Note that b remains a block of any normal subgroup H of G containing N, with defect group $P \cap H$. Therefore, in order to prove (ii), we may assume that G/N has order p. Let S be a simple $kG\bar{b}$-module. By 5.12.6, $\mathrm{Res}^G_N(S)$ is either simple, or $S \cong \mathrm{Ind}^G_N(T)$ for some simple kN-module T. The latter is impossible, because it would imply that N contains the vertices of S, contradicting the fact that P is a vertex of S by 10.3.4. Thus $\mathrm{Res}^G_N(S)$ is simple for every simple $kG\bar{b}$-module S. Every simple $kN\bar{b}$-module T arises in this way: if S is a simple quotient of $\mathrm{Ind}^G_N(T)$, then Frobenius' reciprocity, applied to a surjective kG-homomorphism $\mathrm{Ind}^G_N(T) \to S$ yields a nonzero, hence injective, kN-homomorphism $T \to \mathrm{Res}^G_N(S)$. In particular, b does not annihilate S, and hence S belongs to b. But then $\mathrm{Res}^G_N(S)$ is simple by the previous argument, hence isomorphic to T. Let S and S' be simple $kG\bar{b}$-modules such that $\mathrm{Res}^G_N(S)$ and $\mathrm{Res}^G_N(S')$ are both isomorphic to a simple $kN\bar{b}$-module T. Let i be a primitive idempotent in $kN\bar{b}$ such that iT is nonzero. Then iS and iS' are both nonzero. Since i remains primitive in $kG\bar{b}$ by 5.12.5, it follows that $S \cong S'$. This proves (ii). Let $\chi \in \mathrm{Irr}_K(G, b)$ and $\psi \in \mathrm{Irr}_K(N, b)$ such that $\langle \mathrm{Res}^G_N(\chi), \psi \rangle_N \neq 0$. Let H be the stabiliser of ψ in G. Then, by 3.3.10, there exists $\tau \in \mathrm{Irr}_K(H)$ such that $\chi = \mathrm{Ind}^G_H(\tau)$. However, every $\mathcal{O}$-free $\mathcal{O}G$-module with character χ must have P as a vertex, by 10.3.4. This forces $H = G$; that is, ψ is G-stable. It follows moreover from 3.3.10 that $\mathrm{Res}^G_N(\chi) = a \cdot \psi$ for some positive integer a. In order to show that $a = 1$ we may assume, arguing by induction, that G/N has order p. In that case G/N is cyclic, and therefore, it follows from 5.3.13, every irreducible character in $\mathrm{Irr}_K(N, b)$ extends in p ways to an irreducible character in $\mathrm{Irr}_K(G, b)$. Since $\langle \mathrm{Res}^G_N(\chi), \psi \rangle_N = \langle \chi, \mathrm{Ind}^G_N(\psi) \rangle_G$ and since the degree of $\mathrm{Ind}^G_N(\psi)$ is $p\psi(1)$, it follows that the $p = G/N$ extensions of ψ to G are exactly the irreducible characters in $\mathrm{Irr}_K(G, b)$ whose restriction involves ψ, and hence $a = 1$, and we have $\dim_K(KGb) = p \cdot \dim_K(KNb)$; that is, every irreducible character in $\mathrm{Irr}_K(G, b)$ arises as the extension to G of a character in $\mathrm{Irr}_K(N, b)$. Reverting to the general situation (that is, no longer assuming that N has index p in G), this shows inductively that $\mathrm{Res}^G_N(\chi)$ is irreducible for all $\chi \in \mathrm{Irr}_K(G, b)$, and it also shows that every character ψ in $\mathrm{Irr}_K(N, b)$ arises as the restriction of some χ in $\mathrm{Irr}_K(G, b)$. But then ψ arises as the restriction of at least $|G/N|$ characters in $\mathrm{Irr}_K(G, b)$, obtained from multiplying χ with any of the linear characters of G/N. Since $\dim_K(KGb) = |G/N| \cdot \dim_K(KNb)$, statement (iii) follows. Since N contains all p'-elements of G, it follows from (ii) that the irreducible Brauer characters of in $\mathrm{IBr}_k(G, b)$ are obtained by extending those in $\mathrm{IBr}_k(N, b)$ by zero outside N. Thus the row corresponding to $\chi \in \mathrm{Irr}_K(G, b)$ in the decomposition matrix of $\mathcal{O}Gb$ is equal to the row corresponding to $\mathrm{Res}_N^G(\chi)$ in the decomposition matrix of $\mathcal{O}Nb$, and hence (iv) follows from the previous statement. Since the Cartan matrix of a block is obtained from the decomposition matrix by multiplying it with its transpose, (v) is an immediate consequence of (iv). $\qquad\square$

The next result is a slight generalisation of a theorem due to Erdmann [65].

Theorem 10.3.6 *Let G be a finite group and U a finitely generated indecomposable kG-module with a cyclic vertex Q. Suppose that U has a simple multiplicity module. Then Q is a defect group of the block b of kG to which U belongs.*

Proof Let V be a kQ-source of U. Since Q is cyclic, the isomorphism class of an indecomposable kQ-module is determined by its dimension (cf. 7.1.1). Thus the isomorphism class of V is $N_G(Q)$-stable; that is $N_G(Q, V) = N_G(Q)$. It follows that if M is a multiplicity module with respect to (Q, V), then M is a $k_\alpha N_G(Q)/Q$-module for some class α in $H^2(N_G(Q)/Q; k^\times)$. The $N_G(Q)$-action on $\mathrm{End}_k(M)$ extends the interior action of $C_G(Q)$. Note that Q is contained in $C_G(Q)$ as Q is cyclic. Thus α is the inflation to $N_G(Q)/Q$ of a class in $H^2(N_G(Q)/C_G(Q); k^\times)$ via the canonical surjection $N_G(Q)/Q \to N_G(Q)/C_G(Q)$. The group $N_G(Q)/C_G(Q)$ is a subgroup of $\mathrm{Aut}(Q)$, and as Q is cyclic, it follows that $N_G(Q)/C_G(Q) = Y \times E$ for some abelian p-group Y and a cyclic group E of order dividing $p - 1$. If k is algebraically closed, then $H^2(N_G(Q)/C_G(Q); k^\times)$ is trivial. In general, after possibly replacing k by a large enough field, we may assume that the class α is trivial, and hence that M is a projective simple $kN_G(Q)/Q$-module. (Alternatively, if k is algebraically closed, one could skip this step by replacing $N_G(Q)$ by a suitable central p'-extension.) The restriction of M to $kC_G(Q)$ is projective semisimple. By 10.3.1 there is a (G, b)-Brauer pair of the form (Q, f) such that the image of f in $kC_G(Q)/Q$ does not annihilate M, and such that Q is a defect group of $kC_G(Q)f$. In particular $kC_G(Q)f$ has a unique isomorphism class of simple modules, and thus $N_G(Q, f)$ is the stabiliser in $N_G(Q)$ of this isomorphism class. Denote by D the inverse image of Y in $C_G(Q)$. Then D is a normal subgroup of $N_G(Q)$ of p'-index $|E|$, and $C_G(Q)$ is a normal subgroup of p-power index $|Y|$ of D. It follows from 6.10.6 that $N_G(Q, f) \cap D = C_G(Q)$. Let (P, e) be a maximal (G, b)-Brauer pair containing (Q, f). Then $N_P(Q)$ is contained in $N_G(Q, f) \cap D = C_G(Q)$, hence $N_P(Q) = C_P(Q)$. Since Q is a defect group of $kC_G(Q)f$, it follows that $C_P(Q) = Q$. This implies $N_P(Q) = Q$, hence $P = Q$. $\qquad\square$

Corollary 10.3.7 ([65]) *Let G be a finite group and S a simple kG-module having a cyclic vertex Q. Then Q is a defect group of the block of kG to which S belongs.*

Proof It follows from 5.7.8 that a multiplicity module of S is simple, and hence 10.3.6 implies the result.   □

Corollary 10.3.8 *Let G be a finite group and b a block of kG. Then b has a cyclic defect group if and only if kGb has a simple module with a cyclic vertex.*

Proof This follows from combining 10.3.4 and 10.3.6.   □

10.4 Abelian defect and proper focal subgroup

We assume in this section that K and k are splitting fields for all finite groups and their block algebras. Let G be a finite group and b a block with an abelian defect group P. Let e be a block of $kC_G(P)$ such that (P, e) is maximal (G, b)-Brauer pair. By standard results on coprime group actions, since P is abelian and the inertial quotient $E = N_G(P, e)/C_G(P)$ is a p'-group, it follows that $P = C_P(E) \times [P, E]$. The group $[P, E]$ is then the focal and hyperfocal subgroup of P with respect to the fusion system $\mathcal{F} = \mathcal{F}_{(P,e)}(G, b) = \mathcal{F}_P(P \rtimes E)$. The following result on blocks with abelian defect groups combines two results of A. Watanabe. For b a block of $\mathcal{O}G$ or of kG, we write as before $\ell(G, b) = \ell(\mathcal{O}Gb) = |\mathrm{IBr}_k(G, b)|$ and $\mathbf{k}(G, b) = |\mathrm{Irr}_K(G, b)| = \mathrm{rk}_\mathcal{O}(Z(\mathcal{O}Gb))$.

Theorem 10.4.1 ([223, Theorem 1], [222, Theorem 2]) *Let G be a finite group, b a block of kG, and (P, e_P) a maximal (G, b)-Brauer pair. Suppose that P is abelian, and set $E = N_G(P, e_P)/C_G(P)$. Let (u, c) be a (G, b)-Brauer element contained in (P, e_P) such that $u \in C_P(E)$. Set $H = C_G(u)$.*

(i) We have $\ell(G, b) = \ell(H, c)$ and $\mathbf{k}(G, b) = \mathbf{k}(H, c)$.

(ii) The map sending $z \in Z(kGb)$ to $\mathrm{Br}_{\langle u \rangle}(z)c$ is a k-algebra isomorphism $Z(kGb) \cong Z(kHc)$.

Proof Set $Q = C_P(E)$ and $R = [P, E]$. As mentioned above, we have $P = Q \times R$, and the fusion system $\mathcal{F}$ of b on P determined by the choice of e_P is equal to $\mathcal{F}_P(P \rtimes E)$. That is, $\mathfrak{foc}(\mathcal{F}) = R$. Let Y be a set of representatives of the E-conjugacy classes in R. Then QY is a set of representatives of the E-conjugacy classes in P. For $v \in P$ denote by e_v the unique block of $kC_G(v)$ such that $(v, e_v) \in (P, e_P)$. in particular, $c = e_u$. Thus $\{(v, e_v)|v \in QY\}$ is a set of representatives of the G-conjugacy classes of (G, b)-Brauer elements. We identify $\mathrm{Irr}_K(Q)$ with a subset of $\mathrm{Irr}_K(P)$ by extending every $\lambda \in \mathrm{Irr}_K(Q)$

trivially on R. Equivalently, $\mathrm{Irr}_K(Q)$ is identified with the subset of $\lambda \in \mathrm{Irr}_K(P)$ such that $\mathfrak{foc}(\mathcal{F}) \leq \ker(\lambda)$. With these conventions, the orthogonality relations imply that for any $y \in R$ we have $\sum_{\lambda \in \mathrm{Irr}_K(Q)} \lambda(u^{-1})\lambda(uy) = |Q|$, and for any $v \in Q$ such that $v \neq u$ and any $y \in R$ we have $\sum_{\lambda \in \mathrm{Irr}_K(Q)} \lambda(u^{-1})\lambda(vy) = 0$. In all sums below, λ runs over $\mathrm{Irr}_K(Q)$, v runs over Q, and y runs over the subset Y of R. For $\chi \in \mathrm{Irr}_K(G, b)$, we calculate the sum

$$\sum_{\lambda} \lambda(u^{-1})\langle \lambda * \chi, \chi \rangle$$

in two ways. By 8.10.10, we have $\lambda * \chi \neq \chi$ for $\lambda \neq 1$. Thus the only summand in the above sum that does not vanish is for $\lambda = 1$, and hence we get that

$$\sum_{\lambda} \lambda(u^{-1})\langle \lambda * \chi, \chi \rangle = 1.$$

We now calculate this sum using the formula $\chi = \sum_{u,y} \chi^{(uy, e_{uy})}$ from 8.9.2 (v) and the formula $(\lambda * \chi)^{(vy, e_{vy})} = \lambda(v)\chi^{(vy, e_{vy})}$ from 8.10.3. Using 8.9.2 (vi), the above sum becomes equal to

$$\sum_{\lambda, v, y} \lambda(u^{-1})\lambda(v)\langle \chi^{(vy, e_{vy})}, \chi^{((vy)^{-1}, e_{vy})} \rangle.$$

The summands with $v \neq u$ vanish by the above, and the remaining sum is equal to

$$|Q| \sum_{y} \langle \chi^{(uy, e_{uy})}, \chi^{((uy)^{-1}, e_{uy})} \rangle.$$

Comparing the two ways of calculating the above sum yields

$$\sum_{y} \langle \chi^{(uy, e_{uy})}, \chi^{((uy)^{-1}, e_{uy})} \rangle = \frac{1}{|Q|}.$$

Taking the sum over all $\chi \in \mathrm{Irr}_K(G, b)$ and using 8.9.4 yields that

$$\sum_{y} \ell(C_G(uy), e_{uy}) = \frac{\mathbf{k}(G, b)}{|Q|}.$$

Note that this equation holds for all $u \in Q$. Comparing this expression with the analogous expression with u replaced by 1 yields therefore the equality

$$\sum_{y} \ell(C_G(uy), e_{uy}) = \sum_{y} \ell(C_G(y), e_y).$$

In the two sums, if y is a nontrivial element in Y, then $C_G(uy)$ and $C_G(y)$ are both proper subgroups of G. The blocks e_{uy} and e_y of these groups have again (P, e_P) as maximal Brauer pair, and their inertial quotients are the subgroups of E that centralise uy and y, respectively. In particular, the inertial quotients of these blocks still centralise u. Note that the centralisers of u in the groups $C_G(uy)$ and $C_G(y)$ are both equal to $C_G(\langle u, y \rangle)$. Thus, arguing by induction over the order or G, we may assume that $\ell(C_G(uy), e_{uy}) = \ell(C_G(y), e_y)$ for any non-trivial $y \in Y$. But then the last displayed equality of sums forces that the summands for $y = 1$ must also coincide, and this is precisely the statement

$$\ell(G, b) = \ell(H, c).$$

We show next the corresponding equality for the numbers of ordinary irreducible characters. For this, we use Brauer's formula

$$\mathbf{k}(G, b) = \sum_{v,y} \ell(C_G(vy), e_{vy})$$

from 6.13.12 and the same formula for the block $e_u = c$ of $H = C_G(u)$. Corresponding terms in these two sums coincide by our previously proved equality applied to the blocks $(C_G(vy), e_{vy})$ instead of (G, b). This proves (i).

The Brauer homomorphism $\mathrm{Br}_{\langle u \rangle}$ followed by multiplication with c restricts to an algebra homomorphism $\psi : Z(kGb) \to Z(kHc)$. Since both sides have the same dimension thanks to (i), it suffices to show that ψ is surjective. By 6.17.4, it suffices to show that for any (H, c)-Brauer pair (S, e), the space $\mathrm{Tr}_S^H(kC_H(S)e)$ is contained in the image of ψ. Note that this space is zero if S does not contain u, because u is central in H. Arguing by induction over $|S|$, we may assume $u \in S$. Then $C_H(S) = C_G(S)$, so (S, e) is also a (G, b)-Brauer pair. After replacing (S, e) by an H-conjugate, we may assume that $(S, e) \leq (P, e_P)$; this is an inclusion both of (H, c)-Brauer pairs and of (G, b)-Brauer pairs. Let $a \in kC_G(S)e$. We need to show that $\mathrm{Tr}_S^H(a)c \in \mathrm{Im}(\psi)$. Note that $\mathrm{Tr}_S^H(a)c = \mathrm{Tr}_S^H(aec)$. Since we argue by induction, it suffices to show that $\psi(\mathrm{Tr}_S^G(acb)) - \mathrm{Tr}_S^H(ac)$ belongs to $\sum_T \mathrm{Tr}_T^H((kHc)^T)$, where T runs over the subgroups of P such that $|T| < |S|$. The Mackey formula for relative traces yields

$$\mathrm{Tr}_S^G(ac) = \sum_x \mathrm{Tr}_{H \cap {}^x S}^H({}^x(ac))$$

where x runs over a set of representatives $[H \backslash G / S]$ of the H-S-double cosets in G. Set $U = \langle u \rangle$. Applying Br_U to this sum annihilates all summands such that U is not contained in $H \cap {}^x S$. If U is contained in $H \cap {}^x S$, then ${}^x S \leq H$ because P, hence S, is abelian. Note that $\mathrm{Br}_U(b)c = c$. Therefore

$$\psi(\mathrm{Tr}_S^G(ac)b) = \sum_x \mathrm{Br}_U(\mathrm{Tr}_{{}^x S}^H({}^x(ac)b)c = \sum_x \mathrm{Tr}_{{}^x S}^H(\mathrm{Br}_U({}^x(ac)c)$$

where x runs over the subset of those representatives of H-S-double cosets satisfying $U \leq {}^x S \leq H$. The term with $x = 1$ yields the element $\mathrm{Tr}_S^H(ac)$, so we need to show that the remaining summands are in $\mathrm{Im}(\psi)$. Suppose that

$x \in G \setminus H$. By our induction and 6.17.5, it suffices to show that $\mathrm{Br}_U({}^x(ac)c) \in \ker(\mathrm{Br}_{xS})$. Applying Br_{xS} to this expression yields $\mathrm{Br}_{xS}({}^x(ac)c)$. Since $a = ae$, if this expression is nonzero, then necessarily $\mathrm{Br}_{xS}(({}^xe)c) \neq 0$. Thus ${}^x(S, e) \geq (U, c)$, or equivalently $(U, c)^x \leq (S, e)$. We also have $(U, c) \leq (S, e) \leq (P, e_P)$, and hence conjugation by x yields map in the fusion system $\mathcal{F}$ from U to U^x. Since the fusion systems of (H, c) and (G, b) determined by the choice of (P, e_P) are equal, this forces $x \in H$, a contradiction. The proof is complete. $\qquad\qquad\qquad\qquad\qquad\qquad\qquad\qquad\qquad\qquad\qquad\qquad\qquad\square$

Watanabe's Theorem is applied in the proof of the next result, due to Koshitani and Külshammer.

Theorem 10.4.2 ([108, Theorem]) *Suppose that k is algebraically closed. Let G be finite group and N a normal subgroup such that G/N is a p-group. Let b be a G-stable block of kN. Then b is a block of kG. Let P be a defect group of kGb. Suppose that P is abelian and that $Q = P \cap N$ has a complement R in P. The inclusion $kNb \subseteq kGb$ extends to an isomorphism of k-algebras*

$$kR \otimes_k kNb \cong kGb.$$

Proof By 6.8.11, b remains a block of kG, we have $G = NP$, and Q is a defect group of kNb. Arguing by induction, we may assume that R is cyclic. Since P is abelian, we have $C_G(Q) = RC_H(Q)$. Let Let (Q, e) be a (G, b)-Brauer pair. Since $C_N(Q)$ is normal of p-power index in $C_G(Q) = RC_N(Q)$, it follows that e is contained in $kC_N(Q)$, and hence e is an R-conjugacy class sum of blocks of $kC_N(Q)$. But $k_G(Q)e$ has P as a defect group as P is abelian. Thus $\mathrm{Br}_P(e) \neq 0$, and in particular, $\mathrm{Br}_R(e) \neq 0$. Thus e cannot be a nontrivial R-conjugacy class sum of blocks of $kC_N(Q)$, or equivalently, e remains a block of $kC_N(Q)$ that is R-stable. In particular, (Q, e) is a maximal (N, b)-Brauer pair. The image of R in $\mathrm{Aut}(N)$ is a p-subgroup that fixes Q and the maximal (N, b)-Brauer pair (Q, e). It follows from 6.16.3 that R acts by inner automorphisms on $\mathcal{O}N\hat{b}$, where $\hat{b}$ is the unique block of $\mathcal{O}N$ lifting b. By 10.3.5, we have $|\mathrm{Irr}_K(G, b)| = |G/N| \cdot |\mathrm{Irr}_K(N, b)|$. Equivalently, we have $\dim_k(Z(kGb)) = |G/N| \cdot \dim_k(Z(kHb)$.

Since R acts by inner automorphisms on kNb, for any $x \in R$ there is $s(x) \in (kNb)^\times$ such that $s(x)$ induces the same action as x on kNb, or equivalently, such that $xs(x)^{-1} \in (kGb)^N$. Using 6.8.15, we get that

$$(kGb)^N = \oplus_{x \in R} (xkNb)^N = \oplus_{x \in R} (xs(x)^{-1}kNb)^N = \oplus_{x \in R} xs(x)^{-1}Z(kNb)$$

and hence $\dim_k((kGb)^N) = |G/N| \cdot \dim_k(Z(kNb))$. By the previous argument, this is equal to $\dim_k(Z(kGb))$. Since $Z(kGb) \subseteq (kGb)^N$, it follows that $Z(kGb) = (kGb)^N$. In particular, we have $Z(kNb) \subseteq Z(kGb)$.

Let e be a block of $kC_G(P)$ such that (P, e) is a maximal (G, b)-Brauer pair. Since P is abelian, we have $C_G(P) = PC_N(P)$. The inertial quotient $E = N_G(P, e)/C_G(P) = N_N(P, e)/C_N(P)$ satisfies $[P, E] \subseteq N \cap P = Q$. In particular, Q is E-stable. Since E is a p'-group, it follows that we may choose R to be E-stable, hence that $R \le C_P(E)$.

Let f be a block of $kC_G(R)$ such that (R, f) is a (G, b)-Brauer pair. Since $C_G(R) = RC_N(R)$, it follows that $kC_G(R) \cong kR \otimes_k kC_N(R)$ and that f remains a block of $kC_N(R)$. By Watanabe's Theorem 10.4.1 we have an isomorphism

$$Z(kGb) \cong Z(kC_G(R)f) \cong kR \otimes_k Z(kC_N(R)f).$$

This isomorphism maps $Z(kNb)$ to $Z(kC_N(R)f)$, hence onto $Z(kC_N(R)f)$, as follows from comparing dimensions as above. Let S be the subalgebra of $Z(kGb)$ that is mapped to the image of kR on the right side under this isomorphism. In particular, S is a subalgebra of $Z(kGb)$ isomorphic to kR. The inclusions of S and kNb into kGb yield an algebra homomorphism $S \otimes_k kNb \to kGb$. We need to show that this is an isomorphism. Both sides have the same dimension, so it suffices to show that this map is injective. By 4.12.12 it suffices to show that the induced map $S \otimes_k Z(kNb) \to Z(kGb)$ is injective. Since both sides have the same dimension, it suffices to show that the map $S \otimes_k Z(kNb) \to Z(kGb)$ is surjective. It suffices to show that this holds upon composing this map with the isomorphism $Z(kGb) \cong kR \otimes_k Z(kC_N(R)f)$ from Watanabe's Theorem above. But the composition $S \otimes_k Z(kNb) \to kR \otimes_k Z(kC_N(R)f)$ maps S onto kR and $Z(kNb)$ onto $Z(kC_N(R)f)$. Both sides have the same dimension, hence are isomorphic. $\qquad\qquad\square$

Remark 10.4.3 The isomorphism constructed in the proof of Theorem 10.4.2 is not, in general, an isomorphism of interior P-algebras, since the copy of R on the left side is not necessarily sent to the canonical image of R in kGb. In particular, this isomorphism need not induce a source algebra isomorphism, and it remains an open problem whether this isomorphism lifts to an isomorphism over $\mathcal{O}$. The hypotheses that P is abelian and that Q has a complement in P are both essential; see the remarks at the end of [108].

Remark 10.4.4 The main point of the proof of Theorem 10.4.2 is that it constructs an invertible element s in $Z(kGb)$ of p-power order such that kGb is projective as a module over $\langle s \rangle$. Lifting this element to an invertible element of the same p-power order to the centre over $\mathcal{O}$ is the main obstruction to lifting the isomorphism in Theorem 10.4.2 to an analogous isomorphism over $\mathcal{O}$. Watanabe showed that it suffices to construct a perfect isometry between b as a block of $N \times R$ and as a block of G, because a perfect isometry induces isomorphisms of centres over $\mathcal{O}$, hence sends R to a central p-subgroup of the block b of G

over $\mathcal{O}$. The construction of invertible elements in centres of block algebras over $\mathcal{O}$ is closely related to the Z_p^*-Theorem; see Robinson [188].

10.5 Abelian defect and Frobenius inertial quotient

Let P be a finite abelian p-group, and let E be a p'-subgroup of $\text{Aut}(P)$ acting freely on $P \setminus \{1\}$. Equivalently, the corresponding semidirect product $P \rtimes E$ is a Frobenius group. We refer to [76, Theorem 10.3.1] for the fundamental group theoretic properties of E: the order of E divides $|P| - 1$, and the Sylow subgroups of E are either cyclic or generalised quaternion 2-groups. If E has odd order, then E is metacyclic, and if E has even order, then E has a unique (hence central) involution. In particular, the group $H^2(E; k^\times)$ is trivial. This implies that if $E = N_G(P, e)/C_G(P)$ is the inertial quotient of a block b of a finite group G having (P, e) as a maximal Brauer pair, then the Brauer correspondent of b is Morita equivalent to the untwisted group algebra $\mathcal{O}(P \rtimes E)$. Since P is abelian, the fusion system $\mathcal{F}$ of b on P determined by the choice of the block e of $kC_G(P)$ is equal to that determined by the group $P \rtimes E$. In particular, an endopermutation $\mathcal{O}P$-module V is $\mathcal{F}$-stable if and only if it is E-stable; that is, if and only if ${}^y V \cong V$ for all $y \in E$. The key result of this section, due to Puig, is a stable equivalence of Morita type between $\mathcal{O}Gb$ and $\mathcal{O}(P \rtimes E)$. This situation arises if P is cyclic or a Klein four group, so the material of this section is part of the groundwork for the chapters on blocks with cyclic or Klein four defect groups. As in earlier sections, without further mention, we consider an $\mathcal{O}P$-module V as an $\mathcal{O}\Delta P$-module via canonical isomorphism $P \cong \Delta P$ whenever expedient.

Theorem 10.5.1 ([175, 6.8]) *Let G be a finite group, b a block of $\mathcal{O}G$, P a defect group of b, $i \in (\mathcal{O}Gb)^{\Delta P}$ a source idempotent of b and e the block of $kC_G(P)$ satisfying $\text{Br}_{\Delta P}(i)e \neq 0$. Suppose that P is nontrivial abelian and that k is large enough. Set $E = N_G(P, e)/C_G(P)$ and suppose that E acts freely on $P \setminus \{1\}$. Then there is an indecomposable E-stable endopermutation $\mathcal{O}P$-module V with vertex P and an indecomposable $\mathcal{O}Gb$-$\mathcal{O}(P \rtimes E)$-bimodule X with the following properties.*

(i) *The bimodule X is a direct summand of $\mathcal{O}Gi \otimes_{\mathcal{O}P} \text{Ind}_{\Delta P}^{P \times P}(V) \otimes_{\mathcal{O}P} \mathcal{O}(P \rtimes E)$.*

(ii) *As an $\mathcal{O}(G \times (P \rtimes E))$-module, X has vertex ΔP and source V.*

(ii) *The bimodule X and its dual X^* induce a stable equivalence of Morita type between $\mathcal{O}Gb$ and $\mathcal{O}(P \rtimes E)$.*

The property of E acting freely on $P \setminus \{1\}$ can be read off the structure of the nontrivial Brauer pairs.

Lemma 10.5.2 *Let G be a finite group, b a block of $\mathcal{O}G$ and (P, e) a maximal (G, b)-Brauer pair. Suppose that P is nontrivial abelian. Set $E = N_G(P, e)/C_G(P)$. For any subgroup Q of P denote by e_Q the unique block of $kC_G(Q)$ satisfying $(Q, e_Q) \leq (P, e)$. Then E acts freely on $P \setminus \{1\}$ if and only if e_Q is a nilpotent block of $kC_G(Q)$ for any nontrivial subgroup Q of P.*

Proof Let Q be a nontrivial subgroup of P. Since P is abelian we have $P \subseteq C_G(P) \subseteq C_G(Q)$, and the pair (P, e) is also a maximal $(C_G(Q), e_Q)$-Brauer pair. The corresponding inertial quotient is $E_Q = N_{C_G(Q)}(P, e)/C_G(P) = C_E(Q)$. Thus e_Q is a nilpotent block of $kC_G(Q)$ if and only if $C_E(Q)$ is trivial. Since E acts freely on $P \setminus \{1\}$ if and only if $C_E(Q)$ is trivial for any nontrivial subgroup Q of P, the result follows. $\qquad\square$

Lemma 10.5.3 *Let P be a finite abelian p-group and E a p'-subgroup of $\text{Aut}(P)$. Suppose that E acts freely on $P \setminus \{1\}$. Then, for any nontrivial subgroup Q of P, the unit element 1 of $\mathcal{O}(P \rtimes E)$ remains primitive in $(\mathcal{O}(P \rtimes E))^{\Delta Q}$.*

Proof We may assume $\mathcal{O} = k$. Let Q be a nontrivial subgroup of P. Since E acts freely on $P \setminus \{1\}$, it follows that $C_{P \rtimes E}(Q) = P$, hence $(k(P \rtimes E))(\Delta Q) = kP$, which is a local algebra. Thus $1 = j + j'$ for some primitive idempotent $j \in (k(P \rtimes E))^{\Delta Q}$ satisfying $\text{Br}_{\Delta Q}(j) = 1$ and some idempotent j' satisfying $\text{Br}_{\Delta Q}(j') = 0$. If R is a nontrivial subgroup of Q, then $\text{Br}_{\Delta R}(j) \neq 0$, hence $\text{Br}_{\Delta R}(j) = 1_{kP}$. Thus the first argument, using R instead of Q, implies that $\text{Br}_{\Delta R}(j') = 0$. This holds for any nontrivial subgroup R of Q, and hence 5.8.3 implies that $j' \in (k(P \rtimes E))_1^Q$. Let $u \in P$ and $e \in E$. Then $\text{Tr}_1^Q(ue) = (\sum_{y \in Q} uy(y^{-1})^e)e$. If $y, z \in Q$ such that $y(y^{-1})^e = z(z^{-1})^e$, then $z^{-1}y = (z^{-1}y)^e$. Since E acts freely on $P \setminus \{1\}$, this forces $z = y$ or $e = 1$. If $e = 1$, the above sum is zero. If $e \neq 1$, then the sum $\sum_{y \in Q} uy(y^{-1})^e$ has $|Q|$ distinct terms, hence belongs to the augmentation ideal $I(kP)$. Thus $(k(P \rtimes E))_1^Q \subseteq J(k(P \rtimes E))$, and therefore $j' = 0$, implying the result. $\qquad\square$

Proof of Theorem 10.5.1 We will show that this theorem is a special case of 9.11.8. Set $A = i\mathcal{O}Gi$. As mentioned above, the group $H^2(E; k^\times)$ is trivial, and hence, by 6.15.1, A has a subalgebra $\mathcal{O}(P \rtimes E)$ as interior P-algebra, such that $\mathcal{O}(P \rtimes E)$ is a direct summand of A as an $\mathcal{O}(P \rtimes E)$-$\mathcal{O}(P \rtimes E)$-bimodule. Let Q be a nontrivial subgroup of P, and denote by e_Q the block of $kC_G(Q)$ satisfying $(Q, e_Q) \leq (P, e)$. Since P is abelian, it follows from 6.4.13 that P is a defect group of e_Q and $i_Q = \text{Br}_{\Delta Q}(i)$ is a source idempotent in $(kC_G(Q)e_Q)^{\Delta P}$. Thus $A(\Delta Q)$ is a source algebra of $kC_G(Q)e_Q$. Since E acts freely on $P \setminus \{1\}$, it follows from 10.5.2, that $kC_G(Q)e_Q$ is a nilpotent block. Therefore, by 8.11.5, there is an indecomposable endopermutation kP-module V_Q such that setting

$S_Q = \text{End}_k(V_Q)$, we have an isomorphism of interior P-algebras

$$A(\Delta Q) \cong S_Q \otimes_k kP.$$

As Q runs over the nontrivial subgroups of P, the family of endo-permutation modules V_Q thus obtained satisfies the compatibility conditions of 7.8.2. Thus there is an indecomposable E-stable endopermutation $\mathcal{O}P$-module V with vertex P satisfying

$$V_Q \cong \text{Defres}_{P/Q}^P(V)$$

for any nontrivial subgroup Q of P, or equivalently, satisfying

$$S_Q \cong S(\Delta Q)$$

for any nontrivial subgroup Q of P. Since V is E-stable, it extends by 7.8.4 to an $\mathcal{O}(P \rtimes E)$-module, still denoted V. Consider V as a right $\mathcal{O}(P \rtimes E)$-module via the canonical anti-automorphism of $\mathcal{O}(P \rtimes E)$ and consider the A-$\mathcal{O}(P \rtimes E)$-bimodule

$$U = A \otimes_\mathcal{O} V,$$

with $P \rtimes E$ acting diagonally on the right. Note that since E is a p'-group, the module U is isomorphic to a direct summand of the A-$\mathcal{O}(P \rtimes E)$-bimodule

$$(A \otimes_\mathcal{O} V) \otimes_{\mathcal{O}P} \mathcal{O}(P \rtimes E) \cong A \otimes_{\mathcal{O}P} \text{Ind}_{\Delta P}^{P \times P}(V) \otimes_{\mathcal{O}P} \mathcal{O}(P \rtimes E).$$

Let M be an indecomposable direct summand of U as an A-$\mathcal{O}(P \rtimes E)$-module such that

$$(\text{End}_A(M))(\Delta P) \neq \{0\}.$$

We show that M remains indecomposable as an A-$\mathcal{O}P$-bimodule. It follows from 5.5.9 that all points of P on the interior P-algebra $\text{End}_A(M)$ are local. That is, any indecomposable direct summand M' of M as an A-$\mathcal{O}P$-bimodule satisfies $\text{End}_A(M')(\Delta P) \neq \{0\}$. But by 9.10.5, $A \otimes_\mathcal{O} V$, and hence M, has exactly one indecomposable summand with this property. Therefore M remains indecomposable as an A-$\mathcal{O}P$-bimodule. Moreover, 9.10.5 implies that if Q is a subgroup of P, then the $A(\Delta Q)$-kP-bimodule summand M_Q of $A(\Delta Q) \otimes_k V_Q$ satisfying

$$(\text{End}_A(M))(\Delta Q) \cong \text{End}_{A(\Delta Q)}(M_Q)$$

is indecomposable. The commutative diagram of 9.9.9 implies that M_Q also satisfies an isomorphism

$$(\text{End}_\mathcal{O}(M))(\Delta Q) \cong \text{End}_k(M_Q)$$

so that M_Q is as in 9.11.8. We need to show that if Q is nontrivial, then M_Q induces a Morita equivalence between $A(\Delta Q)$ and $(\mathcal{O}(P \rtimes E))(\Delta Q) \cong kP$. For Q nontrivial, we have $A(\Delta Q) \cong S_Q \otimes_k kP \cong V_Q \otimes_k V_Q^* \otimes_k kP$. As an $A(\Delta P)$-kP-module, this is isomorphic to $V_Q \otimes_k kP \otimes_k V_Q^*$, where the left $A(\Delta Q)$-module structure is given by the natural action of S_Q on V_Q and the left regular action of kP on kP, and where the right action of kP is given by the diagonal action of P on kP and V_Q^*. Thus $A(\Delta Q) \otimes_k V_Q \cong V_Q \otimes_k kP \otimes_k V_Q^* \otimes_k V_Q$. Since $V_Q^* \otimes_k V_Q$ has a trivial direct summand, it follows that $A(\Delta Q) \otimes_k V_Q$ has a summand isomorphic to $V_Q \otimes_k kP$ as an $A(\Delta Q)$-kP-bimodule. The uniqueness of M_Q implies that $M_Q \cong V_Q \otimes_k kP$. Clearly M_Q yields a Morita equivalence between kP and $A(\Delta Q)$. It follows from 9.11.8 that M induces a stable equivalence of Morita type. Thus the $\mathcal{O}Gb$-$\mathcal{O}(P \rtimes E)$-bimodule $X = \mathcal{O}Gi \otimes_A M$ has the properties as stated. $\qquad \square$

By specialising the technicalities leading up to 9.11.8 directly to the situation at hand it is possible to avoid some of the difficulties of the proof of 9.11.8. This yields a shorter proof of 10.5.1 which makes use of the criterion 4.14.12 for when the restriction to a subalgebra is a stable equivalence.

Alternative proof of Theorem 10.5.1 We follow the previous proof of 10.5.1 up to the point where we have constructed a fusion-stable endopermutation $\mathcal{O}P$-module V satisfying $A(Q) \cong S_Q \otimes_k kP$ for any nontrivial subgroup Q of P, where $S = \text{End}_{\mathcal{O}}(V)$ and $S_Q = S(\Delta Q)$. As before, we extend the $\mathcal{O}P$-module structure of V to $\mathcal{O}(P \rtimes E)$ and we consider S as an interior $\mathcal{O}(P \rtimes E)$-algebra. Let e be a primitive idempotent in $(S^{\text{op}} \otimes_{\mathcal{O}} A)^{\Delta P \rtimes E}$ satisfying $\text{Br}_{\Delta P}(e) \neq 0$. Note that P has a unique local point on $S^{\text{op}} \otimes_{\mathcal{O}} A$, and this local point has multiplicity one. It follows from 5.5.9 that all points of P on $A' = e(S^{\text{op}} \otimes_{\mathcal{O}} A)e$ are local. This implies that e remains primitive in $(A')^{\Delta P}$. Thus as an interior P-algebra, A' is primitive, relatively $\mathcal{O}P$-separable and Morita equivalent to A (by 7.4.1 (iii)). Since P is abelian and A' has a $P \times P$-stable $\mathcal{O}$-basis, it follows that for any subgroup Q of P, the interior P-algebra $A'(\Delta Q)$ is primitive. The canonical map $\mathcal{O}(P \rtimes E) \to A'$ is split injective as a bimodule homomorphism, by 6.15.2 applied with $\mathcal{O}(P \rtimes E)$ instead of A. By construction, if Q is nontrivial, then $A'(\Delta Q)$ is of the form $e'(S_Q^{\text{op}} \otimes_k S_Q \otimes_k kP)e'$ for some primitive local idempotent e' in $(S_Q^{\text{op}} \otimes_k S_Q \otimes_k kP)^{\Delta P}$. Using 7.3.10 (iv) applied to $S_Q = S(\Delta Q)$ instead of S we get that $A'(\Delta Q) \cong kP \cong (\mathcal{O}(P \rtimes E))(\Delta Q)$ for any nontrivial subgroup Q of P. It follows now from 4.14.12 that restriction and induction between A' and $\mathcal{O}(P \rtimes E)$ yields a stable equivalence of Morita type. Combining this stable equivalence with the Morita equivalence between A and A' yields a stable equivalence of Morita type between A and $\mathcal{O}(P \rtimes E)$ as stated. $\qquad \square$

The alternative proof of 10.5.1 constructs the stable equivalence between the source algebra A of $\mathcal{O}Gb$ and the algebra $\mathcal{O}(P \rtimes E)$ by breaking it up into two steps. The first step is a Morita equivalence between A and another interior P-algebra A' given by a bimodule with 'endopermutation source'. This is followed by a 'splendid' stable equivalence of Morita type between A' and $\mathcal{O}(P \rtimes E)$, obtained from restriction and induction. The intermediate algebra A' need not be the source algebra of a block, however. This is one of the main reasons for extending the terminology of 'sources' and 'splendid' equivalences to abstract interior P-algebras that are relatively P-separable and have P-stable $\mathcal{O}$-bases (cf. 5.1.19 and 9.8.9).

A consequence of the stable equivalence in 10.5.1 is that the induced partial isometry from 9.3.5 on the group $L^0(G, b)$ of virtual characters which are perpendicular to the group $\mathrm{Pr}_{\mathcal{O}}(G, b)$ of projective characters can be used to show that one inequality of Alperin's weight conjecture holds if the acting group E is abelian. We collect first basic character theoretic facts of $P \rtimes E$.

Proposition 10.5.4 *Let P be a finite nontrivial abelian p-group and E an abelian p'-subgroup of $\mathrm{Aut}(P)$ acting freely on $P \setminus \{1\}$. Suppose that k and K are large enough, and that $\mathrm{char}(K) = 0$. Denote by M the set of irreducible characters of degree 1 and by Λ the set of characters of the form $\mathrm{Ind}_P^{P \rtimes E}(\zeta)$, where ζ runs over a set of representatives of the E-orbits of nontrivial characters in $\mathrm{Irr}_K(P)$.*

- (i) *The characters in Λ are irreducible and have degree $|E|$, and we have $|\Lambda| = \frac{|P|-1}{|E|}$. The characters in M have degree 1, they restrict to the trivial character on P, and we have $|M| = |E|$. All characters of $\mathrm{Irr}_K(P \rtimes E)$ have height zero.*
- (ii) *We have $\mathrm{Irr}_K(P \rtimes E) = \Lambda \cup M$, and this union is disjoint. In particular, $|\mathrm{Irr}_K(P \rtimes E)| = |E| + \frac{|P|-1}{|E|}$.*
- (iii) *The restrictions of the characters in M to the set of p'-elements of $P \rtimes E$ is the set of irreducible Brauer characters $\mathrm{IBr}_k(P \rtimes E)$. In particular, $\ell(k(P \rtimes E)) = |E|$.*
- (iv) *The set $\{\mu + \sum_{\lambda \in \Lambda} \lambda\}_{\mu \in M}$ is the set $\mathrm{IPr}_{\mathcal{O}}(P \rtimes E)$ of the characters of the projective indecomposable $\mathcal{O}(P \rtimes E)$-modules.*
- (v) *The set $\{\lambda - \sum_{\mu \in M} \mu\}_{\lambda \in \Lambda}$ is a $\mathbb{Z}$-basis of the group $L^0(P \rtimes E)$ of generalised characters that vanish on all p'-elements in $P \rtimes E$.*
- (vi) *If λ, $\lambda' \in \Lambda$, then $\lambda - \lambda' \in L^0(P \rtimes E)$.*

Proof Since P is normal in $P \rtimes E$, the simple $k(P \rtimes E)$-modules are acted trivially upon by P, hence they are inflated from the simple kE-modules. Since E is abelian, the simple kE-modules are all 1-dimensional. The statements (i),

(ii) and (iii) are easily deduced from the character table of Frobenius groups described in §3.4. In particular, all characters in $\mathrm{Irr}_K(P \rtimes E)$ have a degree prime to p, hence height zero. One verifies that the characters in the set in (iv) vanish on p-singular elements, hence are in $\mathrm{Pr}_{\mathcal{O}}(P \rtimes E)$. Every character in the set in (iv) involves exactly one character $\mu \in M$ with multiplicity one. Thus a $\mathbb{Q}$-linear combination of these characters is in $\mathrm{Pr}_{\mathcal{O}}(P \rtimes E)$ if and only of this is actually a $\mathbb{Z}$-linear combination. The set in (iv) has $|E|$ elements. Since there are $|E|$ isomorphism classes of simple $k(P \rtimes E)$-modules, hence of projective indecomposable $\mathcal{O}(P \rtimes E)$-modules, it follows that these characters form a $\mathbb{Z}$-basis $\mathrm{Pr}_{\mathcal{O}}(P \rtimes E)$. The sum of these characters is the regular character. The regular character is also equal to the sum of the characters of the projective indecomposable modules, because the simple $k(P \rtimes E)$-modules are 1-dimensional. Since the character of a projective indecomposable $\mathcal{O}(P \rtimes E)$-module is a linear combination with nonnegative integer coefficients of this set, statement (iv) follows. The characters in the set in (v) are perpendicular to those in (iv), hence belong to $L^0(P \rtimes E)$. They are clearly $\mathbb{Z}$-linearly independent. They are in fact a $\mathbb{Z}$-basis of $L^0(P \rtimes E)$, because each character in this set has exactly one of the $\lambda \in \Lambda$ occurring with multiplicity one. This shows (v), and taking the difference of any two of the characters in the set in (v) yields (vi). $\qquad\qquad\square$

The point of statement (vi) in the above result is that an isometry induced by a stable equivalence of Morita type sends $\lambda - \lambda'$ to an element of the form $\chi - \chi'$ for some irreducible characters χ, χ' having the same degree, because this is the only way to write elements whose norm is 2 and that vanish at 1. This is used in the proof of the next result (some parts of which hold in greater generality; see the Remark 10.5.6). We denote by $\ell(A)$ the number of isomorphism classes of simple A-modules, where A is a finite-dimensional k-algebra.

Theorem 10.5.5 *Let G be a finite group, b a block of $\mathcal{O}G$ and (P, e) a maximal (G, b)-Brauer pair. Suppose that P is nontrivial abelian, that k and K are large enough, and that $\mathrm{char}(K) = 0$. Set $E = N_G(P, e)/C_G(P)$ and suppose that E is abelian and acts freely on $P \setminus \{1\}$. We have $|\mathrm{Irr}_K(G, b)| \leq |\mathrm{Irr}_K(P \rtimes E)|$ and $\ell(kG\bar{b}) \leq \ell(kP \rtimes E)$, where $\bar{b}$ is the canonical image of b in kG. If one of these two inequalities is an equality, so is the other. Both equalities hold if $|E| \leq 3$.*

Remark 10.5.6 By results of Brauer [28], Usami [221], Puig and Usami [180], [181] the equality in the statement extends to arbitrary blocks with an abelian defect group and inertial quotient of order at most 4. The inequalities in Theorem 10.5.5 follow also from Sambale's Theorem 6.12.3. We will show later that the equalities $|\mathrm{Irr}_K(G, b)| = |\mathrm{Irr}_K(P \rtimes E)|$ and $\ell(kG\bar{b}) = \ell(kP \rtimes E)$ hold if P is cyclic; see 11.1.3 and 11.10.2.

For the proof of Theorem 10.5.5 we start by proving two lemmas which describe $L^0(G, b)$ and which will also be used in the proof of Theorem 10.5.10 below. With the notation of 10.5.5, by 10.5.1 there is an $\mathcal{O}Gb$-$\mathcal{O}(P \rtimes E)$-bimodule X inducing a stable equivalence of Morita type between $\mathcal{O}Gb$ and $\mathcal{O}(P \rtimes E)$. By 9.3.5, the map $\Phi_X : \mathbb{Z}\mathrm{Irr}_K(P \rtimes E) \to \mathbb{Z}\mathrm{Irr}_K(G, b)$ induced by the exact functor $M \otimes_{\mathcal{O}(P \rtimes E)} -$ on character groups restricts to an isometry

$$L^0(\mathcal{O}(P \rtimes E)) \cong L^0(\mathcal{O}Gb),$$

where the notation is as in 9.3.5. We denote the canonical scalar products on these groups abusively by the same symbol $\langle -, - \rangle$. We use the notation from 10.5.4 for the irreducible characters of $P \rtimes E$.

Lemma 10.5.7 *With the notation and hypotheses above, there is a set of pairwise different characters $\chi_\lambda \in \mathrm{Irr}_K(G, b)$, with $\lambda \in \Lambda$, and a sign $\delta \in \{\pm 1\}$ satisfying*

$$\Phi_X(\lambda - \lambda') = \delta(\chi_\lambda - \chi_{\lambda'}).$$

The set $\{\chi_\lambda\}_{\lambda \in \Lambda}$ is a proper subset of $\mathrm{Irr}_K(G, b)$. If $|\Lambda| = 1$, then the character χ_λ and the sign δ can be chosen arbitrarily, where $\Lambda = \{\lambda\}$. If $|\Lambda| = 2$, then the set $\{\chi_\lambda\}_{\lambda \in \Lambda}$ is unique, while the sign δ depends on a choice of indexation of this set. If $|\Lambda| \geq 3$, then the χ_λ and δ are uniquely determined.

Proof Write $\Phi = \Phi_X$. If $|\Lambda| = 1$, then the first statement holds trivially with any choice of a character χ_λ, where $\Lambda = \{\lambda\}$. If $\lambda \neq \lambda' \in \Lambda$, then $\langle \lambda - \lambda', \lambda - \lambda' \rangle = 2$. Thus $\Phi(\lambda - \lambda') = \chi \pm \chi'$ for some $\chi, \chi' \in \mathrm{Irr}_K(G, b)$. Since $L^0(G, b)$ contains no nonzero character of a KGb-module (because all elements in $L^0(G, b)$ vanish at 1), it follows that

$$\Phi(\lambda - \lambda') = \chi - \chi'.$$

If $|\Lambda| = 2$, set $\chi_\lambda = \chi$, $\chi_{\lambda'} = \chi'$, and $\delta = 1$. Suppose that $|\Lambda| \geq 3$. For any λ, $\lambda' \in \Lambda$, write

$$\Phi(\lambda - \lambda') = \chi_{\lambda,\lambda'} - \chi'_{\lambda,\lambda'}$$

where $\chi_{\lambda,\lambda'}, \chi'_{\lambda,\lambda'} \in \mathrm{Irr}_K(G, b)$. Let $\lambda, \lambda', \lambda''$ be three pairwise different elements in Λ. We have

$$1 = \langle \lambda - \lambda', \lambda - \lambda'' \rangle = \langle \Phi(\lambda - \lambda'), \Phi(\lambda - \lambda'') \rangle.$$

Therefore, the characters $\Phi(\lambda - \lambda')$, with $\lambda' \in \Lambda - \{\lambda\}$, have a common irreducible constituent appearing with the same sign. Thus we may choose notation such that

$$\Phi(\lambda - \lambda') = \delta(\lambda)(\chi_\lambda - \chi_{\lambda,\lambda'})$$

for some $\delta(\lambda) \in \{\pm 1\}$. If $|\Lambda| = 3$, we set $\chi_{\lambda'} = \chi_{\lambda,\lambda'}$, $\chi_{\lambda''} = \chi_{\lambda,\lambda''}$, and it is then clear that $\delta(\lambda) = \delta(\lambda') = \delta(\lambda'')$. We may therefore suppose that $|\Lambda| \geq 4$. Let λ, λ', ν, ν' be four pairwise different characters in Λ. Then

$$\delta(\lambda)(\chi_{\lambda,\lambda'} - \chi_\lambda) = -\Phi(\lambda - \lambda') = \Phi(\lambda' - \lambda) = \delta(\lambda')(\chi_{\lambda'} - \chi_{\lambda',\lambda}).$$

It suffices to show $\chi_\lambda \neq \chi_{\lambda'}$, because this implies $\chi_{\lambda'} = \chi_{\lambda,\lambda'}$ and $\delta(\lambda) = \delta(\lambda')$. If we suppose $\chi_\lambda = \chi_{\lambda'}$, we get the contradiction

$$0 = \langle \lambda - \nu, \lambda' - \nu' \rangle = \langle \Phi(\lambda - \nu), \Phi(\lambda' - \nu') \rangle$$

$$= \langle \delta(\lambda)(\chi_\lambda - \chi_{\lambda,\nu}), \delta(\lambda')(\chi_{\lambda'} - \chi_{\lambda',\nu'}) \rangle$$

$$= \delta(\lambda)\delta(\lambda')(\langle \chi_\lambda, \chi_{\lambda'} \rangle + \langle \chi_{\lambda,\nu}, \chi_{\lambda'\nu'} \rangle) \neq 0.$$

This shows the existence of the set $\{\chi_\lambda\}_{\lambda \in \Lambda}$ and the sign δ with the properties as stated. Since $|\Lambda|$ is the rank of $L^0(\mathcal{O}(P \rtimes E))$, hence by 9.3.5 of $L^0(\mathcal{O}Gb)$, it follows that $\{\chi_\lambda\}_{\lambda \in \Lambda}$ is a proper subset of $\mathrm{Irr}_K(G, b)$. $\qquad \square$

We will need to identify the characters in $\mathrm{Irr}_K(G, b)$ that are not in the set $\{\chi_\lambda\}_{\lambda \in \Lambda}$.

Lemma 10.5.8 *With the notation and hypotheses above, set $N = \mathrm{Irr}_K(G, b) \setminus \{\chi_\lambda\}_{\lambda \in \Lambda}$. The set N is nonempty, and there is a generalised character $\alpha \in \mathbb{Z}\mathrm{Irr}_K(G, b)$ with the following properties.*

(i) We have $\Phi_\chi(\lambda - \sum_{\mu \in M} \mu) = \delta(\chi_\lambda - \alpha)$ for all $\lambda \in \Lambda$.

(ii) The set $\{\chi_\lambda - \alpha\}_{\lambda \in \Lambda}$ is a $\mathbb{Z}$-basis of $L^0(G, b)$.

(iii) The integer $a = \langle \chi_\lambda, \alpha \rangle$ does not depend on the choice of $\lambda \in \Lambda$.

(iv) For any $\chi \in N$ the integer $a_\chi = \langle \chi, \alpha \rangle$ is nonzero.

(v) We have $\alpha = a \sum_{\lambda \in \Lambda} \chi_\lambda - \sum_{\chi \in N} a_\chi \chi$.

(vi) We have $|E| = |\Lambda| a^2 - 2a + \sum_{\chi \in N} a_\chi^2$.

(vii) If $|\Lambda| = 1$, then $a \neq 1$.

(viii) We have $|E| > |\Lambda| a^2 - 2a \geq 0$.

(ix) We have $|\mathrm{Irr}_K(G, b)| = |\mathrm{Irr}_K(P \rtimes E)|$ if and only if $|a_\chi| = 1$ for all $\chi \in N$, and either $a = 0$ or $\{a, |\Lambda|\} = \{1, 2\}$.

Proof It follows from 10.5.7 that the set N is nonempty. Statement (i) is trivial if $\Lambda| = 1$. Assume that $|\Lambda| \geq 2$. Write $\Phi = \Phi_\chi$. For any $\lambda \in \Lambda$ write

$\Phi(\lambda - \sum_{\mu \in M} \mu) = \delta(\chi_\lambda - \alpha_\lambda)$ for some $\alpha_\lambda \in \mathbb{Z}\mathrm{Irr}_K(G, b)$, where δ is from 10.5.7. Thus

$$\delta(\chi_\lambda - \chi_{\lambda'}) = \Phi(\lambda - \lambda') = \Phi\left(\left(\lambda - \sum_{\mu \in M} \mu\right) - \left(\lambda' - \sum_{\mu \in M} \mu\right)\right)$$

$$= \delta(\chi_\lambda - \alpha_\lambda) - \delta(\chi_{\lambda'} - \alpha_{\lambda'})$$

which forces $\alpha_\lambda = \alpha_{\lambda'}$, and hence $\alpha = \alpha_\lambda$ does not depend on $\lambda \in \Lambda$. This shows (i). It follows from 10.5.4(v) that the set $\{\lambda - \sum_{\mu \in M} \mu\}_{\lambda \in \Lambda}$ is a $\mathbb{Z}$-basis of $L^0(\mathcal{O}(P \rtimes E))$, and from 9.3.5 that the set $\{\Phi(\lambda - \sum_{\mu \in M} \mu)\}_{\lambda \in \Lambda}$ is a $\mathbb{Z}$-basis of $L^0(\mathcal{O}Gb)$. Statement (ii) follows. Every irreducible character in $\mathrm{Irr}_K(G, b)$ appears in at least one basis element of $L^0(G, b)$ (because otherwise it would be in $L^0(G, b)^\perp = \mathrm{Pr}_\mathcal{O}(G, b)$, hence of defect zero). Thus every character in the set N has to appear in α. Moreover, the characters χ_λ appear all with the same multiplicity in α, because for any two different $\lambda, \lambda' \in \Lambda$ we have

$$1 = \left\langle \lambda - \sum_{\mu \in M} \mu, \lambda - \lambda' \right\rangle = \langle \chi_\lambda - \alpha, \chi_\lambda - \chi_{\lambda'} \rangle,$$

which implies that

$$\langle \alpha, \chi_\lambda - \chi_{\lambda'} \rangle = 0.$$

The orthogonality relations imply that

$$\alpha = a \sum_{\lambda \in \Lambda} \chi_\lambda + \sum_{\chi \in N} a_\chi \chi$$

for some nonzero integers a_χ, $\chi \in N$, and some integer a. This shows (iii), (iv), and (v). We have

$$|E| + 1 = \left\langle \lambda - \sum_{\mu \in M} \mu, \lambda - \sum_{\mu \in M} \mu \right\rangle = \langle \chi_\lambda - \alpha, \chi_\lambda - \alpha \rangle,$$

hence $|E| = \langle \alpha, \alpha \rangle - 2\langle \chi_\lambda, \alpha \rangle$. Thus statement (v) yields the formula in (vi). If $|\Lambda| = 1$ and $a = 1$, then χ_λ does not occur in the unique basis element $\chi_\lambda - \alpha$ of $L^0(G, b)$, so this is not possible. This shows (vii). Statement (vi) implies that $|E| > |\Lambda|a^2 - 2a$ because N is nonempty. If $|\Lambda|a^2 - 2a < 0$, then $a > 0$, and hence $|\Lambda|a - 2 < 0$. This implies $0 < |\Lambda|a < 2$, hence $|\Lambda|a = 1$. It follows that $|\Lambda| = a = 1$, contradicting (vii). This proves (viii). For statement (ix) note that we have $|\mathrm{Irr}_K(G, b)| = |\mathrm{Irr}_K(P \rtimes E)|$ if and only if $|N| = |E|$. If $a = 0$ and $a_\chi^2 = 1$ for $\chi \in N$, then the formula in (v) forces $|N| = |E|$. Suppose conversely that $|N| = |E|$. Then (viii) forces $a_\chi = \pm 1$ for all $\chi \in N$ and $|\Lambda|a^2 - 2a = 0$. This happens if $a = 0$, or if $|\Lambda|a = 2$. This proves (ix). $\qquad\square$

Remark 10.5.9 The seemingly pathological case $\{a, |\Lambda|\} = \{1, 2\}$ in 10.5.8 (ix) can be eliminated by negating the sign δ in the case $|\Lambda| = 1$ or by exchanging the two characters χ_λ, $\chi_{\lambda'}$ and negating the sign δ if $|\Lambda| = 2$.

Proof of Theorem 10.5.5 Let χ_λ, $\lambda \in \Lambda$, δ, α as in 10.5.7 and 10.5.8. Calculating norms yields

$$|E| + 1 = \left\langle \lambda - \sum_{\mu \in M} \mu, \lambda - \sum_{\mu \in M} \mu \right\rangle = \langle \chi_\lambda - \alpha, \chi_\lambda - \alpha \rangle.$$

Now all characters of $N = \mathrm{Irr}_K(G, b) \setminus \{\chi_\lambda\}_{\lambda \in \Lambda}$ and at least one of the characters $\chi_{\lambda'}$, for some $\lambda' \in \Lambda$, occur in $\chi_\lambda - \alpha$. This forces $|N| + 1 \le |E| + 1$, hence $|N| \le |E|$. The rank of $L^0(H, b)$ is equal to the difference $|\mathrm{Irr}_K(G, b)| - \ell(k G \bar{b})$, hence is invariant under stable equivalences of Morita type. Thus the two inequalities in the statement are in fact equivalent. If $|E| = 1$, then b is nilpotent, hence Morita equivalent to $\mathcal{O}P$, and so the equality holds trivially. Assume that $2 \le |E| \le 3$. Then $0 \le |\Lambda|a^2 - 2a \le 2$ by 10.5.8, where we use that N is nonempty. If $|\Lambda|a^2 - 2a = 0$, then $|E|$ is the sum of the squares a_χ^2, with $\chi \in N$. For $|E| \le 3$ this is only possible with $a_\chi^2 = 1$ for $\chi \in N$ and $|N| = |E|$, whence the last statement in this case. For the remaining cases we show that P must be cyclic of prime order, and hence, as mentioned above, the last statement follows from later results on blocks with cyclic defect groups in 11.1.3 and 11.10.2. If $|\Lambda|a^2 - 2a = 1$, then $a = 1$ and $|\Lambda| = 3$. Since $|P| = 3|E| + 1$ is a power of p, this forces $|E| = 2$ and $|P| = 7 = p$; in particular, P is cyclic. If $|\Lambda|a^2 - 2a = 2$, then $a = 1$ and $|\Lambda| = 4$. Using the inequality $|E| > |\Lambda|a^2 - 2a = 2$ yields $|E| = 3$. This implies $|P| = |\Lambda| \cdot |E| + 1 = 4 \cdot 3 + 1 = 13$; in particular, P is cyclic. $\square$

Alperin's weight conjecture predicts that the equality $|\mathrm{Irr}_K(G, b)| = |\mathrm{Irr}_K(P \rtimes E)|$ should hold in general. The next result shows that if this equality holds, then there is a p-permutation equivalence, hence an isotypy between $\mathcal{O}Gb$ and $\mathcal{O}(P \rtimes E)$. We show further that 'up to a sign' the generalised decomposition numbers can be followed through this isotypy. We use the notation $\mathrm{Irr}_K(P \rtimes E) = \Lambda \cup M$ from above; that is, the characters in M have degree 1 and the characters in Λ have degree $|E|$ and do not have P in their kernel.

Theorem 10.5.10 *Let G be a finite group, b a block of $\mathcal{O}G$, P a defect group of b and $i \in (\mathcal{O}Gb)^{\Delta P}$ a source idempotent of b. Denote by γ the local point of P on $\mathcal{O}Gb$ containing i, and by e the unique block of $kC_G(P)$ satisfying $\mathrm{Br}_{\delta P}(i)e \ne 0$. Suppose that P is abelian, that k and K are large enough, and that $\mathrm{char}(K) = 0$. Set $E = N_G(P, e)/C_G(P)$ and suppose that E is abelian and acts freely on $P \setminus \{1\}$. Suppose that $|\mathrm{Irr}_K(G, b)| = |\mathrm{Irr}_K(P \rtimes E)|$. Let V be an indecomposable* *endopermutation $\mathcal{O}P$-module as in 10.5.1, and let $\delta \in \{\pm 1\}$ and $\{\chi_\lambda\}_{\lambda \in \Lambda}$ as in 10.5.7.*

(i) *For any labelling $\{\chi_\mu\}_{\mu \in M}$ of the complement of $\{\chi_\lambda\}_{\lambda \in \Lambda}$ in $\mathrm{Irr}_K(G, b)$ there are unique signs $\delta(\mu) \in \{\pm 1\}$ for all $\mu \in M$ such that the linear map*

$$\Phi : \mathbb{Z}\mathrm{Irr}_K(P \rtimes E) \to \mathrm{Irr}_K(G, b)$$

defined by $\Phi(\lambda) = \delta\chi_\lambda$ for all $\lambda \in \Lambda$ and $\Phi(\mu) = \delta(\mu)\chi_\mu$ for all $\mu \in M$ is induced by a p-permutation equivalence. In particular, Φ is a perfect isometry.

(ii) *All characters in $\mathrm{Irr}_K(G, b)$ have height zero.*

(iii) *Let u_ϵ be a local pointed (G, b)-element contained in P_γ such that $u \neq 1$. Denote be $\omega(u)$ the sign of the integer $\chi_V(u)$, where χ_V is the character of the endopermutation module V, Then, for all $\mu \in M$ and $\lambda \in \Lambda$ we have*

$$\chi_\mu(u_\epsilon) = \omega(u)\delta(\mu),$$

$$\chi_\lambda(u_\epsilon) = \omega(u)\delta\lambda(u).$$

Proof Let X be the $\mathcal{O}Gb$-$\mathcal{O}(P \rtimes E)$-bimodule as in 10.5.1. That is, X is a direct summand of

$$\mathcal{O}Gi \otimes_{\mathcal{O}P} \mathrm{Ind}_{\Delta P}^{P \times P}(V) \otimes_{\mathcal{O}P} \mathcal{O}(P \rtimes E)$$

where i is a source idempotent in $(\mathcal{O}Gb)^{\Delta P}$, V is an E-stable indecomposable endopermutation $\mathcal{O}P$-module with vertex P, and X induces a stable equivalence of Morita type. The partial isometry $\Phi_X : L^0(P \rtimes E) \cong L^0(G, b)$ induced by the functor $X \otimes_{\mathcal{O}(P \rtimes E)} -$ sends $\lambda - \lambda'$ to $\delta(\chi_\lambda - \chi_{\lambda'})$, where the notation is as in 10.5.7. By 10.5.8 and 10.5.9, we may choose notation such that the integer a in 10.5.8 is zero. Equivalently, if we choose any labelling $\{\chi_\mu\}_{\mu \in M}$ of the complement of $\{\chi_\lambda\}_{\lambda \in \Lambda}$, then by 10.5.8 (v), we have

$$\Phi_X\left(\lambda - \sum_{\mu \in M} \mu\right) = \delta\chi_\lambda - \sum_{\mu \in M} \delta(\mu)\chi_\mu$$

for all $\lambda \in \Lambda$, where $\delta(\mu) \in \{\pm 1\}$. Thus the map $\Phi : \mathrm{Irr}_K(P \rtimes E) \cong \mathrm{Irr}_K(G, b)$ defined by $\Phi(\lambda) = \delta\chi_\lambda$ for all $\lambda \in \Lambda$ and $\Phi(\mu) = \delta(\mu)\chi_\mu$ for all $\mu \in M$ extends Φ_X. Theorem 9.3.7 implies that the difference of the corresponding bicharacters $\chi_\Phi - \chi_X$ is the generalised character ψ of a virtual projective $\mathcal{O}Gb$-$\mathcal{O}(P \rtimes E)$-bimodule and that Φ is a perfect isometry. In order to show that $\chi_\Phi = \chi_X + \psi$ is induced by a p-permutation equivalence it suffices to show that χ_X is the character of a splendid complex. It follows from 7.11.9 that the endopermutation $\mathcal{O}P$-module V has an E-stable endosplit

p-permutation resolution. Therefore, by 9.11.4, X is the homology of a splendid complex, and hence induces a p-permutation equivalence. By 9.2.5, perfect isometries preserve character heights, whence (i). Note that $\mu(u) = 1$, where $\mu \in M$. Thus, setting $\delta(\lambda) = \delta$ for $\lambda \in \Lambda$, the formulae in (ii) are equivalent to $\chi_\lambda(u_\epsilon) = \omega(u)\delta(\lambda)\lambda(u)$ for all $\lambda \in \mathrm{Irr}_K(P \rtimes E)$. In order to prove (ii) we replace the block algebra $\mathcal{O}Gb$ by a source algebra $A = i\mathcal{O}Gi$. The stable equivalence of Morita type with endopermutation source between $\mathcal{O}(P \rtimes E)$ and A is constructed in the alternative proof of 10.5.1 in two steps. The first step is a Morita equivalence between A and $\Lambda' = e(S^{\mathrm{op}} \otimes_\mathcal{O} A)e$, where e belongs to the unique local point of P on A'. The algebra A' is then shown to contain $\mathcal{O}(P \rtimes E)$ as a subalgebra in such a way that induction and restriction between A' and $\mathcal{O}(P \rtimes E)$ yields a stable equivalence of Morita type. By 7.4.3, under the Morita equivalence between A and A', the generalised decomposition numbers of A' are obtained from those of A by multiplying with $\omega(u)$, where $u \in P \setminus \{1\}$. Thus there is no loss in assuming that $V = \mathcal{O}$, hence $A = A'$. Let $\lambda \in \mathrm{Irr}_K(P \rtimes E)$, and let W be an $\mathcal{O}$-free $\mathcal{O}(P \rtimes E)$-module with character λ. Then $A \otimes_\mathcal{O} W$ is an $\mathcal{O}$-free A-module whose character is of the form $\delta(\lambda)\chi_\lambda + \psi_\lambda$ for some $\psi_\lambda \in \mathrm{Pr}_\mathcal{O}(A)$, where we abusively denote the character of A corresponding to χ_λ again by χ_λ. Moreover, $\mathrm{Res}_B^A(A \otimes_B W) \cong W \oplus Y$ for some projective $\mathcal{O}(P \rtimes E)$-module Y. By 5.12.14 the character of Y and the character ψ_λ vanish at u. Thus $\lambda(u) = \delta(\lambda)\chi_\lambda(u)$. Since by 10.5.3 the unit element 1 of $\mathcal{O}(P \rtimes E)$ is primitive in $(\mathcal{O}(P \rtimes E))^{\langle u \rangle}$ it follows that $\mathrm{Br}_{\langle u \rangle}(1) = \mathrm{Br}_{\langle u \rangle}(j) = 1_{kP}$, where $j \in \epsilon$. Thus 5.12.16 implies that $\chi_\lambda(u) = \chi_\lambda(u_\epsilon)$. The result follows. $\qquad\square$

10.6 Blocks of finite p-solvable groups

The character theory of finite p-solvable groups goes back to work of Fong [72]. The algebra structure of blocks of finite p-solvable groups over k has been determined by Külshammer [110], and their source algebras by Puig [179], completing earlier work in [167]. This is based on a version of Fong–Reynolds reduction [72], [184]; we follow the presentation given in [81], [82, §4]. Alperin's weight conjecture holds for blocks of p-solvable finite groups as a consequence of unpublished work by Okuyama. In fact, Alperin's weight conjecture is implied in this case by stronger statements on vertices and Green correspondents of simple modules; see Barker [11], which contains further references and a historic account on this topic. Fong proved in [72] one implication of Brauer's height zero conjecture, namely that if a block of a finite p-solvable group has an abelian defect group, then all characters of that block have height zero. The converse implication was proved by Gluck and Wolf [74], using the classification of finite simple groups. The Alperin–McKay conjecture was proved for blocks of finite p-solvable groups independently by Dade [56] and Okuyama and Wajima [159], and Dade's projective conjecture was proved by Robinson [190].

We assume in this section that k is algebraically closed. For G a finite group, we denote by $O_p(G)$ the largest normal p-subgroup of G, by $O_{p'}(G)$ the largest normal p'-subgroup of G, by $O_{p',p}(G)$ the inverse image in G of $O_p(G/O_{p'}(G))$, and by $O_{p',p,p'}(G)$ the inverse image in G of $O_{p'}(G/O_{p',p}(G))$. A finite group G is called *p-constrained* if $C_G(Q) \subseteq O_{p',p}(G)$, where Q is a Sylow p-subgroup of $O_{p',p}(G)$. Finite p-solvable groups are p-constrained; see 10.6.4 below.

Theorem 10.6.1 ([179]) *Let G be a finite p-solvable group, b a block of $\mathcal{O}G$ with a defect group P, and let $i \in (\mathcal{O}Gb)^P$ be a source idempotent of b. Denote by $\mathcal{F}$ the fusion system of b on P determined by the choice of i. There is an indecomposable endopermutation $\mathcal{O}P$-module V with vertex P whose image in the Dade group $D_{\mathcal{O}}(P)$ has finite order, a finite p-solvable group L having P as a Sylow p-subgroup such that $O_{p'}(L) = \{1\}$ and $C_L(Q) \subseteq Q$, where $Q = O_p(L)$, and a class $\alpha \in H^2(L/Q; k^\times)$ such that, setting $S = \operatorname{End}_{\mathcal{O}}(V)$, we have an isomorphism of interior P-algebras*

$$i\mathcal{O}Gi \cong j(S \otimes_{\mathcal{O}} \mathcal{O}_\alpha L)j$$

for some primitive local idempotent j in $(S \otimes_{\mathcal{O}} \mathcal{O}_\alpha L)^P$. Equivalently, there is a block b' of a finite central p'-extension L' of L such that $\mathcal{O}Gb$ and $\mathcal{O}L'b'$ are Morita equivalent via an indecomposable $\mathcal{O}Gb$-$\mathcal{O}L'b'$-bimodule with vertex ΔP and source V, where we identify P with its preimage in L'. Moreover, the block b' of $\mathcal{O}L'$ is of principal type, and we have

$$\mathcal{F} = \mathcal{F}_P(L', b') = \mathcal{F}_P(L) = N_{\mathcal{F}}(Q).$$

In particular, fusion systems of blocks of finite p-solvable groups are constrained.

Remark 10.6.2 The conditions $O_{p'}(L) = \{1\}$ and $C_L(Q) \subseteq Q$, where $Q = O_p(L)$, ensure that L is bounded in terms of P because $L/Z(Q)$ embeds into $\operatorname{Aut}(Q)$, thanks to the condition $C_L(Q) \subseteq Q$. Since $H^2(L/Q; k^\times)$ is finite, this implies that Donovan's conjecture holds for blocks of finite p-solvable groups. In fact, Puig's conjecture holds for blocks of p-solvable groups because the torsion subgroup of the Dade group $D_{\mathcal{O}}(P)$ is finite, by a result of Puig in [173]. The statement on V being torsion in the Dade group involves the classification of finite simple groups; see the remarks preceding Theorem 7.10.2. There is some redundancy in the statement of 10.6.1: a finite p-solvable group is automatically p-constrained, and hence the inclusion $C_L(Q) \subseteq Q$ is a consequence of the condition $O_{p'}(L) = \{1\}$.

Corollary 10.6.3 *Let G be a finite p-solvable group, b a block of $\mathcal{O}G$ with an abelian defect group P, and let $i \in (\mathcal{O}Gb)^P$ be a source idempotent of b, and E the associated inertial quotient. There is an indecomposable endopermutation $\mathcal{O}P$-module V with vertex P and a class $\alpha \in H^2(E; k^\times)$ such that, setting $S = \text{End}_\mathcal{O}(V)$, we have an isomorphism of interior P-algebras*

$$i\mathcal{O}Gi \cong S \otimes_\mathcal{O} \mathcal{O}_\alpha(P \rtimes E).$$

For the proof of the above theorem and its corollary we will need some basic group theoretic facts on p-solvable groups, collected without proofs in the following lemma.

Lemma 10.6.4 ([76, 3.3]) *Let G be a finite p-solvable group.*

(i) *Let Q be a Sylow p-subgroup of $O_{p',p}(G)$. Then $C_G(Q) \subseteq O_{p',p}(G)$. In particular, we have $Z(G) \subseteq O_{p',p}(G)$, and if G has abelian Sylow p-subgroups, then Q is a Sylow p-subgroup of G.*

(ii) *Let R be a p-subgroup of G. Then $O_{p'}(N_G(R)) = O_{p'}(C_G(R)) = O_{p'}(G) \cap C_G(R)$.*

The proof of (ii) given in [82, 4.2] shows that (ii) holds for finite p-constrained groups. The next lemma is a sufficient criterion for a block to be of principal type.

Lemma 10.6.5 (cf. [81, 3.3]) *Let G be a finite group, Q a Sylow p-subgroup of $O_{p',p}(G)$ satisfying $C_G(Q) \subseteq O_{p',p}(G)$. Let e be a G-stable block of $\mathcal{O}O_{p'}(G)$. Then e remains a block of $\mathcal{O}G$, having as defect groups the Sylow p-subgroups of G, and e is of principal type as a block of $\mathcal{O}G$; that is, $\text{Br}_R(e)$ is a block of $kC_G(R)$ for any p-subgroup R of G.*

Proof Let P be a Sylow p-subgroup of G. The algebra $\mathcal{O}O_{p'}(G)e$ has $\mathcal{O}$-rank prime to p, because its rank is the square of the degree of an irreducible character of the p'-group $O_{p'}(G)$. This algebra has a P-stable basis, because it is a direct factor of $\mathcal{O}O_{p'}(G)$. Thus P fixes at least one element of such a basis, and hence $(\mathcal{O}O_{p'}(G)e)(P) \neq \{0\}$. In particular, we have $\text{Br}_Q(e) \neq 0$. Set $N = O_{p',p}(G)$. By 6.5.4 any block of $\mathcal{O}N$ lies in $\mathcal{O}O_{p'}(G)$. Thus e is a block of $\mathcal{O}N$, having Q as a defect group. Thus $\text{Br}_Q(e)$ is a block of $\mathcal{O}N_N(Q)$, contained in $\mathcal{O}C_N(Q) = \mathcal{O}C_G(Q)$, and $\text{Br}_A(e)$ is $N_G(Q)$-stable. It follows that $\text{Br}_Q(e)$ is a block of $kN_G(Q)$. Let b be a block of $\mathcal{O}G$ such that $be = b$. By 6.8.9, Q is contained in a defect group of b, and hence $\text{Br}_Q(b) \neq 0$. Since $\text{Br}_Q(e)$ is already a block of $\mathcal{O}N_G(Q)$, this forces $\mathrm{Br}_Q(b) = \mathrm{Br}_Q(e)$, hence $b - e \in \ker(\mathrm{Br}_Q)$. That, however, forces $b = e$, because any block $b' \neq b$ satisfying $b'e = b'$ satisfies $\mathrm{Br}_Q(b') \neq 0$ by the previous argument. Since we showed earlier that $\mathrm{Br}_P(e) \neq 0$, it follows that as a block of $\mathcal{O}G$, $b = e$ has P as a defect group. It remains to show that e is of principal type. Let R be a subgroup of P. We have $C_{O_{p'}(G)}(R) = O_{p'}(C_G(R))$ by the second equality in 10.6.4 (ii). The block e remains a block of $\mathcal{O}O_{p'}(G)R$, with R as a defect group. Thus $\mathrm{Br}_R(e)$ is a block of $kN_{O_{p'}(G)R}(R) = kC_{O_{p'}(G)}(R)R = kO_{p'}(C_G(R))R$, hence of $kO_{p'}(C_G(R))$. This block is $C_G(R)$-stable, as e is G-stable. Thus, the first statement applied to the block $\mathrm{Br}_R(e)$ and the group $C_G(R)$ implies that $\mathrm{Br}_R(e)$ remains a block of $kC_G(R)$. $\qquad\square$

This lemma reduces the determination of the source algebras of blocks of finite p-solvable groups to certain blocks of principal type. This reduction is shown to be compatible with the Brauer correspondence.

Proposition 10.6.6 (cf. [81, 3.1]) *Let G be a finite p-solvable group, and let b be a block of $\mathcal{O}G$. There is a subgroup H and an H-stable block e of $\mathcal{O}O_{p'}(H)$ with the following properties.*

(i) *We have $O_{p'}(G) \subseteq H$, the different G-conjugates of e are pairwise orthogonal, e is a block of $\mathcal{O}H$, and $b = \mathrm{Tr}_H^G(e)$.*

(ii) *Any Sylow p-subgroup P of H is a defect group of $\mathcal{O}He$ and of $\mathcal{O}Gb$, any source idempotent $i \in (\mathcal{O}He)^P$ remains a source idempotent in $(OGb)^P$, and we have an equality of source algebras*

$$i\mathcal{O}Hi = i\mathcal{O}Gi.$$

(iii) *The block e of $\mathcal{O}H$ is of principal type.*

(iv) *If P is abelian, then $H = O_{p',p,p'}(H)$.*

(v) *Let c be the block of $\mathcal{O}N_G(P)$ corresponding to b, and let f be the block of $\mathcal{O}N_H(P)$ corresponding to e. Then f is a block of $O_{p'}(N_H(P))$, and we have $c = \mathrm{Tr}_{N_H(P)}^{N_G(P)}(f)$. If $j \in (\mathcal{O}N_H(P)f)^P$ is a source idempotent, then j remains a source idempotent in $(\mathcal{O}N_G(P)c)^P$, and we have an equality of source algebra*

$$j\mathcal{O}N_H(P)j = j\mathcal{O}N_G(P)j.$$

Proof Let d be a block of $\mathcal{O}O_{p'}(G)$ such that $bd \neq 0$. By standard Clifford theoretic results in 6.8.3, we may assume that G stabilises the block d. But in that case, the statements (i) and (ii) follow from 10.6.5 with $H = G$. Let now H, P, e satisfy the conclusions of (i) and (ii). Statement (iii) is a restatement of the second statement in 10.6.5, and (iv) follows from 10.6.4 (i). For (v) we may assume that $\mathcal{O} = k$. Since P is a defect group of kHe and $\mathrm{Tr}_H^G(e) = b$, it follows that $\mathrm{Br}_P(b) = \mathrm{Tr}_{N_H(P)}^{N_G(P)}(\mathrm{Br}_P(e))$. Note that H is the stabiliser of e in G. As kHe is of principal type, it follows further that $\mathrm{Br}_P(e) = f$, and that $N_H(P)$ is the stabiliser of f in $N_G(P)$, whence $c = \mathrm{Tr}_{N_H(P)}^{N_G(P)}(f)$. The statement on source idempotents is a special case of (ii). $\qquad\square$

Proof of Theorem 10.6.1 Set $N = O_{p'}(G)$. By 10.6.6 we may assume that b is a G-stable block of $\mathcal{O}N$. As a block of $\mathcal{O}G$, b is of principal type and has the Sylow p-subgroups of G as defect groups. The G-algebra $T = \mathcal{O}Nb$ is a matrix algebra of rank n^2 for some positive integer n that is prime to p as N is a p'-group. By 5.3.4, the action of a Sylow p-subgroup P of G lifts uniquely to a group homomorphism $P \to T^\times$ such that the image of any element of P in $T^\times$ has determinant 1. Since $P \cap N = \{1\}$, we may choose a set of representatives $\mathcal{R}$ of G/N in G containing P, and then we may choose elements $t_x \in T^\times$ for any $x \in \mathcal{R}$ that act as x on S and that have determinant 1. We then set $t_{xy} = t_x y$ for any $x \in \mathcal{R}$ and any $y \in N$; in particular, $t_y = yb$ for $y \in N$ has order dividing $|N|$, and hence the determinant of t_{xy} is a root of unity of order dividing $|N|n$, which divides $|N|^2$. It follows that with this choice, the values of the 2-cocycle of G defined by $t_x t_y = \alpha(x, y)^{-1} t_{xy}$ for x, y are contained in a finite p'-subgroup of $\mathcal{O}^\times$, hence in the canonical preimage of $k^\times$ in $\mathcal{O}^\times$. Moreover, by the construction of the t_x, this 2-cocycle is the restriction to G of a 2-cocycle of the group $L = G/N$. Since G is p-solvable and $N = O_{p'}(G)$, it follows that L is p-solvable, $O_{p'}(L) = \{1\}$, and $Q = O_p(L)$ satisfies $C_L(Q) = Z(Q)$. By 6.8.13, we have an isomorphism

$$\mathcal{O}Gb \cong T \otimes_{\mathcal{O}} \mathcal{O}_\alpha L$$

sending xe to $t_x \otimes \bar{x}$, where $\bar{x}$ is the image of x in L. By construction, this is an isomorphism of interior P-algebras. The action of P on T stabilises a basis, since T is a direct summand of $\mathcal{O}N$ for the conjugation action by P. Therefore, if we write $T = \mathrm{End}_{\mathcal{O}}(W)$, then W is an endopermutation $\mathcal{O}P$-module of rank n, which is prime to p, and which has thus an indecomposable direct summand of vertex P, or equivalently, which satisfies $T(P) \neq \{0\}$. Since the class of α is contained in $H^2(L; k^\times)$, it follows that there is a central extension L' of L by a finite cyclic p'-group Z, and a block $b' \in Z(\mathcal{O}L')$ such that the canonical map $L' \to L$ induces an algebra isomorphism $\mathcal{O}L'b' \cong \mathcal{O}_\alpha L$. The image of P in L lifts uniquely to L', and by denoting this by P again, the above isomorphisms yield an isomorphism of interior P-algebras

$$\mathcal{O}Gb \cong T \otimes_{\mathcal{O}} \mathcal{O}L'b'.$$

Since b is of principal type by 10.6.6, it follows that for any subgroup R of P the algebra $(\mathcal{O}Gb)(R) \cong T(R) \otimes_k (\mathcal{O}L'b')(R)$ is indecomposable. This forces in particular that the algebra $(\mathcal{O}L'b')(R)$ is indecomposable, and hence that b' is of principal type. A source idempotent $i \in (\mathcal{O}Gb)^P$ corresponds through this isomorphism to a primitive idempotent in $(T \otimes_{\mathcal{O}} \mathcal{O}L'b')^P$, and the projection of W onto any indecomposable direct summand V with vertex P is a primitive local idempotent in T^P. Thus, for some choice of a conjugate of i and a summand V as before, the image of i is a primitive local idempotent in $(S \otimes_{\mathcal{O}} \mathcal{O}L'b')^P$. Since V is obtained from the action of P on a p'-group, it follows from 7.10.2 that V defines a torsion element in $D_{\mathcal{O}}(P)$. The reformulation in terms of Morita equivalences follows from 9.11.9, which also shows that $\mathcal{O}Gb$ and $\mathcal{O}L'b'$ have the same fusion systems on P. $\qquad\square$

Proof of Corollary 10.6.3 It follows from 10.6.6 (v) that $L \cong P \rtimes E$. By 7.4.5 the interior P-algebra $S \otimes_{\mathcal{O}} \mathcal{O}_\alpha(P \rtimes E)$ is primitive. Thus 10.6.3 follows from 10.6.1. $\qquad\square$

Remark 10.6.7 In the proof of 10.6.1 we could have avoided the part showing that α is in $H^2(L; k^\times)$ by first proving the theorem over k, and then using the lifting property 6.4.9.

Theorem 10.6.8 (Puig, 1988) *Let G be a finite p-solvable group, M a simple kG-module, R a vertex of M, and W a kR-source of M. Then W is an endopermutation kR-module.*

Proof We proceed by induction over $|G/Z(G)|$. Let b be the block of kG to which M belongs, let P be a defect group of b containing R, and let $i \in (kGb)^P$ be a source idempotent. We may choose (R, W) such that W is a direct summand of iM as a kR-module. It follows from 10.6.6 that we may assume that b is a G-stable block of $O_{p'}(G)$ and that P is a Sylow p-subgroup of G. With the notation from 10.6.1 there is an interior P-algebra isomorphism

$$ikGi \cong j(S \otimes_k kL'b')j.$$

Through this isomorphism, the simple $ikGi$-module iM corresponds to $j(V \otimes_k N)$ for some simple $kL'b'$-module N. In particular, L' is a central extension of the finite p-solvable group $L = G/O_{p'}(G)$ by a cyclic p'-group Z such that $Q = O_p(L)$ satisfies $C_L(Q) = Z(Q)$. The preimage of Q in L' can be identified with $Z \times Q$. With this identification, Q is normal in L', hence acts trivially on N. Thus N is a simple kL'/Q-module. Since W is a direct summand of $V \otimes_k N$ as a kR-module, it suffices to show that N has an endopermutation source, because V is an indecomposable endopermutation kP-module with vertex P. By the construction of L', we have $Q = O_p(L')$, and Q is a Sylow

p-subgroup of $O_{p',p}(G)$. Moreover, $|(L'/Q)/Z(L'/Q)| \le |L/Q| \le |G/O_{p',p}(G)|$. It follows from 10.6.4 that $Z(G) \subseteq O_{p',p}(G)$ and that if $O_{p',p}(G) = Z(G)$, then $G = Z(G)$ is abelian. In that case every simple kG-module has trivial source. If $Z(G)$ is a proper subgroup of $O_{p',p}(G)$, then $|(L'/Q)/Z(L'/Q)|$ is strictly smaller than $|G/Z(G)|$, and so the result follows by induction. $\square$

Broué's Abelian Defect Conjecture holds for blocks of finite p-solvable groups with an abelian defect. More precisely, there is a splendid Rickard equivalence inducing a Morita equivalence with endopermutation source.

Theorem 10.6.9 ([81, 1.2, 1.3]) *Let G be a finite p-solvable group, and b a block of $\mathcal{O}G$ having an abelian defect group P. Denote by c the block of $\mathcal{O}N_G(P)$ corresponding to b via the Brauer correspondence.*

(i) *There is an indecomposable $\mathcal{O}Gb$-$\mathcal{O}N_G(P)c$-bimodule M with vertex ΔP and an endopermutation module as a source, when regarded as an $\mathcal{O}(G \times N_G(P))$-module, such that M and M^* induce a Morita equivalence between $\mathcal{O}Gb$ and $\mathcal{O}N_G(P)c$.*

(ii) *There is a splendid Rickard complex X of $\mathcal{O}Gb$-$\mathcal{O}N_G(P)c$-bimodules such that X has homology concentrated in degree zero isomorphic to M.*

(iii) *The isomorphism classes of M and of M^* determine is a p-permutation equivalence between the blocks $\mathcal{O}Gb$ and $\mathcal{O}N_G(P)c$. In particular, $\mathcal{O}Gb$ and $\mathcal{O}N_G(P)c$ are isotypic.*

Proof A source algebra of $\mathcal{O}Gb$ is of the form $S \otimes_{\mathcal{O}P} \mathcal{O}_\alpha(P \rtimes E)$ for some $\alpha \in H^2(E; k^\times)$, where E is the inertial quotient of b associated with that source algebra and where $S = \mathrm{End}_\mathcal{O}(V)$ for some indecomposable endopermutation $\mathcal{O}P$-module V. Similarly, a source algebra of $\mathcal{O}N_G(P)c$ is of the form $T \otimes_{\mathcal{O}P} \mathcal{O}_\beta(P \rtimes E)$ for some $\beta \in H^2(E; k^\times)$, where $T = \mathrm{End}_\mathcal{O}(W)$ for some indecomposable endopermutation $\mathcal{O}P$-module W. We need to show that $\beta = \alpha$. By 10.6.6 we may assume that b is a G-stable block of $O_{p'}(G)$, in which case 10.6.6 implies that c is an $N_G(P)$-stable block of $\mathcal{O}O_{p'}(N_G(P))$, where we use that $O_{p'}(N_G(P)) = O_{p'}(C_G(P)) \subseteq O_{p'}(G)$. Note that $G = O_{p',p,p'}(G)$; similarly for $N_G(P)$, and $G = N_G(P)O_{p'}(G)$. Thus $S(P) \cong k \otimes_{\mathcal{O}} T$. It follows from Dade's Fusion Splitting Theorem 7.9.2 that the classes determined by the action of $N_G(P)$ on $\mathcal{O}O_{p'}(G)b$ and on $\mathcal{O}O_{p'}(N_G(P))c$ coincide (as both have values in $k^\times$). Thus these two source algebras are Morita equivalent, as they are both Morita equivalent to $\mathcal{O}_\alpha(P \rtimes E)$. By 9.11.9 this Morita equivalence translates at the block algebra level to a Morita equivalence given by a bimodule M with diagonal vertex ΔP and endopermutation source U, proving (i). The last statement of 9.11.9 implies that U is fusion-stable (that is, E-stable).

By 9.11.10, we may replace U by any other E-stable endopermutation module reducing to $k \otimes_{\mathcal{O}} U$. Therefore, by 7.11.9, we may choose U such that U has an E-stable endosplit p-permutation resolution. It follows from 9.11.5 that there is a splendid Rickard complex with the properties stated in (ii). It follows from (ii) that the isomorphism class $[M]$ of M is equal to $\sum_{m \in \mathbb{Z}} (-1)^m [X_m]$ in the Grothendieck group of $\mathcal{O}Gb$-$\mathcal{O}N_G(P)c$-bimodules. This is a p-permutation equivalence by 9.7.7, hence induces an isotypy by 9.5.4. $\qquad\square$

The following theorem of Navarro and Robinson proves a conjecture of Carlson, Mazza, and Thévenaz in [46].

Theorem 10.6.10 ([156, Theorem]) *Let G be a finite p-solvable group that contains an elementary abelian subgroup of order p^2. Every simple endotrivial kG-module is one-dimensional.*

10.7 Alperin's weight conjecture and alternating sums

Knörr and Robinson [106] reformulated Alperin's weight conjecture in terms of alternating sums indexed by chains of p-subgroups. This reformulation led to a wide range of more precise conjectures, some of which are mentioned in Section 6.12. It also raised the question as to whether there is a complex in the background of which the alternating sum on the right side is the Euler characteristic, leading to speculation whether there is a structural explanation for this conjecture. Symonds showed in [211] that Bredon cohomology yields a complex for the group theoretic version of Alperin's weight conjecture. The adjustments required for a similar cohomological interpretation of the block theoretic version above are described in [133]. Contractible chain complexes modelling the above formulation have been investigated by Boltje in [19]. Further reformulations involving simplicial and cohomological methods include the work by Thévenaz and Webb [216] and Webb [227].

We fix an algebraically closed field k of prime characteristic p. We briefly review some notation from Section 8.13. For G a finite group and b an idempotent in $Z(kG)$, we denote by $\ell(b)$ the number of isomorphism classes of simple kGb-modules, and by $\mathbf{k}(b)$ the number of irreducible characters of G associated with b; that is, $\mathbf{k}(b) = \dim_k(Z(kGb))$. We denote by $w(b)$ the number of isomorphism classes of simple projective kGb-modules; equivalently, $w(b)$ is the number of defect zero blocks occurring in a primitive decomposition of b in $Z(kG)$. This notation is slightly abusive in that it requires keeping track of the finite group G under consideration (because b could be contained in kH for some subgroup H of G). We adopt the convention that for $b = 0$ we set

$\ell(b) = 0$ and that a sum indexed by the empty set is zero. If (Q, e) is a Brauer pair on kG, we denote by $\bar{e}$ the image of e in the algebra $kN_G(Q, e)/Q$; this is a sum of blocks of $kN_G(Q, e)/Q$, and $w(\bar{e})$ is the number of defect zero blocks of $kN_G(Q, e)/Q$ occurring in a decomposition of $\bar{e}$ as a sum of blocks. Alperin's weight conjecture 6.10.2 states that for any finite group G and any block b of kG we have

$$\ell(b) = \sum_{(Q,e)} w(\bar{e}),$$

where (Q, e) runs over a set of representatives of the conjugacy classes of (G, b)-Brauer pairs. We have encountered some variations of the expression on the right side. By 6.10.3, the above equality is equivalent to

$$\ell(b) = \sum_{Q} w(\bar{b}_Q),$$

where $\bar{b}_Q$ is the image in $kN_G(Q)/Q$ of $b_Q = \mathrm{Br}_Q(b)$, regarded as a sum of blocks of $kN_G(Q)/Q$, and where Q runs over a set of representatives of the G-conjugacy classes of p-subgroups of G. If we fix a maximal (G, b)-Brauer pair and denote by $\mathcal{F}$ the associated fusion system on P, then by the material from Section 8.15, the above equality is equivalent to

$$\ell(b) = \sum_{Q} w(k_\alpha \mathrm{Aut}_{\mathcal{F}}(Q))$$

where Q runs over a set of representatives of the $\mathcal{F}$-isomorphism classes of $\mathcal{F}$-centric subgroups of P, and where α is the family of Külshammer–Puig classes (with the obvious notational abuse of using the same letter α for the class at each Q). The key feature of this version is, that the right side is formulated entirely in terms of the three local invariants of a block, namely defect groups, fusion systems, and Külshammer–Puig classes.

Let m be a nonnegative integer. If $\tau = Q_0 < Q_1 < \cdots < Q_m$ is a chain of p-subgroups of G, then we set $b_\tau = \mathrm{Br}_{Q_m}(b)$. Since Br_{Q_m} commutes with the action of $N_G(Q_m)$, hence with the action of $N_G(\tau)$, it follows that b_τ is either 0 or an idempotent in $Z(kN_G(\tau))$. Thus if b_τ is nonzero, then b_τ is a sum of blocks of $kN_G(\tau)$, and the notation $\ell(b_\tau)$ and $\mathbf{k}(\tau)$ refers to b_τ as an idempotent in $Z(kN_G(\tau))$. If $\sigma = (Q_0, e_0) < (Q_1, e_1) < \cdots < (Q_m, e_m)$ is a chain of Brauer pairs on kG, then, by Lemma 8.13.11, the block e_m of $kC_G(Q_m)$ remains a block of $kN_G(\sigma)$. We set $e_\sigma = e_{Q_m}$, and regard e_σ as a block of $kN_G(\sigma)$. Again, the notation $\ell(e_\sigma)$ and $\mathbf{k}(e_\sigma)$ refers to e_σ as a block of $kN_G(\sigma)$. We denote by $\mathcal{P}_G$ the simplicial complex of chains $Q_0 < Q_1 < \cdots < Q_m$ of nontrivial p-subgroups of G and by $\mathcal{P}_G/G$ a set of representatives of the G-conjugacy classes of such chains. We denote by $\mathcal{N}_G$ the subcomplex of those chains where all Q_i are normal in the maximal term Q_m, and by $\mathcal{E}_G$ the subcomplex of those chains where Q_m is elementary abelian. As before, we denote by $\mathcal{N}_G/G$ and $\mathcal{E}_G/G$ sets of representatives of G-conjugacy classes in $\mathcal{N}_G$ and $\mathcal{E}_G$. For b a block of kG, we denote by $\mathcal{P}_{G,b}$ the simplicial complex of chains of nontrivial (G, b)-Brauer pairs. We use the analogous notation $\mathcal{N}_{G,b}$, $\mathcal{E}_{G,b}$, as well as the notation $\mathcal{P}_{G,b}/G$, $\mathcal{N}_G/G$, $\mathcal{N}_{G,b}/G$ for sets of representatives of G-conjugacy classes.

Theorem 10.7.1 (cf. [106, Theorem 3.8, Theorem 4.6]) *With the notation above, the following are equivalent.*

(i) *For any finite group G and any block b of kG, Alperin's weight conjecture holds; that is, we have*

$$\ell(b) = \sum_{(Q,e)} w(\bar{e}),$$

where (Q, e) runs over a set of representatives of the conjugacy classes of (G, b)-Brauer pairs.

(ii) *For any finite group and any block b of kG with a nontrivial defect group, we have*

$$\ell(b) = \sum_{\tau \in \mathcal{N}_G/G} (-1)^{|\tau|}\ell(b_\tau).$$

(iii) *For any finite group and any block b of kG with a nontrivial defect group, we have*

$$\ell(b) = \sum_{\sigma \in \mathcal{N}_{G,b}/G} (-1)^{|\sigma|}\ell(e_\tau).$$

(iv) *For any finite group and any block b of kG with a nontrivial defect group, we have*

$$\mathbf{k}(b) = \sum_{\tau \in \mathcal{P}_G/G} (-1)^{|\tau|}\mathbf{k}(b_\tau).$$

(v) *For any finite group and any block b of kG with a nontrivial defect group, we have*

$$\mathbf{k}(b) = \sum_{\sigma \in \mathcal{P}_{G,b}/G} (-1)^{|\sigma|}\mathbf{k}(e_\sigma).$$

Moreover, the statements in (iv), (v) are equivalent to the analogous statements with $\mathcal{P}_G$, $\mathcal{P}_{G,b}$ replaced by $\mathcal{N}_G$, $\mathcal{N}_{G,b}$ or by $\mathcal{E}_G$, $\mathcal{E}_{G,b}$, respectively.

Remark 10.7.2 The three versions of Alperin's weight conjecture counting weights as stated at the beginning of this section are equivalent for each block.

The equivalences in the Knörr–Robinson reformulation above are, however, logically different in that the proofs do not show that if one of the equations holds for a particular block, then all equations hold for that block – they show that if one of the equations holds for all blocks, then so do the other.

Since Alperin's weight conjecture holds trivially for blocks of defect zero, we could reformulate statement (i) in Theorem 10.7.1 for blocks with a non-trivial defect group, and in that case, it would suffice to have the sum in (i) run over a set of representatives of nontrivial (G, b)-Brauer pairs. The equivalence between (i) and (ii) introduces alternating sums indexed by chains of p-subgroups, but (ii) still counts simple modules rather than irreducible characters. The step from there to counting characters requires new ideas. The key result for this step is the virtual projectivity result 10.7.8 below. The equivalences between (ii) and (iii) as well as between (iv) and (v) are both straightforward consequences of Proposition 8.13.12, which describes the passage between chains of p-subgroups and chains of Brauer pairs. The last statement, allowing for the replacement of $\mathcal{P}_G, \mathcal{P}_{G,b}$ by $\mathcal{N}_G, \mathcal{N}_{G,b}$ or by $\mathcal{E}_G, \mathcal{E}_{G,b}$ is an immediate consequence of Proposition 8.13.13. Therefore, in order to prove Theorem 10.7.1, we will focus on the equivalence of (i) and (iii), and the equivalence of (iii) and (v) with $\mathcal{N}_G$ instead of $\mathcal{P}_G$.

We start with an easy observation, which organises chains of p-subgroups of positive lengths in $\mathcal{N}_G$ in terms of their initial subgroup and chains of shorter lengths in a quotient of the normaliser of the initial subgroup.

Lemma 10.7.3 *Let G be a finite group and m a positive integer. Let $\sigma = Q_0 < Q_1 < \cdots < Q_m$ and $\tau = R_0 < R_1 < \cdots < R_m$ be chains of length m of nontrivial p-subgroups of G in $\mathcal{N}_G$ such that $Q_0 = R_0 = Q$. Then $\sigma /Q = Q_1/Q < Q_2/Q < \cdots < Q_m/Q$ and $\tau /Q = R_1/Q < R_2/Q < \cdots < R_m/Q$ are chains of nontrivial p-subgroups of $N_G(Q)/Q$ of length $m - 1$. Moreover, σ and τ are G-conjugate if and only if σ /Q and τ /Q are $N_G(Q)/Q$-conjugate.*

Proof This is a trivial exercise. □

Lemma 10.7.4 *Let G be a finite group, m a positive integer and $\sigma = Q_0 < Q_1 < \cdots < Q_m$ a chain of nontrivial p-subgroups of G in $\mathcal{N}_G$. Set $Q = Q_0$. Let b be an idempotent in $Z(kG)$ such that $\mathrm{Br}_Q(b) \neq 0$ and denote by c the image of $\mathrm{Br}_Q(b)$ in $kN_G(Q)/Q$. Set $\bar{\sigma} = Q_1/Q < Q_2/Q < \cdots < Q_m/Q$. The image of $b_\sigma = \mathrm{Br}_{Q_m}(b)$ in $kN_G(Q)/Q$ is equal to $c_{\bar{\sigma}} = \mathrm{Br}_{Q_m/Q}(c)$. In particular, we have*

$$\ell(b_\sigma) = \ell(c_{\bar{\sigma}}).$$

Proof Note that since Q is normal in Q_m by the assumptions on σ, we have $b_\sigma = \mathrm{Br}_{Q_m}(b) = \mathrm{Br}_{Q_m}(\mathrm{Br}_Q(b))$. Thus the statement that the image of b_σ in $kN_G(Q)/Q$

is equal to $c_{\bar{\sigma}}$ is the special case of 6.5.18 applied to $N_G(Q)$, $\mathrm{Br}_Q(b)$, and Q_m instead of G, b, R, respectively. Therefore, the algebra $kN_G(Q)/Qc_{\bar{\sigma}}$ is the image of $kN_G(Q)b_\sigma$ under the canonical algebra homomorphism $kN_G(Q) \to kN_G(Q)/Q$. Since the kernel of this homomorphism is contained in the radical, the last statement follows. $\square$

The equivalence of (i) and (ii) in Theorem 10.7.1 will be a consequence of the following result.

Lemma 10.7.5 ([106, 3.6, 3.7]) *Let G be a finite group and b a block of kG. For Q a nontrivial p-subgroup of G, set $N_Q = N_G(Q)/Q$ and denote by c_Q the image of $\mathrm{Br}_Q(b)$ in kN_Q. We have*

$$\sum_{\sigma \in \mathcal{P}_G/G} (-1)^{|\sigma|}\ell(b_\sigma) = \sum_{Q} \ell(c_Q) - \sum_{\tau \in \mathcal{P}_G/G} (-1)^{|\tau|}\ell((c_Q)_\tau),$$

where Q runs over a set of representatives of the G-conjugacy classes of nontrivial p-subgroups of G. If b has defect zero, then this sum is zero.

Proof The chains of length zero in the sum on the left side yield the terms $\ell(b_Q)$, where $b_Q = \mathrm{Br}_Q(b)$, regarded as a (possibly empty) sum of blocks of $N_G(Q)$, and clearly $\ell(b_Q) = \ell(c_Q)$, since the kernel of the canonical algebra homomorphism $kN_G(Q) \to kN_Q$ is in the radical. The chains of positive lengths in the sum of the left side are partitioned according to their initial term as in 10.7.3. Since in that process, chains are shortened by one term, we have to adjust the signs, accounting for the minus signs in front of the sums $\sum_{\tau \in \mathcal{P}_G/G}(-1)^{|\tau|}\ell((c_Q)_\tau)$ on the right side. Finally, the fact that $\ell((c_Q)_\tau)$ shows up on the right side follows from 10.7.4. If b has defect zero, then $\mathrm{Br}_Q(b) = 0$ for any nontrivial p-subgroup Q of G, whence the last statement. $\square$

We reformulate the equivalence of (i) and (ii) in Theorem 10.7.1 in a slightly more precise way.

Theorem 10.7.6 *The following statements are equivalent.*

  (i) *For any finite group G and any block b of kG, Alperin's weight conjecture holds.*
 (ii) *For any finite group G and any block b of kG with a nontrivial defect group we have*

$$\ell(b) = \sum_{\tau \in \mathcal{N}_G/G} (-1)^{|\tau|}\ell(b_\tau).$$

(iii) For any finite group G and any idempotent b of Z(kG), we have

$$\ell(b) - w(b) = \sum_{\tau \in \mathcal{N}_G/G} (-1)^{|\tau|} \ell(b_\tau).$$

Proof If b is a block of defect zero, then $\ell(b) = w(b) = 1$ and by 10.7.5, the sum $\sum_{\tau \in \mathcal{N}_G/G}(-1)^{|\tau|}\ell(b_\tau)$ is zero. If b is a block of positive defect, then $w(b) = 0$. Thus if b is an arbitrary idempotent in $Z(kG)$, then $w(b)$ counts the number of defect zero blocks in a primitive decomposition of b in $Z(kG)$. The equivalence of (ii) and (iii) is an immediate consequence. Suppose that (iii) holds. Let G be a finite group and b a block of kG of positive defect. Since (iii) holds, if follows that the left side in 10.7.5 is $\ell(b)$. The right side is, again by (iii), equal to the sum of the $w(c_Q)$, and that is the statement of Alperin's weight conjecture for the block b. Thus (iii) implies (i). Suppose that (i) holds. We show (iii) by induction over $|G|$. Note that for Q a nontrivial p-subgroup of G, the group $N_Q = N_G(Q)/Q$ in the statement of 10.7.5 has order strictly smaller than $|G|$. Thus, by (iii) applied to the sum of blocks c_Q of kN_Q, the right side in 10.7.5 is equal to the sum of the $w(c_Q)$. But since Alperin's weight conjecture is assumed to hold, this implies that the left side in 10.7.5 is equal to $\ell(b)$. This shows that (iii) holds also for b. $\qquad\square$

Let G be a finite group. For U a finitely generated kG-module, we denote by $[U]$ the isomorphism class of U, regarded as an element of the Grothendieck group $\mathcal{A}(kG)$ of finitely generated kG-modules with respect to split exact sequences. As a consequence of the Krull–Schmidt Theorem, $\mathcal{A}(kG)$ is a free abelian group having as a basis the set of isomorphism classes of finitely generated indecomposable kG-modules. An element X in $\mathcal{A}(kG)$ can be written in the form $X = [U] - [U']$ for some finitely generated kG-modules U, U'. After cancelling isomorphic summands in U, U', we may assume that U, U' have no nonzero isomorphic direct summands, and then U, U' are uniquely determined by X. The elements in $\mathcal{A}(kG)$ are called *virtual modules*. A virtual module X is called projective if it can be written in the form $X = [U] - [U']$ with U, U' finitely generated projective kG-modules.

Definition 10.7.7 Let G be a finite group and b a block of kG. For any chain $\sigma \in \mathcal{P}_G$, consider $kN_G(\sigma)b_\sigma$ as a (possibly zero) module for $kN_G(\sigma)$ with respect to the conjugation action of $N_G(\sigma)$. We define the virtual kG-module $L(b)$ by setting

$$L(b) = \sum_{\sigma \in \mathcal{P}_G/G} (-1)^{|\sigma|} \left[\mathrm{Ind}_{N_G(\sigma)}^G (kN_G(\sigma)b_\sigma) \right].$$

This is slightly different from the virtual module $L(B)$ defined in [106, §3], since we consider only chains of nontrivial p-subgroups in the definition of $L(b)$. In particular, $L(b)$ is zero if b has defect zero, and thus the results below take slightly different forms from their equivalent versions in [106]. By 8.13.13, replacing $\mathcal{P}_G$ by $\mathcal{N}_G$ or $\mathcal{E}_G$ in Definition 10.7.7 does not change $L(b)$.

Theorem 10.7.8 ([106, Theorem 4.2]) *Let G be a finite group and b a block of kG. Consider kGb as a kG-module with respect to the conjugation action of G. The virtual kG-module $[kGb] - L(b)$ is projective.*

We need to show that for Q a nontrivial p-subgroup of G, the indecomposable summands in the terms of $[kGb] - L(b)$ cancel each other. For that purpose, we use the following terminology. We say for short that *two finite-dimensional kG-modules U, V have isomorphic summands with vertex Q*, if in direct sum decompositions of U and V, the direct sum of all summands with vertex Q of U is isomorphic to the direct sum of all summands with vertex Q of V, or equivalently, if U and V become isomorphic upon deleting all summands that do not have Q as a vertex. By 5.5.21 and 5.11.4, two finite-dimensional p-permutation kG-modules U, V have isomorphic summands with vertex Q if and only if the $kN_G(Q)$-modules $U(Q)$ and $V(Q)$ have isomorphic summands with vertex Q. Applied to kGb, with the conjugation action by G, this yields immediately the following result.

Lemma 10.7.9 *Let G be a finite group and b an idempotent in $Z(kG)$. Let Q be a p-subgroup of G. Consider kGb as a kG-module with respect to the conjugation action of G. The three $kN_G(Q)$-modules $\operatorname{Res}_{N_G(Q)}^G(kGb)$, $kC_G(Q)\operatorname{Br}_Q(b)$, and $kN_G(Q)\operatorname{Br}_Q(b)$ have isomorphic summands with vertex Q.*

Proof Since $kC_G(Q)b_Q = (kGb)(Q) = (kN_G(Q)b_Q)(Q)$, this follows from 5.5.21 and 5.11.4. □

Proof of Theorem 10.7.8 Let Q be a nontrivial p-subgroup of G. We need to show that the indecomposable summands with vertex Q involved in the terms of $[kGb] - L(b)$ cancel. Set $N = N_G(Q)$. By 5.11.4 or by 5.5.20 it suffices to show that all summands with vertex Q in the restriction to N of $[kGb] - L(b)$ cancel. Consider a chain $\sigma = Q_0 < Q_1 < \cdots < Q_m$ in $\mathcal{N}_G$. Set $G_\sigma = N_G(\sigma)$ and $B_\sigma = kG_\sigma b_\sigma$. Mackey's formula yields

$$\operatorname{Res}_N^G(\operatorname{Ind}_{G_\sigma}^G(B_\sigma)) \cong \oplus_{x \in [N \backslash G / G_\sigma]} \operatorname{Ind}_{N \cap {}^x G_\sigma}^N(B_{{}^x\sigma}),$$

where we use that ${}^x B_\sigma = B_{{}^x\sigma}$. A summand on the right side can only contribute a summand with vertex Q if Q is contained in $G_{{}^x\sigma} = {}^x G_\sigma$. Thus we need to consider only G-conjugacy class representatives of chains σ in $\mathcal{P}_G$ with the property $Q \leq G_\sigma$. This inclusion is equivalent to requiring that σ is Q-stable. Denote by $\mathcal{P}_G^Q$ the set of Q-stable chains in $\mathcal{P}_G$. The group N acts by conjugation on this set, since Q is normal in N.

We verify that if σ runs over $\mathcal{P}_G/G$ and for each σ, x runs over the elements in $[N\backslash G/G_\sigma]$ such that $Q \leq G_{x\sigma}$, then ${}^x\sigma$ runs over $\mathcal{P}_G^Q/N$. If $\tau \in \mathcal{N}_G^Q/N$, then there is a unique $\sigma \in \mathcal{N}_G/G$ and some $x \in G$ such that $\tau = {}^x\sigma$. We need to show that if x, $y \in G$ such that ${}^x\sigma$, ${}^y\sigma \in \mathcal{P}_G^Q$, then ${}^x\sigma$ and ${}^y\sigma$ are N-conjugate if and only if x, y belong to the same double coset in $[N\backslash G/G_\sigma]$. Now ${}^x\sigma$ and ${}^y\sigma$ are N-conjugate if and only if ${}^y\sigma = {}^{nx}\sigma$ for some $n \in N$, hence if and only if $y^{-1}nx \in G_\sigma$ for some $n \in N$, or equivalently, if and only if $yG_\sigma = nxG_\sigma$ for some $n \in N$, which is clearly equivalent to $y \in NxG_\sigma$.

By 10.7.9, the summands of $\mathrm{Res}_N^G(kGb)$ with vertex Q are exactly the summands of kNb_Q with vertex Q. It suffices therefore to check that

$$[kNb_Q] - \sum_{\sigma \in \mathcal{P}_G^Q/N} (-1)^{|\sigma|} \mathrm{Ind}_{N \cap G_\sigma}^N (B_\sigma)$$

involves no summand with vertex Q. On the right side, the chain Q of length zero yields the term $[kNb_Q]$, so we need to show that the contributions of summands with defect Q from all chains other than the chain Q of length zero cancel. We pair up the chains in $\mathcal{P}_G^Q$ as follows. Let $\sigma = Q_0 < Q_1 < \cdots < Q_m$ be a Q-stable chain different from the chain Q of length zero; that is, we have $m \geq 1$, or $m = 0$ and Q_0 is not G-conjugate to Q but normalised by Q_0. If $Q < Q_0$, we form σ' by adding Q to σ as the minimal term of σ'. If $Q = Q_0$ and hence $m \geq 1$, form σ' by deleting the minimal group Q_0 from σ. If Q is not contained in Q_0, let j be maximal such that Q is not contained in Q_j, so that $Q_j < QQ_j$. If $QQ_j = Q_{j+1}$, form σ' by deleting the term Q_{j+1} from σ. If $QQ_j < Q_{j+1}$ or if $j = m$, form σ' by adding the group QQ_j to σ; so in that case, QQ_j is either between Q_j and Q_{j+1}, or $QQ_j = QQ_m$ is the new maximal group of σ', added to the top end of σ. An easy verification shows that $\sigma = (\sigma')'$ for all chains σ in $\mathcal{P}_G^Q$ other than Q, and that the lengths of σ and σ' differ by 1. Moreover, for $x \in N$, we clearly have $(\sigma^x)' = (\sigma')^x$; in particular, we have $N \cap G_\sigma = N_N(\sigma) = N_N(\sigma') = N \cap G_{\sigma'}$, and the pairing thus constructed induces a pairing on $\mathcal{P}_G^Q/N$ with the chain Q removed. It suffices therefore to show that for σ in $\mathcal{P}_G^Q$ different from Q, the summands with vertex Q of the modules $\mathrm{Ind}_{N \cap G_\sigma}^N (B_\sigma)$ and of $\mathrm{Ind}_{N \cap G_{\sigma'}}^N (B_{\sigma'})$ coincide. We may assume that $|\sigma| = |\sigma'| + 1$. It suffices to observe that B_σ and $B_{\sigma'}$ have isomorphic summands with vertex Q as modules over $N \cap G_\sigma = N \cap G_{\sigma'}$. Note that this group is the normaliser of Q in the groups G_σ and $G_{\sigma'}$, so in order to compare summands with vertex Q, we may (using 10.7.9) apply the Brauer construction with respect to Q to B_σ and $B_{\sigma'}$. This yields isomorphic modules. Indeed, since σ is the longer of the two chains, its maximal term contains Q, and hence the unit element b_σ of B_σ is contained in $kC_G(Q)$. The unit element $b_{\sigma'}$ satisfies therefore $\mathrm{Br}_Q(b_{\sigma'}) = b_\sigma$. The result follows. $\qquad\square$

Theorem 10.7.8 is the key ingredient for the passage from counting simple modules to counting characters, in conjunction with the earlier character theoretic result 6.5.16.

Theorem 10.7.10 ([106, 4.5]) *Let G be a finite group and b a block of kG. We have*

$$\mathbf{k}(b) - \ell(b) = \sum_{\sigma \in \mathcal{P}_G/G} (-1)^{|\sigma|}(\mathbf{k}(b_\sigma) - \ell(b_\sigma)).$$

Proof Let γ be the generalised character of the virtual module $[kGb] - L(b)$. Since this virtual module is projective by 10.7.8, it follows that γ vanishes on p'-elements. In other words, γ is equal to the generalised Brauer character γ' of $[kGb] - L(b)$ extended by zero on p-singular elements. Thus we have

$$\langle \gamma, 1 \rangle = \langle \gamma', 1 \rangle'.$$

Applying 6.5.16 to the terms in $[kGb] - L(b)$ and using Frobenius' reciprocity for the terms in $L(b)$ yields

$$\langle \gamma, 1 \rangle = \mathbf{k}(b) - \sum_{\sigma \in \mathcal{P}_G/G} (-1)^{|\sigma|}\mathbf{k}(b_\sigma),$$

$$\langle \gamma', 1 \rangle' = \ell(b) - \sum_{\sigma \in \mathcal{P}_G/G} (-1)^{|\sigma|}\ell(b_\sigma).$$

The result follows. $\qquad\square$

Proof of Theorem 10.7.1 As mentioned before, the equivalence between (i) and (ii) follows from 10.7.6, and the equivalences between (ii) and (iii) as well as between (iv) and (v) follow from 8.13.12. The equivalence between (ii) and (iv) follows from 10.7.10. The last statement on the passage between the different set $\mathcal{P}_G$, $\mathcal{N}_G$, $\mathcal{E}_G$ and their analogues for Brauer pairs follows from 8.13.13. $\qquad\square$

The fact that $L(b)$ is virtually projective implies that its Tate cohomology vanishes (cf. 4.13.7 and 4.13.10 for notation and basic facts). This has the following consequence.

Theorem 10.7.11 ([116, Theorem 1]) *Let G be a finite group and b a block of kG. Set $B = kGb$. For any chain $\sigma = Q_0 < Q_1 < \cdots < Q_m$ in $\mathcal{P}_G$ set*

$B_\sigma = kN_G(\sigma)\mathrm{Br}_{Q_m}(b)$. *For any integer n we have*

$$\dim_k(\widehat{HH}^n(B)) = \sum_{\sigma \in \mathcal{P}_G/G} (-1)^{|\sigma|}\dim_k(\widehat{HH}^n(B_\sigma)).$$

Proof Since B, regarded as a virtual kG-module with the conjugation action by G, is isomorphic to $L(b)$ plus a virtual projective module, it follows that for any integer n we have

$$\hat{H}^n(G; B) \cong \hat{H}^n(G; L(b)) = \sum_{\sigma \in \mathcal{P}_G/G} (-1)^{|\sigma|}\hat{H}^n(G; \mathrm{Ind}_{N_G(\sigma)}^G(B_\sigma)).$$

Applying the appropriate version of 4.13.11, we get that this expression is equal to the virtual k-vector space

$$\sum_{\sigma \in \mathcal{P}_G/G} (-1)^{|\sigma|}\hat{H}^n(N_G(\sigma); B_\sigma).$$

The result follows from 4.13.12. □

Since $\widehat{HH}^*(B)$ is invariant under Morita equivalences, it follows from Proposition 8.13.12 that this theorem can be rewritten in terms of chains of Brauer pairs as follows.

Theorem 10.7.12 ([116, Theorem 2]) *Let G be a finite group and b a block of kG. Set $B = kGb$. For any chain of nontrivial B-Brauer pairs $\sigma = (Q_0, e_0) < (Q_1, e_1) < \cdots < (Q_m, e_m)$ set $C_\sigma = kN_G(\sigma)e_m$. For any integer n we have*

$$\dim_k(\widehat{HH}^n(B)) = \sum_{\sigma \in \mathcal{P}_{G,b}/G} (-1)^{|\sigma|}\dim_k(\widehat{HH}^n(C_\sigma)).$$

Since $HH^0(B) \cong Z(B)$ has dimension $\mathbf{k}(b)$, since $\widehat{HH}^0(B) \cong \underline{Z}(B)$ and since by 2.16.12, we have dimension $\dim_k(Z^{\mathrm{pr}}(B)) = \dim_k(B_1^G)$, it follows from Theorem 10.7.1 that Alperin's weight conjecture is equivalent to the statement $\dim_k(HH^0(B)) = \sum_{\sigma \in \mathcal{P}_G/G}(-1)^{|\sigma|}\dim_k(HH^0(B_\sigma))$ for all B. By Theorem 10.7.11, this equality holds with HH replaced by $\widehat{HH}$. Therefore, since by the above we have $\dim_k(HH^0(B)) - \dim_k(\widehat{HH}^0(B)) = \dim_k(B_1^G)$, it follows that Alperin's weight conjecture for all blocks is equivalent to the equality

$$\dim_k(B_1^G) = \sum_{\sigma \in \mathcal{P}_G/G} (-1)^{|\sigma|}\dim_k((B_\sigma)_1^{N_G(\sigma)})$$

for all blocks B. See [106, Theorem 4.6] and the remarks at the end of [106, Section 4].

11

Blocks with Cyclic Defect Groups

The structure theory of blocks with cyclic defect groups is well understood and illustrates a great variety of methods in block theory. Brauer described in [25] the ordinary and modular characters of blocks with a defect group of prime order. Dade extended these results in [53] to blocks with arbitrary cyclic defect groups, using a technique due to Thompson [218]. Janusz [91] and Kupisch [117] gave descriptions of the modular indecomposable representations. J. A. Green [77] showed that a block with cyclic defect groups has a periodic long exact sequence of projective indecomposable modules, which is a structural link between modular and characteristic zero representations. Gabriel and Riedtmann [73] showed that the structure of the block algebra does not in fact depend so much on the presence of a finite group, but rather on a purely algebra theoretic property: the block algebra over a field of prime characteristic of a block with cyclic defect groups is stably equivalent to a symmetric serial algebra. Rickard proved in [185], that this stable equivalence lifts to a derived equivalence. This derived equivalence has been lifted in [125] over a valuation ring, and Rouquier has given in [196] an explicit description of a splendid Rickard complex. In particular, Broué's abelian defect group conjecture holds for blocks with cyclic defect groups. The source algebras of blocks with cyclic defect groups are classified in [127], implying in particular, that Puig's finiteness conjecture holds in this case.

Throughout Chapter 11, the word 'module' means 'finitely generated module'. Let $\mathcal{O}$ be a complete discrete valuation ring with residue field $k = \mathcal{O}/J(\mathcal{O})$ of prime characteristic p and a quotient field K of characteristic zero.

374

11.1 The structure of blocks with cyclic defect groups

We describe in this section the main results on the structure of blocks with cyclic defect groups and their source algebras. Proofs are given in subsequent sections. We assume in this section that k is a splitting field for all finite groups and algebras involved.

Blocks with a cyclic defect group satisfy Alperin's weight conjecture, Broué's abelian defect conjecture, Brauer's height zero conjecture, and the finiteness conjectures of Donovan and Puig. The proofs of these conjectures follow from more precise structural results. We start with an elementary observation on the local structure.

Theorem 11.1.1 *Let G be a finite group and b a block of $\mathcal{O}G$ with a nontrivial cyclic defect group P and inertial quotient E. The group E is cyclic of order dividing $p - 1$, and E acts freely on $P \setminus \{1\}$.*

This follows from elementary group theoretic considerations: the automorphism group of a nontrivial cyclic p-group P is a direct product of a cyclic group of order $p - 1$ acting freely on $P \setminus \{1\}$ and an abelian p-group. We will restate this in a slightly more precise way in 11.2.1. It follows that E is determined by its order, and hence E does not depend on the choice of a block e of $kC_G(P)$ such that (P, e) is a maximal (G, b)-Brauer pair. Since E acts freely on $P \setminus \{1\}$ it follows from 10.5.1 that there is a stable equivalence of Morita type between $\mathcal{O}Gb$ and $\mathcal{O}(P \rtimes E)$, given by a bimodule with endopermutation source:

Theorem 11.1.2 *Let G be a finite group and b a block of $\mathcal{O}G$ with a nontrivial cyclic defect group P and inertial quotient E. There is a stable equivalence of Morita type between $\mathcal{O}Gb$ and $\mathcal{O}(P \rtimes E)$ given by an $\mathcal{O}Gb$-$\mathcal{O}(P \rtimes E)$-bimodule M and its dual M^* such that, as an $\mathcal{O}(G \times (P \rtimes E))$-module, M has vertex ΔP and an indecomposable endopermutation $\mathcal{O}\Delta P$-module W as a source with the properties that the subgroup of order p of P acts trivially on W and that P acts with determinant 1 on W.*

This will be proved in 11.2.5. The endopermutation $\mathcal{O}P$-module W is uniquely characterised, up to isomorphism, by the properties in 11.1.2, but the bimodule M itself is not since one can compose the stable equivalence induced by M with any even power of the Heller operator. This does not affect W because W has period 2. This stable equivalence is one of the corner stones for the structure theory of blocks with cyclic defect groups. Thanks to the precise description of the indecomposable $k(P \rtimes E)$-modules in §11.3, it is possible to show that this stable equivalence preserves the numbers of isomorphism classes of simple modules and irreducible characters:

Theorem 11.1.3 *Let G be a finite group and b a block of $\mathcal{O}G$ with a nontrivial cyclic defect group P and inertial quotient E. We have $|\mathrm{IBr}_k(G, b)| = |E|$. If K is large enough for G, then $|\mathrm{Irr}_K(G, b)| = |E| + \frac{|P|-1}{|E|}$. In particular, Alperin's weight conjecture holds for blocks with cyclic defect groups.*

This will be proved in 11.10.2. The equality $|\mathrm{IBr}_k(G, b)| = |E|$ has the following immediate consequence:

Corollary 11.1.4 *A block with a cyclic defect group is nilpotent if and only if it has a unique isomorphism class of simple modules.*

The next results describe parts of the structure of projective indecomposable modules of blocks with a cyclic defect group. The short exact sequences in statement (i) of the next result are due to Green [77], statement (ii) is due to Janusz [91], and (iii) is from [123, 15.5].

Theorem 11.1.5 *Let A be a block algebra or source algebra of a block b with a nontrivial cyclic defect group P and inertial quotient E. Let I be a set of representatives of the conjugacy classes of primitive idempotents in A. There are unique permutations ρ, σ of I and, for any $i \in I$, unique $\mathcal{O}$-pure submodules U_i, V_i of Ai, such that the following hold for any $i \in I$.*

(i) There are exact sequences of A-modules

$$0 \longrightarrow U_{\sigma(i)} \longrightarrow A\sigma(i) \longrightarrow V_i \longrightarrow 0$$

$$0 \longrightarrow V_{\rho(i)} \longrightarrow A\rho(i) \longrightarrow U_i \longrightarrow 0.$$

(ii) The images $\bar{U}_i$ and $\bar{V}_i$ in $\bar{A}\bar{i} = k \otimes_{\mathcal{O}} Ai$ of U_i and V_i, respectively, are uniserial, and we have

$$\bar{U}_i \cap \bar{V}_i = \mathrm{soc}(\bar{A}\bar{i}),$$

$$\bar{U}_i + \bar{V}_i = \mathrm{rad}(\bar{A}\bar{i}).$$

(iii) The A-modules U_i, V_i are indecomposable with P as a vertex, the endopermutation $\mathcal{O}P$-module W from 11.1.2 is a source of U_i, and $\Omega_P(W)$ is a source of V_i, for all $i \in I$.

Moreover, the permutations $\rho \circ \sigma$ and $\sigma \circ \rho$ are transitive cycles on I.

This will be proved in 11.11.3. All indecomposable endopermutation $\mathcal{O}P$-modules with determinant 1 on which Z acts trivially arise as a source W as in 11.1.5; this follows from examples due to Dade in [53]. There is some redundancy in the last statement of 11.1.5: since $\sigma \circ \rho$ and $\rho \circ \sigma$ are conjugate it follows that if one is a transitive cycle, so is the other. The transitivity of $\sigma \circ \rho$ implies that if there exists $i \in I$ such that $i = \rho(i) = \sigma(i)$, then $I = \{i\}$ consists of a single element. This holds if and only if the block b is a nilpotent block. Splicing the short exact sequences in 11.1.5 (i) together yields a periodic long exact sequence of projective indecomposable A-modules

$$\cdots \longrightarrow A\sigma(\rho(i)) \longrightarrow A\rho(i) \longrightarrow Ai \longrightarrow A\sigma^{-1}(i) \longrightarrow \cdots.$$

This sequence has period $2|I|$. Every indecomposable projective A-modules arises, up to isomorphism, exactly twice in any subsequence of $2|I|$ consecutive terms because $\sigma \circ \rho$ is a transitive cycle on I. This is Green's 'walk around the Brauer tree' in [77].

Corollary 11.1.6 *With the notation from 11.1.5, for any $i \in I$ we have $\Omega_A(U_i) \cong V_{\rho(i)}$ and $\Omega_A(V_i) \cong U_{\sigma(i)}$; in particular, we have $\Omega_A^2(U_i) \cong U_{\sigma(\rho(i))}$ and $\Omega_A^2(V_i) \cong V_{\rho(\sigma(i))}$.*

Proof This is an immediate consequence of the short exact sequences in 11.1.5. $\qquad\qquad\qquad\qquad\qquad\qquad\qquad\qquad\qquad\qquad\qquad\qquad\square$

The next theorem shows that unless $|P| = p$ and $|E| = p - 1$, there is a unique distinguished orbit of either ρ or σ, called the *exceptional orbit*. This orbit can be characterised in terms of the $\mathcal{O}$-ranks of the endomorphism algebras of the modules U_i, V_i from 11.11.3.

Theorem 11.1.7 *With the notation from 11.1.5, the following hold.*

(i) *The rational number $m = \frac{|P|-1}{|I|}$ is an integer.*

(ii) *If $m > 1$, then there is exactly one orbit v of either ρ or σ on I, called the exceptional orbit, which is characterised by the following two properties.*

(iii) *If the ρ-orbit of an element $i \in I$ is not exceptional, then the $\mathcal{O}$-rank of $\mathrm{End}_A(U_i)$ is 1, or equivalently, the character of U_i is irreducible; similarly, if the σ-orbit of an element $i \in I$ is nonexceptional, then the $\mathcal{O}$-rank of $\mathrm{End}_A(V_i)$ is 1, or equivalently, the character of V_i is absolutely irreducible,*

(iv) *If the exceptional orbit v is a ρ-orbit and $i \in v$, then the $\mathcal{O}$-rank of $\mathrm{End}_A(U_i)$ is equal to m; if v is a σ-orbit and $i \in v$, then the $\mathcal{O}$-rank of $\mathrm{End}_A(V_i)$ is equal to m.*

This will be proved in 11.11.3. This follows also from the description of the composition series of $k \otimes_{\mathcal{O}} U_i$, $k \otimes_{\mathcal{O}} V_i$ in 11.1.8. By 11.1.3 we have $|I| = |\mathrm{IBr}_k(A)| = |E|$. Since $|E|$ divides $p - 1$, it follows that the case $m = 1$ is equivalent to $|P| = p$ and $|E| = p - 1$. In that case there is no exceptional orbit. The Loewy layers of the projective indecomposable $k \otimes_{\mathcal{O}} A$-modules and the composition series of the uniserial submodules $k \otimes_{\mathcal{O}} U_i$ and $k \otimes_{\mathcal{O}} V_i$ of $k \otimes_{\mathcal{O}} Ai$ are as follows.

Theorem 11.1.8 *Let A be a block algebra or source algebra of a block with a nontrivial cyclic defect group P and inertial quotient E. Let I be a set of representatives of the conjugacy classes of primitive idempotents in A. Set* $m = \frac{|P|-1}{|I|}$. *If $m > 1$, denote by v the exceptional orbit of either ρ or σ. For $i \in I$ let U_i, V_i be the submodules of Ai as in 11.1.5, and set $S_i = Ai/J(A)i$. Set $\bar{A} = k \otimes_{\mathcal{O}} A$, $\bar{U}_i = k \otimes_{\mathcal{O}} U_i$, $\bar{V}_i = k \otimes_{\mathcal{O}} V_i$, and for $i \in I$ denote by $\bar{i}$ the canonical image of i in $\bar{A}$. Identify $\bar{U}_i$, $\bar{V}_i$ to their canonical images in $\bar{A}\bar{i}$. For any $i \in I$ the following hold.*

(i) *The $\bar{A}$-modules $\bar{U}_i$ and $\bar{V}_i$ are the unique uniserial submodules of $\bar{A}\bar{i}$ satisfying*

$$\bar{U}_i \cap \bar{V}_i = \mathrm{soc}(\bar{A}\bar{i}),$$

$$\bar{U}_i + \bar{V}_i = \mathrm{rad}(\bar{A}\bar{i}).$$

(ii) *We have $\mathrm{rad}(\bar{A}\bar{i})/\mathrm{soc}(\bar{A}\bar{i}) = \bar{U}_i/\mathrm{soc}(\bar{A}\bar{i}) \oplus \bar{V}_i/\mathrm{soc}(\bar{A}\bar{i})$. The two summands in this decomposition have no isomorphic composition factor, and any uniserial submodule of $\bar{A}\bar{i}$ is contained in $\bar{U}_i$ or in $\bar{V}_i$.*

(iii) *Every proper submodule of $\bar{A}\bar{i}$ is equal to $\bar{U}_i' + \bar{V}_i'$ for some submodule $\bar{U}_i'$ of $\bar{U}_i$ and some submodule $\bar{V}_i'$ of $\bar{V}_i$; in particular, $\bar{A}\bar{i}$ has only finitely many submodules.*

(iv) *The simple composition factors of the unique composition series of $\bar{U}_i$ are $S_{\rho(i)}, S_{\rho^2(i)}, \ldots, S_{\rho^a(i)} \cong S_i$, where a is the length $|i^\rho|$ of the ρ-orbit i^ρ of i if i^ρ is not exceptional, and $a = m|i^\rho|$ if i^ρ is the exceptional orbit.*

(v) *The simple composition factors of the unique composition series of $\bar{V}_i$ are $S_{\sigma(i)}, S_{\sigma^2(i)}, \ldots, S_{\sigma^b(i)} \cong S_i$, where b is the length $|i^\sigma|$ of the σ-orbit i^σ of i if i^σ is not exceptional, and $b = m|i^\sigma|$ if i^σ is the exceptional orbit.*

This will be proved in 11.6.3 and 11.6.4 in the more general context of symmetric algebras stably equivalent to a serial symmetric algebra, where we also show that this theorem determines the quiver with relations of a cyclic block.

There is another way to encode the Morita equivalence class of $\bar{A}$ combinatorially, namely in terms of the *Brauer tree*. A graph is called a *tree* if it is connected and has no simple cycles, or equivalently, if any two vertices are connected by a unique simple path. A finite connected graph with n vertices is a tree if and only if it has $n - 1$ edges. The two permutations ρ, σ of the set I satisfy the following combinatorial properties.

Theorem 11.1.9 *Let A be a block algebra or source algebra of a block with a nontrivial cyclic defect group P and inertial quotient E. Let I be a set of representatives of the conjugacy classes of primitive idempotents in A, and let ρ, σ be the two permutations of I as in 11.1.5. Set $m = \frac{|P|-1}{|I|}$. If $m > 1$, denote by v the exceptional orbit of either ρ or σ.*

(i) *The intersection of a ρ-orbit and of a σ-orbit on I is either empty or contains exactly one element in I. In particular, if $|I| \geq 2$, then $\rho \neq \sigma$.*

(ii) *Let $\Gamma(A)$ be the graph whose set of vertices is the disjoint union of the sets of ρ-orbits abd σ-orbits on I and whose set of edges is labelled by the set I, such that the edge labelled i links the ρ-orbit and the σ-orbit containing i. Then $\Gamma(A)$ is a tree.*

The first statement of 11.1.9 will be proved in 11.6.4, and the second statement will be proved in 11.7.2. The datum $(I, \{\rho, \sigma\})$, for $m = 1$, or $(I, \{\rho, \sigma\}, m, v)$ for $m > 1$, is called the *Brauer tree of A* or of the corresponding block. Any two permutations ρ, σ on I determine a graph as in statement (ii), but in general this graph need not be a tree. The specification of I together with $\{\rho, \sigma\}$ is more precise than just the abstract tree $\Gamma(A)$ in that it specifies cyclic orders at each vertex of the tree given by the action of ρ and σ, respectively, on the ρ-orbit or σ-orbit of an element $i \in I$. This is equivalent to prescribing a realisation of the tree $\Gamma(A)$ in the plane with the counterclockwise order on the set of edges emanating from a fixed vertex. In addition, if the integer $m = \frac{|P|-1}{|I|}$ is greater than 1, then the Brauer tree keeps track of a distinguished vertex v, called the *exceptional vertex* of the Brauer tree, corresponding to the unique exceptional orbit of either ρ or σ, and m is called the *exceptional multiplicity* of the Brauer tree. If $m = 1$, it is sometimes convenient to choose an exceptional vertex at random, whenever this does not introduce any confusion, in order to avoid having to make the distinction between Brauer trees with or without an exceptional vertex. Not every tree arises as the Brauer tree of a block with a cyclic defect group. See for instance [70] for results on possible Brauer trees and [89] for extensive calculations of Brauer trees for sporadic finite simple groups. Both the quiver with relations and the Brauer tree of A with the exceptional multiplicity (if any) determine the basic algebras, hence the Morita equivalence class, of $\bar{A}$ in a combinatorial way. Since $|I|$ is bounded by the order of P, there are at most finitely many Brauer trees arising from blocks with a fixed cyclic defect group P, and we obtain Donovan's conjecture in this case:

Corollary 11.1.10 *Donovan's conjecture holds for blocks with cyclic defect groups.*

Broué's conjecture holds for blocks with cyclic defect groups. Over k this was first shown by Rickard [185], lifted over $\mathcal{O}$ in [125], and a splendid Rickard complex was constructed by Rouquier [196], extending prior results from [194].

Theorem 11.1.11 ([196]) *Let G be a finite group and b a block of $\mathcal{O}G$ with a nontrivial cyclic defect group P and inertial quotient E. There is a splendid Rickard complex of $\mathcal{O}Gb$-$\mathcal{O}(P \rtimes E)$-bimodules. In particular, $\mathcal{O}Gb$ and $\mathcal{O}(P \rtimes E)$ are derived equivalent, and they are isotypic, provided that K is large enough.*

This will be proved in 11.12.3. The derived equivalence between $\mathcal{O}(P \rtimes E)$ and $\mathcal{O}Gb$ yields in particular a perfect isometry between $\mathcal{O}(P \rtimes E)$ and $\mathcal{O}Gb$, which can be used to calculate the generalised decomposition numbers of $\mathcal{O}Gb$. In order to show Theorem 11.1.11, we will need to proceed in reverse order; that is, we will first construct a perfect isometry, extending the partial isometry given by the stable equivalence of Morita type in 11.1.2. The character theory of the Frobenius group $P \rtimes E$ is well understood; see 10.5.4. Write $\mathrm{Irr}_K(P \rtimes E) = M \cup \Lambda$ as a disjoint union with M consisting of all irreducible characters of $P \rtimes E$ that have P in their kernel, and λ consisting of all irreducible characters of the form $\mathrm{Ind}_P^{P \rtimes E}(\zeta)$ for some nontrivial $\zeta \in \mathrm{Irr}_K(P)$. Note that $|\Lambda| = \frac{|P|-1}{|E|} = m$; if $m > 1$, then m is the exceptional multiplicity. The perfect isometry that we will describe now yields the generalised decomposition numbers. Moreover, the ordinary decomposition numbers determine again the Brauer tree as an abstract tree with an exceptional vertex, but they do not determine the cyclic order of the edges at each vertex. For $|P| = p$, the decomposition numbers go back to work of Brauer [25], for general cyclic P the decomposition numbers are due to Dade [53], and the perfect isometry was noted in [122]. The next three theorems will be proved in §11.10 together with 11.11.3.

Theorem 11.1.12 *Let G be a finite group and b a block of $\mathcal{O}G$ with a nontrivial cyclic defect group P and inertial quotient E. Suppose that K is large enough for G. There is a perfect isometry between $\mathbb{Z}\mathrm{Irr}_K(P \rtimes E) \cong \mathbb{Z}\mathrm{Irr}_K(G, b)$ which sends $\mu \in M$ to $\delta(\mu)\chi_\mu$ for some $\delta(\mu) \in \{\pm 1\}$ and some $\chi_\mu \in \mathrm{Irr}_K(G, b)$ and which sends $\lambda \in \Lambda$ to $\delta\chi_\lambda$ for some $\delta \in \{\pm 1\}$ that is independent of λ, and some $\chi_\lambda \in \mathrm{Irr}_K(G, b)$. In particular, we have*

$$\mathrm{Irr}_K(G, b) = \{\chi_\mu\}_{\mu \in M} \cup \{\chi_\lambda\}_{\lambda \in \Lambda},$$

and $|\mathrm{Irr}_K(G, b)| = |E| + \frac{|P|-1}{|E|}$.

Since perfect isometries induce height preserving character bijections and isomorphisms of centres, this yields a proof of Brauer's height zero conjecture for blocks with cyclic defect groups.

Corollary 11.1.13 *Let G be a finite group and b a block of $\mathcal{O}G$ with a nontrivial cyclic defect group P and inertial quotient E. All irreducible characters in $\mathrm{Irr}_K(G, b)$ have height zero, and we have an isomorphism of centres $Z(\mathcal{O}(P \rtimes E)) \cong Z(\mathcal{O}Gb)$.*

The perfect isometry above and the endopermutation $\mathcal{O}P$-module W provide a way to calculate the generalised decomposition numbers as follows.

Theorem 11.1.14 *Let $A = \mathcal{O}Gb$ be a block algebra of a block b of a finite group G with a nontrivial cyclic defect group P, let γ be a local point of P on A, denote by $E = N_G(P_\gamma)/C_G(P)$ the corresponding inertial quotient. Suppose that K is large enough for G. For $u \in P \setminus \{1\}$, denote by $\omega(u)$ the sign of the nonzero integer $\chi_W(u)$, where χ_W is the character of the indecomposable endopermutation $\mathcal{O}P$-module W which is a source of the modules U_i. There is a unique local point ϵ of u on A such that $u_\epsilon \in P_\gamma$, and then we have*

$$\chi_\mu(u_\epsilon) = \delta(\mu)\omega(u)$$

$$\chi_\lambda(u_\epsilon) = \delta\omega(u)\lambda(u)$$

for all $\mu \in M$ and $\lambda \in \Lambda$.

The connections between the perfect isometry above, the characters of the modules U_i, V_i and the ordinary decomposition numbers of a block with cyclic defect groups are described in the next result.

Theorem 11.1.15 *Let $A = \mathcal{O}Gb$ be a block algebra of a block b of a finite group G with a nontrivial cyclic defect group P and inertial quotient E. Set $m = \frac{|P|-1}{|E|}$. Suppose that K is large enough for G.*

(i) *Let i be a primitive idempotent in A. The character of the projective indecomposable A-module Ai is equal to either $\chi_\mu + \chi_{\mu'}$ for some μ, $\mu' \in M$ such that $\delta(\mu) \neq \delta(\mu')$, or to $\chi_\mu + \sum_{\lambda \in \Lambda} \chi_\lambda$, where $\mu \in M$ such that $\delta(\mu) = \delta$.*

(ii) *If the character of Ai is equal to $\chi_\mu + \chi_{\mu'}$ for some μ, $\mu' \in M$, then neither i^ρ nor i^σ is exceptional, and the notation can be chosen such that χ_μ is the character of U_i and $\chi_{\mu'}$ is the character of V_i.*

(iii) *Suppose that $m > 1$. If i^ρ is exceptional, then $\sum_{\lambda \in \Lambda} \chi_\lambda$ is the character of U_i and χ_μ is the character of V_i; if i^σ is exceptional, then $\sum_{\lambda \in \Lambda} \chi_\lambda$ is the character of V_i and χ_μ is the character of U_i.*

(iv) *The ordinary decomposition numbers of A are either* 0 *or* 1.

(v) *The set of* $|I| + 1$ *characters* $\{\chi_\mu\}_{\mu \in M} \cup \{\sum_{\lambda \in \Lambda} \chi_\lambda\}$ *corresponds bijectively to the set of vertices of the Brauer tree in such a way that there is an edge in the Brauer tree between two vertices if and only if the sum of the two characters labelling these vertices is the character of a projective indecomposable A-module. In particular, as an abstract tree, the Brauer tree of A is determined by the matrix of ordinary decomposition numbers.*

The ordinary decomposition numbers do not determine the cyclic order of the edges emanating from a fixed vertex of the Brauer tree. In particular, they do not determine the Morita equivalence class of the block. With some extra effort, the structural information on the projective indecomposable A-modules in 11.1.8 can be used to determine the Morita equivalence class of A; this extends earlier work of Plesken in [165] on group rings over p-adic integers.

Theorem 11.1.16 ([125, Proposition 3.10]) *Let A be a block algebra or source algebra of a block with a nontrivial cyclic defect group P. Let I be a set of representatives of the conjugacy classes of primitive idempotents in A. Let ρ, σ be the permutations of I satisfying 11.1.5, with exceptional vertex v, if the integer $m = \frac{|P|-1}{|I|}$ is greater than 1. The $\mathcal{O}$-algebra A is Morita equivalent to the basic $\mathcal{O}$-algebra generated by the set $I \cup \{r, s\}$, with the relations*

$$1 = \sum_{i \in I} i,$$

$$i^2 = i, \, (i \in I),$$

$$ij = 0, \, (i, j \in I, i \neq j),$$

$$rs = sr = 0,$$

$$ir^{|i^\rho|} + is^{|i^\sigma|} = |P|i, \quad i \in I, i^\rho \neq v, i^\sigma \neq v,$$

$$ir^{m|i^\rho|} + is^{|i^\sigma|} = \sum_{t=0}^{m-1} \lambda_t ir^{t|i^\rho|}, \quad i \in I, i^\rho = v, i^\sigma \neq v,$$

$$ir^{|i^\rho|} + is^{m|i^\sigma|} = \sum_{t=0}^{m-1} \lambda_t is^{t|i^\sigma|}, \quad i \in I, i^\rho \neq v, i^\sigma = v,$$

where the λ_t are coefficients in $J(\mathcal{O})$ which are determined by P and E, with $\lambda_0 = |P|$. In particular, the Morita equivalence class of A is determined by the Brauer tree $(I, \{\rho, \sigma\}, v, m)$.

Since I is bounded in terms of $|P|$, we get Donovan's conjecture for blocks with a cyclic defect group over $\mathcal{O}$. Over k, Donovan's conjecture goes back to work of Janusz [91] and Kupisch [117]. Reducing the relations in 11.1.16 modulo $J(\mathcal{O})$ yields a presentation of $k \otimes_{\mathcal{O}} A$ such that in the last three relations, the right side is zero (because the images in k of $|P|$ and the λ_t are zero). They correspond to those described in 11.8.6. Thus Donovan's conjecture holds for block algebras with cyclic defect groups over $\mathcal{O}$.

Corollary 11.1.17 *There are only finitely many Morita equivalence classes of block algebras over $\mathcal{O}$ with a fixed cyclic defect group P.*

It may well happen that two source algebras are Morita equivalent but not isomorphic as interior P-algebras. This phenomenon arises whenever two blocks are Morita equivalent via a bimodule with a nontrivial endopermutation source. For blocks with cyclic defect groups, this is the only way to obtain Morita equivalent but not source algebra equivalent blocks, and therefore, the source algebras of a block with a cyclic defect group P are determined by the Brauer tree together with an endopermutation $\mathcal{O}P$-module. The next theorem shows that the isomorphism class of a source algebra A of a block with a nontrivial cyclic defect group P is determined by the isomorphism class of the datum (I, ρ, σ, v, W) as in 11.1.5. If $(I', \rho', \sigma', v, W')$ is the datum associated with another source algebra A' of a block with defect group P as in 11.1.5, then an isomorphism $(I, \rho, \sigma, v, W) \cong (I', \rho', \sigma', v, W')$ consists of an isomorphism of $\mathcal{O}P$-modules $W \cong W'$ and a bijection $\tau : I \cong I'$ satisfying $\tau \circ \rho = \rho' \circ \tau$ and $\tau \circ \sigma = \sigma' \circ \tau$, and $\tau(v) = v'$ in case there are exceptional orbits v, v'. If $|I| = 1$, then A is the source algebra of a nilpotent block, hence completely determined the isomorphism class of an indecomposable endopermutation $\mathcal{O}P$-module with vertex P as described in 8.11.5. The next theorem describes the information required to determine the source algebras for $|I| \geq 2$.

Theorem 11.1.18 ([127, §2]) *Let A be a source algebra of a block with a nontrivial cyclic defect group P. Suppose that $|I| \geq 2$. Then the isomorphism class of A as an interior P-algebra is determined by the datum (I, ρ, σ, v, W), where the notation is as in 11.1.5.*

This theorem shows that in order to parametrise the source algebras in the case $|I| \geq 2$, we need to distinguish between ρ and σ, while for the Morita equivalence class it is sufficient to consider the unordered pair $\{\rho, \sigma\}$. Distinguishing between ρ and σ amounts to introducing a parity on the set of vertices, labelling the vertices of the Brauer tree with either ρ or σ in such a way that neighbouring vertices (connected by an edge i) have different labels. The passage from (I, ρ, σ, v, W) to (I, σ, ρ, v, W) amounts to modifying the algebra

A in such a way that the modules U_i have as common source $\Omega(W)$ instead of W, or equivalently, the U_i and V_i are exchanged. By 6.4.5, the source algebras of a block need not all be isomorphic as interior P-algebras, but only isomorphic 'up to automorphisms' of P. The above theorem implicitly states that the source algebras of a block with a cyclic defect group P are all isomorphic. We state this explicitly as follows.

Theorem 11.1.19 *Let A be a source algebra of a block with a nontrivial cyclic defect group P, identified to its image in $A^{\times}$. Every group automorphism of P extends to an algebra automorphism of A.*

The theorems 11.1.18 and 11.1.19 will be proved in §11.15. There are only finitely many possible data $(I, \rho, \sigma, \upsilon, W)$ for a fixed cyclic P, and hence Puig's conjecture holds for blocks with cyclic defect groups:

Corollary 11.1.20 ([127, Theorem 2.7.(iii)]) *Let P be a finite cyclic p-group. There are only finitely many isomorphism classes of interior P-algebras that arise as source algebras of blocks with P as a defect group.*

Remark 11.1.21 The proof of 11.1.18 below is different from that given in [127] in that it does not use Weiss' criterion [228]. The use of Weiss' criterion in the context of blocks with cyclic defect groups goes back to work of L. L. Scott [204]. It leads to more precise statements on conjugacy classes of p-subgroups in source algebras of blocks with cyclic defect groups; see for instance [127]. See the Remark 11.14.4 below for more details and references.

11.2 Local structure

Blocks with a cyclic defect group are special cases of blocks with an abelian defect group and a Frobenius inertial quotient. We state this, and restate some of the consequences from §10.5, notably the fact that there is a stable equivalence of Morita type between a block with a cyclic defect group and its Brauer correspondent, given by a bimodule with 'diagonal' vertex and endopermutation source. We assume in this section that k is a splitting field for all finite groups and their subgroups that arise.

A finite cyclic p-group P has a unique subgroup Z of order p, and the study of a block of a finite group G with a cyclic defect group P passes through the corresponding blocks of $C = C_G(Z)$ and $H = N_G(Z)$. More precisely, if b is a block of $\mathcal{O}G$ with P as a defect group, denoting by c the block of $\mathcal{O}H$ corresponding to b, we will show that $\mathcal{O}Hc$ is Morita equivalent to a semidirect product group algebra $\mathcal{O}(P \rtimes E)$ for some cyclic subgroup E of $\mathrm{Aut}(P)$ of order $p-1$, and that $\mathcal{O}Gb$ is stably equivalent to $\mathcal{O}Hc$. The characters of $\mathcal{O}Hc$ correspond to those of $\mathcal{O}(P \rtimes E)$, and up to certain signs, the generalised decomposition matrix of $\mathcal{O}Hc$ is determined by that of $\mathcal{O}(P \rtimes E)$. The partial isometry induced by the stable equivalence between $\mathcal{O}Hc$ and $\mathcal{O}Gb$ can be extended to a perfect isometry, yielding the generalised decomposition matrix for $\mathcal{O}Gb$. For the algebra structure, including the ordinary decomposition matrix of $\mathcal{O}Gb$, one needs to use the fact that $kH\bar{c}$ is a *serial* algebra. We first collect basic statements on the local structure of a block with cyclic defect groups.

Theorem 11.2.1 *Let G be a finite group and b a block of $\mathcal{O}G$. Let (P, e) be a maximal (G, b)-Brauer pair, and suppose that P is cyclic. Let γ be the unique local point of P on $\mathcal{O}Gb$ such that $\mathrm{Br}_P(\gamma) \subseteq kC_G(P)e$.*

(i) *The inertial quotient $E = N_G(P, e)/C_G(P)$ of b is cyclic of order dividing $p - 1$.*

(ii) *The group E acts freely on $P \setminus \{1\}$.*

(iii) *For any Brauer pair (Q, f) contained in (P, e) we have $N_G(Q, f) = N_G(P, e)C_G(Q)$.*

(iv) *For any Brauer pair (Q, f) contained in (P, e) we have $N_G(Q, f)/C_G(Q) \cong E$.*

(v) *For any nontrivial Brauer pair (Q, f) contained in (P, e), the block f of $kC_G(Q)$ is nilpotent; in particular, f has a unique irreducible Brauer character φ_Q.*

(vi) *For any nontrivial subgroup Q of P there is a unique local point δ of Q on $\mathcal{O}Gb$ such that $Q_\delta \leq P_\gamma$.*

Proof By [76, I.3.10], the automorphism group $\mathrm{Aut}(P)$ of the cyclic p-group P is isomorphic to the abelian group $(\mathbb{Z}/p^a\mathbb{Z})^\times$, hence isomorphic to the direct product of a cyclic group of order $p - 1$ and a finite abelian p-group. The cyclic subgroup of order $p - 1$ of $\mathrm{Aut}(P)$ acts freely on $P \setminus \{1\}$. This implies (i) and (ii). Let (Q, f) be a Brauer pair such that $(Q, f) \leq (P, e)$. By 8.5.7 we have $N_G(Q, f) \subseteq N_G(P, e)C_G(Q)$. Since P is cyclic, we have $N_G(P) \subseteq N_G(Q)$, hence $N_G(P, e) \subseteq N_G(Q, f)$ because f is uniquely determined by e. This implies (iii), and (iv) follows from (iii). Since E acts freely on $P \setminus \{1\}$ it follows that the block f of $C_G(Q)$ has a trivial inertial quotient, hence is nilpotent by 8.11.3. It follows from 8.11.5 that $kC_G(Q)f$ has a unique irreducible Brauer character φ_Q, whence (v). Moreover, $kC_G(Q)f$ has a unique conjugacy class of primitive idempotents. Thus, if δ is a local point of Q on $\mathcal{O}Gb$ such that $Q_\delta \subseteq P_\gamma$, then the uniqueness of the inclusion of Brauer pairs implies that $\mathrm{Br}_Q(\delta)$ is the unique conjugacy class of primitive idempotents in $kC_G(Q)f$, whence (vi). $\square$

Once a cyclic defect group P is fixed, the inertial quotient E is uniquely determined, as a subgroup of $\mathrm{Aut}(P)$, by its order, because the p'-part of $\mathrm{Aut}(P)$ is a unique cyclic subgroup of order $p - 1$ and $\mathrm{Aut}(P)$ is abelian. We denote by $P \rtimes E$ the semidirect product corresponding to the action of E on P. We determine the source algebras of blocks with a cyclic defect group P in the cases where the unique subgroup of order p is central and normal, respectively.

Theorem 11.2.2 *Let G be a finite group and b a block of $\mathcal{O}G$ with a nontrivial cyclic defect group P and inertial quotient E. Denote by Z the unique subgroup of order p of P. Set $C = C_G(Z)$ and $H = N_G(Z)$. We have $N_G(P) \subseteq H$. Let c be the block of $\mathcal{O}H$ corresponding to b through the Brauer correspondence. Then $c \in (\mathcal{O}C)^H$. Let f be a block of $\mathcal{O}C$ such that $fc = f$. Then P is a defect group of f. Let $j \in (\mathcal{O}Cf)^P$ be a source idempotent of f.*

 (i) *The block f is nilpotent with P as a defect group. There is up to isomorphism a unique indecomposable endopermutation $\mathcal{O}P$-module W with determinant 1 on which Z acts trivially such that*

$$jOCj \cong S \otimes_\mathcal{O} \mathcal{O}P$$

 as interior P-algebras, where $S = \mathrm{End}_\mathcal{O}(W)$. In particular, $\mathcal{O}Cf$ and $\mathcal{O}P$ are Morita equivalent.

 (ii) *The idempotent j remains a source idempotent for the block c of $\mathcal{O}H$, the $\mathcal{O}P$-module structure of W extends to an $\mathcal{O}(P \rtimes E)$-module structure, and there is an isomorphism*

$$jOHj \cong S \otimes_\mathcal{O} \mathcal{O}(P \rtimes E)$$

 as interior P-algebras. In particular, $\mathcal{O}Hc$ and $\mathcal{O}(P \rtimes E)$ are Morita equivalent.

Proof Since Z is the unique subgroup of order p we have $N_G(P) \subseteq H$. Thus, by the Brauer correspondence 6.7.1, there is a unique block c of $\mathcal{O}H$ with P as a defect group such that $\mathrm{Br}_P(b) = \mathrm{Br}_P(c)$. It follows from 6.2.6 that c is contained in $(\mathcal{O}C)^H$, hence a sum of blocks of $\mathcal{O}C$. Any defect group of c is conjugate to P in H, hence contains Z, and hence is contained in C. It follows from 6.8.9 that f has P as a defect group. For any choice of a maximal Brauer pair it follows from 11.2.1 that the inertial quotient of f acts freely on $P \setminus \{1\}$. Since Z is in the centre of C it follows that the inertial quotient of f acts trivially on P. Thus the inertial quotient of f is trivial. It follows from 8.11.3 that f is nilpotent with P as a defect group. Since Z is central in C, the elements of Z act trivially on any simple $kC\bar{f}$-module. The statement on the structure of the source algebra $jOCj$ follows from 8.11.5. The endopermutation kP-module

$k \otimes_{\mathcal{O}} W$ is a source of a simple $k C \bar{f}$-module, and hence Z acts trivially on $k \otimes_{\mathcal{O}} W$. But then Z acts trivially on the unique lift W of $k \otimes_{\mathcal{O}} W$ with determinant 1. A block algebra is Morita equivalent to any of its source algebras by 6.4.6. This proves (i). The last statement of 6.2.6 implies that j remains a source idempotent of $\mathcal{O}Hc$. Setting $N = N_G(Z, f)$, it follows from the above that f remains a block of $\mathcal{O}N$ that, by 6.8.3, is source algebra equivalent to $\mathcal{O}Hc$. More precisely, we have $j\mathcal{O}Hj = j\mathcal{O}Nj$. By 11.2.1 (iv) we have $N/C \cong E$, and by 1.2.10 we have $H^2(E; k^{\times}) = \{0\}$. The block $\mathcal{O}N_G(Z, f)f$ is a p'-extension of the nilpotent block $\mathcal{O}C_G(Z)f$. The statement (ii) on the source algebras of $\mathcal{O}N_G(Z)f$ follows therefore from 8.12.4 (iv). $\qquad \square$

Theorem 11.2.2 implies in particular that kHc is Morita equivalent to $k(P \rtimes E)$. We will see in the next section that every indecomposable $k(P \rtimes E)$-module has a unique composition series. The key tool to determine the structure of $\mathcal{O}Gb$ using the precise knowledge of the structure of $\mathcal{O}Hc$ from 11.2.2 is a splendid stable equivalence of Morita type between the two block algebras, thanks to the fact that H is strongly embedded in G with respect to P.

Lemma 11.2.3 *Let G be a a finite group and P a nontrivial cyclic p-subgroup of G. Denote by Z the unique subgroup of order p of P and set $H = N_G(Z)$. For any $x \in G \setminus H$ we have $P \cap {}^x P = \{1\}$. In particular, H contains $N_G(Q)$ for any nontrivial subgroup Q of P.*

Proof Let $x \in G$ such that $P \cap {}^x P \neq \{1\}$. Then P and ${}^x P$ contain the unique subgroup Z of order p of P. But the unique subgroup of order p of ${}^x P$ is also equal to ${}^x Z$. It follows that $Z = {}^x Z$, hence $x \in H$. $\qquad \square$

Theorem 11.2.4 *Let G be a finite group and b a block of $\mathcal{O}G$ with a cyclic defect group P. Denote by Z the unique subgroup of order p of P and set $H = N_G(Z)$. Let c be the block of $\mathcal{O}H$ corresponding to b through the Brauer correspondence. Then $\mathrm{Br}_Q(b) = \mathrm{Br}_Q(c)$ for any nontrivial subgroup Q of P, and the $\mathcal{O}Gb$-$\mathcal{O}Hc$-bimodule $M = b\mathcal{O}Gc$ and its dual $M^* \cong c\mathcal{O}Gb$ induce a splendid stable equivalence of Morita type between $\mathcal{O}Gb$ and $\mathcal{O}Hc$.*

Proof By 11.2.3 this is a special case of 9.8.6. $\qquad \square$

As a consequence, we obtain a proof of 11.1.2, which we restate for convenience.

Corollary 11.2.5 *Let G be a finite group and b a block of $\mathcal{O}G$ with a nontrivial cyclic defect group P and inertial quotient E. There is a stable equivalence of Morita type between $\mathcal{O}Gb$ and $\mathcal{O}(P \rtimes E)$ given by a bimodule with vertex ΔP and an endopermutation $\mathcal{O}P$-module W as a source. Moreover, this stable* *equivalence can be chosen in such a way that W has determinant* 1 *and such that the subgroup Z of order p of P acts trivially on W.*

Proof This follows from combining the stable equivalence in 11.2.4 and the Morita equivalence in 11.2.2 (ii).   □

While stable equivalences of Morita type are abundant in block theory, one of the major hurdles to put them to good use is that it is not known in general whether they preserve the numbers of ordinary irreducible characters. It will require some detailed knowledge of the module category of $k(P \rtimes E)$ to show that for blocks with cyclic defect groups the stable equivalence in 11.2.4 extends to a perfect isometry, hence preserves in particular the numbers of irreducible characters. A key observation needed in order to determine the ordinary decomposition numbers is that there is no nonprojective indecomposable $\mathcal{O}$-free $\mathcal{O}Gb$-module U such that $k \otimes_{\mathcal{O}} U$ remains indecomposable and such that the character of U belongs to the group $\mathrm{Proj}_{\mathcal{O}}(G, b)$ spanned by the characters of the projective indecomposable $\mathcal{O}Gb$-modules. This property relies on the classification of the indecomposable $k(P \rtimes E)$-modules in §11.4.

11.3 Selfinjective serial algebras

Definition 11.3.1 Let A be a finite-dimensional k-algebra. An A-module U is called *uniserial* if it has a unique composition series. The algebra A is called *serial*, if every indecomposable A-module is uniserial.

The structure theory of selfinjective serial algebras goes back to work of Nakayama [153], and these algebras are also frequently called *Nakayama algebras*. Serial symmetric algebras are in particular selfinjective, and they arise in the context of blocks with cyclic defect groups because of the following result.

Theorem 11.3.2 *Let P be a finite cyclic p-group, and let E be a p'-subgroup of* Aut(P).

(i) *The algebra $k(P \rtimes E)$ is indecomposable, symmetric, and serial.*
(ii) *The algebra $k(P \rtimes E)$ has $|E|$ isomorphism classes of simple modules, all of which have dimension* 1, *and the projective indecomposable $k(P \rtimes E)$-modules all have dimension $|P|$. In particular, $k(P \rtimes E)$ is split basic.*

For $|E| = 1$, the above theorem implies that kP is uniserial, providing an alternative proof of 7.1.1 (v). In order to prove Theorem 11.3.2, we develop some general properties of serial algebras and their modules. With the exception of the above theorem, all results in this section remain valid without any hypothesis on the characteristic of k.

Lemma 11.3.3 *Let A be a finite-dimensional k-algebra and let U be a finite-dimensional A-module. The following are equivalent.*

(i) *The module U is uniserial.*

(ii) *The series $U \supset J(A)U \supset J(A)^2U \supset \cdots \supset J(A)^mU = \{0\}$ is a composition series of U, where m is the smallest positive integer satisfying $J(A)^mU = \{0\}$.*

(iii) *Every submodule of U is equal to $J(A)^sU$ for some integer $s \geq 0$.*

(iv) *For any two submodules V, W of U we have $V \subseteq W$ or $W \subseteq V$.*

(v) *The right A-module $U^* = \text{Hom}_k(U, k)$ is uniserial.*

Moreover, if U is uniserial, so is every submodule and every quotient module of U.

Proof The module $J(A)U = \text{rad}(U)$ is the intersection of all maximal submodules of U. Thus the quotient $U/J(A)U$ is simple if and only if $J(A)U$ is the unique maximal submodule in U. It follows that U is uniserial if and only if the sequence given by the submodules $J(A)^kU$ is a composition series of U, or equivalently, if $J(A)^sU/J(A)^{s+1}U$ is simple for $0 \leq s \leq m - 1$. This implies the equivalence of the statements (i), (ii), (iii). Clearly (iii) implies (iv). If U is not uniserial, then there is $s \geq 0$ such that the semisimple module $J(A)^sU/J(A)^{s+1}U$ is not simple, hence equal to a direct sum $X \oplus Y$ for some nonzero semisimple modules X, Y. Then the inverse images U, V of X, Y in $J(A)^sU$ do not satisfy (iv), whence the equivalence of the first four statements. The equivalence of (i) and (v) follows immediately from the properties of the k-duality functor as described in 2.9.2. For the last statement one checks that the properties (iii) or (iv) pass to submodules and quotient modules. $\square$

If I is a set of representatives of the conjugacy classes of primitive idempotents of a finite-dimensional k-algebra A (that is, I is a set of representatives of the points of A), then, by 4.7.17, the set $\{Ai\}_{i\in I}$ is a set of representatives of the isomorphism classes of the finite-dimensional projective indecomposable A-modules, and, setting $S_i = Ai/J(A)i$, the set $\{S_i\}_{i\in I}$ is a set of representatives of the isomorphism classes of simple A-modules. We denote by $\ell(U)$ the length of a composition series of a finite-dimensional A-module. The following theorem describes the structure of serial selfinjective algebras.

Theorem 11.3.4 *Let A be a finite-dimensional nonsimple indecomposable selfinjective serial k-algebra. Denote by I a set of representatives of the conjugacy classes of primitive idempotents in A, and set $S_i = Ai/J(A)i$ for any $i \in I$.*

(i) *There is a unique transitive cycle π on the set I such that $J(A)i/J(A)^2i \cong S_{\pi(i)}$ for any $i \in I$.*

(ii) *The projective indecomposable A-modules Ai have all the same composition length q, and for any $i \in I$, the composition factors of Ai from the top to the bottom are S_i, $S_{\pi(i)}$, $S_{\pi^2(i)}$, ..., $S_{\pi^{q-1}(i)}$.*

(iii) *For any $i \in I$, there is a short exact sequence of A-module*

$$0 \longrightarrow S_{\pi^q(i)} \longrightarrow A\pi(i) \longrightarrow J(A)i \longrightarrow 0.$$

(vi) *If A is symmetric, then $|I|$ divides $q - 1$ and for any $i \in I$ we have $S_{\pi^{q-1}(i)} \cong S_i$.*

The hypothesis 'nonsimple' in the above theorem is mainly there to avoid trivialities; it ensures that none of the simple modules S_i is projective, or equivalently, that $J(A)i \neq \{0\}$ for any $i \in I$.

Proof As Ai is uniserial, $J(A)i/J(A)^2i$ is simple for any $i \in I$, by 11.3.3. Thus $J(A)i/J(A)^2i \cong S_{\pi(i)}$ for a uniquely determined $\pi(i) \in I$. In other words, $A\pi(i)$ is a projective cover of $J(A)i$. By induction, we show that $A\pi^s(i)$ is a projective cover of $J(A)^si$, where $i \in I$ and s is a positive integer such that $J(A)^si \neq \{0\}$. For $s = 1$ we are done, so assume $s > 1$. By induction, there is a surjective homomorphism $\varphi : A\pi^{s-1}(i) \to J(A)^{s-1}i$. By the first argument applied to $\pi^{s-1}(i)$ instead of i, there is also a surjective homomorphism $\psi : A_{\pi^s(i)} \to J(A)\pi^{s-1}(i)$. Now $J(A)\pi^{s-1}(i)$ is the unique maximal submodule of $A\pi^{s-1}(i)$, thus it is mapped by φ onto the unique maximal submodule of $J(A)^{s-1}i$, which is $J(A)^si$. It follows that $\varphi \circ \psi$ is a surjective homomorphism from $A_{\pi^s(i)}$ onto $J(A)^si$. This shows that the composition series $Ai \supset J(A)i \supset J(A)^2i \supset \cdots \supset J(A)^qi = \{0\}$ of Ai has consecutive composition factors S_i, $S_{\pi(i)}$, ..., $S_{\pi^{q-1}(i)}$, where $q = q(i)$ is the composition length of Ai (which at this stage may still depend on i). As A is selfinjective, then the projective indecomposable A-modules are precisely the injective indecomposable A-modules. Any simple A-module S_j, where $j \in J$, has therefore an injective envelope of the form Ai for some $i \in I$. Since $\mathrm{soc}(Ai) \cong S_{\pi^{q-1}}(i)$ we have $\pi^{q-1}(i) = j$. In particular, π is surjective, thus a permutation as I is a finite set. We have to show that π is a transitive cycle on I. Let $i, j \in I$. If Ai, Aj have a common composition factor, then there are integers s, t such that $S_{\pi^s(i)} \cong S_{\pi^t(j)}$; thus i, j belong to the same π-orbit in that case. Note that if $\mathrm{Hom}_A(Ai, Aj)$ is nonzero, then Ai, Aj have a common composition factor isomorphic to S_i. Brauer's Theorem 4.10.5 implies that π is transitive as

A is an indecomposable algebra. As $A\pi(i)$ is a projective cover of $J(A)i$, where $i \in I$, there is a surjective A-homomorphism $\tau : A\pi(i) \to J(A)i$. Now τ cannot be injective, because in that case $J(A)i \cong A\pi(i)$ would be an injective module, thus would split off as direct summand of Ai, which is impossible. Thus $\ell(A\pi(i)) \geq \ell(J(A)i) + 1 = \ell(Ai)$. Since π is a transitive cycle, we have equality and $\ker(\tau)$ is simple. Thus $\ker(\tau) \cong \mathrm{soc}(A\pi(i)) \cong S_{\pi^q(i)}$. From this we get the short exact sequence as stated in (iii), and that the Ai have all the same composition length q. Furthermore, if A is symmetric, then $\mathrm{soc}(Ai) \cong S_i$, and hence $\pi^{q-1}(i) = i$. Therefore the order $|I|$ of the cycle π divides $q - 1$, which concludes the proof. $\qquad\square$

Corollary 11.3.5 *With the notation and hypotheses of 11.3.4, for any $i \in I$ we have $\Omega^2(S_i) \cong S_{\pi^q(i)}$. If A is symmetric, we have $\Omega^2(S_i) \cong S_{\pi(i)}$.*

Proof Let $i \in I$. The obvious short exact sequence $0 \to J(A)i \to Ai \to S_i \to 0$ implies that $\Omega(S_i) \cong J(A)i$. By the short exact sequence in 11.3.4 we have $\Omega(J(A)i) \cong S_{\pi^q(i)}$, from which 11.3.5 follows. $\qquad\square$

Proposition 11.3.6 (cf. [153, Theorem 11]) *A finite-dimensional k-algebra A is serial if and only if every indecomposable finite-dimensional projective or injective A-module is uniserial.*

Proof Assume that every indecomposable finite-dimensional projective or injective A-module is uniserial, and let U be a finite-dimensional indecomposable A-module. We have to show that U is uniserial. Let V be a uniserial submodule of U of maximal dimension with this property. Since every simple submodule of U is uniserial, it follows that V is nonzero. Let I be a minimal injective envelope of V. Note that $\mathrm{soc}(I) \cong \mathrm{soc}(V)$ is simple, as V is uniserial; thus I is indecomposable. Since I is injective, the injection $V \to I$ extends to an A-homomorphism $\pi : U \to I$. Set $X = \ker(\pi)$. Then U/X is isomorphic to a submodule of I. By the assumptions, I is uniserial, and hence U/X is so, too. Thus $S = (U/X)/J(A)(U/X)$ is simple. Denote by P a minimal projective cover of U/X. Then P is indecomposable, as $P/J(A)P \cong S$, and hence P is uniserial by the assumptions. Since P is projective, any surjective map $P \to U/X$ lifts to a map $\mu : P \to U$ satisfying $U = X + \mathrm{Im}(\mu)$. Note that $X \cap V = \{0\}$; this is because π extends an injective map $V \to I$. Thus $(V + X)/X \cong V$, and hence $\dim_k(V) \leq \dim_k(U/X) \leq \dim_k(\mathrm{Im}(\mu))$. But $\mathrm{Im}(\mu)$ is a quotient of the uniserial module P, thus itself uniserial. The maximality of V implies that $V \cong U/X$, and therefore $V \oplus X = U$. Since U is indecomposable, we have $X = \{0\}$, and hence $U = V$ is uniserial. The converse is trivial. $\qquad\square$

Any indecomposable injective module over a finite-dimensional k-algebra A is isomorphic to the k-dual of a projective indecomposable right A-module. The property of a module to be uniserial is invariant under k-duality. Thus 11.3.6 can be formulated as follows: A is serial if and only if for any primitive idempotent i in A the A-module Ai and the right A-module iA are uniserial.

Lemma 11.3.7 *A finite-dimensional selfinjective k-algebra A is serial if and only if $J(A)i/J(A)^2i$ is simple or zero for any primitive idempotent i in A.*

Proof If A is serial, for any primitive idempotent $i \in A$, the projective indecomposable A-module Ai is uniserial, and thus $J(A)i/J(A)^2i$ is either simple or zero by 11.3.3. Conversely, let $i \in A$. If $J(A)i$ is zero, then Ai is simple, and hence Ai is trivially uniserial. If $J(A)i/J(A)^2i$ is simple, then $J(A)i$ has a projective cover that is indecomposable, thus isomorphic to Aj for some primitive idempotent $j \in A$. Thus $J(A)^{s+1}i$ is a quotient of $J(A)^s j$ for any positive integer s. Therefore, if $J(A)^s j/J(A)^{s+1} j$ is simple, then $J(A)^{s+1}i/J(A)^{s+2}i$ is either simple or zero. An easy induction argument implies that Ai is uniserial. Thus every projective indecomposable A-module is uniserial. Since A is selfinjective, the classes of projective and injective modules coincide. Thus 11.3.7 follows from 11.3.6. $\square$

The above lemma is going to be used in the following theorem, due to Nakayama, which characterises symmetric serial algebras and implies also Theorem 11.3.2

Theorem 11.3.8 *Let A be a selfinjective indecomposable k-algebra. The following are equivalent.*

(i) There is an element $t \in J(A)$ such that $J(A) = tA$ or $J(A) = At$.
(ii) The algebra A is serial, and all projective indecomposable A-modules occur with the same multiplicity in a decomposition of A as left A-module.

If the two above conditions hold, we have $At = tA$.

Proof If A is simple, then the result holds trivially with $t = 0$. Assume that A is not simple. Suppose that $J(A) = At$ for some $t \in A$. Let $a \in A$. If $ta = 0$, then $J(A)a = Ata = \{0\}$. As A is symmetric, we have $aJ(A) = \{0\}$. In particular, $at = 0$. Therefore we have a well-defined linear map from tA to At that sends ta to at; in particular, we have $\dim_k(tA) \geq \dim_k(At)$. But since $t \in J(A)$, we have also $tA \subset J(A) = At$. Together we get the equality $tA = At = J(A)$. In a completely analogous way one shows that if $J(A) = tA$, then $J(A) = At$. Let now J be a primitive decomposition of 1_A in A. Then $A = \oplus_{j \in J} Aj$, and therefore

$$J(A) = \oplus_{j \in J} J(A)j = \sum_{j \in J} Ajt.$$

We are going to show that the last sum is also direct. Let $a_j \in Aj$ for any $j \in J$, such that $\sum_{j \in J} a_j t = 0$. Thus the sum $\sum_{j \in J} a_j$ belongs to the left annihilator of $J(A) = tA$, thus also to the right annihilator. In particular, $ta_i = t(\sum_{j \in J} a_j)i = 0$ for any $i \in J$. Again, a_i belongs to the right annihilator of $J(A)$, thus also to its left annihilator, and whence $a_i t = 0$ for any $i \in J$. This shows that

$$J(A) = \oplus_{j \in J} J(A)j = \oplus_{j \in J} Ajt.$$

Moreover, $J(A)j$ is indecomposable because its socle is simple, and Ajt is indecomposable since it is a quotient of Aj, and hence its semisimple quotient is simple. The Krull–Schmidt Theorem implies, that there is a permutation π on J such that

$$J(A)j \cong A\pi(j)t$$

for any $j \in J$. That is, $J(A)j$ is a quotient of $A\pi(j)$, and thus $J(A)j/J(A)^2 j$ is either simple or zero. Lemma 11.3.7 implies that A is serial. Moreover, if $i, j \in J$ such that $Ai \cong Aj$, then $A\pi(i) \cong A\pi(j)$. Therefore the multiplicity of $A\pi(i)$ in A is greater or equal to the multiplicity of Ai in A. By 11.3.4, π permutes transitively the isomorphism classes of finite-dimensional projective indecomposable A-modules, and thus the multiplicities have all to be equal. This shows that (i) implies (ii). Suppose that (ii) holds. Choose a system of representatives I of the conjugacy classes of primitive idempotents in A in such a way, that the elements of I are pairwise orthogonal (this may be done, for instance, by choosing I as a subset of a primitive decomposition of 1_A in A). Set $e = \sum_{i \in I} i$. Then e is an idempotent in A and we have $Ae = \oplus_{i \in I} Ai$. The hypothesis on A, that its indecomposable direct summands occur all with the same multiplicity, means that $A \cong (Ae)^m$ as left A-modules, for some positive integer m. Thus $A/J(A) \cong (Ae/J(A)e)^m$. Now this is, by 11.3.4, isomorphic to $(J(A)e/J(A)^2 e)^m \cong J(A)/J(A)^2$. Let $t \in J(A)$ such that $t + J(A)^2$ is the image of $1 + J(A)$ under this chain of isomorphisms. Since $1 + J(A)$ generates $A/J(A)$ as a left A-module, it follows that $t + J(A)^2$ generates $J(A)/J(A)^2$ as a left A-module. Thus $At + J(A)^2 = J(A)$. By Nakayama's Lemma 1.10.4 we have $At = J(A)$. This completes the proof. □

Proof of Theorem 11.3.2 Set $A = k(P \rtimes E)$. The algebra A is symmetric by 2.11.2 and indecomposable as an algebra by 6.2.9. The canonical group homomorphism $P \rtimes E \to E$ with P as kernel induces a surjective algebra homomorphism $A \to kE$, having as its kernel the ideal $I(kP)A$ generated by the augmentation ideal of kP. Since kE is semisimple by Maschke's Theorem, we have $J(A) \subset I(kP)A$. But we also have $I(kP)A \subset J(A)$ by 1.11.8. Let y be a generator of P. For any positive integer k we have $y^k - 1 = (y-1)(y^{k-1} + \cdots + y + 1)$; thus $I(kP) = (y-1)kP$. Together we get $J(A) = (y-1)A$. Thus A is serial by 11.3.8. Since E is a cyclic group of order dividing $p - 1$, it follows that the group algebra kE decomposes as a left module as a direct sum of $|E|$ pairwise nonisomorphic simple modules of dimension 1. Thus A has $|E|$ isomorphism classes of simple modules, all of dimension 1, and hence A is basic. Moreover, A decomposes as a left module into a direct sum of $|E|$ pairwise nonisomorphic projective indecomposable modules. Any of these is projective, hence free, as a kP-module, and therefore has dimension divisible by $|P|$. By comparing dimensions, it follows that the projective indecomposable A-modules all have dimension $|P|$. $\square$

The algebra structure of a split serial selfinjective algebra can be described in terms of its quiver and relations:

Theorem 11.3.9 *Let A be a split indecomposable non simple selfinjective serial k-algebra. Let e be the number of isomorphism classes of simple A-modules and let q be the composition length of the projective indecomposable A-modules.*

(i) The Ext-*quiver Q of A is a cyclic graph with e vertices:*

(ii) Let J be the ideal in the path algebra kQ generated by all paths of lengths q in Q. If A is split basic, then there is a k-algebra isomorphism

$$A \cong kQ/J;$$

in particular, the isomorphism class of A is uniquely determined by the integers e and m.

Proof Let I be a system of representatives of the conjugacy classes of primitive idempotents in A. The vertices of Q are in bijection with the set I, and for any $i \in I$, the quotient $J(A)i/J(A)^2i$ is simple and thus isomorphic to $S_{\pi(i)}$ for a unique $\pi(i) \in I$. Thus there is a unique arrow in Q starting at the vertex i, and this arrow goes to the vertex $\pi(i)$. By 11.3.4, π is a transitive cycle on I, which proves (i). Moreover, again by 11.3.4, there is a homomorphism $A\pi(i) \to Ai$ with image $J(A)$. Thus there is an element $t_i \in \pi(i)Ai$ such that $J(A)i = At_i$. Since $t_i \in J(A)$, any product of at least q of theses elements is zero because all projective indecomposable A-modules have length q. Thus there is an algebra homomorphism $kQ \to A$ mapping any vertex i to i as element in A and any arrow $i \to \pi(i)$ to t_i. By 4.9.14, this algebra homomorphism is surjective, and its kernel contains J. Since A is split basic, every simple A-module has dimension 1, and thus A has dimension em. The paths of length smaller than q determine a k-basis of kQ/J, and thus the dimension of kQ/J is also eq. This shows that the algebra homomorphism above induces an isomorphism $kQ/J \cong A$. $\qquad\qquad\square$

Even if an indecomposable selfinjective serial algebras is not split, all of its simple modules have isomorphic endomorphism algebras.

Proposition 11.3.10 *Let A be a selfinjective serial k-algebra and let U be an indecomposable A-module. Set $S = \mathrm{soc}(U)$ and $T = U/\mathrm{rad}(U)$. We have isomorphisms*

$$\mathrm{End}_A(S) \cong \mathrm{End}_A(U)/\mathrm{Hom}_A(U, \mathrm{rad}(U)) \cong \mathrm{End}_A(T).$$

Proof Denote by $\pi_U : P_U \to U$ a minimal projective cover of U and by $\sigma : U \to T$ the canonical surjection. Let $\beta : U \to T$ be an A-homomorphism. Since P_U is projective and σ is surjective, there is a homomorphism $\delta : P_U \to U$ such that $\sigma \circ \delta = \beta \circ \pi_U$. As P_U is uniserial, we have $\ker(\pi_U) \subset \ker(\delta)$. Thus there is $\alpha \in \mathrm{End}_A(U)$ such that $\alpha \circ \pi_U = \delta$, Composing this on the left by σ yields $\sigma \circ \alpha \circ \pi_U = \sigma \circ \delta = \sigma \circ \beta \circ \pi_U$. As π_U is surjective, it follows that $\sigma \circ \alpha = \beta$. In other words, the map σ_* sending $\alpha \in \mathrm{End}_A(U)$ to $\beta \in \mathrm{Hom}_A(U, T)$ is surjective. Clearly $\ker(\sigma_*) = \mathrm{Hom}_A(U, \mathrm{rad}(U))$. This shows the second isomorphism in the statement. Dually, the canonical injection $\rho : S \to U$ induces a map $\rho^* : \mathrm{End}_A(U) \to \mathrm{Hom}_A(S, U)$ mapping $\alpha \in \mathrm{End}_A(U)$ to $\alpha \circ \rho$. By dualising the above arguments one sees easily that ρ^* is surjective with kernel $\mathrm{Hom}_A(U, \mathrm{rad}(U))$ again, which yields the first isomorphism. $\qquad\square$

Corollary 11.3.11 *Let A be an indecomposable selfinjective serial k-algebra. There is a finite-dimensional division algebra D over k such that $\mathrm{End}_A(S) \cong D$ for any simple A-module S.*

Proof If A is simple, it has only one isomorphism class of simple modules. Thus we may assume that A is not simple. Let I be a set of representatives of the conjugacy classes of primitive idempotents in A. For any $i \in I$ set $S_i = Ai/J(A)i$. Choose some $i \in I$ and denote by U the unique quotient of length 2 of Ai. Then $U/\mathrm{rad}(U) \cong S_i$ and $\mathrm{soc}(U) \cong S_{\pi(i)}$, where π is the transitive cycle on I determined by 11.3.4. We have $\mathrm{End}_A(S_i) \cong \mathrm{End}_A(S_{\pi(i)})$ by 11.3.10, and hence 11.3.11 follows from the transitivity of π. $\qquad \square$

11.4 Modules over selfinjective serial algebras

Let k be a field. A striking feature of a selfinjective serial k-algebra is that it has only finitely many isomorphism classes of indecomposable modules. These are classified as follows.

Theorem 11.4.1 *Let A be an indecomposable non simple selfinjective serial k-algebra. Let I be a set of representatives of the conjugacy classes of primitive idempotents in A, and set $S_i = Ai/J(A)i$ for any $i \in I$. Denote by π the cyclic permutation on I satisfying $J(A)i/J(A)^2i \cong S_{\pi(i)}$ and let q be the composition length of the projective indecomposable A-modules Ai, where $i \in I$.*

   (i) *For any $i \in I$ and any integer a such that $1 \le a \le q$ there is, up to isomorphism, a unique indecomposable A-module $U_{i,a}$ satisfying $U_{i,a}/\mathrm{rad}(U_{i,a}) \cong S_i$ and $\ell(U_{i,a}) = a$.*
  (ii) *For any $i \in I$ and any integer a such that $1 \le a \le q$, the composition factors from top to bottom of the unique composition series of $U_{i,a}$ are S_i, $S_{\pi(i)}, \ldots, S_{\pi^{a-1}(i)}$.*
 (iii) *The set $\{U_{i,a}\}_{i\in I, 1\le a\le q}$ is a complete set of representatives of the isomorphism classes of indecomposable A-modules; in particular, there are $|I| \cdot q$ isomorphism classes of indecomposable A-modules.*

Proof Let U be an indecomposable A-module. By the hypotheses on A, the module U is uniserial. In particular, $U/\mathrm{rad}(U)$ is simple. Thus there is a unique $i \in I$ such that $U/\mathrm{rad}(U) \cong S_i = Ai/J(A)i$. Therefore, Ai is a projective cover of U, and hence $U \cong Ai/V$ for some submodule V of Ai. Since Ai is uniserial, V is uniquely determined by U. If $a = \ell(U)$, then $\ell(V) = q - a$. All statements follow from 11.3.4. $\qquad \square$

A selfinjective serial algebra is a direct product of finitely many indecomposable selfinjective serial algebras. Thus the last statement of the preceding theorem has the following consequence:

Corollary 11.4.2 *Any selfinjective serial k-algebra has finite representation type.*

We prove next a technical lemma, which is going to be used in the section below in order to determine the structure of symmetric algebras stably equivalent to symmetric serial algebras.

Lemma 11.4.3 *Let A be a selfinjective serial k-algebra, let U, V be indecomposable A-modules and let φ, $\psi : U \to V$ be A-homomorphisms. The following are equivalent:*

(i) $\mathrm{Im}(\varphi) = \mathrm{Im}(\psi)$.
(ii) $\ker(\varphi) = \ker(\psi)$.
(iii) There is an automorphism α of U such that $\varphi = \psi \circ \alpha$.
(iv) There is an automorphism β of V such that $\psi = \beta \circ \varphi$.

Proof Since U, V are uniserial, any submodule of U or V is completely determined by its composition length. As $\ell(\mathrm{Im}(\varphi)) = \ell(U) - \ell(\ker(\varphi))$, we get the equivalence of (i) and (ii). Again since U is uniserial, any automorphism α of U maps every submodule to itself; thus $\ker(\varphi) = \ker(\varphi \circ \alpha)$. Thus (iii) implies (ii). Similarly, any automorphism β of V preserves any submodule of V; thus $\mathrm{Im}(\psi) = \mathrm{Im}(\beta \circ \psi)$. Hence (iv) implies (i). Suppose that (ii) holds. As V is uniserial, it has a uniserial injective envelope I_V. We factor φ in the obvious way

$$U \twoheadrightarrow U/\ker(\varphi) \hookrightarrow V \hookrightarrow I_V$$

and ψ similarly. Since $\ker(\varphi) = \ker(\psi)$ and I_V is injective, there is an endomorphism ϵ of I_V such that $\epsilon \circ \varphi = \psi$ (where we identify V to its image in I_V). The ϵ restricts to an endomorphism β of V since I_V is uniserial. As β restricts to the identity on $U/\ker(\varphi) = U/\ker(\psi)$, it follows that β is injective, hence an automorphism. Thus (ii) implies (iii). Suppose finally that (i) holds. As U is uniserial, it has a uniserial projective cover $\pi_U : P_U \to U$. Since π_U is surjective, we have $\mathrm{Im}(\varphi \circ \pi_U) = \mathrm{Im}(\varphi)$ and $\mathrm{Im}(\psi \circ \pi_U) = \mathrm{Im}(\psi)$. Thus, as P_U is projective, there is an endomorphism δ of P_U such that $\psi \circ \pi_U \circ \delta = \varphi \circ \pi_U$. As P_U is uniserial, δ induces an endomorphism α on U such that $\alpha \circ \pi_U = \pi_U \circ \delta$. Then $\varphi \circ \pi_U = \psi \circ \pi_U \circ \delta = \psi \circ \alpha \circ \pi_U$. As π_U is surjective, it follows from this equality that we have $\varphi = \psi \circ \alpha$. Then α has to be surjective as $\mathrm{Im}(\varphi) = \mathrm{Im}(\psi)$, thus α is an automorphism of U. This shows that (i) implies (iv) and concludes the proof. $\square$

Thanks to the precise description of the module category of a selfinjective serial algebra, we can get a very detailed understanding of its stable category. We start by looking at the effect of the Heller operator on modules; we will show, that every self stable equivalence of Morita type is the composition of a power of the Heller operator and a Morita equivalence.

Proposition 11.4.4 *Let A be an indecomposable non simple selfinjective serial k-algebra and let I be a set of representatives of the conjugacy classes of primitive idempotents in A. Let q be the length of the projective indecomposable A-modules. Denote by π the transitive cycle on I satisfying $J(A)i/J(A)^2 i \cong A\pi(i)/J(A)\pi(i)$ for any $i \in I$. For any $i \in I$ and any integer a such that $1 \le a \le q$, denote by $U_{i,a}$ the quotient of Ai of composition length a. We have $\Omega(U_{i,a}) \cong U_{\pi^a(i),q-a}$ and $\Omega^2(U_{i,a}) \cong U_{\pi^q(i),a}$. In particular, if A is symmetric, then $\Omega^2(U_{i,a}) \cong U_{\pi(i),a}$.*

Proof By the definition of Ω, there is a short exact sequence

$$0 \longrightarrow \Omega(U_{i,a}) \longrightarrow Ai \longrightarrow U_{i,a} \longrightarrow 0.$$

As the last composition factor of $U_{i,a}$ is $S_{\pi^{a-1}(i)}$, the top composition factor of $\Omega(U_{i,a})$ is $S_{\pi^a(i)}$ by 11.3.4 and 11.4.1. Since $q = \ell(Ai)$, we have $\ell(\Omega(U_{i,a})) = q - a$. This shows the first isomorphism. Applying Ω again yields the second isomorphism. Again by 11.3.4, if A is symmetric, then the order of π divides $q - 1$, so $\pi^q(i) = \pi(i)$, implying the last statement. $\qquad\square$

Corollary 11.4.5 *Let A be an indecomposable non simple symmetric serial k-algebra, and let I be a set of representatives of the conjugacy classes of primitive idempotents in A. Denote by q the length of the projective indecomposable A-modules. If $q \ge 3$, then every simple A-module has period $2|I|$. If q is odd, then every indecomposable nonprojective A-module has period $2|I|$.*

Proof Suppose that $q \ge 3$. Let S be a simple A-module. Then applying an odd power of Ω to S yields a module of length $q - 1$, which is different from 1 by the assumption on q. Thus 11.4.4 implies that S has period $2|I|$. Suppose that q is odd. Let U be an indecomposable nonprojective A-module of length a. Since q is odd, we have $q - a \ne a$, and hence any odd power of Ω applied to U yields a module that is not isomorphic to U. The result follows from 11.4.4. $\qquad\square$

The hypotheses on q are necessary in this Corollary. If $A = k[x]/(x^q)$ is the local uniserial algebra of length q with $q = 2^n$ for some positive integer n, then A uniserial module of length 2^{n-1} has period 1.

For the understanding of the stable category, it is crucial to be able to detect homomorphisms that factor through a projective module. We have the following criterion, due to Alperin [4, Lemma 3]:

Proposition 11.4.6 *Let A be an indecomposable selfinjective serial k-algebra and let U, V be indecomposable A-modules and* $\alpha : U \to V$ *a nonzero A-homomorphism. The following are equivalent:*

  (i) The homomorphism α *factors through a projective A-module.*
  (ii) We have $\ell(\Omega(V)) \leq \ell(\ker(\alpha))$.
  (iii) We have $\ell(\Omega^{-1}(U)) \leq \ell(\mathrm{coker}(\alpha))$.

In particular, if there is a nonzero homomorphism $U \to V$ *that factors through a projective A-module, then* $\ell(\Omega(V)) < \ell(U)$.

Proof Let $\pi : P \to V$ be a projective cover of V. Denote by X the inverse image of $\mathrm{Im}(\alpha)$ in P; that is, π induces an isomorphism $P/X \cong V/\mathrm{Im}(\alpha)$. In particular, we have $\ell(\Omega(V)) = \ell(P) - \ell(V) = \ell(X) - \ell(\mathrm{Im}(\alpha)) = \ell(X) - \ell(U) + \ell(\ker(\alpha))$. Using an injective envelope $U \to I$ of U and the fact that $\ell(P) = \ell(I)$ by 11.3.4, we get in the same way that $\ell(\Omega^{-1}(U)) = \ell(X) - \ell(U) + \ell(\mathrm{coker}(\alpha))$. Thus both inequalities in (i) and (ii) are equivalent to $\ell(X) \leq \ell(U)$. The homomorphism α factors through a projective module if and only if it factors through $\pi : P \to V$, hence if and only if there is a homomorphism $\tau : U \to P$ satisfying $\alpha = \pi \circ \tau$, which happens if and only if there is a homomorphism $\tau : U \to P$ satisfying $\mathrm{Im}(\tau) = X$. Since $U, X, \mathrm{Im}(\alpha)$ all have isomorphic quotients, such a τ exists indeed if and only if $\ell(U) \geq \ell(X)$. $\square$

Corollary 11.4.7 *Let A be an indecomposable selfinjective serial k-algebra and let U, V be indecomposable A-modules and* $\alpha : U \to V$ *an A-homomorphism that does not factor through a projective module. If* $U \cong J(A)i \neq 0$ *for some primitive idempotent* $i \in A$ *then* α *is surjective, and if* $V \cong J(A)i \neq 0$ *for some primitive idempotent* $i \in A$ *then* α *is injective.*

Proof If $U \cong J(A)i$ then $\ell(\Omega^{-1}(U)) = 1$, and if $V \cong J(A)i$, then $\ell(\Omega(V)) = 1$, so in both cases the conclusion follows from 11.4.6. $\square$

Proposition 11.4.6 can be exploited to determine the dimension of the stable homomorphism spaces between any two finitely generated modules over a selfinjective serial algebra.

Proposition 11.4.8 *Let A be a split indecomposable nonsimple selfinjective serial k-algebra and let U be an indecomposable A-module. We have* $\underline{\mathrm{End}}_A(U) \cong k$ *if and only if* $\ell(U) \leq |I|$ *or* $\ell(\Omega(U)) \leq |I|$.

Proof If $\ell(U) \leq |I|$, then by 11.4.1 the composition factors of U are pairwise nonisomorphic, and hence $\mathrm{End}_A(U) \cong k$; in particular, $\underline{\mathrm{End}}_A(U) \cong k$. If $\ell(\Omega(U)) \leq |I|$, then $\underline{\mathrm{End}}_A(U) \cong \underline{\mathrm{End}}_A(\Omega(U)) \cong k$, where we use the fact that Ω induces an equivalence on $\underline{\mathrm{mod}}(A)$. Suppose conversely that $\underline{\mathrm{End}}_A(U) \cong k$.

After replacing U by $\Omega(U)$, if necessary, we may assume that $\ell(U) \leq \ell(\Omega(U))$. Let α be a nonzero endomorphism of U. If α factors through a projective A-module, then 11.4.6 implies that $\ell(\Omega(U)) \leq \ell(\ker(\alpha)) < \ell(U)$, a contradiction. This shows that no nonzero endomorphism of U factors through a projective A-module, and hence $\mathrm{End}_A(U) \cong \underline{\mathrm{End}}_A(U) \cong k$. It follows from 11.4.1 that the composition factors of U are pairwise nonisomorphic, and hence that $\ell(U) \leq |I|$. The result follows. $\qquad\square$

11.5 Picard groups of selfinjective serial algebras

For A a finite-dimensional k-algebra we denote as before by $\mathrm{Aut}_0(A)$ the group of all automorphisms of A that stabilise the isomorphism classes of all finitely generated A-modules.

Proposition 11.5.1 *Let A be an indecomposable nonsimple selfinjective serial k-algebra. Let $\alpha \in \mathrm{Aut}(A)$. There is an indecomposable A-module U such that $_\alpha U \cong U$ if and only if $\alpha \in \mathrm{Aut}_0(A)$.*

Proof Note that $_\alpha U$ and U have the same composition length for any A-module U. Suppose that U is an indecomposable A-module such that $_\alpha U \cong U$. Since U is uniserial, its socle S is simple and satisfies $_\alpha S \cong S$. It follows from 2.14.3 and 11.3.5 that α stabilises any simple A-module. By 11.4.1, any indecomposable A-module is determined by its socle and composition length. Thus $\alpha \in \mathrm{Aut}_0(A)$. The converse is trivial. $\qquad\square$

Proposition 11.5.1 shows that an automorphism that stabilises the isomorphism class of one indecomposable A-module automatically stabilises the isomorphism classes of all indecomposable modules. The analogous statement for self stable equivalences is not true in general: if A is split local uniserial of dimension 4, then Ω stabilises the unique isomorphism class of 2-dimensional indecomposable modules, but not the unique isomorphism class of simple modules. The following proposition is a sufficient criterion for a self stable equivalence of Morita type of A to stabilise the isomorphism classes of all indecomposable A-modules.

Proposition 11.5.2 *Let A be an indecomposable nonsimple selfinjective serial k-algebra. Suppose that $A/J(A)$ is separable. Let M and N be indecomposable A-A-bimodules inducing a stable equivalence of Morita type on A. Suppose that $M \otimes_A S \cong S$ in $\underline{\mathrm{mod}}(A)$ for some simple A-module S. Then M and N induce a Morita equivalence satisfying $M \otimes_A U \cong U$ in $\mathrm{mod}(A)$ for any finitely generated A-module U; equivalently, we have $M \cong A_\alpha$ for some $\alpha \in \mathrm{Aut}_0(A)$.*

Proof Let I be a set of representatives of the conjugacy classes of primitive idempotents in A, and set $S_i = Ai/J(A)i$ for all $i \in A$. Denote by π the transitive cycle from 11.3.5 satisfying $\Omega^2(S_i) \cong S_{\pi(i)}$ for all $i \in I$. Since $M \otimes_A -$ commutes with Ω on $\underline{\mathrm{mod}}(A)$, it follows that $M \otimes_A S \cong S$ in $\underline{\mathrm{mod}}(A)$ for any simple A-module. By 4.14.9 we have $M \otimes_A S \cong S$ in $\mathrm{mod}(A)$ for any simple A-module S, and therefore, by 4.14.10, the bimodules M and N induce a Morita equivalence. As in the proof of the previous proposition, it follows from the structure of indecomposable A-modules that $M \otimes_A -$ stabilises the isomorphism classes of all finitely generated A-modules, and hence that $M \cong A_\alpha$ for some $\alpha \in \mathrm{Aut}_0(A)$ by 2.8.16 (v). $\qquad\square$

Proposition 11.5.3 *Let A be an indecomposable nonsimple selfinjective serial k-algebra. Suppose that $A/J(A)$ is separable. Let n be an integer.*

(i) *If n is even, then the bimodule $\Omega^n_{A \otimes_k A^{\mathrm{op}}}(A)$ induces a Morita equivalence on A.*

(ii) *If n is odd, and if the composition length of the projective indecomposable A-modules is at least 3, then the bimodule $\Omega^n_{A \otimes_k A^{\mathrm{op}}}(A)$ does not induce a Morita equivalence.*

Proof Let I be a set of representatives of the conjugacy classes of primitive idempotents in A, and set $S_i = Ai/J(A)i$ for all $i \in A$. Denote by π the transitive cycle from 11.3.5 satisfying $\Omega^2(S_i) \cong S_{\pi(i)}$ for all $i \in I$. Suppose first that n is even. Write $n = 2t$ for some integer t. It follows that $\Omega^n_{A \otimes_k A^{\mathrm{op}}}(A) \otimes_A S_i \cong S_{\pi^t(i)}$. Thus $\Omega^n_{A \otimes_k A^{\mathrm{op}}}(A)$ induces a stable equivalence of Morita type which permutes the isomorphism classes of simple A-modules. It follows from 4.14.10 that $\Omega^n_{A \otimes_k A^{\mathrm{op}}}(A)$ induces a Morita equivalence. Suppose that n is odd. Write $n = 2t + 1$. Let S be a simple A-module. Then $S' = \Omega^{2t}(S)$ is simple. Thus the composition length of $\Omega^n(S) \cong \Omega(S')$ is $q - 1$ where q is the composition length of the projective indecomposable A-modules. Since $q \geq 3$ we have $q - 1 \geq 2$, hence $\Omega^n(S)$ is not simple, and therefore Ω^n is not induced by a Morita equivalence. $\qquad\square$

Theorem 11.5.4 *Let A be an indecomposable nonsimple selfinjective serial k-algebra. Suppose that A is split. Let M and N be indecomposable A-A-bimodules that induce a stable equivalence of Morita type on A. There is an integer n and an automorphism $\alpha \in \mathrm{Aut}_0(A)$ such that*

$$M \cong (\Omega^n_{A \otimes_k A^{\mathrm{op}}}(A))_\alpha$$

as A-A-bimodules. In particular, we have $M \otimes_A U \cong \Omega^n(U)$ in $\underline{\mathrm{mod}}(A)$ for any A-module U.

Proof We show first that the stable equivalence induced by M sends any simple module to a Heller translate of a simple module. Let S be a simple A-module, and let i be a primitive idempotent in A such that $S \cong$ soc(Ai). Denote by q the composition length of the projective indecomposable A-modules. If V is an indecomposable A-module such that $\underline{\mathrm{Hom}}_A(S, V) \neq \{0\}$, then soc$(V) \cong S$. Thus V is isomorphic to a submodule of Ai. It follows that up to isomorphism, there are exactly $q - 1$ nonprojective indecomposable A-modules V satisfying $\underline{\mathrm{Hom}}_A(S, V) \neq \{0\}$. If V, V' are submodules of Ai, then $V \subseteq V'$ or $V' \subseteq V$, because Ai is uniserial. It follows from 4.13.3 that at least one of the spaces $\underline{\mathrm{Hom}}_A(V, V')$ or $\underline{\mathrm{Hom}}_A(V', V)$ is nonzero. The image of S under any stable equivalence must have the analogous property. Let U be the up to isomorphism unique direct summand of $M \otimes_A S$ that is not projective. Since $\underline{\mathrm{End}}_A(U) \cong \underline{\mathrm{End}}_A(S) \cong k$, it follows from 11.4.8 that $\ell(U) \leq |I|$ or $\ell(\Omega(U)) \leq |I|$. After replacing M by $\Omega_{A \otimes_k A^{\mathrm{op}}}(M)$, if necessary, we may assume that $\ell(U) \leq |I|$ and that U is not isomorphic to $\Omega(S')$ for any simple A-module S'. We will show that then U is simple. Arguing by contradiction, suppose that $2 \leq \ell(U) \leq |I|$. In particular, this forces $|I| \geq 2$. Let T be the simple quotient of U. Then T is not isomorphic to soc(U) because of the assumption $2 \leq \ell(U) \leq |I|$ and the description of the composition series of U as in 11.4.1. Since U is not isomorphic to $\Omega(S')$ for any simple A-module S', it follows from 11.4.1 that U is contained in a nonprojective indecomposable A-module W such that $\ell(W) = \ell(U) + 1$. The top composition factor of W is not isomorphic to the top composition factor T of U. Thus $\mathrm{Hom}_A(W, T) = \{0\}$. Since soc$(W) = $ soc(U), we also have $\mathrm{Hom}_A(T, W) = \{0\}$. However, by the first argument, at least one of these two spaces has to be nonzero, a contradiction. This shows that U is simple. It follows from 11.3.5 that after replacing M by $\Omega^n(M)$ for some integer n we may assume that $M \otimes_A S \cong S$ in $\underline{\mathrm{mod}}(A)$. Then 11.5.2 implies that $M \cong A_\alpha$ for some $\alpha \in \mathrm{Aut}_0(A)$. The result follows. $\square$

Exercise 11.5.5 Let A be an indecomposable nonsimple selfinjective serial k-algebra such that A is split. The purpose of this exercise is to reformulate the above theorem in terms of stable Picard groups. For the notation, see 4.15.1, 2.8.16, 2.17.16.

(a) Show that $\mathrm{StPic}(A)$ is generated by the images of $\Omega_{A \otimes_k A^{\mathrm{op}}}(A)$ and of $\mathrm{Out}_0(A)$ in $\mathrm{StPic}(A)$.

(b) Show that if the composition length of the projective indecomposable A-modules is at least 3, then the Picard group $\mathrm{Pic}(A)$ is generated by the images of $\Omega^2_{A \otimes_k A^{\mathrm{op}}}(A)$ and $\mathrm{Out}_0(A)$; deduce that $\mathrm{Pic}(A)$ is normal of index two in $\mathrm{StPic}(A)$, and that the nontrivial class of $\mathrm{StPic}(A)/\mathrm{Pic}(A)$ is represented by the bimodule $\Omega_{A \otimes_k A^{\mathrm{op}}}(A)$.

(c) Show that if the projective indecomposable A-modules have length 2, then $\mathrm{StPic}(A) = \mathrm{Pic}(A)$.

Exercise 11.5.6 Let A be an indecomposable nonsimple symmetric serial k-algebra. Denote by q the composition length of the projective indecomposable A-modules, and by e the number of isomorphism classes of simple A-modules. Set $m = \frac{q-1}{e}$. Show the Cartan matrix of A is equal to the $e \times e$ matrix of the form

$$
\begin{pmatrix}
m+1 & m & \cdots & m & m \\
m & m+1 & \cdots & m & m \\
\cdots & \cdots & \cdots & \cdots & \cdots \\
m & m & \cdots & m+1 & m \\
m & m & \cdots & m & m+1
\end{pmatrix}
$$

and that its determinant is equal to $em + 1$.

11.6 Symmetric algebras stably equivalent to a serial algebra

Let A, B be indecomposable symmetric non simple algebras over a field k. Suppose that B is serial and that A, B are stably equivalent. By work of Gabriel and Riedtmann [73] these hypotheses are sufficient to determine the algebra structure of A. We follow the presentation from [123]. Let I and J be sets of representatives of the conjugacy classes of primitive idempotents in A and B, respectively. For any $i \in I$ we set $S_i = Ai/J(A)i$ and for any $j \in J$ we set $T_j = Bj/J(B)j$. Thus $\{S_i\}_{i \in I}$ and $\{T_j\}_{j \in J}$ are sets of representatives of the isomorphism classes of simple modules over A and B, respectively. We fix k-linear equivalences $\mathcal{F} : \underline{\mathrm{mod}}(A) \cong \underline{\mathrm{mod}}(B)$ and $\mathcal{G} : \underline{\mathrm{mod}}(B) \cong \underline{\mathrm{mod}}(A)$ between the stable categories of finitely generated modules over A and B, respectively. Since $\mathcal{F}$ is an equivalence, it maps every non projective indecomposable A-module to a B-module, which is indecomposable as an object of the stable category; that is, which is a direct sum of an indecomposable nonprojective B-module and some projective B-module. Thus we may choose the notation in such a way that for any indecomposable nonprojective A-module U, the B-module $\mathcal{F}(U)$ is again indecomposable nonprojective, and for any indecomposable nonprojective B-module V, the A-module $\mathcal{G}(V)$ is again indecomposable nonprojective. The fact that $\mathcal{F}$ is a k-linear equivalence means that for any two A-modules U, U', the functors $\mathcal{F}$ induces a linear isomorphism

$$
\underline{\mathrm{Hom}}_A(U, U') \cong \underline{\mathrm{Hom}}_B(\mathcal{F}(U), \mathcal{F}(U')),
$$

which is compatible with composition of morphisms in the stable categories of A, B, and which is in particular an algebra isomorphism for $U = U'$. A similar statement holds for $\mathcal{G}$. We will frequently make use of the basic facts on stable equivalences from §4.13. The first result shows that A and B have the same number of isomorphism classes of simple modules. The proof we give here is due to J. A. Green [77]; it shows more precisely, that the choice of a stable equivalence $\mathcal{F}$ determines two bijections between the sets of isomorphism classes of simple modules over A and B.

Proposition 11.6.1 *There are unique maps γ, $\delta : I \to J$ such that, for any $i \in I$, we have isomorphisms $\mathrm{soc}(\mathcal{F}(S_i)) \cong T_{\gamma(i)}$ and $\mathcal{F}(S_i)/\mathrm{rad}(\mathcal{F}(S_i)) \cong T_{\delta(i)}$. The maps γ and δ are bijections. In particular, we have $|I| = |J|$; that is, A and B have the same number of isomorphism classes of simple modules.*

Proof Since B is serial, for any $i \in I$ the B-module $\mathcal{F}(S_i)$ is uniserial, hence $\mathrm{soc}(\mathcal{F}(S_i)) \cong T_{\gamma(i)}$ for a uniquely determined element $\gamma(i) \in J$. Similarly, $\mathcal{F}(S_i)/\mathrm{rad}(\mathcal{F}(S_i)) \cong T_{\delta(i)}$ for a uniquely determined element $\delta(i) \in J$. Let $j \in J$ and let $i \in I$ such that S_i is isomorphic to a summand of the semisimple A-module $\mathrm{soc}(\mathcal{G}(T_j))$. Thus, by 4.13.4, we have $0 \neq \underline{\mathrm{Hom}}_A(S_i, \mathcal{G}(T_j)) \cong \underline{\mathrm{Hom}}_B(\mathcal{F}(S_i), T_j)$, hence $j = \delta(i)$. This shows that δ is surjective; a similar argument shows that γ is surjective. Let i, $i' \in I$ such that $\delta(i) = \delta(i')$. This means that the uniserial modules $\mathcal{F}(S_i)$, $\mathcal{F}(S_{i'})$ have the same projective cover Bj, where $j = \delta(i) = \delta(i')$. Choose notation such that $\ell(\mathcal{F}(S_i)) \geq \ell(\mathcal{F}(S_{i'}))$, where ℓ denotes the composition length. Then, since the involved modules are uniserial, there is a surjective B-homomorphism from $\mathcal{F}(S_i)$ onto $\mathcal{F}(S_{i'})$. It follows from 4.13.3 and 4.13.4 that $0 \neq \underline{\mathrm{Hom}}_B(\mathcal{F}(S_i), \mathcal{F}(S_{i'})) \cong \underline{\mathrm{Hom}}_A(S_i, S_{i'}) \cong \mathrm{Hom}_A(S_i, S_{i'})$, and hence $i = i'$. This shows that δ is injective; a similar argument yields that γ is injective. $\qquad\Box$

Proposition 11.6.2 *For any simple A-module S and any simple B-module T we have $\mathrm{End}_A(S) \cong \mathrm{End}_B(T)$. In particular, A is split if and only if B is split.*

Proof With the notation of 11.6.1, let $i \in I$, set $S = S_i$ and $T = T_{\delta(i)}$. We have $\mathrm{End}_A(S) \cong \underline{\mathrm{End}}_A(S) \cong \underline{\mathrm{End}}_B(\mathcal{F}(S))$. The surjection $\mathcal{F}(S) \to T$ induces, by 11.3.10, a surjection $\mathrm{End}_B(\mathcal{F}(S)) \to \mathrm{End}_B(T)$. The right side is a division algebra, and hence this map induces a surjection $\underline{\mathrm{End}}_B(\mathcal{F}(S)) \to \mathrm{End}_B(T)$. But $\underline{\mathrm{End}}_B(\mathcal{F}(S))$ is also a division algebra, so the last map is an isomorphism. Thus $\mathrm{End}_A(S) \cong \mathrm{End}_B(T)$. By 11.3.11, the isomorphism class of the division algebra $\mathrm{End}_B(T)$ does not depend on T, whence the result. $\qquad\Box$

We state now two main results on the structure of projective indecomposable A-modules. We keep the notation γ, δ for the two bijections between I and

J introduced in 11.6.1. We denote by π the transitive cycle on J satisfying $\Omega^2(T_j) \cong T_{\pi(j)}$ for all $j \in J$, from 11.3.4.

Theorem 11.6.3 *For any $i \in I$ there are unique submodules U_i, V_i of Ai such that $U_i \cong \mathcal{G}(T_{\delta(i)})$ and $V_i \cong \mathcal{G}(\Omega_B(T_{\gamma(i)}))$. These submodules have moreover the following properties.*

(i) *The submodules U_i and V_i of Ai are uniserial.*

(ii) *We have $U_i \cap V_i = \mathrm{soc}(Ai)$.*

(iii) *We have $U_i + V_i = \mathrm{rad}(Ai)$.*

(iv) *We have $J(A)i/\mathrm{soc}(Ai) = U_i/\mathrm{soc}(Ai) \oplus V_i/\mathrm{soc}(Ai)$, and the modules $U_i/\mathrm{soc}(Ai)$ and $V_i/\mathrm{soc}(Ai)$ have no common composition factor.*

(v) *Every proper submodule of Ai is equal to $U_i' + V_i'$ for some submodule U_i' of U_i and some submodule V_i' of V_i. In particular, Ai has only finitely many submodules.*

(vi) *Every uniserial submodule of Ai is contained in U_i or V_i.*

(vii) *The set of submodules $\{U_i, V_i\}$ of Ai is uniquely determined by the properties (i), (ii), and (iii).*

Theorem 11.6.4 *With the notation from 11.6.3, there are unique permutations ρ, σ on I such that, for any $i \in I$, we have $U_i/\mathrm{rad}(U_i) \cong S_{\rho(i)}$ and $V_i/\mathrm{rad}(V_i) \cong S_{\sigma(i)}$. Moreover, the following hold.*

(i) *The simple composition factors of U_i are $S_{\rho(i)}, S_{\rho^2(i)}, \ldots, S_{\rho^a(i)} \cong S_i$, where $a = \ell(U_i)$. The number a depends only on the ρ-orbit of i and is equal to a multiple of the length of the ρ-orbit of i. Similarly, the simple composition factors of V_i are $S_{\sigma(i)}, S_{\sigma^2(i)}, \ldots, S_{\sigma^b(i)} \cong S_i$, where $b = \ell(V_i)$. The number b depends only on the σ-orbit of i and is equal to a multiple of the length of the σ-orbit of i. Moreover, at least one of a, b is equal to the length of the ρ-orbit, σ-orbit of i, respectively.*

(ii) *The permutation $\sigma \circ \rho$ is a transitive cycle on I; a ρ-orbit and a σ-orbit have at most one common element. In particular, if $|I| \geq 2$, then $\rho \neq \sigma$.*

(iii) *For any $i \subset I$, there are short exact sequences*

$$0 \longrightarrow U_{\sigma(i)} \longrightarrow A_{\sigma(i)} \xrightarrow{\alpha_{\sigma(i)}} V_i \longrightarrow 0$$

$$0 \longrightarrow V_{\rho(i)} \longrightarrow A_{\rho(i)} \xrightarrow{\beta_{\rho(i)}} U_i \longrightarrow 0 .$$

(iv) *For any $i \in I$ we have $\Omega_A(V_i) \cong U_{\sigma(i)}$ and $\Omega_A(U_i) \cong V_{\rho(i)}$.*

(v) *We have $\rho = \gamma^{-1} \circ \delta$ and $\sigma = \delta^{-1} \circ \pi \circ \gamma$.*

Theorem 11.3.4, describing the structure of symmetric serial algebras, is the special case of Theorem 11.6.4 in which one of the permutations ρ, σ is the identity.

Corollary 11.6.5 *Suppose that the projective indecomposable A-modules have length at least 3. Then the $2|I|$ modules U_i, V_i, with $i \in I$, are pairwise nonisomorphic and have period $2|I|$.*

Proof A projective indecomposable A-module of length at least 3 has at least two nonisomorphic indecomposable nonprojective submodules, so B has at least two nonisomorphic indecomposable nonprojective modules, and hence B does not have a projective indecomposable module of length 2. Since the modules U_i and V_i correspond to the simple B-modules and their Heller translates, the result follows from 11.4.5. $\qquad\Box$

For the proof of the two theorems above we will need a series of intermediate steps.

Lemma 11.6.6 *Let U, V, W be indecomposable nonprojective A-modules. Suppose that the B-module $\mathcal{F}(U)$ is simple, that the socles of the A-modules U, V, W are simple and that $\ell(\mathcal{F}(V)) \leq \ell(\mathcal{F}(W))$.*

 (i) If V, W are isomorphic to submodules of U, then W is isomorphic to a submodule of V.
 (ii) If U is isomorphic to a submodule of V and of W, then V is isomorphic to a submodule of W.

Proof Let $\alpha : V \to U$ and $\beta : W \to U$ be injective A-homomorphisms. In particular, α, β do not factor through a projective module. Thus the B-homomorphisms $\mathcal{F}(\alpha)$, $\mathcal{F}(\beta)$ from $\mathcal{F}(V)$ and $\mathcal{F}(W)$ to $\mathcal{F}(U)$ do not factor through a projective B-module, hence are both surjective as $\mathcal{F}(U)$ is simple. Since $\ell(\mathcal{F}(V)) \leq \ell(\mathcal{F}(W))$ there is a surjective B-homomorphism $\varphi : \mathcal{F}(W) \to \mathcal{F}(V)$. By 11.4.3 we can choose φ in such a way that $\mathcal{F}(\alpha) \circ \varphi = \mathcal{F}(\beta)$. Thus the A-homomorphism $\beta - \alpha \circ \mathcal{G}(\varphi)$ from W to U factors through a projective A-module. It follows from 4.13.6 that $\mathcal{G}(\varphi)$ is injective, whence (i). Similarly, let $\gamma : U \to V$ and $\delta : U \to W$ be injective A-homomorphisms. By the same arguments as before, $\mathcal{G}(\gamma)$, $\mathcal{G}(\delta)$ do not factor through a projective module, hence are injective as $\mathcal{F}(U)$ is simple. Thus $\mathcal{F}(V)$, $\mathcal{F}(W)$ have isomorphic socles, and since they are uniserial, there is an injective homomorphism $\psi : \mathcal{F}(V) \to \mathcal{F}(W)$. By 11.4.3 we can choose ψ in such a way that $\mathcal{F}(\delta) \circ \psi = \mathcal{F}(\gamma)$. Thus $\delta \circ \mathcal{G}(\psi) - \gamma$ factors through a projective A-module, and 4.13.6 implies again that $\mathcal{G}(\psi)$ is injective. $\qquad\Box$

Proposition 11.6.7 *Let U be an indecomposable A-modules and S a simple A-module. Then S appears with multiplicity at most 1 in a decomposition of $\mathrm{soc}(U)$ and a decomposition of $U/\mathrm{rad}(U)$.*

Proof This is trivial if U is projective, so we may assume that U is non-projective. Let α, β be injective homomorphisms from S to U. By 4.13.4, none of α, β factors through a projective module. Thus $\mathcal{F}(\alpha), \mathcal{F}(\beta)$ are two homomorphisms from $\mathcal{F}(S)$ to $\mathcal{F}(U)$ that do not factor through a projective module. Since $\mathcal{F}(S)$ is uniserial we may choose notation such that $\ker(\mathcal{F}(\beta)) \subseteq \ker(\mathcal{F}(\alpha))$. Since $\mathcal{F}(U)$ is also uniserial we have $\mathrm{Im}(\alpha) \subseteq \mathrm{Im}(\beta)$. Thus there is an endomorphism δ of $\mathcal{F}(S)$ satisfying $\mathcal{F}(\beta) \circ \delta = \mathcal{F}(\alpha)$. Then $\mathcal{G}(\delta)$ is an endomorphism of S such that $\beta \circ \mathcal{G}(\delta) - \alpha$ factors through a projective module. Since S is simple this forces $\beta \circ \mathcal{G}(\delta) - \alpha = 0$; in particular, $\mathrm{Im}(\alpha) = \mathrm{Im}(\beta)$ is the unique submodule of U isomorphic to S. A dual argument shows the corresponding statement for $U/\mathrm{rad}(U)$. $\qquad\square$

Proposition 11.6.8 *Let U be an indecomposable A-module such that $\mathrm{soc}(U)$ is simple. Any two different submodules of U are non-isomorphic.*

Proof Let V, W be two isomorphic submodules of U. If $\ell(V) = \ell(W) = 1$ then $U = V = \mathrm{soc}(U)$. Suppose that $\ell(V) \geq 2$. Arguing by induction we may assume that every maximal submodule of V is also a maximal submodule of W. If V has two different maximal submodules, their sum is equal to both V and W, thus we may assume that V, W have a unique maximal submodule M. If $V \neq W$ then $(V + W)/M \cong V/M \oplus W/M$. Note that $V + W$ is indecomposable since its socle is simple. Since $V/M \cong W/M$ this contradicts 11.6.7, whence the result. $\qquad\square$

Proposition 11.6.9 *For any $j \in J$ the A-modules $\mathcal{G}(T_j)$ and $\mathcal{G}(\Omega_B(T_j))$ are uniserial. Moreover, for any $i \in I$, we have $S_i \cong \mathrm{soc}(\mathcal{G}(T_{\delta(i)})) \cong \mathrm{soc}(\mathcal{G}(\Omega_B(T_{\gamma(i)})))$.*

Proof By the definition of the maps δ, γ, we have $\underline{\mathrm{Hom}}_A(S_i, \mathcal{G}(T_j)) \cong \underline{\mathrm{Hom}}_B(\mathcal{F}(S_i), T_j)$, and this is nonzero if and only if $j = \delta(i)$. Similarly, $\underline{\mathrm{Hom}}_A(S_i, \mathcal{G}(\Omega_B(T_j))) \cong \underline{\mathrm{Hom}}_B(\mathcal{F}(S_i), \Omega_B(T_j))$ is nonzero if and only if $j = \gamma(i)$, because $\Omega_B(T_j)$ is equal to the maximal submodule $J(B)j$ of Bj, and any homomorphism from an indecomposable B-module to $J(B)j$ that does not factor through a projective module is necessarily injective, by 11.4.7. This shows that the socle of $\mathcal{G}(T_j)$ is simple, as claimed. Let U, V be submodules of $\mathcal{G}(T_j)$ such that $\ell(\mathcal{F}(U)) \geq \ell(\mathcal{F}(V))$. In order to show that $\mathcal{G}(T_j)$ is uniserial it suffices, by 11.3.3, to show that U contains V. Since $\mathcal{F}(\mathcal{G}(T_j)) \cong T_j$ is simple, there is, by 11.6.6, an injective homomorphism $\pi : U \rightarrow V$. Since the socle of

$\mathcal{G}(T_j)$ is simple it follows from 11.6.8 that $U = \mathrm{Im}(\pi)$, hence that $U \subseteq V$ as required.   □

Proof of Theorem 11.6.3 Let $i \in I$. By 11.6.9 there are submodules U_i, V_i of Ai such that $U_i \cong \mathcal{G}(T_{\delta(i)})$ and $V_i \cong \mathcal{G}(\Omega_B(T_{\gamma(i)}))$. By 11.6.8, U_i, V_i are unique. It follows from 11.6.9 that U_i is uniserial. It follows also from 11.6.9, but applied to the stable equivalence $\mathcal{G} \circ \Omega_B$, that V_i is uniserial. This shows (i). Set $X = U_i \cap V_i$. We have $S_i \cong \mathrm{soc}(Ai) \subseteq X$. If $\ell(\mathcal{F}(X)) \geq \ell(\mathcal{F}(\mathrm{soc}(Ai)))$ then X is isomorphic to a submodule of $\mathrm{soc}(Ai)$ by 11.6.6 (i) hence $X = \mathrm{soc}(Ai)$. If $\ell(\mathcal{F}(X)) \leq \ell(\mathcal{F}(\mathrm{soc}(Ai)))$ then $\ell(\Omega_B^{-1}(\mathcal{F}(X))) \geq \ell(\Omega_B^{-1}(\mathcal{F}(\mathrm{soc}(Ai))))$, hence again $X = \mathrm{soc}(Ai)$ by 11.6.6 applied with $\Omega_B^{-1} \circ \mathcal{F}$ instead of $\mathcal{F}$. This proves (ii). Set $Y = U_i + V_i$. Since U_i, V_i are indecomposable non-projective we have $Y \subseteq J(A)i$. Note that $J(A)i = \Omega_A(S_i)$, hence $\mathcal{F}(J(A)i) \cong \Omega_B(\mathcal{F}(S_i))$. If $\ell(\mathcal{F}(Y)) \geq \ell(\mathcal{F}(J(A)i))$ then $J(A)i$ is isomorphic to a submodule of Y by 11.6.6 (ii), hence $X = J(A)i$. If $\ell(\mathcal{F}(Y)) \geq \ell(J(A)i)$ then again by 11.6.6 applied with $\Omega_B^{-1} \circ \mathcal{F}$ instead of $\mathcal{F}$ we get $Y = J(A)i$. This proves (iii). The equality $J(A)i/\mathrm{soc}(Ai) = U_i/\mathrm{soc}(Ai) \oplus V_i/\mathrm{soc}(Ai)$ follows from (ii) and (iii). Suppose that $U_i/\mathrm{soc}(Ai)$ and $V_i/\mathrm{soc}(Ai)$ have a common composition factor S'. Then there are (necessarily uniserial) submodules $U' \subseteq U_i$ and $V' \subseteq V_i$ which properly contain $\mathrm{soc}(Ai) = U_i \cap V_i$ and whose top composition factor is S'. But then $W = U' + V'$ is an indecomposable module (because its socle is simple) such that S' occurs with multiplicity at least 2 in $W/\mathrm{rad}(W)$, contradicting 11.6.7. This proves (iv). Any proper nontrivial submodule of Ai contains $\mathrm{soc}(Ai)$ and is contained in $J(Ai) = U_i + V_i$. Statement (v) follows from (iv) and 1.8.5, applied to $U_i/\mathrm{soc}(Ai) \oplus V_i/\mathrm{soc}(Ai)$. Let U be a proper uniserial submodule of Ai. Then U is a uniserial submodule of $J(A)i = U_i + V_i$. Its image in $J(A)i/\mathrm{soc}(Ai)$ is uniserial, hence contained in $U_i/\mathrm{soc}(Ai)$ or $V_i/\mathrm{soc}(Ai)$ by (iv). Statement (vi) follows. Let U, V be submodules of Ai such that (i), (ii), and (iii) hold for U, V instead of U_i, V_i. By (ii), the modules U, V contain $\mathrm{soc}(Ai)$. By (i) and (vi), the modules U, V are submodules of U_i or V_i. Using (ii) and (iii) we get that $U/\mathrm{soc}(Ai) \oplus V/\mathrm{soc}(Ai) = J(A)i/\mathrm{soc}(Ai) = U_i/\mathrm{soc}(Ai) \oplus V_i/\mathrm{soc}(Ai)$. Statement (vii) follows.   □

Proof of Theorem 11.6.4 Since U_i, V_i are uniserial for any $i \in I$ there are uniquely determined elements $\rho(i)$, $\sigma(i)$ in I such that $U_i/\mathrm{rad}(U_i) \cong S_{\rho(i)}$ and $V_i/\mathrm{rad}(V_i) \cong S_{\sigma(i)}$. We consider the bijections γ, δ from 11.6.1. For i, $i' \in I$ we have $\mathrm{Hom}_A(U_i, S_{i'}) \cong \underline{\mathrm{Hom}}_B(T_{\delta(i)}, \mathcal{F}(S_{i'})) \neq 0$ if and only if $\gamma(i') = \delta(i)$. This shows that $\rho = \gamma^{-1} \circ \delta$; in particular, ρ is indeed a permutation of I. Let π be the unique transitive cycle on J from 11.3.5 such that $\Omega_B^2(T_j) \cong T_{\pi(j)}$ for all $j \in J$. For i, $i' \in I$ we have $\mathrm{Hom}_A(V_i, S_{i'}) \cong \underline{\mathrm{Hom}}_B(\Omega_B(T_{\gamma(i)}, \mathcal{F}(S_{i'}))) \cong \underline{\mathrm{Hom}}_B(\Omega_B^2(T_{\gamma(i)}, \Omega_B(\mathcal{F}(S_{i'})))) \cong$

$\underline{\mathrm{Hom}}_B(T_{\pi(\gamma(i))}, \Omega_B(\mathcal{F}(S_{i'})))$, and this is nonzero if and only if $T_{\pi(\gamma(i))} \cong$ $\mathrm{soc}(\Omega_B(\mathcal{F}(S_{i'}))) \cong \mathcal{F}(S_{i'})/\mathrm{rad}(\mathcal{F}(S_{i'})) \cong T_{\delta(i')}$, hence if and only if $i' =$ $\delta^{-1}(\pi(\gamma(i)))$. This shows that $\sigma = \delta^{-1} \circ \pi \circ \gamma$ is a permutation of I, proving (v). Moreover, $\sigma \circ \rho = \delta^{-1} \circ \pi \circ \delta$ is then a transitive cycle on I because π is a transitive cycle on J. We have $\Omega_A^{-1}(V_{\rho(i)}) \cong \Omega_A^{-1}(\mathcal{G}(\Omega_B(T_{\gamma(\rho(i))}))) \cong$ $\mathcal{G}(T_{\gamma(\rho(i))}) \cong U_i$ because $\gamma \circ \rho = \delta$. Similarly, $\Omega_A^{-1}(U_{\sigma(i)}) \cong V_i$, whence the short exact sequences in (iii). Statement (iv) is an obvious consequence of (iii). We have $\alpha_{\sigma(i)}(V_{\sigma(i)}) = \mathrm{rad}(V_i)$, by the exact sequences in (iii) and by 11.6.3. Similarly, $\beta_{\rho(i)}(U_{\rho(i)}) = \mathrm{rad}(U_i)$. This yields the composition factors of U_i and of V_i as stated in (i). Since the kernels and cokernels of the restrictions of the maps $\alpha_{\sigma(i)}$, $\beta_{\rho(i)}$ are all simple we get that $\ell(U_i) = \ell(U_{\rho(i)})$ and $\ell(V_i) = V_{\sigma(i)}$. Thus a and b depend only on the ρ-orbit and σ orbit of i, respectively. Since $S_i \cong \mathrm{soc}(Ai) = \mathrm{soc}(U_i) = \mathrm{soc}(V_i)$ we have $\rho^a(i) = i = \sigma^b(i)$, and hence a and b are multiples of the lengths of the ρ-orbit and σ-orbit of i, respectively. At least one of a or b has to be equal to that orbit length because otherwise S_i would be a composition factor of both $U_i/\mathrm{soc}(Ai)$, $V_i/\mathrm{soc}(Ai)$, contradicting 11.6.3 (iii). For the same reason the ρ-orbit of i and the σ-orbit of i have no common element other than i. Thus if $\rho = \sigma$, then all orbits have length 1; that is, $\rho = \sigma = \mathrm{Id}_I$. Since $\sigma \circ \rho$ is transitive on I, this forces $|I| = 1$. This completes the proof of (ii), whence the result. $\qquad \square$

For a complete determination of the structure of the projective indecomposable A-modules we need to determine the multiples of the lengths of the ρ-orbits and σ-orbits that can actually occur in the statement of 11.6.4 (i). The answer to that is surprisingly simple: all but possibly one of these multiplicities are equal to 1. This will be shown in the next section.

11.7 The Brauer tree

The structure of an algebra A as in the previous section can be described in terms of a tree, the *Brauer tree of A*, with some additional combinatorial information (namely an *exceptional vertex* of the Brauer tree and a cyclic order on each set of edges of this tree having a common vertex). The edges of the Brauer tree will be in bijection with the isomorphism classes of simple A-modules. As in the previous section, A and B are stably equivalent indecomposable nonsimple symmetric k-algebras with B serial, I and J are sets of representatives of the conjugacy classes of primitive idempotents, and we set $S_i = Ai/J(A)i$ for $i \in I$ and $T_j = Bj/J(B)j$ for $j \in J$. We denote by $\mathcal{F} : \underline{\mathrm{mod}}(A) \to \underline{\mathrm{mod}}(B)$ and $\mathcal{G} : \underline{\mathrm{mod}}(B) \to \underline{\mathrm{mod}}(A)$ k-linear equivalences that are inverse to each other, with notation chosen as before in such a way that if U (resp. V) is an indecomposable nonprojective A-module (resp. B-module), then $\mathcal{F}(U)$ (resp. $\mathcal{G}(V)$) is an indecomposable nonprojective B-module (resp. A-module). We denote by U_i, V_i the uniserial modules of Ai as in 11.6.3 and by ρ, σ the permutation of I as in 11.6.4.

Definition 11.7.1 For any $i \in I$ denote by i^ρ the ρ-orbit in I containing i and by i^σ the σ-orbit in I containing i. We define a graph $\Gamma(A)$ as follows. The vertex set of $\Gamma(A)$ consists of the disjoint union of the sets of ρ-orbits and σ-orbits in I, and for any $i \in I$ there is exactly one edge, labelled again i, between the ρ-orbit i^ρ and the σ-orbit i^σ. Moreover, we set $m(i^\rho) = \ell(U_i)/|i^\rho|$, called the *multiplicity of i^ρ*, and we set $m(i^\sigma) = \ell(V_i)/|i^\sigma|$, called the *multiplicity of i^σ*.

A vertex of $\Gamma(A)$ that is a ρ-orbit is called of *type ρ*, otherwise of *type σ*. There is no edge between vertices of the same type. By 11.6.4 (ii) a ρ-orbit and a σ-orbit have at most one common element, and thus there is at most one edge between two vertices of different type. The multiplicities $m(i^\rho)$, $m(i^\sigma)$ are integers, by 11.6.4 (i), and for any fixed $i \in I$, at least one of these numbers is equal to 1. We can be more precise:

Theorem 11.7.2 *The graph $\Gamma(A)$ is a tree, and all but possibly one of the multiplicities of the ρ-orbits and σ-orbits in I are equal to* 1.

We collect the technicalities of the proof, following the arguments from [124, §2].

Lemma 11.7.3 ([124, Proposition (2.13)]) *Let $i, j \in I$.*

(i) *If S_j is a composition factor of $U_i/\mathrm{soc}(Ai)$ then $\ell(\Omega_B(\mathcal{F}(S_j))) < \ell(\mathcal{F}(S_i))$. In particular, if $m(i^\rho) > 1$ then $\ell(\Omega_B(\mathcal{F}(S_i))) < \ell(\mathcal{F}(S_i))$.*

(ii) *If S_j is a composition factor of $V_i/\mathrm{soc}(Ai)$ then $\ell(\Omega_B^{-1}(\mathcal{F}(S_j))) > \ell(\mathcal{F}(S_i))$. In particular, if $m(i^\sigma) > 1$ then $\ell(\Omega_B^{-1}(\mathcal{F}(S_i))) > \ell(\mathcal{F}(S_i))$.*

Proof If S_j is a composition factor of $U_i/\mathrm{soc}(Ai)$ there is a submodule W of length at least 2 of U_i satisfying $W/\mathrm{rad}(W) \cong S_j$. Thus we have a diagram of A-modules

$$
\begin{array}{ccc}
 & U_i & \\
 & \uparrow{\scriptstyle\iota} & \\
S_i \xrightarrow{\alpha} & W & \xrightarrow{\beta} S_j
\end{array}
$$

in which α, ι are injective, β is surjective, and $\beta \circ \alpha = 0$ because $\ell(W) \geq 2$. Applying $\mathcal{F}$ yields thus a diagram of B-modules

$$T_{\delta(i)}$$

$$\uparrow \iota'$$

$$\mathcal{F}(S_i) \xrightarrow{\alpha'} \mathcal{F}(W) \xrightarrow{\beta'} \mathcal{F}(S_j)$$

in which the morphism $\iota' \circ \alpha'$ does not factor through a projective module, and hence both α', ι' are surjective as $T_{\delta(i)}$ is simple and $\mathcal{F}(W)$ is uniserial. Now β' does not factor through a projective module, hence is nonzero. Since α' is surjective it follows that $\beta' \circ \alpha'$ is nonzero. But $\beta \circ \alpha = 0$, hence $\beta' \circ \alpha'$ factors through a projective module. Thus 11.4.6 implies that $\ell(\Omega_B(\mathcal{F}(S_j))) < \ell(\mathcal{F}(S_i))$. If $m(i^\rho) > 1$ then S_i is a composition factor of $U_i/\mathrm{soc}(Ai)$, hence the second statement in (i) follows from the first. A similar argument, using $\Omega_B^{-1} \circ \mathcal{F}$ instead of $\mathcal{F}$, yields (ii). $\qquad\square$

Proof of Theorem 11.7.2 Since $\sigma \circ \rho$ is a transitive cycle on I the graph $\Gamma(A)$ is connected. Since a ρ-orbit and a σ-orbit have at most one element in common, the graph $\Gamma(A)$ has no multiple edges. We need to show that $\Gamma(A)$ has no circular subgraph. Consider three consecutive edges in $\Gamma(A)$

$$i_1^\sigma \overset{i_1}{\rule{2cm}{0.4pt}} i_1^\rho \overset{i_2}{\rule{2cm}{0.4pt}} i_2^\sigma \overset{i_3}{\rule{2cm}{0.4pt}} i_3^\rho$$

such that $i_1 \neq i_2$, $i_2 \neq i_3$, $i_1^\rho = i_2^\rho$ and $i_2^\sigma = i_3^\sigma$. By 11.7.3 we have

$$\ell(\mathcal{F}(S_{i_1})) > \ell(\Omega_B(\mathcal{F}(S_{i_2}))) = \ell(\Omega_B^{-1}(\mathcal{F}(S_{i_2}))) > \ell(\mathcal{F}(S_{i_3})).$$

Given two neighbouring vertices of $\Gamma(B)$, exactly one is a ρ-orbit, and the other is a σ-orbit. Thus a circular path in $\Gamma(A)$ would contain an even number of vertices and edges, hence would be a sequence of edges $i = i_1, i_2, \ldots, i_{2n+1} = i$ for some positive integer n and some $i \in I$. This would yield a contradiction $\ell(\mathcal{F}(S_i)) > \ell(\mathcal{F}(S_i))$. Similarly, if there were two different orbits of ρ or σ in I with multiplicities greater than 1, going back and forth along the path in $\Gamma(A)$ linking these two orbits yields a contradiction in conjunction with 11.7.3. $\qquad\square$

Definition 11.7.4 If one of the multiplicities of the ρ-orbits and σ-orbits is greater than one we call this the *exceptional multiplicity* of A and the corresponding vertex in the tree $\Gamma(A)$ is called the *exceptional vertex* of $\Gamma(A)$.

The datum $(I, \{\rho, \sigma\})$ together with a pair (v, m) consisting of the exceptional vertex v and the exceptional multitplicity m (if any) is called the *Brauer tree of A*.

Any pair of permutations on I will define a graph as in 11.7.1, but this will not always be a tree. Keeping the permutations ρ, σ as part of the definition of the Brauer tree amounts to specifying a cyclic order on any set of edges emanating from a common vertex of $\Gamma(A)$, or equivalently, amounts to realising the tree $\Gamma(A)$ in the plane. In addition, the Brauer tree can have one distinguished vertex v, and in that case, there is a multiplicity $m > 1$ attached to this vertex.

Remark 11.7.5 The description of the composition series of the projective indecomposable B-modules implies that the Brauer tree $\Gamma(B)$ of B is equal to $(J, \{\mathrm{Id}_J, \pi\}, v, m)$; here the exceptional vertex v is the unique π-orbit J and $m = \frac{q-1}{e}$, where q is the common length of the projective indecomposable B-modules and where $e = |J|$ is the number of isomorphism classes of simple B-modules, with the convention that there is no exceptional vertex if $m = 1$. The underlying tree of $\Gamma(B)$ is a star, the exceptional vertex (if any) is the centre of this star, and the transitive cycle π corresponds to a realisation of this star in the plane.

As an immediate consequence of Theorem 11.7.2 and the description of the composition series of the uniserial submodules U_i and V_i of the projective indecomposable A-modules Ai in 11.6.4, we obtain the Cartan matrix of A.

Theorem 11.7.6 *Denote by $C = (c_{ii'})_{i,i' \in I}$ the Cartan matrix of A; that is, $c_{ii'}$ is the number of composition factors isomorphic to S_i in a composition series of Ai', for i, $i' \in I$. Denote by v the exceptional vertex in the Brauer tree of A, if any, and by m the corresponding exceptional multiplicity.*

(i) For $i \in I$ we have

$$c_{ii} = \begin{cases} 2 & \text{if } i^\rho \neq v \text{ and } i^\sigma \neq v \\ m+1 & \text{if } i^\rho = v \text{ or } i^\sigma = v. \end{cases}$$

(ii) For i, $i' \in I$ such that $i \neq i'$ we have

$$c_{ii'} = \begin{cases} 1 & \text{if } i^\rho = (i')^\rho \neq v \text{ or } i^\sigma = (i')^\sigma \neq v \\ m & \text{if } i^\rho = (i')^\rho = v \text{ or } i^\sigma = (i')^\sigma = v \\ 0 & \text{if } i^\rho \neq (i')^\rho \text{ and } i^\sigma \neq (i')^\sigma. \end{cases}$$

Proof It follows from 11.6.3 that the composition factors of Ai are the same as of $U_i \oplus V_i$, counting multiplicities. The result follows from the composition series of the modules U_i, V_i in 11.6.4, together with the fact that there is at most one vertex with an exceptional multiplicity, by 11.7.2. $\qquad\square$

The alternative in the statement (ii) of Theorem 11.7.6 covers all possible Cartan invariants that are not in the diagonal, because if $i \neq i'$, then necessarily $i^\rho \neq (i')^\rho$ or $i^\sigma \neq (i')^\sigma$.

Remark 11.7.7 If one of A or B has an exceptional vertex, then so does the other, and in that case the exceptional multiplicities of A and B are equal. One way to see this is as follows. As in 11.5.6, one shows first that the determinant of the Cartan matrix of A is equal to $em + 1$, where $e = |I|$ and m is the exceptional multiplicity of A, with the convention $m = 1$ if there is no exceptional vertex. This is done, for instance, in detail in Alperin [4, §24]. One then goes on to show that the determinants of the Cartan matrices of A and B are equal. If the stable equivalence between A and B is of Morita type, this follows immediately from 4.14.13, and this is an assumption that in the context of blocks with cyclic defect groups is satisified. We will see that for blocks with cyclic defect groups, we obtain the equality of the exceptional multiplicities far more easily from character theoretic results, as stated in 11.1.15 for instance.

Proposition 11.7.8 *With the notation above, let i, $i' \in I$. Suppose that A is split. Denote by m the exceptional multiplicity, if any.*

(i) *If $i^\rho = (i')^\rho$ is nonexceptional, then $\dim_k(\mathrm{Hom}_A(U_i, U_{i'})) = 1$.*

(ii) *If $i^\rho = (i')^\rho$ is exceptional, then $\dim_k(\mathrm{Hom}_A(U_i, U_{i'})) = m$.*

(iii) *If $i^\rho \neq (i')^\rho$ then $\mathrm{Hom}_A(U_i, U_{i'}) = \{0\}$.*

(iv) *If $i^\sigma = (i')^\sigma$ is nonexceptional, then $\dim_k(\mathrm{Hom}_A(V_i, V_{i'})) = 1$.*

(v) *If $i^\sigma = (i')^\sigma$ is exceptional, then $\dim_k(\mathrm{Hom}_A(V_i, V_{i'})) = m$.*

(vi) *If $i^\sigma \neq (i')^\sigma$ then $\mathrm{Hom}_A(V_i, V_{i'}) = \{0\}$.*

Proof We prove (i) and (ii) simultaneously. Suppose that $i^\rho = (i')^\rho$. It follows from the description of the composition series of U_i in 11.6.4 that $S_{\rho(i)}$ is the top composition factor of U_i, and that $S_{\rho(i)}$ is a composition factor of $U_{i'}$, with multiplicity either 1 or m, depending on whether i^ρ is nonexceptional or exceptional. Thus it suffices to show that for any submodule W of $U_{i'}$ with top composition factor $S_{\rho(i)}$, there is a homomorphism $U_i \twoheadrightarrow U_{i'}$ with image equal to W. There is certainly a homomorphism $\mu : A\rho(i) \to U_{i'}$ with image W; this follows from the projectivity of $A\rho(i)$. The top composition factor of $V_{\rho(i)}$ is $S_{\sigma(\rho(i))}$. If this is different from $\rho(i)$, then this is not a composition factor of $U_{i'}$, because the ρ-orbit and the σ-orbit of i' have no other common element but i'. Thus the kernel of β contains $V_{\rho(i)}$, and hence β factors through the surjective map $A\rho(i) \to U_i$, inducing a homomorphism $U_i \to U_{i'}$ with image W as required.

This shows (i) and (ii). If the ρ-orbits of i and of i' are different, then U_i and $U_{i'}$ have no common composition factors, implying (iii). The statements (iv), (v) and (vi) are proved analogously.   □

11.8 Split basic Brauer tree algebras

We describe in this section Brauer tree algebras in terms of generators and relations. Let A and B be stably equivalent indecomposable nonsimple symmetric k-algebras with B serial, let I and J be sets of representatives of the conjugacy classes of primitive idempotents, and we set $S_i = Ai/J(A)i$ for $i \in I$ and $T_j = Bj/J(B)j$ for $j \in J$. Denote by q the common length of the projective indecomposable B-modules, and set $m = \frac{q-1}{|J|}$. If $m > 1$, then m is the exceptional multiplicity of B, and the exceptional orbit is in that case all of J, the orbit of the transitive cycle π from 11.3.4.

Fix stable equivalences $\mathcal{F} : \underline{\text{mod}}(A) \to \underline{\text{mod}}(B)$ and $\mathcal{G} : \underline{\text{mod}}(B) \to \underline{\text{mod}}(A)$ that are inverse to each other, with notation chosen as before in such a way that if U (resp. V) is an indecomposable nonprojective A-module (resp. B-module), then $\mathcal{F}(U)$ (resp. $\mathcal{G}(V)$) is an indecomposable nonprojective B-module (resp. A-module). We denote by U_i, V_i the uniserial modules of Ai as in 11.6.3 and by ρ, σ the permutation of I as in 11.6.4. We apply previous results to determine a basic algebra of A in terms of generators and relations. We use the same letter I for the generators corresponding to a primitive decomposition of the unit element in the basic algebra.

Theorem 11.8.1 *Suppose that A is split basic. Then A is isomorphic to the k-algebra with generating set $I \cup \{r, s\}$ and the following list of relations.*

$$1 = \sum_{i \in I} i, \quad i^2 = i \ (i \in I), \quad ij = 0 \ (i, j \in I, i \neq j),$$

$$rs = 0 = sr,$$

$$ir^{|i^\rho|} + is^{|i^\sigma|} = 0, \quad i \in I, i^\rho \neq v, i^\sigma \neq v,$$

$$ir^{m|i^\rho|} + is^{|i^\sigma|} = 0, \quad i \in I, i^\rho = v, i^\sigma \neq v,$$

$$ir^{|i^\rho|} + is^{m|i^\sigma|} = 0, \quad i \in I, i^\rho \neq v, i^\sigma = v.$$

In particular, the Morita equivalence class of A is determined by the Brauer tree $(I, \{\rho, \sigma\}, v, m)$ or $(I, \{\rho, \sigma\})$ depending on whether there is an exceptional vertex or not.

Proof Choose a set I of pairwise orthogonal representatives of the conjugacy classes of primitive idempotents of A. Then $1 = \sum_{i \in I} i$, since A is assumed to be basic. Thus the relations $i^2 = i$ for $i \in I$, $ii' = 0$ for $i, i' \in I$ such that $i \neq i'$, and $\sum_{i \in I} i = 1$ are trivially satisfied in A. For any $i, i' \in I$, any homomorphism $Ai \to Ai'$ is given by right multiplication with an element in iAi'. In particular, for any $i \in I$, there are elements $r_i \in iA\rho^{-1}(i)$ and $s_i \in iA\sigma^{-1}(i)$ such that $\beta_i(a) = ar_i$ and $\alpha_i(a) = as_i$ for all $a \in Ai$. Set $r = \sum_{i \in I} r_i$ and $s = \sum_{i \in I} s_i$. Since the elements of I are pairwise orthogonal, we have $ir = r_i = r\rho^{-1}(i)$ and $is = s_i = s\sigma^{-1}(i)$, or equivalently, $\rho(i)r = ri$ and $\sigma(i)s = si$ for all $i \in I$. We have $rs = 0 = sr$ by the exactness of the sequences in 11.6.4. For the remaining relations, consider first the case where none of i^ρ and i^σ are exceptional. By considering the composition series of the U_i, V_i, one sees that right multiplication by $ir^{|i^\rho|}$ is an endomorphism of Ai that sends Ai onto $\mathrm{soc}(Ai)$. Similarly, right multiplication by $is^{|i^\sigma|}$ is an endomorphism sending Ai onto $\mathrm{soc}(Ai)$. Since k is a splitting field for A, it follows that there is a nonzero scalar $\lambda_i \in k^\times$ satisfying

$$ir^{|i^\rho|} + \lambda_i is^{|i^\sigma|} = 0.$$

A similar argument shows that if $i^\rho = v$, then

$$ir^{m|i^\rho|} + \lambda_i is^{|i^\sigma|} = 0$$

for some $\lambda_i \in k^\times$, and that if $i^\sigma = v$, then

$$ir^{|i^\rho|} + \lambda_i is^{m|i^\sigma|} = 0$$

for some $\lambda_i \in k^\times$. It remains to show that the λ_i can all be chosen to be equal to 1. Using the fact that $\Gamma(A)$ is a tree, this can be achieved as follows. Choose a vertex at the end of the tree. This is an orbit of length 1 of either ρ or of σ. We show this in the case where this is a ρ-orbit; the case where this is a σ-orbit is strictly analogous. That is, we fix $i \in I$ such that $\rho(i) = i$. Then i^ρ is a vertex of the Brauer tree that is at an end of this tree. If i^σ is nonexceptional, replace s_i by $\lambda_i s_i$, and if i^σ is exceptional, replace s_i by $\mu_i s_i$, where $\mu_i^m = \lambda_i$. Thus we may assume that for this particular i we have $\lambda_i = 1$. Let $j \in I$. There is a unique shortest path $i = i_0, i_1, \ldots, i_r = j$ in the Brauer tree such that i_s and i_{s+1} have a common vertex in the Brauer tree, for $0 \leq s \leq r - 1$. We argue by induction over r. If $r = 0$, then $j = i$, so there is nothing to prove. Suppose that $r > 0$ is odd. Then the vertex furthest away from i^ρ in this path is j^ρ. Replace r_j by $\lambda_j^{-1} r_j$ or by $\mu_j^{-1} r_j$, depending on whether j^ρ is not exceptional or exceptional, where as before $\mu_j^m = \lambda_j$. This choice yields $\lambda_j = 1$ and has no impact on the relations involving edges that are nearer to i. Similarly, if $r > 0$ is even, then the vertex furthest away from i^ρ is j^σ, and we replace s_j by $\lambda_j s_j$ or $\mu_j s_j$, as before. The uniqueness of a path in the Brauer tree from i to j ensures that in any orbit of either ρ or σ we change at most one of the r_i or s_i. It follows that we may choose r, s such that all relations as in the statement hold in A. Thus A is a quotient of the algebra C generated by $I \cup \{r, s\}$ and the relations as stated. By comparing the dimensions of iCi' and iAi, where i, $i' \in I$ (and where we abusively use the same notation for the images of I in C and in A), one sees that the canonical algebra homomorphism $C \to A$ is an isomorphism. $\square$

For the classification of the source algebras of blocks with cyclic defect groups we will need to compare two algebras that are stably equivalent to the same serial symmetric algebra. Let A' be another indecomposable nonsimple symmetric algebra that is stably equivalent to B, with fixed inverse stable equivalences $\mathcal{F}' : \underline{\mathrm{mod}}(A') \cong \underline{\mathrm{mod}}(B)$ and $\mathcal{G}' : \underline{\mathrm{mod}}(B) \cong \underline{\mathrm{mod}}(A')$. Denote by I' a set of representatives of the conjugacy classes of primitive idempotents in A'. Let ρ', σ' be the permutations of I' satisfying 11.6.4 with A' and I' instead of A and I. If $m > 1$ we assume that both A and A' have an exceptional orbit, with the same exceptional multiplicity m. (As mentioned in 11.7.7, this assumption is automatically satisfied, but we will prove this only in the case of blocks with cyclic defect groups.) We denote by $U'_{i'}$, $V'_{i'}$ the uniserial submodules of $A'i'$ satisfying 11.6.3.

Proposition 11.8.2 *Suppose that A and A' are split basic. Let $\tau : I \to I'$ be a bijective map satisfying $\tau \circ \rho = \rho' \circ \tau$ and $\tau \circ \sigma = \sigma' \circ \tau$. Suppose that if A and A' have exceptional orbits v, v', respectively, then $\tau(v) = v'$. Then there is a k-algebra isomorphism $\alpha : A' \cong A$ satisfying $_\alpha Ai \cong A'\tau(i)$ for all $i \in I$. Any such isomorphism satisfies $_\alpha U_i \cong U'_{\tau(i)}$ and $_\alpha V_i \cong V'_{\tau(i)}$ for all $i \in A$.*

Proof The description of A and A' in terms of generating sets $I \cup \{r, s\}$ and $I' \cup \{r', s'\}$ in 11.8.1 implies immediately that there is an algebra isomorphism $\alpha : A' \cong A$ sending $\tau(i) \in I'$ to i and r', s' to r, s, respectively. The map α^{-1} sends $i \in I$ to $\tau(i)$ and induces hence an isomorphism of A-modules $_\alpha Ai \cong A'\alpha^{-1}(i) = A'\tau(i)$. Right multiplication by $\rho(i)r = ri$ on $A\rho(i)$ is a homomorphism from $A\rho(i)$ to Ai with image U_i. It follows that $_\alpha U_i \cong U'_{\tau(i)}$. If $\beta : A' \cong A$ is another isomorphism satisfying $_\beta Ai \cong A'\tau(i)$ for all $i \in I$, then $\beta \circ \alpha^{-1}$ stabilises the isomorphism class of Ai, hence of S_i, for all $i \in I$. Since the uniserial submodules U_i, V_i of Ai are determined by their composition factors, it follows that $\beta \circ \alpha^{-1}$ stabilises their isomorphism classes as well. Thus $_\beta U_i \cong _\alpha U_i$ and $_\beta V_i \cong _\alpha V_i$ for all $i \in I$, implying the result. $\square$

Any automorphism of A induces an automorphism of the Brauer tree, hence a permutation of I that normalises $\{\rho, \sigma\}$. In most cases – including blocks with cyclic defect groups – this permutation actually centralises ρ and σ. In order to show this, we will need the following elementary observation regarding automorphisms of trees.

Lemma 11.8.3 *Let T be a finite tree. Let τ be an automorphism of T; that is, τ permutes the sets of vertices and edges of T preserving adjacency. Then τ fixes one vertex or one edge.*

Proof We argue by induction over the number of vertices of T. We may assume that T has at least two edges and that T is not a star because the central vertex of a star with at least 2 edges is fixed by any tree automorphism. Thus not every edge of T has a vertex at the end of the tree. The edges having a vertex at the end of the tree are permuted by τ. Removing the vertices at the end of the tree and the edges having a vertex at the end of the tree yields therefore a tree T' with fewer vertices than T, and τ restricts to an automorphism of T'. By induction, τ fixes a vertex or an edge of T', hence of T. $\qquad\square$

Theorem 11.8.4 *Suppose that A is split basic. Assume that if the Brauer tree of A has no exceptional vertex, then $|I|$ is either 1 or even. Let τ be a permutation of I. The following are equivalent.*

(i) There exists an automorphism α of A such that $_{\alpha}Ai \cong A\tau(i)$ for all $i \in I$.
(ii) The permutation τ centralises ρ and σ, and τ stabilises the exceptional orbit, if any.

Proof It follows from 11.8.2 that (ii) implies (i). If $|I| = 1$, then the converse implication holds trivially with $\alpha = \mathrm{Id}_A$. We assume that $|I| \geq 2$. Suppose that $\alpha \in \mathrm{Aut}(A)$ satisfies $_{\alpha}Ai \cong A\tau(i)$ for all $i \in I$. Note that then also $_{\alpha}S_i \cong S_{\tau(i)}$. By 11.6.4, we have $\Omega(V_i) \cong U_{\sigma(i)}$ and $\Omega(U_i) \cong V_{\rho(i)}$. By 2.14.3 we have $\Omega(_{\alpha}U) \cong {_{\alpha}\Omega(U)}$ for any indecomposable A-module U. By 11.6.3, the set of uniserial submodules $\{U_i, V_i\}$ of Ai is uniquely determined by the properties $U_i \cap V_i = \mathrm{soc}(Ai)$ and $U_i + V_i = \mathrm{rad}(Ai)$. Thus either $_{\alpha}U_i \cong U_{\tau(i)}$ or $_{\alpha}U_i \cong V_{\tau(i)}$. We will show that the first case leads to the conclusion as in (ii), and that the second case cannot arise. Suppose that $_{\alpha}U_i \cong U_{\tau(i)}$. Then

$$_{\alpha}V_{\rho(i)} \cong {_{\alpha}\Omega(U_i)} \cong \Omega(_{\alpha}U_i) \cong \Omega(U_{\tau(i)}) \cong V_{\rho(\tau(i))}.$$

Thus $\tau(\rho(i)) = \rho(\tau(i))$. Repeating this argument by applying Ω, and using the fact that $\sigma \circ \rho$ is transitive on I, it follows that τ centralises ρ and σ. Moreover, if there is an exceptional vertex, then τ must preserve this, because U_i and $U_{\tau(i)}$ have the same composition lengths, similarly for V_i. Thus in this case statement (ii) holds. Suppose finally that $_{\alpha}U_i \cong V_{\tau(i)}$. The same reasoning as before, applying repeatedly Ω, yields that $\tau \circ \sigma = \rho \circ \tau$ and $\tau \circ \rho = \sigma \circ \tau$. Thus τ exchanges the ρ-orbits and σ-orbits. In particular, τ stabilises no orbit, and hence the Brauer tree of A has no exceptional vertex. Since $|I| \geq 2$, the hypotheses imply that $|I|$ is even. By 11.8.3, τ fixes an edge i of the Brauer tree, exchanging the two vertices i^ρ and i^σ adjacent to this edge. Then τ fixes no other edge, and τ^2 fixes i and the two vertices i^ρ, i^σ. But then, since τ^2 centralises ρ and σ, it follows that τ^2 is the identity on I. This means that τ fixes exactly the edge i but no other edge, and hence τ pairs up the edges in $I \setminus \{i\}$, implying that $|I|$ is odd, a contradiction. Thus this case cannot arise, completing the proof. □

We will show later that if the Brauer tree of a block algebra with cyclic defect groups has no exceptional vertex, then its vertex set I satisfies $|I| = p - 1$, hence either $|I| = 1$ or $|I|$ is even; in particular, the hypotheses of 11.8.4 are satisfied for basic algebras of blocks with cyclic defect groups. We conclude this section with a description of the quiver of A, which is easily obtained from combining the above results.

Lemma 11.8.5 *For any $i \in I$ the following hold.*

(i) *The projective indecomposable A-module Ai is uniserial if and only of one of U_i or V_i is simple.*

(ii) *The module U_i is simple if and only if $\rho(i) = i$ and the ρ-orbit $i^\rho = \{i\}$ is not exceptional.*

(iii) *The module V_i is simple if and only if $\sigma(i) = i$ and the σ-orbit $i^\sigma = \{i\}$ is not exceptional.*

(iv) *Both U_i and V_i are simple if and only if $I = \{i\}$ and none of the two vertices is exceptional. In that case, a basic algebra of A is local of dimension 2.*

(v) *If $\rho(i) \neq i$ and $\sigma(i) \neq i$, or if $\rho(i) \neq i$ and $i^\sigma = v$, or if $i^\rho(i) = v$ and $\sigma(i) \neq i$, then $\mathrm{rad}(Ai)/\mathrm{rad}^2(Ai) \cong S_{\rho(i)} \oplus S_{\sigma(i)}$.*

(vi) *If $\sigma(i) = i$ and $i^\sigma \neq v$, then $\mathrm{rad}(Ai)/\mathrm{rad}^2(Ai) \cong S_{\rho(i)}$.*

(vii) *If $\rho(i) = i$ and $i^\rho \neq v$, then $\mathrm{rad}(Ai)/\mathrm{rad}^2(Ai) \cong S_{\sigma(i)}$.*

Proof All statements follow from the description of the composition series of the modules U_i, V_i in 11.6.4. For (iv) one needs to observe that if both U_i, V_i are simple, then $\rho(i) = i = \sigma(i)$ and there is no exceptional orbit. Since $\rho \circ \sigma$ is transitive on I, this forces $I = \{i\}$. □

Lemma 11.8.5 determines the quiver of A, and the relations of this quiver are determined by using 11.8.1. Summarising the above, if $|I| \geq 2$, then the quiver of A is a union of cycles, indexed by the orbits of ρ and σ which are either nontrivial or exceptional, and if $|I| = 1$, the quiver has one vertex and one loop. The relations follow from the composition series of the uniserial modules U_i and V_i, or also from 11.8.1. The details are as follows.

Theorem 11.8.6 *Suppose that A is split. Let $(I, \{\rho, \sigma\}, v, m)$ be the Brauer tree of A.*

(i) *The quiver of A is the quiver Q having the set I as set of vertices, with the following arrows.*
   For any $i \in I$ such that i^ρ is nontrivial we have an arrow $\rho_i : i \to \rho(i)$; that is, we have a cyclic subgraph

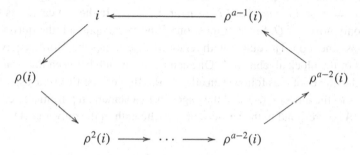

where $a = |i^\rho|$. We denote by ρ_{ii} the simple cyclic path of length a starting and ending at i in this graph.
   For any $i \in I$ such that i^σ is nontrivial we have an arrow $\sigma_i : i \to \sigma(i)$; that is, we have a cyclic subgraph

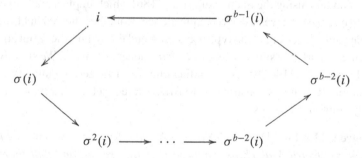

where $b = |i^\sigma|$. We denote by σ_{ii} the simple cyclic path of length b starting and ending at i in this graph.
   If $i \in I$ such that $\rho(i) = i$ and either i^ρ is exceptional or $\sigma(i) = i$, then there is a loop denoted ρ_{ii} at i. If $i \in I$ such that $\sigma(i) = i$ and i^σ is exceptional, then there is a loop denoted σ_{ii} at i.

(ii) *Let N be the ideal in the quiver algebra kQ generated by the following list of elements:*
   $\rho_{ii} + \sigma_{ii}$ *if $i \in I$ such that $\rho(i) \neq i$ and $\sigma(i) \neq i$;*
   $\rho_{ii}^m + \sigma_{ii}$ *if $i \in I$ such that i^ρ is exceptional and $\sigma(i) \neq i$;*
   $\rho_{ii} + \sigma_{ii}^m$ *if $i \in I$ such that $\rho(i) \neq i$ and i^σ exceptional;*

$\sigma_i \sigma_{ii}^m$ if $\rho(i) = i$ and i^σ is exceptional;

$\rho_i \rho_{ii}^m$ if i^ρ is exceptional and $\sigma(i) = i$;

ρ_{ii}^2 if $\rho(i) = i = \sigma(i)$ and none of i^ρ, i^σ is exceptional.

We have an algebra isomorphism $A \cong kQ/N$.

The quiver Q of A is a union of simple cyclic paths corresponding to the nontrivial orbits of ρ and of σ, together with a loop at the edge i if $\rho(i) = i$ and $i^\rho = v$, or if $\sigma(i) = i$ and $i^\sigma = v$, or if $I = \{i\}$. In the last case, this loop is the only arrow of Q. Since at most one orbit is exceptional, the quiver Q has at most one loop. The quiver with relations determines the Morita equivalence class of the block algebra kGb. One can easily switch between the Brauer tree and the quiver of A as follows: an edge of the Brauer tree $\Gamma(A)$ corresponds to a vertex of the quiver $Q(A)$, and the edges in $\Gamma(A)$ emanating from a fixed vertex of $\Gamma(A)$ correspond to the vertices of a cyclic path in the quiver $Q(A)$.

11.9 The derived category of a Brauer tree algebra

Stably equivalent Brauer tree algebras are derived equivalent. This fact, due to Rickard, is proved in [185] by constructing explicitly a one-sided tilting complex, and then using the main result from [186] which implies that a one-sided tilting complex induces a derived equivalence. Rouquier showed in [196] that a stable equivalence of Morita type between A and the symmetric serial algebra B lifts in a remarkably simple way to a 2-term Rickard complex. We slightly modify the proofs in [196, 10.3.4], avoiding characteristic zero arguments, thereby extending Rouquier's construction to Brauer tree algebras that may not necessarily come from blocks.

Theorem 11.9.1 (cf. [196, 10.3.8]) *Let A and B be split indecomposable nonsimple symmetric k-algebras. Suppose that B is serial, and that there is an indecomposable A-B-bimodule M such that M and ist dual M^* induce a stable equivalence of Morita type between A and B. Then there is a finitely generated projective $A \otimes_k B^{op}$-module P and an $A \otimes_k B^{op}$-homomorphism $\pi : P \to M$ such that the complex*

$$X = \cdots \longrightarrow 0 \longrightarrow P \xrightarrow{\pi} M \longrightarrow 0 \longrightarrow \cdots$$

with M in degree zero is a Rickard complex. In particular, the derived categories of mod(A) *and* mod(B) *are equivalent as triangulated categories.*

We keep the notation and hypotheses from Theorem 11.9.1 throughout this section. The functors $M \otimes_B -$ and $M^* \otimes_A -$ induce inverse equivalences $\underline{\mathrm{mod}}(B) \cong \underline{\mathrm{mod}}(A)$. By 4.14.9, if T is a simple B-module, then the A-module $\mathcal{G}(T) = M \otimes_B T$ is indecomposable nonprojective, and by 11.6.9, the A-module $\mathcal{G}(T)$ is uniserial. Similarly, if S is a simple A-module, then the B-module $\mathcal{F}(S) = M^* \otimes_A S$ is indecomposable nonprojective; in particular, $\mathcal{F}(S)$ is uniserial. We use further the following notation. Denote by I a set of representatives of the conjugacy classes of primitive idempotents in A, and by J a set of representatives of the conjugacy classes of primitive idempotents in B. Set $S_i = Ai/J(A)i$ for all $i \in I$ and $T_j = Bj/J(B)j$ for all $j \in J$. By 11.6.1 there are unique bijections $\delta, \gamma : I \to J$ such that $T_{\delta(i)}$ is isomorphic to the unique simple quotient of $\mathcal{F}(S_i)$ and such that $T_{\gamma(i)}$ is isomorphic to the unique simple submodule of $\mathcal{F}(S_i)$. By 11.6.3, for any $i \in I$ there are unique uniserial submodules U_i and V_i of Ai isomorphic to $\mathcal{G}(T_{\delta(i)})$ and $\mathcal{G}(\Omega(T_{\gamma(i)}))$, respectively. By 11.6.4, there are unique permutations ρ and σ of I such that the top composition factors of U_i and V_i are isomorphic to $S_{\rho(i)}$ and $S_{\sigma(i)}$, respectively. In particular, $A\rho(i)$ is a projective cover of $U_i \cong M \otimes_B T_{\delta(i)}$. By 4.5.11, the projective indecomposable right B-module $\delta(i)B$ is a projective cover of the simple right B-module $T^*_{\delta(i)}$. It follows from the description of projective covers of bimodules in 4.5.12, that a projective cover of the $A \otimes_k B^{\mathrm{op}}$-module M has the form

$$P_M = \oplus_{i \in I} A\rho(i) \otimes_k \delta(i)B$$

together with a surjective $A \otimes_k B^{\mathrm{op}}$-homomorphism $\pi : P_M \to M$.

The permutations ρ and σ determine a tree whose vertices are the ρ-orbits and σ-orbits, with exactly one edge labelled i linking the ρ-orbit i^ρ and the σ-orbit i^σ of i, as in 11.7.1 and subsequent results. Denote by v the exceptional vertex with exceptional multiplicity m; if there is no exceptional vertex, we choose one randomly and set $m = 1$. Note that there is a unique edge $\rho(i)$ that links the ρ-orbit $i^\rho = \rho(i)^\rho$ and the σ-orbit $\rho(i)^\sigma$. Since there is a unique minimal path from v to any other vertex in the Brauer tree, it follows that we have a well-defined notion of distance from v – this is the number of edges of a minimal path from v to any other vertex. In particular, of the two adjacent vertices i^ρ and $\rho(i)^\sigma$, exactly one is further away from v than the other.

The construction of Rouquier's bimodule complex is based on the following partition of I into two subsets. We denote by I_0 the set of all $i \in I$ such that the vertex i^ρ of the Brauer tree is further away from the exceptional vertex than the vertex $\rho(i)^\sigma$. In particular i^ρ is nonexceptional in that case. We set $I_1 = I \setminus I_0$; that is, I_1 consists of all $i \in I$ such that $\rho(i)^\sigma$ is further away from v than i^ρ. In particular, $\rho(i)^\sigma$ is nonexceptional in that case. We set

$$P = \oplus_{i \in I_1} A\rho(i) \otimes_k \delta(i)B.$$

Thus P is a direct summand of the projective cover P_M of M. Denote the restriction of the surjective map $\pi : P_M \to M$ to P again by π, and set

$$X = \quad \cdots \quad\longrightarrow\quad 0 \quad\longrightarrow\quad P \quad\overset{\pi}{\longrightarrow}\quad M \quad\longrightarrow\quad 0 \quad\longrightarrow\quad \cdots$$

with M in degree zero. Then $X^* \otimes_A X$ is a chain complex with at most three nonzero terms of the form

$$M^* \otimes_A P \quad\longrightarrow\quad M^* \otimes_A M \oplus P^* \otimes_A P \quad\longrightarrow\quad P^* \otimes_A M$$

and these three terms are in the degrees 1, 0, -1, respectively. This complex is selfdual, and the bimodules $M^* \otimes_A P, P^* \otimes_A P, P^* \otimes_A M$ are projective. The only nonprojective term is $M^* \otimes_B M$, which has B as its unique nonprojective indecomposable direct summand, up to isomorphism, as M and M^* induce a stable equivalence of Morita type. We keep this notation. The plan is to show that the complex X just constructed satisfies the conclusion of Theorem 11.9.1.

Lemma 11.9.2 *Let $i \in I$.*

(i) *If $i \in I_0$, then $X \otimes_B T_{\delta(i)} \cong U_i$, regarded as a complex concentrated in degree zero. In that case, the vertex i^ρ is nonexceptional.*

(ii) *If $i \in I_1$, then $X \otimes_B T_{\delta(i)} \cong A\rho(i) \to U_i$, viewed as a 2-term complex with U_i in degree 0. In that case $\rho(i)^\sigma$ is nonexceptional.*

Proof We have $\delta(i)B \otimes_B T_{\delta(i)} \cong k$ and $\delta(i)B \otimes_B T_{\delta(i')} \cong \{0\}$ for i, $i' \in I$ such that $i \neq i'$. Thus both isomorphisms for $X \otimes_B T_{\delta(i)}$ follow immediately from the construction of X. The statements on i^ρ and $\rho(i)^\sigma$ being nonexceptional have been noted earlier. □

The shift $C[-1]$ of a chain complex C by -1 is defined by $C[-1]_n = C_{n-(-1)} = C_{n+1}$, with differential multiplied by -1; this amounts to shifting the complex C to the right by one position and multiplying the differential by a sign.

Lemma 11.9.3 *Let i, $i' \in I$. Every chain map $X \otimes_B T_{\delta(i)} \to X \otimes_B T_{\delta(i')}[-1]$ is zero.*

Proof If $i' \in I_0$, this is trivial, since in that case the complex $X \otimes_B T_{\delta(i')}[-1]$ is concentrated in degree -1, while $X \otimes_B T_{\delta(i)}$ is always zero in degree -1.

Suppose that $i' \in I_1$. Thus

$$X \otimes_B T_{\delta(i')}[-1] \cong A\rho(i') \to U_{i'}$$

with $A\rho(i)$ in degree zero, and where the map $A\rho(i') \to U_{i'}$ is surjective, with kernel $V_{\rho(i')}$. If $i \in I_0$, then a chain map $X \otimes_B T_{\delta(i)} \to X \otimes_B T_{\delta(i')}[-1]$ is given by a commutative diagram of the form

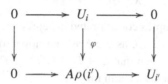

for some A-homomorphism $U_i \to A\rho(i')$. Being a chain map forces $\mathrm{Im}(\varphi) \subseteq V_{\rho(i')}$. The composition factors of U_i are labelled by the ρ-orbit of i starting with $\rho(i)$, and the composition factors of $V_{\rho(i')}$ are labelled by the σ-orbit of $\rho(i')$. Since a ρ-orbit and a σ-orbit have at most one common element, it follows that $\mathrm{Im}(\varphi)$ is zero or simple. If it is simple, then the top composition factor $S_{\rho(i)}$ of U_i is isomorphic to the bottom composition factor $S_{\rho(i')}$, but that forces $i = i'$, contradicting the choice of $i \in I_0$ and $i' \in I_1$. Thus $\varphi = 0$. Finally, if $i \in I_1$, then a chain map $X \otimes_B T_{\delta(i)} \to X \otimes_B T_{\delta(i')}[-1]$ is given by a commutative diagram of the form:

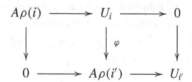

Since the map $A\rho(i) \to U_i$ in this diagram is surjective, it follows that $\varphi = 0$, which concludes the proof. $\qquad\square$

Lemma 11.9.4 *The complex of $B \otimes_k B^{\mathrm{op}}$-modules $X^* \otimes_A X$ is split, with homology concentrated in degree zero, isomorphic to $B \oplus Q$ for some projective $B \otimes_k B^{\mathrm{op}}$-module Q.*

Proof In order to show that $X^* \otimes_A X$ is split with homology concentrated in degree zero, it suffices to show that the second map $M^* \otimes_A M \oplus P^* \otimes_A P \to P^* \otimes_A M$ is surjective. Indeed, if this map is surjective, then it is split surjective as $P^* \otimes_A M$ is projective as a $B \otimes_k B^{\mathrm{op}}$-module. Moreover, the surjectivity of this map is equivalent to having zero homology in degree -1. Since $X^* \otimes_A X$ is selfdual, it follows that also the first map $M^* \otimes_A P \to M^* \otimes_A M \oplus P^* \otimes_A P$

is split injective, implying that the homology in degree 1 is zero as well. Arguing by contradiction, suppose that the map $M^* \otimes_A M \oplus P^* \otimes_A P \to P^* \otimes_A M$ is not surjective. Then its image is a proper submodule of $P^* \otimes_A M$, hence contained in a maximal submodule of $P^* \otimes_A M$. Thus it is contained in the kernel of a surjective map $P^* \otimes_A M \to L$ for some simple $B \otimes_k B^{op}$-module L. This map defines therefore a chain map

$$X^* \otimes_A X \to L[-1]$$

with $L[-1]$ the chain complex concentrated in degree -1, equal to L. This chain map is not homotopic to zero, as there are no nonzero homotopies. Since B is split, there are i, $i' \in I$ such that $L \cong T_{\delta(i)} \otimes_k T_{\delta(i')}^*$. The adjunction between $X^* \otimes_A -$ and $X \otimes_B -$ yields a chain map

$$X \to X \otimes_B T_{\delta(i)} \otimes_k T_{\delta(i')}^*[-1] \cong \operatorname{Hom}_k(T_{\delta(i')}, X \otimes_B T_{\delta(i)})[-1]$$

which is not homotopic to zero. The tensor-Hom adjunction with the k-B-bimodule $T_{\delta(i')}^*$ yields a chain map

$$X \otimes_B T_{\delta(i')} \to X \otimes_B T_{\delta(i)}[-1]$$

which is not homotopic to zero. This, however, contradicts Lemma 11.9.3. Thus $X^* \otimes_B X$ is indeed split with homology concentrated in degree zero. Since the terms in degree 1 and -1 of $X^* \otimes_A X$ are projective, it follows that the homology of $X^* \otimes_A X$ in degree zero is a direct summand of the bimodule $M^* \otimes_A M \oplus P^* \otimes_A P$. This bimodule has exactly one indecomposable nonprojective summand, and this summand is isomorphic to B, because M and M^* induce a stable equivalence of Morita type. Thus the homology of $X^* \otimes_A X$ in degree zero is isomorphic to $B \oplus Q$ for some projective $B \otimes_k B^{op}$-module Q as claimed.   □

In order to finish the proof of Theorem 11.9.1, we will need some earlier calculations in 11.7.8 which follow from the composition series of the submodules U_i, V_i of Ai. Let i, $i' \in I$. If i, i' belong to the same ρ-orbit, then $\dim_k(\operatorname{Hom}_A(U_i, U_{i'}))$ is equal to either 1 or m, depending on whether this ρ-orbit is nonexceptional or exceptional. If i, i' do not belong to the same ρ-orbit, then $\operatorname{Hom}_A(U_i, U_{i'}) = \{0\}$. The dimensions of the spaces iAi' are the Cartan invariants of A, described in 11.7.6. We will use without further comment the fact from 2.9.3 that for any two finite-dimensional A-modules U, V, the dimensions of $\operatorname{Hom}_A(U, V)$ and of $V^* \otimes_A U$ are equal.

Proof of Theorem 11.9.1 It remains to show that the complex of $A \otimes_k A^{op}$-modules $X \otimes_B X^*$ and the complex of $B \otimes_k B^{op}$-modules $X^* \otimes_A X$ are homotopy equivalent to A and B, respectively. By 4.14.16, it suffices to show that $X^* \otimes_A X \simeq B$. By 11.9.4 we have $X^* \otimes_A X \simeq B \oplus Q$ for some projective $B \otimes_k B^{\mathrm{op}}$-module Q, and $X^* \otimes_A X$ is a split complex. Thus, for any $i, i' \in I$, we have a homotopy equivalence of complexes of k-vector spaces

$$T^*_{\delta(i')} \otimes_B X^* \otimes_A X \otimes_B T_{\delta(i)} \simeq T^*_{\delta(i')} \otimes_B T_{\delta(i)} \oplus T^*_{\delta(i')} \otimes_B Q \otimes_B T_{\delta(i)}.$$

If Q is nonzero, then $T^*_{\delta(i')} \otimes_B Q \otimes_B T_{\delta(i)}$ is nonzero for some $i, i' \in I$. Thus it suffices to show that $T^*_{\delta(i')} \otimes_B Q \otimes_B T_{\delta(i)} = \{0\}$ for all $i, i' \in I$. This is equivalent to showing that there is a homotopy equivalence of complexes of k-vector spaces

$$T^*_{\delta(i')} \otimes_B X \otimes_A X \otimes_R T_{\delta(i)} \simeq T^*_{\delta(i')} \otimes_B T_{\delta(i)}$$

for all $i, i' \in I$. The right side is zero if $i \neq i'$, and isomorphic to k if $i = i'$. The left side is a complex consisting of three terms in degree $-1, 0, 1$, such that the first map is injective and the second map is surjective. Indeed, the complex $X^* \otimes_A X$ has these properties, and since $X^* \otimes_A X$ is split, these properties are preserved upon tensoring by $T^*_{\delta(i')} \otimes_B -$ and $- \otimes_B T_{\delta(i)}$. Thus we need to show that the complex $T^*_{\delta(i')} \otimes_B X \otimes_A X \otimes_B T_{\delta(i)}$ is acyclic for $i \neq i'$ and that it has homology k in degree 0 if $i = i'$.

We start with the case $i = i'$. There are two subcases, depending on whether $i \in I_0$ or $i \in I_1$. Suppose first that $i \in I_0$. By 11.9.2 we have an isomorphism of complexes

$$T^*_{\delta(i)} \otimes_B X^* \otimes_A X \otimes_B T_{\delta(i)} \cong U_i^* \otimes_A U_i.$$

Since i^ρ is nonexceptional, it follows that the right side is one-dimensional as required. Suppose next that $i \in I_1$. Then

$$T^*_{\delta(i)} \otimes_B X^* \otimes_A X \otimes_B T_{\delta(i)} \cong (U_i^* \to \rho(i)A) \otimes_A (A\rho(i) \to U_i)$$

$$\cong U_i^*\rho(i) \to U_i^* \otimes_A U_i \oplus \rho(i)A\rho(i) \to \rho(i)U_i,$$

where the left map is injective and the right map is surjective. The left term, right term, and the summand $U_i^* \otimes_A U_i$ of the middle term all have the same dimension which is either 1 or m, depending on whether i^ρ is nonexceptional or exceptional. The remaining term $\rho(i)A\rho(i)$ has dimension 2 or $m + 1$, depending on whether i^ρ is nonexceptional or exceptional. In both cases the homology of this complex is one-dimensional and concentrated in degree zero. This concludes the argument in the case where $i = i'$.

Assume now that $i \neq i'$. We need to show that the complex $T^*_{\delta(i')} \otimes_B X^* \otimes_A X \otimes_B T_{\delta(i)}$ is acyclic. There are four subcases to consider, according to whether i, i' belong to I_0 or I_1.

Suppose first that both i and i' belong to I_0. Then

$$T^*_{\delta(i')} \otimes_B X^* \otimes_A X \otimes_B T_{\delta(i)} \cong U^*_{i'} \otimes_A U_i.$$

Since both $i, i' \in I$, the ρ-orbits of i, i' are both further away from v than the σ-orbits of $\rho(i), \rho(i')$, respectively. But then i, i' must belong to two different ρ-orbits; indeed, if i, i' are in the same ρ-orbit, then $\rho(i), \rho(i')$ are in the same ρ-orbit. The uniqueness of the paths from v to these ρ-orbits forces that then also the σ-orbits of $\rho(i), \rho(i')$ coincide. That, however, forces $i = i'$, a contradiction. Thus i, i' belong to different ρ-orbits, and hence $U^*_{i'} \otimes_A U_i$ is zero as required.

Suppose next that $i \in I_1$ and $i \in I_0$. Then

$$T^*_{\delta(i')} \otimes_B X^* \otimes_A X \otimes_B T_{\delta(i)} \cong U^*_{i'} \otimes_A (A\rho(i) \to U_i) \cong U^*_{i'}\rho(i) \to U^*_{i'} \otimes_A U_i,$$

and the map in the last complex is injective. If i, i' belong to different ρ-orbits, then this complex is zero. If i, i' belong to the same ρ-orbit, then both terms in this complex have the same dimension 1 or m, depending on whether this ρ-orbit is nonexceptional or exceptional. Thus the map in this complex is an isomorphism, and hence this complex is acyclic. The case $i \in I_0$ and $i' \in I_1$ follows similarly.

Suppose finally that both i, i' belong to I_1. Then

$$T^*_{\delta(i')} \otimes_B X^* \otimes_A X \otimes_B T_{\delta(i)} \cong (U^*_{i'} \to \rho(i')A) \otimes_A (A\rho(i) \to U_i)$$

$$\cong U^*_{i'}\rho(i) \to U^*_{i'} \otimes_A U_i \oplus \rho(i')A\rho(i) \to \rho(i')U_i.$$

Since $i, i' \in I_1$ and $i \neq i'$, it follows as above that the σ-orbits of $\rho(i)$ and of $\rho(i')$ are different. Thus, if also the ρ-orbits of i and i' are different, then this complex is zero. If i, i' belong to the same ρ-orbits, then the four spaces $U^*_{i'}\rho(i)$, $U^*_{i'} \otimes_A U_i$, $\rho(i')A\rho(i)$, $\rho(i')U_i$ all have dimension 1 or m, depending on whether this ρ-orbit is nonexceptional or exceptional, and hence again, this complex is acyclic. This concludes the proof of Theorem 11.9.1. □

11.10 Irreducible characters and decomposition numbers

We prove in this section the Theorems 11.1.12, 11.1.14, and 11.1.15. Let G be a finite group, b a block of $\mathcal{O}G$ with a nontrivial cyclic defect group P and inertial quotient E. Denote by Z the subgroup of order p of P and set $H = N_G(Z)$. Denote by c the block of $\mathcal{O}H$ with P as a defect group which corresponds to b through the Brauer correspondence. We fix a block e of $kC_G(P)$ such that

(P, e) is a maximal (G, b)-Brauer pair. Since $C_G(P) \subseteq H$, it follows that (P, e) is also a maximal (H, c)-Brauer pair. We assume in this section that the field K is a splitting field for all involved block algebras. Since $\mathcal{O}Hc$ is Morita equivalent to $\mathcal{O}(P \rtimes E)$, the set $\mathrm{Irr}_K(H, c)$ is indexed by the set $\mathrm{Irr}_K(P \rtimes E)$. As E acts freely on $P \setminus \{1\}$, the group $P \rtimes E$ is a Frobenius group, and the set of irreducible characters is a disjoint union

$$\mathrm{Irr}_K(P \rtimes E) = \Lambda \cup M,$$

where Λ is the set of characters of the form $\mathrm{Ind}_P^{P \rtimes E}(\zeta)$, with ζ running over a set of representatives of the E-conjugacy classes of nontrivial irreducible characters of P, and where M consists of the irreducible characters of $P \rtimes E$ with P in their kernel. The characters in Λ have degree $|E|$, and the characters in M have degree 1 and are obtained from extending the irreducible characters of E trivially to $P \rtimes E$. We have $|\Lambda| = \frac{|P|-1}{|E|}$ and $|M| = |E|$. The set

$$\left\{ \mu + \sum_{\lambda \in \Lambda} \lambda \right\}_{\mu \in M}$$

is the set of characters of the projective indecomposable $\mathcal{O}(P \rtimes E)$-modules; in particular, this set is a basis of the abelian subgroup $\mathrm{Pr}_{\mathcal{O}}(P \rtimes E)$ of $\mathbb{Z}\mathrm{Irr}_K(P \rtimes E)$. The set

$$\left\{ \lambda - \sum_{\mu \in M} \mu \right\}_{\lambda \in \Lambda}$$

is a basis of $L^0(P \rtimes E)$. Indeed, the elements of this set are orthogonal to the characters of the projective indecomposable $\mathcal{O}(P \rtimes E)$-modules, they are $\mathbb{Z}$-linearly independent, and they are in fact a $\mathbb{Z}$-basis because each element has exactly one of the $\lambda \in \Lambda$ occurring with multiplicity 1. The ordinary decomposition numbers can be read off the above description of the characters of the projective indecomposable $\mathcal{O}(P \rtimes E)$-modules. The remaining part of the generalised decomposition matrix of $\mathcal{O}(P \rtimes E)$ is equal to the matrix

$$(\lambda(u))_{\lambda, u}$$

with $\lambda \in \mathrm{Irr}_K(P \rtimes E)$ and u running over a set of representatives of the E-conjugacy classes of elements in $P \setminus \{1\}$. This follows from the fact that for $u \in P \setminus \{1\}$ we have $C_{P \rtimes E}(u) = P$, hence $C_{P \rtimes E}(u)$ has a unique irreducible Brauer character, and 1 is the unique idempotent in $\mathcal{O}P$. The structure of the source algebras of $\mathcal{O}Hc$ implies that $\mathcal{O}Hc$ is Morita equivalent to $\mathcal{O}(P \rtimes E)$ via a Morita equivalence having an endopermutation $\mathcal{O}P$-module W as a source. Thus the characters of $\mathcal{O}Hc$ are labelled by those of $P \rtimes E$ in such a way that corresponding characters give rise to the same ordinary and generalised decomposition numbers, up to signs determined by the character of W. More precisely:

Theorem 11.10.1 *With the notation above, the following hold.*

(i) *There is a perfect isometry $\mathbb{Z}\mathrm{Irr}_K(P \rtimes E) \cong \mathrm{Irr}_K(H, c)$ sending $\lambda \in \mathrm{Irr}_K(P \rtimes E)$ to $\eta_\lambda \in \mathrm{Irr}_K(H, c)$; that is, all signs of this isometry are positive.*

(ii) *The set $\{\sum_{\mu \in M} \eta_\mu - \eta_\lambda\}_{\lambda \in \Lambda}$ is a $\mathbb{Z}$-basis of $L^0(H, c)$.*

(iii) *The set $\{\eta_\mu + \sum_{\lambda \in \Lambda} \eta_\lambda\}_{\mu \in M}$ is a $\mathbb{Z}$-basis of $\mathrm{Pr}_\mathcal{O}(H, c)$.*

(iv) *For any local pointed element u_ϵ on $\mathcal{O}Hc$ associated with (P, e) and any $\lambda \in \mathrm{Irr}_K(P \rtimes E)$ we have*

$$\eta_\lambda(u_\epsilon) = \omega(u)\lambda(u),$$

where $\omega(u)$ is the sign of the character value $\chi_W(u)$.

Proof Statements (i) follows trivially from the fact that $\mathcal{O}Hc$ is Morita equivalent to the source algebra $S \otimes_\mathcal{O} \mathcal{O}(P \rtimes E)$, where $S = \mathrm{End}_\mathcal{O}(W)$, with W as above (or as in 11.2.2). The statements (ii) and (iii) follow immediately from the descriptions of $L^0(P \rtimes E)$ and $\mathrm{Pr}_\mathcal{O}(P \rtimes E)$ above. Statement (iv) follows from (i) this and 7.4.3. □

The passage from $\mathcal{O}Hc$ to $\mathcal{O}Gb$ is made via the splendid stable equivalence of Morita type from 11.2.4.

Theorem 11.10.2 *With the notation above, the following hold.*

(i) *We have $\ell(\mathcal{O}Gb) = \ell(\mathcal{O}Hc) = |E|$ and $|\mathrm{Irr}_K(G, b)| = |\mathrm{Irr}_K(H, c)| = |E| + \frac{|P|-1}{|E|}$.*

(ii) *There is a perfect isometry $\mathbb{Z}\mathrm{Irr}_K(H, c) \cong \mathbb{Z}\mathrm{Irr}_K(G, b)$ which extends the partial isometry $L^0(H, c) \cong L^0(G, b)$ induced by the stable equivalence $b\mathcal{O}Gc \otimes_{\mathcal{O}Hc} -$. This isometry sends η_λ to $\delta(\lambda)\chi_\lambda$ for some $\delta(\lambda) \in \{\pm 1\}$ and $\chi_\lambda \in \mathrm{Irr}_K(G, b)$, for all $\lambda \in \Lambda$, and moreover, if $\lambda, \lambda' \in \Lambda$, then $\delta(\lambda) = \delta(\lambda')$.*

(iii) *Set $\delta = \delta(\lambda)$, where $\lambda \in \Lambda$. The set $\{\sum_{\mu \in M} \delta(\mu)\chi_\mu - \delta\chi_\lambda\}_{\lambda \in \Lambda}$ is a $\mathbb{Z}$-basis of $L^0(G, b)$.*

(iv) *Set $\delta = \delta(\lambda)$, where $\lambda \in \Lambda$. The set $\{\delta(\mu)\chi_\mu + \sum_{\lambda \in \Lambda} \delta\chi_\lambda\}_{\mu \in M}$ is a $\mathbb{Z}$-basis of $\mathrm{Pr}_\mathcal{O}(G, b)$.*

(v) *Any local pointed element u_ϵ on $\mathcal{O}Gb$ associated with (P, e) such that $u \neq 1$ corresponds to a unique local pointed element $u_{\epsilon'}$ on $\mathcal{O}Hc$ associated* *with (P, e), and we have*

$$\chi_\lambda(u_\epsilon) = \delta(\lambda)\eta_\lambda(u_{\epsilon'}).$$

Proof Since $kG\bar{b}$ is stably equivalent to the symmetric serial algebra $kH\bar{c}$, it follows from 11.6.1 that $\ell(\mathcal{O}Gb) = \ell(\mathcal{O}Hc)$. Since $\mathcal{O}Hc$ and $\mathcal{O}(P \rtimes E)$ are Morita equivalent, we have $\ell(\mathcal{O}Hc) = \ell(\mathcal{O}(P \rtimes E)) = |E|$. By 9.3.5 we have $L^0(H, c) \cong L^0(G, b)$, hence both sides have the same rank over $\mathbb{Z}$. Again since $\mathcal{O}Hc$ and $\mathcal{O}(P \rtimes E)$ are Morita equivalent, we have $L^0(H, c) \cong L^0(P \rtimes E)$. By the remarks at the beginning of this section, the rank of this group is $\frac{|P|-1}{|E|}$. Since $|\mathrm{Irr}_K(G, b)| = \ell(\mathcal{O}Gb) + \mathrm{rk}_\mathbb{Z}(L^0(G, b))$, it follows that $|\mathrm{Irr}_K(G, b)| = |\mathrm{Irr}_K(H, c)|$ as stated in (i). Theorem 10.5.10 (i) yields the perfect isometry as stated in (ii). A perfect isometry preserves the groups generated by characters of projective modules, as well as the groups that are orthogonal to these. Thus (iii) and (iv) are an immediate consequence of the perfect isometry in (ii) and the bases of $L^0(H, c)$ and $\mathrm{Pr}_\mathcal{O}(H, c)$ as described in 11.10.1. Let $u \in P \setminus \{1\}$. Then 11.2.4 implies that $(\mathcal{O}Gb)(\langle u \rangle) = kC_G(u)\mathrm{Br}_{\Delta\langle u \rangle}(b) = (\mathcal{O}Hc)(\langle u \rangle)$, and hence we get a bijection between the local points of $\langle u \rangle$ on $\mathcal{O}Gb$ and on $\mathcal{O}Hc$ as stated. Let ϵ be a local point of $\langle u \rangle$ on $\mathcal{O}Gb$ associated with (P, e) (that is, $u_\epsilon \in P_\gamma$, where γ is the unique local point of P on $\mathcal{O}Gb$ such that $\mathrm{Br}_{\Delta P}(\gamma)e \neq \{0\}$). Let ϵ' be the corresponding local point of $\langle u \rangle$ on $\mathcal{O}Hc$ (that is, $\mathrm{Br}_{\langle u \rangle}(\epsilon) = \mathrm{Br}_{\langle u \rangle}(\epsilon')$). Let U be an $\mathcal{O}$-free $\mathcal{O}Gb$-module with character χ_λ, for some $\lambda \in \mathrm{Irr}_K(P \rtimes E)$, then the character of the $\mathcal{O}Hc$-module cU is of the form $\delta(\lambda)\eta_\lambda + \psi_\lambda$ for some $\psi \in \mathrm{Pr}_\mathcal{O}(H, c)$. Let $j \in \epsilon$ and $j' \in \epsilon'$. By 5.12.16 we have $\chi_\lambda(uj) = \chi_\lambda(uj') = \delta(\lambda)\eta_\lambda(uj')$ where we use the fact that $j' \in \mathcal{O}Hc$ and that $\psi_\lambda(uj') = 0$ by 5.12.14. This proves (v). $\qquad \square$

Combining 11.10.1 and 11.10.2 yields proofs of 11.1.12 and of 11.1.14. In order to determine the ordinary decomposition numbers we need the following observation.

Lemma 11.10.3 *With the notation above, let U be an $\mathcal{O}$-free $\mathcal{O}Gb$-module such that $k \otimes_\mathcal{O} U$ is indecomposable. Then the character of U is not contained in $\mathrm{Pr}_\mathcal{O}(G, b)$; equivalently, the character of U does not vanish on all p-singular elements in G.*

Proof Arguing by induction, suppose that the character of U is in $\mathrm{Pr}_\mathcal{O}(G, b)$. Through the stable equivalence between $\mathcal{O}Gb$ and $\mathcal{O}(P \rtimes E)$ given by induction and restriction (truncated by the block idempotents) the module U corresponds to an indecomposable $\mathcal{O}(P \rtimes E)$-module V. The character of V belongs to $\mathrm{Pr}_\mathcal{O}(P \rtimes E)$. By 4.14.6, the module $k \otimes_\mathcal{O} V$ remains indecomposable and nonprojective. Thus $k \otimes_\mathcal{O} V$ is a uniserial nonprojective module, and hence its dimension is strictly less than $|P|$. But then the character of V cannot be in $\mathrm{Pr}_{\mathcal{O}}(P \rtimes E)$ because any ordinary character in $\mathrm{Pr}_{\mathcal{O}}(P \rtimes E)$ has degree divisible by $|P|$. This contradiction proves the result. $\qquad \square$

For the following result we make use of Thompson's technique using pure submodules, as described in 4.16.4.

Proposition 11.10.4 *With the notation above, let U be a projective indecomposable $\mathcal{O}Gb$-module, and let ψ be the character of U. Suppose there are nonzero ordinary characters ψ_1 and ψ_2 such that $\psi = \psi_1 + \psi_2$. Then none of ψ_1 or ψ_2 is contained in $\mathrm{Pr}_{\mathcal{O}}(G, b)$.*

Proof By 4.16.4, there is an $\mathcal{O}$-pure submodule U_1 of U having ψ_1 as character. Then $k \otimes_{\mathcal{O}} U_1$ is a submodule of the projective indecomposable $kG\bar{b}$-module $k \otimes_{\mathcal{O}} U$. It follows that the socle of $k \otimes_{\mathcal{O}} U_1$ is simple, and hence in particular, $k \otimes_{\mathcal{O}} U_1$ is indecomposable. The preceding lemma implies that ψ_1 is not in $\mathrm{Pr}_{\mathcal{O}}(G, b)$. The result follows. $\qquad \square$

We restate and prove the first statement from 11.1.15. We keep the notation above; in particular, we denote by $\delta(\mu)$ and δ the signs from the perfect isometry in 11.10.2.

Theorem 11.10.5 *Let ψ be the character of a projective indecomposable $\mathcal{O}Gb$-module. Then either $\psi = \chi_\mu + \chi_{\mu'}$ for some $\mu, \mu' \in M$ such that $\delta(\mu) \neq \delta(\mu')$, or $\psi = \chi_\mu + \sum_{\lambda \in \Lambda} \lambda$ for some $\mu \in M$ such that $\delta(\mu) = \delta$. In particular, all ordinary decomposition numbers of $\mathcal{O}Gb$ are either 1 or 0.*

Proof The characters $\chi_\mu + \chi_{\mu'}$, with $\delta(\mu) \neq \delta(\mu')$, and the characters $\chi_\mu + \sum_{\lambda \in \Lambda} \chi_\lambda$ belong all to the group $\mathrm{Pr}_{\mathcal{O}}(G, b)$, because they are easily obtained as $\mathbb{Z}$-linear combinations of the basis described in 11.10.2 (iv). Using this basis we can write ψ in the form

$$\psi = \sum_{\mu \in M} a_\mu \delta(\mu) \chi_\mu + \left(\sum_{\mu \in M} a_\mu \right) \sum_{\lambda \in \Lambda} \delta \chi_\lambda$$

for some coefficients $a_\mu \in \mathbb{Z}$. Since ψ is an actual character, we have $a_\mu \delta(\mu) \geq 0$ for all $\mu \in M$ and $(\sum_{\mu \in M} a_\mu) \delta \geq 0$. Thus, if $\sum_{\mu \in M} a_\mu = 0$, then there are at least two different nonzero coefficients a_μ, $a_{\mu'}$ with a different sign, which implies that $\delta(\mu) \neq \delta(\mu')$. Thus $\psi = \chi_\mu + \chi_{\mu'} + \psi'$ for some ordinary character ψ'. But since $\chi_\mu + \chi_{\mu'}$ is in $\mathrm{Pr}_{\mathcal{O}}(G, b)$, it follows from 11.10.4 that $\psi' = 0$, hence $\psi = \chi_\mu + \chi_{\mu'}$. Similarly, if $\sum_{\mu \in M} a_\mu \neq 0$, then there is at least one $\nu \in M$ such that a_ν is nonzero and has the same sign as the nonzero integer $\sum_{\mu \in M} a_\mu$. This implies that $\delta(\nu) = \delta$, and hence that $\psi = \chi_\nu + \sum_{\lambda \in \Lambda} \lambda + \psi'$

for some ordinary character ψ'. As before, it follows from 11.10.4 that $\psi' = 0$, and hence that $\psi = \chi_\nu + \sum_{\lambda \in \Lambda} \lambda$ as claimed. □

This gives rise to a graph having as vertices the characters χ_μ with $\mu \in M$ and the character $\sum_{\lambda \in \Lambda} \lambda$, with an edge between χ_μ and $\chi_{\mu'}$ whenever $\chi_\mu + \chi_{\mu'}$ is the character of a projective indecomposable $\mathcal{O}Gb$-module, and an edge between χ_μ and $\sum_{\lambda \in \Lambda} \lambda$ whenever $\chi_\nu + \sum_{\lambda \in \Lambda} \lambda$ is the character of a projective indecomposable $\mathcal{O}Gb$-module. This is a tree, as it is connected (as a consequence of Brauer's Theorem 4.10.5) and the fact that it has $|I| + 1$ vertices and $|I|$ edges none of which is a loop. We will see in the next section that this graph is the abstract tree underlying the Brauer tree (this is the last statement in 11.1.15).

Remark 11.10.6 Theorem 11.1.14 is a special case of Theorem 10.5.10 (ii), but for P cyclic, the proof is is easier: the intermediate algebra A' constructed in the alternative proof of 10.5.1, needed for the proof of 10.5.10 (ii), can be avoided by considering instead the source algebras of the corresponding block of the normaliser of the unique subgroup of order p as above.

11.11 Walking around the Brauer tree

Let G be a finite group, b a block of $\mathcal{O}G$ with a nontrivial cyclic defect group, let Z be the subgroup of order p of P, and set $H = N_G(Z)$. Denote by c the unique block of $\mathcal{O}H$ with defect group P such that $\mathrm{Br}_Z(b) = \mathrm{Br}_Z(c)$. Let E be the inertial quotient of b and of c. Set $A = \mathcal{O}Gb$ and $B = \mathcal{O}Hc$. Let I be a set of representatives of the conjugacy classes of primitive idempotents in A and J a set of representatives of the conjugacy classes of primitive idempotents in B. The image of I in $\bar{A} = k \otimes_\mathcal{O} A$ is still a set of representatives of the conjugacy classes of primitive idempotents in $\bar{A}$, and a similar statement holds for the image of J in $\bar{B} = k \otimes_\mathcal{O} B$. We set $S_i = Ai/J(A)i$ for $i \in I$ and $T_j = Bj/J(B)j$ for $j \in J$.

By 11.2.4, the A-B-bimodule $bOGc$ and its dual, which is isomorphic to $cOGb$, induce a stable equivalence of Morita type between A and B. In particular, if U is an indecomposable A-module that is not relatively $\mathcal{O}$-projective, then the B-module $c\mathrm{Res}_H^G(U) \cong cOGb \otimes_A U$ has, up to isomorphism, a unique indecomposable direct summand that is not relatively $\mathcal{O}$-projective, and we will denote this indecomposable summand by $\mathcal{F}(U)$. Similarly, if V is an indecomposable B-module that is not relatively $\mathcal{O}$-projective, then the A-module $b\mathrm{Ind}_H^G(V) = bOGc \otimes_B V$ has up to isomorphism a unique indecomposable direct summand that is not relatively $\mathcal{O}$-projective, and we will denote this indecomposable summand by $\mathcal{G}(V)$. Since H contains the normaliser of any nontrivial subgroup of P, the module $\mathcal{F}(U)$ is the Green correspondent of U and the module $\mathcal{G}(V)$ is the Green correspondent of V. Since the operators $\mathcal{F}$ and $\mathcal{G}$ are induced by a stable equivalence, it follows that $\mathcal{G}(\mathcal{F}(U)) \cong U$ for any indecomposable non relatively $\mathcal{O}$-projective A-module U and $\mathcal{F}(\mathcal{G}(V)) \cong V$ for any indecomposable non relatively $\mathcal{O}$-projective B-module V.

We denote further by W the indecomposable endopermutation $\mathcal{O}P$-module from 11.2.2; that is, Z acts as identity on W, and $S \otimes_{\mathcal{O}} \mathcal{O}(P \rtimes E)$ is a source algebra of $\mathcal{O}Hc$, where $S = \mathrm{End}_{\mathcal{O}}(W)$. The algebra $\bar{B}$ is serial symmetric algebra, and we can apply the results from §11.3. We write abusively Ω for the Heller translates of A and of B.

Proposition 11.11.1 *With the notation above, the following hold.*

(i) *There is a unique transitive cyclic permutation π of J such that $\Omega^2(T_j) \cong T_{\pi(j)}$ for any $j \in J$.*

(ii) *For any $j \in J$ there is up to isomorphism a unique $\mathcal{O}$-free B-module W_j such that $k \otimes_{\mathcal{O}} W_j \cong T_j$ and such that W is a source of W_j.*

(iii) *We have $\Omega^2(W_j) \cong W_{\pi(j)}$ for any $j \in J$.*

(iv) *The $\mathcal{O}P$-module $\Omega_P(W)$ is a source of $\Omega(W_j)$.*

Proof Let T be a simple $kP \rtimes E$-module. Then $\dim_k(T) = 1$, and the Brauer character φ of T is the restriction to the p'-elements of a unique ordinary irreducible character μ which belongs to the set of nonexceptional characters M having P in their kernel. Thus there is up to isomorphism a unique $\mathcal{O}(P \rtimes E)$-module Y of $\mathcal{O}$-rank 1 such that $k \otimes_{\mathcal{O}} Y \cong T$ and such that P acts trivially on Y. Through the Morita equivalence between $\mathcal{O}(P \rtimes E)$ and the source algebra $S \otimes_{\mathcal{O}} \mathcal{O}(P \rtimes E)$ of B, the module Y corresponds to the module of the form $W \otimes Y$. This module, restricted to P, has W as a summand since P acts trivially on Y. Thus P is a vertex and W a source of the B-module W_j corresponding to $W \otimes_{\mathcal{O}} Y$ through the standard Morita equivalence between $S \otimes_{\mathcal{O}} \mathcal{O}(P \rtimes E)$ and B. This Morita equivalence commutes with Heller translates, and hence all remaining statements follow from the analogous statements on the serial symmetric algebra $k(P \rtimes E)$ from 11.3.4, its corollary 11.3.5, and the uniqueness of the modules W_j lifting the simple modules T_j. $\qquad\square$

Corollary 11.11.2 *With the notation above, the following hold.*

(i) *The trivial $\mathcal{O}(P \rtimes E)$-module $\mathcal{O}$ has period $2|E|$.*

(ii) *The $\mathcal{O}(P \rtimes E)$-$\mathcal{O}(P \rtimes E)$-bimodule $\mathcal{O}(P \rtimes E)$ has period $2|E|$.*

(iii) *There is an $\mathcal{O}$-algebra automorphism ϵ of $\mathcal{O}(P \rtimes E)$ of order $|E|$ which acts as identity on P such that $\Omega^2_{\mathcal{O}(P \rtimes E) \otimes_{\mathcal{O}} \mathcal{O}(P \rtimes E)^{op}}(\mathcal{O}(P \rtimes E)) \cong {}_\epsilon\mathcal{O}(P \rtimes E)$.*

Proof The statement (i) on the period of the trivial $\mathcal{O}(P \rtimes E)$-module $\mathcal{O}$ follows from 11.11.1. By 2.4.5, we can identify the $\mathcal{O}(P \rtimes E)$-$\mathcal{O}(P \rtimes E)$-bimodule $\mathcal{O}(P \rtimes E)$ with $\mathrm{Ind}^{P \rtimes E \times P \rtimes E}_{\Delta(P \rtimes E)}(\mathcal{O})$. Since induction commutes with the Heller operator, it follows that $\mathcal{O}(P \rtimes E)$ has also period $2|E|$. This shows (ii). By 11.5.3, the bimodule $\Omega^2_{\mathcal{O}(P \rtimes E) \otimes_{\mathcal{O}} \mathcal{O}(P \rtimes E)^{op}}(\mathcal{O}(P \rtimes E))$ induces a Morita equivalence on $\mathcal{O}(P \rtimes E)$. This Morita equivalence sends the trivial $\mathcal{O}(P \rtimes E)$-module $\mathcal{O}$ to an $\mathcal{O}(P \rtimes E)$-module Y of $\mathcal{O}$-rank 1, isomorphic to $\Omega^2(\mathcal{O})$. Thus P acts trivially on Y, and if y is a generator of E, then y acts on Y has multiplication by some primitive $|E|$-th root of unity ζ. There is a unique automorphism ϵ of $\mathcal{O}(P \rtimes E)$ sending $u \in P$ to u and y to ζy, and clearly ϵ has order $|E|$. Thus $Y \cong {}_\epsilon\mathcal{O}$. Since Ω commutes with induction, we have isomorphisms $\Omega^2_{\mathcal{O}(P \rtimes E) \otimes_{\mathcal{O}} \mathcal{O}(P \rtimes E)^{op}}(\mathcal{O}(P \rtimes E)) \cong \mathrm{Ind}^{P \rtimes E \times P \rtimes E}_{\Delta(P \rtimes E)}(Y) \cong {}_\epsilon\mathcal{O}(P \rtimes E)$. Statement (iii) follows. $\square$

By 11.6.1 and 11.6.3 there are unique bijections $\gamma : I \to J$ and $\delta : I \to J$, and for any $i \in I$, there are unique uniserial submodules $\bar{U}_i$, $\bar{V}_i$ of $\bar{A}\bar{i}$ such that $\bar{U}_i \cong \mathcal{G}(T_{\delta(i)})$ and $\bar{V}_i \cong \mathcal{G}(\Omega(T_{\gamma(i)}))$. The simple $\bar{B}$-modules T_j lift canonically to $\mathcal{O}$-free B-modules W_j, and hence the $\bar{U}_i$, $\bar{V}_i$ have canonical lifts to $\mathcal{O}$-free A-modules. By 11.6.4 there are unique permutations ρ and σ of I such that we have short exact sequences of $\bar{A}$-modules

$$0 \longrightarrow \bar{U}_{\sigma(i)} \longrightarrow \bar{A}_{\sigma(i)} \longrightarrow \bar{V}_i \longrightarrow 0$$

$$0 \longrightarrow \bar{V}_{\rho(i)} \longrightarrow \bar{A}_{\rho(i)} \longrightarrow \bar{U}_i \longrightarrow 0.$$

The next result shows that these short exact sequences lift over $\mathcal{O}$, proving parts of 11.1.5 and of 11.1.15.

Theorem 11.11.3 ([77]) *With the notation above, the following hold for any $i \in I$.*

(i) *There are unique $\mathcal{O}$-pure submodules U_i and V_i of Ai such that $U_i \cong \mathcal{G}(W_{\delta(i)})$ and $V_i \cong \mathcal{G}(\Omega(W_{\gamma(i)}))$.*

(ii) *The group P is a vertex of U_i and of V_i, the module W is a source of U_i, and its Heller translate $\Omega_P(W)$ is a source of V_i.*

(iii) *We have short exact sequences*

$$0 \longrightarrow U_{\sigma(i)} \longrightarrow A_{\sigma(i)} \longrightarrow V_i \longrightarrow 0,$$

$$0 \longrightarrow V_{\rho(i)} \longrightarrow A_{\rho(i)} \longrightarrow U_i \longrightarrow 0.$$

(iv) *The characters of U_i and V_i are disjoint, and they belong to the set $\{\chi_\mu\}_{\mu \in M} \cup \{\sum_{\lambda \in \Lambda} \chi_\lambda\}$. The character of U_i depends only on the ρ-orbit of i, and the character of V_i depends only on the σ-orbit of i.*

(v) *The character of Ai is equal to the character of $U_i \oplus V_i$, and any A-endomorphism of Ai sends U_i to U_i and V_i to V_i.*

(vi) *The canonical maps $\mathrm{End}_A(U_i) \to \mathrm{End}_{\bar{A}}(\bar{U}_i)$ and $\mathrm{End}_A(V_i) \to \mathrm{End}_{\bar{A}}(\bar{V}_i)$ are surjective.*

(vii) *If i^ρ is nonexceptional, then the character of U_i is equal to χ_μ for some $\mu \in M$, and equal to $\sum_{\lambda \in \Lambda} \chi_\lambda$ otherwise. Similarly, if i^σ is nonexceptional, then the character of V_i is equal to χ_μ for some $\mu \in M$, and equal to $\sum_{\lambda \in \Lambda} \chi_\lambda$ otherwise.*

Proof Denote by π the transitive cycle on J from 11.3.4 such that $\Omega^2(T_j) \cong T_{\pi(j)}$ for $j \in J$. The uniqueness property of the lifts W_j implies that then also $\Omega^2(W_j) \cong W_{\pi(j)}$. Thus we have

$$\Omega(V_i) \cong \Omega(\mathcal{G}(\Omega(W_{\gamma(i)}))) \cong \mathcal{G}(\Omega^2(W_{\gamma(i)})) \cong \mathcal{G}(W_{\pi(\gamma(i))}) \cong U_{\sigma(i)},$$

where the last isomorphism uses the equality $\pi \circ \gamma = \delta \circ \sigma$, from 11.6.4 (v). Similarly, we have

$$\Omega(U_i) \cong \Omega(\mathcal{G}(W_{\delta(i)})) \cong \mathcal{G}(\Omega(W_{\delta(i)})) \cong V_{\rho(i)},$$

where the last isomorphism uses the equality $\gamma \circ \rho = \delta$, again from 11.6.4 (v). This yields the short exact sequences in (iii). Since the U_i, V_i arise as kernels of homomorphisms starting at Ai, they are $\mathcal{O}$-pure in Ai. By 11.10.5 the character of Ai is a sum of pairwise different irreducible characters. Thus any $\mathcal{O}$-pure submodule of Ai is uniquely determined by its character, hence by its isomorphism class. This shows the uniqueness statement in (i). The stable equivalence $\mathcal{G}$ is splendid and coincides with the Green correspondence; in particular, $\mathcal{G}$ preserves vertices and sources. Thus $\mathcal{G}(W_{\delta(i)})$ and $\mathcal{G}(\Omega(W_{\gamma(i)}))$ have P as a vertex, and their sources are W and $\Omega_P(W)$, respectively. This shows (ii). Let $i \in I$. We use again the description of the character of Ai as in 11.10.5. Suppose that the character of Ai is equal to $\chi_\mu + \chi_{\mu'}$, for some $\mu, \mu' \in M$. Then a nonzero proper $\mathcal{O}$-pure submodule of Ai must have as character either χ_μ or $\chi_{\mu'}$. In particular, the characters of U_i, V_i are in $\{\chi_\mu, \chi_{\mu'}\}$. Suppose next that the character of Ai is equal to $\chi_\mu + \sum_{\lambda \in \Lambda} \chi_\lambda$. It suffices to show that if some χ_λ appears in the character of U_i or V_i, then $\sum_{\lambda \in \Lambda} \chi_\lambda$ appears in that character. Denote by θ the character of U_i, and suppose that there is $\lambda \in \Lambda$ such that $\langle \theta, \chi_\lambda \rangle \neq 0$. Note that the character of $\mathcal{F}(U_i) \cong W_{\delta(i)}$ is equal to $\eta_{\mu'}$ for some $\mu' \in M$. Let λ, λ' be two different elements in Λ. Then $\chi_\lambda - \chi_{\lambda'} \in L^0(G, b)$ is orthogonal to the characters of projective A-modules. Using the perfect isometry from 11.10.2, we get that

$$\langle \theta, \chi_\lambda - \chi_{\lambda'} \rangle = \delta \langle \eta_{\mu'}, \eta_\lambda - \eta_{\lambda'} \rangle = 0.$$

Thus all χ_λ appear with the same multiplicity in the character of U_i. A similar argument shows that this is true for the character of V_i. If $\rho(i) \neq i$ or $\sigma(i) \neq i$, then the composition series of $\bar{U}_i$ and of $\bar{V}_i$ are different. Thus the characters of U_i and of V_i are different in that case, and therefore their sum must be the character of Ai. If $\rho(i) = i = \sigma(i)$, then $I = \{i\}$. In that case, (iii) implies that $V_i \cong Ai/U_i$, hence the character of Ai is the sum of the characters of U_i and V_i. Since the decomposition numbers are all 0 or 1, this forces again that the characters of U_i and V_i are different. The short exact sequences in (iii) imply that the character of Ai is also equal to that of the modules $U_i \oplus V_{\sigma^{-1}(i)}$ and $U_{\rho^{-1}} \oplus V_i$. This shows that V_i and $V_{\sigma^{-1}(i)}$ have the same character, and that U_i and $U_{\rho^{-1}(i)}$ have the same character. The fact that the decomposition numbers are all 0 or 1 implies further that every endomorphism of Ai preserves any $\mathcal{O}$-pure submodule, hence in particular U_i and V_i. This completes the proof of (iv) and (v). The algebra $\bar{A}\bar{i}$ is symmetric, hence selfinjective. Thus any endomorphism of $\bar{U}_i$ extends to an endomorphism of $\bar{A}\bar{i}$. Any endomorphism of $\bar{A}\bar{i}$ lifts to an endomorphism of Ai. Any endomorphism of Ai preserves U_i and V_i by the uniqueness properties of these modules. This implies that any endomorphism of $\bar{U}_i$ lifts to an endomorphism of U_i. A similar argument for V_i yields (vi). Statement (vii) is an immediate consequence of the previous statements in conjunction with the composition series of the uniserial modules $\bar{U}_i$ and $\bar{V}_i$ from 11.6.4. □

Proposition 11.11.4 *With the notation above, set* $m = \frac{|P|-1}{|E|}$. *The Brauer tree of $\mathcal{O}Gb$ has an exceptional vertex if and only if* $m > 1$, *in which case m is the exceptional multiplicity. In particular, the Brauer tree of $\mathcal{O}Gb$ has no exceptional vertex if and only if $|P| = p$ and $|E| = p - 1$.*

Proof By 11.11.3 (vi), we have $\mathrm{rk}_{\mathcal{O}}(\mathrm{End}_A(U_i)) = \dim_k(\mathrm{End}_{\bar{A}}(\bar{U}_i))$. It follows from 11.7.8 that the Brauer tree of $\mathcal{O}Gb$ has an exceptional vertex if and only if for some $i \in I$, one of $\mathrm{End}_A(U_i)$ or $\mathrm{End}_A(V_i)$ has $\mathcal{O}$-rank greater than one. Suppose that $\mathrm{End}_A(U_i)$ has $\mathcal{O}$-rank greater than one. Then the character of U_i is not irreducible, hence equal to $\sum_{\lambda \in \Lambda} \chi_\lambda$ by 11.11.3 (vii). But then the $\mathcal{O}$-rank of $\mathrm{End}_A(U_i)$ is equal to $|\Lambda| = m$. Again by 11.7.8, this shows that m is the exceptional multiplicity. A similar argument using V_i concludes the proof. $\qquad\square$

11.12 A splendid Rickard complex for blocks with cyclic defect

The main result of this section, due to Rouquier [196], describes a splendid Rickard complex between a block algebra with a nontrivial cyclic defect group P and the algebra $\mathcal{O}(P \rtimes E)$, where E is the inertial quotient of the block. In particular, Broué's abelian defect group conjecture holds for blocks with cyclic defect groups. Splendid Rickard complexes have been defined with respect to a choice of source idempotents. If we do not specify such a choice in the results below, then we implicitly assert that the complex is splendid for some choice. It turns out in that the complex is then splendid for any such choice. This follows from Theorem 11.1.19, proved in §11.15 below, which states that all source algebras of a block with a cyclic defect group P are isomorphic as interior P-algebras.

The proof we present follows [131, §5]. Let G be a finite group, b a block of $\mathcal{O}G$ with a nontrivial cyclic defect group P and inertial quotient E. Denote by Z the unique subgroup of order p of P and set $H = N_G(Z)$. Denote by c the unique block of $\mathcal{O}H$ with P as a defect group corresponding to b through the Brauer correspondence. Set $A = \mathcal{O}Gb$ and $B = \mathcal{O}Hc$. We assume in this section that k is a splitting field for all involved block algebras. We noted earlier that the inertial quotient E is determined as a subgroup of $\mathrm{Aut}(P)$ by its order, and hence all block algebras and all of their source algebras in this section have the same fusion system $\mathcal{F} = \mathcal{F}_P(P \rtimes E)$ of $P \rtimes E$ on P.

The starting point is the splendid stable equivalence of Morita type between A and B, given by induction and restriction. More precisely, by 11.2.4, the A-B-bimodule $b\mathcal{O}Gc$ and its dual induce a stable equivalence of Morita type between A and B. Since P is nontrivial, the blocks A and B are nonprojective as bimodules. It follows from 4.14.2 that $b\mathcal{O}Gc$ has up to isomorphism a unique indecomposable direct summand, denoted M. Then M and its dual induce a stable equivalence of Morita type between A and B. We keep this notation.

Theorem 11.12.1 ([196, 10.2]) *There is a direct summand (Q, π) of a projective cover of M as an A-B-bimodule such that the 2-term complex*

$$X = Q \xrightarrow{\ \pi\ } M$$

with M in degree zero is a splendid Rickard complex.

Proof The bimodule M has ΔP as a vertex, and the bimodule Q is projective. Thus X is splendid. The algebra B is Morita equivalent to $\mathcal{O}(P \rtimes E)$, and $k(P \rtimes E)$ is indecomposable symmetric serial. Thus for $\mathcal{O} = k$ the result follows from 11.9.1. Since projective modules lift uniquely, up to isomorphism, from k to $\mathcal{O}$, it follows for general $\mathcal{O}$ that there is a complex X of the form as stated with the property that $k \otimes_{\mathcal{O}} X$ is a splendid Rickard complex. Using general lifting theorems one can conclude that X itself is a Rickard complex, but in this case one can see this directly, retracing some of the arguments used in the proof of 11.9.1, as follows. The complex $X \otimes_B X^*$ has three terms, of which only the degree zero term is nonprojective, having $M \otimes_B M^*$ as a direct summand, hence having A as a direct summand. The right map in the 3-term complex $k \otimes_{\mathcal{O}} X \otimes_B X^*$ is surjective by 11.9.1, hence the right map in the 3-term complex $X \otimes_B X^*$ is surjective by Nakayama's Lemma, and therefore this map is split surjective, as the right term is a projective bimodule. Since $X \otimes_B X^*$ is selfdual, it follows that the left map is split injective. Thus $X \otimes_B X^*$ is a split complex, with homology in degree zero isomorphic to A plus possibly a projective summand. However, this projective summand is zero, as it is zero upon taking residues modulo $J(\mathcal{O})$. Thus $X \otimes_B X^* \simeq A$. It follows from 4.14.16 that X is a splendid tilting complex. $\qquad\square$

The next step applies earlier results, showing that under suitable hypotheses, a Morita equivalence with endopermutation source is induced by a splendid Rickard equivalence. By 11.2.2, the source algebras of $\mathcal{O}Hc$ are isomorphic to $S \otimes_{\mathcal{O}} \mathcal{O}(P \rtimes E)$, where $S = \operatorname{End}_{\mathcal{O}}(W)$ for some indecomposable endopermutation $\mathcal{O}P$-module W with vertex P and determinant 1. Moreover, W is unique up to isomorphism, hence E-stable. In terms of Morita equivalences, this can be restated as follows: for any choice of a source idempotent j in $(\mathcal{O}Hc)^{\Delta P}$ there is an indecomposable direct summand N of the $\mathcal{O}Hc$-$\mathcal{O}(P \rtimes E)$-bimodule

$$\mathcal{O}Hj \otimes_{\mathcal{O}P} \operatorname{Ind}_{\Delta P}^{P \times P}(W) \otimes_{\mathcal{O}P} \mathcal{O}(P \rtimes E)$$

with vertex ΔP and source W such that N and its dual induce a Morita equivalence between $\mathcal{O}Hc$ and $\mathcal{O}(P \rtimes E)$. By 7.11.9, the module W has a split p-permutation resolution Y_W.

Theorem 11.12.2 *There is a splendid Rickard complex Y of $\mathcal{O}Hc$-$\mathcal{O}(P \rtimes E)$-bimodules that is isomorphic to a direct summand of the complex*

$$\mathcal{O}Hj \otimes_{\mathcal{O}P} \operatorname{Ind}_{\Delta P}^{P \times P}(Y_W) \otimes_{\mathcal{O}P} \mathcal{O}(P \rtimes E)$$

such that Y has homology concentrated in degree zero and isomorphic to N.

Proof This is a special case of 9.11.5. $\qquad\square$

Combining the two previous results yields a proof of Theorem 11.1.11.

Theorem 11.12.3 *The complex $X \otimes_B Y$ is a splendid Rickard complex of $A\text{-}\mathcal{O}(P \rtimes E)$-bimodules. In particular, Broué's abelian defect group conjecture holds for blocks with cyclic defect groups.*

Proof The block algebras A, B, $\mathcal{O}(P \rtimes E)$ all have the same fusion system on P. Thus $X \otimes_B Y$ is a splendid Rickard complex by 9.7.9. The source algebras of the Brauer correspondent of b are isomorphic to $\mathcal{O}(P \rtimes E)$, so this yields a splendid Rickard complex between A and its Brauer correspondent. $\qquad\square$

11.13 The $\mathcal{O}$-algebra structure of blocks with cyclic defect groups

In this section we prove Theorem 11.1.16, which describes the $\mathcal{O}$-algebra structure of basic algebras of blocks with cyclic defect groups, following [125]. We assume that k is large enough, but we do not need this hypothesis for K.

Let G be a finite group, b a block of $\mathcal{O}G$ with a nontrivial cyclic p-group P as a defect group. Set $A = \mathcal{O}Gb$. Let I be a set of representatives of the conjugacy classes of primitive idempotents in A. We use the notation from 11.1.5. In particular, for $i \in I$, we denote by $\alpha_i : Ai \to A\sigma^{-1}(i)$ and $\beta_i : Ai \to A\rho^{-1}(i)$ some A-homomorphisms that make the sequences

$$0 \longrightarrow U_i \longrightarrow Ai \overset{\alpha_i}{\longrightarrow} V_{\sigma^{-1}(i)} \longrightarrow 0$$

$$0 \longrightarrow V_i \longrightarrow Ai \overset{\beta_i}{\longrightarrow} U_{\rho^{-1}(i)} \longrightarrow 0.$$

from 11.1.5 exact. We abusively denote the homomorphism $Ai \to V_{\sigma^{-1}(i)}$ induced by α_i again by α_i; similarly for β_i. The exactness of these sequences implies that $\alpha_i \circ \beta_{\rho(i)} = 0$ and $\beta_i \circ \alpha_{\sigma(i)} = 0$ for all $i \in I$. For any two elements $i, i' \in I$ belonging to the same σ-orbit, we define $\alpha_{i,i'} : Ai \to Ai'$ as the composition

$$\alpha_{i,i'} = \alpha_{\sigma^{-t}(i)} \circ \cdots \circ \alpha_{\sigma^{-1}(i)} \circ \alpha_i$$

where t is the smallest nonnegative integer such that $\sigma^{-(t+1)}(i) = i'$. By the above, we have $\ker(\alpha_{i,i'}) = U_i$ and $\operatorname{Im}(\alpha_{i,i'}) \subseteq V_{i'}$. In particular, we have $\beta_{i'} \circ \alpha_{i,i'} = 0$. Similarly, if i, i' belong to the same ρ-orbit, we define $\beta_{i,i'} : Ai \to Ai'$ by

$$\beta_{i,i'} = \beta_{\rho^{-t}(i)} \circ \cdots \circ \beta_{\rho^{-1}(i)} \circ \beta_i$$

where now t is the smallest nonnegative integer such that $\rho^{-(t+1)}(i) = i'$. As before, we have $\ker(\beta_{i,i'}) = V_i$ and $\text{Im}(\beta_{i,i'}) \subseteq V_{i'}$; in particular, we have $\alpha_{i'} \circ \beta_{i,i'} = 0$. If i, i', i'' belong to the same σ-orbit such that $\sigma^{-t}(i) = i'$ and $\sigma^{-s}(i) = i''$ for some integers t, s such that $0 \leq t < s < |i^\sigma|$, then

$$\alpha_{i,i''} = \alpha_{i',i''} \circ \alpha_{i,i'},$$

and similar formulae hold for the $\beta_{i,i'}$. For any $i \in I$, the map $\alpha_{i,i}$ is the endomorphism of Ai that is the composition of the $\alpha_{i'}$ with i' running over the σ-orbit of i in reverse order. The induced endomorphism of $\bar{A}\bar{i}$ sends $\bar{A}\bar{i}$ to its socle, provided that i^σ is not the exceptional orbit. A similar statement holds for $\beta_{i,i}$. We use this notation to describe $\mathcal{O}$-bases of the spaces of A-homomorphisms between projective indecomposable A-modules. We start with nonisomorphic projective indecomposable A-modules.

Lemma 11.13.1 ([125, 3.1]) *Let i, i' be two different elements in I.*

(i) *If i and i' do not belong to a common orbit of either ρ or σ, then $\text{Hom}_A(Ai, Ai') = \{0\}$.*

(ii) *If i and i' belong to the same σ-orbit, then the set $\{\alpha_{i,i'}\alpha_{i,i}^t\}_{0 \leq t \leq n-1}$ is an $\mathcal{O}$-basis of $\text{Hom}_A(Ai, Ai')$, where $n = 1$ is i^σ is nonexceptional, and $n = m$ otherwise.*

(iii) *If i and i' belong to the same ρ-orbit, then the set $\{\beta_{i,i'}\beta_{i,i}^t\}_{0 \leq t \leq n-1}$ is an $\mathcal{O}$-basis of $\text{Hom}_A(Ai, Ai')$, where $n = 1$ is i^σ is nonexceptional, and $n = m$ otherwise.*

Proof By Nakayama's Lemma, it suffices to show that the images of the given sets in $\text{Hom}_{\bar{A}}(\bar{A}\bar{i}, \bar{A}\bar{i'})$ are k-bases. This follows easily from the composition series of the uniserial modules $\bar{U}_i$ and $\bar{V}_i$ as described in 11.1.8. $\qquad\square$

We consider next endomorphisms of projective indecomposable A-modules.

Lemma 11.13.2 ([125, 3.2]) *Let $i \in I$.*

(i) *If i does not belong to the exceptional orbit, then the sets $\{\text{Id}_{Ai}, \alpha_{i,i}\}$ and $\{\text{Id}_{Ai}, \beta_{i,i}\}$ are $\mathcal{O}$-bases of $\text{End}_A(Ai)$.*

(ii) *If i^σ is exceptional, then the set $\{\alpha_{i,i}^t\}_{0 \leq t \leq m}$ is an $\mathcal{O}$-basis of $\text{End}_A(Ai)$.*

(iii) *If i^ρ is exceptional, then the set $\{\beta_{i,i}^t\}_{0 \leq t \leq m}$ is an $\mathcal{O}$-basis of $\text{End}_A(Ai)$.*

Proof As in the proof of the previous lemma, this follows from combining Nakayama's Lemma and the composition series of the uniserial modules $\bar{U}_i$ and $\bar{V}_i$. $\qquad\square$

By 11.1.5 the submodules U_i and V_i of a projective indecomposable A-module Ai are pure, and by 11.1.15 their characters are disjoint. Thus any automorphism of Ai preserves these submodules. This means that we can compose the morphisms α_i and β_i with any automorphisms of the Ai without affecting the properties described in the two previous lemmas. The next two lemmas will be used to describe a suitable choice of the endomorphisms of Ai depending on wether i belongs to a nonexceptional orbit or to the exceptional orbit.

Lemma 11.13.3 ([125, 3.4])   *For any $i \in I$ we have* $\underline{\mathrm{End}}_A(U_i) \cong \underline{\mathrm{End}}_A(V_i) \cong$ $\mathcal{O}/|P|\mathcal{O}$.

Proof The modules V_i are obtained by taking Heller translates of the U_i, and since taking Heller translates is a stable equivalence of Morita type, it follows that $\underline{\mathrm{End}}_A(U_i) \cong \underline{\mathrm{End}}_A(V_i)$ for any $i \in I$. Through the stable equivalence between A and $\mathcal{O}(P \rtimes E)$, the modules U_i correspond to the $\mathcal{O}(P \rtimes E)$-modules of rank 1 whose restriction to $\mathcal{O}P$ is the trivial $\mathcal{O}P$-module. Thus, if W is an $\mathcal{O}(P \rtimes E)$-module such that the restriction of W to $\mathcal{O}P$ is the trivial $\mathcal{O}P$-module $\mathcal{O}$, then $\underline{\mathrm{End}}_A(U_i) \cong \underline{\mathrm{End}}_{\mathcal{O}(P \rtimes E)}(W)$. Thus it suffices to show that $\underline{\mathrm{End}}_{\mathcal{O}(P \rtimes E)}(W) \cong \mathcal{O}/|P|\mathcal{O}$. Since $\mathrm{End}_{\mathcal{O}(P \rtimes E)}(W) \cong \mathcal{O}$, it suffices to show that $|P|$ is the smallest positive integer with the property that $|P| \cdot \mathrm{Id}_W$ factors through a projective $\mathcal{O}(P \rtimes E)$-module. Clearly $|P| \cdot \mathrm{Id}_W = \mathrm{Tr}_1^{P \rtimes E}(\frac{1}{|E|}\mathrm{Id}_W)$, so $|P| \cdot \mathrm{Id}_W$ factors through a projective $\mathcal{O}(P \rtimes E)$-module by Higman's criterion. Restricting to $\mathcal{O}P$, it suffices to show that $|P|$ is the smallest positive integer such that $|P| \cdot \mathrm{Id}_{\mathcal{O}}$ factors through a projective $\mathcal{O}P$-module. Suppose that $\gamma \in \mathrm{End}_{\mathcal{O}P}(\mathcal{O})$ factors through a projective $\mathcal{O}P$-module. Then γ factors through the canonical projective cover $\eta : \mathcal{O}P \to \mathcal{O}$ sending $y \in P$ to 1. Moreover, any $\mathcal{O}P$-homomorphism $\tau : \mathcal{O} \to \mathcal{O}P$ sends 1 to $\lambda \sum_{y \in P} y$. Thus if $\gamma = \eta \circ \tau$, then $\gamma(1) = \lambda \eta(\sum_{y \in P} y) = |P|\lambda$, which shows that γ is a scalar multiple of $|P| \cdot \mathrm{Id}_{\mathcal{O}}$, whence the result. $\square$

Lemma 11.13.4   *There is an element $z \in (\mathcal{O}P)^E$ such that the set $\{z^t\}_{1 \leq t \leq m}$ is an $\mathcal{O}$-basis of the annihilator of $\sum_{x \in P} x$ in $(\mathcal{O}P)^E$. If z' is another such element, then $z' = zw$ for some invertible element $w \in (\mathcal{O}P)^E$. Moreover, we may choose z such that*

$$\sum_{x \in P} x + z^m = \sum_{t=0}^{m-1} \lambda_t z^t$$

for some $\lambda_t \in J(\mathcal{O})$, $0 \leq t \leq m - 1$, and then $\lambda_0 = |P|$.

Proof Let y be a generator of P. Set $z = \prod_{e \in E}(^e y - 1)$. Any factor in this product is in the augmentation ideal of $\mathcal{O}P$, hence any positive power of z

annihilates $\sum_{x\in P} x$, and we have $z \in J(\mathcal{O}P)^{|E|} \cap (\mathcal{O}P)^E \subseteq J((\mathcal{O}P)^E)$, hence $z^t \in J((\mathcal{O}P)^E)^t$ for any positive integer t. Denote by $\bar{z}$ the image of z in kP. Then $\bar{z}kP = J(kP)^{|E|}$, hence $\bar{z}^m kP = J(kP)^{|E|m} = J(kP)^{|P|-1} \neq \{0\}$. Thus $\bar{z}^m \neq 0$. Let $t \geq 0$. If $\bar{z}^t \neq 0$, then $\bar{z}^{t+1}(kP)^E$ is a proper subspace of $\bar{z}^t(kP)^E$. Since $\dim_k((kP)^E) = m+1$, it follows that $\dim_k(\bar{z}^t(kP)^E/\bar{z}^{t+1}(kP)^E) = 1$ for $0 \leq t \leq m$. Thus the set $\{\bar{z}^t\}_{1\leq t\leq m}$ is a k-basis of the annihilator of $\sum_{x\in P} x$ in $(kP)^E$. Using Nakayama's Lemma implies that the set $\{z^t\}_{1\leq t\leq m}$ is an $\mathcal{O}$-basis of the annihilator of $\sum_{x\in P} x$ in $(\mathcal{O}P)^E$. If z' is any element in $(\mathcal{O}P)^E$ that annihilates $\sum_{x\in P} x$, then $z' = \sum_{t=1}^m \mu_t z^t$ for some $\mu_t \in \mathcal{O}$. Moreover, $\{(z')^t\}_{1\leq t\leq m}$ is then an $\mathcal{O}$-basis of the annihilator of $\sum_{x\in P} x$ in $(\mathcal{O}P)^E$ if and only if $\mu_1 \in \mathcal{O}^\times$. In that case, the element $w = \sum_{t=1}^m \mu_t z^{t-1}$ is contained in $\mu_1(1 + J((\mathcal{O}P)^E))$, hence invertible in $(\mathcal{O}P)^E$, and we have $z' = zw$. For the last statement, write

$$\sum_{x\in P} x = \sum_{t=0}^m \lambda_t z^t$$

for some coefficients λ_t; this is possible since $|P| \cdot 1 - \sum_{x\in P} x$ annihilates $\sum_{x\in P} x$, hence is a linear combination of the z^t with $1 \leq t \leq m$. This shows also that $\lambda_0 = |P|$. The algebra $(kP)^E$ is symmetric local, hence has a unique minimal ideal, which is generated by the images in $(kP)^E$ of z^m and of $\sum_{x\in P} x$. This shows that $\lambda_m \in \mathcal{O}^\times$ and $\lambda_t \in J(\mathcal{O})$ for $1 \leq t \leq m-1$. Since m is prime to p, there exists an m-th root ζ of $-\lambda_m$ in $\mathcal{O}$. If we replace z by ζz, then $\lambda_m = -1$. The result follows. $\qquad\square$

We describe now our choices of the homomorphisms α_i and β_i in case that i^σ and i^ρ are nonexceptional, respectively.

Lemma 11.13.5 ([125, 3.5]) *We may choose the homomorphisms α_i, β_i in such a way that for any $i \in I$ we have $\alpha_i \circ \alpha_{i,i} = |P|\alpha_i$ if i^σ is nonexceptional, and $\beta_i \circ \beta_{i,i} = |P|\beta_i$ if i^ρ is nonexceptional. Any choice with these properties satisfies moreover $\alpha_{i,i} + \beta_{i,i} = |P|\mathrm{Id}_{Ai}$ if both i^σ and i^ρ are nonexceptional.*

Proof Let $i \in I$. Suppose that i^σ is nonexceptional. Then $\mathrm{End}_A(V_i) \cong \mathcal{O}$, as there is no repetition in the composition series of $\bar{V}_i$. Let φ be an endomorphism of V_i that factors through a projective A-module. Then φ factors through the projective cover $\alpha_{\sigma(i)} : A\sigma(i) \to V_i$. Since V_i is $\mathcal{O}$-pure in Ai, any homomorphism $V_i \to A\sigma(i)$ extends to a homomorphism $Ai \to A\sigma(i)$. It follows that φ factors through the restriction $\alpha_{i,i}|_{V_i}$ of $\alpha_{i,i}$ to V_i. It follows from 11.13.3 that $\{|P|\mathrm{Id}_{V_i}\}$ and $\{\alpha_{i,i}|_{V_i}\}$ are both $\mathcal{O}$-bases of the space $\mathrm{End}_A^{\mathrm{pr}}(V_i)$. This forces $\alpha_{i,i}|_{V_i} = \tau_i|P|\mathrm{Id}_{V_i}$ for some $\tau_i \in \mathcal{O}^\times$. As $\alpha_i \circ \alpha_{i,i} = \alpha_{\sigma^{-1}(i),\sigma^{-1}(i)} \circ \alpha_i$ it follows that τ_i depends only on the σ-orbit i^σ of i. Thus if we replace α_i by $\tau_i^{-1}\alpha_i$, this amounts to also modifying $\alpha_{i,i}$ by τ_i^{-1}, and hence we get that $\alpha_i \circ \alpha_{i,i} = |P|\alpha_i$

for this particular i, hence for any i in that σ-orbit, hence for any $i \in I$ such that i^σ is nonexceptional. The same argument shows that we may choose the β_i such that $\beta_i \circ \beta_{i,i} = |P|\beta_i$ for any $i \in I$ such that i^ρ is nonexceptional. If i^σ and i^ρ are both nonexceptional, then by 11.13.2 there are λ_i, $\mu_i \in \mathcal{O}$ such that $\alpha_{i,i} + \beta_{i,i} = \lambda_i \mathrm{Id}_{Ai} + \mu_i \alpha_{i,i}$. Composing with β_i yields $\lambda_i = |P|$ because $\beta_i \circ \alpha_{i,i} = 0$. Composing with α_i implies $\mu_i = 0$, whence the result. $\qquad \square$

In order to describe a suitable choice of the α_i, β_i, with i belonging to the exceptional orbit, we need the following notation and lemma. Denote by $\varphi : Z(\mathcal{O}(P \rtimes E)) \cong Z(A)$ the isomorphism from 11.1.13 obtained from the perfect isometry $\mathbb{Z}\mathrm{Irr}_K(P \rtimes E) \cong \mathbb{Z}\mathrm{Irr}_K(A)$ as described in 11.1.12. That is, φ is induced by the isomorphism $Z(K(P \rtimes E)) \cong Z(K \otimes_{\mathcal{O}} A)$ sending $e(\lambda)$ to $e(\chi_\lambda)$ and $e(\mu)$ to $e(\chi_\mu)$, for $\lambda \in \Lambda$ and $\mu \in M$. The algebra $(\mathcal{O}P)^E$ is a subalgebra of $Z(\mathcal{O}(P \rtimes E))$. Thus, for any $i \in I$, there is an algebra homomorphism $\varphi_i : (\mathcal{O}P)^E \to \mathrm{End}_A(Ai)$ sending $c \in (\mathcal{O}P)^E$ to the endomorphism of Ai given by left multiplication with $\varphi(c)$ on Ai.

Lemma 11.13.6 ([125, 3.6]) *With the notation above, let $i \in I$.*

(i) *If i^σ is exceptional, then φ_i is an isomorphism, and we have $\varphi_i(\sum_{x \in P} x) = \beta_{i,i}$.*

(ii) *If i^ρ is exceptional, then φ_i is an isomorphism, and we have $\varphi_i(\sum_{x \in P} x) = \alpha_{i,i}$.*

Proof An easy calculation in $K(P \rtimes E)$ shows that $\sum_{x \in P} x = |P| \sum_{\mu \in M} e(\mu)$. Thus $\varphi(\sum_{x \in P} x) = |P| \sum_{\mu \in M} e(\chi_\mu)$. Suppose that i^σ is exceptional. Then the character of V_i is equal to $\sum_{\lambda \in \Lambda} \chi_\lambda$. Thus $\varphi_i(\sum_{x \in P} x)$ annihilates V_i. Equivalently, $\varphi_i(\sum_{x \in P} x)$ is an endomorphism of Ai that annihilates V_i. Since i^σ is exceptional, it follows that i^ρ is not exceptional, and hence the character of U_i, which is equal to the character of Ai/V_i, is irreducible. Thus the space of endomorphisms of Ai that annihilate V_i has $\mathcal{O}$-rank 1. By construction, the endomorphism $\beta_{i,i}$ annihilates V_i. This implies that $\varphi_i(\sum_{x \in P} x) = \tau \beta_{i,i}$ for some $\tau \in K$. As $(\sum_{x \in P} x)^2 = |P| \sum_{x \in P} x$ and $\beta_{i,i}^2 = |P|\beta_{i,i}$ we have $\tau = 1$. This implies that the image in $\mathrm{End}_{\bar{A}}(\bar{A}\bar{i})$ of $\varphi(\sum_{x \in P} x)$ is nonzero (because it sends $\bar{A}\bar{i}$ onto its socle). The image of $\sum_{x \in P} x$ in $(kP)^E$ generates the unique minimal ideald in $(kP)^E$, and hence the homomorphism $(kP)^E \to \mathrm{End}_{\bar{A}}(\bar{A}\bar{i})$ is injective. Both sides have dimension $m + 1$, and so this is an isomorphism. But then φ_i is surjective by Nakayama's Lemma, and hence φ_i is an isomorphism, again as both sides have the same $\mathcal{O}$-rank $m + 1$. This concludes the proof of (i), and the same argument proves (ii). $\qquad \square$

If i does not belong to the exceptional orbit, then the arguments of the preceding lemma can be used to show that the image of φ_i is $\mathcal{O} \cdot \mathrm{Id}_{A_i}$, but we will not need this here. For the next lemma, which describes a choice of the α_i, β_i in case i belongs to the exceptional orbit, we need the following notation from 11.13.4. Let $z \in (\mathcal{O}P)^E$ such that $\{z^t\}_{1 \leq t \leq m}$ is an $\mathcal{O}$-basis of the annihilator of $\sum_{x \in P} x$ in $(\mathcal{O}P)^E$ and such that $\sum_{x \in P} x + z^m = \sum_{t=0}^{m-1} \lambda_t z^t$, with coefficients $\lambda_t \in J(\mathcal{O})$ for $0 \leq t \leq m-1$. By 11.13.4 this implies that $\lambda_0 = |P|$.

Lemma 11.13.7 ([125, 3.8]) *Let $i \in I$.*

(i) *If i^σ is exceptional, then we may choose α_i such that $\alpha_{i,i}^m + \beta_{i,i} = \sum_{t=0}^{m-1} \lambda_t \alpha_{i,i}^t$. In that case and with this choice we have $\alpha_i \circ \alpha_{i,i} = \sum_{t=0}^{m-1} \lambda_t \alpha_i \circ \alpha_{i,i}^t$.*

(ii) *If i^σ is exceptional, then we may choose β_i such that $\alpha_{i,i} + \beta_{i,i}^m = \sum_{t=0}^{m-1} \lambda_t \beta_{i,i}^t$. In that case and with this choice we have $\beta_i \circ \beta_{i,i} = \sum_{t=0}^{m-1} \lambda_t \beta_i \circ \beta_{i,i}^t$.*

Proof Suppose that i^σ is exceptional. By 11.13.1 we have

$$\alpha_i \circ \alpha_{i,i}^m = \sum_{t=0}^{m-1} \mu_t^i \alpha_i \circ \alpha_{i,i}^t$$

for some coefficients $\mu_t^i \in \mathcal{O}$, where $0 \leq t \leq m-1$. Precomposing with $\alpha_{\sigma(i)}$ yields

$$\alpha_i \circ \alpha_{i,i}^m \circ \alpha_{\sigma)(i)} = \sum_{t=0}^{m-1} \mu_t^i \alpha_i \circ \alpha_{i,i}^t \alpha_{\sigma(i)}$$

and combining this with the equalities $\alpha_{i,i}^t \circ \alpha_{\sigma(i)} = \alpha_{\sigma(i)} \circ \alpha_{\sigma(i),\sigma(i)}^t$ yields

$$\alpha_i \circ \alpha_{\sigma(i)} \circ \alpha_{\sigma(i),\sigma(i)}^m = \sum_{t=0}^{m-1} \mu_t^i \alpha_i \circ \alpha_{\sigma(i)} \circ \alpha_{\sigma(i),\sigma(i)}^t.$$

As $\ker(\alpha_i) \cap \mathrm{Im}(\alpha_{\sigma(i)}) = U_i \cap V_i = \{0\}$, it follows that

$$\alpha_{\sigma(i)} \circ \alpha_{\sigma(i),\sigma(i)}^m = \sum_{t=0}^{m-1} \mu_t^i \alpha_{\sigma(i)} \circ \alpha_{\sigma(i),\sigma(i)}^t.$$

This shows that the μ_t^i do not depend on the choice of i in the exceptional σ-orbit. By 11.13.6 there is a unique element $z' \in (\mathcal{O}P)^E$ such that $\varphi(z') = \alpha_{i,i}$, and we have $\varphi_i(\sum_{x \in P} x) = \beta_{i,i}$. $\qquad\square$

Proof of Theorem 11.1.16 We use the notation of 11.1.16 with A a block algebra with a nontrivial cyclic defect group P. Choose a set I of pairwise orthogonal representatives of the conjugacy classes of primitive idempotents of A.
Then $e = \sum_{i \in I} i$ is an idempotent in A, and the algebra eAe is a basic algebra
of A, having e as its unit element. Thus the relations $i^2 = i$ for $i \in I$, $ii' = 0$
for $i, i' \in I$ such that $i \neq i'$, and $\sum_{i \in I} i = 1_{eAe}$ are trivially satisfied in eAe.
For any $i, i' \in I$, any homomorphism $Ai \to Ai'$ is given by right multiplica-
tion with an element in iAi'. In particular, for any $i \in I$, there are elements
$r_i \in iA\rho^{-1}(i)$ and $s_i \in iA\sigma^{-1}(i)$ such that $\beta_i(a) = ar_i$ and $\alpha_i(a) = as_i$ for all
$a \in Ai$. Set $r = \sum_{i \in I} r_i$ and $s = \sum_{i \in I} s_i$. Since the elements of I are pairwise
orthogonal, we have $ir = r_i = r\rho^{-1}(i)$ and $is = s_i = s\sigma^{-1}(i)$, or equivalently,
$\rho(i)r = ri$ and $\sigma(i)s = si$ for all $i \in I$. We have $rs = 0 = sr$ by the exactness of
the sequences in 11.1.5, restated at the beginning of this section. The elements
I, r, s satisfy the remaining relations as stated in 11.1.16 as an immediate consequence of 11.13.5 and 11.13.7. Thus eAe is a quotient of the $\mathcal{O}$-algebra C
with generators $I \cup \{r, s\}$ and relations as stated in 11.1.16. Using abusively the
same letters $i \in I$ and r, s for the generators of C and their images in eAe, it
follows that the canonical surjective algebra homomorphism $C \to eAe$ sends
iCi' onto iAi', for $i, i' \in I$. Suppose that $i \neq i'$. If i, i' are not in the same orbit of
either ρ or σ, then the relations as well as 11.13.1 imply that $iCi' = \{0\} = iAi'$.
If i, i' are in the same σ-orbit, say $i' = \sigma - t - 1(i)$ with $t \geq 0$ smallest possible,
then $\alpha_{i,i'}$ is given by right multiplication with is^{t+1}. In particular, $\alpha_{i,i}$ is given by
right multiplication with is^b, where $b = |i^\sigma|$. Set $n = 1$ if i^σ is nonexceptional,
and $n = m$ otherwise. It follows from 11.13.1 (i) and the relations in C that
the image in iCi' of the set $\{is^{1+t+ub}\}_{0 \leq u \leq n-1}$ generates iCi' as an $\mathcal{O}$-module,
and that the image in iAi' of this set is an $\mathcal{O}$-basis. Since iAi' is a canonical
quotient of iCi', this implies that iCi' is $\mathcal{O}$-free of the same rank as iAi'. A similar argument, using 11.13.1 (ii) and 11.13.2 shows this if $i \neq i'$ belong to the
same ρ-orbit and if $i = i'$, respectively. Combining these cases implies that the
canonical map $C \to eAe$ is an isomorphism. $\qquad\square$

11.14 Picard groups of blocks with cyclic defect groups

We determine the Picard group of a block with a nontrivial cyclic defect group
P in terms of its Brauer tree and its inertial quotient E. We suppose that k and
K are large enough for all block algebras in this section. If the Brauer tree has
an exceptional vertex, then the exceptional multiplicity is equal to $\frac{|P|-1}{|E|}$. In
what follows, we suppress specifying the exceptional multiplicity, if any, in the
Brauer tree. Let G be a finite group, b a block of $\mathcal{O}G$ with a nontrivial cyclic

defect group P and inertial quotient E. Set $A = \mathcal{O}Gb$ and $\bar{A} = k \otimes_{\mathcal{O}} A$. We use the notation as in 11.11.3. As before, we denote by $\mathrm{Aut}_0(A)$ the subgroup of all $\alpha \in \mathrm{Aut}(A)$ that stabilise the isomorphism classes of all finitely generated $k \otimes_{\mathcal{O}} A$-modules. We denote by $\mathrm{Out}_0(A)$ the image of $\mathrm{Aut}_0(A)$ in $\mathrm{Out}(A)$. We start with a sufficient criterion to detect self stable equivalences of Morita type of A that are in fact Morita equivalences induced by automorphisms in $\mathrm{Aut}_0(A)$.

Proposition 11.14.1 *Let N be an indecomposable A-A-bimodule such that N and its dual N^* induce a stable equivalence of Morita type. Suppose that there exists $i \in I$ such that $N \otimes_A U_i \cong U_i$ in $\underline{\mathrm{mod}}(A)$. Then $N \otimes_A U \cong U$ for any finitely generated A-module, or equivalently, $N \cong A_\alpha$ for some $\alpha \in \mathrm{Aut}_0(A)$.*

Proof By 11.6.3, the modules $k \otimes_{\mathcal{O}} U_i$ correspond to the simple $k(P \rtimes E)$-modules through a stable equivalence of Morita type between A and $\mathcal{O}(P \rtimes E)$, given by an A-$\mathcal{O}(P \rtimes E)$-bimodule denoted M, and its dual. Thus $M^* \otimes_A N \otimes_A M$ and its dual determine a stable equivalence of Morita type of $\mathcal{O}(P \rtimes E)$ that stabilises a simple $k(P \rtimes E)$-module. It follows from 11.5.2 that this stable equivalence of Morita type stabilises the isomorphism classes of all finitely generated $k(P \rtimes E)$-modules. But then $N \otimes_A -$ stabilises the isomorphism classes of all indecomposable nonprojective A-modules, hence is a Morita equivalence induced by an automorphism in $\mathrm{Aut}_0(A)$. $\qquad\square$

Theorem 11.14.2 *Let G be a finite group, b a block of $\mathcal{O}G$ with a nontrivial cyclic defect group P and inertial quotient E. Set $A = \mathcal{O}Gb$ and $\bar{A} = k \otimes_{\mathcal{O}} A$. The bimodule ΩA generates a cyclic central subgroup $C_{2|E|}$ of order $2|E|$ in $\mathrm{StPic}(A)$, and the following hold.*

(i) We have $\mathrm{StPic}(A) \cong \mathrm{StPic}(\mathcal{O}(P \rtimes E)) \cong C_{2|E|} \times \mathrm{Out}_0(\mathcal{O}(P \rtimes E))$.
(ii) If $|P| \geq 3$, then $\mathrm{StPic}(\bar{A}) \cong \mathrm{StPic}(k(P \rtimes E)) \cong C_{2|E|} \times \mathrm{Out}_0(k(P \rtimes E))$.
(iii) If $|P| = p = 2$, then $\mathrm{StPic}(\bar{A}) = \mathrm{Pic}(\bar{A}) \cong \mathrm{Aut}(kP) \cong k^\times$.

Proof Since there is a stable equivalence of Morita type between A and $\mathcal{O}(P \rtimes E)$, it follows from 2.17.17 that we may assume that $A = \mathcal{O}(P \rtimes E)$. It follows from 11.5.4 that $\mathrm{StPic}(A)$ is generated by ΩA and by $\mathrm{Out}_0(A)$, and that a similar statement holds over k. Moreover, ΩA generates a central subgroup of $\mathrm{StPic}(A)$, by 2.17.8. The period of $\mathcal{O}(P \rtimes E)$ is $2|E|$, and if $|P| \geq 3$, then the period of $k(P \rtimes E)$ is also $2|E|$. This shows (i) and (ii). Suppose that $|P| = 2$. Then A is Morita equivalent to kP, hence has period 1. Denote by t the nontrivial element of P. Any automorphism of kP fixes 1 and stabilises $J(kP) = k(1 - t)$. Thus, for any $\lambda \in k^\times$ there is a unique automorphism of kP

sending $1 - t$ to $\lambda(1 - t)$, and any automorphism arises in this way, whence the last isomorphism in (iii). $\qquad\square$

Theorem 11.14.3 ([128, Theorem 4.7]) *Let G be a finite group, b a block of $\mathcal{O}G$ with a nontrivial cyclic defect group and inertial quotient E. Denote by $\bar{b}$ the image of b in kG. Suppose that the block algebra $\mathcal{O}Gb$ is split. Let $(I, \{\rho, \sigma\}, v)$ or $(I, \{\rho, \sigma\})$ be the Brauer tree of $\mathcal{O}Gb$, depending on whether $\mathcal{O}Gb$ has an exceptional vertex v or not. Denote by F the group of permutations of I that centralise $\{\rho, \sigma\}$ and that stabilise the exceptional vertex v, if any. Then F is cyclic of order dividing $|E|$, and we have group isomorphisms*

$$\mathrm{Pic}(\mathcal{O}Gb) \cong F \times \mathrm{Out}_0(\mathcal{O}(P \rtimes E))$$

and

$$\mathrm{Pic}(kG\bar{b}) \cong F \times \mathrm{Out}_0(k(P \rtimes E)).$$

The group F is generated by the smallest power of the Heller operator that induces a Morita equivalence on $\mathcal{O}Gb$, or equivalently, that permutes the isomorphism classes of simple $kG\bar{b}$-modules.

Proof Suppose first that $|I| = 1$, or equivalently, that E is trivial. Then $\mathcal{O}Gb$ is Morita equivalent to $\mathcal{O}P$. Since the isomorphism class of any indecomposable kP-module is determined by its dimension, it follows that $\mathrm{Out}_0(\mathcal{O}P) = \mathrm{Out}(\mathcal{O}P)$. The algebra $\mathcal{O}P$ is basic, and hence $\mathrm{Out}(\mathcal{O}P) \cong \mathrm{Pic}(\mathcal{O}P) \cong \mathrm{Pic}(\mathcal{O}Gb)$. Similarly, we have $\mathrm{Out}_0(kP) = \mathrm{Out}(kP) \cong \mathrm{Pic}(kG\bar{b})$. Thus the theorem holds if $|I| = 1$. Assume that $|I| \geq 2$. Then E is nontrivial, hence p is odd. It follows that if the Brauer tree of b has no exceptional vertex, then $|I| = |E| = p - 1 = |P| - 1$, hence $|I|$ is even. Let A be a basic algebra of $\mathcal{O}Gb$. By 4.9.7 we have $\mathrm{Pic}(\mathcal{O}Gb) \cong \mathrm{Out}(A)$. By 11.8.4, every automorphism of A induces a permutation on I that centralises ρ and σ, and that stabilises the exceptional vertex, if any. The Picard group of $\mathcal{O}Gb$ is a subgroup of $\mathrm{StPic}(\mathcal{O}Gb)$. By 2.17.17 and 4.15.1 there is an isomorphism $\mathrm{StPic}(\mathcal{O}Gb) \cong \mathrm{StPic}(\mathcal{O}(P \rtimes E))$ that induces an isomorphism $\mathrm{Out}_0(\mathcal{O}Gb) \cong \mathrm{Out}_0(\mathcal{O}(P \rtimes E))$. Thus this isomorphism sends $\mathrm{Pic}(\mathcal{O}Gb)$ to a subgroup of $\mathrm{StPic}(\mathcal{O}(P \rtimes E)) \cong E \times \mathrm{Out}_0(\mathcal{O}(P \rtimes E))$. This implies that $\mathrm{Pic}(\mathcal{O}Gb) \cong F \times \mathrm{Out}_0(\mathcal{O}(P \rtimes E))$ for some subgroup F of E. In order to show that F is as stated, we need to show that every permutation τ of I that centralises ρ and σ and that stabilises the exceptional vertex, if any, is induced by a self Morita equivalence of $\mathcal{O}Gb$, or equivalently, by an automorphism of A. Over k this follows from 11.8.4. There are two ways to see that this remains true over $\mathcal{O}$. One can use the description of A in 11.1.16 that the explicit construction of α in 11.8.4 lifts verbatim over $\mathcal{O}$. Alternatively, since $\mathrm{StPic}(A)$ is generated by $\mathrm{Out}_0(A)$ and the Heller operator, it follows that over k, the group F is generated by the smallest power of the Heller operator that yields a Morita equivalence. Since the Heller operator lifts over $\mathcal{O}$ and since the lift of a Morita equivalence is again a Morita equivalence, it follows directly that F lifts to a subgroup of $\mathrm{Pic}(\mathcal{O}Gb)$. This proves in particular also the last statement, whence the result. $\qquad\square$

Remark 11.14.4 Using Weiss' criterion from [228] (and its extensions to more general coefficient rings in [193], [177, Appendix 1]) one can calculate $\mathrm{Out}_0(\mathcal{O}(P \rtimes E))$ more precisely. By [127, Proposition 4.3], if $E \neq \{1\}$, then $\mathrm{Out}_0(\mathcal{O}(P \rtimes E)) \cong \mathrm{Aut}(P)/E$. If $E = \{1\}$, then it follows from results of Higman [85] that $\mathrm{Out}(\mathcal{O}P) \cong P \rtimes \mathrm{Aut}(P)$. Weiss' criterion can be used to show that in fact any Morita equivalence between two blocks with cyclic defect groups is given by a bimodule with endopermutation source (this follows from [127, 2.6, 2.7], for instance). See [20] for more precise statements.

Calculating the analogue of Picard groups for derived categories is more difficult; see Rouquier and Zimmermann [197], Zimmermann [230, §6.12], and Yekutieli [229].

11.15 The source algebras of blocks with cyclic defect groups

Proof of Theorem 11.1.19 Let G be a finite group and b a block of $\mathcal{O}G$ with a nontrivial cyclic defect group P and inertial quotient E. Denote by Z the unique subgroup of order p of P, and set $H = N_G(Z)$. Let c be the block of $\mathcal{O}H$ corresponding to b, and let $j \in (\mathcal{O}Hc)^P$ be a source idempotent of c. By 11.2.2 we have an isomorphism of interior P-algebras

$$jOHj \cong S \otimes_\mathcal{O} \mathcal{O}(P \rtimes E),$$

where $S = \mathrm{End}_\mathcal{O}(W)$ for some indecomposable endopermutation $\mathcal{O}P$-module W on which Z acts trivially and which has determinant 1. Denote by $\bar{j}$ the image of j in kH. The standard Morita equivalence between block algebras and almost source algebras 6.4.6 implies that $\bar{j}kH\bar{j}$ is Morita equivalent to $kH\bar{c}$, hence symmetric indecomposable and nonsimple. The algebra $\bar{j}kH\bar{j}$ is also Morita equivalent to $k(P \rtimes E)$, hence serial by 11.3.2. Let φ be an automorphism of P. Since $\mathrm{Aut}(P)$ is abelian, it follows that φ extends trivially to a group automorphism of $P \rtimes E$, hence to an algebra automorphism of $\mathcal{O}(P \rtimes E)$, still denoted φ, which induces the identity on $\mathcal{O}E$. Any automorphism of P stabilises any indecomposable kP-module. The fact that W is up to isomorphism the unique lift of $k \otimes_{\mathcal{O}} W$ which has determinant 1 implies that any automorphism of P stabilises W. Thus φ extends to an algebra automorphism of S. It follows that φ extends via the structural map $P \to (S \otimes_{\mathcal{O}} \mathcal{O}(P \rtimes E))^{\times}$ to an algebra automorphism, still denoted φ, of $S \otimes_{\mathcal{O}} \mathcal{O}(P \rtimes E)$. This automorphism is the identity on $1 \otimes \mathcal{O}E$, hence stabilises the isomorphism classes of the simple $\bar{S} \otimes_k k(P \rtimes E)$-modules, where $\bar{S} = k \otimes_{\mathcal{O}} S = \mathrm{End}_k(k \otimes_{\mathcal{O}} W)$. It follows from 11.5.1 that $\varphi \in \mathrm{Aut}_0(j\mathcal{O}Hj)$; that is, φ stabilises the isomorphism classes of all finitely generated $\bar{j}kH\bar{j}$-modules. By 11.2.4 the bimodule $b\mathcal{O}Gc$ and its dual induce a stable equivalence of Morita type between $\mathcal{O}Gb$ and $\mathcal{O}Hc$. It follows again from the standard Morita equivalences in 6.4.6 that the $j\mathcal{O}Gbj$-$j\mathcal{O}Hj$-bimodule $j\mathcal{O}Gbj$ and its dual induce a stable equivalence of Morita type between $j\mathcal{O}Gbj$ and $j\mathcal{O}Hj$. By 4.14.1, as a right module, $j\mathcal{O}Gbj$ is a progenerator for $j\mathcal{O}Hj$, so we may identify $j\mathcal{O}Hj$ with its image $j\mathcal{O}Hbj$ in $j\mathcal{O}Gbj$, obtained from multiplying $j\mathcal{O}Hj$ by b. Thus we have a stable equivalence of Morita type between $j\mathcal{O}Gbj$ and $j\mathcal{O}Hbj$ induced by restriction and induction. Thus 4.15.2 implies that φ extends to an automorphism ψ of $j\mathcal{O}Gbj$ in $\mathrm{Aut}_0(j\mathcal{O}Gbj)$. Since $\mathrm{Br}_P(j)$ is a primitive idempotent in $(j\mathcal{O}Gbj)(P) = (j\mathcal{O}Hj)(P)$ it follows that P has a unique local point γ on $j\mathcal{O}Gbj$. Let $i \in \gamma$. Then $\psi(i) \in \gamma$, hence $\psi(i) = cic^{-1}$ for some $c \in (j\mathcal{O}Gbj)^{\times}$. Define ψ' by $\psi'(a) = c^{-1}ac$ for all $a \in j\mathcal{O}Gbj$. Then ψ' fixes i, hence induces an automorphism of A. Since ψ and ψ' differ by conjugation with a P-fixed element, it follows that they induce the same automorphism of P on the images of P in $j\mathcal{O}Hj$ and $i\mathcal{O}Gi$. The result follows. $\qquad\square$

Proof of Theorem 11.1.18 Let G, G' be finite groups, b a block of $\mathcal{O}G$, and b' a block of $\mathcal{O}G'$. Suppose that b and b' have a common nontrivial cyclic defect group P. Suppose that $\mathcal{O}Gb$ and $\mathcal{O}G'b'$ give rise to isomorphic data (I, ρ, σ, v, W) and $(I, \rho', \sigma', v', W')$. That is, we have $W \cong W'$, and there is a bijection $\tau : I \cong I'$ satisfying $\tau \circ \rho = \rho' \circ \tau$ and $\tau \circ \sigma = \sigma' \circ \tau$, as well as $\tau(v) = v'$, provided there are exceptional vertices. By 11.1.19 all source algebras of $\mathcal{O}Gb$ are isomorphic as interior P-algebras, and all source algebras of $\mathcal{O}G'b'$ are isomorphic as interior P-algebras. By 9.7.4, in order to show that $\mathcal{O}Gb$ and $\mathcal{O}G'b'$ have isomorphic source algebras, it suffice to show that there is a splendid Morita equivalence between $\mathcal{O}Gb$ and $\mathcal{O}G'b'$. In view of 6.4.9 we may assume that $\mathcal{O} = k$. Denote by Z the subgroup of order p of P. Set $H = N_G(Z)$ and $H' = N_{G'}(Z)$. Let c be the block of kH corresponding to b, and let c' be the block of kH' corresponding to b'. By 11.2.4 the bimodule $bkGc$ and its dual induce a stable equivalence of Morita type. This is a splendid stable equivalence. A similar statement holds for $b'kG'c'$. Since $W \cong W'$, it follows from 11.2.2 that source algebras of $B = kHc$ and $B' = kH'c'$ are isomorphic, or equivalently, that there is a splendid stable Morita equivalence between B and B'. All four blocks b, b', c, c' have the defect group P in common and the same fusion system, which is equal to that of $P \rtimes E$. Thus tensoring the three splendid stable equivalences between $A = kGb$, B, B', and $A' = kG'b'$ together yields a stable equivalence of Morita type between A and A', given by an indecomposable A-A'-bimodule M, which has vertex ΔP and trivial source when viewed as a $k(G \times G')$-module. Since the trivial kP-module has period 2, it follows that $\Omega^{2t}(M)$ is still a splendid stable equivalence of Morita type, for any integer t.

Denote by I a set of representatives of the conjugacy classes of primitive idempotents in A. Similarly, denote by I' a set of representatives of the conjugacy classes of primitive idempotents in A'. By 11.6.3, the simple modules of B and B' correspond to the uniserial submodules U_i of Ai and $U'_{i'}$ of $A'i'$ through the stable equivalence between A (resp. A') and B (resp. B'). Thus the stable equivalence induced by M sends the submodule U_i of Ai to the submodule $U'_{i'}$ of $A'i'$ for some $i' \in I'$ depending on $i \in I$. It follows from the assumptions and 11.8.2 that A and A' have isomorphic basic algebras. More precisely, there is a Morita equivalence given by an A-A'-bimodule N and its dual satisfying $N \otimes_A U_i \cong U'_{\tau(i)}$ for all $i \in I$. Thus $M \otimes_{A'} N^*$ is an A-A-bimodule inducing a self stable equivalence of A. By construction, this stable equivalence permutes the isomorphism classes of the U_i. Since the group generated by Ω^2 permute these isomorphism classes transitively, we may assume, after possibly replacing M by $\Omega^{2t}(M)$ for some integer t, that the stable equivalence given by $M \otimes_{A'} N^*$ stabilises one of the U_i. But then 11.14.1 implies that the unique indecomposable direct bimodule summand of $M \otimes_{A'} N^*$ is of the form A_α for some $\alpha \in \mathrm{Aut}_0(A)$. After replacing N by $_{\alpha^{-1}}N$ we may assume that the unique indecomposable nonprojective bimodule summand of $M \otimes_{A'} N^*$ is isomorphic to A. Tensoring with $- \otimes_A N$ on the right implies that $N \cong M$. This shows that M is both splendid and induces a Morita equivalence. Thus A and A' have isomorphic source algebras. $\qquad\square$

12

Blocks with Klein Four Defect Groups

A block of a finite group over an algebraically closed field of characteristic 2 with a Klein four defect group P is Morita equivalent to either kP, kA_4 or the principal block algebra of kA_5. This result, due to Erdmann, implies that Donovan's conjecture holds for blocks with a Klein four defect group. We will see in this chapter that this lifts to a complete discrete valuation ring $\mathcal{O}$ with k as residue field and that more precisely the source algebras are isomorphic to the unique primitive interior P-algebras with P as defect associated with $S \otimes_{\mathcal{O}} \mathcal{O}P$, $S \otimes_{\mathcal{O}} \mathcal{O}A_4$ or $S \otimes_{\mathcal{O}} \mathcal{O}A_5$, where $S = \mathrm{End}_{\mathcal{O}}(\Omega_P^n(\mathcal{O}))$ for some integer n. Using the classification of finite simple groups it has been shown in [51] that $n = 0$, implying in particular that Puig's conjecture holds for blocks with a Klein four defect group. Throughout this Chapter, $\mathcal{O}$ is a complete discrete valuation ring with an algebraically closed residue field $k = \mathcal{O}/J(\mathcal{O})$ of characteristic 2 and a quotient field K.

12.1 The structure of blocks with Klein four defect groups

Blocks with a Klein four defect group satisfy Alperin's weight conjecture, Broué's abelian defect conjecture, Brauer's height zero conjecture, and the finiteness conjectures of Donovan and Puig. As in the case of blocks with cyclic defect groups, the above mentioned conjectures follow from far more precise structural results. Let P be a Klein four group. Then P is isomorphic to a Sylow 2-subgroup of A_4 and of A_5. Thus, if we identify P with a Sylow 2-subgroup of A_4 or A_5, then $\mathcal{O}A_4$ becomes an interior P-algebra; similarly for A_5. It does not matter which identification we choose because any automorphism of a Sylow 2-subgroup of A_4 or A_5 extends to an automorphism of A_4 or A_5, hence to an algebra automorphism of $\mathcal{O}A_4$ or the principal block algebra of $\mathcal{O}A_5$, respectively. Thus all source algebras of a given block with a Klein four defect group P are isomorphic as interior P-algebras. The automorphism group of a Klein four group is isomorphic to S_3, and hence the inertial quotient E of a block b with a Klein four defect group P is either trivial or cyclic of order 3. If E is trivial, then b is nilpotent.

Theorem 12.1.1 ([51, Theorem 1.1]) *Let A be a source algebra of a block of a finite group algebra over $\mathcal{O}$ having a Klein four group P as a defect group. Then, as an interior P-algebra, A is isomorphic to either $\mathcal{O}P$ or $\mathcal{O}A_4$ or the principal block algebra of $\mathcal{O}A_5$. In particular, Puig's conjecture holds for blocks with a Klein four defect group.*

This implies that a block algebra having a Klein four defect group P is Morita equivalent to either $\mathcal{O}P$, $\mathcal{O}A_4$, or $\mathcal{O}A_5b_0$, where b_0 is the principal block of $\mathcal{O}A_5$, and that the matrix of all decomposition numbers is equal to that of $\mathcal{O}P$, $\mathcal{O}A_4$, or $\mathcal{O}A_5b_0$, respectively. The proof of Theorem 12.1.1 in [51] invokes the classification of finite simple groups and is beyond the scope of this book. We will prove a slightly weaker result that still yields the Morita equivalence classes and the decomposition numbers, except for some signs in the generalised decomposition numbers. The source algebras are determined 'up to a power of the Heller operator'.

Theorem 12.1.2 ([126, Theorem 1.1]) *Let A be a source algebra of a block of a finite group algebra over $\mathcal{O}$ having a Klein four group P as a defect group. Then there is an integer n such that, setting $S = \mathrm{End}_{\mathcal{O}}(\Omega_P^n(\mathcal{O}))$, the interior P-algebra A is isomorphic to either $S \otimes_{\mathcal{O}} \mathcal{O}P$ or $S \otimes_{\mathcal{O}} \mathcal{O}A_4$ or $i(S \otimes_{\mathcal{O}} \mathcal{O}A_5b_0)i$, where b_0 is the principal block of $\mathcal{O}A_5$ and where i is primitive local in $(S \otimes_{\mathcal{O}} \mathcal{O}A_5b_0)^P$.*

The unit element of $S \otimes_{\mathcal{O}} \mathcal{O}A_4$ remains primitive in $(S \otimes_{\mathcal{O}} \mathcal{O}A_4)^P$. If $n \neq 0$, then the unit element of $S \otimes_{\mathcal{O}} \mathcal{O}A_5b_0$ is not primitive in $(S \otimes_{\mathcal{O}} \mathcal{O}A_5b_0)^P$, but equal to $i + i'$, where i is primitive local and $i' \in (S \otimes_{\mathcal{O}} \mathcal{O}A_5)_1^P$, whence the slightly more technical statement. Any automorphism of P stabilises the Heller translates $\Omega^n(\mathcal{O})$, and any group automorphism of P extends to A_4 and A_5. Thus 12.1.2 is still precise enough to imply that all source algebras of a given block with defect group P are isomorphic as interior P-algebras. The proof of Theorem 12.1.2 in §12.3 below simplifies the original proof in [126]. Theorem 12.1.2 does not imply Puig's conjecture, because it does not rule out that there could be infinitely many possibilities for the integer n in the statement. The Morita equivalence classes of these algebras are independent of this integer.

Corollary 12.1.3 ([126, Corollary 1.4]) *Let G be a finite group and B a block algebra of $\mathcal{O}G$ having a Klein four defect group P. Then B is Morita equivalent to either $\mathcal{O}P$ or $\mathcal{O}A_4$ or the principal block algebra of $\mathcal{O}A_5$.*

Corollary 12.1.4 (Erdmann [66]) *Donovan's conjecture holds for blocks with a Klein four defect group.*

Corollary 12.1.5 (Brauer [28, §7]) *A block with a Klein four defect group has four ordinary irreducible characters, all of height zero, and either one or three irreducible Brauer characters.*

Not all irreducible Brauer characters need to have height zero: it follows from 12.2.9 or 12.2.10 that if B is Morita equivalent to the principal block of $\mathcal{O}A_5$, then two of the three irreducible Brauer characters have height 1.

Notation 12.1.6 Let G be a finite group, b a block of $\mathcal{O}G$, and let (P, e) be a maximal (G, b)-Brauer pair. Suppose that P is a Klein four group. Denote by $\hat{e}$ the block of $\mathcal{O}C_G(P)$ that lifts the block e of $kC_G(P)$. Let c be the block of $\mathcal{O}N_G(P)$ with P as a defect group such that $\mathrm{Br}_{\Delta P}(b) = \mathrm{Br}_{\Delta P}(c)$; that is, c is the Brauer correspondent of b. Set $H = N_G(P, e)$ and $E = H/C_G(P)$; that is, E is the inertial quotient of b with respect to the maximal (G, b)-Brauer pair (P, e). The $\mathcal{O}N_G(P)c$-$\mathcal{O}He$-bimodule $\mathcal{O}N_G(P)e$ and its dual induce a splendid Morita equivalence between $\mathcal{O}N_G(P)c$ and $\mathcal{O}He$; in particular, $\mathcal{O}N_G(P)c$ and $\mathcal{O}He$ are source algebra equivalent. Let f be a primitive idempotent in $(\mathcal{O}Gb)^{\Delta H}$ such that $\mathrm{Br}_{\Delta P}(f) = e$. Let $j \in \mathcal{O}C_G(P)\hat{e}$ be a primitive idempotent. Then j is primitive in $(\mathcal{O}He)^{\Delta P}$, hence a source idempotent of e as a block of $\mathcal{O}H$, and thus also a source idempotent of the block c of $\mathcal{O}N_G(P)$. Set $i = jf$. By 6.15.1, i is a source idempotent of $\mathcal{O}Gb$ in $(\mathcal{O}Gb)^{\Delta P}$. Multiplication by f induces a unitary algebra homomorphism

$$j\mathcal{O}Hj \to i\mathcal{O}Gi,$$

and by 6.14.1 we have an isomorphism of interior P-algebras

$$j\mathcal{O}Hj \cong \mathcal{O}(P \rtimes E),$$

where we use that $H^2(E; k^\times)$ is trivial as E is cyclic. Set $A = i\mathcal{O}Gi$. We identify $\mathcal{O}(P \rtimes E)$ as a subalgebra of A through the previous maps.

12.2 A stable equivalence for blocks with Klein four defect groups

The results on blocks with a Frobenius inertial quotient in Theorem 10.5.1 imply that a block with a Klein four defect group is stably equivalent to its Brauer correspondent via a bimodule with endopermutation source. Since endopermutation modules of Klein four groups over k are Heller translates of the trivial module, it follows that there is even a splendid stable equivalence between such a block and its Brauer correspondent. The considerable technicalities in Theorem 10.5.1 become significantly easier in the special case of blocks with a Klein four defect group, which is what we describe in this section. We use the notation from 12.1.6. In particular, $A = i\mathcal{O}Gi$ is a source algebra of a block b of $\mathcal{O}G$ having a Klein four defect group P and inertial quotient E. We identify $\mathcal{O}(P \rtimes E)$ to a subalgebra of A. We will show that restriction and induction between A and $\mathcal{O}(P \rtimes E)$ yields a stable equivalence of Morita type.

Lemma 12.2.1 *Let Q be a nontrivial subgroup of P. Then $A(\Delta Q) \cong kP$.*

Proof Since P is abelian, it follows from 6.4.13 that $i_Q = \mathrm{Br}_{\Delta Q}(i)$ is a source idempotent in $kC_G(Q)$ of a block e_Q having P as a defect group. Since Q is nontrivial, it follows from 10.5.2 that e_Q is a nilpotent block. By the structure theorem on nilpotent blocks we have $i_Q kC_G(Q)i_Q \cong \mathrm{End}_k(V) \otimes_k kP$ where V is an indecomposable endopermutation kP-module with vertex P. Since Q is in the centre of $C_G(Q)$, it follows that Q acts trivially on V. Therefore, V is an indecomposable endopermutation kP/Q-module with vertex P/Q. But P/Q has order at most 2, so $V \cong k$.   □

Theorem 12.2.2 *We have $A = \mathcal{O}(P \rtimes E) \oplus Y$ for some projective $\mathcal{O}(P \rtimes E)$-$\mathcal{O}(P \rtimes E)$-bimodule Y.*

Proof Lemma 12.2.1 implies that for any nontrivial subgroup Q of P we have $A(\Delta Q) \cong kP \cong (\mathcal{O}(P \rtimes E))(\Delta Q)$, and hence $Y(\Delta Q) = \{0\}$. Since A and $\mathcal{O}(P \rtimes E)$ have the same fusion system on P, it follows from 9.8.5 that Y is projective.   □

Corollary 12.2.3 *The restriction functor*

$$\mathrm{Res}^A_{\mathcal{O}(P \rtimes E)} = {}_{\mathcal{O}(P \rtimes E)}A \otimes_A -$$

and the induction functor

$$\mathrm{Ind}^A_{\mathcal{O}(P \rtimes E)} = A \otimes_{\mathcal{O}(P \rtimes E)} -$$

induce a splendid stable equivalence of Morita type between A and $\mathcal{O}(P \rtimes E)$.

Proof This follows from 12.2.2 and 9.8.4.   □

Corollary 12.2.4 *The $\mathcal{O}Gb$-$\mathcal{O}He$-bimodule $\mathcal{O}Gf$ and its $\mathcal{O}$-dual $f\mathcal{O}G$ induce a splendid stable equivalence of Morita type between $\mathcal{O}Gb$ and $\mathcal{O}He$.*

Proof We have $i(\mathcal{O}Gf)j = i\mathcal{O}Gi = A_{\mathcal{O}(P \rtimes E)}$. Thus 12.2.4 follows from 12.2.3 combined with the standard Morita equivalences between block algebras and their source algebras.   □

Corollary 12.2.5 *We have*

$$|\mathrm{Irr}_K(G, b)| = |\mathrm{Irr}_K(A)| = |\mathrm{Irr}_K(\mathcal{O}(P \rtimes E))| = 4,$$

$$|\mathrm{IBr}_k(G, b)| = |\mathrm{Irr}_k(k \otimes_\mathcal{O} A)| = |\mathrm{Irr}_k(k(P \rtimes E))| = |E|.$$

Proof This is a special case of 10.5.5. Since the calculations in this case are simpler than in the general case, we sketch the arguments. If $|E| = 1$, then the block b is nilpotent, hence $A \cong \mathrm{End}_\mathcal{O}(\Omega^n(\mathcal{O})) \otimes_\mathcal{O} \mathcal{O}P$ for some integer n. In that case the equalities hold trivially. Suppose that $|E| = 3$. Let $T_1 = k$, T_2, T_3 be three pairwise nonisomorphic $k(P \rtimes E)$-modules; these are obtained from inflating the three simple kE-modules to $k(P \rtimes E)$. Since P is contained in the kernel of any simple $k(P \rtimes E)$-module, they are a set of representatives of the isomorphism classes of simple $k(P \rtimes E)$-modules. Each T_i lifts uniquely to an irreducible character η_i of $P \rtimes E$. If $\zeta : P \to \mathcal{O}^\times$ is a nontrivial linear character of P, then $\eta_0 = \mathrm{Ind}_P^{P \rtimes E}(\zeta)$ is irreducible, and we have

$$\mathrm{Irr}_K(P \rtimes E) = \{\eta_0, \eta_1, \eta_2, \eta_3\}.$$

The set

$$\{\eta_i + \eta_0 | 1 \le i \le 3\}$$

is the set of characters of the projective indecomposable $\mathcal{O}(P \rtimes E)$-modules, and the set

$$\left\{\sum_{i=1}^3 \eta_i - \eta_0\right\}$$

is a basis of $L^0(P \rtimes E)$. The unique element in this basis has norm square 4. Since a stable equivalence of Morita type between A and $\mathcal{O}(P \rtimes E)$ induces an isometry $L^0(P \rtimes E) \cong L^0(A)$, it follows that the image of this element in $L^0(A)$ has norm square 4. This image cannot be of the form $\pm 2\chi$ for some $\chi \in \mathrm{Irr}_K(A)$ because $L^0(A)$ does not contain any actual nonzero character. Thus this image is of the form $\sum_{i=0}^3 \delta_i \chi_i$ for some signs $\delta_i \in \{\pm 1\}$ and where $\mathrm{Irr}_K(A) = \{\chi_i | 0 \le i \le 3\}$. This shows that $|\mathrm{Irr}_K(G, b)| = 4$, and then $|\mathrm{IBr}_k(G, b)| = 4 - \mathrm{rk}_\mathbb{Z}(L^0(G, b)) = 3$ as required. $\qquad\square$

We use the stable equivalence between A and $\mathcal{O}(P \rtimes E)$ to determine the restrictions to $k(P \rtimes E)$ of the simple $k \otimes_\mathcal{O} A$-modules. We have $P \rtimes E \cong A_4$, and we have classified in 7.2.12 the indecomposable kA_4-modules with a one-dimensional stable endomorphism ring. Set $\bar{A} = k \otimes_\mathcal{O} A$.

Lemma 12.2.6 *Let S be a simple $\bar{A}$-module. Then $\mathrm{Res}_{k(P \rtimes E)}^{\bar{A}}(S)$ is an inde-composable nonprojective $k(P \rtimes E)$-module, which is either isomorphic to*

$\Omega^n_{P \rtimes E}(T)$ *for some integer n and some simple* $k(P \rtimes E)$-*module T, or which has dimension* 2.

Proof Since restriction from $\bar{A}$ to $k(P \rtimes E)$ induces a stable equivalence, it follows that $\mathrm{Res}^{\bar{A}}_{k(P \rtimes E)}(S)$ is a direct sum of an indecomposable nonprojective $k(P \rtimes E)$-module V satisfying $\underline{\mathrm{End}}_{k(P \rtimes E)}(V) \cong k$, and a projective $k(P \rtimes E)$-module. It follows further from 6.4.11 that $\mathrm{Res}^{\bar{A}}_{kPE}(S)$ has no nonzero projective direct summand. It follows from 7.2.12 that either $V \cong \Omega^n_{P \rtimes E}(T)$ for some integer n and some simple $k(P \rtimes E)$-module T, or $\dim_k(V) = 2$, whence the result. $\qquad\square$

Lemma 12.2.7 *Let n be a positive integer, and let T be a simple* $k(P \rtimes E)$-*module. Then every simple* $k(P \rtimes E)$-*module is isomorphic to a composition factor of* $\Omega^n_{P \rtimes E}(T)$.

Proof If $|E| = 1$ this is trivial. Suppose that $|E| = 3$. Let T_1, T_2, T_3 be three pairwise nonisomorphic simple $k(P \rtimes E)$-modules. Set $M = \Omega^n_{P \rtimes E}(T)$ for some simple $k(P \rtimes E)$-module T. Arguing by contradiction, suppose that all simple composition factors of M are isomorphic to T_2 or T_3. Suppose first that $n > 0$. By 7.2.10 (ii) we have $\mathrm{Res}^{k(P \rtimes E)}_{kP}(M) \cong \Omega^n_P(k)$, and by 4.11.8 we have $\mathrm{rad}(M) = \mathrm{soc}(M)$. It follows from 7.2.3 that $\dim_k(\mathrm{soc}(M)) = \dim_k(M/\mathrm{soc}(M)) - 1$, or equivalently, $\dim_k(M) = 2\dim_k(\mathrm{soc}(M)) + 1$. Let U_i be a projective cover of T_i, where $1 \le i \le 3$. Then U_i is also an injective envelope of T_i. By 7.2.10 we have $\mathrm{rad}(U_i)/\mathrm{soc}(U_i) \cong T_j \oplus T_l$, where $\{i, j, l\} = \{1, 2, 3\}$, and hence U_i has exactly two submodules, denoted T_i^j, T_i^l, of dimension two, with top composition factor T_j, T_l, respectively, and socle T_i. Since M is indecomposable nonprojective, it follows that M is a submodule of a direct sum of $\dim_k(\mathrm{soc}(M))$ modules of the form $\mathrm{rad}(U_i)$, where $i \in \{2, 3\}$. Since M has no submodule isomorphic to T_1, M is in fact a submodule of a direct sum of $\dim_k(\mathrm{soc}(M))$ modules isomorphic to T_3^2 or T_2^3. Such a direct sum would have dimension $2\dim_k(\mathrm{soc}(M))$, but $\dim_k(M) = 2\dim_k(\mathrm{soc}(M)) + 1$, a contradiction. The case $n < 0$ is proved either similarly, or by using the fact that $\Omega^n(T)$ is dual to $\Omega^{-n}(T')$, where T' is dual to T. $\qquad\square$

Remark 12.2.8 In cohomological terms, Lemma 12.2.7 states that for any two simple $k(P \rtimes E)$-modules T, T' and any nonzero integer n, at least one of $\widehat{\mathrm{Ext}}^n_{k(P \rtimes E)}(T, T')$ or $\widehat{\mathrm{Ext}}^n_{k(P \rtimes E)}(T', T)$ is nonzero. Indeed, T' is a composition factor of $\Omega^n_{P \rtimes E}(T)$ if and only if T' is a submodule or a quotient of $\Omega^n_{P \rtimes E}(T)$, because the Loewy length of indecomposable nonprojective $k(P \rtimes E)$-modules is at most 2.

Proposition 12.2.9 *Suppose that* $|E| = 3$. *Let* $\{T_1, T_2, T_3\}$ *be a set of representatives of the isomorphism classes of simple* $k(P \rtimes E)$-*modules, and for any* $i, j \in \{1, 2, 3\}$, $i \neq j$, *denote by* T_j^i *an indecomposable 2-dimensional* $k(P \rtimes E)$-*module with composition series* T_i, T_j, *from top to bottom. Write* $\Omega = \Omega_{P \rtimes E}$. *Then there is a set of representative* $\{S_1, S_2, S_3\}$ *of the isomorphism classes of simple* $\bar{A}$-*modules such that exactly one of the two following statements holds.*

(i) There is an integer n such that for $1 \leq i \leq 3$ *we have*

$$\mathrm{Res}^{\bar{A}}_{k(P \rtimes E)}(S_i) \cong \Omega^n(T_i).$$

(ii) There is an integer n such that

$$\mathrm{Res}^{\bar{A}}_{k(P \rtimes E)}(S_1) \cong \Omega^n(T_1),$$

$$\mathrm{Res}^{\bar{A}}_{k(P \rtimes E)}(S_2) \cong \Omega^n(T_3^2),$$

$$\mathrm{Res}^{\bar{A}}_{k(P \rtimes E)}(S_3) \cong \Omega^n(T_2^3).$$

Proof By 6.15.1 (iv), at least one of the simple $\bar{A}$-modules has odd dimension. Choose notation such that $\dim_k(S_1)$ is odd. It follows from 12.2.6 that we may choose notation such that $\mathrm{Res}^{\bar{A}}_{k(P \rtimes E)}(S_1) \cong \Omega^n(T_1)$ for some integer n. We consider first the case where S_2 has odd dimension. Then $\mathrm{Res}^{\bar{A}}_{k(P \rtimes E)}(S_2) \cong \Omega^m(T)$ for some integer m and some simple $k(P \rtimes E)$-module T. Since $\mathrm{Hom}_{\bar{A}}(S_i, S_j)$ is zero if $i \neq j$, we get that

$$\{0\} = \underline{\mathrm{Hom}}_{k(P \rtimes E)}(\Omega^n(T_1), \Omega^m(T)) \cong \underline{\mathrm{Hom}}_{k(P \rtimes E)}(T_1, \Omega^{m-n}(T)).$$

Thus $\Omega^{m-n}(T)$ has no submodule isomorphic to T_1. Similarly, $\Omega^{m-n}(T)$ has no quotient isomorphic to T_1. Since indecomposable nonprojective $k(P \rtimes E)$-modules have Loewy length at most 2, this implies that T_1 is not isomorphic to a composition factor of $\Omega^{m-n}(T)$. Thus 12.2.7 forces $m = n$. But then T and T_1 are not isomorphic. Choose notation such that $T = T_2$. If also $\dim_k(S_3)$ is odd, then this argument forces $\mathrm{Res}^{\bar{A}}_{k(P \rtimes E)}(S_3) \cong \Omega^n(T_3)$, which is the situation described in (i). Suppose that $\dim_k(S_3)$ is even. Since Ω permutes the isomorphism classes of 2-dimensional indecomposable $k(P \rtimes E)$-modules, we may write $\mathrm{Res}^{\bar{A}}_{k(P \rtimes E)}(S_3) \cong \Omega^n(T_j^i)$ for some $i, j \in \{1, 2, 3\}$, $i \neq j$. Then, as above, we have

$$\{0\} = \underline{\mathrm{Hom}}_{k(P \rtimes E)}(\Omega^n(T_1), \Omega^n(T_j^i)) \cong \underline{\mathrm{Hom}}_{k(P \rtimes E)}(T_1, T_j^i).$$

Thus $j \neq 1$. Similarly,

$$\{0\} = \underline{\mathrm{Hom}}_{k(P \rtimes E)}(\Omega^n(T_j^i), \Omega^n(T_1)) \cong \underline{\mathrm{Hom}}_{k(P \rtimes E)}(T_j^i, T_1),$$

hence $i \neq 1$. Thus $\{i, j\} = \{2, 3\}$. But this argument, with S_2 and T_2 implies that also $\{i, j\} = \{1, 3\}$, which is impossible. This shows that if $\dim_k(S_2)$ is odd, then so is $\dim_k(S_3)$, and hence that (i) holds. By 12.2.6 it remains to consider the case where $\dim_k(S_2) = \dim_k(S_3) = 2$. In that case, the above arguments imply that (ii) holds, up to possibly exchanging the notation for S_2 and S_3. $\square$

The above result applies in particular to the principal block b_0 of A_5.

Corollary 12.2.10 *Consider $P \rtimes E \cong A_4$ as a subgroup of A_5. Let b_0 be the principal block of $\mathcal{O}A_5$, and set $A' = \mathcal{O}A_5 b_0$. Then A' is the unique source algebra of $\mathcal{O}A_5 b_0$, and $\bar{A}' = k \otimes_\mathcal{O} A'$ has three pairwise nonisomorphic simple modules $S_1' = k$, S_2, S_3, such that*

$$\mathrm{Res}^{\bar{A}'}_{k(P \rtimes E)}(S_1') = T_1,$$

$$\mathrm{Res}^{\bar{A}'}_{k(P \rtimes E)}(S_2') = T_3^2,$$

$$\mathrm{Res}^{\bar{A}'}_{k(P \rtimes E)}(S_3') = T_2^3.$$

Proof The algebra A' has three isomorphism classes of simple modules, namely the trivial module, and two 2-dimensional simple modules whose restrictions to P remain indecomposable. Thus 6.4.17 implies that A' is its own source algebra. The rest follows from 12.2.9. $\square$

One can also prove this corollary by a direct verification.

12.3 The source algebras of blocks with Klein four defect

Proof of Theorem 12.1.2 We use the notation as in 12.1.6. If $|E| = 1$, then b is nilpotent, and hence, by 8.11.5, we have an isomorphism of interior P-algebras $A \cong \mathrm{End}_\mathcal{O}(V) \otimes_\mathcal{O} \mathcal{O}P$ for some indecomposable endopermutation $\mathcal{O}P$-module V with vertex P and determinant 1. It follows either from the classification of indecomposable kP-modules in 7.2.1 or from Dade's classification of endopermutation modules over abelian p-groups in 7.8.1, that we may choose $V \cong \Omega_P^n(\mathcal{O})$ for some integer n. This proves Theorem 12.1.2 if $|E| = 1$.

Suppose now that $|E| = 3$. We need to consider the two possibilities according to 12.2.9. By 6.4.9 may assume $\mathcal{O} = k$.

Suppose first that, for some integer n, we have

$$\mathrm{Res}^A_{k(P \rtimes E)}(S_i) \cong \Omega^n(T_i)$$

for $1 \leq i \leq 3$. As an A-$k(P \rtimes E)$-bimodule, A remains indecomposable, and induces a stable equivalence of Morita type. Thus its dual, which induces the induction functor $A \otimes_{k(P \rtimes E)} -$, sends simple modules to indecomposable modules, hence satisfies

$$A \otimes_{k(P \rtimes E)} T_i \cong \Omega^{-n}(S_i).$$

It follows that the A-$k(P \rtimes E)$-bimodule $M = \Omega^n_{A \otimes_k k(P \rtimes E)}(A)$ induces a stable equivalence of Morita type that sends T_i to S_i. Therefore, by 4.14.10, the bimodule M and its dual induce a Morita equivalence. By 9.11.3, this bimodule is isomorphic to a direct summand of

$$A \otimes_{kP} \mathrm{Ind}^{P \times P}_{\Delta P}(\Omega^n_P(k)) \otimes_{kP} k(P \rtimes E).$$

Thus 9.11.9 implies that $A \cong e(S \otimes_k k(P \rtimes E))e$ for some primitive local idempotent e in $(S \otimes_k k(P \rtimes E))^{\Delta P}$, where $S = \mathrm{End}_k(\Omega^n_P(k))$. The unit element of $S \otimes_{kP} k(P \rtimes E)$ remains however primitive in $(S \otimes_k k(P \rtimes E))^{\Delta P}$ because the restrictions to P of the simple $S \otimes_k k(P \rtimes E)$-modules remain indecomposable (they are isomorphic to $\Omega^n_P(k)$ as the simple $k(P \rtimes E)$-modules are one-dimensional). Thus $e = 1_{S \otimes_k k(P \rtimes E)}$, and the result follows in this case.

Suppose finally that, for some integer n, we have

$$\mathrm{Res}^A_{k(P \rtimes E)}(S_1) \cong \Omega^n(T_1),$$

$$\mathrm{Res}^A_{k(P \rtimes E)}(S_2) \cong \Omega^n(T_3^2),$$

$$\mathrm{Res}^A_{k(P \rtimes E)}(S_3) \cong \Omega^n(T_2^3).$$

Consider $P \rtimes E \cong A_4$ as a subgroup of A_5. Let b_0 be the principal block of kA_5, and set $A' = kA_5 b_0$. By 12.2.10, the algebra A' has three pairwise nonisomorphic simple modules $S_1' = k, S_2, S_3$, such that

$$\mathrm{Res}^{A'}_{k(P \rtimes E)}(S_1') = T_1,$$

$$\mathrm{Res}^{A'}_{k(P \rtimes E)}(S_2') = T_3^2,$$

$$\mathrm{Res}^{A'}_{k(P \rtimes E)}(S_3') = T_2^3.$$

The restriction functors from A and A' to $k(P \rtimes E)$, respectively, are both stable equivalences of Morita type. Composing one with the inverse of the other yields a stable equivalence of Morita type from A' to A sending S_i' to $\Omega^{-n}_A(S_i)$, for $1 \le i \le 3$. This stable equivalence of Morita type is given by a direct bimodule summand M of $A \otimes_{k(P \rtimes E)} A'$. Thus M is a summand of $A \otimes_{kP} A'$. As in the previous case, the bimodule $\Omega^n_{A \otimes_k (A')^{\mathrm{op}}}(M)$ yields a stable equivalence of Morita type sending S_i' to S_i, hence, by 4.14.10, a Morita equivalence given by a bimodule with vertex ΔP and source $\Omega^n_P(k)$. The result follows again from 9.11.9. This concludes the proof of Theorem 12.1.2. $\square$

Remark 12.3.1 The structure of the source algebras in Theorem 12.1.2 implies that in order to calculate Picard groups of blocks with a Klein four defect group P, it suffices to calculate the Picard groups of $\mathcal{O}P$, $\mathcal{O}A_4$ and the principal block algebra of $\mathcal{O}A_5$. Using Weiss' criterion, it follows from Theorem 12.1.2, that every Morita equivalence between two blocks with a Klein four defect group is given by a bimodule with an endotrivial source. Theorem 12.1.1 (which requires the classification of finite simple groups) implies that in fact any Morita equivalence between two blocks with a Klein four defect group is given by a bimodule with a linear source, and even trivial source, if the considered blocks are not nilpotent. See [20] for precise statements.

12.4 Derived categories of blocks with Klein four defect

Broué's abelian defect group conjecture holds for blocks with a Klein four defect group.

Theorem 12.4.1 *Let G be a finite group and b a block of $\mathcal{O}G$ with a Klein four defect group P. Let c be the corresponding block of $\mathcal{O}N_G(P)$. There is a splendid Rickard complex of $\mathcal{O}Gb$-$\mathcal{O}N_G(P)c$-bimodules.*

We prove first the following special case, due to Rickard. The proof follows Rouquier's construction of Rickard complexes in the cyclic block case.

Theorem 12.4.2 ([187, §3]) *Set $A = \mathcal{O}A_5 b_0$, where b_0 is the principal block idempotent of $\mathcal{O}A_5$. Set $B = \mathcal{O}A_4$, and identify B with its image in A via multiplication by b_0. Set $M = {}_B A$. There is a projective B-A-bimodule Q and a homomorphism $\pi : Q \to M$ such that the 2-term complex*

$$X = Q \xrightarrow{\ \pi\ } M$$

is a Rickard complex.

Proof By 4.14.16 it suffices to show that $X^* \otimes_B X \simeq A$. Using the same arguments for lifting Rickard complexes as in the proof of 11.12.1 we may assume that $\mathcal{O} = k$. We follow the strategy of the proof of 11.9.1. Let $\{T_1, T_2, T_3\}$ be a set of representatives of the isomorphism classes of simple B-modules, such that T_1 is the trivial module. For $i, j \in \{1, 2, 3\}$ such that $i \neq j$ denote by T_j^i a uniserial B-module with composition series T_i, T_j, from top to bottom. For $1 \leq i \leq 3$ denote by Q_i a projective cover of T_i. Note that Q_i is also a projective cover of T_j^i for $j \neq i$. By 12.2.10 there is a set of representatives of the isomorphism classes of simple A-modules $\{S_1, S_2, S_3\}$ with S_1 the trivial module, such that the restrictions of S_1, S_2, S_3 to B are isomorphic to T_1, T_3^2, T_2^3, respectively. For $1 \leq i \leq 3$ denote by R_i a projective cover of the simple right A-module S_i^*. By 4.5.12, a projective cover of M is isomorphic to $\oplus_{i=1}^3 Q_i \otimes_{\mathcal{O}} R_i$. We set

$$Q = \oplus_{i=2}^3 Q_i \otimes_{\mathcal{O}} R_i$$

and denote by $\pi : Q \to M$ the restriction to Q of a surjective bimodule homomorphism from this projective cover to M. Thus we have

$$X \otimes_A S_1 \cong T_1,$$

$$X \otimes_A S_2 \cong (Q_2 \to T_3^2),$$

$$X \otimes_A S_3 \cong (Q_3 \to T_2^3),$$

where the maps ending at T_3^2 and T_2^3 are surjective. For any $i, j \in \{1, 2, 3\}$, any chain map

$$X \otimes_A S_i \to X \otimes_A S_j[-1]$$

is zero. Indeed, for $j = 1$ this is trivial, for $i = 1$ this follows from the fact that T_1 is not isomorphic to a submodule of Q_2 or Q_3. For $\{i, j\} = \{2, 3\}$, such a chain map would be given by a commutative diagram of the form:

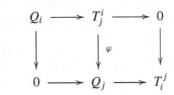

The surjectivity of the map $Q_i \to T_j^i$ in this diagram forces $\varphi = 0$. The exact same argument as in the proof of 11.9.4 shows that this implies that $X^* \otimes_B X$ is a split complex of A-A-bimodules, with homology concentrated in degree zero, and isomorphic to $A \oplus U$ for some projective A-A-bimodule U. Arguing as in the proof of 11.9.1, it suffices to show that $U = \{0\}$. Since $X^* \otimes_B X \simeq A \oplus U$ this is equivalent to showing that

$$S_i^* \otimes_A X^* \otimes_B X \otimes_A S_j \simeq S_i^* \otimes_A S_j$$

where $i, j \in \{1, 2, 3\}$. The right side is zero if $i \neq j$ and k if $i = j$. The left side is a complex with three terms in degree $1, 0, -1$, such that the first map is injective and the second map is surjective, as $X^* \otimes_B X$ has these properties and is split. Thus is suffices to show that the left side is acyclic for $i \neq j$ and has homology concentrated in degree 0 isomorphic to k if $i = j$. If $i = j = 1$, then the left side is $T_1^* \otimes_B T_1 \cong k$. If $i = 1$ and $j = 2$, then the left side is

$$T_1^* \otimes_B (Q_2 \to T_3^2) \cong T_1^* \otimes_B T_3^2 \cong \{0\}$$

and a similar argument shows that the left side is zero if $i = 1$ and $j = 3$. A dual argument shows that the left side is zero if $i \neq 1$ and $j = 1$. For $i = j = 2$, the left side is isomorphic to

$$((T_3^2)^* \to Q_2^*) \otimes_B (Q_2 \to T_3^2).$$

This yields a 3-term complex in which the left and right term are k and the middle term is k^3, hence a complex that is homotopy equivalent to k in degree zero. The analogous statement holds for $i = j = 3$. For $i = 2$ and $j = 3$ we get the complex

$$((T_3^2)^* \to Q_2^*) \otimes_B (Q_3 \to T_2^3).$$

This yields a 3-term complex in which the left and right term are k and the middle term is k^2, hence an acyclic complex. This concludes the proof. $\quad\square$

Proof of Theorem 12.4.1 If $\mathcal{O}Gb$ is either nilpotent or Morita equivalent to $B = \mathcal{O}A_4$, then $\mathcal{O}Gb$ and $\mathcal{O}N_G(P)c$ are Morita equivalent via a bimodule with endopermutation source $\Omega_P^n(\mathcal{O})$ for some integer n. This endopermutation $\mathcal{O}P$-module has an endosplit p-permutation resolution, hence gives rise to a splendid Rickard complex by 9.11.5. If $\mathcal{O}Gb$ is Morita equivalent to $A = \mathcal{O}A_5b_0$, then a source algebra of $\mathcal{O}Gb$ is Morita equivalent to A via a bimodule with source $\Omega^n(\mathcal{O})$. Again by 9.11.5, there is a splendid Rickard complex of $\mathcal{O}Gb$-A-bimodules. Tensoring this with a splendid Rickard complex of A-B-bimodules obtained from 12.4.2 yields a splendid Rickard complex of $\mathcal{O}Gb$-B-bimodules. Since B is a source algebra of $\mathcal{O}N_G(P)c$, combining this with the standard Morita equivalence yields a Rickard complex of $\mathcal{O}Gb$-$\mathcal{O}N_G(P)c$-bimodules. $\quad\square$

Remark 12.4.3 The proof of Theorem 12.4.1 shows – modulo using the classification of finite simple groups via Theorem 12.1.1 – that there is a splendid Rickard complex of $\mathcal{O}Gb$-$\mathcal{O}N_G(P)c$-bimodules of length at most 2. By making use of Theorem 12.1.2 instead – hence avoiding the classification of finite simple groups – the above proof of Theorem 12.4.1 yields no such bound, as there is no bound on the length of the endosplit p-permutation resolutions of the Heller translates $\Omega_P^n(\mathcal{O})$.

Remark 12.4.4 The source algebras of the remaining 2-blocks of tame representation type – that is, blocks with generalised quaternion, dihedral or semidihedral defect groups – remain largely out of reach, despite the fact that the k-algebra structures of these blocks are very well understood, thanks to Erdmann's seminal work on tame blocks (see [67] and the references therein). Erdmann's work relies significantly on further methods from the representation theory of finite-dimensional algebras which have not been introduced in this book, most notably the theory of almost split sequences, due to Auslander and Reiten.

Appendix

A.1 The tensor product

The tensor product of two finite-dimensional vector spaces U, V over a field k is a *pair* consisting of a k-vector space W and an embedding $U \times V \to W$ such that any bilinear map from $U \times V$ to some other k-vector space X extends uniquely to a k-linear map from W to X. Such a space W always exists: if $m = \dim_k(U)$ and $n = \dim_k(V)$, then we can take for W any k-vector space with dimension mn. An embedding $U \times V \to W$ with the above property can be specified as follows: choose a k-basis B of U, a k-basis C of V, and then identify $B \times C$ to a k-basis, say D, of W via some bijection $\beta : B \times C \cong D$. Then any bilinear map $\lambda : U \times V \to X$ extends to the unique linear map $\mu : W \to X$ defined by $\mu(\beta(b, c)) = \lambda(b, c)$ for all $(b, c) \in B \times C$. This characterises the tensor product as a solution of a universal problem. The ideas behind this construction extend to much more general situations where k is replaced by any algebra A over some commutative ring k, where U is a right A-module, V a left A-module and the resulting tensor product W, denoted by $U \otimes_A V$, is a k-module. Moreover, this construction is functorial in both U and V. Let k be a commutative ring.

Definition A.1.1 Let A be a k-algebra, U a right A-module, V a left A-module and W a k-module. An *A-balanced map from $U \times V$ to W* is a k-bilinear map $\beta : U \times V \to W$ satisfying $\beta(ua, v) = \beta(u, av)$ for all $u \in U$, $v \in V$ and $a \in A$.

If $A = k$, then an A-balanced map is a k-bilinear map.

Definition A.1.2 Let A be a k-algebra, U a right A-module and V a left A-module. A *tensor product of U and V over A* is a pair (W, β) consisting of a k-module W and an A-balanced map $\beta : U \times V \to W$ such that for any further pair (W', β') consisting of an k-module W' and an A-balanced map $\beta' : U \times V \to W'$ there is a unique k-linear map $\gamma : W \to W'$ such that $\beta' = \gamma \circ \beta$.

The next result establishes the existence of tensor products. We may then speak of "the" tensor product, because any solution of a universal problem, if it exists, is unique up to unique isomorphism.

Theorem A.1.3 *Let A be a k-algebra, U a right A-module and V a left A-module. There exists a tensor product (W, β) of U and V over A, and if (W', β') is another tensor product of U and V over A, there is a unique k-linear isomorphism $\gamma : W \to W'$ such that $\beta' = \gamma \circ \beta$.*

Proof Let M be the free k-module having as basis a set of symbols $u \otimes v$ indexed by the elements $(u, v) \in U \times V$. Let I be the k-submodule of M generated by the set of all linear combinations of these symbols of the form $ua \otimes v - u \otimes av$, $(u + u') \otimes v - u \otimes v - u' \otimes v$, $u \otimes (v + v') - u \otimes v - u \otimes v'$, $r(u \otimes v) - (ru) \otimes v$, $r(u \otimes v) - u \otimes (rv)$, where $u, u' \in U$, $v, v' \in V$, $a \in A$ and $r \in k$. Set $W = M/I$ and define $\beta : U \times V \to W$ to be the unique map sending $(u, v) \in U \times V$ to the image of the symbol $u \otimes v$ in W. It follows from the definition of the submodule I of M that β is an A-balanced map. Given any further k-module W' together with an A-balanced map $\beta' : U \times V \to W'$ there is a unique map $M \to W'$ mapping the symbol $u \otimes v$ to $\beta'(u, v)$. Since β' is k-balanced, this map has I in its kernel and induces hence a unique map $\gamma : W \to W'$ mapping the image of $u \otimes v$ in W to $\beta'(u, v)$. Thus γ is the unique k-linear map from W to W' satisfying $\beta' = \gamma \circ \beta$. This proves the existence of a tensor product (W, β) of U and V over A. The uniqueness is a formal routine exercise: if (W', β') is another tensor product, there are unique k-linear maps $\gamma : W \to W'$ and $\delta : W' \to W$ such that $\beta' = \gamma \circ \beta$ and $\beta = \delta \circ \beta'$. Thus $\beta = \delta \circ \gamma \circ \beta$. But also $\beta = \text{Id}_W \circ \beta$. Thus $\delta \circ \gamma = \text{Id}_W$ by the universal property of (W, β). Similarly, $\gamma \circ \delta = \text{Id}_{W'}$. Thus γ and δ are the uniquely determined isomorphisms between (W, β) and (W', β'). $\square$

With the notation of the previous theorem, we denote a tensor product (W, β) of U and V over A by $U \otimes_A V = W$, and $u \otimes v = \beta(u, v)$, for all $(u, v) \in U \times V$. The property that β is k-bilinear takes the following form: for $u, u' \in U$, $v, v' \in V$ and $r \in k$ we have $(u + u') \otimes v = (u \otimes v) + (u' \otimes v)$, we have $u \otimes (v + v') = (u \otimes v) + (u \otimes v')$, and we have $r(u \otimes v) = (ru) \otimes v = u \otimes (rv)$. The property that β is A-balanced reads then $ua \otimes v = u \otimes av$, for all $a \in A$. An element in $U \otimes_A V$ of the form $u \otimes v$ is called an *elementary tensor*. Not all elements in $U \otimes_A V$ are elementary tensors, but they are finite k-linear combinations of elementary tensors.

Proposition A.1.4 *Let A be a k-algebra, U a right A-module and V a left A-module. The set of elementary tensors $\{u \otimes v | (u, v) \in U \times V\}$ generates $U \otimes_A V$ as a k-module.*

Proof By the construction of $U \otimes_A V$ in the proof of A.1.3 the set of images $u \otimes v$ in $U \otimes_A V$ generates $U \otimes_A V$ as a k-module. One can see this also using the universal property: if we take for W the submodule of $U \otimes_A V$ generated by the set of elementary tensors then W, together with the map $U \times V \to W$ sending (u, v) to $u \otimes v$ is easily seen to be a tensor product of U and V over A. Thus the inclusion $W \subseteq U \otimes_A V$ must be an isomorphism. $\qquad \square$

When using the notation $u \otimes v$ for elementary tensors it is important to keep track of the algebra A over which the tensor product is taken – confusion could arise if B is a subalgebra of A, in which case the tensor products $U \otimes_A V$ and $U \otimes_B V$ are both defined, while the elementary tensors would be denoted by the same symbol $u \otimes v$. In those cases it is important to specify the meaning of $u \otimes v$, which could be done, for instance by naming the structural map $\beta : U \times V \to U \otimes_A V$ explicitly, as in the definition of the tensor product. The following string of results describes the basic formal properties of tensor products: compatibility with bimodule structures, with algebra structures, functoriality, associativity and additivity.

Proposition A.1.5 *Let A, B, C be k-algebras, let U be an A-B-bimodule and V a B-C-bimodule. Then the tensor product $U \otimes_B V$ has a unique structure of A-C-bimodule satisfying $a \cdot (u \otimes v) \cdot c = (au) \otimes (vc)$ for any $a \in A$, $c \in C$, $u \in U$ and $v \in V$.*

Proof Let $a \in A$. The map $U \times V \to U \otimes_B V$ sending $(u, v) \in U \times V$ to $au \otimes v$ is clearly B-balanced. Thus there is a unique k-linear map $\varphi_a : U \otimes_B V \to U \otimes_B V$ such that $\varphi(u \otimes v) = au \otimes v$. If a' is another element in A we have $\varphi_{a'} \circ \varphi_a = \varphi_{a'a}$ because this is true on the tensors $u \otimes v$. Thus $a \cdot (u \otimes v) = (au) \otimes v$ defines a unique left A-module structure on $U \otimes_B V$. In a similar way one sees that $(u \otimes v) \cdot c = u \otimes (vc)$ defines a unique right C-module structure on $U \otimes_B V$. For any $r \in k$ and $(u, v) \in U \times V$ we have $(ru) \otimes v = u \cdot (r1_B) \otimes v = u \otimes (r1_B) \cdot v = u \otimes (rv)$, and hence the left and right k-module structure of $U \otimes_B V$ coincide. $\qquad \square$

Proposition A.1.6 *Let A, B be k-algebras, let U be an A-module and V a B-module. There is a unique k-algebra structure on $A \otimes_k B$ satisfying*

$(a \otimes b)(a' \otimes b') = aa' \otimes bb'$ *for all* a, $a' \in A$, b, $b' \in B$, *and there is a unique* $A \otimes_k B$-*module structure on* $U \otimes_k V$ *satisfying* $(a \otimes b) \cdot (u \otimes v) = au \otimes bv$ *for all* $a \in A$, $b \in B$, $u \in U$ *and* $v \in V$.

Proof Let $a \in A$ and $b \in B$. The map sending $(u, v) \in U \times V$ to $au \otimes bv \in U \otimes_k V$ is k-balanced, hence extends uniquely to an k-endomorphism $\lambda_{a,b}$ of $U \otimes_k V$ mapping $u \otimes v$ to $au \otimes bv$. Then the map sending $(a, b) \in A \times B$ to $\lambda_{a,b} \in \text{End}_k(U \otimes_k V)$ is k-balanced, hence extends uniquely to a map

$$\lambda : A \otimes_k B \longrightarrow \text{End}_k(U \otimes_k V)$$

sending $a \otimes b$ to $\lambda_{a,b}$. Consider first the case where $U = A$ and $V = B$. We use in this case λ to define a multiplication μ on $A \otimes B$ as follows: for $x, y \in A \otimes_k B$, we set $\mu(x, y) = \lambda(x)(y)$. By construction, if $x = a \otimes b$ and $y = a' \otimes b'$, then $\mu(x, y) = aa' \otimes bb'$. This shows that μ defines a distributive and associative multiplication: as usual, it suffices to check this on tensors and there it is clear. Having shown that $A \otimes_k B$ has the algebra structure as claimed, we observe that for general U, V, the map λ is an k-algebra homomorphism; as before, one sees this by checking on tensors. Thus $U \otimes_k V$ gets in this way an $A \otimes_k B$-module structure, and this is exactly the structure as stated. $\quad\square$

Proposition A.1.7 *Let* A, B, C *be* k-*algebras, let* U, U' *be* A-B-*bimodules, and let* V, V' *be* B-C-*bimodules. For any* A-B-*bimodule homomorphism* $\varphi : U \to U'$ *and any* B-C-*bimodule homomorphism* $\psi : V \to V'$ *there is a unique* A-C-*bimodule homomorphism* $\varphi \otimes \psi : U \otimes_B V \to U' \otimes_B V'$ *mapping* $u \otimes v$ *to* $\varphi(u) \otimes \psi(v)$ *for all* $u \in U$ *and* $v \in V$.

Proof The map sending $(u, v) \in U \times V$ to $\varphi(u) \otimes \psi(v)$ is B-balanced and extends hence uniquely to a map $\varphi \otimes \psi$ as stated. $\quad\square$

Proposition A.1.8 *Let* A, B, C, D *be* k-*algebras, let* U *be an* A-B-*bimodule, let* V *be a* B-C-*bimodule and let* W *be a* C-D-*bimodule. There is a unique isomorphism of* A-D-*bimodules*

$$U \otimes_B (V \otimes_C W) \cong (U \otimes_B V) \otimes_C W$$

mapping $u \otimes (v \otimes w)$ *to* $(u \otimes v) \otimes w$ *for all* $u \in U$, $v \in V$ *and* $w \in W$.

Proof For any $u \in U$ the map $V \times W \to (U \otimes_B V) \otimes_C W$ sending (v, w) to $(u \otimes v) \otimes w$ is C-balanced, hence extends to a unique map $V \otimes_C W \to (U \otimes_B V) \otimes_C W$ sending $v \otimes w$ to $(u \otimes v) \otimes w$. This works for all $u \in U$, and hence we get a map $U \times (V \otimes_C W) \to (U \otimes_B V) \otimes_C W$ mapping $(u, v \otimes w)$ to $(u \otimes v) \otimes w$. This map now is B-balanced, and hence extends uniquely to a map $\Phi : U \otimes_B (V \otimes_C W) \to (U \otimes_B V) \otimes_C W$ sending $u \otimes (v \otimes w)$ to $(u \otimes v) \otimes w$. In a completely analogous way one shows that there is a unique map $\Psi : (U \otimes_B V) \otimes_C W \to U \otimes_B (V \otimes_C W)$ sending $(u \otimes v) \otimes w$ to $u \otimes (v \otimes w)$. Then Φ and Ψ are inverse to each other because they are so on tensors. Finally, both Φ, Ψ are A-D-bimodule homomorphisms because they are compatible with the A-D-bimodule structure on tensors. $\qquad\square$

Proposition A.1.9 *Let A, B, C be k-algebras, let $\{U_i\}_{i \in I}$ be a family of A-B-bimodules indexed by some set I, and let V be a B-C-bimodule. We have a canonical isomorphism of A-C-bimodules*

$$(\oplus_{i \in I} U_i) \otimes_B V \cong \oplus_{i \in I}(U_i \otimes_B V).$$

Proof The proof consists of playing off the universal properties of direct sums and of the tensor product. In order to keep notation minimal, we write here $\oplus$ for the direct sum indexed by the set I. The direct sum $\oplus U_i$ is, by definition, an A-B-bimodule coming along with canonical homomorphisms $\iota_i : U_i \to \oplus U_i$ with the universal property that for any further A-B-bimodule M endowed with homomorphisms $\iota'_i : U_i \to M$ there is a unique homomorphism of A-B-bimodules $\alpha : \oplus U_i \to M$ such that $\iota'_i = \alpha \circ \iota_i$ for all $i \in I$. Similarly, the right side in the isomorphism of the statement is a direct sum, hence comes with canonical A-C-bimodule homomorphisms $\sigma_i : U_i \otimes_B V \to \oplus(U_i \otimes_B V)$ fulfilling the analogous universal property. In order to show that the left side in the statement is isomorphic to the right side, we construct maps $\tau_i : U_i \otimes_B V \to (\oplus U_i) \otimes_B V$ and show that they fulfill the same universal property. For any $i \in I$ we have a map $U_i \times V \to (\oplus U_i) \otimes_B V$ mapping (u_i, v) to $\iota_i(u) \otimes v$, where $u_i \in U_i$ and $v \in V$. This map is B-balanced, hence extends uniquely to a map $\tau_i : U_i \otimes_B V \to (\oplus U_i) \otimes_B V$ sending $u_i \otimes v$ to $\iota_i(u_i) \otimes v$. Let now N be any further A-C-bimodule endowed with A-C-bimodule homomorphisms $\tau'_i : U_i \otimes_B V \to N$ for all $i \in I$. Given $v \in V$ we have a map $U_i \to N$ sending u_i to $\tau'_i(u_i \otimes v)$, thus a unique map $\oplus U_i \to M$ sending $\iota_i(u_i)$ to $\tau'_i(u_i \otimes v)$. Since this holds for all $v \in V$, we get a map $\oplus U_i \times V \to N$ sending $(\iota_i(u_i), v)$ to $\tau'_i(u_i \otimes v)$. This map is B balanced and induces hence a unique map $\beta : (\oplus U_i) \otimes_B V \to N$ sending $\iota_i(u_i) \otimes v$ to $\tau'_i(u_i \otimes v)$. Thus the map β is the unique map satisfying $\beta \circ \tau_i = \tau'_i$ for all $i \in I$. This shows that the left side in the statement, endowed with the family of maps τ_i, is a direct sum of the module $U_i \otimes_B V$, hence canonically isomorphic to the right side. $\qquad\square$

Of course, the obvious analogue of the above result holds, too: if U is an A-B-bimodule and $\{V_i\}_{i \in I}$ a family of B-C-bimodules, then we have a canonical isomorphism of A-C-bimodules $U \otimes_B (\oplus_{i \in I} V_i) \cong \oplus_{i \in I}(U \otimes_B V_i)$; this is proved just in the same way. Combining the above statements shows that taking tensor products is a covariantly functorial construction, and just as for the functors using homomorphism spaces briefly discussed at the end of the last section, this functor has certain exactness properties – it is *right exact*:

Proposition A.1.10 *Let A, B be k-algebras and M an A-B-bimodule. There is a unique k-linear covariant functor $M \otimes_B -: \mathrm{Mod}(B) \to \mathrm{Mod}(A)$ sending any B-module V to the A-module $M \otimes_B V$ and sending any homomorphism of B-modules $\varphi : V \to V'$ to the homomorphism of A-modules $\mathrm{Id}_M \otimes \varphi : M \otimes_B V \to M \otimes_B V'$. Moreover, for any exact sequence of B-bimodules of the form $W \to V \to U \to 0$, the induced sequence of A-modules $M \otimes_B W \to M \otimes_B V \to M \otimes_B U \to 0$ is exact.*

Proof The fact that $M \otimes_B -$ is a covariant functor follows immediately from the preceding statements. For the exactness property observe first that the map $M \otimes_B V \to M \otimes_B U$ is surjective because its image contains all elementary tensors $m \otimes u$ thanks to the fact that the map $V \to U$ is surjective. We need to show the exactness at $M \otimes_B V$. Let $I \subseteq M \otimes_B V$ be the image of the map $M \otimes_B W \to M \otimes_B V$. This is contained in the kernel of the map $M \otimes_B V \to M \otimes_B U$, and hence induces a surjective map $\varphi : (M \otimes_B V)/I \to M \otimes_B U$. We need to show that φ is injective. For this it suffices to construct a map $\psi : M \otimes_B U \to (M \otimes_B V)/I$ such that $\psi \circ \varphi$ is the identity on $(M \otimes_B V)/I$. Let $m \in M$ and $u \in U$. Choose $v \in V$ in the preimage of u. Define a map $M \times U \to (M \otimes_B V)/I$ by sending (m, u) to the image $(m \otimes v) + I$. One checks that this does not depend on the choice of v, and that the resulting map is B-balanced, hence induces a map $\psi : M \otimes_B U \to (M \otimes_B V)/I$. By construction the composition $\psi \circ \varphi$ is the identity map. Alternatively, as in the case of the proof of the exactness properties of homomorphism space functors in A.1.13, the exactness statement follows from general abstract nonsense on adjoint functors 2.3.6 and 2.2.4. $\square$

The functor $M \otimes_B -$ need not be *exact*; that is, it need not preserve injective homomorphisms – see the example A.1.11 (e) below. A right B-module M is called *flat* if the functor $M \otimes_B -$ from $\mathrm{Mod}(B)$ to $\mathrm{Mod}(k)$ is exact. Again, there is an obvious analogue for right modules: there is a unique k-linear covariant functor $- \otimes_A M : \mathrm{Mod}(A^{\mathrm{op}}) \to \mathrm{Mod}(B^{\mathrm{op}})$ sending a right A-module U to the right B-module $U \otimes_A M$ and sending a homomorphism of right A-modules $\varphi : U \to U'$ to the homomorphism of right B-modules $\varphi \otimes \mathrm{Id}_M : U \otimes_A M \to U' \otimes_A M$.

Examples A.1.11

(a) Let A be a k-algebra. Then A can be considered as an A-A-bimodule via multiplication in A. For any left A-module U we have a canonical isomorphism of left A-modules $A \otimes_A U \cong U$ mapping $a \otimes u$ to au, where $a \in A$ and $u \in U$. The existence of such a map follows from the fact that the map $A \times U \to U$ sending (a, u) to au is trivially A-balanced. The inverse of this map sends $u \in U$ to $1_A \otimes u$. In conjunction with the above corollary this shows that the functor $A \otimes_A -$ on $\mathrm{Mod}(A)$ is isomorphic to the identity functor on $\mathrm{Mod}(A)$. Similarly, for any right A-module V we have a canonical isomorphism of right A-modules $V \otimes_A A \cong V$ sending $v \otimes a$ to va, where $a \in A$ and $v \in V$.

(b) The remarks at the beginning of this section show that if k is a field and U, V are finite-dimensional k-vector spaces, then $U \otimes_k V$ is a k-vector space of finite dimension $\dim_k(U) \cdot \dim_k(V)$.

(c) There is a canonical isomorphism of $\mathbb{Q}$-vector spaces $\mathbb{Q} \otimes_\mathbb{Z} \mathbb{Z} \cong \mathbb{Q}$ mapping $q \otimes n$ to qn; the inverse maps q to $q \otimes 1$. For any positive integer n we have $\mathbb{Q} \otimes_\mathbb{Z} \mathbb{Z}/n\mathbb{Z} = \{0\}$ because if $q \in \mathbb{Q}$ and $c + n\mathbb{Z} \in \mathbb{Z}/n\mathbb{Z}$ then $q \otimes (c + n\mathbb{Z}) = \frac{q}{n} \otimes (nc + n\mathbb{Z}) = 0$ since $nc + n\mathbb{Z} = 0_{\mathbb{Z}/n\mathbb{Z}}$. In other words, tensoring a finitely generated abelian group A with $\mathbb{Q}$ yields a vector space over $\mathbb{Q}$ whose dimension is the rank of the free part of A and which annihilates all torsion in A. This reasoning extends to the more general situation of an integral domain $\mathcal{O}$ with quotient field K: tensoring any torsion $\mathcal{O}$-module M by K yields zero, while tensoring a free $\mathcal{O}$-module of finite rank n yields a K-vector space of dimension n.

(d) If n, m are coprime positive integers then $(\mathbb{Z}/m\mathbb{Z}) \otimes_\mathbb{Z} (\mathbb{Z}/n\mathbb{Z}) = \{0\}$. Indeed, there are integers a, b such that $am + bn = 1$.

Thus for $c, d \in \mathbb{Z}$ we have $(c + m\mathbb{Z}) \otimes_\mathbb{Z} (d + n\mathbb{Z}) = ((cam + cbn) + m\mathbb{Z}) \otimes (d + n\mathbb{Z}) = (cam + m\mathbb{Z}) \otimes (d + n\mathbb{Z}) + (c + m\mathbb{Z}) \otimes (bnd + n\mathbb{Z})$, which is zero because $cam + m\mathbb{Z}$ and $bnd + n\mathbb{Z}$ are zero in $\mathbb{Z}/m\mathbb{Z}$ and $\mathbb{Z}/n\mathbb{Z}$, respectively.

(e) Let A be a k-algebra, U a right A-module and $\varphi : V \to V'$ a homomorphism of left A-modules. If φ is surjective then the induced map $\mathrm{Id}_U \otimes \varphi$ from $U \otimes_A V$ to $U \otimes_A V'$ is surjective as well by A.1.10. It is not true, in general, that if φ is injective then $\mathrm{Id}_U \otimes \varphi$ is injective. Here is a general source of examples for this phenomenon: let I be a nonzero ideal in A whose square I^2 is zero. Denote by $\varphi : I \to A$ the inclusion map; this is in particular a homomorphism of left A-modules. Since I is an ideal, we may consider I also as right A-modules. Tensoring by $I \otimes_A -$ yields a map $\mathrm{Id}_I \otimes \varphi : I \otimes_A$

$I \to I \otimes_A A$. This map is zero: if $a, b \in I$, then the image of $a \otimes b$ in $I \otimes A$ can be written in the form $a \otimes b = a \otimes b \cdot 1_A = ab \otimes 1$, and this is zero as $ab \in I^2 = \{0\}$ by the assumptions. Simple examples of this format arise for $A = \mathbb{Z}/4\mathbb{Z}$ and $I = 2\mathbb{Z}/4\mathbb{Z}$, with $k = \mathbb{Z}$. One checks that $I \otimes_A I \cong I \otimes_\mathbb{Z} I$ is nonzero, with exactly two elements.

The tensor product is closely related to functors obtained from taking homomorphism spaces. Let A, B be k-algebras, M an A-B-bimodule and U an A-module. Then $\operatorname{Hom}_A(M, U)$ becomes a B-module via $(b \cdot \mu)(m) = \mu(mb)$, where $m \in M$, $b \in B$ and $\mu \in \operatorname{Hom}_A(M, U)$. This construction is covariant functorial: if $\alpha : U \to V$ is a homomorphism of A-modules, then the induced map $\operatorname{Hom}_A(M, U) \to \operatorname{Hom}_A(M, V)$ sending $\mu \in \operatorname{Hom}_A(M, U)$ to $\alpha \circ \mu$ is easily seen to be a B-homomorphism. We denote by $\operatorname{Hom}_A(M, -)$ the functor from $\operatorname{Mod}(A)$ to $\operatorname{Mod}(B)$ obtained in this way. Similarly, $\operatorname{Hom}_A(U, M)$ becomes a right B-module via $(v \cdot b)(u) = v(u)b$, where $u \in U$, $b \in B$ and $v \in \operatorname{Hom}_A(U, M)$. This construction is now contravariant functorial: if $\alpha : U \to V$ is a homomorphism of A-modules then the induced map $\operatorname{Hom}_A(V, M) \to \operatorname{Hom}_A(U, M)$ sending $v \in \operatorname{Hom}_A(V, M)$ to $v \circ \alpha$ is a homomorphism of right B-modules. We denote by $\operatorname{Hom}_A(-, M)$ the contravariant functor from $\operatorname{Mod}(A)$ to $\operatorname{Mod}(B^{\text{op}})$ obtained in this way. The covariant functor $\operatorname{Hom}_A(M, -)$ commutes with direct products, whereas the contravariant functor $\operatorname{Hom}_A(-, M)$ sends direct sums to direct products. Finite direct products coincide with direct sums, so both functors $\operatorname{Hom}_A(M, -)$ and $\operatorname{Hom}_A(-, M)$ commute with finite direct sums.

Proposition A.1.12 *Let A, B be k-algebras, M an A-B-bimodule, and let $\{U_i\}_{i \in I}$ be a family of A-modules. We have a canonical isomorphisms*

$$\operatorname{Hom}_A\left(M, \prod_{i \in I} U_i\right) \cong \prod_{i \in I} \operatorname{Hom}_A(M, U_i),$$

$$\operatorname{Hom}_A(\oplus_{i \in I} U_i, M) \cong \prod_{i \in I} \operatorname{Hom}_A(U_i, M).$$

Proof The first isomorphism is induced by the family of canonical projections $\prod_{i \in I} U_i \to U_j$, where $j \in I$. The second isomorphism is induced by the family of canonical injections $U_j \to \oplus_{i \in I} U_i$, where $j \in I$. $\qquad\square$

These functors $\operatorname{Hom}_A(M, -)$ and $\operatorname{Hom}_A(-, M)$ have furthermore the following exactness properties.

Proposition A.1.13 *Let A, B be k-algebras and M be an A-B-bimodule.*

(i) *If $0 \to U \to V \to W$ is an exact sequence of A-modules, then the induced sequence of B-modules $0 \to \operatorname{Hom}_A(M, U) \to \operatorname{Hom}_A(M, V) \to \operatorname{Hom}_A(M, W)$ is exact.*

(ii) *If $W \to V \to U \to 0$ is an exact sequence of A-modules, then the induced sequence of right B-modules $0 \to \operatorname{Hom}_A(U, M) \to \operatorname{Hom}_A(V, M) \to \operatorname{Hom}_A(W, M)$ is exact.*

Proof Straightforward verification. Alternatively, this follows from general abstract nonsense on adjoint functors 2.3.6 and 2.2.4. □

In other words, the functor $\operatorname{Hom}_A(M, -)$ is *left exact*. It is not exact, in general, because it need not preserve surjective homomorphisms. This leads to the consideration of *projective modules*; see 1.12.3 and 1.12.4. Similarly, the functor $\operatorname{Hom}_A(-, M)$ need not be exact – this leads to the consideration of *injective modules*; see 1.12.16. The single most important general statement in module theory is arguably Theorem 2.2.4 – used in the proof of A.1.13 – stating that *the functor $M \otimes_B -$ is left adjoint to the functor* $\operatorname{Hom}_A(M, -)$. The left exactness of $\operatorname{Hom}_A(M, -)$ and the right exactness of $M \otimes_B -$ are formal consequences of this adjunction.

A.2 Triangulated categories

A triangulated category is an additive category with an additional structure of *exact triangles*, which should be thought of as a replacement for short exact sequences. This concept has been developed independently by J. L. Verdier, and, in a topological context, by D. Puppe. The two main examples of triangulated categories that we consider here - stable categories of selfinjective algebras and homotopy categories of chain complexes – both arise in a similar way: as additive quotient categories of abelian categories in such a way, that the short exact sequences of the abelian category induce the exact triangles in the quotient category. Given an additive category $\mathcal{C}$ and an additive functor $\Sigma : \mathcal{C} \to \mathcal{C}$ on $\mathcal{C}$, we call a *triangle in $\mathcal{C}$* a sequence of the form

$$X \xrightarrow{f} Y \xrightarrow{g} Z \xrightarrow{h} \Sigma(X)$$

where X, Y, Z are objects in C and f, g, h are morphisms in C. The triangles in C form the objects of a category: a *morphism of triangles* is a triple (u, v, w) of morphisms in C making the diagram

$$
\begin{array}{ccccccc}
X & \xrightarrow{f} & Y & \xrightarrow{g} & Z & \xrightarrow{h} & \Sigma(X) \\
\downarrow u & & \downarrow v & & \downarrow w & & \downarrow \Sigma(u) \\
X' & \xrightarrow{f'} & Y' & \xrightarrow{g'} & Z' & \xrightarrow{h'} & \Sigma(X')
\end{array}
$$

commutative, where the two rows are triangles in C.

Definition A.2.1 A *triangulated category* is a triple $(C, \Sigma, \mathcal{T})$ consisting of an additive category C, a covariant additive self equivalence $\Sigma : C \to C$, and a class $\mathcal{T}$ of triangles in C – called *exact* or sometimes also *distinguished triangles* in C – fulfilling the axioms T1, T2, T3, T4 below.

T1: For any object X in C, the triangle $0 \longrightarrow X \xrightarrow{\mathrm{Id}_X} X \longrightarrow 0$ is exact (i.e., belongs to the class $\mathcal{T}$), for any morphism $f : X \to Y$ in C there is an exact triangle of the form

$$
X \xrightarrow{f} Y \xrightarrow{g} Z \xrightarrow{h} \Sigma(X)
$$

for some object Z in C and some morphisms g, h, any triangle in C that is isomorphic to an exact triangle is itself exact (i.e., the class $\mathcal{T}$ is closed under isomorphisms).

T2: Any commutative diagram in C of the form

$$
\begin{array}{ccccccc}
X & \xrightarrow{f} & Y & \xrightarrow{g} & Z & \xrightarrow{h} & \Sigma(X) \\
\downarrow u & & \downarrow v & & & & \\
X' & \xrightarrow{f'} & Y' & \xrightarrow{g'} & Z' & \xrightarrow{h'} & \Sigma(X')
\end{array}
$$

whose rows are exact triangles, can be completed to a commutative diagram

$$
\begin{array}{ccccccc}
X & \xrightarrow{\,f\,} & Y & \xrightarrow{\,g\,} & Z & \xrightarrow{\,h\,} & \Sigma(X) \\
\downarrow{\scriptstyle u} & & \downarrow{\scriptstyle v} & & \downarrow{\scriptstyle w} & & \downarrow{\scriptstyle \Sigma(u)} \\
X' & \xrightarrow{\,f'\,} & Y' & \xrightarrow{\,g'\,} & Z' & \xrightarrow{\,h'\,} & \Sigma(X')
\end{array}
$$

for some morphism w.

T3: If the triangle $X \xrightarrow{\,f\,} Y \xrightarrow{\,g\,} Z \xrightarrow{\,h\,} \Sigma(X)$ in $\mathcal{C}$ is exact, so is the triangle

$$
Y \xrightarrow{\,g\,} Z \xrightarrow{\,h\,} \Sigma(X) \xrightarrow{\,-\Sigma f\,} \Sigma(Y).
$$

T4: Given any sequence of two composable morphisms $X \xrightarrow{\,f\,}$ $Y \xrightarrow{\,g\,} Z$ in $\mathcal{C}$ there is a commutative diagram in $\mathcal{C}$ whose first two rows and middle two columns are exact triangles:

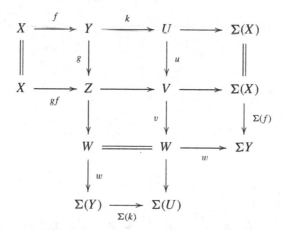

Remark A.2.2 The axiom T4 describes in which way the three triangles over f, g, $g \circ f$ are connected. This axiom is called the *octahedral axiom* for the following reason: if we rewrite a triangle $X \xrightarrow{\,f\,} Y \xrightarrow{\,g\,} Z \xrightarrow{\,h\,} \Sigma(X)$

in the form

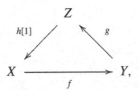

where [1] means that h "is of degree 1", then the diagram in T4 takes the following form:

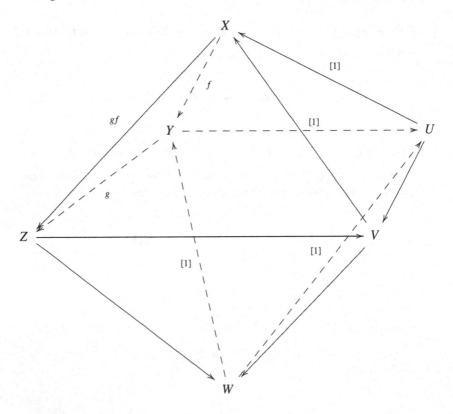

Proposition A.2.3 *Let* $X \xrightarrow{f} Y \xrightarrow{g} Z \xrightarrow{h} \Sigma(X)$ *be an exact triangle in a triangulated category* $\mathcal{C}$, *and let* U *be an object in* $\mathcal{C}$.

(i) *We have* $g \circ f = 0$ *and* $h \circ g = 0$.

(ii) *Given any morphism* $j : Y \to U$ *there is a morphism* $i : Z \to U$ *satisfying* $i \circ g = j$ *if and only if* $j \circ f = 0$.

(iii) *Given any morphism* $q : U \longrightarrow Y$ *there is a morphism* $p : U \to X$ *satisfying* $f \circ p = q$ *if and only if* $g \circ q = 0$.

Proof By T1 the triangle $0 \longrightarrow X \Longrightarrow X \longrightarrow 0$ is exact, and hence, by T3, the triangle $X \Longrightarrow X \longrightarrow 0 \longrightarrow \Sigma(X)$ is exact. Applying T2 yields the existence of a commutative diagram

$$
\begin{array}{ccccccc}
X & = & X & \longrightarrow & 0 & \longrightarrow & \Sigma(X) \\
\| & & \downarrow{\scriptstyle f} & & \downarrow & & \| \\
X & \xrightarrow{f} & Y & \xrightarrow{g} & Z & \longrightarrow & \Sigma(X)
\end{array}
$$

which shows that $g \circ f = 0$. The same argument, after turning the triangle by means of T3, shows that $h \circ g = 0$, whence (i). If $j \circ f = 0$, then it follows from T1 and T2 that we have a commutative diagram

$$
\begin{array}{ccccccc}
X & \xrightarrow{f} & Y & \xrightarrow{g} & Z & \xrightarrow{h} & \Sigma(X) \\
\downarrow & & \downarrow{\scriptstyle j} & & \downarrow{\scriptstyle i} & & \downarrow \\
0 & \longrightarrow & U & = & U & \longrightarrow & 0
\end{array}
$$

for some morphism i, which means that $i \circ g = j$. Conversely, if there is a morphism $i : Z \to U$ such that $i \circ g = j$, then $j \circ f = i \circ g \circ f = 0$, since $g \circ f = 0$ by (i). This proves (ii). The last statement is proved by applying a dual argument to the exact triangle $Y \xrightarrow{g} Z \xrightarrow{h} \Sigma(X) \xrightarrow{-\Sigma(f)} \Sigma(Y)$. □

Lemma A.2.4 *Let* $(\mathcal{C}, \Sigma, \mathcal{T})$ *be a triangulated category, and let*

$$
\begin{array}{ccccccc}
& & Y & \xrightarrow{g} & Z & & \\
& & \downarrow{\scriptstyle v} & & \downarrow{\scriptstyle 0} & & \\
X' & \xrightarrow{f'} & Y' & \xrightarrow{g'} & Z' & \xrightarrow{h'} & \Sigma(X') \\
\downarrow{\scriptstyle 0} & & \downarrow{\scriptstyle v'} & & & & \\
X'' & \xrightarrow{f''} & Y'' & & & &
\end{array}
$$

be a commutative diagram in $\mathcal{C}$ *whose middle row is an exact triangle. We have* $v' \circ v = 0$.

Proof Since $g' \circ v = 0$ there is, by A.2.3, a morphism $p : Y \to X'$ such that $f' \circ p = v$. Similarly, since $v' \circ f' = 0$ there is a morphism $i : Z' \to Y''$ such that $v' = i \circ g'$. Together we obtain $v' \circ v = i \circ g' \circ f \circ p = 0$, since $g' \circ f' = 0$ by A.2.3. $\qquad\square$

Proposition A.2.5 *Let* $(\mathcal{C}, \Sigma, \mathcal{T})$ *be a triangulated category and let*

$$
\begin{array}{ccccccc}
X & \xrightarrow{\ f\ } & Y & \xrightarrow{\ g\ } & Z & \xrightarrow{\ h\ } & \Sigma(X) \\
\downarrow{\scriptstyle u} & & \downarrow{\scriptstyle v} & & \downarrow{\scriptstyle w} & & \downarrow{\scriptstyle \Sigma(u)} \\
X' & \xrightarrow{\ f'\ } & Y' & \xrightarrow{\ g'\ } & Z' & \xrightarrow{\ h'\ } & \Sigma(X')
\end{array}
$$

be a commutative diagram in $\mathcal{C}$ *whose rows are exact triangles. If* u, v *are isomorphisms, so is* w.

Proof By applying T2 to u^{-1} and v^{-1} we may assume that $X = X'$, $Y = Y'$, $Z = Z'$, $f = f'$, $g = g'$, $h = h'$, $u = \mathrm{Id}_X$ and $v = \mathrm{Id}_Y$. Thus we are down to considering the endomorphism $(\mathrm{Id}_X, \mathrm{Id}_Y, w)$ of the exact triangle $X \xrightarrow{\ f\ } Y \xrightarrow{\ g\ } Z \xrightarrow{\ h\ } \Sigma(X)$, and we have to show that w is an automorphism. Clearly $(\mathrm{Id}_X, \mathrm{Id}_Y, \mathrm{Id}_Z)$ is an endomorphism of this triangle, too, thus taking the difference of the two endomorphisms yields an endomorphism $(0, 0, \mathrm{Id}_Z - w)$. Using T3 and A.2.4 shows that $(\mathrm{Id}_Z - w)^2 = 0$, or equivalently, $\mathrm{Id}_Z = w \circ (2\mathrm{Id}_Z - w)$, which implies that w is invertible with inverse $2\mathrm{Id}_Z - w$. $\qquad\square$

Corollary A.2.6 *Let* $(\mathcal{C}, \Sigma, \mathcal{T})$ *be a triangulated category. If the triangle*

$$
Y \xrightarrow{\ g\ } Z \xrightarrow{\ h\ } \Sigma(X) \xrightarrow{\ -\Sigma f\ } \Sigma(Y)
$$

is exact in $\mathcal{C}$, *so is the triangle*

$$
X \xrightarrow{\ f\ } Y \xrightarrow{\ g\ } Z \xrightarrow{\ h\ } \Sigma(X).
$$

Proof By T1, there is an exact triangle in $\mathcal{C}$ of the form $X \xrightarrow{\ f\ } Y \xrightarrow{\ g'\ } Z' \xrightarrow{\ h'\ } \Sigma(X)$. We turn this triangle three times, and the first triangle in the statement twice; this yields two exact triangles in $\mathcal{C}$

$$\Sigma X \xrightarrow{-\Sigma f} \Sigma Y \xrightarrow{-\Sigma g'} \Sigma Z' \xrightarrow{-\Sigma h'} \Sigma^2(X)$$

$$\Sigma(X) \xrightarrow{-\Sigma f} \Sigma(Y) \xrightarrow{-\Sigma g} \Sigma Z \xrightarrow{-\Sigma h} \Sigma^2(X)$$

and by Proposition A.2.5, these two triangles are isomorphic. Since Σ is an equivalence, it follows that the triangles

$$X \xrightarrow{f} Y \xrightarrow{g'} Z' \xrightarrow{h'} \Sigma(X)$$

$$X \xrightarrow{f} Y \xrightarrow{g} Z' \xrightarrow{h} \Sigma(X)$$

are isomorphic, and as the first one is exact, so is the second by T1. $\qquad\square$

Corollary A.2.7 *Let $(\mathcal{C}, \Sigma, \mathcal{T})$ be a triangulated category, and let*

$$X \xrightarrow{f} Y \xrightarrow{g} Z \xrightarrow{h} \Sigma(X)$$

be an exact triangle in $\mathcal{C}$.

(i) *f is an isomorphism if and only if $Z = 0$.*
(ii) *g is an isomorphism if and only if $X = 0$.*
(iii) *h is an isomorphism if and only if $Y = 0$.*

Proof If f is an isomorphism, then, by A.2.5 and T1, the given exact triangle is isomorphic to the exact triangle $X \xrightarrow{\quad\quad} X \longrightarrow 0 \longrightarrow \Sigma(X)$, thus $Z = 0$. Conversely, if $Z = 0$, turning the triangle by T3 shows that $\Sigma(f)$ is an isomorphism, and hence f is so, too, as Σ is an equivalence. This shows (i), and the other statements follow from (i) with T3 and the fact, that Σ is an equivalence. $\qquad\square$

Corollary A.2.8 *Let $(\mathcal{C}, \Sigma, \mathcal{T})$ be a triangulated category and let*

$$X \xrightarrow{f} Y \xrightarrow{g} Z \xrightarrow{h} \Sigma(X)$$

be an exact triangle in C. For any object U in C, the functors $\text{Hom}_C(U, -)$ *and* $\text{Hom}_C(-, U)$ *induce long exact sequences of abelian groups*

$$\cdots \to \text{Hom}_C(U, \Sigma^n(X)) \to \text{Hom}_C(U, \Sigma^n(Y)) \to \text{Hom}_C(U, \Sigma^n(Z))$$

$$\to \text{Hom}_C(U, \Sigma^{n+1}(X)) \to \cdots$$

$$\cdots \to \text{Hom}_C(\Sigma^{n+1}(X), U) \to \text{Hom}_C(\Sigma^n(Z), U) \to \text{Hom}_C(\Sigma^n(Y), U)$$

$$\to \text{Hom}_C(\Sigma^n(X), U) \to \cdots .$$

Proof By A.2.3, the sequence $\text{Hom}_C(U, X) \to \text{Hom}_C(U, Y) \to \text{Hom}_C(U, Z)$ is exact. Turning the triangle by means of T3 and its converse A.2.6 yields the first of the two long exact sequences. An analogous argument shows the exactness of the second sequence. $\qquad\square$

Proposition A.2.9 *Let* $(C, \Sigma, \mathcal{T})$ *be a triangulated category and let* $f : X \to Y$ *be a morphism in C.*

(i) *f is an epimorphism if and only if f has a right inverse.*
(ii) *f is a monomorphism if and only if f has a left inverse.*
(iii) *f is an isomorphism if and only if f is both an epimorphism and a monomorphism.*

Proof If f has a right inverse, f is trivially an epimorphism. Conversely, suppose that f is an epimorphism. Consider an exact triangle of the form

$$X \xrightarrow{\ f\ } Y \xrightarrow{\ g\ } Z \xrightarrow{\ h\ } \Sigma(X).$$

Since $g \circ f = 0$ and since f is an epimorphism, we have $g = 0$. Thus $g \circ \text{Id}_Y = 0$. Thus there is $r : Y \to X$ such that $f \circ r = \text{Id}_Y$, which shows that r is a right inverse of f. This shows (i), and a dual argument shows statement (ii). If f is both an epimorphism and a monomorphism, then it has a right inverse r and a left inverse l, and then $l = l \circ f \circ r = r$, which shows that f is an isomorphism. The converse in (iii) is trivial. $\qquad\square$

This shows that epimorphisms and monomorphisms in a triangulated category are all split. As a consequence, a triangulated category is abelian if and only if it is semisimple. There are, however, nontrivial examples of proper abelian subcategories of triangulated categories arising in the context of central p-extensions of finite groups.

A.3 The *k*-stable category is triangulated

Let k be a commutative ring and let A be a k-algebra such that the classes of relatively k-projective A-modules and relatively k-injective A-modules coincide. We show that then the k-stable category $\underline{\text{Mod}}(A)$ is triangulated. We begin by defining the class of exact triangles in $\underline{\text{Mod}}(A)$. We write Σ and Ω instead of Σ_A and Ω_A and denote relatively k-projective covers and relatively k-injective envelopes of A-modules as in the previous section. Morphisms in $\underline{\text{Mod}}(A)$ are denoted by underlined letters $\underline{f}$ with the implicit understanding that f is a representative of $\underline{f}$ in $\text{Mod}(A)$. We need to construct for any morphism $\underline{f} : U \to V$ a third object W and morphism $\underline{g} : V \to W$ and $\underline{h} : W \to \Sigma(U)$. We cannot simply take for W the cokernel of a representative f of $\underline{f}$, because this would not be independent of the choice of f. It turns out that if we can find a k-split injective representative f of $\underline{f}$, then the cokernel W of f is well-defined. The problem here is that the morphism $\underline{f}$ may not always have a k-split injective representative. We can force this by adding a k-injective envelope (I, ι) of U to V; for any representative f of $\underline{f}$ the A-homomorphism

$$U \xrightarrow{\ \ \binom{f}{\iota}\ \ } V \oplus I$$

is k-split injective because ι is, and in the k-stable category $\underline{\text{Mod}}(A)$ the module I is zero, so V and $V \oplus I$ are canonically isomorphic, and modulo identification via this isomorphism, f and $\binom{f}{\iota}$ represent the same morphism $\underline{f}$ in $\underline{\text{Mod}}(A)$. In this way exact triangles in $\underline{\text{Mod}}(A)$ are induced by short exact sequences in $\text{Mod}(A)$. The following definition makes this precise.

Definition A.3.1 A triangle $U \xrightarrow{\underline{f}} V \xrightarrow{\underline{g}} W \xrightarrow{\underline{h}} \Sigma(U)$ in $\underline{\text{Mod}}(A)$ is *exact* if it is isomorphic to a triangle for which there exists a commutative diagram of A-modules of the form

$$
\begin{array}{ccccccccc}
0 & \longrightarrow & U & \xrightarrow{\ f\ } & V & \xrightarrow{\ g\ } & W & \longrightarrow & 0 \\
& & \Big\| & & \Big\downarrow{\scriptstyle v} & & \Big\downarrow{\scriptstyle h} & & \\
0 & \longrightarrow & U & \xrightarrow[\iota_U]{} & I_U & \xrightarrow[\kappa_U]{} & \Sigma(U) & \longrightarrow & 0
\end{array}
$$

such that the first row is k-split exact.

The second row in this diagram is automatically k-split exact, by the definition of relatively k-injective envelopes.

Theorem A.3.2 *The k-stable category* $\underline{\mathrm{Mod}}(A)$, *together with the functor* Σ *and the class of exact triangles defined in A.3.1 is triangulated. If k is Noetherian then* $\underline{\mathrm{mod}}(A)$ *is a full triangulated subcategory of* $\underline{\mathrm{Mod}}(A)$.

With the notation of A.3.1, a trivial verification shows that the sequence

$$0 \longrightarrow V \xrightarrow{\binom{g}{v}} W \oplus I_U \xrightarrow{(h,-\kappa_U)} \Sigma(U) \longrightarrow 0$$

is then also k-split exact. This observation can be used to give an alternative characterisation of exact triangles which will be useful for the proof of Theorem A.3.2:

Proposition A.3.3 *Suppose we have two k-split exact sequences of A-modules of the form*

$$0 \longrightarrow U \xrightarrow{\binom{f}{i}} V \oplus P \xrightarrow{(g,-p)} W \longrightarrow 0$$

$$0 \longrightarrow V \xrightarrow{\binom{g}{j}} W \oplus Q \xrightarrow{(h,-q)} X \longrightarrow 0.$$

Then the following hold.

(i) The sequence

$$0 \longrightarrow U \xrightarrow{\binom{i}{jf}} P \oplus Q \xrightarrow{(hp,-q)} X \longrightarrow 0$$

is k-split exact.

(ii) If P, Q are relatively k-injective, then the sequence in (i) determines a unique isomorphism $X \cong \Sigma(U)$ in $\underline{\mathrm{Mod}}(A)$, *and the triangle*

$$U \xrightarrow{f} V \xrightarrow{g} W \xrightarrow{h} X \cong \Sigma(U) \text{ is exact.}$$

(iii) Up to isomorphism, any exact triangle in $\underline{\mathrm{Mod}}(A)$ *arises as in (ii).*

Proof The composition of $\binom{i}{jf}$ and $(hp, -q)$ is equal to $hpi - qjf$. The exactness hypotheses on the first two sequences in the statement implies that $gf = pi$ and $hg = jq$. Thus $hpi - qjf = hgf - hgf = 0$. Let $x \in P$, $y \in Q$ such that $(x, y) \in \ker(hp, -q)$. Then $hp(x) = q(y)$. Thus $(p(x), y) \in \ker(h, -q)$, which by the assumptions is equal to the image of $\binom{g}{j}$. So there is $v \in V$ such that $g(v) = p(x)$ and $j(v) = y$. This implies that $(v, x) \in \ker(g, -p)$, and so there is $u \in U$ such that $f(u) = v$ and $i(u) = x$. But then $jf(u) = y$, hence (x, y) belongs to the image of $\binom{i}{jf}$. This shows that the sequence in (i) is exact. Again, by the assumptions there are k-linear retractions $r : V \oplus P \to U$

and $s : W \oplus Q \to V$ of $\binom{f}{i}$ and $\binom{g}{j}$, respectively. One checks that the map $(r|_P, r|_V \circ S|_Q) : P \oplus Q \to U$ is a k-linear retraction of $\binom{i}{jf}$. This proves (i). Suppose now that P, Q are relatively k-injective. Then there is a unique isomorphism $X \cong \Sigma(U)$ in $\underline{\mathrm{Mod}}(A)$ obtained from a commutative diagram of A-modules with k-split exact rows

$$
\begin{array}{ccccccccc}
0 & \longrightarrow & U & \longrightarrow & P \oplus Q & \longrightarrow & X & \longrightarrow & 0 \\
& & \| & & \downarrow{\scriptstyle v} & & \downarrow & & \\
0 & \longrightarrow & U & \longrightarrow & I_U & \longrightarrow & \Sigma(U) & \longrightarrow & 0.
\end{array}
$$

Moreover, the following diagram is commutative

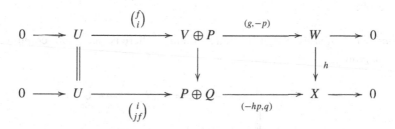

where the vertical morphism in the middle is equal to j on V and Id_P on P. Since $V \cong V \oplus P$ in $\underline{\mathrm{Mod}}(A)$ this shows that the triangle in (ii) is exact. Statement (iii) follows from (ii) applied to the case where P is zero and $Q = I_U$. □

The proof of Theorem A.3.2 requires further the following lemma about lifting commutative diagrams in $\underline{\mathrm{Mod}}(A)$ to commutative diagrams in $\mathrm{Mod}(A)$.

Lemma A.3.4 *Suppose we have a commutative diagram in $\underline{\mathrm{Mod}}(A)$ of the form:*

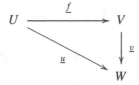

Suppose that $\underline{f}$ has a k-split injective representative f. Then, for any representative u of $\underline{u}$ there is a representative v of $\underline{v}$ making the following diagram in

Mod(A) *commutative:*

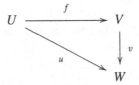

Proof Let u, v be any representatives of $\underline{u}$, $\underline{v}$, respectively. The commutativity $\underline{u} = \underline{v}\underline{f}$ means that $vf - u$ factors through a relatively k-injective A-module I; say $vf - u = ba$ for some A-homomorphisms $a : U \to I$ and $b : I \to W$. Since f is k-split injective, there is a homomorphism $c : V \to I$ such that $a = cf$. Thus $vf - u = bcf$, or equivalently, $(v - bc)f = u$. Since bc factors through I, the homomorphism $v - bc$ represents still $\underline{v}$. Replacing v by $v - bc$ proves the result. $\qquad\square$

Proof of Theorem A.3.2 Let U be an A-module. There is an obvious commutative diagram

$$0 \longrightarrow 0 \longrightarrow U = U \longrightarrow 0$$
$$0 \longrightarrow 0 \longrightarrow 0 \longrightarrow 0 \longrightarrow 0$$

which shows that

$$0 \longrightarrow U \xrightarrow{\operatorname{Id}_U} U \longrightarrow 0$$

is an exact triangle. Let $f : U \to V$ be a homomorphism of A-modules and let $i : U \to P$ be a relatively k-injective envelope of U. By the above, the A-homomorphism

$$\binom{f}{i} : U \to V \oplus P$$

is k-split injective. Taking its cokernel yields a k-split exact sequence of the form

$$0 \longrightarrow U \xrightarrow{\binom{f}{i}} V \oplus P \xrightarrow{(g,-p)} W \longrightarrow 0.$$

Applying this argument to the homomorphism g instead of f yields a k-split exact sequence of the form

$$0 \longrightarrow V \xrightarrow{\binom{g}{j}} W \oplus Q \xrightarrow{(h,-q)} X \longrightarrow 0$$

where $j : V \to Q$ is a relatively k-injective envelope of V. Thus, by A.3.3, there is an exact triangle of the form

$$U \xrightarrow{f} V \xrightarrow{g} W \xrightarrow{h} \Sigma(U)$$

which proves that the axiom T1 holds. Suppose next we have two exact triangles in $\underline{\mathrm{Mod}}(A)$ given by two pairs of k-split exact sequences

$$0 \longrightarrow U \xrightarrow{f} V \xrightarrow{g} W \longrightarrow 0$$

$$0 \longrightarrow V \xrightarrow{\binom{g}{j}} W \oplus P \xrightarrow{(h,-p)} X \longrightarrow 0$$

$$0 \longrightarrow U' \xrightarrow{f'} V' \xrightarrow{g'} W' \longrightarrow 0$$

$$0 \longrightarrow V' \xrightarrow{\binom{g'}{j'}} W' \oplus P' \xrightarrow{(h',-p')} X' \longrightarrow 0$$

where P, P' are relatively k-injective A-modules. Suppose we have two A-homomorphisms $u : U \to U'$ and $v : V \to V'$ making the diagram

$$
\begin{array}{ccc}
U & \xrightarrow{\;f\;} & V \\
\underline{u} \downarrow & & \downarrow \underline{v} \\
U' & \longrightarrow & V' \\
& \underline{f'} &
\end{array}
$$

in $\underline{\mathrm{Mod}}(A)$ commutative. Thanks to Lemma A.3.4 we can choose u, v such that the diagram

$$
\begin{array}{ccc}
U & \xrightarrow{\;f\;} & V \\
u \downarrow & & \downarrow v \\
U' & \longrightarrow & V' \\
& f' &
\end{array}
$$

is commutative in $\mathrm{Mod}(A)$. Then there is a unique A-homomorphism $w : W \to W'$ making the diagram

$$
\begin{array}{ccccccccc}
0 & \longrightarrow & U & \xrightarrow{\ f\ } & V & \xrightarrow{\ g\ } & W & \longrightarrow & 0 \\
 & & \downarrow{\scriptstyle u} & & \downarrow{\scriptstyle v} & & \downarrow{\scriptstyle w} & & \\
0 & \longrightarrow & U' & \xrightarrow{\ f'\ } & V' & \xrightarrow{\ g'\ } & W' & \longrightarrow & 0
\end{array}
$$

commutative. We are going to show that $(\underline{u}, \underline{v}, \underline{w})$ is a morphism of triangles. As P' is relatively k-injective and $\binom{g}{j} : V \to W \oplus P$ is k-split injective, there is a commutative diagram of A-modules

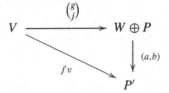

for some A-homomorphisms $a : W \to P'$ and $b : P \to P'$. Setting

$$
\omega = \begin{pmatrix} w & 0 \\ a & b \end{pmatrix} : W \oplus P \to W' \oplus P'
$$

the square

$$
\begin{array}{ccc}
V & \xrightarrow{\binom{g}{j}} & W \oplus P \\
\downarrow{\scriptstyle v} & & \downarrow{\scriptstyle \omega} \\
V' & \xrightarrow{\binom{g'}{j'}} & W' \oplus P'
\end{array}
$$

is commutative, and hence there is a unique A-homomorphism $x : X \to X'$ making the diagram

$$
\begin{array}{ccccccccc}
0 & \longrightarrow & V & \xrightarrow{\binom{g}{j}} & W \oplus P & \xrightarrow{(h,-p)} & X & \longrightarrow & 0 \\
 & & \downarrow{\scriptstyle v} & & \downarrow{\scriptstyle \omega} & & \downarrow{\scriptstyle x} & & \\
0 & \longrightarrow & V' & \xrightarrow{\binom{g'}{j'}} & W' \oplus P' & \xrightarrow{(h',-p')} & X' & \longrightarrow & 0
\end{array}
$$

commutative. In order to see that $\underline{x}$ gets identified to $\Sigma(\underline{u})$ through the isomorphisms $X \cong \Sigma(U)$ and $X' \cong \Sigma(U')$ in $\underline{\mathrm{Mod}}(A)$, it suffices to observe that the diagram

$$
\begin{array}{ccccccccc}
0 & \longrightarrow & U & \xrightarrow{\ jf\ } & P & \xrightarrow{\ -p\ } & X & \longrightarrow & 0 \\
 & & \big\downarrow{\scriptstyle u} & & \big\downarrow{\scriptstyle b} & & \big\downarrow{\scriptstyle x} & & \\
0 & \longrightarrow & U' & \xrightarrow{\ j'f'\ } & P' & \xrightarrow{\ -p'\ } & X' & \longrightarrow & 0
\end{array}
$$

is commutative, with k-split exact rows. This completes the proof of axiom T2. Consider three k-split exact sequences of A-modules of the form

$$
0 \longrightarrow U \xrightarrow{\ f\ } V \xrightarrow{\ g\ } W \longrightarrow 0
$$

$$
0 \longrightarrow V \xrightarrow{\ \binom{g}{j}\ } W \oplus P \xrightarrow{\ (h,-p)\ } X \longrightarrow 0
$$

$$
0 \longrightarrow W \xrightarrow{\ \binom{h}{l}\ } X \oplus Q \xrightarrow{\ (t,-q)\ } Y \longrightarrow 0
$$

with P, Q relatively R-injective. One checks that the diagram

$$
\begin{array}{ccccccccc}
0 & \longrightarrow & U & \xrightarrow{\ jf\ } & P & \xrightarrow{\ -p\ } & X & \longrightarrow & 0 \\
 & & \big\downarrow{\scriptstyle f} & & \big\downarrow{\scriptstyle \binom{\mathrm{Id}_P}{0}} & & \big\downarrow{\scriptstyle -t} & & \\
0 & \longrightarrow & V & \xrightarrow[\binom{j}{lg}]{} & P \oplus Q & \xrightarrow[(tp,-q)]{} & Y & \longrightarrow & 0
\end{array}
$$

is commutative with k-split exact rows. This shows that $\underline{t}$ gets identified to $-\Sigma(\underline{f})$ under the canonical isomorphisms $X \cong \Sigma(U)$ and $Y \cong \Sigma(V)$ in $\underline{\mathrm{Mod}}(A)$. Thus axiom T3 holds. Suppose we have a commutative diagram in $\underline{\mathrm{Mod}}(A)$ of the form:

$$
\begin{array}{ccc}
U & \xrightarrow{\ \underline{f}\ } & V \\
 & {\scriptstyle \underline{h}}\searrow & \big\downarrow{\scriptstyle \underline{g}} \\
 & & W \ .
\end{array}
$$

By adding, if necessary, relatively R-projective direct summands to V, W we may assume that $\underline{f}$, $\underline{g}$, $\underline{h}$ have R-split injective representatives f, g, h, respectively. By A.3.4 we may choose these representatives in such a way that the diagram

$$
\begin{array}{ccc}
U & \xrightarrow{\ f\ } & V \\
 & h \searrow & \downarrow g \\
 & & W
\end{array}
$$

becomes commutative in $\mathrm{Mod}(A)$. Consider the three exact triangles defined by the three pairs of k-split exact sequences

$$0 \longrightarrow U \xrightarrow{\ f\ } V \xrightarrow{\ a\ } W' \longrightarrow 0 \qquad 0 \longrightarrow V \xrightarrow{\binom{a}{ig}} W' \oplus P \xrightarrow{(x,-p)} X \longrightarrow 0$$

$$0 \longrightarrow V \xrightarrow{\ g\ } W \xrightarrow{\ b\ } U' \longrightarrow 0 \qquad 0 \longrightarrow W \xrightarrow{\binom{b}{i}} U' \oplus P \xrightarrow{(y,-q)} Y \longrightarrow 0$$

$$0 \longrightarrow U \xrightarrow{\ h\ } W \xrightarrow{\ c\ } V' \longrightarrow 0 \qquad 0 \longrightarrow W \xrightarrow{\binom{c}{i}} V' \oplus P \xrightarrow{(z,-r)} Y \longrightarrow 0.$$

Let v, w be the unique A-homomorphisms making the diagram

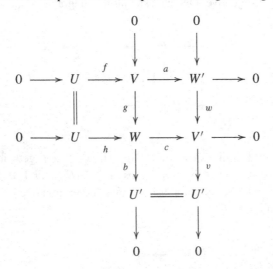

commutative. The two rows and the left column in this diagram are k-split exact by assumption. The right column is easily checked to be exact, and it is also

k-split: one gets a section for v by taking a section for b and composing it with c. By A.3.3 we have three k-split exact sequences

$$0 \longrightarrow U \xrightarrow{igf} P \xrightarrow{-p} X \longrightarrow 0$$

$$0 \longrightarrow V \xrightarrow{ig} P \xrightarrow{-q} Y \longrightarrow 0$$

$$0 \longrightarrow U \xrightarrow{i} P \xrightarrow{-r} Z \longrightarrow 0.$$

Since P is relatively k-injective, we can identify $\Sigma(U)$, $\Sigma(V)$, $\Sigma(W)$ to X, Y, Z, respectively, through these exact sequences. If we consider the two columns in the previous diagram, it follows from the proof of axiom T2 that the triple $(\underline{a}, \underline{b}, \underline{\mathrm{Id}}_{U'})$ defines a morphism from the exact triangle

$$V \xrightarrow{\underline{g}} W \xrightarrow{\underline{h}} U' \xrightarrow{\underline{y}} \Sigma(V)$$

to the exact triangle

$$W' \xrightarrow{\underline{w}} V' \xrightarrow{\underline{v}} U' \longrightarrow \Sigma(W').$$

It remains to show that the diagram

$$
\begin{array}{ccc}
V & \xrightarrow{\underline{z}} & Z = \Sigma(U) \\
\downarrow{\underline{v}} & & \downarrow{\Sigma(\underline{f})} \\
U' & \xrightarrow[\underline{y}]{} & Y = \Sigma(V)
\end{array}
$$

is commutative in $\underline{\mathrm{Mod}}(A)$. This is equivalent to showing that the diagram

$$
\begin{array}{ccc}
V' & \xrightarrow{z} & Z \\
\downarrow{v} & & \downarrow{s} \\
U' & \xrightarrow[y]{} & Y
\end{array}
$$

is commutative, where s is the unique A-homomorphism making the diagram

$$
\begin{array}{ccccccccc}
0 & \longrightarrow & U & \xrightarrow{\;ih\;} & P & \xrightarrow{\;-r\;} & Z & \longrightarrow & 0 \\
 & & \downarrow{\scriptstyle f} & & \| & & \downarrow{\scriptstyle s} & & \\
0 & \longrightarrow & V & \xrightarrow[\;ig\;]{} & P & \xrightarrow[\;-q\;]{} & Y & \longrightarrow & 0
\end{array}
$$

commutative. Thus $\underline{s}$ gets identified with $\Sigma(f)$ through the isomorphisms $Z \cong \Sigma(U)$ and $Y \cong \Sigma(V)$. It suffices therefore to show that the diagram

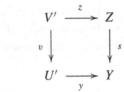

is commutative in $\mathrm{Mod}(A)$. Since $c : W \to V'$ is an epimorphism, it suffices to prove the equality $szc = yvc$. Now $zc = ri$ and $yb = qi$; thus together, $szc = sri = qi = yb = yvc$ as required. This completes the proof of the octahedral axiom T4 and shows that $\underline{\mathrm{Mod}}(A)$ is a triangulated category. If k is Noetherian, the above proof shows also that all constructions and commutative diagrams above can be performed within $\underline{\mathrm{mod}}(A)$, which completes the proof of Theorem A.3.2.   □

A.4 Homotopy categories are triangulated

If $\mathcal{C}$ is an additive category, then the chain homotopy category $K(\mathcal{C})$ together with the shift functor [1] is triangulated. To see this we need to define a suitable class of exact triangles.

Definition A.4.1 Let $\mathcal{C}$ be an additive category and let $f : (X, \delta) \to (Y, \epsilon)$ be a chain map of complexes over $\mathcal{C}$. The *mapping cone of f* is the complex $C(f)$ over $\mathcal{C}$ given by $C(f)_n = X_{n-1} \oplus Y_n$ with differential Δ

$$
\Delta_n = \begin{pmatrix} -\delta_{n-1} & 0 \\ f_{n-1} & \epsilon_n \end{pmatrix} : X_{n-1} \oplus Y_n \to X_{n-2} \oplus Y_{n-1}
$$

for any integer n. The mapping cone comes along with canonical chain maps $i(f) : Y \to C(f)$ given by the canonical monomorphims $Y_n \hookrightarrow X_{n-1} \oplus Y_n$ and

$p(f) : C(f) \to X[1]$ given by the canonical projections $X_{n-1} \oplus Y_n \twoheadrightarrow X_{n-1} = X[1]_n$ for any integer n.

One checks that $\Delta \circ \Delta = 0$. The cone $C(X)$ of a chain complex X as defined in 1.18.9 is the mapping cone of the identity chain map Id_X. Associated with any chain map $f : X \to Y$ we have a triangle in $(\mathrm{Ch}(\mathcal{C}), [1])$ given by the 'mapping cone sequence'

$$X \xrightarrow{\;f\;} Y \xrightarrow{\;\iota(f)\;} C(f) \xrightarrow{\;p(f)\;} X[1]$$

and we denote by $\mathcal{T}$ the class of all triangles in $K(\mathcal{C})$ isomorphic to the image of a triangle in $\mathrm{Ch}(\mathcal{C})$ of this form. We are going to show that this yields a structure of a triangulated category for $K(\mathcal{C})$.

Theorem A.4.2 *Let $\mathcal{C}$ be an additive category. Then $K(\mathcal{C})$, endowed with the shift functor [1] and the class $\mathcal{T}$ of triangles induced by mapping cone sequences, is a triangulated category. Moreover, the categories $K^+(\mathcal{C})$, $K^-(\mathcal{C})$, $K^b(\mathcal{C})$ are full triangulated subcategories in $K(\mathcal{C})$.*

Axiom T1 holds trivially: by definition, any chain map is part of a mapping cone sequence, and the mapping cone of the zero map $0 \to X$ is just X again, so the mapping cone sequence degenerates to $0 \to X \to X \to 0$. The following proposition shows, that T2 holds:

Proposition A.4.3 *Let*

$$\begin{array}{ccc} X & \xrightarrow{\;f\;} & Y \\ {\scriptstyle u}\downarrow & & \downarrow{\scriptstyle v} \\ X' & \xrightarrow{\;f'\;} & Y' \end{array}$$

be a commutative diagram of chain complexes over an additive category $\mathcal{C}$. Then there is a chain map $w : C(f) \to C(f')$ making the diagram

$$\begin{array}{ccccccc} X & \xrightarrow{\;f\;} & Y & \xrightarrow{\;i(f)\;} & C(f) & \xrightarrow{\;p(f)\;} & X[1] \\ {\scriptstyle u}\downarrow & & {\scriptstyle v}\downarrow & & \downarrow{\scriptstyle w} & & \downarrow{\scriptstyle u[1]} \\ X' & \xrightarrow{\;f'\;} & Y' & \xrightarrow{\;i(f')\;} & C(f') & \xrightarrow{\;i(f')\;} & X'[1] \end{array}$$

homotopy commutative.

Proof Since $v \circ f \sim f' \circ u$, there is a homotopy $h : X \to Y'$ such that $v \circ f - f' \circ u = \epsilon' \circ h + h \circ \delta$, where δ and ϵ' are the differentials of X and Y', respectively. Set

$$w_n = \begin{pmatrix} u_{n-1} & 0 \\ h_{n-1} & v_n \end{pmatrix} : X_{n-1} \oplus Y_n \to X'_{n-1} \oplus Y'_n$$

for any $n \in \mathbb{Z}$. A straightforward verification shows that $w = (w_n)_{n \in \mathbb{Z}}$ is a chain map from $C(f)$ to $C(f')$ which makes the middle and right square in the above diagram commutative. $\square$

The next proposition describes in which way the mapping cones $C(f)$, $C(i(f))$, $C(p(f))$ are connected, implying in particular, that the axiom T3 holds for the class $\mathcal{T}$ in $K(\mathcal{C})$:

Proposition A.4.4 *Let $f : X \to Y$ be a chain map of chain complexes over an additive category $\mathcal{C}$. Denote by $q(f) : C(i(f)) \to X[1]$ the graded map given by the canonical projections $q(f)_n = (0, \mathrm{Id}_{X_{n-1}}, 0) : Y_{n-1} \oplus (X_{n-1} \oplus Y_n) \to X_{n-1}$, for any $n \in \mathbb{Z}$. Denote by $s(f) : C(p(f)) \to Y[1]$ the graded map given by $s(f)_n = (0, \mathrm{Id}_{Y_{n-1}}, f_{n-1}) : X_{n-2} \oplus Y_{n-1} \oplus X_{n-1} \to Y_{n-1}$, for any $n \in \mathbb{Z}$. Then $q(f)$ and $s(f)$ are homotopy equivalences making the following diagram of chain complexes homotopy commutative:*

$$
\begin{array}{ccccccc}
& & Y & \xrightarrow{i(f)} & C(f) & \xrightarrow{i(i(f))} & C(i(f)) & \xrightarrow{p(i(f))} & Y[1] \\
& & \| & & \| & & \downarrow{\scriptstyle q(f)} & & \| \\
X & \xrightarrow{f} & Y & \xrightarrow{i(f)} & C(f) & \xrightarrow{p(f)} & X[1] & \xrightarrow{-f[1]} & Y[1] & \xrightarrow{-i(f)[1]} & C(f)[1] \\
& & & & \| & & \| & & \uparrow{\scriptstyle -s(f)} & & \| \\
& & & & C(f) & \xrightarrow{p(f)} & X[1] & \xrightarrow{i(p(f))} & C(p(f)) & \xrightarrow{p(p(f))} & C(f)[1]
\end{array}
$$

Proof The verification, that both $q(f)$, $s(f)$ are chain maps, is straightforward. We construct homotopy inverses $r(f)$, $t(f)$ of $q(f)$, $s(f)$, respectively, as follows. Set

$$r(f)_n = \begin{pmatrix} -f_{n-1} \\ \mathrm{Id}_{X_{n-1}} \\ 0 \end{pmatrix} : X_{n-1} \to Y_{n-1} \oplus X_{n-1} \oplus Y_n$$

for any integer n. Then $r(f)$ is a chain map satisfying $q(f) \circ r(f) = \mathrm{Id}_{X[1]}$. In order to show that $r(f) \circ q(f) \sim \mathrm{Id}_{C(i(f))}$, we define the homotopy h on $C(i(f))$ by

$$
h_n = \begin{pmatrix} 0 & 0 & \mathrm{Id}_{Y_n} \\ 0 & 0 & 0 \\ 0 & 0 & 0 \end{pmatrix} : Y_{n-1} \oplus X_{n-1} \oplus Y_n \to Y_n \oplus X_n \oplus Y_{n+1}
$$

for any integer n. If Δ denotes the differential of $C(i(f))$, we have $\Delta \circ h + h \circ \Delta = \mathrm{Id}_{C(i(f))} - r(f) \circ q(f)$. This shows also the homotopy commutativity of the upper part of the diagram, since $q(f) \circ i(i(f)) = p(f)$ and $p(i(f)) \circ r(f) = -f[1]$. We proceed similarly for $t(f)$. Set

$$
t(f)_n = \begin{pmatrix} 0 \\ \mathrm{Id}_{Y_{n-1}} \\ 0 \end{pmatrix} : Y_{n-1} \to X_{n-2} \oplus Y_{n-1} \oplus X_{n-1}
$$

for any integer n. Clearly $t(f)$ is a chain map satisfying $s(f) \circ t(f) = \mathrm{Id}_{Y[1]}$. In order to show that $t(f) \circ s(f) \sim \mathrm{Id}_{C(p(f))}$, we define the homotopy k on $C(p(f))$ by

$$
k_n = \begin{pmatrix} 0 & 0 & \mathrm{Id}_{X_{n-1}} \\ 0 & 0 & 0 \\ 0 & 0 & 0 \end{pmatrix} : X_{n-2} \oplus Y_{n-1} \oplus X_{n-1} \to X_{n-1} \oplus Y_n \oplus X_n
$$

for any integer n. If Π denotes the differential of $C(p(f))$, we have $\Pi \circ k + k \circ \Pi = \mathrm{Id}_{C(p(f))} - t(f) \circ s(f)$. This shows also the homotopy commutativity of the lower part of the diagram, since $s(f) \circ p(f) = f[1]$ and $p(p(f)) \circ t(f) = i(f)[1]$. $\qquad\square$

It remains to show, that the octahedral axiom T4 holds:

Proposition A.4.5 *Given two composable chain maps* $X \xrightarrow{\ f\ }$ $Y \xrightarrow{\ g\ } Z$ *of complexes over an additive category* C, *there are chain maps* u, v *making the following diagram of chain complexes* *commutative*

$$
\begin{array}{ccccccc}
X & \xrightarrow{\ f\ } & Y & \xrightarrow{\ i(f)\ } & C(f) & \xrightarrow{\ p(f)\ } & X[1] \\
\| & & \downarrow{\scriptstyle g} & & \downarrow{\scriptstyle u} & & \| \\
X & \xrightarrow[\ gf\ ]{} & Z & \xrightarrow[\ i(gf)\ ]{} & C(gf) & \xrightarrow[\ p(gf)\ ]{} & X[1] \\
& & \downarrow{\scriptstyle i(g)} & & \downarrow{\scriptstyle v} & & \downarrow{\scriptstyle f[1]} \\
& & C(g) & = & C(g) & \xrightarrow[\ p(g)\ ]{} & Y[1] \\
& & \downarrow{\scriptstyle p(g)} & & \downarrow{\scriptstyle w} & & \\
& & Y[1] & \xrightarrow[\ i(f)[1]\ ]{} & C(f)[1] & &
\end{array}
$$

and there is a homotpy equivalence $t(u) : C(g) \to C(u)$ *such that the diagram*

$$
\begin{array}{ccccccc}
C(f) & \xrightarrow{\ u\ } & C(gf) & \xrightarrow{\ v\ } & C(g) & \xrightarrow{\ w\ } & C(f)[1] \\
\| & & \| & & \downarrow{\scriptstyle t(u)} & & \| \\
C(f) & \xrightarrow[\ u\ ]{} & C(gf) & \xrightarrow[\ i(u)\ ]{} & C(u) & \xrightarrow[\ p(u)\ ]{} & C(f)[1]
\end{array}
$$

is homotopy commutative.

Proof For any $n \in \mathbb{Z}$ set

$$
u_n = \begin{pmatrix} \mathrm{Id}_{X_{n-1}} & 0 \\ 0 & g_n \end{pmatrix} : X_{n-1} \oplus Y_n \to X_{n-1} \oplus Z_n,
$$

$$
v_n = \begin{pmatrix} f_{n-1} & 0 \\ 0 & \mathrm{Id}_{Z_n} \end{pmatrix} : X_{n-1} \oplus Z_n \to Y_{n-1} \oplus Z_n,
$$

$$
w_n = \begin{pmatrix} 0 & 0 \\ \mathrm{Id}_{Y_{n-1}} & 0 \end{pmatrix} : Y_{n-1} \oplus Z_n \to X_{n-2} \oplus Y_{n-1}.
$$

A straightforward verification shows that $u = (u_n)_{n \in \mathbb{Z}}$, $v = (v_n)_{n \in \mathbb{Z}}$ and $w = (w_n)_{n \in \mathbb{Z}}$ are chain maps which make the first diagram in the statement commutative. For $t(u)$ we take the morphism given by the obvious split monomorphisms

$$
Y_{n-1} \oplus Z_n \longrightarrow (X_{n-2} \oplus Y_{n-1}) \oplus (X_{n-1} \oplus Z_n)
$$

and we define a morphism $s(u) : C(u) \to C(g)$ given by the projections

$$(X_{n-2} \oplus Y_{n-1}) \oplus (X_{n-1} \oplus Z_n) \xrightarrow{\hspace{3cm}} Y_{n-1} \oplus Z_n$$

for any $n \in \mathbb{Z}$. Then $s(u) \circ t(u) = \mathrm{Id}_{C(g)}$, and it remains to show that $t(u) \circ s(u) \sim \mathrm{Id}_{C(u)}$. For this we consider on $C(u)$ the homotopy h given, for any $n \in \mathbb{Z}$, by the map

$$h_n : (X_{n-2} \oplus Y_{n-1}) \oplus (X_{n-1} \oplus Z_n) \to (X_{n-1} \oplus Y_n) \oplus (X_n \oplus Z_{n+1}),$$

where h_n is zero on the summands X_{n-2}, Y_{n-1}, Z_n, and h_n maps X_{n-1} identically to its canonical image in $C(u)_{n+1}$. Then in the second diagram in the statement, the left and middle square are commutative. Clearly $p(v) \circ s(u) = p(u)$; thus the right square is homotopy commutative, as $s(u)$ is a homotopy inverse to $t(u)$. ☐

This completes the proof of Theorem A.4.2. We note some immediate consequences.

Corollary A.4.6 *Let C be an additive category, let $f : X \to Y$ be a chain map of complexes over C, and consider the mapping cone sequence*

$$X \xrightarrow{f} Y \xrightarrow{i(f)} C(f) \xrightarrow{p(f)} X[1].$$

(i) We have $i(f) \circ f \sim 0$.
(ii) f is a homotopy equivalence if and only if $C(f) \simeq 0$.
(iii) $i(f)$ is a homotopy equivalence if and only if $X \simeq 0$.
(iv) $p(f)$ is a homotopy equivalence if and only if $Y \simeq 0$.
(v) If two of X, Y, $C(f)$ are homotopic to zero, so is the third.

Proof Statement (i) follows from A.2.3 (i), but one can see this also directly: the canonical monomorphisms $X_n \hookrightarrow X_n \oplus Y_{n+1}$ define a homotopy $h : X \longrightarrow C(f)$ through which $i(f) \circ f$ becomes homotopic to the zero map. The statements (ii), (iii), (iv) are all particular cases of A.2.7. Finally, (v) follows from (ii) and (iii). ☐

Corollary A.4.7 *Let C be an additive category, $f : X \to Y$ a chain map of complexes over C, and let U be a complex over C.*

(i) The covariant functor $\mathrm{Hom}_{K(C)}(U, -)$ induces a long exact sequence

$$\cdots \to \mathrm{Hom}_{K(C)}(U, X[n]) \to \mathrm{Hom}_{K(C)}(U, Y[n]) \to \mathrm{Hom}_{K(C)}(U, C(f)[n])$$

$$\to \mathrm{Hom}_{K(C)}(U, X[n+1]) \to \cdots.$$

(ii) The contravariant functor $\mathrm{Hom}_{K(\mathcal{C})}(-, U)$ *induces a long exact sequence*

$$\cdots \to \mathrm{Hom}_{K(\mathcal{C})}(X[n+1], U) \to \mathrm{Hom}_{K(\mathcal{C})}(C(f)[n], U)$$
$$\to \mathrm{Hom}_{K(\mathcal{C})}(Y[n], U) \to \mathrm{Hom}_{K(\mathcal{C})}(X[n], U) \to \cdots.$$

Proof This is a particular case of A.2.8. $\qquad\square$

Corollary A.4.8 *Let A be an algebra over a commutative ring k and let* $f : X \to Y$ *be a chain map of complexes of A-modules. Taking homology induces a long exact sequence of A-modules*

$$\cdots \to H_n(X) \to H_n(Y) \to H_n(C(f)) \to H_{n-1}(X) \to \cdots.$$

Proof By 1.18.4 we have that $H_n(X) \cong \mathrm{Hom}_{K(\mathrm{Mod}(A))}(A[n], X) \cong \mathrm{Hom}_{K(\mathrm{Mod}(A))}(A, X[-n])$, thus the statement follows from the first of the two long exact sequences in the previous corollary applied to $U = A$. $\qquad\square$

Corollary A.4.9 *Let A be an algebra over a commutative ring k and let* $f : X \to Y$ *be a chain map of complexes of A-modules. The following are equivalent.*

 (i) *f is a quasi-isomorphism.*
 (ii) *C(f) is acyclic.*
 (iii) *For any bounded below complex P of projective A-modules, the map f induces an isomorphism* $\mathrm{Hom}_{K(\mathrm{Mod}(A))}(P, X) \cong \mathrm{Hom}_{K(\mathrm{Mod}(A))}(P, Y)$.
 (iv) *For any bounded above complex I of injective A-modules, the map f induces an isomorphism* $\mathrm{Hom}_{K(\mathrm{Mod}(A))}(Y, I) \cong \mathrm{Hom}_{K(\mathrm{Mod}(A))}(X, I)$.

Proof The equivalence of (i), (ii) follows from the long exact homology sequence in the previous corollary, and the equivalence with (iii), (iv) follows then from the long exact sequences in A.4.7, together with the characterisation 1.18.7 of acyclic complexes, using the fact that $\mathrm{Mod}(A)$ has enough projective and injective objects. $\qquad\square$

We have two ways of producing long exact sequences: via mapping cone sequences and via short exact sequences of complexes. Both approaches are equivalent in the sense that we can view mapping cone sequences as being induced by short exact sequences of complexes in the same way we defined triangles in a stable module category using short exact sequences of modules. One can use this to give an alternative proof of the fact that homotopy categories of complexes over abelian categories are triangulated, following the lines of the proof for stable categories. The advantage of the approach in the present section is that it holds for additive categories.

Theorem A.4.10 *Let A be an algebra over a commutative ring k and let*

$$0 \longrightarrow X \xrightarrow{\ f\ } Y \xrightarrow{\ g\ } Z \longrightarrow 0$$

be a short exact sequence of chain complexes A-modules. The maps
$s_n = (0, g_n) : X_{n-1} \oplus Y_n \to Z_n$ *induce a quasi-isomorphism* $s : C(f) \to Z$
making the diagram of chain complexes of A-modules

$$
\begin{array}{ccccc}
X & \xrightarrow{\ f\ } & Y & \xrightarrow{\ i(f)\ } & C(f) \\
\| & & \| & & \big\downarrow{\scriptstyle s} \\
0 \longrightarrow X & \xrightarrow{\ f\ } & Y & \xrightarrow{\ g\ } & Z \longrightarrow 0
\end{array}
$$

commutative, and we have an isomorphism of long exact sequences

$$
\begin{array}{ccccccccc}
\cdots \longrightarrow & H_n(X) & \longrightarrow & H_n(Y) & \longrightarrow & H_n(C(f)) & \longrightarrow & H_{n-1} & \longrightarrow \cdots \\
& \| & & \| & & \big\downarrow{\scriptstyle H_n(s)} & & \\
\cdots \longrightarrow & H_n(X) & \longrightarrow & H_n(Y) & \longrightarrow & H_n(Z) & \longrightarrow & H_{n-1} & \longrightarrow \cdots
\end{array}
$$

where the first row is from A.4.8 and the second row from 1.17.5. Moreover,
if the first exact sequence of chain complexes is degreewise split, then s is a
homotopy equivalence.

Proof The commutativity of the diagram is, as usual, a straightforward verification. One can verify either directly that s is a quasi-isomorphism (which,
effectively, would yield another proof of Theorem 1.17.5), or use the 5-Lemma.
Denote by δ, ϵ, ζ, Δ the differentials of X, Y, Z, $C(f)$, respectively. Suppose
now that the first exact sequence of chain complexes in the statement is degree
wise split; that is, there are graded morphisms $u : Y \to X$ and $v : Z \to Y$ satisfying $\mathrm{Id}_Y = fu + vg$. Note that then $f = fuf$, whence $uf = \mathrm{Id}_X$ as f is a
monomorphism; in particular, u is a retraction for f. Similarly, $gv = \mathrm{Id}_Z$; in particular, v is a section for g. The morphism $v\zeta - \epsilon v : Z \to Y$ is graded of degree
-1, and it is actually a chain map from Z to $Y[1]$, since $(-\epsilon)(v\zeta - \epsilon v) =$
$-\epsilon v\zeta = (v\zeta - \epsilon v)\zeta$. This chain map satisfies $g(v\zeta - \epsilon v) = gv\zeta - \zeta gv = 0$,
as $gv = \mathrm{Id}_Z$. Thus this map factors through f. Let $r : Z \to X$ be the graded
morphism of degree -1 such that $fr = v\zeta - \epsilon v$. Since the right side is a chain
map and f is a monomorphism, r itself can be viewed as a chain map from

$Z[-1]$ to X. Consider the associated triangle

$$Z[-1] \xrightarrow{\ r\ } X \xrightarrow{\ i(r)\ } C(r) \xrightarrow{\ p(r)\ } Z.$$

A straightforward verification shows that $C(r) \cong Y$ via the inverse chain maps given by the morphisms $(v_n, f_n) : Z_n \oplus X_n \to Y_n$ and $\binom{g_n}{u_n} : Y_n \to Z_n \oplus X_n$ for any integer n. Thus $C(i(r)) \cong C(f)$. By A.4.4, we have also a homotopy equivalence $q(r) : C(i(r)) \to Z$. Together we find a homotopy equivalence $C(f) \simeq Z$, and this is easily seen to be the chain map s as defined. $\qquad\square$

This yields another proof of 1.18.18, as well as the following consequence.

Corollary A.4.11 *With the notation and hypotheses of A.4.10, if $Y \simeq 0$ then $Z \simeq X[1]$.*

Bibliography

[1] J. L. Alperin, *Sylow intersections and fusion*, J. Algebra **6** (1967), 222–241.

[2] J. L. Alperin, *The Green correspondence and normal subgroups*, J. Algebra **104** (1986), 74–77.

[3] J. L. Alperin, *Weights for finite groups*, Proc. Symp. Pure Math. **47** (1987), 369–379.

[4] J. L. Alperin, *Local representation theory*, Cambridge Studies in Advanced Mathematics **11**, Cambridge University Press (1986).

[5] J. L. Alperin and M. Broué, *Local methods in block theory*, Ann. Math. **110** (1979), 143–157.

[6] J. L. Alperin and P. Fong, *Weights for symmetric and general linear groups*, J. Algebra **131** (1990), 2–22.

[7] J. L. Alperin, M. Linckelmann, and R. Rouquier, *Source algebras and source modules*, J. Algebra **239** (2001), 262–271.

[8] J. An, 2-*Weights for general linear groups*, J. Algebra **149** (1992), 500–527.

[9] J. An, *Uno's invariant conjecture for the general linear and unitary groups in nondefining characteristics*, J. Algebra **284** (2005), 462–479.

[10] M. Aschbacher, R. Kessar, and B. Oliver, *Fusion systems in algebra and topology*, London Math. Soc. Lecture Notes Series **391**, Cambridge University Press (2011).

[11] L. Barker, *On p-soluble groups and the number of simple modules associated with a given Brauer pair*, Quart. J. Math. Oxford **48** (1997), 133–160.

[12] L. Barker, *On contractibility of the orbit space of a G-poset of Brauer pairs*. J. Algebra **212** (1999), 460–465.

[13] V. A. Bašev, *Representations of the group $\mathbb{Z}_2 \times \mathbb{Z}_2$ in a field of characteristic 2* (Russian). Dokl. Akad. Nauk. SSSR **141** (1961), 1015–1018.

[14] D. J. Benson, *Representations and cohomology, Vol. I: Cohomology of groups and modules*, Cambridge studies in advanced mathematics **30**, Cambridge University Press (1991).

[15] D. J. Benson, *Representations and cohomology, Vol. II: Cohomology of groups and modules*, Cambridge studies in advanced mathematics **31**, Cambridge University Press (1991).

[16] D. J. Benson and R. Kessar, *Blocks inequivalent to their Frobenius twists*. J. Algebra **315** (2007), 588–599.

497

[17] T. R. Berger, *Irreducible modules of solvable groups are algebraic*, Proc. Conf. Finite Groups, Utah (1976), 541–553.

[18] T. R. Berger, *Solvable groups and algebraic modules*, J. Algebra **57** (1979), 387–406.

[19] R. Boltje, *Alperin's weight conjecture and chain complexes*, J. London Math. Soc. **68** (2003) 83–101.

[20] R. Boltje, R. Kessar, and M. Linckelmann, *On Morita equivalences with endo-permutation source.* Preprint (2017).

[21] R. Boltje and B. Külshammer, *The ring of modules with endo-permutation source*, Manuscripta Math. **120** (2006), 359–376.

[22] R. Boltje and B. Xu, *On p-permutation equivalences: between Rickard equivalences and isotypies*, Trans. Amer. Math. Soc. **360** (2008), 5067–5087.

[23] S. Bouc, *Modules de Möbius.* C. R. Acad. Sci. Paris Sér. I, **299** (1984), 49–52.

[24] S. Bouc, *The Dade group of a p-group*, Invent. Math., **164** (2006), 189–231.

[25] R. Brauer, *Investigations on group characters*, Ann. Math. **42** (1941), 936–958.

[26] R. Brauer, *Number theoretical investigations on groups of finite order*, Proceedings of the international symposium on algebraic number theory, Tokyo and Nikko, Science Council of Japan, Tokyo (1995), 55–62.

[27] R. Brauer, *On blocks and sections in finite groups II*, Amer. J. Math. **90** (1968), 895–925.

[28] R. Brauer, *Some applications of the theory of blocks of characters of finite groups IV*, J. Algebra **17** (1971), 489–521.

[29] R. Brauer, *On 2-blocks with dihedral defect groups*, Symp. Math. XIII, Academic Press (1974), 367–393.

[30] R. Brauer, *Notes on representations of finite groups, I*, J. London Math. Soc. (2), **13** (1976), 162–166.

[31] R. Brauer and W. Feit, On the number of irreducible characters of finite groups in a given block, Nat. Acad. Sci. USA **45** (1959), 361–365.

[32] R. Brauer and C. J. Nesbitt, *On the modular characters of groups.* Ann. Math. **42** (1941), 556–590.

[33] C. Broto, R. Levi, and B. Oliver, *The homotopy theory of fusion systems*, J. Amer. Math. Soc. **16** (2003), 779–856.

[34] C. Broto, N. Castellana, J. Grodal, R. Levi, and B. Oliver, *Subgroup families controlling p-local finite groups*, Proc. London Math. Soc. (3) **91** (2005), 325–354.

[35] M. Broué, *Projectivité relative, blocs, groupes de défaut.* Thèse de Doctorat d'Etat es-sciences, Université Paris VII (1975).

[36] M. Broué, *Radical, hauteurs, p-sections et blocs*, Ann. Math. **107** (1978), 89–107.

[37] M. Broué, *Brauer coefficients of p-subgroups associated with a p-block of a finite group*, J. Algebra **56** (1979), 365–383.

[38] M. Broué, *Isométries parfaites, types de blocs, catégories dérivées*, Astérisque **181–182** (1990), 61–92.

[39] M. Broué, *Equivalences of blocks of group algebras*, in: Finite dimensional algebras and related topics, Kluwer (1994), 1–26.

[40] M. Broué and L. Puig, *Characters and local structure in G-algebras*, J. Algebra **63** (1980), 306–317.

[41] M. Broué and L. Puig, *A Frobenius theorem for blocks*, Invent. Math. **56** (1980), 117–128.

[42] K. S. Brown, *Cohomology of groups*. Graduate Texts in Math. **87** (1982), Springer, New York.

[43] M. Cabanes, *Extensions of p-groups and construction of characters*, Comm. Algebra **15** (1987), 1297–1311.

[44] M. Cabanes, *Brauer morphism between modular Hecke algebras*, J. Algebra **115** (1988), 1–31.

[45] J. F. Carlson, *A characterization of endotrivial modules over p-groups*, Manuscripta Math. **97** (1998), 303–307.

[46] J. F. Carlson, N. Mazza, and J. Thévenaz, *Endotrivial modules for p-solvable groups*, Trans. Amer. Math. Soc. **363** (2011), 4979–4996.

[47] J. Chuang and R. Kessar, *Symmetric groups, wreath products, Morita equivalences, and Broué's Abelian Defect Group Conjecture*, Bull. London Math. Soc. **34** (2002), 174–185.

[48] J. Chuang and R. Rouquier, *Derived equivalences for symmetric groups and sl₂-categorification*, Ann. Math. **167** (2008), 245–298.

[49] G. Cliff, W. Plesken, and A. Weiss, *Order-theoretic properties of the center of a block*, in: The Arcata Conference on Representations of Finite Groups (editor: P. Fong), Proc. Sympos. Pure Math. **47**, Amer. Math. Soc, Providence RI (1987), 413–420.

[50] D. A. Craven, *The theory of fusion systems*, Cambridge Studies in Advanced Mathematics **131**, Cambridge University Press (2011).

[51] D. Craven, C. Eaton, R. Kessar, and M. Linckelmann, *The structure of blocks with a Klein four defect group*, Math. Z. **208** (2011), 441–476.

[52] C. W. Curtis and I. Reiner, *Methods of representation theory* Vol. II, John Wiley and Sons, New York, London, Sydney (1987).

[53] E. C. Dade, *Blocks with cyclic defect groups*, Ann. Math. **84** (1966), 20–48.

[54] E. C. Dade, *Block extensions*, Illinois J. Math. **17** (1973), 198–272.

[55] E. C. Dade, *Endo-permutation modules over p-groups, I, II*, Ann. Math. **107** (1978), 459–494, **108** (1978), 317–346.

[56] E. C. Dade, *A correspondence of characters*, Proc. Symp. Pure Math. **37** (1980), 401–404.

[57] E. C. Dade, *Extending endo-permutation modules*. Preprint (1982).

[58] E. C. Dade, *Counting characters in blocks, I*, Invent. Math. **109** (1992), 187–210.

[59] E. C. Dade, *Counting characters in blocks, II*, J. Reine Angew. Math. **448** (1994), 97–190.

[60] J. Dieudonné, *Sur la réduction canonique des couples de matrices*. Bull. Soc. Math. France **74** (1946), 130–146.

[61] O. Düvel, *On Donovan's conjecture*, J. Algebra **272** (2004), 1–16.

[62] C. W. Eaton, *The equivalence of some conjectures of Dade and Robinson*, J. Algebra **271** (2004), 638–651.

[63] C. Eaton, R. Kessar, B. Külshammer, and B. Sambale, *2-blocks with abelian defect groups*, Adv. Math. **254** (2014), 706–735.

[64] C. Eaton and A. Moretó, *Extending Brauer's height zero conjecture to blocks with nonabelian defect groups*, Int. Math. Res. Not. **2014** (2014), 5581–5601.

[65] K. Erdmann, *Blocks and simple modules with cyclic vertices*, Bull. London Math. Soc. **9** (1977), 216–218.

[66] K. Erdmann, *Blocks whose defect groups are Klein four groups: a correction*, J. Algebra **76** (1982), 505–518.

[67] K. Erdmann, *Blocks of tame representation type and related algebras*, Lecture Notes Math. **1428**, Springer Verlag, Berlin Heidelberg (1990).

[68] Y. Fan and L. Puig, *On blocks with nilpotent coefficient extensions*, Alg. Repr. Theory **2** (1999).

[69] W. Feit, *Irreducible modules of p-solvable groups*, Proc. Symp. Pure Math. **37** (1980), 405–412.

[70] W. Feit, *Possible Brauer trees*, Illinois J. Math. **28** (1984), 43–56.

[71] W. Feit and J. G. Thompson, *Solvability of groups of odd order*, Pacific J. Math. **13** (1963), 775–1029.

[72] P. Fong, *On the characters of p-solvable groups*, Trans. Amer. Math. Soc. **98** (1961), 263–284.

[73] P. Gabriel and Ch. Riedtmann, *Group representations without groups*, Comment. Math. Helvet. **54** (1979), 240–287.

[74] D. Gluck and T. R. Wolf, *Brauer's height zero conjecture for p-solvable groups*. Trans. Amer. Math. Soc. 282 (1984), 137–152.

[75] D. M. Goldschmidt, *A conjugation family for finite groups*, J. Algebra **16** (1970), 138–142.

[76] D. Gorenstein, *Finite groups*, Chelsea Publishing Company, New York (1980).

[77] J. A. Green, *Walking around the Brauer tree*, J. Austr. Math. Soc. **17** (1974), 197–213.

[78] J. Grodal, *Higher limits via subgroup complexes*. Annals of Math. **55** (2002), 405–457.

[79] M. E. Harris, *On the p-deficiency class of a finite group*, J. Algebra **94** (1985), 411–424.

[80] M. E. Harris and R. Knörr, *Brauer correspondences for covering blocks of finite groups*, Comm. Algebra **13** (1985), 1213–1218.

[81] M. E. Harris and M. Linckelmann, *Splendid derived equivalences for blocks of finite p-solvable groups*, J. London Math. Soc. (2) **62** (2000), 85–96.

[82] M. E. Harris and M. Linckelmann, *On the Glauberman and Watanabe correspondences for blocks of finite p-solvable groups*, Trans. Amer. Math. Soc. **354** (2002), 3435–3453.

[83] A. Heller and I. Reiner, *Indecomposable representations*. Ill. J. Math. **5** (1961), 314–323.

[84] L. Héthelyi, B. Külshammer, and B. Sambale, *A note on Olsson's Conjecture*, J. Algebra **398** (2014), 364–385.

[85] G. Higman, *Units in group rings*, D. Phil. Thesis, Oxford Univ. (1939).

[86] G. Hiss, *Morita equivalences between blocks of finite Chevalley groups*, Proc. Representation Theory of Finite and Algebraic Groups, eds: n. Kawanaka, G. Michler, and K. Uno, Osaka University, Osaka (2000), 128–136.

[87] G. Hiss and R. Kessar, *Scopes reduction and Morita equivalence classes of blocks in finite classical groups*, J. Algebra **230** (2000), 378–423.

[88] G. Hiss and R. Kessar, *Scopes reduction and Morita equivalence classes of blocks in finite classical groups II*, J. Algebra **283** (2005), 522–563.

[89] G. Hiss and K. Lux, *Brauer trees of sporadic simple groups*, Oxford Science Publications, The Clarendon Press, Oxford University Press, New York (1989).

[90] I. M. Isaacs and G. Navarro, *New refinements of the McKay conjecture for arbitrary finite groups*, Ann. Math. **156** (2002), 333–344.

[91] G. Janusz, *Indecomposable modules for finite groups*, Ann. Math. **89** (1969), 209–241.

[92] D. L. Johnson, *Indecomposable representations of the group (p, p) of fields of characteristic p*, J. London Math. Soc. **1** (1969) 43–50.

[93] T. Jost, *Morita equivalences for blocks of finite general linear groups*, Manuscripta Math. **91** (1996), 121–144.

[94] R. Kessar, *A remark on Donovan's conjecture*, Archiv Math. (Basel) **82** (2004), 391–394.

[95] R. Kessar, *On isotypies between Galois conjugate blocks*. Buildings, finite geometries and groups, Springer Proc. Math. **10**, 153–162 (2012), Springer, New York.

[96] R. Kessar and M. Linckelmann, *On blocks with Frobenius inertial quotient*, J. Algebra **249** (2002), 127–146.

[97] R. Kessar and M. Linckelmann, *Fusion systems with one weight I*, unpublished notes (2007).

[98] R. Kessar and M. Linckelmann, *ZJ-theorems for fusion systems*, Trans. Amer. Math. Soc. **360** (2008), 3093–3106.

[99] R. Kessar and M. Linckelmann, *On stable equivalences and blocks with one simple module*, J. Algebra **323** (2010), 1607–1621.

[100] R. Kessar and M. Linckelmann, *Bounds for Hochschild cohomology of block algebras*, J. Algebra **337** (2011), 318–322.

[101] R. Kessar, M. Linckelmann, and G. Navarro, *A characterisation of nilpotent blocks*. Proc. Amer. Math. Soc. **143** (2015), 5129–5138.

[102] R. Kessar and G. Malle, *Quasi-isolated blocks and Brauer's height zero conjecture*, Ann. Math. **178** (2013), 321–386.

[103] R. Kessar and R. Stancu, *A reduction theorem for fusion systems of blocks*, J. Algebra **319** (2008), 806–823.

[104] R. Knörr, *On the vertices of irreducible modules*, Ann. Math. **110** (1979), 487–499.

[105] R. Knörr, *A remark on covering blocks*, J. Algebra **103** (1986), 208–210.

[106] R. Knörr and G. R. Robinson, *Some remarks on a conjecture of Alperin*, J. London Math. Soc. **39** (1989), 48–60.

[107] S. Koshitani, *Conjectures of Donovan and Puig for principal 3-blocks with abelian defect groups*. Comm. Algebra **31** (2003), 2229–2243.

[108] S. Koshitani and B. Külshammer, *A splitting theorem for blocks*. Osaka J. Math. **33** (1996), 343–346.

[109] S. Koshitani and M. Linckelmann, *The indecomposability of a certain bimodule given by the Brauer construction*, J. Algebra **285** (2005), 726–729.

[110] B. Külshammer, *On p-blocks of p-solvable groups*, Comm. Algebra **9** (1981), 1763–1785.

[111] B. Külshammer, *Symmetric local algebras and small blocks of finite groups*, J. Algebra **88** (1984), 190–195.

[112] B. Külshammer, *Crossed products and blocks with normal defect groups*, Comm. Algebra **13** (1985), 147–168.

[113] B. Külshammer, *A remark on conjectures in modular representation theory*, Arch. Math. **49** (1987), 396–399.

[114] B. Külshammer, *Roots of simple modules*, Canad. Math. Bull. **49** (1) (2006), 96–107.

[115] B. Külshammer and L. Puig, *Extensions of nilpotent blocks*, Invent. Math. **102** (1990), 17–71.

[116] B. Külshammer and G. R. Robinson, *An alternating sum for Hochschild cohomology of a block*. J. Algebra **249** (2002), 220–225.

[117] H. Kupisch, *Unzerlegbare Moduln endlicher Gruppen mit zyklischer p-Sylow Gruppe*, Math. Z. **108** (1969), 77–104.

[118] R. Levi and B. Oliver, *Construction of 2-local finite groups of a type studied by Solomon and Benson*, Geom. Topol. **6** (2002), 917–990 (electronic).

[119] R. Levi and B. Oliver, *Correction to: Construction of 2-local finite groups of a type studied by Solomon and Benson*, Geom. Topol. **9** (2005), 2395–2415 (electronic).

[120] A. Libman, *The gluing problem does not follow from homological properties of* $\Delta_p(G)$, Homology, Homotopy Appl. **12** (2010), 1–10.

[121] A. Libman, *The gluing problem in the fusion systems of the symmetric, alternating and linear groups*, J. Algebra **341** (2011), 209–245.

[122] M. Linckelmann, *Le centre d'un bloc à groupes de défaut cycliques*, C.R.A.S. **306**, série I (1988), 727–730.

[123] M. Linckelmann, *Variations sur les blocs à groupes de défaut cycliques*, Thèse, Univ. Paris VII (1988).

[124] M. Linckelmann, *Modules in the sources of Green's exact sequences for cyclic blocks*, Invent. Math. **97**, 129–140.

[125] M. Linckelmann, *Derived equivalence for cyclic blocks over a p-adic ring*, Math. Z. **207** (1991), 293–304.

[126] M. Linckelmann, *The source algebras of blocks with a Klein four defect group*, J. Algebra **167** (1994), 821–854.

[127] M. Linckelmann, *The isomorphism problem for cyclic blocks and their source algebras*, Invent. Math. **125**, (1996), 265–283.

[128] M. Linckelmann, *Stable equivalences of Morita type for self-injective algebras and p-groups*, Math. Z. **223** (1996) 87–100.

[129] M. Linckelmann, *On derived equivalences and local structure of blocks of finite groups*, Turkish J. Math. **22** (1998), 93–107.

[130] M. Linckelmann, *Transfer in Hochschild cohomology of blocks of finite groups*, Algebras Representation Theory **2** (1999), 107–135.

[131] M. Linckelmann, *On splendid derived and stable equivalences between blocks of finite groups*, J. Algebra **242** (2001), 819–843.

[132] M. Linckelmann, *Fusion category algebras*, J. Algebra **277** (2004), 222–235.

[133] M. Linckelmann, *Alperin's weight conjecture in terms of equivariant Bredon cohomology*, Math. Z. **250** (2005), 495–513.

[134] M. Linckelmann, *Introduction to fusion systems*, in: (eds. M. Geck, D. Testerman, and J. Thévenaz) *Group representation theory*, EPFL Press, Lausanne (2007), 79–113.

[135] M. Linckelmann, *Blocks of minimal dimension*, Arch. Math. **89** (2007), 311–314.

[136] M. Linckelmann, *Trivial source bimodule rings for blocks and p-permutation equivalences*, Trans. Amer. Math. Soc. **361** (2009), 1279–1316.

[137] M. Linckelmann, *The orbit space of a fusion system is contractible*, Proc. London Math. Soc. **98** (2009), 191–216.

[138] M. Linckelmann, *On $H^2(\mathcal{C}; k^\times)$ for fusion systems*, Homotopy, Homology and Applications **11** (2009), 203–218.

[139] M. Linckelmann, *On dimensions of block algebras*, Math. Res. Lett. **16** (2009), 1011–1014.

[140] M. Linckelmann, *On automorphisms and focal subgroups of blocks*. Proc. Edinburgh Math. Soc., to appear.

[141] M. Linckelmann and N. Mazza, *The Dade group of a fusion system*, J. Group Theory **12** (2009), 55–74.

[142] M. Linckelmann, *On stable equivalences with endopermutation source*, J. Algebra **434** (2015), 27–45.

[143] M. Linckelmann and L. Puig, *Structure des p'-extensions des blocs nilpotents*, C. R. Acad. Sc. Paris **304**, Série I (1987), 181–184.

[144] W. Lück, *Transformation groups and algebraic K-theory*. Springer Lecture Notes in Mathematics, Springer Verlag **1408** (1989).

[145] G. Malle and G. Navarro, *Blocks with equal height zero degrees*, Trans. Amer. Math. Soc. **363** (2011), 6647–6669.

[146] G. Malle and G. R. Robinson, *On the number of simple modules in a block of a finite group*. J. Algebra **475** (2017), 423–438.

[147] A. Marcuş, *On equivalences between blocks of group algebras: reduction to simple components*. J. Algebra **184** (1996), 372–396.

[148] A. Marcuş, *Broué's Abelian defect group conjecture for alternating group*, Proc. Amer. Math. Soc., **132** (2003), 7–14.

[149] N. Mazza, *Modules d'endo-permutation*, Ph.D. Thesis, Lausanne, 2003.

[150] M. Murai, *On a minimal counterexample to the Alperin-McKay conjecture*, Proc. Japan Acad. Ser. A Math. Sci. **87** (2011), 192–193.

[151] H. Nagao, *On a conjecture of Brauer for p-solvable groups*, J. Math. Osaka City Univ. **13** (1962), 35–38.

[152] H. Nagao and Y. Tsushima, *Representations of finite groups*, Academic Press, San Diego (1989).

[153] T. Nakayama, *On Frobeniusean algebras. I*, Ann. Math. **40** (1939), 611–633.

[154] G. Navarro, *The McKay conjecture and Galois automorphisms*. Ann. Math. **160** (2004), 1129–1140.

[155] G. Navarro and G. R. Robinson, *Blocks with p-power character degrees*, Proc. Amer. Math. Soc. **133** (2005), 2845–2851.

[156] G. Navarro and G. R. Robinson, *On endo-trivial modules for p-solvable groups*, Math. Z. **270** (2012), 983–987.

[157] G. Navarro and P. Tiep, *A reduction theorem for the Alperin weight conjecture*, Invent. Math. **184** (2011), 529–565.

[158] T. Okuyama, *Module correspondences in finite groups*, Hokkaido Math. J. **10** (1981), 299–318.

[159] T. Okuyama and S. Wajima, *Character correspondence and p-blocks of p-solvable groups*, Osaka J. Math. **17** (1980), 801–806.

[160] T. Okuyama and Y. Tsushima, *Local properties of p-block algebras of finite groups*, Osaka J. Math. **20** (1983), 33–41.

[161] J. Olsson, *On 2-blocks with quaternion and quasidihedral defect groups.* J. Algebra **36** (1975), 212–241.

[162] S. Park, *Realizing a fusion system by a single finite group*, Archiv Math. (Basel) **94** (2010), 405–410.

[163] S. Park, *The gluing problem for some block fusion systems*, J. Algebra **323** (2010), 1690–1697.

[164] C. Picaronny and L. Puig, *Quelques remarques sur un thème de Knörr*, J. Algebra **109** (1987), 69–73.

[165] W. Plesken, *Group rings of finite groups over p-adic integers*, Lecture Notes in Mathematics **1026** (1983), Springer Verlag, Berlin, ii+151 pp.

[166] L. Puig, *Structure locale dans les groupes finis*, Bull.SocMath.France, Mémoire **47** (1976).

[167] L. Puig, *Local block theory in p-solvable groups*, Proc. Symp. Pure Math. **37** (1980), 385–388.

[168] L. Puig, *Local extensions in endo-permutation modules split: a proof of Dade's theorem*, Publ. Math. Univ. Paris VII **2** (1984), 199–205.

[169] L. Puig, *Pointed groups and construction of characters.* Math. Z. **176** (1981), 265–292.

[170] L. Puig, *Nilpotent blocks and their source algebras*, Invent. Math. **93** (1988), 77–116.

[171] L. Puig, *Local fusion in block source algebras*, J. Algebra **104** (1986), 358–369.

[172] L. Puig, *Sur les P-algèbres de Dade*, unpublished manuscript (1988).

[173] L. Puig, *Affirmative answer to a question of Feit*, J. Algebra **131** (1990), no. 2, 513–526.

[174] L. Puig, *Pointed groups and construction of modules*, J. Algebra **116** (1988), 7–129.

[175] L. Puig, *Une correspondance de modules pour les blocs à groupes de défaut abéliens*, Geom. Dedicata **37** (1991), no. 1, 9–43.

[176] L. Puig, *On Joanna Scopes' criterion of equivalence for blocks of symmetric groups*, Algebra Colloq. **1** (1994), 25–55.

[177] L. Puig, *On the local structure of Morita and Rickard equivalences between Brauer blocks*, Progress in Math. **178**, Birkhäuser Verlag, Basel (1999).

[178] L. Puig, *The hyperfocal subalgebra of a block*, Invent. Math. **141** (2000), 365–397.

[179] L. Puig, *Block source algebras in p-solvable groups*, Michigan Math. J. **58** (2009), 323–338.

[180] L. Puig and Y. Usami, *Perfect isometries for blocks with abelian defect groups and Klein four inertial quotients.* J. Algebra **160** (1995), 192–225.

[181] L. Puig and Y. Usami, *Perfect isometries for blocks with abelian defect groups and cyclic inertial quotients of order 4.* J. Algebra **172** (1995), 205–213.

[182] D. Quillen, *Homotopy properties of the poset of non-trivial p-subgroups of a group.* Adv. in Math. **28** (1978), 101–128.

[183] K. Ragnarsson, *Classifying spectra of saturated fusion systems*, Algebr. Geom. Topol. **6** (2006), 195–252.

[184] W. F. Reynolds, *Blocks and normal subgroups of finite groups*, Nagoya Math. **22** (1963), 15–32.

[185] J. Rickard, *Derived categories and stable equivalence*, J. Pure Applied Algebra **61** (1989), 303–317.

[186] J. Rickard, *Morita theory for derived categories*, J. London Math. Soc. **39** (1989), 436–456.

[187] J. Rickard, *Splendid equivalence: derived categories and permutation modules*, Proc. London Math. Soc. **72** (1996), 331–358.

[188] G. R. Robinson, *The Z_p^*-theorem and units in blocks*. J. Algebra **134** (1990), 353–355.

[189] G. R. Robinson, *Local structure, vertices, and Alperin's conjecture*, Proc. London Math. Soc. **72** (1996), 312–330.

[190] G. R. Robinson, *Dade's projective conjecture for p-solvable groups*, J. Algebra **229** (2000), 234–248.

[191] G. R. Robinson, *More bounds on norms of generalized characters with applications to p-local bounds and blocks*, Bull. London Math. Soc. **37** (2005) 555–565.

[192] G. R. Robinson, *On the focal defect group of a block, characters of height zero, and lower defect group multiplicities*, J. Algebra **320** (2008), 2624–2628.

[193] K. W. Roggenkamp, *Subgroup rigidity of p-adic group rings (Weiss arguments revisited)*. J. London Math. Soc. **46** (1992), 432–448.

[194] R. Rouquier, *From stable equivalences to Rickard equivalences for blocks with cyclic defect groups*, Proc. 'Groups 1993, Galway-Saint-Andres', volume 2, London Math. Soc. Series **212**, 512–523, Cambridge University Press (1995).

[195] R. Rouquier, *Block theory via stable and Rickard equivalences*, in: Modular representation theory of finite groups (Charlottesville, VA 1998), (M. J. Collins, B. J. Parshall, L. L. Scott), DeGruyter, Berlin, 2001, 101–146.

[196] R. Rouquier, *The derived category of blocks with cyclic defect groups*, in: Derived Equivalences for Group Rings (S. König, A. Zimmermann), Lecture Notes in Math. **1685**, Springer Verlag, Berlin-Heidelberg, 1998, 199–220.

[197] R. Rouquier and A. Zimmermann, *Picard groups for derived module categories*. Proc. London Math. Soc. **87** (2003), 197–225.

[198] A. Ruiz and A. Viruel, *The classification of p-local finite groups over the extraspecial group of order p^3 and exponent p*, Math. Z. **248** (2004), no. 1, 45–65.

[199] A. Salminen, *On the sources of simple modules in nilpotent blocks*, J. Algebra **319** (2008), 4559–4574.

[200] A. Salminen, *Endopermutation modules arising from the action of a p-group on a defect zero block*, J. Group Theory **12** (2009), 201–207.

[201] B. Sambale, *Blocks of finite groups and their invariants*. Lecture Notes in Mathematics **2127** (2014), xiv+243 pp.

[202] B. Sambale, *Cartan matrices and Brauer's k(B)-conjecture III*. Manuscripta Math. **146** (2015), 505–518.

[203] J. Scopes, *Cartan matrices and Morita equivalence for blocks of the symmetric groups*, J. Algebra **142** (1991), 441–455.

[204] L. L. Scott, *Defect groups and the isomorphism problem*, SMF Astérisque **181/182** (1990), 257–262.

[205] L. L. Scott, unpublished notes (1990).

[206] J. Słomińska, *Homotopy colimits on EI-categories*, Lecture Notes in Mathematics, Springer Verlag, Berlin **1474** (1991), 273–294.

[207] B. Späth, *A reduction theorem for the Alperin-McKay conjecture*, J. Reine Angew. Math. **680** (2013), 153–189.

[208] B. Späth, *A reduction theorem for the blockwise Alperin weight conjecture*, J. Group Theory **16** (2) (2013), 159–220.

[209] R. Stancu, *Almost all generalized extraspecial p-groups are resistant*, J. Algebra **249** (2002), no. 1, 120–126.

[210] R. Stancu, *Control of the fusion in fusion systems*, J. Algebra and its Applications **5** (2006), no. 6, 817–837.

[211] P. Symonds, *The Bredon cohomology of subgroup complexes*, J. Pure and Applied Algebra **199** (2005), 261–298.

[212] J. Thévenaz, *Permutation representations arising from simplicial complexes*. J. Combin. Theory A **46** (1987), 121–155.

[213] J. Thévenaz, *Duality in G-algebras*, Math. Z. **200** (1988) 47–85.

[214] J. Thévenaz, *Endo-permutation modules, a guided tour*, in: (eds. M. Geck, D. Testerman, and J. Thévenaz) *Group representation theory*, EPFL Press, Lausanne (2007), 115–147.

[215] J. Thévenaz, *G-algebras and modular representation theory*, Oxford Science Publications, Clarendon, Oxford (1995).

[216] J. Thévenaz and P. J. Webb, *A Mackey functor version of a conjecture of Alperin*. Astérisque **181–182** (1990), 263–272.

[217] A. Turull, *Strengthening the McKay conjecture to include local fields and local Schur indices*, J. Algebra **319** (2008), 4853–4868.

[218] J. Thompson, *Vertices and sources*, J. Algebra **6** (1967), 1–6.

[219] K. Uno, *Conjectures on character degrees for the simple Thompson group*, Osaka J. Math. **41** (2004), 11–36.

[220] J. M. Urfer, *Modules d'endo-p-permutation*, Ph.D. Thesis, Lausanne, 2006.

[221] Y. Usami, *On p-blocks with abelian defect groups and inertial index 2 or 3, I.* J. Algebra **119** (1988), 123–146.

[222] A. Watanabe, *Note on a p-block of a finite group with abelian defect groups*, Osaka J. Math. **26** (1989), 829–836.

[223] A. Watanabe, *Notes on p-blocks of characters of finite groups*, J. Algebra **136** (1991), 109–116.

[224] A. Watanabe, *On nilpotent blocks of finite groups*, J. Algebra **163** (1994), 128–134.

[225] P. J. Webb, *Subgroup complexes*. Proc. Symp. Pure Math. **47** (1987), Amer. Math. Soc., Providence.

[226] P. J. Webb, *A split exact sequence of Mackey functors*. Comment. Math. Helvet. **66** (1991), 34–69.

[227] P. J. Webb, *Standard stratifications of EI categories and Alperin's weight conjecture*. J. Algebra **320** (2008), 4073–4091.

[228] A. Weiss, *Rigidity of p-adic p-torsion*, Ann. Math. **127** (1988), 317–332.

[229] A. Yekutieli, *Dualizing complexes, Morita equivalence, and the derived Picard group of a ring.* J. London Math. Soc. **60** (1999), 723–746.

[230] A. Zimmermann, *Representation theory. A homological algebra point of view.* Algebra and Applications **19**, Springer, Cham, Heidelberg, New York, Dordrecht, London, 2014.

Index

Printed in the United States
By Bookmasters